Date Due

FEB 2 8 2001

LOCALIZATION AND NEUROIMAGING IN NEUROPSYCHOLOGY

FOUNDATIONS
OF
NEUROPSYCHOLOGY

A Series of Textbooks, Monographs, and Treatises

Series Editor

LAIRD S. CERMAK

Memory Disorders Research Center, Boston Veterans Administration Medical Center, Boston, Massachusetts, USA

LOCALIZATION AND NEUROIMAGING IN NEUROPSYCHOLOGY

Edited by

313 256481

Andrew Kertesz

Department of Clinical Neurological Sciences
Lawson Research Institute
St. Joseph's Health Center
London, Ontario
Canada

Academic Press

A Division of Harcourt Brace & Company

San Diego   New York   Boston   London   Sydney   Toyko   Toronto

Using functional magnetic resonance imaging (FMRI), activity during a task requiring monitoring of semantic features of auditory words is visualized. For further details, see Chapter 7.

This book is printed on acid-free paper. ∞

Academic Press, Inc.

525 B Street, Suite 1900, San Diego, California 92101-4495

United Kingdom Edition published by
Academic Press Limited
24–28 Oval Road, London NW1 7DX

Library of Congress Cataloging-in-Publication Data

Localization and neuroimaging in neuropsychology / edited by Andrew
    Kertesz.
        p.   cm. -- (Foundations of neuropsychology)
    Includes bibliographical references and index.
    ISBN 0-12-405045-X
    1. Brain--Imaging.   2. Brain mapping.   3. Neuropsychiatry.
    I. Kertesz, Andrew.   II. Series: Foundations of neuropsychology San
    Diego, Calif.)
        [DNLM: 1. Nervous System Diseases--diagnosis.   2. Diagnostic
    Imaging.   WL 141 L811 1994]
    RC473.B7L63   1994
    616.8'04754--dc20
    DNLM/DLC
    for Library of Congress   93-37924
                                                                          CIP

PRINTED IN THE UNITED STATES OF AMERICA
94   95   96   97   98   99   MM   9   8   7   6   5   4   3   2   1

Contents

Chapter 4 Localizing the Neural Generators of Event-Related
              Brain Potentials

Diane Swick, Marta Kutas, and Helen J. Neville

Chapter 5 Use of Positron Emission Tomography to Study Aphasia

E. Jeffrey Metter and Wayne R. Hanson

Chapter 6 Functional Activation and Cognition: The ^{15}O PET
              Subtraction Method

Howard Chertkow and Daniel Bub

Chapter 7 Human Brain Mapping with Functional Magnetic
Resonance Imaging

Jeffrey R. Binder and Stephen M. Rao

Chapter 8 Anatomical Asymmetries and Cerebral Lateralization

Andrew Kertesz and Margaret A. Naeser

Chapter 21   Neuroimaging in Dementia

Kevin F. Gray and Jeffrey L. Cummings

Contributors

Numbers in parentheses indicate the pages on which the authors' contributions begin.

Martin L. Albert (429), Department of Neurology, Boston University School of Medicine, Veterans Administration Medical Center, Boston, Massachusetts 02130

Santiago Arroyo (57), Department of Neurology, Johns Hopkins University, School of Medicine, Baltimore, Maryland 21205

Marlene Behrmann (331), Department of Psychology, Carnegie Mellon University, Pittsburgh, Pennsylvania 15313

Jeffrey R. Binder (185), Department of Neurology, Medical College of Wisconsin, Milwaukee, Wisconsin 53226

Sandra E. Black (331), Department of Medicine (Neurology), Sunnybrook Health Science Centre, Toronto, Ontario, Canada M4N 3M5

Daniel Bub (151), Department of Neurology and Neurosurgery, Montreal Neurological Institute, McGill University, Montreal, Quebec, Canada H3A 2B4

Stefano Cappa (545), Clinica Neurologica Università di Brescia-II, Neurologia Spedali Civili, 25125 Brescia, Italy

Laird S. Cermak (599), Memory Disorders Research Center, Veteran's Administration Medical Center, Boston, Massachusetts 02130

Howard Chertkow (151), Department of Neurology and Neurosurgery, Lady Davis Institute for Medical Research, McGill University, Montreal, Quebec, Canada H3T 1E2, and Research Centre, Centre Hospitalier Côte-des-Neiges, Montreal, Quebec, Canada H3W 1W5

Jeffrey L. Cummings (621), Neurobehavior Unit, West Los Angeles Veteran's Administration Medical Center, Los Angeles, California 90073

Leslie J. Gonzalez Rothi (407), Audiology and Speech Pathology Service, Department of Veterans Affairs Medical Center, Gainesville, Florida 32608

Barry Gordon (57), Department of Neurology, Johns Hopkins University, School of Medicine, and The Zanvyl Krieger Mind/Brain Institute, Johns Hopkins University, Baltimore, Maryland 21205

Kevin F. Gray (621), Department of Psychiatry, and Biobehavioral Sciences, University of California at Los Angeles, School of Medicine, Los Angeles, California 90024

Wayne R. Hanson (123), Audiology and Speech Pathology Service, Veterans Administration Medical Center, Sepulveda, California 91343

John Hart (57), Department of Neurology, Johns Hopkins University, School of Medicine, and The Zanvyl Krieger Mind/Brain Institute, Johns Hopkins University, Baltimore, Maryland 21205

Kenneth M. Heilman (407, 495), Department of Veterans Affairs Medical Center, and Department of Neurology, Neurology Service, College of Medicine, University of Florida, Gainesville, Florida 32610

Janet Jankowiak (429), Department of Neurology, Boston Veteran's Administration Medical Center, Boston, Massachusetts 02130

Terry L. Jernigan (599), Departments of Psychiatry and Radiology, and Department of Veteran's Affairs Medical Center, University of California at San Diego, La Jolla, California 92093

Andrew Kertesz (1, 213, 525, 567), Department of Clinical Neurological Sciences, Lawson Research Institute, St. Joseph's Health Center, London, Ontario, Canada N6A 4V2

Andrew Kirk (525), Department of Medicine (Neurology), University of Saskatchewan, Saskatoon, Saskatchewan, Canada S7N 0X0

Marta Kutas (73), Department of Cognitive Science, University of California at San Diego, La Jolla, California 92093

Ronald P. Lesser (57), Department of Neurology, Johns Hopkins University, School of Medicine, and The Zanvyl Krieger Mind/Brain Institute, Johns Hopkins University, Baltimore, Maryland 21205

E. Jeffrey Metter (123), The Baltimore Longitudinal Study of Aging, Gerontology Research Center, National Institute of Aging, Baltimore, Maryland 21224

Margaret A. Naeser (213, 245), Department of Neurology and Aphasia Research Center, Boston University School of Medicine, and Veteran's Administration Medical Center, Boston, Massachusetts 02130

Helen J. Neville (73), Neuropsychology Lab, The Salk Institute, San Diego, California 92186

George A. Ojemann (35), Department of Neurological Surgery, University of Washington School of Medicine, Seattle, Washington 98195

Adele S. Raade (407), Department of Communicative Disorders, Boston University, 635 Commonwealth Avenue, Boston, Massachusetts 02215

Stephen M. Rao (185), Department of Neurology, Neuropsychology Section, Medical College of Wisconsin, Milwaukee, Wisconsin 53226

Steven Z. Rapcsak (297), Neurology Service, Veteran's Administration Medical Center, Tucson, Arizona 85723

David P. Roeltgen (377), Department of Neurology, Hahnemann University, Philadelphia, Pennsylvania 19102

Alan B. Rubens (297), Department of Neurology, University of Arizona, Tucson, Arizona 85721

Justine Sergent (473), Department of Neurology and Neurosurgery, Montreal Neurological Institute, McGill University, Montreal, Quebec, Canada H3A 2B4

Diane Swick (73), Department of Neurology and Center for Neuroscience, University of California at Davis, Veteran's Administration Medical Center, Martinez, California 94553

Edward Valenstein (495), Department of Neurology, and Center for Neuropsychological Studies, University of Florida, Gainesville, Florida 32610

Claus-W. Wallesch (545), Neurologische Klinik, Klinikum der Albert-Ludwigs Universität, D-7800 Freiburg, Germany

Robert T. Watson (495), Department of Neurology, Center for Neuropsychological Studies, and College of Medicine, University of Florida, Gainesville, Florida 32610

Preface and Overview

Progress in neuroimaging and neuropsychology necessitates a frequent update of continuously forthcoming data. Advances in neuroimaging in the past 10 years have been nothing short of spectacular. These changes have been closely followed by theoretical and empirical progress in neuropsychology and cognitive science. Some readers may be familiar with my previous book *Localization in Neuropsychology*. Although continuity between these complementary volumes exists, this book is substantially different, representing up-to-date ideas and information. An attempt is made to be comprehensive and cohesive in theory, methodology, and content. The first chapter is intended as an exposition of the theory of localization and a brief survey of the methodologies. In comparison to its predecessor, this chapter is much expanded and updated. The first eight chapters focus on methodologies. The next nine chapters emphasize syndrome and symptom-oriented content, summarizing whatever localization information is available in specific functional areas. In addition, some chapters focus on the localization of lesions in certain anatomical structures, and others focus on more diffusely distributed functions or lesions, such as memory and dementia. Even in these areas previously thought to be nonlocalizable, an increasing amount of work has been done with neuroimaging. The time of publication is more propitious than ever because of the emergence of functional activation by positron emission tomography (PET) and magnetic resonance imaging (MRI), both of which are represented in this book by several of the leaders in their respective fields.

The chapter on *cortical stimulation* offers a different perspective from that of *lesion localization*. Ojemann provides not only new data, but also a new model for *language organization*. Separate systems for naming, reading, speech production, perception, and recent verbal memory are postulated to be arranged like a mosaic in the cortex, requiring widespread activation during function. In addition to a critical summary of the stimulation data, this chapter also draws attention to new developments in the field, including data about optical reflectance of the surface tissue, which changes with function. Cerebral organization of language in children, sex differences, naming in multiple languages, changes in the electrocorticogram, and the relationships of production to perception and of naming to reading are also discussed. The chapter by Lesser et al. on *subdural electrode stimulation* describes a relatively recent technique that has become successful clinically and provides a great deal of scientific information. The technique is less invasive than open cortical stimulation, although anatomical verification is less direct. The extent of functional and anatomical fractionation achieved by the stimulation method is astonishing.

Event-related potentials (ERPs) are time-locked to brief cognitive events. In contrast to their relatively low spatial resolution, ERPs provide physiological information about covert, cognitive, and linguistic processing. Swick et al. describe the recently improved accuracy in localization and the expanded knowledge of the underlying cognitive mechanisms that have been achieved with this technique. The authors discuss the approaches used to localize ERP components and survey the effect of lesions and pharmacologicals on evoked potentials in animals and humans. They also summarize magnetoencephalography integrated with magnetic resonance imaging as an adjunct to localizing current dipoles in the brain, complementary to ERPs. Current source density maps further contribute to the topography of scalp potentials and are also useful in differentiating among various cognitive processes.

As did Pierre Marie almost a century ago, Metter and Hanson, on the basis of *positron emission tomography* studies, suggest that temporoparietal dysfunction underlies all aphasic deficits. Additional prefrontal hypometabolism modifies the clinical picture. Subcortical damage appears to have a direct effect on language, as well as an indirect effect associated with prefrontal hypometabolism. Small lesions without distant hypometabolism are asymptomatic. If the deficits are related to the distant hypometabolism, as the results imply, the issue of persisting hypometabolism in the face of substantial recovery must be further studied.

Functional activation with ^{15}O PET scanning opens up a new chapter on the study of cognition. Chertkow and Bub, with a careful approach, explore the methodological pitfalls of the so-called subtraction technique and provide insight into the exciting field of functional cortical mapping with PET scanning. An interesting example is the presentation of independent evidence for frontal and temporoparietal sites of semantic processing. Although the field is still young, this chapter examines the results with an analytical and questioning approach that will be very important in sorting out empirical evidence from artifact. There are three other chapters in this book that present the results of PET activation, underlying the critical importance of this methodology in functional localization.

Brain mapping with functional magnetic resonance imaging is the latest of the revolutionary technologies, utilizing noninvasive physiological changes in blood flow to detect cerebral functional activation. The chapter by Binder and Rao introduces the principles of magnetic resonance signal changes produced by neuronal activation and changes in oxyhemoglobin that provide the basis of the technique. These authors present the application of the methodology to specific cognitive domains, such as human motor control and language, and describe what may become the most important functional imaging technique of the future, not only because it may become more widely available but because its capabilities are different, aiming at a briefer stimulus-response period than the other activation methods.

Anatomical and functional asymmetries and their biological significance are reviewed by Kertesz and Naeser. Some of these asymmetries can be correlated *in vivo* using neuroimaging. Potentially important handedness and sex effects are described, but individual variability often exceeds group differences. The practical use of determining functional laterality or predicting recovery through neuroimaging asymmetries awaits further study. *Recovery from aphasia* is related to many factors. Naeser investigates lesion location, while controlling some of the other parameters. This chapter gives examples of detailed computerized tomography analyses of groups of patients with an aphasic syndrome or the structural correlations of the recovery of a symptom, and the prediction of the efficacy of a certain therapy. The sophisticated analysis of the interesting syndrome of *transcortical aphasia* by Rapcsak and Rubens incorporates modern empirical evidence to confirm the theoretical considerations in the initiation of speech and movement in general. The roles of medial and dorsolateral frontal lesions, subcortical structures, and their various combinations in the production of the clinical syndromes are outlined authoritatively.

Alexia is one of the modular deficits that can be studied both in isolation and in relationship to other deficits. Black and Behrmann combine a thorough review of the classical lesion evidence, the anatomy of the visual system, and the burgeoning literature on various models of reading with the results of recent ^{15}O PET imaging studies and the information processing approach in cognitive psychology. Included are their own important studies of neglect dyslexia. The detailed classification of *agraphia* and its association with lesions provides insight into the complexity of writing and its disturbances. Roeltgen attempts to integrate the neurological and neurolinguistic approach using recent advances in these areas. The degree of convergence is encouraging and points out the value of both directions of study. A special section on the impairment of the written language system in Japanese adds further interest. Difficulty with purposeful movement is commonly labeled *apraxia* and left hemisphere dominance for action has been recognized since Liepmann. Rothi et al. follow a multilevel regional approach, describing apraxic syndromes, according to the type of action that is affected, such as limb apraxia and buccofacial apraxia, in addition to the lesion location, such as callosal apraxia or subcortical apraxia.

Freud's definition of *agnosia* as the failure to recognize objects without primary visual or intellectual impairment or aphasia has stood the test of time despite ongoing debate about accompanying deficits. Jankowiak and Albert describe this fascinating clinical syndrome and summarize the current knowledge of the lesions that produce it. A complex but regular combination of features corresponds to lesion localization to an extent that is matched only in the dominant hemisphere language area. These authors also complete their survey with evidence from PET activation of various posterior head regions on object identification tasks. To illustrate the uniqueness of *face recognition*, Sergent invites us to participate in an experiment of matching monkey faces. She analyzes the components of face processing, and presents the results of tests on four patients with different types of involvement, as well as the analysis of their lesions. She then shows the results of PET activation studies of face and object processing on these patients and on normal individuals. This converging evidence of the structural and functional dissociation of face and object processing, and of the distinct operations within face recognition, establishes the specific roles of the right fusiform and lingual gyri in face perception, and of the parahippocampal gyrus in retrieving identity.

Directed attention and its deficit, neglect, are of considerable biological significance. Heilman et al. define the various terms used in the *neglect* literature and review the anatomy of neglect and the comparative anatomy of attention. These authors propose a highly interactive network that mediates spatially direct attention and motor intention. These networks can be disrupted at separate levels. In addition to the contribution of widely distributed cortical structures, the subcortical nuclei play a pivotal role. The role of the frontal lobes in the intentional network is particularly emphasized. Recent advances in the fractionation of the neglect syndrome are also included. *Constructional impairment*, a combination of visuoperceptual function and executive motor function, is an important and frequent clinical syndrome that has been under scrutiny for considerable time. This extremely complex syndrome covers the range of nonverbal cortical processing from gnostic to practic function. The controversial and, at times, conflicting literature is summarized by Kirk and Kertesz, who include examples from their own clinical studies using localizing information concerning the contributions from various hemispheric structures to the clinical deficit.

Cappa and Wallesch contribute their comprehensive treatment of the role of *subcortical structures in cognitive deficits*. The basal ganglia have an obviously important coordinating role in motor, oculomotor, nonmotor cognitive, limbic, and language functions. In fact, few aspects of cognition exist in which these structures have not been implicated either directly or indirectly. The anatomical, physiological, and pharmacological factors in the production of functional deficits, particularly aphasia, apraxia, and neglect, are reviewed with attention to recent developments in neuroscience. *Frontal lesions* produce dramatic changes in personality, executive function, attention, motor behavior, motivation, memory, and language. Kertesz reviews these symptoms in relationship to the lesion sites that produce them. Frontal lobes are often affected by head injury and tumor, and recently have been considered to be important in schizophrenia. In addition, certain dementias show a predilection for the frontal lobes. This large region of the human brain is no longer terra incognita, due to a great deal of convergent research into its functions and their localization.

The *structures underlying memory*, particularly recent memory, encoding, and retrieval, are considered bilateral, and part of the diencephalic limbic system. Recent advances, especially the visualization of small lesions with MRI, are reviewed by Jernigan and Cermak, who also present their semi-automated analysis of the structural changes on MRI in Korsakoff's and Alzheimer's disease and the multivariate analysis of the anatomical correlates of memory processing experiments such as priming and recognition. These latter studies suggest that data from less discrete and milder impairment and from more diffuse damage can be informative. Cummings and Gray cover the tremendously expanded field of *neuroimaging in dementia*, an area that was practically nonexistent until recently, except as a method of excluding focal conditions that may simulate dementia. The evidence, as clearly documented in this chapter, suggests that specific structural changes in certain types of dementia are diagnostic, as well as reflective of the stage and the extent of the illness. This chapter demonstrates that even degenerative disease can be fruitfully imaged and that some features can be localized, while the nature and the location of the changes are correlated with the pathology and the symptoms of the disease.

Finally, special acknowledgments are due to my administrative assistant, Bonita Caddel, who kept the project going and typed a substantial portion of it, as well as helped with the indexing, to my colleagues in my lab (Pat McCabe and Marsha Polk), who did some proofreading and provided constructive comments, and thanks to the editorial staff at Academic Press, Nikki Fine and Diane Scott, for all their help.

Andrew Kertesz

Chapter 1

Localization and Function: Old Issues Revisited and New Developments

Andrew Kertesz

I. INTRODUCTION

Advances in integrating function with structure in the nervous system have been accelerated by technological developments in neuroimaging and by theoretical and empirical developments in cognitive neuropsychology and behavioral neurology. The result is an exploding knowledge base that deserves an update. This chapter addresses some of the fundamental issues in function and its localizability in the brain.

Most neuroscientists have no difficulty accepting the idea that certain functions are related to certain structures in the brain. Some would, however, qualify that only physiological functions, not psychological ones, are structure bound (Bullock, 1965). Others would consider the gap between function and structure too large even to attempt integration and remain content with functional analysis, casting aside the relevance of structure (Mehler, Morton, & Jusczyk, 1984). Some of the fundamental issues of functional–anatomical relationships can be presented as the following questions: (1) What is considered a neuropsychological function? (2) How are these functions interrelated? (3) To what extent are functions localizable? (4) What are the anatomical structures necessary to carry out such function in the normal brain? (5) How are functions localized biochemically and physiologically? (6) What are the connections of functionally related structures? (7) What other structures can compensate if certain parts of the brain are damaged?

Central to all these issues is the idea that, after brain damage, the observed behavior does not reflect the simple equation of normal function

1

missing but represents a new functional state of reorganization. Underlying the complex phenomena of reorganized function are the various issues of substitution, brain redundancy, vicarious function, diaschisis, regeneration, regrowth, and retraining. A somewhat wishful concept of "transparency" assumes that a lesion results in a relatively transparent modification of the normal system; therefore, deficit analysis reveals normal function (Caramazza, 1986). Most serious investigators accept Jackson's (1878) warning that only lesions and not functions can be localized. Goldstein's (1948) rejoinder was that the question is not "Where is a definite function localized?" but "How does a lesion modify the function of the brain so a definite symptom comes to the forefront?" Localizing lesions can and does, however, provide information about function, although this information is often indirect. Sophisticated lesion analysis, including remote metabolic and transsynaptic events integrated with deficit analysis based on knowledge of normal function and functional activation, goes a long way toward clarifying functional–structural relationships.

II. FUNDAMENTAL ISSUES

A. What Is a Function? Holistic vs. Modular Organization

Defining a function is difficult and the definition is arbitrary at the best of times. Clinicians, physiologists, and psychologists have different concepts of the same behavior. Certain psychological concepts of function may not be appropriate to describe actual brain function or connectivity. On the other hand, anatomy and physiology may not provide even the questions, let alone the answers, about behavior. Functional analysis may proceed on theoretical and empirical lines, but theories of function are influenced by anatomical, physiological, and behavioral evidence.

A unitary theory of brain function was formulated by physiologists who observed that functions were interconnected and ablation of one part of the cortex did not cause permanent deficit. Flourens and others called this aspect of brain function the "sensorium commune" in the 19th century, implying that all functions were interrelated in some fashion. Golgi, who perfected the silver stain for nerve tissue, thought that the filaments of the nervous system were interconnected to form a unified network or syncytium. The reflex physiology of Sechenov and the association psychology of James contributed to the notion of a complex network functioning without clear anatomical boundaries. Another argument against strict localization of function was forwarded by Lashley (1929), who considered large regions of the brain to consist of more or less equipotential components. Equipotentiality explained the evidence for the considerable amount of substitution after lesions. After Lashley, a whole generation of behaviorists were interested only in studying perception and behavior without paying attention to the "black box" of the enigmatic brain in which localization did not seem to be relevant.

However, anatomists such as Rolando, Gall, Spurzheim, Romberg, and others described the cortical gyri and developed techniques to preserve the brain and to trace cortical connections and long tracts. These investigators also began to assign function to them—at times on the basis of empirical evidence, other times in a speculative fashion—contradicting the holistic view of intelligence and behavior. Phrenology was a fanciful mosaicist view of cerebral function, in some instances reduced to examining the external landmarks on the skull. Scientific support for the localization of function was provided by the physiologists of the second half of the 19th century, who observed the close relationship between stimulated cortical points and movements in animals (Fritsch & Hitzig, 1870) and extirpation of occipital cortex and loss of vision (Munk, 1881), and by clinicians who correlated language impairment with the left perisylvian brain region (Broca, 1861; Wernicke, 1874). Cajal (1890) recognized the individuality of nerve cells or neurons, despite their interconnection by synapses. After the recognition of the distinct anatomy of various neurons, their differing function also was investigated, mainly through electrophysiology and cortical and intracellular recording.

In the clinical sciences, evidence accrued that brain damage may result in a selective loss of psychological function, even more often than in global impairment. The dissociation of various functions, especially the appearance of double dissociation in which the loss of function appears in one direction in one case and in another direction in the other, indicated a great deal of modularity in how the brain is organized and suggested localization of function to certain areas of the brain that can be damaged while sparing function in other areas. For example, it became evident that pure alexia can occur without agraphia (Dejerine, 1892). Cognitive research since has utilized the description of various functional dissociations caused by circumscribed or localized lesions. The number of dissociated cognitive functions described in the last 20 years has increased enormously, and has encouraged investigators to think of the brain as very highly modular in nature. This highly fractionated, functional mosaic has relatively poor correspondence to brain structure. Few scientists today view brain structure as having a strict pinpoint localization corresponding to a functional mosaic although, of all the techniques, electrical stimulation of the cortical surface comes close to this concept (see Chapter 2).

Some of the cortical maps being constructed by functional activation studies (see Petersen, Fox, Posner, Mintun, & Raichle, 1988 and Chapters 6, 7, and 11 of this volume) resemble the cortical maps of behavioral neurologists based on lesioned patients, although interesting differences exist (see subsequent discussion and chapters).

Opponents of cerebral localization have been pointing out the numerous difficulties with the concept of centers of function. The evidence for widespread cerebral activation in cortical functions has been accumulating physiologically (Barlow, 1981; Mountcastle, 1978). The reticular activating system, the ubiquitous thalamocortical projections, and the multiplicity and redundancy of limbic and cortical connections indicate that central nervous system activity is complex and that much of the brain may be activated even for simpler acts of cognition. These physiological notions have been confirmed by the multiple areas that are shown by the cerebral blood flow and regional glucose metabolism techniques to be activated in reasonably well-defined stages of the mental activity (Ingvar & Schwartz, 1974; Raichle, Herscovitch, Mintun, Martin, & Power, 1984). Indirect measures of function, such as EEG, event-related averaged potentials, or the distribution of neurotransmitters, also support the diffuse and interactive nature of cerebral activity.

Recent advances in cognitive modeling are important in the interpretation of functional deficits. These changes are based on the principles of information processing and are permeated by the computer metaphor. Fractionation, or staging of function, and cognitive modeling with box and arrow diagrams, however, have many methodological and conceptual problems. Models that are consistent with known physiology and anatomy rarely are constructed. Complex behaviors often are fractionated further, although a function that is considered separate may have the same underlying mechanism. Most of these new "modules of function" are defined in terms of computer science or artificial intelligence. These constructs may have little, if any, relationship to a real psychological function, or even less to a physiological function (Crick, 1989). Alternative theories often are elaborated to interpret functions and are changed to fit the empirical facts. Reduction of a complex behavior to its components is fraught with the hazard of losing the meaning and biological significance of the behavior to the organism. From the perspective of localization, more complex behaviors are likely to have widespread input and will be affected by lesions in many areas. Minutely fractionated functions, on the other hand, are likely to be components of many complex behaviors and are less likely to be localizable, if at all.

One of the major distinctions between certain functions that are considered to be more modular than others that involve "central" or diffusely distributed cognition is the extent to which they are bound to certain brain structures (Fodor, 1983). Some hypothetical function such as phonemic assembly or the "phonological output buffer" (which is impaired when the deficit phenomenon of phonemic paraphasia is elicited) is impaired most consistently by lesions in the left inferior parietal cortex. Other language functions, such as word retrieval, may be distributed in a diffuse fashion through the language area, since they seem to be affected by lesions from many locations within the perisylvian cortex of the left hemisphere. Widespread distribution of certain cognitive functions, such as attention and memory, indicate their contribution to many other processes. Others such as problem solving, social skills, or judgment defy efforts to localize them, although from time-to-time the prefrontal cortex is appointed to be responsible for some of these functions (see Chapter 19). Some functions await definitions that are commonly agreed on. Large neuronal networks are obviously necessary in much of the performance of complex cognitive tasks, and even individuals who are interested in functional–anatomical correlations are reluctant to assign gyri or smaller structures to such functions.

Fractionation of function often implies (although not necessarily) that a serial activation of the processes takes place. However, some physiological and computational considerations suggest that mental processes operate in parallel and not in series (Hinton & Anderson, 1981; McClelland & Rumelhart, 1986). Discrete, manipulable, modular units of cognition presumably have similar counterparts in a neural organization but, to date, little concrete evidence is available that minute fractionation of function through cognitive modeling has a counterpart in physiological organization. Doubt has been expressed over whether the human world of perception and processing of input can be analyzed into independent elements (Dreyfus, 1979). Attempts to segregate perception from central processing have not always proven to be successful. For instance expectancy and previous knowledge, in other words, the precept, greatly influence how stimuli are perceived.

The analysis of behavior after a lesion should be supplanted by studies in normal individuals to confirm conclusions about the function of a lesioned organism. The functional analysis of normal cognitive systems supplies an important background in which to test the damaged or reorganized functions in patients. A detailed analysis of a function is necessary to the understanding of complex ways in which a lesion can affect it. The accurate up-to-date measurement of deficit is an essential prerequisite for meaningful localization. Much of the older clinicopathological literature is handicapped by a rather rudimentary description of the psychological deficits.

Theories of normal psychological processes, on the other hand, may benefit from considering the ways in which these processes can be disrupted. Clinical and anatomical constraints should be helpful in cognitive modeling to bring it closer to reality. It is obvious that complex cellular systems that interact must be interconnected, and this occurs most readily if cells are close together. Functional contiguity, however, may be even more important than geographic contiguity in the cortex.

The behavior observed after a lesion, however, often is not analyzable in terms of normal function. As long as caution is used in the interpretation of pathological behavior as a model of brain function, the gap between them can be bridged with the help of some extrapolation and converging evidence from other methodologies.

B. How Are Functions Related? Symptoms vs. Syndromes: Single Case vs. Group Studies

The analytical examination of isolated behaviors, defined by current theory, has gained support against the approach based on syndromes, or behaviors affected together by certain lesions in the brain. Clinicians generally attempt to deal with syndromes since they represent the total deficit picture in the patient, although they may undertake symptom analysis that goes beyond the syndrome. Isolated or pure psychological phenomena, representing single modules of function, are difficult to define and even more difficult to localize yet this route often is chosen by experimental psychologists. Often one function is emphasized only to establish a theoretical point and other symptoms are ignored. Drawing the line between a pure function and a syndrome of related functions may be impossible. In the clinicopathological paradigm of localization, it is common to start with syndromes or with lesions in a certain location and determine the number of deficits related to them.

The major advantage of syndromes is that they usually are connected to a localizable lesion, whereas symptoms are less likely to be. The syndrome vs function controversy is exemplified by a series of articles on the Gertsmann syndrome (Benton, 1961; Strub & Geschwind, 1983). The combination of agraphia, acalculia, finger agnosia, and right–left confusion has caught on among clinicians as a useful syndrome, signifying a dominant parietal lobe lesion (Gerstmann, 1940). The rationale and justification, as well as its appropriateness, have been questioned and the discussion that resulted can be applied to other syndromes in localization studies. Intuitive clinical taxonomy can and should be re-examined with modern statistical methods. Clinical taxonomies of aphasias also have been attacked for being arbitrary and based on a prejudicial conceptual framework, but numerical taxonomy with cluster analysis has shown that grouping of patients based on performance scores is not only clinically but also statistically valid (Kertesz, 1979). Different taxonomies have been used for clinical localization than for functional analysis, but to increase our knowledge of the brain and its function, integration of the two levels is most desirable.

The taxonomy issue is a crucial one in many aspects of localization. Many discrepancies simply are related to the same terminology applied to different behaviors, for example, the different definitions of transcortical motor aphasia. Sometimes the opposite occurs, and similar behaviors are grouped differently; therefore, the conclusions are contradictory. One example is the use of the term "jargon" for the stereotypies of global aphasia rather than restricting the term to fluent Wernicke's aphasia. These disagreements may come about because a phenomenon is defined poorly and described only qualitatively. Even when standardized tests are used, classifications differ because the criteria for each syndrome or deficit are different. Despite some of these differences, as long as the behavioral criteria are defined clearly on the basis of standardized measurements, various groups can be compared with reasonable efficiency. Correlation with lesion sites has been successful and convincingly consistent for many syndromes of higher cerebral function.

A considerable amount of cognitive theory has been elaborated on the basis of single case reports in the last 15 years. The dangers of generalizing from single case reports for or against localizing certain functions, or the fractionation of function, are great. The human brain is complex, not only functionally but also in terms of individual variability in many biological factors. In many examples in the literature, a single case (or a few selected cases) is documented to show that a lesion in a certain location does or does not produce the symptoms expected. However, important biological factors such as the time elapsed from the date of injury may be ignored or considered inadequately in the interpretation of the behavior observed. Therefore, the wrong conclusions may be drawn. Nevertheless, since experimental series of lesions are often not available to investigate the function of an area, single well-documented case reports that are followed with several examinations and that have good lesion localization can be very informative. Special statistical techniques controlling for other factors that influence behavior have been developed to make single case studies more scientifically acceptable, but these cases also need controls or groups of similar subjects for comparison (Kazdin, 1982).

Group studies, on the other hand, may provide information about the function that is commonly affected when the same or similar lesion is seen. Such studies, however, may obscure individual differences in specific functions. Detailed exploration of behavior or experimentation is also difficult to carry out in large groups. However, group studies are useful for generalizing the findings and for proving that the observations are not by chance or caused by some factor other than the one that is being examined. Group studies and single case studies make important contributions, provided scientific criteria are met. Advocating one at the expense of the other does not seem to be productive.

C. What Function Is Localizable?

Some early investigators, such as Goltz (1881), have concluded that the cerebral cortex is the organ of intelligence that is impaired permanently by large lesions, but in which smaller focal lesions cannot cause permanent or selective deficits. Subsequent stimulation and lesion studies have contradicted this idea to some extent and have provided a great deal of new information (Phillips, Zeki, & Barlow, 1984). Some of the discrepancies were attributed to species differences. The extent to which a function can be localized in the brain is variable. For certain functions, relatively little argument exists concerning localization to a certain cytoarchitectonic area. These regions are the primary sensory and motor areas, including the cortical areas for special sense such as hearing and vision, which are clearly responsible for perceiving the environment and initiating, coordinating, and maintaining movements. Head (1926) emphasized the distinction between the localizable functions of motion, sensation, and vision—which have a clear relationship to body parts, muscles, sensory inputs, and stimuli—and "disorders of speech or similar higher grade functions that have no such relation to parts of the body or their projection in space."

The correlation of the columnar organization of visual cortex with visual feature analysis (Hubel & Wiesel, 1968), a similar organization in the somatosensory cortex (Mountcastle, 1957), and the elaboration of visual maps in the cortex (Van Essen, 1979; Zeki, 1983) are prime examples of the neurophysiological knowledge of the primary and secondary neocortex. However, much of the actual matching of cortical physiology with functions has been controversial (Phillips et al., 1984). Even the localization of primary motor functions in the prerolandic cortex has been the subject of a great deal of argument, especially concerning the extent to which part of a movement can be localized. Stimulation and ablation studies indicate that fairly complex functional units seem to be affected together (Zulch, Creutzfeldt, & Galbraith, 1975).

Lesions in animals were less successful in reproducing the selectivity of certain cortical columnar organizations. For instance, a large lesion of the striate cortex, which is known from electrophysiological studies to have a preponderance of orientation-selective cells, leads only to a slight increase in the threshold for the discrimination of orientations (Pasik & Pasik, 1980). In other words, behavioral studies do not always replicate physiological ones, indicating that physiological function may differ from reorganized function after a lesion.

D. What Are the Anatomical Structures that Are Necessary for Certain Functions?

The anatomical definition of cortical areas that may have functional significance is difficult and depends on the techniques used. Until recently, cytoarchitectonics was the major methodology. Definition of most of the areas has been based on subtle differences that resulted in little agreement among the investigators. Brodmann (1909) viewed the cytoarchitectonic divisions as *"organs of the brain."* Traditional stains have been supplemented with new techniques to study the distribution of cellular enzymes, such as cytochrome oxidase, and of neural transmitters (Livingstone & Hubel, 1984). Microelectrode mapping allows representations to be revealed with a considerable amount of accuracy, producing a functionally distinct subdivision of cortex of the sensory surface, or motor maps by movements. However, higher order or tertiary association areas may be relatively unresponsive under ordinary stimulation conditions, and cortex with complex organization may be difficult to map. Integration of the new methods suggests rather sharp functional boundaries (Kaas, 1987). The new methods confirm that large complex mammalian brains have more distinct cortical areas than smaller, primitive brains, although a few basic areas of the cortex are present in all mammals. The evolutionary advance in brain organization is accompanied by increases in the number of unimodal sensory fields, not only by increases in the multimodal association cortex as was traditionally thought.

Primary areas were thought to have fairly uniform cytoarchitecture, but recent investigations indicate that even the primary areas such as the striate cortex are not uniform in function (Zeki, 1983). Similarly, the primary auditory area maps show a great deal of parallel organization that would be expected in secondary or tertiary cortices (Seldon, 1985). When the striate cortex (designated V1) is stained for cytochrome oxidase, a metabolic architecture reveals blobs of highly stained cells that respond to color and selectively form-responsive cells located between these blobs (Livingstone & Hubel, 1988). Secondary visual cortex (V2) has a similar metabolic architecture consisting of "thick stripes" sensitive to direction and motion and "thin stripes" sensitive to color. Cells sensitive to form are distributed in the interstripe regions. Separation of various forms of the stimuli may

occur even at the geniculate level, where the magnocellular portion inner-
vates layer 4B of V1; from here signals pass through the stripes of V2 to
V5 to respond to motion. The color system, in turn, receives input from
the parvocellular layers of the lateral geniculate nucleus. Form is processed by two systems, one linked to color and the other to motion. Perception is likely to be based on the coactivation of several cortical fields,
modified by a "top down" feedback mechanism (Zeki and Shipp, 1988).
Some of this cortical selectivity has been confirmed with lesion studies
and positron emission tomography (PET) activation in humans. Viewing
patches of color increases blood flow in the fusiform gyrus (V4) and viewing motion increases blood flow in the lateral convexity of the occipital
lobe (V5) (Fox, Mintun, Raichle, Miezin, Allman, & Van Essen, 1986).

Functional heterogeneity within a cortical field allows parallel processing of information. Cerebral maps probably are not fixed passive representations, since the work of several researchers showed a great deal of
plasticity in the somatosensory cortex (Merzenich, Kaas, Wall, Nelson, Sur,
& Felleman, 1983). Representation in the cortex is probably dynamic,
changing with time and antecedent activity involving attentional mechanisms of the moment. The search for certain cortical modules that would
be unifying in their function, despite the changeability and plasticity of
cerebral maps, provides certain principles of cortical organization that
represent the anatomical and functional basis of a psychological function.
The anatomy of a cortical module as a piece of neural tissue subserving a
function is conceptualized as similar to the printed circuits in a computer
(Szentagothai, 1978). As many different modules as sensory systems may
exist. The reality of such cortical modules has been established to some
extent in the "barrels" of the somesthetic cortex of rats responding to the
stimulation of whiskers (Woolsey & Van Der Loos, 1970). On the other
hand, the "orientation" columns in the visual cortex do not have an obvious anatomical demarcation; they are defined only by electrophysiological
responses to functional stimuli (Hubel & Wiesel, 1968).

Arguments for the so-called "grandmother cell" that would respond to
one type of stimulus exclusively (Barlow, 1972) have been replaced by the
concepts of networks responding to certain stimulations, that are also variable and show a great deal of plasticity. Recent considerations of cortical
function suggest that more information about a stimulus exists in an assembly of cells than in any one of them because of the possible patterns of
covariation among the cells (MacKay, 1978). From the perceptual network,
the associate areas integrate information which then is assigned a psychological percept such as a color, shape, or an individual's face according to
the memories or precepts that are available to modify the perception. The
integration of this perceptual process with "top down" processing seems

to be the function not only of the secondary and tertiary association areas but also of the primary cortex.

Integration of signals from cells from several cortical areas representing different attributes, all of which are responding to the same object, occurs by temporal synchrony. This "binding" is accomplished by re-entry inputs from one area to other areas. The re-entrant system is more diffuse, collecting from all areas of primary perceptual and secondary association areas, in contrast to the forward propagating system described earlier that is highly specialized (Zeki & Shipp, 1988). The integration of external information is achieved by certain networks carrying out several neuropsychological functions that are considered to be distinct, such as perception and comprehension, at the same time. This hypothesis explains how multiple sensory maps that have been demonstrated by physiologists would integrate sensation or how sensation and motor function, or even perception and memory, could be coordinated.

Large areas of the brain that are commonly called tertiary association areas seem less clearly related to any particular function, and show a great deal of evidence of plasticity and interchangeability. Some of these tertiary association areas have, nevertheless, at least a hemispheric functional specialization. The unique role of some left hemisphere tertiary association areas in language has been well established since the time of Broca, although some evidence exists for oral and visual comprehension of concrete nouns and automatic speech output in the right hemisphere, which may not be as speechless as once was thought. Polysensory cortical areas or tertiary association areas are likely to mediate cross-modal matching, although some studies indicate that subcortical structures such as the amygdala have important cross-modal function (Murray & Mishkin, 1985).

The temporal tertiary association areas integrate auditory, visual, and even tactile sensory information. Anatomical evidence indicates that each sensory system projects through a series of cortical fields from primary sensory areas to the insular region of the temporal lobe, which is connected closely with the limbic areas important in the consolidation of memory (Turner, Mishkin, & Knapp, 1980). Areas of the temporal neocortex project topographically to the inferior convexity and orbital prefrontal cortex through the arcuate fasciculus in the macaque and the human (Jones & Powell, 1970; Moran, Mufson, & Mesulam, 1987; Pandya & Kuypers, 1979). The dominant (usually) left temporal lobe is crucial for comprehension of syntax and phonology.

The frontal cortex has been much studied in animals and humans. The extensive connections of this complex association cortex have been reviewed by Goldman-Rakic (1987); the functional studies in humans have been reviewed by Stuss and Benson (1984, see also Ch. 19.) Nauta (1971) suggested that the prefrontal cortex has a unique relationship among the cortical areas of interoceptive and exteroceptive sensory domains, and plays the role of synthesizing inner and outer sensory worlds. The prefrontal cortex appears to be the ultimate target of sensory cascades, as shown by the cortico-cortical connections using the Nauta method (Jones & Powell, 1970; Pandya & Kuypers, 1979). Various synthetic functions such as "spatial memory," "response inhibition," "short-term memory," "polymodal integration," "planning," and "the temporal structuring behavior" were allocated to the prefrontal cortex (Fuster, 1980; Jacobsen, 1936). Goldman-Rakic (1987) argued that the prefrontal cortex is necessary for regulating behavior guided by representations or internalized models of reality.

The left parietal polymodal tertiary association area is considered to have a specific integrative function for visual and auditory language processes (Geschwind, 1965; Wernicke, 1874). The parietal lobes are important polysensory areas bilaterally. The simultaneous injection of distinguishable anterograde tracers in the principal sulcus and the posterior parietal cortex of monkeys revealed that these two areas are mutually interconnected with as many as 15 other cortical areas such as the anterior cingulate, posterior cingulate, and supplementary motor cortex; the ventral and dorsal premotor areas; the orbital prefrontal, prearcuate, frontal opercular, insular, superior temporal, and parahippocampal cortex; the presubiculum and caudomedial lobule; and the medial prestriate cortex (Selemon & Goldman-Rakic, 1988).

Micro-organization of the cortex is constantly in a state of fluidity. Receptors activate a variable amount of cortical space. The extent of each receptor field is influenced by the competition between inputs. Increasing use probably increases cortical space and decreasing use probably decreases the area. Such features would account for learning, improvement in perceptual motor skills with practice, and a considerable amount of compensation that occurs after injury (Kaas, 1987). The topographical order is most clearly evident in the primary and secondary visual, somatosensory, auditory, and motor cortices. Not every structure in the brain has the same degree of specificity; each species is different to the extent that certain functions are related to certain areas of the brain, as well as in the extent to which other areas can compensate for the loss of function. In addition to the interspecies differences are substantial differences that may be common to certain biological groups, such as gender or age groups, but these group differences often are overshadowed by interindividual differences. The precision and stability of the wiring diagram is counterbalanced by the plasticity in the nervous system that is influenced by environmental change, aging, and biochemical environment.

E. What Are the Connections of Functionally Related Structures?

Intercortical and intercerebral connections follow a certain pattern that distinguishes between primary and other cortical areas, according to Flechsig (1901), who specifically established, on the basis of myelogenetic studies, that the secondary and tertiary association areas are interconnected and that the primary ones are connected only with their secondary cortices. Although some evidence exists that modifies Flechsig's rule, this idea remains an important principle of cortical connectivity. Flechsig also established that primary areas mature and myelinate first, and secondary and tertiary areas mature later with the acquisition of experience.

Anatomical connections between cortical areas were confirmed physiologically, first with strychnine stimulation and electrical recording (Dusser de Barenne & McCulloch, 1938). Cerebral connectivity has been traced since by a variety of molecules transported in axons or by action potentials that can be recorded intracellularly or very near the cell surface. Each of these techniques has revealed a topographical map that represents evidence for localized activity in the cerebral cortex (Goldman-Rakic, 1988). Each cortical field has complex connections with neighboring areas and subcortical regions, as well as across lobes and even hemispheres. Even simple stimuli delivered to a receptor surface activate an array of interacting locations in a multitude of cortical areas and subcortical nuclei that are related to a certain modality.

The type of connections exhibits a certain degree of regularity. The *"feed forward"* connections terminate in layers III and IV, which contain the stellate neurons that initiate processing in an area. From there, connections then terminate in the upper and lower layers, and these *"feedback"* connections modulate the outflow of information (Maunsell & Van Essen, 1983). Only a portion of the arborization seems to be activated at any particular time. Connections at all levels appear to be superabundant, contributing to the plasticity of the cortex. The importance of cortical connections and the mechanisms of disconnections in the production of the neurobehavioral syndromes was placed in modern perspective by Geschwind (1965). This kind of connectionism has been advanced further by the concepts of cortical fields and their multiple activation, in which the covariance of these fields, rather than their simple connections, will produce functional and behavioral changes (Phillips et al., 1984).

The term *"connectionism"* has been used in an entirely different sense in brain modeling and artificial intelligence (McClelland and Rumelhart,

1986). Neural networks refer to computer models that consist of processing units that share some of the interconnecting properties of neurons. Each unit receives excitatory and inhibitory inputs from a number of other units and, if the strength of the signal exceeds a given threshold, that unit sends signals to other units and provides feedback. *NET-talk* is a model developed by Sejnowski and Rosenberg (1987) that processes English pronunciation using an algorithm called *back propagation* and a three-layered architecture: an input layer, an output layer in the middle, and a "hidden" layer. An interesting feature is a degree of redundancy or plasticity that allows considerable "lesioning" of the network without impairing its function. Neural network modeling has been criticized for being unrealistic. Nevertheless, computer modeling may be a valuable source of hypotheses about how real neural networks function (Churchland and Sejnowski, 1992).

F. How Are Functions Localized?

The biochemical basis of the individuality of neurons has been expanded greatly. Approximately 50 neurotransmitters have been identified to date that have functional differences. An even more recent discovery is the classification of neuronal receptors and the receptor regulating systems, such as G proteins or excitatory amino acids (Goodman, 1990). The individuality of neuronal function is diversified by selective gene expression within seemingly homogeneous populations of nerve cells. Although every cell contains the same set of genes, individual cells express or activate only a small subset, resulting in an astounding variation among individual cells. Immediate early genes (IEGs) that are activated rapidly by bursts of action potentials may provide a link to persistent changes in gene expression, which may underlie memory. Impulse activity increases the expression of genes that encode trophic factors or proteins that promote the survival of neurons. Hebb (1949) suggested the formation of neuronal ensembles altered by pre- and postsynaptic activity as a basis for learning or localized function (Hebb's law). Long-term potentiation (LTP) is an alteration in the responsivity of postsynaptic cells that was described initially in the hippocampus (Bliss & Lomo, 1973). The neurotransmitter study of the persisting increase of synaptic efficacy that would follow functional activity has been explored extensively. The most common excitatory neurotransmitter related to LTP is glutamate, according to current neurochemical research (Madison, Malenka, & Nicoll, 1991). Postsynaptic depression also has been described (Desmond & Levy, 1983). This area of study has become one of the ways in which functional localization can be connected to structures of the nervous system.

G. How Is Localization of Function Related to Altered Function after Lesions?

One of the major problems in localization is whether or not the performance observed should be attributed to a damaged area that functions without some of its lost components or to other functionally related or even unrelated structures that compensate for the damaged one. One example is the difficulty in constructional tasks observed with right hemisphere lesions. One cannot be certain whether this difficulty is related to poor performance of the damaged right hemisphere or to performance of the left hemisphere, which performs without the usual right-sided input for the task. At times after a lesion, not only a deficit or negative symptoms but also new behavior or positive symptoms are observed. Positive symptoms may represent elements of neural activity that had been controlled or suppressed by the structure destroyed by the lesion. Sometimes a second lesion even promotes recovery by this mechanism (Irle, 1987).

The sudden removal of large areas of functioning brain, in stroke or in experimental lesions, may produce distant effects in functionally connected neural structures. This phenomenon also has been known as *diaschisis* since the time of von Monakow (1914). Slowly growing tumors, on the other hand, often displace tissues with relatively little functional deficit. Rapidly expanding tumors produce distant effects by edema, hemorrhage, and vascular occlusion. Therefore, in many instances, lesions of the same size or location will not produce the same deficit unless the study is controlled strictly for etiology. This principle often has been violated in studies of functional localization in neuropsychology. Occasionally, different etiologies will produce the same deficit but with different localization. One example is the appearance of transcortical sensory aphasia and Wernicke's aphasia in cases of Alzheimer's disease with diffuse neuronal degeneration (at least not localizable with current methods to the same extent as a focal infarct).

Other biological factors may cause similar lesions to produce different deficits. A major difference is the age of the organism. When a child sustains brain damage, the plasticity of the brain allows for compensation by the homologous hemisphere or other structures as long as the lesion is not a progressive one. Hemispherectomy in a young child (Basser, 1962) or damage to the prepubertal brain results in almost complete recovery, indicating the possibility of hormonal influence on plasticity. Certain aphasic syndromes, such as fluent or Wernicke's aphasia, appear to be more common in older individuals. Whether this is phenomenon related to anatomical differences in the vasculature affected or to continuing functional lateralization remains controversial (Brown & Jaffe, 1975; Kertesz & Sheppard, 1981).

Sex differences in cerebral organization are considered to be significant by some investigators. Language is assumed to be more bilaterally distributed in women; therefore, lesions may have a different effect in different sexes (McGlone, 1977). However, epidemiological studies of aphasia tend not to support this contention (Kertesz & Sheppard, 1981). Although important psychological sex differences may exist, evidence from clinical studies is not supportive of substantial sex differences in intra- or interhemispheric brain organization.

Handedness is considered an important factor in cerebral dominance for certain functions. Left-handers may have somewhat different cerebral organization, not just mirroring right-handers but including important qualitative differences that may result in certain lesions producing different deficits in left- than in right-handers. Gloning, Gloning, Haub, and Quatember (1969) suspected that left-handers are likely to become aphasic regardless of which hemisphere is damaged, but we found fewer left-handers among our aphasics than expected from the general population, even in the acute stage (Kertesz, 1979). In the chronic stage, one would expect to see fewer left-handed aphasics because of the suggested better recovery from the deficit (Gloning et al., 1969).

Individual differences in brain organization may be associated with structural differences as well. Anatomical differences between individuals are often observable, but are difficult to quantitate. Some differences follow a pattern, such as the 65:15 ratio of larger planum temporale on the left vs the right (Geschwind & Levitsky, 1968). Perhaps this accounts for the variable extent superior temporal lobe lesions affect behavior. Hemispheric anatomical asymmetries are just a beginning in our consideration of the anatomical variables in localization. The complex gyral pattern of the human brain also shows a great deal of individual variation. The importance of this variation becomes evident when a neurosurgeon is called on to stimulate or excise a portion of the cortex.

Deficits caused by a lesion are often unstable. The early deficit that may be related to edema, cellular reaction, transient ischemia, and so on is followed by a great deal of spontaneous recovery in trauma or stroke. The chronic deficit is related not only to a loss of function but to compensatory changes by the whole brain or homologous areas or neighboring areas during subsequent stages of recovery. The deficit with acute lesion cannot be considered in the same category as the recovered state, although the lesion persists. These two instances lead to conflicting conclusions about localization. Therefore, in every attempt to correlate deficit with lesion, the time from onset is crucial. Failure to consider this variable is a major source of confusion in this field of study.

The criticism that a performance observed after a lesion has little to do with the function of the lesioned area applies to other methods of localization, such as electrical stimulation of the brain, the local application of neurotransmitters, or recording from single units since these units are connected to the whole system and the response is the function of other structures as well (Glassman, 1978). To a certain extent, this consideration remains important in the interpretation of any method, but does not justify the elimination of any approach in the search for causality, an important principle in science.

Behavior in a complex nervous system may be disturbed because the whole system or only a component of it is affected with similar results, provided that the impaired component is crucial enough in the performance of a function. However, the same phenomenon, therefore, can be observed from lesions at different locations. The possibility of the organism using alternative structures for a function, or the "redundancy" of organization, also would account for variously located lesions producing the same deficit. That the same function is represented at various levels of the nervous system is the well-known principle of "hierarchial" representation, and probably accounts for some of the recovery or reorganization, rather than the "take-over" by completely unrelated structures or "vicarious functioning."

The general principles developed by many researchers in the anatomical correlations of complex network functions were summarized by Mesulam (1981):

> (1) Components of a single complex function are represented within distinct but interconnected sites which collectively constitute an integrated network for that function; (2) individual cortical areas contain the neural substrate for components of several complex functions and may therefore belong to several partially overlapping networks; (3) lesions confined to a single cortical region are likely to result in multiple deficits; (4) severe and lasting impairments of an individual complex function usually require the simultaneous involvement of several components in the relevant network; and (5) the same complex function may be impaired as a consequence of a lesion in one of several cortical areas, each of which is a component of an integrated network of that function.

These principles are general enough to explain many of the findings in lesion localization studies, and are also applicable to stimulation studies.

III. METHODOLOGY OF FUNCTION–LESION CORRELATION

A. Clinicoanatomical Correlation

This well-established method is further advanced by the development of new micromorphological techniques. The study of human cytoarchitectonics and myeloarchitectonics was overtaken by experimental neuroanatomy in animals, bolstered by autoradiography and axonal tracers such as horseradish peroxidase (HRP). Advances in pigment cytoarchitectonics, and the application of silver stains to anterograde and retrograde degeneration of tracts in human lesions, opened new opportunities to study human cerebral connectivity and cortical organization in postmortem anatomical material (Galaburda & Mesulam, 1983; Sanides, 1962). The accuracy of postmortem localization makes the method the standard to which others are compared. Autopsy examination of the human brain allows topographical localization of gyral and sulcal structures and fiber tracts, and the accurate determination of the extent of lesions. In addition to gross topography, the myelin and cellular stains allow microscopic analysis of neuronal structures and lesions.

The major disadvantage of the autopsy method is that the brain usually becomes available only long after the patient has been examined clinically. Only occasionally is it possible to perform the postmortem examination of the brain shortly after a detailed clinical examination or some experimental analysis (Landis, Regard, Bliestle, & Kleihues, 1988). Patients who die after a cerebral event usually are affected so severely that they could not be examined in the acute premortem illness in any detail. If a patient dies after a progressively declining course, as in the case of a brain tumor, by the time of autopsy the brain is distorted severely and will not reflect the original state that was studied previously. Although many circumscribed lesions have been autopsied, the corresponding clinical examinations often have been scanty and incomplete. Unfortunately, the reverse is also true. When the clinical examination has been conducted in great length, the patient often recovers; when death eventually occurs for some other reason, permission to carry out a postmortem examination is not granted, the patient dies in another location, or, even if the brain is available, aging and new lesions are added and the opportunity to establish correlation is lost.

Naturally occurring lesions may or may not respect anatomical boundaries. Interpretation in terms of the structures involved is a difficult methodology and requires considerable experience. A lesion may destroy an actual area that subserves a function, it may disconnect two areas that are part of the network, or it may change behavior by damaging longitudinal pathways that transverse the area affected by pathology (the transit effect). *Diaschisis* refers to the remote suppressive effect on functionally connected areas that otherwise remain intact anatomically. This consideration, of course, applies to lesion localization by other modalities as well, and increases the complexity of interpretation enormously.

B. Neurosurgical Resection

Neurosurgical resection of tumors and surgery for epilepsy, and at times operation on penetrating head injuries, provides a traditional measure of localization but has many problems. Tumor resection is not the most suitable material, because slowly grown tumors compress the brain in an insidious fashion and often there is a great deal of compensation for the slowly incurred deficit. In fact, this may result in false negative conclusions concerning the role of certain areas in the brain. Malignant tumors infiltrate the brain tissue beyond resection; determining how far they actually exert their influence is often difficult. The distant effects of edema, vascular insufficiency, and displacement of tissues are difficult to quantitate on surgical exploration. Epilepsy surgery often is performed in scarred brains, and the removed tissue may not be functional at all. The extent of the excision is difficult to judge because of the limitation of the exposure and the variability and distortion of the surface anatomy.

Hemispherectomies have been used to eradicate malignant tumors in adults (Smith, 1966) and to excise severely scarred epileptogenic tissue in infants and children. Some of these cases have been studied extensively. Left hemispherectomies in adults produce global aphasia, but the individuals retain emotional and automatic utterances, articulated even at a sentence level, and some comprehension of single nouns. The limitations on studying adult hemispherectomies are the rarity of the operation (most surgeons no longer consider this treatment for tumors) and the usually relentless spread of the neoplasm into the other hemisphere. Infantile hemispherectomies have been well studied, indicating almost complete recovery of function on either side. However, sophisticated language measures have indicated less than normal development in verbal IQ and in grammatical competence in children who had a left hemispherectomy in their infancy (Dennis & Whitaker, 1976).

C. Callosotomy

Performed for intractable epilepsy, callosotomy provides an opportunity to study the function of each hemisphere independently (Sperry, Gazzaniga, & Bogen, 1969). In this respect, the method is not a surgical resection but a disconnection of two functional areas. The results have illuminated many right hemisphere functions, including a certain capacity of comprehending language (a more detailed review is presented by Gazzaniga, 1970). Some of the operations have been less than complete, as recent magnetic resonance imaging (MRI) studies have indicated, which may have led to some false conclusions concerning function. On the basis of one left hemispherectomy and two commissurotomies, Zaidel (1976) estimated the vocabulary of the right hemisphere to be at or above the level of a 10-year-old child. Gazzaniga (1983), in subsequent studies, found that the normal right hemisphere has less language than originally was thought. The performance of these patients after commissurotomy could be attributable to early left hemisphere damage resulting in the reorganization of language functions or their transfer to the right hemisphere.

More recent experiments with split-brain patients indicate that some degree of intrahemispheric transfer takes place outside the corpus callosum. Even in split-brain patients, the left hemisphere dominance exerts itself and interprets nonverbal responses from the right hemisphere into whatever verbal framework the left hemisphere experiences have constructed. Gazzaniga and LeDoux (1978) postulated that a number of mental systems—emotional, motivational, and perceptual—are monitored by the verbal system. This "verbal self looks out and sees what the person is doing and from that knowledge it interprets reality." This concept of verbal self control is somewhat similar to Freud's theory of the ego as it is applied to language dominance. Thus, the left hemisphere not only interprets the word in terms of language but also integrates mental functions in a purposeful consciousness, which is the essence of our existence.

D. Cortical Stimulation

Cortical stimulation and cortical recording during epilepsy surgery make a major contribution to cerebral localization. Initially, these methods were used by physiologists to map the function of the cerebral cortex in animals; subsequently, neurosurgeons such as Otto Foerster, in the 1920s, tried to identify cortical function with stimulation in preparation for their excisions. Sensory, motor, and language areas could be spared using this technique. Penfield and Roberts (1959) mapped the language areas, and Van Buren, Fedio, and Frederick (1978), Ojemann and Whitaker (1978), and Ojemann and Mateer (1979) added additional information concerning cortical localization. These studies showed that, indeed, Broca's area produces the most consistent naming difficulty. These investigators also found a rather selective pattern in which several millimeters made a great deal of difference in determining the functional alterations that were obtained on stimulation. For instance, one site would interfere with one language but not the other, and a very small change in the location of the electrodes would reverse the situation. Language appeared to be organized concentrically around the Sylvian fissure rather than following the anteroposterior dichotomy of the lesion studies. Right hemisphere stimulation disrupted face recognition, labeling of emotional expression, and line orientation (Mateer, 1983).

Recording from neurons in the superior temporal gyrus showed activation on listening to verbal material, particularly to phonological distinctions. Differences in meaning or inflection were associated with changes in the pattern of neuronal discharge from the middle and inferior temporal gyrus (Creutzfeldt, Ojemann, & Lettich, 1989). Medial temporal lobe recording produced stimulus-specific firing on sight of a particular word or face during epilepsy surgery. This region seemed to encode distinct stimuli in a given context, with a remarkable degree of specificity (Heit, Smith, & Halgren, 1988). Implanted subdural electrodes also have been used to record and to stimulate activity in ambulant patients (see also Chapter 3 in this volume). In addition to the perisylvian language cortex, stimulation of the inferior basal region produced speech arrest more on the left than on the right (Lüders, Lesser, Dinner, Morris, Wyllie, & Godoy, 1988).

Transcranial magnetic stimulation of human brain has been reported to eliciting motor function (Barker, Jalinous, & Freeston, 1985). The threshold for excitation of motor cortex is reduced markedly by slight voluntary contraction of the target muscle. The mechanism of facilitation is neural activity at the spinal and cortical level. Dendrites and presynaptic terminals, or even cell bodies and efferent axons, are assumed to be stimulated by magnetic stimuli increasing in strength (Mills, Murray, & Hess, 1987). Magnetic stimulation of language cortex also has been tried, but to date only negative phenomena such as anomia have been obtained.

The advantage of the method is that it can be applied in the awake cooperative patient who can perform selected functions during stimulation. In this respect, the procedure resembles other so-called functional methods of localization, such as CBF and PET. However, stimulation interferes with functions, interrupts or alters them, and, less frequently, elicits positive phenomena. In this respect, the procedure is more like the lesion method, but stimulation involves a much smaller area, allowing greater resolution for mapping. Several functions can be examined by repeated stimulation of the same area. Limitations of the technique include the logistics of a lengthy surgical procedure, difficulty reproducing stimulation on the same spot, and the brevity of tasks that can be employed during the short period of stimulation (maximum about 12 sec).

The subjects usually have epilepsy or tumors with reorganized brains, and the conclusions may not always be generalizable. Recent advances and a summary of the method are found in Chapter 2.

E. Electroencephalography

Electroencephalography is a physiological method of localization that utilizes the small electrical potentials generated by neuronal activity that can be detected by surface electrodes on the scalp or on the brain surface during operations. Electroencephalograms (EEGs) have been used for more than 60 years. However, the method suffers from relatively low specificity, since the electrical potential changes often present a summation of remote effects. The localization of cognitive function in aphasia with EEG has been reported by Marinesco, Sager, and Kreindler (1936) and Tikofsky, Kooi, and Thomas (1960). Galin and Ornstein (1972) observed suppression of the alpha rhythm in the dominant hemisphere during verbal tasks, and in the nondominant hemisphere during spatial tasks. An attempt to assess intercortical connections was made by measuring the degree of correlation between pairs of electrodes, called *cortical coupling* (Callaway & Harris, 1974). Sophisticated computer-based statistical methodology has been used to display functional alterations and their location.

F. Event-Related Potentials

Determination of event-related potentials (ERP) became possible through averaging computers. Differences in ERP lateralization were observed using verbal and nonverbal stimuli (Buchsbaum & Fedio, 1970), words and nonsense syllables (Shelburne, 1972), and contextual meaning (Brown, Marsh, & Smith, 1973; Teyler et al., 1973). Long latency ERPs are sensitive to task relevance and expectancy. The 300-msec positive component is called *P300*, and is considered endogenous and less dependent on stimulus characteristics. A potential that precedes movement or cognition, called the *contingent negative variation,* has been correlated with hemispheric dominance and has been useful in various cognitive studies (Low, Wada, & Fox, 1973). Another example of negative potentials associated with processing is the *N400 potential,* which is related to the occurrence of context incongruity in serial semantic tasks (Kutas & Hillyard, 1980; Neville, Kutas, & Schmidt, 1982).

Cerebral ERPs are technically difficult to acquire because of the many artifacts that interfere with the study. Some of these technical difficulties have been discussed by Desmedt (1977). Another problem with the method is the time restriction on the stimuli. The stimulation must be very short and must have a definite onset, duration, and offset to be connected with a cerebral event. In a way, this method is the opposite of CBF studies (see subsequent discussion), which require more sustained cerebral activity to be analyzed. The major advantage of ERP measurement is that it reflects a physiological event connected with actual cerebral processing, although the resolution of localization is poor and the potentials are usually over a large area of the scalp and the brain. Advances are updated and summarized in Chapter 4.

G. Computerized Tomography

Computerized tomography (CT) represents a significant breakthrough in the radiology of the central nervous system. The latest generation scanners even show some grey and white matter differentiation and facilitate seeing changes in density, such as brain edema. Contrast enhancement with radio-opaque material, such as organic iodine, increases the visualization of vascular structures and the increased vascularity around an acute infarct. However, CT does not show lesions as early as MRI can, and the early changes tend to be less distinct than the ones obtained several weeks poststroke. One of the reasons for negative localization with a clear-cut clinical syndrome could be related to the timing of the CT scan. Hemorrhages are quite dramatic; brain tumors also are shown superiorly on CT scans. Old infarcts are quite distinct, with sharp margins and lower densities than the surrounding brain. Enlargement of the ventricles and sulci also provides an accurate measurement of atrophy.

The correlation of lesions with behavioral studies began as soon as CT equipment became available. These studies differ in the quality of the scans, the sophistication of localization, the number of patients included, the actual neuropsychological measurements used, and the definition of the syndromes. One of the most important considerations, which is missing from some of the studies, is the time from onset. Some patients with large lesions may have recovered considerably, which may lead to the conclusion that the area involved plays no role in a function. Mixing acute and chronic patients is especially misleading. CT studies nevertheless have contributed a great deal to our knowledge about localization of lesions in syndromes of language and nonverbal cognitive impairment.

The major advantage of a CT scan is its ability to localize lesion in vivo with reasonably high anatomical resolution, especially in the late model scanners which show grey and white matter distinctly. CT scanning is especially useful in chronic lesions in which the edges of infarcts are well outlined and in horizontal cuts, which usually are oriented 15° above the orbitomeatal line. CT scans are now available in most major centers, and the method has become standard for neurological investigation. The scanning time is brief, with no associated discomfort. A clinical indication is needed for scanning, because the necessary repeated frequent exposures to X rays are considered harmful. Cortical landmarks may not be seen, and one must use ventricular and bony landmarks for anatomical orientation. Variations in head position, head size, ventricular size, extent of atrophy, and cerebral asymmetries are important to consider in accurate localization. The variation in the sulcal and gyral pattern, and the depth of the cortex in the central portions of the brain, is often underestimated in CT studies.

H. Magnetic Resonance Imaging

MRI is the latest imaging modality, and it provides the most accurate localization of lesions without invasive radiation. The technique uses the inherent magnetic properties of spinning atomic nuclei by placing the structure to be imaged in a large magnet and applying short-wave radiofrequency pulses to produce a resonance signal that can be quantified and computerized. Superior anatomical detail can be achieved, with excellent grey and white matter differentiation and an accurate outline of the edge of the brain from cerebrospinal fluid (CSF) spaces. The brain can be imaged in coronal and sagittal sections, in addition to the horizontal one that is the usual plane obtained in other modalities. This imaging flexibility, combined with anatomical accuracy, already has established MRI as a useful clinical and research tool. The apparent lack of biohazard allows the study of normal individuals without clinical indications, as well as a more frequent imaging of patients than is possible with other modalities that use ionizing radiation (Sweetland, Kertesz, Prato, & Nantau, 1987).

Various pulse sequences can be used to probe the metabolic and molecular environment of various regions of the brain, making this technique much more dynamic than the CT scan. Currently, two major pulse sequences are used. One, called *inversion recovery* (IR), emphasizes grey and white matter differences and provides excellent anatomical detail. The second, called *spin echo* (SE), is utilized to detect edema and other metabolic changes associated with lesions. SE sequences are often superior to CT scanning in the early detection of cerebral infarct, thus reducing the false negative rate obtained on CT in the first 2–3 days of a stroke. The technique is particularly suited to detecting demyelinating plaques and some other degenerative diseases as well as early strokes.

Functional imaging with MRI is becoming a promising investigative tool. First, contrast material (gadolinium) injection provided some measure of cerebral blood flow, but recent diffusion and perfusion imaging is capable of detecting changes in cerebral blood flow that are associated with normal function. MRI has the potential to provide functional information because several parameters—such as changes in proton density, longitudinal and horizontal relaxation times, chemical shifts, magnetic susceptibility, and effect of flow on signal—are influenced by function. Positive enhancement with gadolinium, similar to contrast enhancement in CT, reflects a change in the function of the blood–brain barrier. The signal loss that accompanies the transit of the contrast agent through a region of interest is proportional to the tissue blood volume. Maps of circulation obtained this way are similar to those from PET studies of blood volume obtained with ^{18}F fluoro-deoxyglucose, and have been used to detect changes in cerebral blood volume induced during task activation paradigms such as visual stimulation.

Even more recently, an approach without contrast agents demonstrated real-time changes in magnetic resonance (MR) signal in response to functional activation. This method uses the decrease in arterio–venous oxygenation difference that accompanies the regional increase in blood flow. A reduction of deoxyhemoglobin concentration produces regional signal intensity enhancement in the area of activation, so oxyhemoglobin acts as the body's own contrast agent that measures regional tissue oxygen consumption. An example of a new type of pulse sequence used for detecting cerebral blood flow is the dynamic FLASH (fast low angle shot) technique. This approach exploits the sensitivity of the gradient echo sequence to the varying concentrations of paramagnetic deoxyhemoglobin in the region of activation. The activation occurs 6–9 sec after the psychophysical stimulus is applied. After the stimulation is switched off, the MR signal returns to basal value in a similar period of time. Further, a decrease in the basal MR signal was noted in the activated area after 60–90 sec of persistent activation, which was considered to be related to an autoregulatory adaptation of increased overall brain activity associated with information processing (Belliveau et al. 1991).

Preliminary MRI studies of brain activation included the visual systems, sensorimotor, and language processing areas. Some of the functional activation maps of the visual cortex, for example, have been obtained at spatial resolution almost two orders of magnitude better than that of PET studies. These new MRI approaches allow a noninvasive correlation of brain anatomy with function at a high level of spatial resolution and within a brief temporal interval. From the small amount of data available, a remarkable amount of intersubject reproducibility can be seen. Interindividual differences are, however, also detected, reflecting the differences in the anatomy of the cortex investigated. These techniques eliminate the averaging of results and the use of geometrically standardized brain as employed with PET data (see Chapter 7).

Magnetic resonance spectroscopy (MRS) is sensitive to water nuclei (protons), lactate, phosphate compounds, and some amino acids; this sensitivity can be used for localization of function. Although the chemical specificity is high, the sensitivity and resolution is low, and the relationship to function is at a biochemical rather than an anatomical level.

I. Cerebral Blood Flow

The CBF technique utilizes the physiological and pathological alterations in regional blood supply. The technique is based on estimating the clearance of radioactive isotopes from various regions of the brain using surface detectors. The values are expressed as percentages of the hemispheric average. Increases of blood flow are assumed to be associated with increased neuronal and, therefore, functional activity. A sustained repetitive task at least 3–5 min in duration, is required for activation. The color-coded images of CBF have become popular in illustrating that the brain is activated in multiple locations when, for instance, a person reads or that the right hemisphere also "lights up" when a person speaks. The resting pattern shows precentral high and postcentral low flows (Ingvar & Schwartz, 1974). Simple repetitive movements of the mouth, hand, or foot augment CBF in the contralateral sensorimotor area, supporting the topography of the cortical "homunculus."

The important advantage of the technique is that it reflects physiological metabolic change that accompanies psychological function and can be used to study normal processes. Note, however, that these changes are not measuring neuronal events directly. These techniques are best suited to the study of sustained repetitive acts of cognition. Lesions such as infarcts that produce a deficit may appear as areas of low flow with an area of "luxury perfusion" or high flow surrounding them. The major disadvantage is a relatively poor resolution of the noninvasive xenon inhalation method: it only measures blood flow on the surface. The more accurate intra-arterial injection is used rarely because it is invasive and requires the indications of an angiogram. A more recent modification of the technique combines ^{133}Xe inhalation with computerized tomography, achieving three-dimensional representations in slices similar to those in PET scanning.

A less expensive and more available measure of functional activation is single photon emission computerized tomography (SPECT), but the resolution is much lower than that of PET. The correlation to function is not as direct and the technique is more widely used to measure steady state rather than activation of function. SPECT tracers such as radioactive iso-topes of iodine and technetium are combined with lipophilic amines and cross the blood–brain barrier easily. Some of them leave the brain slowly; their distribution represents a record of regional cerebral blood flow at the time of injection. Activation must precede and coincide with the injection. Imaging can be carried out several hours later without loss of information. This flexibility is a useful feature in a clinical setting. SPECT is used most often to detect chronic alteration of cerebral blood flow associated with lesions or degenerative disease (Alavi & Hirsch, 1991).

J. Positron Emission Tomography

PET measures oxygen and glucose metabolism using a positron emit-ting isotope and a computerized tomographic scanner. This method also can be used to study regional blood flow. This complex and expensive technique is available only in a few centers equipped with a particle accel-erator (cyclotron) and a team of nuclear physicists, radiopharmacists, com-puter experts, isotope specialists, clinicians, and experimental psycholo-gists. The positron-labeled metabolites, such as [^{18}F]fluoro-deoxyglucose, must be given immediately at their source because of their short half-life. The nature of tracer kinetics requires that the physiological activity studied must be sustained for 20–40 min. Metabolic scanning with PET has confirmed the differences in hemispheric activation between verbal and visuospatial task performance (Mazziotta & Phelps, 1984). A bolus administration of ^{15}O has shortened the period of observation and allowed repeated measurements in the same subjects in one session (Raichle et al., 1984). A new device, the "super-PET—ru1," is capable of a temporal res-olution of less than 1 min (Ter-Pogossian, Ficke, Mintun, Herscovitch, Fox, & Raichle, 1984).

PET studies have included measurements of cerebral blood flow, oxy-gen utilization, glucose metabolism, dopamine, opiate, serotonin, acetyl-choline, glutamate, and gamma aminobutyric acid as well as the imaging of various other physiological processes. Although the resolution power of the PET scanners has been 1–2 cm, devices have been built with resolu-tion of 2.5 mm. Some of these studies suggest that even subtle task differ-ences can result in the recruitment of different cortical areas and that the connection between the areas is a dynamic balance of stimulation and inhibition, the interpretation of which may be quite difficult. Some of the activation studies have shown somewhat unexpected localization, such as semantic processing in the medial frontal regions on a word association task. Anatomical information from MRI can be combined with functional information from the PET scan. The integration of standard regions of interests on the PET scan with MRI anatomical templates provides a visual image of the statistically significant changes that occur on activation (Sergent, Zuck, Terriah, & MacDonald, 1992).

The complexity of some of the functions measured may result in conflicting results (Petersen et al., 1988; Wise, Chollet, Hadar, Friston, Hoffner, & Frackowiak, 1991). The activation paradigm appears to be critical for the results. The processing components, the stimuli, and the responses must be defined exactly; otherwise comparisons will be nonspecific and the activation will involve different processes. Foci of activation may represent only parts of large-scale networks that are converging in one particular area. The interpretation of the activated area is difficult because cerebral blood flow may not represent stimulation or inhibition equally. Changes in neural activity may happen long before any change in the blood flow occurs; therefore, most of the PET paradigms require sustained activation. The spatial analysis of these results still is limited, and much information depends on the statistical analysis that is used to display the data. Some of the PET activation studies access cognitive processes, such as "inner speech," that may not be available for analysis of explicit responses.

REFERENCES

Alavi, A., & Hirsch, L. J. (1991). Studies of central nervous system disorders with single photon emission computed tomography and positron emission tomography: Evolution of the past two decades. *Seminars in Nuclear Medicine, 21*, 58–81.

Barker, A. T., Jalinous, R., & Freeston, I. L. (1985). Non-invasive magnetic stimulation of the human motor cortex. *Lancet, 2*, 1106–1107.

Barlow, H. B. (1972). Single units and sensation: A neuron doctrine for perceptual psychology? *Perception, 1*, 371–394.

Barlow, H. B. (1981). The Ferrier lecture: Critical limiting factors in the design of the eye and visual cortex. *Proceedings of the Royal Society, London, 212*, 1–34.

Basser, L. S. (1962). Hemiplegia of early onset and the faculty of speech with special reference to the effects of hemispherectomy. *Brain, 85*, 427–460.

Belliveau, J. W., Kennedy, D. N., McKinstry, R. C., Buchbinder, B. R., Weisskoff, R. M., Cohen, M. S., Vevea, J. M., Brady, T. J., & Rosen, B. R. (1991). Functional mapping of the human visual cortex by magnetic resonance imaging. *Science, 254*, 716–719.

Benton, A. L. (1961). The fiction of the "Gerstmann syndrome." *Journal of Neurology, Neurosurgery and Psychiatry, 24*, 176–181.

Bliss, T. V., & Lomo, T. (1973). Long-lasting potentiation of synaptic transmission in the dentate area of the anaesthetized rabbit following stimulation of the perforant path. *Journal of Physiology (London), 232*, 331–356.

Broca, P. (1861). Remarques sur le siege de la faculte du langage articule suivies d'une observation d'amphemie (perte de la parole). *Bulletin et Memoires de la Societe Anatomique de Paris, 36*, 330–357.

Brodmann, K. (1909). *Vergleichende Lokalisationslehre der Grosshirnrinde*. Leipzig: Barth.

Brown, J., & Jaffe, J. (1975). Hypothesis on cerebral dominance. *Neuropsychologia, 13*, 107–110.

Brown, W. S., Marsh, J. T., & Smith, J. C. (1973). Contextual meaning effects on speech-evoked potentials. *Behavioral Biology, 9*, 755–761.

Buchsbaum, M., & Fedio, P. (1970). Hemispheric differences in evoked potentials to verbal and nonverbal stimuli in the left and right visual fields. *Physiology and Behavior, 5*, 207–210.

Bullock, T. H. (1965). Physiological bases of behavior. In J. A. Moore (Ed.), *Ideas in modern biology* (pp. 32–56). New York: Natural History Press.

Cajal, S. R. (1890). Reprinted in J. DeFelipe & E. G. Jones (Eds.), *Cajal on the cerebral cortex*. Oxford: Oxford University Press, 1988.

Callaway, E., & Harris, P. R. (1974). Coupling between cortical potentials from different areas. *Science, 183*, 873–875.

Caramazza, A. (1986). On drawing inferences about the structure of normal cognitive systems from the analysis of patterns of impaired performance: The case of single-patient studies. *Brain and Cognition, 5*, 41–66.

Churchland, P. S., & Sejnowski, T. J. (1992). *The computational brain*. Cambridge, Massachusetts: MIT Press.

Creutzfeldt, O., Ojemann, G., & Lettich, E. (1989). Neuronal activity in the human lateral temporal lobe. I. Responses to speech. *Experimental Brain Research, 77*, 451–475.

Crick, F. (1989). The recent excitement about neural networks. *Nature (London), 337*, 129–132.

Dejerine, J. (1892). Des différentes variétés de céité verbale. *Memoires de la Societé Biologique*, 1–30.

Dennis, M., & Whittaker, M. A. (1976). Language acquisition following hemidecortication: Linguistic superiority of the left over the right hemisphere. *Brain and Language, 2*, 472–482.

Desmedt, J. E. (1977). Some observations on the methodology of cerebral evoked potentials in man. In J. E. Desmedt (Ed.), *Attention, voluntary contraction and event-related cerebral potentials* (Vol. 1, pp. 12–29). Basel: Karger.

Desmond, N. L., & Levy, W. B. (1983). Synaptic correlates of associative potentiation/depression: an ultrastructural study in the hippocampus. *Brain and Research, 265*, 21–30.

Dreyfus, H. L. (1979). *What computers can't do: The limits of artificial intelligence* (2d Ed.). New York: Harper & Row.

Dusser de Barenne, J. G., & McCulloch, W. S. (1938). The direct functional interrelation of sensory cortex and optic thalamus. *Journal of Neurophysiology, 1*, 176–186.

Flechsig, P. (1901). Developmental (myelogenetic) localisation of the cerebral cortex in the human subject. *Lancet, ii*, 1027–1029.

Fodor, J. A. (1983). *The modularity brain*. Cambridge, Massachusetts: MIT Press.

Fox, P. T., Mintun, M. A., Raichle, M. E., Miezin, F. M., Allman, J. M., & Van Essen, D. C. (1986). Mapping human visual cortex with positron emission tomography. *Nature, 323*, 806–809.

Fritsch, G. T., & Hitzig, E. (1870). Über die elektrische Erregbarkeit des Grosshirns. *Archiv fuer die Anatomie und Physiologie, Leipzig*, 300–322.

Fuster, J. M. (1980). *The prefrontal cortex*. New York: Raven Press.

Galaburda, A. M., & Mesulam, M. M. (1983). Neuroanatomical aspects of cerebral localization. In A. Kertesz (Ed.), *Localization in neuropsychology* (pp. 21–61). New York: Academic Press.

Galin, D., & Ornstein, R. (1972). Lateral specialization of cognitive mode: An EEG study. *Psychophysiology, 9*, 412–418.

Gazzaniga, M. S. (1970). *The bisected brain*. New York: Century & Crofts.

Gazzaniga, M. S. (1983). Right hemisphere language following brain bisection: A 20-year perspective. *American Psychologist, 38*, 342–346.

Gazzaniga, M. S., & LeDoux, J. E. (1978). *The integrated mind.* New York: Plenum Press.

Gerstmann, J. (1940). Syndrome of finger agnosia, disorientation for right and left, agraphia and acalculia. *Archives of Neurology and Psychiatry, 444*, 398–408.

Geschwind, N. (1965). Disconnexion syndromes in animals and man. *Brain, 88*, 237–294, 585–644.

Geschwind, N., & Levitsky, W. (1968). Human brain: Left-right asymmetries in temporal speech regions. *Science, 161*, 186–187.

Glassman, R. B. (1978). The logic of the lesion experiment and its role in the neural sciences. In S. Finger (Ed.), *Recovery from brain damage* (pp. 3–31). New York: Plenum Press.

Gloning, I., Gloning, K., Haub, G., & Quatember, R. (1969). Comparison of verbal behaviour in right-handed and non-right-handed patients with anatomically verified lesions of one hemisphere. *Cortex, 5*, 43–52.

Goldman-Rakic, P. S. (1987). Circuitry of the prefrontal cortex and the regulation of behavior by representational memory. *Handbook of Physiology, 5*, 373–417.

Goldman-Rakic, P. S. (1988). Topography of cognition: Parallel distributed networks in primate association cortex. *Annual Review of Neuroscience, 11*, 137–156.

Goldstein, K. (1948). *Language and language disturbances.* New York: Grune & Stratton.

Goltz, F. (1881). In W. MacCormac (Ed.), *Transactions of the 7th International Medical Congress,* (Vol. 1, pp. 218–228). London: Kolkmann.

Goodman, R. H. (1990). Regulation of neuropeptide gene expression. *Annual Review of Neuroscience, 13*, 111–127.

Head, H. (1926). *Aphasia and kindred disorders of speech.* Cambridge: Cambridge University Press.

Hebb, D. O. (1949). *Organization of behavior.* New York: Wiley.

Heit, G., Smith, M. E., & Halgren, E. (1988). Neural encoding of individual words and faces by the human hippocampus and amygdala. *Nature (London), 333*, 773–775.

Hinton, G. E., & Anderson, J. A. (1981). *Parallel models of associative memory.* Hillsdale, New Jersey: Erlbaum Associates.

Hubel, D., & Wiesel, T. (1968). Receptive fields and functional architecture of monkey striate cortex. *Journal of Physiology, 195*, 215–243.

Ingvar, D. H., & Schwartz, M. S. (1974). Blood flow patterns induced in the dominant hemisphere by speech and reading. *Brain, 97*, 273–388.

Irle, E. (1987). Lesion size and recovery of function: Some new perspectives. *Brain Research Reviews, 12*, 307–320.

Jackson, H. J. (1878). On affections of speech from disease of the brain. *Brain, 1*, 304–330.

Jacobsen, C. C. (1936). Studies of cerebral function in primates. *Comparative Psychology and Monographs, 13*, 1–68.

Jones, E. G., & Powell, T. P. (1970). An anatomical study of converging sensory pathways within the cerebral cortex of the monkey. *Brain, 93*, 793–820.

Kaas, J. H. (1987). The organization of neocortex in mammals: Implications for theories of brain function. *Annual Review of Psychology, 38*, 129–151.

Kazdin, A. E. (1982). *Single case research designs.* New York: Oxford University Press.

Kertesz, A. (1979). *Aphasia and associated disorders: Taxonomy, localization and recovery.* New York: Grune & Stratton.

Kertesz, A., & Sheppard, A. (1981). The epidemiology of aphasic and cognitive impairment in stroke—Age, sex, aphasia type and laterality differences. *Brain, 104*, 117–128.

Kutas, M., & Hillyard, S. A. (1980). Reading between the lines event-related brain potentials during nature sentence processing. *Brain and Language, 11*, 354–373.

Landis, T., Regard, M., Bliestle, A., and Kleihues, P. (1988). Prosopagnosia and agnosia for noncanonical views. *Brain, 111*, 1287–1297.

Lashley, K. S. (1929). *Brain mechanisms and intelligence*. Chicago: University of Chicago Press.

Livingstone, M. S., & Hubel, D. H. (1984). Anatomy and physiology of a color system in the primate visula cortex. *Journal of Neuroscience, 4*, 309–356.

Livingstone, M. S., & Hubel, D. H. (1988). Segregation of form, color, movement, and depth: Anatomy, physiology, and perception. *Science, 240*, 740–749.

Low, W., Wada, J. A., & Fox, M. (1973). Electroencephalographic localization of the conative aspects of language production in the human brain. *Transactions of the American Neurological Association, 98*, 129–133.

Lüders, H., Lesser, R. P., Dinner, D. S., Morris, H. H., Wyllie, E., & Godoy, J. (1988). Localization of cortical function: New information from extraoperative monitoring of patients with epilepsy. *Epilepsia, 29*, S56–S65.

MacKay, D. M. (1978). The dynamics of perception. In P. A. Buser & A. Rougeul-Buser (Eds.), *Cerebral correlates of conscious experience* (pp. 53–68). Amsterdam: Elsevier.

Madison, D. V., Malenka, R. C., & Nicoll, R. A. (1991). Mechanisms underlying long-term potentiation of synaptic transmission. *Annual Review of Neuroscience, 14*, 379–397.

Marinesco, G., Sager, O., & Kreindler, A. (1936). Etudes electroencephalographiques; Electroencephalogrammes dans l'aphasie. *Bulletin de l'Academie de Medecine (Paris), 116*, 182.

Mateer, C. (1983). Localization of language and visuospatial functions by electrical stimulation. In A. Kertesz (Ed.), *Localization of neuropsychology* (pp. 153–183). New York: Academic Press.

Maunsell, J. H. R., & Van Essen, D. C. (1983). The connections of the middle temporal visual area (MT) and their relationship to a cortical hierarchy in the macaque monkey. *Journal of Neuroscience, 3*, 2563–2586.

Mazziotta, J. C., & Phelps, M. E. (1984). Human sensory stimulation and deprivation: Positron emission tomographic results and strategies. *Neurology, 14*, 40–60.

McClelland, J. L., & Rumelhart, D. E. (1986). *Parallel distributed processing: Explorations in the microstructure of cognition. Vol. 2: Psychological and biological models*. London: MIT Press.

McGlone, J. (1977). Sex differences in the cerebral organization of verbal functions in patients with unilateral brain lesions. *Brain, 100*, 775–793.

Mehler, J., Morton, J., & Jusczyk, P. W. (1984). On reducing language to biology. *Cognitive Neuropsychology, 1*, 1.

Merzenich, M. M., Kaas, J. H., Wall, J., Nelson, R. J., Sur, M., & Felleman, D. (1983). Topographic reorganisation of somatosensory cortical areas 3b and 1 in adult monkeys following restricted deafferentation. *Neuroscience, 8*, 33–55.

Mesulam, M. M. (1981). A cortical network for directed attention and unilateral neglect. *Annals of Neurology, 10*, 309–325.

Mills, K. R., Murray, N. M., & Hess, C. W. (1987). Magnetic and electrical transcranial brain stimulation: Physiological mechanisms and clinical applications. *Neurosurgery, 20*, 164–168.

Moran, M. A., Mufson, E. J., & Musulam, M. M. (1987). Neural inputs into the temporopolar cortex of the thesus monkey. *Journal of Comparative Neurology, 256*, 88–103.

Mountcastle, V. B. (1957). Modality and topographic properties of single neurons of cat's somatic sensory cortex. *Journal of Neurophysiology, 20*, 408–434.

Mountcastle, V. B. (1978). An organizing principle for cerebral function: the unit module and the distributed system. In G. M. Edelman and V. B. Mountcastle (Eds.), *The mindful brain*. Cambridge, Massachusetts: MIT Press.

Munk, H. (1881). *Ueber die Funktionen der Grosshirnrinde. Gesammelte Mitteilungen aus den Jahren 1877–1880*. Berlin: Hirschwald.

Murray, E. A., & Mishkin, M. (1985). Amygdalectomy impairs crossmodal association in monkeys. *Science, 228*, 604–606.

Nauta, W. J. H. (1971). The problem of the frontal lobe: A reinterpretation. *Journal of Psychiatric Research, 8*, 167–187.

Neville, H. J., Kutas, M., & Schmidt, A. (1982). Event-related potential studies of cerebral specialization during reading. I. Studies of normal adults. *Brain and Language, 16*, 300–315.

Ojemann, G. A., & Mateer, C. (1979). Human language cortex: Localization of memory, syntax and sequential motor-phoneme identification systems. *Science, 250*, 1401–1403.

Ojemann, G. A., & Whitaker, H. A. (1978). Language localization and variability. *Brain and Language, 6*, 239–260.

Pandya, D. N., and Kuypers, H. G. J. M. (1979). Cortico-cortical connections in the rhesus monkey. *Brain Research, 13*, 13–36.

Pasik, T., & Pasik, P. (1980). Extrageniculostriate vision in primates. In S. Lessell and J. T. W. van Dalin (Eds.), *Neuro-ophthalmology, 1980* (Vol. 1, pp. 95–119). Amsterdam: Excerpta Medica.

Penfield, W., & Roberts, L. (1959). *Speech and brain mechanisms*. Princeton, New Jersey: Princeton University Press.

Petersen, S. E., Fox, P. T., Posner, M. I., Mintun, M., & Raichle, M. E. (1988). Positron emission tomographic studies of the cortical anatomy of single-word processing. *Nature (London), 331*, 585–589.

Phillips, C. G., Zeki, S., & Barlow, H. B. (1984). Localization of function in the cerebral cortex. *Brain, 107*, 327–361.

Raichle, M. E., Herscovitch, P., Mintun, A. M., Martin, W. R. W., & Power, W. (1984). Dynamic measurements of local blood flow and metabolism in the study of higher cortical function in humans with positron emission tomography. *Neurology, 14*, 48–49.

Sanides, F. (1962). *Die Architektonik des menschlichen Stirnhirns*. Berlin: Springer-Verlag.

Sejnowski, T. J., & Rosenberg, C. R. (1987). Parallel networks that learn to pronounce English text. *Complex Systems, 1*, 145–168.

Seldon, H. L. (1985). The anatomy of speech perception: human auditory cortex. In A. Peters & E. G. Jones (Eds.), *Cerebral cortex*. Vol. 4: Association and Auditory Cortices (pp. 273–327). New York: Plenum Press.

Selemon, L. D., and Goldman-Rakic, P. S. (1988). Common cortical and subcortical targets of the dorsolateral, prefrontal and posterior parietal cortices in the rhesus monkey: Evidence for a distributed neural network subserving spatially guided behaviour. *Journal of Neuroscience, 8*, 4049–4068.

Sergent, J., Zuck, E., Terriah, S., & MacDonald, B. (1992). Distributed neural network underlying musical sight-reading and keyboard performance. *Science, 257*, 106–109.

Shelburne, S. A. (1972). Visual evoked responses to word and nonsense syllable stimuli. *Cortex, 12*, 325–336.

Smith, A. (1966). Speech and other functions after left (dominant) hemispherectomy. *Journal of Neurology, Neurosurgery and Psychiatry, 29*, 467–471.

Sperry, R. W., Gazzaniga, M. S., & Bogen, J. E. (1969). Interhemispheric relationship: the neocortical commissures; syndromes of hemispheric disconnection. In P. J. Vinken & G. W. Bruyn (Eds.), *Handbook of clinical neurology* (Vol. 4, pp. 273–290). Amsterdam: North-Holland.

Strub, R. L., & Geschwind, N. (1983). Localization in Gerstmann Syndrome. In A. Kertesz (Ed.), *Localization in neuropsychology* (pp. 295–321). New York: Academic Press.

Stuss, D. T., & Benson, D. F. (1984). Neuropsychological studies of the frontal lobes. *Psychological Bulletin, 95*, 3–28.

Sweetland, J., Kertesz, A., Prato, F. S., & Nantau, K. (1987). The effect of magnetic resonance imaging on human cognition. *Magnetic Resonance Imaging, 5,* 129–135.

Szentagothai, J. (1978). The neuron network of the cerebral cortex: A functional interpretation. *Proceedings of the Royal Society, B, 201,* 219–248.

Ter-Pogossian, M. M., Ficke, D. C., Mintun, M. A. Herscovitch, P., Fox, P. T., & Raichle, M. E. (1984). Dynamic cerebral positron emission tomographic studies. *Annals of Neurology (Suppl.), 15,* S46–47.

Teyler, T. J., Roemer, R. A., Harrison, T. F., et al. (1973). Human scalp-recorded evoked-potential correlates of linguistic stimuli. *Bulletin of the Psychonomic Society, 1,* 333–324.

Tikofsky, R. S., Kooi, K. A., & Thomas, M. H. (1960). Electroencephalographic findings and recovery from aphasia. *Neurology, 10,* 154–156.

Turner, B. H., Mishkin, M., & Knapp, M. (1980). Organization of the amygdalopetal projections from modality specific cortical association areas in the monkey. *Journal of Comparative Neurology, 191,* 515.

Van Buren, J., Fedio, P., & Frederick, G. (1978). Mechanism and localization of speech in the parietotemporal cortex. *Neurosurgery, 2,* 233–239.

Van Essen, D. C. (1979). Visual areas of the mammalian cerebral cortex. *Annual Review of Neuroscience, 2,* 227–263.

von Monakow, C. (1914). *Die Lokalisation in Grosshirn und der Abbau der Funktionen durch cortikale Herde.* Wiesbaden: Bergmann.

Wernicke, C. (1874). *Der aphasische symptomenkomplex.* Breslau: Cohn & Weigart. Reprinted in *Boston Studies on the Philosophy of Science* (Vol. IV). Dordrecht: Reidel.

Wise, R. J., Chollet, F., Hadar, U., Friston, K., Hoffner, E., & Frackowiak, R. (1991). Distribution of cortical neural networks involved in word comprehension and word retrieval. *Brain, 114,* 1803–1817.

Woolsey, T., & Van Der Loos, H. (1970). The structural organization of layer IV in the somatosensory region (S1) of mouse cerebral cortex. *Brain Research, Amsterdam, 17,* 205–242.

Zaidel, E. (1976). Auditory vocabulary of the right hemisphere following brain bisection or hemidecortication. *Cortex, 12,* 187–211.

Zeki, S. M. (1983). The distribution of wavelength and orientation selective cells in different areas of monkey visual cortex. *Proceedings of the Royal Society, B, 217,* 449–470.

Zeki, S., & Shipp, S. (1988). The functional logic of cortical connections. *Nature (London) 335,* 311–317.

Zulch, K. J., Creutzfeldt, O., & Galbraith, G. C. (1975). *Cerebral localization.* New York: Springer-Verlag.

Chapter 2

Cortical Stimulation and Recording in Language

George A. Ojemann

I. INTRODUCTION

The investigation of cortical language organization during neurosurgical operations provides a perspective different from that derived from the study of language changes with brain lesions that constitutes traditional aphasiology. Not surprisingly, then, intraoperative investigations have led to a model of the organization of language in dominant hemisphere perisylvian cortex that differs from the traditional one, a model summarized in another review (Ojemann, 1991a). The major features of the model derived from intraoperative investigations include the following:

1. Compartmentalization of language function into largely separate systems: separate systems have been identified for naming (including separate systems for naming in different languages), reading, recent verbal memory, and speech production and perception; additional separate systems are likely
2. Several localized essential areas in each system, often 1–2 cm^2 in surface extent, as well as widely dispersed neurons (even in nondominant hemisphere) specific to a particular system
3. Parallel activation of all components of the system during the appropriate aspects of language

Evidence for this model and some of its other features are among the topics included in this chapter.

Intraoperative language investigations derive two types of data. In one type, the link with language is made when alteration of activity in a local brain area disrupts some language measure, thus identifying function of that brain area to be essential to that aspect of language. Brain lesions are related to a language in the same way. In intraoperative investigations, activity in a local brain area usually is altered by surface electrical stimulation, which blocks local function most likely from local depolarization blockade. One study using imaging of the intrinsic changes in reflectance that occur when neurons depolarize has shown that, when various parameters including electrodes and currents level are typical for this type of stimulation in patients, the area of depolarization is confined to the tissue between the electrodes (Haglund, Ojemann, & Blasdel, 1993a). A second type of intraoperative data correlates changes in physiological measures with some aspects of language. At present, physiological measures derived intraoperatively include surface electrocorticographic recordings and extracellular recordings of single neuron activity. These data can indicate participation of an area in the aspect of language investigated, but will not indicate whether that participation is essential for language. Scalp-recorded evoked potentials and imaging measures of metabolic changes such as those derived from positron emission tomography (PET) provide similar data. Except for the single neuron studies, which are unique to intraoperative investigation, the different perspective is not the type of data derived as much as the ability intraoperatively to map the essential or participatory roles of multiple local cortical areas with a finer resolution than can be derived currently from lesion, imaging, or scalp electrophysiological studies and, in the case of electrical stimulation, the ability to alter function only transiently, so effects during different parts of a behavioral measure can be determined, for example, differences between effects on memory input, storage, or retrieval (Ojemann, 1983).

These methods for the investigation of cortical language organization during neurosurgical operations are applied in several different settings. One setting is craniotomy conducted under local anesthesia. The brain is insensitive to touch or pain. Thus, with appropriate blocking of scalp and dural sensation with local anesthetics and use of an ultra-short-acting anesthetic such as propofol for removal of the skull, both the physiological recordings needed to plan the operation and the investigation of language can be conducted on a reasonably comfortable cooperative patient. I use this setting to obtain all types of data on language organization in adults and adolescents. Another setting involves placement of arrays of subdural recording electrodes, usually in grids of electrodes 1 cm apart, at craniotomy under general anesthesia, with leads from the electrodes through the scalp. These electrodes are left in place chronically for a period of a few days to few weeks; stimulation and electrocorticographic recording are carried out outside the operating room. Some investigators have derived all data on language organization from this setting; I usually reserve it for patients, such as young children, who cannot cooperate in an awake operation. A few language investigations, including single neuron recording, have been carried out in a third setting, that of chronically implanted depth electrodes. In all these settings, the patient's underlying disease state determines the locations of the craniotomy or electrode placement and, thus, the area available for investigation of language.

All these settings, of course, occur in special patient populations. These patients have underlying diseases that provide the indications for the surgical procedures, usually medically intractable epilepsy or tumors and other structural lesions near functionally important brain areas. This problem of special populations occurs in nearly all studies of human language. For example, classical aphasiology was derived largely from the special population of patients with vascular disease. Because the influence of the disease on language organization is largely unknown, language organization in "normals" most likely represents those features supported by a convergence of findings from the different populations.

All intraoperative techniques are most suited to investigation of intrahemispheric organization of language, rather than of lateralization. Rarely is surgical access to homologous areas on both sides available. Thus, lateralization issues usually can be studied with intraoperative techniques only by comparing findings from different patients with right or left hemispheral exposure. The intracarotid amobarbital perfusion test (Wada and Rasmussen, 1960) is a much better method of assessing lateralization, especially when modified to determine the effect of both right and left carotid perfusion in the same patient (Woods, Dodrill, & Ojemann, 1988). This type of assessment is obtained regularly in my patients before craniotomy under local anesthesia. About 15% of them have some evidence of involvement of the right hemisphere with speech. In one-third to one-half of those patients, language is exclusively in the right hemisphere; in the remainder, language representation is bilateral (Woods et al., 1988). In this largest series of bilateral carotid assessments reported, no statistically significant relationship was seen between left-handedness and involvement of the right hemisphere in language, once patients who were left-handed because the right hand was weak from a left frontal or parietal brain lesion were excluded. Except as noted, the remainder of this chapter is confined to a discussion of cortical language organization in the setting of the usual pattern of left language dominance.

II. CORTICAL LANGUAGE ORGANIZATION DERIVED FROM STUDY OF OBJECT NAMING

A. Essential Areas

Object naming is the aspect of language most extensively investigated with intraoperative techniques, a phenomenon that is partly historical.

Naming was the language measure used by Penfield when he introduced the technique of electrical stimulation mapping (Penfield & Roberts, 1959). In addition, naming deficits are a feature of all types of aphasia, and therefore would seem to constitute a good screening test for language localization.

Electrical stimulation of the lateral cortical surface in a quiet patient usually evokes responses from primary cortices. With exposure of perisylvian areas, localized face and tongue movements from motor cortex and localized face and tongue tingling sensations from somatosensory cortex commonly are evoked at small currents; buzzing from primary auditory cortex is reported much less often, even with large stimulating currents. Spontaneous speech is almost never evoked. If the stimulating current is kept below the threshold for after-discharge, the "experiential" and "interpretative" responses described by Penfield (Mullan & Penfield, 1959; Penfield & Perot, 1963) are also very rare. However, if the patient repetitively names common objects, stimulation at currents below afterdischarge thresholds at some sites that evoked nothing in the quiet patient will alter naming. At some of those sites, the naming alteration will be an arrest of speech output, whereas at other sites speech will be present but the object name will be either incorrect or omitted (anomia). Errors occur repeatedly each time some sites are stimulated, whereas at other sites errors occur only on some stimulations, although if the error rate of naming in the absence of stimulation is low, and it usually is, these occasional errors with stimulation also can be shown to be statistically significant.

Figure 1 shows the location of sites at which stimulation altered naming in one patient when assessed at craniotomy under local anesthesia using bipolar stimulation with electrodes separated by 5 mm. Sites at which naming was altered repeatedly in that patient are only 1–2 cm² in surface extent. The pattern illustrated by Fig. 1 is encountered frequently. Usually several essential areas are separated by cortex not related to naming. In a series of 117 patients with left dominant perisylvian cortical stimulation mapping during naming, 67% had at least 2 separate sites, usually one in the posterior inferior frontal lobe and the other temporo-parietally, whereas 24% had at least 3 (Ojemann, Ojemann, Lettich, & Berger, 1989a). Moreover, each of those areas is usually quite localized. In the 117-patient series, only 16% showed a contiguous area of repeated naming errors greater than 6 cm² (Ojemann et al., 1989a). Few patients, then, have any one essential language area as large as the traditional Broca's or Wernicke's area. The transition from a site of repeated naming errors to surrounding cortex in which no naming changes are evoked is sometimes abrupt (Fig. 1); in other patients, the area in which repeated naming errors

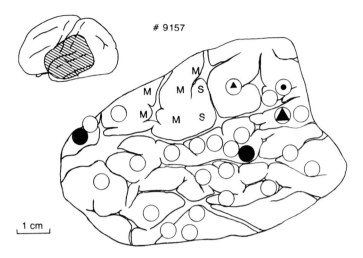

Figure 1 Localization of sites identified by the electrical stimulation mapping technique as essential for naming in English and Spanish, in a 44-year-old ambidextrous male, left dominant by intracarotid amytal assessment. Each circle represents a site at which stimulation effects were assessed, using biphasic square wave pulses each of 1-msec duration at 60 Hz delivered in a bipolar manner through electrodes separated by 5 mm at a current of 3 mA. Filled circles represent sites of repeated naming errors in English; the small circle represents a site at which a single naming error in English was evoked in one of three samples. No errors in naming the same object pictures in Spanish were evoked at any of these sites. The large triangle identifies a site at which repeated naming errors in Spanish occurred; the small triangle represents the site of a single error in three trials on naming in Spanish. No errors in naming in English were evoked in these sites. The baseline error rates in the absence of stimulation were 1.2% for English and 4.8% for Spanish. M and S identify motor and sensory responses from face rolantic cortex. Note the highly localized nature of the sites essential for naming in each language, and the separation of sites essential for naming of the same pictures in the two different languages. Other patients showing highly localized areas essential for naming in one language are discussed by Ojemann (1983, 1988, 1991b) and by Ojemann et al. (1989a). Other patients with localization of areas essential for naming in two languages are discussed by Ojemann and Whitaker (1978) and Ojemann (1983,1991b).

are evoked is surrounded by a region of occasional errors, indicating a more gradual transition (Whitaker & Ojemann, 1978).

Ojemann and Dodrill (Ojemann, 1983) investigated the question of how essential to language these localized sites are at which stimulation repeatedly evokes naming errors. In that study, changes in the Wepman aphasia battery 1 mo after anterior temporal lobectomies were determined for resections that either did or did not come within 2 cm of sites with repeated naming errors. Subtle but definite deficits in the aphasia battery were present when resections encroached on those sites related to naming, deficits that were not present when the resection did not come within 2 cm of those sites. Moreover, no relationship was found between errors on the aphasia battery and the size of the resection, the degree of postoperative seizure control, or the preoperative language performances. Thus, these localized sites of repeated naming errors are essential for language. Indeed, stimulation mapping during naming has become a standard neurosurgical technique for identifying brain that must be preserved in a resection to avoid an aphasia (Berger, Ojemann, & Lettich, 1990; Ojemann et al., 1989a).

Substantial variation in the exact location of these sites is evident in this patient population. In the 117-patient series, 90 patients had mapping of both frontal and temporo-parietal language areas. Of those patients, 15% had all essential language areas in temporo-parietal cortex; none could be identified in frontal lobe despite extensive mapping. In another 17%, all essential language areas were in frontal cortex; none were identified in temporo-parietal cortex and, in several of these cases, resections in classical Wernicke's area did not alter language even acutely (Ojemann et al., 1989a). Figure 2 shows the variation in location of essential areas for naming when individual maps of all 117 patients are aligned by motor cortex and the end of the Sylvian fissure. When dominant hemisphere perisylvian cortex is divided into the zones illustrated in Fig. 2, only the zone immediately in front of face motor cortex contains essential language areas in a substantial majority of patients (79%). Elsewhere in perisylvian cortex, including all of the traditional Wernicke's area, the probability of finding essential language areas in any one zone is rather low, in 36% of patients or less. Essential language areas also extend well beyond the traditional language areas. For example, 15% of patients had essential areas in anterior superior temporal gyrus, in front of the level of rolandic cortex. This variability in location of essential language areas was similar in patients with lesions that clearly began in early life and those with lesions that likely began in adulthood (Ojemann et al., 1989a). This variability is greater than the variability in gross gyral morphology in the perisylvian region (although that is considerable), and is much greater than any described variability in location of specific cytoarchitectonic areas. The traditional maps of Broca's and Wernicke's areas, then, are really artifacts produced by pooling the multiple, relatively small, localized essential areas for language in individual patients that show substantial variability in their location across patients.

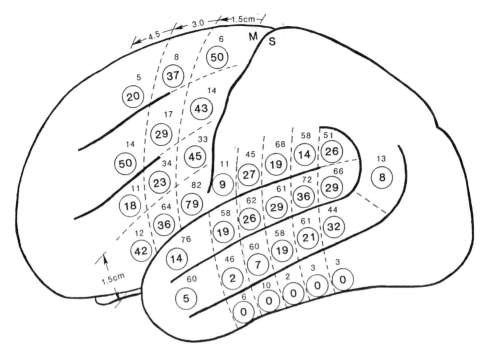

Figure 2 Variability in language localization in 117 patients, all left dominant for language. Individual maps were aligned to motor cortex (M) and to the end of the sylvian fissure. Cortex was divided into zones identified by the dash lines. The upper number in each zone is the number of subjects with a site in that zone at which effects of stimulation during naming were assessed. The lower number in a circle in each zone is the percentage of those patients with sites with significant evoked naming errors. Reproduced with permission from Ojemann et al. (1989a).

A somewhat unexpected finding has been the importance to language of the surface cortex on the tops of gyri. The human brain, of course, is folded extensively, with well over half the cortex buried in sulci. However, surface stimulation of the dominant perisylvian area identifies at least one site essential to language in nearly all patients. For example, the 117-patient series was derived from naming stimulation mapping of 119 consecutive nonaphasic patients. Moreover, these surface sites indicate whether a resection that includes buried cortex will or will not result in language deficits. Thus, essential language areas are not distributed randomly across buried perisylvian cortex, nor do they extend far away from surface sites in sulci. The few occasions on which language representation on banks of sulci has been assessed support this idea, since surface language areas were found to extend either only a short distance into an adjacent sulcus or not at all (Ojemann et al., 1989a).

The surgical exposure occasionally includes posterior superior frontal cortex. As evident from Fig. 2, errors in naming also were evoked rather frequently from sites there. These sites are likely to represent the lateral extension of the language representation in the supplementary motor area. Penfield and his associates (Penfield & Roberts, 1959) evoked changes in naming from the medial face of the left frontal lobe. From those findings and the effects of excisions, Penfield proposed a language role for the supplementary motor area. However, in contrast to left perisylvian language areas in which at least large lesions are associated with permanent language deficits, language deficits of extensive left posterior–frontal–supplementary motor excisions, although often acutely dramatic with mutism, recover rapidly, often within a week. Fried et al. (1991) reported detailed stimulation mapping during naming through chronically implanted electrodes over medial frontal lobe. These researchers described a motor homonucleus in supplementary motor area beginning at leg motor cortex, with face more anterior, and evoked naming changes anterior to that. My findings with mapping in the same area are somewhat different; arrests of naming are evoked from interhemispheric fissure sites immediately anterior to leg motor cortex, and from those sites contiguous with sites of evoked naming changes on lateral superior frontal gyrus (Ojemann, 1992).

Stimulation of chronic subdural electrode grids during naming in adult patients generally has shown somewhat more widespread changes than observed with direct intraoperative stimulation at craniotomy. The reason for this phenomenon is not entirely clear, but subdural grid stimulation also often shows wider sensory and motor representation than observed with stimulation at craniotomy. On the other hand, considerable variability in the location of sites with evoked naming changes is also present with subdural grid stimulation (Lesser, Lüders, Dinner, Hahn, & Cohen, 1984). In addition, using that technique, Lüders et al. (1986) evoked changes in multiple language measures, including naming, from left inferior basal temporal cortex as far anterior as 3.5 cm behind the temporal tip in some patients. They conclude that in some, but not all, patients this area also subserves language, although excision of these sites of evoked language change usually has not been associated with later language deficits, again in contrast to perisylvian cortex.

The subdural grid technique is the only method for stimulation mapping in young children. With that technique, I have investigated localization of essential naming sites in 9 children from 4 to 10 years of age (Berger et al., 1989; Ojemann et al., 1989a; Ojemann, Berger, & Lettich, 1989c). Sites of evoked errors for naming easily recognized objects surprisingly are localized in these young children; changes are confined to one electrode or, at most, two adjacent electrodes on a grid in which electrodes are separated by 1 cm. Both frontal and temporo-parietal sites are present in 4-year-olds (including the youngest patient tested to date, 4 years and 3 days old), although changes with temporo-parietal stimulation have been confined to a single electrode. In studies conducted to date, the youngest patients have shown single frontal and temporo-parietal sites. Multiple separate temporo-parietal sites have not been identified in patients younger than 8 years old. Substantial variability in location of naming sites is evident even in this small series. These findings suggest that highly localized essential areas for language are present in quite young children. At present, the only suggestion of changes in language localization with age is that additional localized essential areas may appear, since similar highly localized essential sites have been seen in the oldest patients studied, at age 70–80.

Comparison of mapping through electrodes chronically implanted in adults over lateral temporal cortex for diagnostic purposes with mapping at craniotomy some months later has provided the opportunity to address the question of the stability of essential language areas over this interval, in the absence of any intervening brain lesion. Investigators generally assume that such localization is stable, although of course this hypothesis cannot be tested with brain lesions and, in fact, limited data are available to substantiate that assumption. When I have had the opportunity to compare localizations separated by a few months, the sites with and without evoked naming changes generally have been similar (Ojemann, 1983).

The patterns of location of essential areas for naming also have been related to patient sex and preoperative verbal ability as measured by the verbal IQ (VIQ). In the part of the 117-patient series, with VIQs below the mean (99), males were significantly more likely than females to have essential sites for naming in supramarginal gyrus. Moreover, females were overrepresented in those patients in which no temporo-parietal essential language sites were identified. Together, these findings indicate that less temporo-parietal language representation occurs in some females, likely accounting for the different effects in males and females of temporo-parietal brain lesions reported by Kimura (1983). Significantly different patterns for location of essential sites for naming also were identified in patients with high VIQs compared with those with low VIQs. Temporal cortex was the site of the greatest difference; patients with low VIQs were likely to have essential naming sites in superior temporal gyrus whereas those with high VIQs were more likely to have such sites in middle temporal gyrus (Ojemann et al., 1989a). Thus, the cortical organization of language seems to differ in individuals with different verbal abilities (insofar as these abilities are measured by the VIQ).

B. Physiologic Correlates of Essential Sites in the Electrocorticogram

Changes in the electrocorticogram (ECoG; the EEG recorded directly from brain) recorded during naming that differentiate sites essential for naming from surrounding cortex have been identified (Fried, Ojemann, & Fetz, 1981; Ojemann, Fried, & Lettich, 1989b). Frontal sites independently identified as essential for naming by stimulation mapping showed slow potentials of 1-sec duration that were less evident at surrounding sites not related to naming. These potentials were present when naming was done silently and when it was done aloud, but were less evident when the same visual stimulus used for naming was given with the instruction to match a spatial feature. In temporo-parietal lobes, essential naming sites were differentiated from surrounding cortex by the appearance of local desynchronization (replacement of 7–12 Hz ECoG activity by low voltage fast activity) during naming. This change was quantitated by spectral density measures and was shown statistically to appear at naming sites during naming and not at those sites during a spatial task using the same visual stimuli or at surrounding nonnaming sites during naming (Ojemann et al., 1989b).

These ECoG changes that differentiate essential naming sites from surrounding cortex during naming occur in parallel at frontal and temporal sites, appearing shortly after the visual stimulus and lasting essentially the entire time (about 1 sec) required for naming. These changes provide direct evidence for the parallel processing that has been inferred from PET and lesion studies (Damasio, 1990; Raichle, 1990). No physiological evidence for the serial processing from posterior to anterior language cortex, as postulated in the classical model of cortical organization of language, has been demonstrated in these studies. Instead, when combined with the stimulation mapping localization, the new data suggest that language in human association cortex is organized like other functions in primate association cortex—into parallel distributed systems.

ECoG changes similar to those that differentiate essential sites for naming from surrounding cortex during naming can be evoked in nonhuman animals by activating the thalamocortical activating system. Thus, this system may act on a moment-to-moment basis to select the cortical areas appropriate to a particular aspect of language. Evidence that the thalamus

plays a role in language has been derived from both stimulation (Ojemann, 1975) and lesion (Reynolds, Turner, Harris, Ojemann, & Davis,
1979) studies, including independent evidence that the thalamic role in
language might involve activity of the thalamocortical activating system
(Ojemann, 1975).

C. Physiologic Correlates in Neuronal Activity at Nonessential Sites

Microelectrode recording of neuronal activity is invasive, so in my studies this procedure has been restricted to cortex that is to be removed
subsequently at the operation. Of course, this material never includes essential language areas. Thus, the recordings of neuronal activity during
naming are all at nonessential sites. Nevertheless, changes in neuronal
activity that are specific to naming and not to spatial control tasks in
response to the same visual stimuli are recorded occasionally in nonessential areas of the dominant hemisphere, as illustrated in Fig. 3, and even
from nondominant temporal cortex (J. Ojemann et al., 1991). The most
common change has been a sustained shift in the level of activity, either
increased activity or, as in Fig. 3, decreased activity. This shift often occurs
with the initiating of naming and lasts throughout its duration, thus occurring in parallel with the physiological events at essential sites. The
entire system related to a language function then, both the multiple essential areas and the many individual widely dispersed neurons, seems to be
activated in parallel.

III. ORGANIZATION OF DIFFERENT LANGUAGE FUNCTIONS

A. Naming in Multiple Languages

When similar intraoperative investigations of multiple language functions are undertaken, separation of functions is encountered frequently,
whether of essential areas or of the behavioral correlates of changes in
neuronal activity. Some of the most convincing examples of this separation
are from investigations in bilingual patients. Stimulation mapping during
naming the same object pictures in two different languages in the same
patient has shown some separation of sites essential for each language in
nearly all subjects in which this has been studied (Black & Ronner, 1987;
Ojemann, 1983; Ojemann & Whitaker, 1978; Rapport, Tan, & Whitaker,
1983). Figure 1 shows another example in which sites with repeated naming errors in English and Spanish are each quite localized, but to different
nearby brain areas. The sole exception seems to be a recent patient of mine

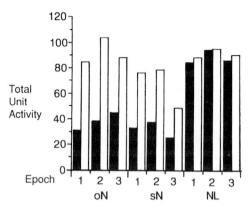

Figure 3 Neuronal activity during overt naming (oN) and silent naming (sN) of the same objects in English (■) and Spanish (□), as well as when these object pictures were presented in a line matching task (NL), in recordings from the anterior superior temporal gyrus in a 27-year-old female. Activity during each task is divided into three epochs of 1.3 sec each. Each bar represents the summation of activity recorded during six different trials of presentation of the same object pictures in English or Spanish naming or in the line matching task. Note the marked reduction in activity during overt and silent naming in English compared with Spanish, a reduction in activity that lasts throughout the duration of the tasks although all the overt naming responses occur in the second epochs. Thus, this example demonstrates differential effects of naming on activity of a single neuronal population with naming in one language compared with another and demonstrates that change being sustained throughout the duration of the task (T. F. Schwartz and G. A. Ojemann, unpublished observations). A somewhat different analysis of this same activity, with similar findings, can be found in a report by Ojemann et al. (1990).

who was just learning (and was not fluent in) the second language (G.A. Ojemann and D. Cawthon, unpublished data). In addition to a partial separation of essential areas for two languages, regardless of when acquired, the area essential for a language acquired later seems to be slightly larger than that for the first language, perhaps with a tendency for the area essential to the second language to be located further from the perisylvian area (D. Cawthon and G. A. Ojemann, unpublished data).

Separation of essential areas is also evident when stimulation effects on oral languages and manual communication systems are compared. Sites essential for the manual communication system—finger spelling—but not oral language were identified in the anterior temporal lobe of a hearing patient; an anterior temporal resection that included those sites was associated with a disproportionate deficit in finger spelling compared with oral language (Mateer et al., 1982). That patient also had a parietal site in which both languages were altered. In a hearing patient fluent in American Sign Language (ASL), separate temporal lobe sites were identified for naming in only the oral language and only ASL, as was a site in which naming in both was intact but oral interpretation of signs was disturbed (Haglund, Ojemann, Lettich, Bellugi, & Corina, 1993b).

Neuronal populations in nonessential areas also usually change activity in response to naming in only one of two languages. All five neuronal populations (in 3 patients) recorded during bilingual naming have demonstrated differential responses in the two oral languages, as shown in Fig. 3 (Ojemann, Cawthon, & Lettich, 1990). A neuronal population recorded at a nonessential site in anterior temporal lobe in the patients fluent in ASL increased activity only in response to naming in oral language but not in response to naming by signing or in response to spatial matching tasks to those same visual stimuli (Haglund et al., 1993b).

B. Production and Perception

In the traditional model of the organization of language cortex, the major functional separation is between language production, related to posterior frontal lobe, and language perception, related to posterior temporal lobe. Intraoperative stimulation mapping confirms the presence of a posterior inferior frontal area essential for all speech output, a final motor pathway for speech (Ojemann, 1983). Stimulation at these sites not only evokes an arrest of all speech, but disrupts the ability to mimic single speech gestures. This area is the most consistent site related to language in population studies, although even it demonstrates some individual variability in exact locations (Ojemann et al., 1989a). Intraoperative recording of ECoG changes with naming indicates activity in that area throughout the language process, even without overt speech production.

In addition, stimulation mapping identified essential areas for speech production more widely in perisylvian cortex, more anteriorly in inferior frontal gyrus, and in portions of superior temporal gyrus and anterior parietal operculum. With stimulation of these sites, the ability to mimic single speech gestures is intact, but mimicry of sequences of speech gestures is disturbed (Ojemann & Mateer, 1979). Thus, intraoperative investigations indicate that the neural mechanisms essential to speech production are distributed beyond posterior inferior frontal lobe, in perisylvian cortex of the dominant hemisphere. This result corresponds with the report that lesions that produce permanent motor aphasias must include dominant hemisphere perisylvian superior temporal and anterior parietal areas, as well as posterior frontal cortex (Mohr, 1976).

Temporal cortical microelectrode recording during speech production demonstrated changes in neuronal populations that were distributed widely in superior and middle temporal gyri of both hemispheres (Creutzfeldt, Ojemann, & Lettich, 1989a). Some neuronal populations in the nondominant hemisphere related to speech production show greater changes in activity with speech production from visuospatial rather than lexical tasks, suggesting that those populations may provide a link between the visuospatial function and language (J. Ojemann, G. Ojemann, & Lettich, 1992).

Intraoperative investigations also suggest that the relationship between speech production and perceptions is complex and sometimes overlapping. Stimulation mapping during perception of stop consonants evoked changes from many of the same dominant perisylvian sites in which speech production was altered, although stimulation was applied only during the period of stop consonant presentation and not during a response period (Ojemann, 1983; Ojemann & Mateer, 1979). Thus, common dominant perisylvian cortical sites are essential to speech production and perception. This combination is not predicated by the classical model of language organization, but is predicted by the psycholinguistically devised motor therapy of speech perception, which postulates that creation of a motor model is involved in decoding speech sounds (Liberman, Cooper, Shankweiler, & Studdert-Kennedy, 1967), the two functions stimulation mapping often relates to the same cortical sites. Alternatively, those sites could be related to some other function that is common to speech production and perception, such as precise timing (Calvin, 1983; Tallal, 1983). Evoked potentials to the speech sounds "pa" and "ba" also have been recorded from posterior frontal cortex (as well as from superior temporal gyrus), providing further evidence for speech perception occurring in cortex regions known to have a role in speech production (Mateer & Cameroon, 1989).

Temporal lobe microelectrode recordings suggest an even more complex relationship between speech production and perception. A few recordings show the changes predicted by the stimulation mapping findings: neuronal populations that change activity in the same way with speech production and perception (Ojemann, 1991b). These neurons have been restricted to left superior temporal gyrus. One such neuron seems to have specific patterns of activity in response to perception of specific words, and similar patterns during production of those words (Ojemann,

1991b). More commonly, however, neuronal activity changes with only speech production or only perception or changes in opposite directions with perception and production (Creutzfeldt et al., 1989a, Creutzfeldt, Ojemann, & Lettich, 1989b; Ojemann, 1991b). This differential effect was present even though the same word was enunciated in both tasks—in one by another individual, in the other by the patient. Most of the neurons involved in perceiving the speech of others, then, are not active during generation of one's own speech. Neuronal activity changes with only speech production or only speech perception were recorded widely in the superior and middle gyri of either temporal lobe.

Some neuronal populations, including some recorded in nondominant temporal cortex, seem to have specific patterns of activity in response to perception of specific words (Creutzfeldt et al., 1989b). Populations have been identified that become active only on the second syllable or subsequent syllables during perception of multisyllabic words (Creutzfeldt et al., 1989b). Some populations may alter activity in response to specific phonemes. One dominant hemisphere population altered activity with normal speech and normal speech backward but not with speech with degraded frequency responses, suggesting a population specifically tuned to speech frequencies. These findings suggest highly specific subdivisions in the neuronal mechanism of speech perception, involving neurons distributed over both hemispheres.

Prosody has not been studied systematically intraoperatively. However, stimulation of left face sensory cortex often is associated with a change in articulation, so that speech has a foreign accent. Changes in rate of articulation of speech sounds, including acceleration, has been evoked with stimulation of some left temporal cortex sites (Smith 1980).

C. Naming Compared with Reading

Partial separation of mechanisms for reading from those for naming is evident both in the location of essential areas and in the activity of neurons in nonessential areas. Assessment of the effects of stimulation during naming and sentence reading in 55 patients demonstrated changes in only one of the two functions at 77% of the sites with an evoked change in either one or both (Ojemann, 1989). Separation was particularly evident at temporo-parietal sites. The type of error evoked at sites related to reading alone differed from that at sites common to reading and naming. At the latter sites, stimulation usually evoked slow effortful reading. At some sites related to reading along, however, reading during stimulation was often fluent, but with errors on verb endings, pronouns, prepositions, and conjunctions (Ojemann, 1983), suggesting that these sites were essential for the syntactic aspects of reading. Errors of that type were evoked from both frontal and temporo-parietal lobes.

   The individual patterns of location of temporo-parietal sites essential for only reading or only naming differed in patients with high or low preoperative VIQs. The 20 of the 55 patients with the highest VIQs were likely to have sites essential for only reading in superior temporal gyrus and sites for only naming in middle temporal gyrus, whereas the 20 patients with lowest VIQs were likely to have the reverse pattern, a difference that was highly statistically significant (Ojemann, 1989). This finding of a relationship between VIQ and location of sites essential for naming alone or reading alone is quite robust statistically, and is supported by the finding mentioned earlier in a larger series, including 62 more patients, that sites essential to naming in superior gyrus are significantly more likely in patients with lower VIQs. However, explaining the finding is difficult. If the higher VIQs reflect higher reading skills, and if grapheme–phoneme linkage is important to skill in reading as has been suggested (Liberman, 1989), then the higher reading skill with superior temporal gyrus location of sites essential for reading may reflect the close relationship to primary and secondary auditory cortex, with a presumed role in phoneme identification. Some evidence suggests a visual association area in human middle temporal gyrus (G. Ojemann, J. Ojemann, Hoglund, Holmes, & Lettich, 1992). If so, the more optimal placement of naming sites in middle temporal gyrus may be a consequence of a closer link to visual information.

   Single neuronal recordings at nonessential sites have identified changes in activity with word reading that are not present with spatial matching tasks in response to the same visual stimuli (Ojemann, 1989; Ojemann, Creutzfeldt, Lettich, & Haglund, 1988). Changes with word reading have been recorded more frequently than changes with object naming in the same patients (Ojemann et al., 1988), suggesting that reading requires activity of more neurons than does naming. This result may explain why, in some cases of dyslexia in which deficit in all language functions are found, reading is involved more severely. Neurons that change activity with both naming and reading have been recorded rarely; instead, changes, when present, are usually in response to word reading or to naming alone but not to both, another example of the separation of functions at the level of neurons in nonessential cortex (Ojemann, 1989).

D. Recent Verbal Memory

   Sites essential for recent verbal memory have been identified with stimulation mapping in lateral temporal cortex (Ojemann, 1978, 1983; Ojemann

& Dodrill, 1985). Including these sites in a resection increased the likelihood of a recent verbal memory deficit after anterior temporal lobectomy (whereas no relationship of memory deficits to extent of mesial resection was seen in those patients) (Ojemann & Dodrill, 1987). Moreover, sparing lateral cortical memory sites (in addition to medial temporal studies) avoided a memory deficit after resection in patients in which such a deficit was expected based on performance on memory measures during the intracarotid amobarbital perfusion test (Ojemann & Dodrill, 1985). Thus, recent verbal memory depends on dominant hemisphere lateral temporal cortex, as well as on the hippocampal formation.

Sites essential for recent verbal memory also have been identified in parietal and frontal cortex. Language usually is not altered by stimulation at cortical sites at which recent verbal memory changes are evoked (Ojemann, 1978). Moreover, recent verbal memory is usually intact at sites at which language is disturbed by stimulation. These findings have been interpreted as a biological basis for the behaviorally observed separation between episodic memory and generalized word memory. When localization of sites essential for naming, reading, recent verbal memory, and orofacial movement mimicry were identified by stimulation mapping in 14 patients, the perisylvian cortex essential for language and orofacial movements was surrounded by sites essential for only one language function; these in turn were surrounded frontally, parietally, and anterior temporally by sites essential for only recent verbal memory (Ojemann, 1983; Ojemann & Mateer, 1979). Some evidence exists for lateralized motor and memory systems in primates (Dewson, 1977; Trevathen, 1974), raising the possibility that phylogenetically specialized language mechanisms develop between these two systems.

The recent verbal memory measure used in these studies was an input–storage–retrieval paradigm, patterned after that of Peterson and Peterson (1959); each trial of the measure has the name of an object as the information entered into recent verbal memory (input phase) that is stored over the distraction provided by reading a sentence or naming other objects (storage phase), and then is retrieved at a visual cue (retrieval phase). On different trials of this measure, the effect of stimulation during the input phase alone, the storage phase alone, or the retrieval phase alone was assessed (Ojemann, 1978, 1983). At temporo-parietal sites, most alterations in memory followed stimulation during the input or storage phase whereas at frontal sites, most memory changes occurred with stimulation at the time of retrieval. Thus, different phases of memory seem to have separate mechanisms.

Microelectrode recording in dominant anterior lateral temporal cortex demonstrated significant changes during recent verbal memory measures in 12 of 17 neuronal populations (Ojemann et al., 1988). Control measures in this study included use of the same visual cues in language and spatial matching tasks. Thus, a large proportion of lateral cortical neurons are active during recent verbal memory. Half these neuronal populations also change activity with one of the several language functions measured, so separation between language and recent memory may not be as complete at a neuronal level as in identification of essential sites. The most common pattern during recent verbal memory was an increase in frequency of neuronal activity at entry of information into memory and again at retrieval, an increase that persisted well beyond the time required to identify the information entering memory or the time required to retrieve it. In several populations, this increase in activity was associated with increases in specific intervals between neuronal discharges, suggesting some patterning to the activity (Ojemann et al., 1990). Preliminary findings suggest that, in some neuronal populations, this increased activity diminishes when the same item is retrieved again (Haglund, Ojemann, Schwartz, & Lettich, 1993c). The role of these sustained increases in neuronal activity during the input and retrieval phases of recent verbal memory is unknown, but in conjunction with the importance of temporal cortex in memory storage, the hypothesis that such activity represents the memory "engram" is not totally unreasonable.

IV. SUMMARY

Intraoperative investigation of the cortical organization of language in the dominant hemisphere indicates the presence of multiple, at least partially separate, systems for different aspects of language: a system for naming, in large part separate from a system for naming in another language, but also partly separate from other systems for reading or recent verbal memory. Each system consists of multiple essential areas and widely dispersed neurons. Essential areas are often quite localized, but with considerable individual variation. Patterns of the localization of these essential areas differ between males and females and with verbal ability. The widely dispersed neurons often change activity only with one system. The entire system, both essential areas and dispersed neurons, is activated in parallel.

On the other hand, these investigations suggest a complex relationship between language perception and production. In part these two functions have a common biological substrate. When neuronal mechanisms for these two functions are separated, however, the system for speech perception seems to be inhibited during speech production. The common biological substrate for language perception and production is identified most often in perisylvian cortex; more specialized systems related to single language or memory functions most often surround this perisylvian area frontally and temporo-parietally.

REFERENCES

Berger, M., Kincaid, J., Ojemann, G., & Lettich, E. (1989). Brain mapping techniques to maximize resections, safety and seizure control in children with brain tumors. *Neurosurgery, 25,* 786–792.

Berger, M., Ojemann, G., & Lettich, E. (1990). Neurophysiological monitoring during astrocytoma surgery. In H. Winn & M. Mayberg (Eds.), *The role of surgery in brain tumor management.* Philadelphia: Saunders.

Black, P., & Ronner, S. (1987). Cortical mapping for defining the limits of tumor resection. *Neurosurgery, 20,* 914–919.

Calvin, W. (1983). Timing sequences as a foundation for language. *The Behavioral and Brain Sciences, 6,* 210–211.

Creutzfeldt, O., Ojemann, G., & Lettich, E. (1989a). Neuronal activity in human lateral temporal lobe. I. Responses to speech. *Experimental Brain Research, 77,* 451–475.

Creutzfeldt, O., Ojemann, G., & Lettich, E. (1989b). Neuronal activity in human lateral temporal lobe. II. Responses to the subjects own voice. *Experimental Brain Research, 77,* 476–489.

Damasio, A. (1990). Synchronous activation in multiple cortical regions: A mechanism for recall. *Seminars in Neuroscience, 2,* 287–296.

Dewson, J. (1977). Preliminary evidence of hemispheric asymmetry of auditory function in monkeys. In S. Harnad (Ed.), *Lateralization in the nervous system.* New York: Academic Press.

Fried, I., Katz, A., McCarthy, G., Sass, K., Williamson, P., Spencer, S., & Spencer, D. (1991). Functional organization of human supplementary motor cortex studied by electrical stimulation. *Journal of Neuroscience, 11,* 3656–3666.

Fried, I., Ojemann, G. A., & Fetz, E. E. (1981). Language-related potentials specific to human language cortex. *Science, 212,* 353–356.

Haglund, M., Ojemann, G., & Blasdel, G. (1993a). Optical imaging of bipolar cortical stimulation. *Journal of Neurosurgery, 78,* 785–793.

Haglund, M., Ojemann, G., Lettich, E., Bellugi, U., & Corina, D. (1993b). Dissociation of cortical and single unit activity in spoken and signed languages. *Brain and Language, 44,* 19–27.

Haglund, M., Ojemann, G., Schwartz, T., & Lettich, E. (1993c). Neuronal activity in human lateral temporal cortex during serial retrieval from short-term memory. *Journal of Neuroscience,* in press.

Kimura, D. (1983). Sex differences in cerebral organization for speech and praxic functions. *Canadian Journal of Psychology, 37,* 19–35.

Lesser, R., Lueders, H., Dinner, D., Hahn, J., & Cohen, L. (1984). The location of speech and writing functions in the frontal language area. Results of extraoperative cortical stimulation. *Brain, 107,* 275–291.

Liberman, A. M. (1989). Reading is hard just because listening is easy. In G. Lennerstrand, L. Lundberg, & C. VonEuler (Eds.), *Brain and reading.* New York: Stockton Press.

Liberman, A. M., Cooper, F. S., Shankweiler, D. P., & Studdert-Kennedy, M. (1967). Perception of the speech code. _Psychology Review, 74,_ 431–461.

Lüders, H., Lesser, R., Hahn, J. et al. (1986). Basal temporal language area demonstrated by electrical stimulation. _Neurology, 36,_ 505–510.

Mateer, C., & Cameron, P. (1989). Electrophysiological correlates of language stimulation mapping and evoked potential studies. In F. Boller & J. Grafman (Eds.), _Handbook of neuropsychology_ (Vol. 2, pp. 91–116). Amsterdam: Elsevier.

Mateer, C. A., Polen, S. B., Ojemann, G. A., et al. (1982). Cortical localization of finger spelling and oral language. A case study. _Brain and Language, 17,_ 46–57.

Mohr, J. P. (1976). Broca's area and Broca's aphasia. In _Studies in neurolinguistics._ New York: Academic Press.

Mullan, S., & Penfield, W. (1959). Illusions of comparative interpretation and emotion produced by epileptic discharge and by electrical stimulation in temporal cortex. _Archives of Neurology and Psychiatry, 81,_ 269–284.

Ojemann, G. A. (1975). Language and the thalamus: Object naming and recall during and after thalamic stimulation. _Brain and Language, 2,_ 101–120.

Ojemann, G. (1978). Organization of short-term verbal memory in language areas of human cortex: Evidence from electrical stimulation. _Brain and Language, 5,_ 331–348.

Ojemann, G. A. (1983). Brain organization for language from the perspective of electrical stimulation mapping. _The Behavioral and Brain Sciences, 6,_ 189–230.

Ojemann, G. A. (1988). Effect of cortical and subcortical stimulation on human language and verbal memory. In F. Plum (Ed.). _Language communication and the brain_ (pp. 101–115). New York: Raven Press.

Ojemann, G. A. (1989). Some brain mechanisms for reading. In C. Von Euler, I. Lundberg, & G. Lennerstrand (Eds.), _Brain and reading_ (pp. 47–59). New York: Macmillan.

Ojemann, G. (1991a). Cortical organization of language. _Journal of Neuroscience, 11,_ 2281–2287.

Ojemann, G. (1991b). Cortical organization of language and verbal memory based on intraoperative investigations. _Progress in Sensory Physiology, 12,_ 193–230.

Ojemann, G. (1992). Localization of language in frontal cortex. In P. Chauvel, A. Delgado-Eschueta, E. Halgren, & J. Bancaud (Eds.), _Frontal lobe seizures and epilepsies._ New York: Raven Press.

Ojemann, G., Berger, M., & Lettich, E. (1989c). Resective surgery for epilepsy in young children: Method for foci near eloquent areas. _Epilepsia, 30,_ 642.

Ojemann, G., Cawthon, D., & Lettich, E. (1990). Localization and physiological correlates of language and verbal memory in human lateral temporo-parietal cortex. In A. Scheibel & A. Wechsler (Eds.), _Neurobiology of higher cognitive functions_ (pp. 185–202). New York: Guildford Press.

Ojemann, G. A., Creutzfeldt, O., Lettich, E., & Haglund, M. (1988). Neuronal activity in human lateral temporal cortex related to short-term verbal memory, naming and reading. _Brain, 111,_ 1383–1403.

Ojemann, G. A., & Dodrill, C. B. (1985). Verbal memory deficits after left temporal lobectomy for epilepsy. _Journal of Neurosurgery, 62,_ 101–107.

Ojemann, G., & Dodrill, C. B. (1987). Intraoperative techniques for reducing language and memory deficits with left temporal lobectomy. In P. Wolf, M. Dam, D. Janz, & F. Dreifuss (Eds.), _Advances in Epileptology_ (Vol. 16, pp. 327–330). New York: Raven Press.

Ojemann, G. A., Fried, I., & Lettich, E. (1989b). Electrocorticographic (ECoG) correlates of language. I. Desynchronization in temporal language cortex during object naming. _EEG and Clinical Neurophysiology, 73,_ 453–463.

Ojemann, G. A., & Mateer, C. (1979). Human language cortex: Localization of memory, syntax and sequential motor-phoneme identification systems. _Science, 205,_ 1401–1403.

Ojemann, G., Ojemann, J. Haglund, M., Holmes, M., & Lettich, E. (1992). Visually related activity in human temporal cortical neurons. In B. Gulyas, D. Ottoson, & P. Roland (Eds.), *Functional organization of the human visual cortext.*

Ojemann, G., Ojemann, J., Lettich, E., & Berger, M. (1989a). Cortical language localization in left-dominant hemisphere. *Journal of Neurosurgery, 71*, 316–326.

Ojemann, G. A., & Whitaker, H. A. (1978). The bilingual brain. *Archives of Neurology, 35*, 409–412.

Ojemann, J., Ojemann, G., & Lettich, E. (1992). Neuronal activity related to faces and matching in human right nondominant temporal cortex. *Brain, 115*, 1–13.

Penfield, W., & Perot, P. (1963). The brains record of auditory and visual experience—A final summary and discussion. *Brain, 86*, 595–696.

Penfield, W., & Roberts, L. (1959). *Speech and brain mechanisms.* Princeton: Princeton University Press.

Peterson, L., & Peterson, M. (1959). Short-term retention of individual verbal items. *Journal of Experimental Psychology, 58*, 193–198.

Raichle, M. (1990). Exploring the mind with dynamic imaging. *Seminars in Neuroscience, 2*, 307–315.

Rapport, R. L., Tan, C. T., Whitaker, H. A. (1983). Language function and dysfunction among Chinese- and English-speaking polyglots: Cortical stimulation, Wada testing, and clinical studies. *Brain and Language, 18*, 342–366.

Reynolds, A., Turner, P., Harris, A., Ojemann, G., & Davis, L. (1979). Left thalamic hemorrhage with dysphasia: A report of five cases. *Brain and Language, 7*, 62–73.

Smith, B. (1980). Cortical stimulation and speech timing: A preliminary observation. *Brain and Language, 10*, 89–97.

Tallal, P. (1983). A precise timing mechanism may underline a common speech perception and production area in the perisylvian cortex of the dominant hemispheres. *The Behavioral and Brain Sciences, 6*, 219–220.

Trevathen, C. (1974). Functional relations of disconnected hemispheres with the brain stem, and with each other: monkey and man. In M. Kinsbourne & W. Smith (Eds.), *Hemispheric disconnection and cerebral function.* Springfield, Illinois: Thomas.

Wada, J. & Ramussen, T. (1960). Intracarotid injections of sodium amytal for the lateralization of cerebral speech dominance. *Journal of Neurosurgery, 17*, 266–282.

Whitaker, H. A., & Ojemann, G. A. (1978). Graded localization of naming from electrical stimulation mapping of left cerebral cortex. *Nature (London), 270*, 50–51.

Woods, R., Dodrill, G., & Ojemann, G. (1988). Brain injury, handedness and speech lateralization in a series of amobarbital studies. *Annals of Neurology, 23*, 510–518.

Chapter 3

Use of Subdural Electrodes for the Study of Language Functions

Ronald P. Lesser, Santiago Arroyo, John Hart, and Barry Gordon

I. INTRODUCTION

Language processing currently is postulated to be divided into independent components or modules. The main objective of current models has been to specify the nature of these processing components and their interconnections. [The mechanisms of processing within each component often are left unspecified, but see McClelland & Rumelhart (1981).] Evidence for these components and their interrelationships derives from studies of language breakdown in brain-injured individuals via two methodologies. The cognitive neuropsychological method seeks to identify relationships through very detailed models of the components, but typically does not try to reference them to neuroanatomical structures. The classical functional–neuroanatomical approach (our term) attempts to determine functional relationships and to relate them to specific brain regions.

Many of the models that have been proposed are models of normal language functioning. Researchers generally assume that acquired disorders of language can be explained by damage to one or another component(s) of normal architectures, that is, pathology does not add any new functions. This assumption and these models—whether the functional–neuroanatomical models or the cognitive neuropsychological ones—have been extremely successful in accounting for many otherwise obscure patterns of patient performance (Geschwind, 1965a,b; Patterson, Marshall, & Coltheart, 1985).

Despite successes, these models and explanations have many limitations (cf. Ellis, 1985a,b; Henderson, 1981,1982; Marshall, 1986) and a need exists to improve the methods by which the components of a model are isolated and through which their interconnections are established (Marshall, 1986; Patterson, 1981; Safran, 1982; Seidenberg, 1985; Seidenberg,

Bruck, Fornarolo, & Backman, 1985; Shallice, 1981). Components have been isolated operationally by finding one task, or combinations of tasks, that serves as a marker. For example, when a subject is asked to decide whether a string of letters is a word or not (the lexical decision task), only the visual word identification process is logically necessary, so failure on this task can be considered indicative of an impairment of that stage (e.g., Katz & Sevush, 1987). Sometimes, however, several tasks must be used that have different input and/or output requirements but share a (hypothetical) single process. Failure on all relevant tasks (given appropriate performance on other tasks) is considered evidence that this one process is impaired. For example, detecting whether words rhyme and detecting whether pictures rhyme is assumed to tap the common phonologic output stage of the reading and naming models.

Once components have been identified, the next basic question is whether any two components are independent or are somehow associated in the same processing chain. To establish independence, researchers ideally try to find double dissociations of the components, that is, one component is intact (in some cases) even when the other is not (in other cases), and vice versa. Single dissociations give less convincing evidence of the same independence. To establish the functional relatedness of two processes, the researcher tries to establish that their performances covary.

This logic is quite accepted. The problems in practice are with satisfying the basic assumptions (that the patient under study had been normal, and that his or her language processes have not changed because of the pathology), with knowing what aspects of performance are relevant and impaired, and with being able to compare patients. Many of these problems can be traced to the fact that, despite the goal of the understanding of normal language, the bulk of these experiments are done on patients who have suffered lesions (infarctions, hemorrhages, tumors, trauma) that are relatively large and capriciously located, who have essentially fixed deficits, and who are tested many weeks or years after the onset of their illness. A different approach is possible with direct electrical stimulation of the brain.

Electrical cortical stimulation has been used clinically for more than 50 years for the localization of language (and other) functions in patients in whom cortical resections are being considered. This technique has been employed most frequently for intractable epilepsy (Ajmone-Marsan, 1980; Fedio & Van Buren, 1974; Goldring, 1978; Goldring & Gregore, 1984; Jefferson, 1935; Laxer, Needleman, & Rosenbaum, 1984; Lesser, Hahn, Lüders, Rothner, & Erenberg, 1981; Lesser et al., 1987b; Lüders et al., 1982; Ojemann, 1983a; Penfield & Jasper, 1954; Penfield & Perot, 1963; Penfield & Roberts, 1959; Rapport, Tan, & Whitaker, 1983; Rasmussen & Milner,

1975; Schomer, Erba, Blume, Spiers, & Ives, 1984; Serafetinides, 1966; Van Buren, Ajmone-Marsan, & Mutsuga, 1975; Van Buren, Fedio, & Frederick, 1978; Wyler, Ojemann, Lettich, & Ward, 1984), but increasingly also has been used for tumors (Morris et al., 1986) and other nonepileptic lesions. Direct cortical electrical stimulation mapping of language at the time of cortical exposure during surgery (Jefferson, 1935; Fedio & Van Buren, 1974; Ojemann, 1983a; Penfield & Jasper, 1954; Penfield & Perot, 1963; Penfield & Roberts, 1959; Rapport et al., 1983; Rasmussen & Milner, 1975; Serafetinides, 1966; Van Buren et al., 1975) necessarily has severe time limitations, which limits the amount of language testing that can be accomplished (60 min or less). Recent technical advances, coupled with clinical needs, have led to the implantation of arrays of subdural electrodes. The patient then can be fully awake and comfortable while the electrodes are in place, for periods of days to weeks. Several clinicians are now using such subdural arrays (Ajmone-Marsan, 1980; Goldring, 1978; Goldring & Gregore, 1984; Laxer et al., 1984; Lesser, Lüders, Dinner, Hahn, & Cohen, 1984; Lesser et al., 1981; Lüders et al., 1982, 1986a,b; J. C. Morris, Cole, Banker, & Wright, 1984a; H. H. Morris, Lüders, Lesser, Dinner, & Hahn, 1984; Schomer et al., 1984; Van Buren et al., 1975; Wyler et al., 1984), taking advantage of the systematic unhurried stimulation studies this technique allows.

Essentially, the technique involves having a small (e.g., 0.5–3 mm in diameter) electrode resting directly on a cortical gyrus while the patient is awake and responsive. The patient then performs a task. Current is applied through the electrode at some point during performance. Levels of current that do not produce seizures or after-discharges (as previously determined at rest) nonetheless may interfere with task performance. The impairment typically lasts the duration of the current application, and (usually) ends after the current is turned off. Hence, cortical stimulation acts as a reversible cortical lesion under these circumstances. The anatomical extent of the electrical "lesion" often appears to be only several millimeters in diameter (Lesser et al., 1984, 1986; H. H. Morris et al., 1984; Ojemann, 1983a).

The cortical stimulation technique offers the unique advantage over lesion studies of validating the prelesion (prestimulation) level of performance on a processing component or task by testing without current (or with current in another region). Therefore, individual differences in functioning can be determined and taken into account because the subject serves as his or her own control. This internal control can help address two criticisms of lesion studies: such studies may be reporting rare but normal variations as pathological performance and, by their nature, they may allow only relatively gross impairments to be admissible as evidence.

In addition, the technique allows both cortical stimulation and baseline (nonstimulation) testing to be repeated, which can help establish the reliability of any purported impairments.

Single case lesion studies have been championed to meet the problems that are inherent in averaging over groups with unknown premorbid status and unknown functional impairments (Schwartz, 1984; Shallice, 1979; Smits, Gordon, Witte, Rasmusson, & Zarzecki, 1991). However, establishing double dissociations in a single case with a fixed lesion is impossible. Cortical stimulation studies can be single case studies as well (although admittedly with less extensive testing possible, because of time constraints). However, because of the reversibility of the stimulation "lesion", and the possibility of finding different impairments at different stimulation sites, these studies have the potential of showing double dissociations in the same individual. Additionally, data from single case stimulation studies can be summed for group comparisons to filter out idiosyncratic effects.

II. ELECTRODE CHARACTERISTICS AND PLACEMENT

Subdural electrode materials should be made of a good electrical conductor, inert to normal tissues and resistant to corrosion. Both stainless steel and platinum–iridium compound have these properties and have been used successfully. However, theoretically, platinum–iridium may have some advantages because of its higher stability for stimulation procedures and its lack of magnetic properties (Mortimer, Shealy, & Wheeler, 1970; White & Gross, 1974). Electrode contacts are of different diameters, varying from 2 to 5 mm, with center-to-center distances of 0.5–2 cm between electrodes. They are embedded in a thin (0.5 mm) transparent silastic plate in which the electrodes are laid out in a rectangular array. Silastic serves three purposes. First, it keeps the electrodes in a determined position; second, it allows visual inspection of the cortical surface under the electrodes, thus permitting visualization of the actual relationships of the electrodes to the underlying brain convolutions and vessels; third, silastic increases the flexibility of the plate, making possible its general adaptation to the overall brain surface. The electrodes are arranged in a variety of arrays to be applicable to different placement locations. The most usual designs are 8×8, 6×8, and 2×8, called grids, and 1×8 and 1×4, called strips.

Subdural electrodes are put in place with the patient under general anesthesia. The approach to their placement differs depending on the expected region of the patient's seizures and on the number of electrodes thought to be necessary in each case. Placement of subdural grids is done through a craniectomy. Scalp and bone flaps are tailored depending on the area of interest; then the dura is opened and the electrodes are put in place under direct visual inspection. Special attention is directed to avoiding positioning the electrodes over cortical veins because electrodes so placed could produce painful sensations when later stimulated (Lesser et al., 1985). The grid is fixed in place by suturing the corners to the dura and by stitching the dura to the cable. The dura is closed with continuous suture; we usually employ human graft dura to avoid excessive pressure on the grid and the subjacent brain. The craniectomy is sealed with silastic adhesive, and the scalp incision is closed in two layers. The cables from the grid exit through the craniectomy. They are tunneled under the scalp and exit at a certain distance from the craniectomy opening. Strict antiseptic procedures are used in caring for the external surgical incisions.

Subdural strips are inserted through a blur hole and are slipped into the subarachnoid space under the temporal lobe or over the convexity of the hemisphere. Frequently, bilateral strips are used (Spencer, 1989; Spencer, Spencer, Mattson, & Williamson, 1984; Spencer, Spencer, Williamson, & Mattson, 1990; Wyler, Richey, & Hermann, 1989; Wyler et al., 1984a&b) for lateralization within both hemispheres. Although language testing can be performed through strips, the limited number of electrodes results in considerably more limited data about language (or other) functioning in the area of interest.

These electrodes are placed for clinical purposes, generally to locate areas of epileptogenesis in patients with epilepsy and (especially in the case of grids) to determine the relationship of these areas to regions important to language, motor, or other functions. Therefore their arrangement differs among patients. In patients with temporal lobe seizure foci, unilateral or bilateral strips or grids can be used. For example, in our research center, two unilateral grids usually are inserted in the case of a unilateral temporal lobe focus of unknown extent over the "language dominant" hemisphere: a 6×8 or 8×6 grid over the convexity of the brain covering the temporal and inferior temporal regions and a 2×8 grid under the basal temporal region.

III. INTERPRETATION OF THE RESULTS OF STIMULATION

Electrical stimulation of the human brain identifies areas that are, at least acutely, essential for a given task. Being essential, however, does not mean that these sites are the only areas that can generate the function that has been impaired. Several observations support this assertion. First, some portions of functionally relevant areas can be resected without an evident functional deficit (Uematsu et al., 1991a; (Uematsu, Lesser, & Gordon, 1991b). Second, when an electrode is stimulated for a prolonged period, the deficit elicited at the beginning of the stimulation might, in some cases, fade away at the end of it (Lesser et al., 1986). Third, partial lesions, for example, those caused by strokes, in functionally important cortical areas do not necessarily produce measurable or permanent functional deficits (Jenkins & Merzenich, 1987). Fourth, some hours after a motor nerve has been transected, stimulation of the cortical area that, before transection, was responsible for the movement of muscles innervated by that nerve now influences another group of muscles (Donoghue & Sanes, 1988; Donoghue, Suner, & Sanes, 1990). Fifth, neurons neighboring a lesioned area gain new receptive fields after the acute lesion (Jenkins & Merzenich, 1987). The quick time course over which these changes sometimes occur suggests that new connections have not grown. Instead, existing synaptic connections have been altered in their effects and functional reorganization has occurred consequently by modification of the inhibitory and excitatory connections between the cells involved (Jacobs & Donoghue, 1991; Jenkins & Merzenich, 1987; Smits et al., 1991).

Consequently, deficits found during cortical stimulation may reflect an acute interference with the neural connections. Because testing occurs immediately, reorganization is unlikely to occur. Findings during stimulation therefore do not necessarily reflect the same kinds of pathology found in patients with long standing structural lesions.

IV. STIMULATION PARAMETERS

Electrical stimulation should be delivered with a current-regulated patient-isolated stimulator. The stimulus should be of alternating polarity to avoid progressive polarization of the electrodes, minimizing the release of ions at the metal–solution interface and thus diminishing the possibility of neural damage (Lilly, 1961; Mortimer et al., 1970). We use 0.3-msec pulses of alternating polarity in trains of 50 pulses per second and 1–5 sec in duration. We start at a low intensity, 0.5–1 mA, and increase stepwise at 0.5- to 1-mA intervals until one of three end points is achieved: functional changes appear, after-discharges are recorded, or the intensity limits of the stimulators (15–17.5 mA) are reached (Gordon et al., 1990; Lesser et al., 1984,1987a,b). In addition, electrode stimulation may evoke painful or unpleasant sensations because of stimulation of fifth cranial nerve twigs that accompany the pial vasculature (Lesser et al., 1985). When this occurs, stimulation in general cannot be performed at that site.

After-discharges induced by electrical stimulation are defined as rhythmical spikes occurring in response to stimulation, once the stimulation is finished. The presence of after-discharges depends on the intensity of the stimulus and on the excitability of the subjacent cortex. After-discharges may propagate and disrupt functions in remote areas. Further, stimulation may trigger subclinical or clinical seizures that are of poor localizing value. Because of these considerations, in general the goal is to obtain functional changes in the absence of after-discharges.

V. SAFETY

Patients with subdural electrodes should be considered in a high risk category with respect to electrical safety, and should be isolated from ground and from the recording instrument. Also, several concerns exist about the safety of subdural electrode stimulation. First, could it be injurious to brain tissues? Second, could it kindle seizures? We briefly review these issues here.

Animal studies have shown that damage to the cortex occurs when certain stimulation settings have been exceeded. Variables that best correlate with neural tissue damage from electrical stimulation are charge per phase and charge density per phase (Agnew & McCreery, 1987; Agnew, Yuen, & McCreery, 1983; McCreery & Agnew, 1983; Yuen, Agnew, Bullara, Jacques, & McCreery, 1981). These parameters determine the current passed for the duration of the stimulus pulse and per area of tissue. For example, if the current or the duration of the stimulus is increased, a higher charge per phase or charge density per phase is attained. On the contrary, if the area of tissue stimulated by the electrode or the electrode diameter is increased, a lower charge density per phase is achieved. In animal experiments, when charge per phase is over 0.4 μC/ph and charge density per phase is over 40 $\mu C/cm^2 \cdot ph$, neural damage occurs. Several mechanisms are believed to be involved in tissue damage. First, neural excitability produced by stimulation can alter the brain mechanisms to maintain the extracellular–intracellular potassium and calcium homeostasis (Agnew & McCreery, 1987; Agnew et al., 1983; McCreery & Agnew, 1983). Second, stimulation induces changes in transference across the electrode–tissue interface that may produce electrochemical toxic products or electrode dissolution products (White & Gross, 1974).

These parameters are based on continuous chronic stimulation experiments in animals. By comparison, electrodes are left in people for a relatively short period of time and stimulation occurs for only a brief portion of the period of implantation. We have analyzed the effects of electrical stimulation after chronic intracranial subdural recording and stimulation. Nonspecific findings such as mild diffuse gliosis of cortical layers 1 and 2 and an increase in mononuclear inflammatory cells in the subarachnoid space were shown. These findings are most likely related to a reaction to a foreign body, that is, to the placement of the silastic plate. No pathological features correlated with the presence of the electrodes or with the amount of electrical stimulation given (Gordon et al., 1990). Further, neural damage typically reported as caused by electrical stimulation (i.e., vacuolation and degeneration of the neurons) (Yuen et al., 1981) was not observed.

Although being concerned about the possibility of a kindling effect of electrical stimulation in humans is reasonable, no reports exist that this has ever actually happened over more than 50 years of experience with stimulation procedures (Lesser et al., 1987a). Kindling is achieved by repeated cortical stimulation in daily sessions, resulting in the development of permanent lowering of the after-discharge threshold both at the point of the stimulation and at a distant areas (Sato, Racine, & McIntyre, 1990). The kindling phenomenon occurs only with certain stimulus parameters and with daily stimulus repetition for a prolonged period of time over the same area. These conditions never occur with electrical stimulation through subdural electrodes. In addition, kindling the cortex is much more difficult than kindling subcortical structures (Racine, 1978). Moreover, in primates kindling becomes more difficult than in rats (Wada, Mizuguchi, & Osawa, 1978).

Induction of single seizures is a well-known risk of electrical stimulation of the cortex. To minimize their appearance, stimulation intensity at each site should begin at a low level and increase gradually.

VI. EFFECTS OF STIMULATION

In the cases of patients with chronic fixed structural lesions (e.g., after strokes) and of patients with chronic epilepsy coming to electrical cortical stimulation, we can question how representative their language processing mechanisms are of language processing mechanisms in normal individuals, given the possibility of long-term language reorganization and other changes (Coltheart, 1980; Henderson, 1982; Ojemann, 1979,1983a; Ojemann & Mateer, 1979; Ojemann & Whitaker, 1978; Rasmussen & Milner, 1977). First, however, examination of sufficient patients should reveal invariances and patterns that provide an indication of normal functions and arrangements. Second, some patients coming to cortical stimulation have relatively recently acquired lesions (e.g., tumors; H. H. Morris et al., 1984); thus, stimulation typically includes sites separate from the lesion itself. Such patients offer the potential of studying neuroanatomic language distributions when long-term disruptions and reorganization due to epilepsy and its concomitants are less likely. Third, since current application is acute and brief, no appreciable chance exists for functional reorganization to occur during the time of stimulation. Fourth, although how electrical stimulation affects cortical functioning is not clear, how destructive lesions produce their effects is not clear either. If these arguments are accepted, stimulation studies should contribute to knowledge of normal language functions.

Electrical stimulation with the 3-mm surface electrode we have described is likely to affect both cortical neurons and subcortical U fibers and other fiber pathways (Landau, Bishop, & Clare, 1965; Ranck, 1975). Therefore, cortical stimulation is likely to have both local and more remote effects that are not necessarily more complex, nor less well understood, than the effects of destructive lesions (e.g., Astrup, Siesjo, & Symon, 1981; Kass, 1987).

The functional–neuroanatomical and the cognitive neuropsychological approaches both have relied on the fractionations of behavior that accidental cerebral lesions and other forms of pathology appear to cause. Electrical cortical stimulation also has been shown to fractionate behaviors and to isolate processing components, although this effect has not been tested as extensively to date. These observations (which are reviewed in the next section), the nature of cortical stimulation, and the well-known existence of fractionation in the case of cerebral lesions make it very likely that fractionation of other types of behaviors also will be observed with cortical stimulation.

The two great advantages of the fractionation that may be observed with electrical stimulation over that caused by cerebral lesions are that it will be reversible and that it is under experimental control. Reversibility is important in part because it helps establish the existence of a true deficit for that patient, something that must be inferred in lesion studies. However, of even greater potential importance is the fact that deficits can be induced under experimental control while the effects on other functions are determined. Hence, with this particular technique, the investigator may be able to determine whether or not language components are necessarily associated in the same patient, and perhaps even determine some information about their dynamic interrelationships, which is not possible in lesion studies.

VII. FUNCTIONAL FRACTIONATIONS OBSERVED

Electrical cortical stimulation frequently produces dissociations in closely related functions such as reading and naming (e.g., Lesser et al.,

1984,1986; H. H. Morris et al., 1984; Ojemann, 1983a,b). For example, in bilingual subjects, a number of investigators have reported that naming can be accomplished in one language but not in the other during stimulation at the same site (Ojemann, 1983a; Ojemann & Whitaker, 1978; Penfield & Jasper, 1954; Rapport et al., 1983).

In another study, H. H. Morris et al. (1984) were able to fractionate the components of Gerstmann's syndrome by stimulating different sites within and just inferior to the angular gyrus. The tasks they used were naming of pictures, objects, colors, and shapes, and reading of connected paragraphs. Stimulation was begun after the patients had started reading, or before naming began, and resulted in deficits within a few seconds. Elementary motor deficits were excluded; the patient was able to make rapid alternating movements of the tongue and repeat syllables when he or she was unable to name objects or to read. Of significance here is that anomia alone occurred at three sites, whereas alexia alone occurred at one site. This result could be evidence that more unique, earlier stages of processing of each task were being affected, although this possibility was not pursued.

Lesser et al. (1986) also were able to induce impairments of auditory comprehension, responsive naming, and reading in various combinations (but without additional systematic exploration of their interrelationships) by stimulating the posterior superior temporal gyrus. In one case (number 210), recognition testing after a responsive naming task and after current had been turned off showed better performance than testing after naming during stimulation. Whether this result was due to partial residual semantic knowledge or to recovery of completely intact representations of word forms was not assessed definitively. Lesser and his colleagues (1984) also have demonstrated, during stimulation of anterior speech areas, that speech arrest could be produced with or without impairments of rapid finger movements or writing, but writing was never impaired without also producing difficulties with finger movements or speech. In the context presented earlier, these findings suggest that writing in these patients somehow depends on speech and finger movement, but not vice versa.

VIII. ANATOMICAL FRACTIONATION

The same studies just reviewed also help support the notion that cortical stimulation produces very anatomically delimited effects. For the most part, areas of speech involvement obey the classical neuroanatomical boundaries (Lesser et al., 1984,1986; Lüders et al., 1986a; H. H. Morris et al., 1984), supporting at least that degree of neuroanatomical resolution.

An even finer degree of spatial specificity is suggested by the routine findings that electrodes 1 cm apart could give very different patterns of results, comparable to the results of others (Ojemann, 1983a).

In addition, stimulation of a circumscribed area of the basal temporal lobe produces speech arrest or slowing of speech, as well as a confrontation naming deficit (Burnstine et al., 1990; Kluin, Abou-Khalil, & Hood, 1988; Lüders et al., 1986a,1989,1991). This effect occurs in response to stimulation of the fusiform (Lüders et al., 1989,1991), inferior temporal, and parahippocampal (Burnstine et al., 1990) gyri. The length of the area varies, beginning as anterior as 11 mm from the temporal tip and extending to 74 mm posteriorly (Burnstine et al., 1990; Lüders et al., 1989,1991).

IX. POSSIBLE LIMITATIONS

Several possible difficulties arise in the interpretation of the data from electrical stimulation studies (see, for example, the commentary by Ojemann, 1983a), including atypicality of the subject population and lack of patient subgroups of sufficient size (e.g., of the tumor patients); effects of ongoing seizures and spike discharges; unappreciated changes in stimulation threshold; learning effects of repeated trials; changes in the stimulation threshold with repeated trials; differences between electrical stimulation effects and lesion effects in general and between electrical stimulation effects and the state of the chronic postlesion brain in particular; the possibility that the experimental variance will be too great, or the statistical power insufficient, with the current N values; and the possibility that some of the performance effects are due to memory failures and report failures, rather than to processing defects in the tasks themselves.

In our (and others') use of the technique, we have found it important to try to anticipate these problems. For example, the single-item technique, with immediate responses, minimizes any memory burden or memory effects and facilitates the choice of the optimal time for presenting each trial. Patients undergoing cortical stimulation should have continuous concurrent electroencephalogram (EEG) monitoring; trials that overlap with frequent spikes or with frank seizures can be noted and discarded. If some of the dissociations discussed earlier can be produced reliably in more than one patient, then the finding cannot be attributed to premorbid idiosyncracies or to peculiarities of the technique itself. However, negative findings (that is, the inability to reproduce any of the classic disassociations) obviously will be much more difficult to interpret. The question then will be whether cortical stimulation produces its own pattern of dissociations that perhaps has no relation to the stages and the pathways derived from chronic, stable, lesioned patients. Even so, this finding would be significant in its own right.

Cortical stimulation is a technique that has been utilized for over half a century for the purpose of assisting in answering surgically related localization questions in patients with seizures and with structural lesions. In addition, because of the reversible nature of the effects produced, stimulation offers the possibility of developing new insights into the mechanisms underlying human language.

ACKNOWLEDGMENTS

This investigation has been supported in part by the National Institutes of Health (Grant No. 1 RO1 NS26553 from the National Institute of Neurological Disorders and Stroke), The Whittier Foundation, The Seaver Foundation, and The McDonnel-Pew Program in Cognitive Neuroscience. S. Arroyo has been supported by a grant from the Fondo de Investigaciones Sanitarias, Spain.

REFERENCES

Agnew, W. F., & McCreery, D. B. (1987). Considerations for safety in the use of extracranial stimulation for motor evoked potentials. *Neurosurgery, 20,* 143–147.

Agnew, W. F., Yuen, T. G. H., & McCreery, D. B. (1983). Morphologic changes after prolonged electrical stimulation of the cat's cortex at defined charge densities. *Experimental Neurology, 79,* 397–411.

Ajmone-Marsan, C. (1980). Depth electrography and electrocorticography. In M. J. Aminoff (Ed.), *Electrodiagnosis in clinical neurology* (pp. 167–196). New York: Churchill Livingstone.

Astrup, J., Siesjo, B. K., & Symon, L. (1981). Thresholds in cerebral ischemia: The ischemic penumbra. *Stroke, 12,* 723–725.

Burnstine, T. H., Lesser, R. P., Hart, J., Uematsu, S., Zinreich, S. J., Krauss, G. L., Fisher, R. S., Vining, E. P., & Gordon, B. (1990). Characterization of the basal temporal language area in patients with left temporal lobe epilepsy. *Neurology, 40,* 966–970.

Coltheart, M. (1980). Deep dyslexia: A right hemisphere hypothesis. In M. Coltheart, K. Patterson, & J. C. Marshall (Eds.), *Deep dyslexia* (pp. 325–380). London: Routledge & Kegan Paul.

Donoghue, J. P., & Sanes, J. N. (1988). Organization of adult motor cortex representation patterns following neonatal forelimb nerve injury in rats. *Journal of Neurosciences, 8,* 3221–3232.

Donoghue, J. P., Suner, S., & Sanes, J. N. (1990). Dynamic organization of primary motor cortex output to target muscles in adult rats. II. Rapid reorganization following motor cortex nerve lesions. *Experimental Brain Research, 79,* 492–503.

Ellis, A. W. (1985a). The production of spoken words: A cognitive neuropsychological perspective. In A. W. Ellis (Ed.), *Progress in the psychology of language,* (Vol. 2, pp. 107–145). Hillsdale, New Jersey: Erlbaum Associates.

Ellis, A. W. (1985b). The cognitive neuropsychology of developmental (and acquired) dyslexia: A critical review. *Cognitive Neuropsychology, 2*, 169–205.

Fedio, P., & Van Buren, J. (1974). Memory deficits during electrical stimulation of the speech cortex in conscious man. *Brain and Language, 1*, 29–42.

Geschwind, N. (1965a). Disconnexion syndromes in animals and man. *Brain, 88*, 585–644.

Geschwind, N. (1965b). Disconnexion syndromes in animals and man. *Brain, 88*, 237–294.

Goldring, S. (1978). A method for surgical management of focal epilepsy, especially as is relates to children. *Journal of Neurosurgery, 49*, 344–356.

Goldring, S., & Gregore, E. M. (1984). Surgical management of epilepsy using epidural recordings to localize the seizure focus. Review of 100 cases. *Journal of Neurosurgery, 40*, 447–466.

Gordon, B., Lesser, R. P., Rance, N. E., Hart, J., Webber, R. S., Uematsu, S., & Fisher, R. S. (1990). Parameters for direct cortical electrical stimulation in the human: Histopathologic confirmation. *Electroencepahlography and Clinical Neurophysiology, 75*, 371–377.

Henderson, L. (1981). Information processing approaches to acquired dyslexia. *Quarterly Journal of Experimental Psychology, 33A*, 507–522.

Henderson, L. (1982). *Orthography and word recognition in reading*. New York: Academic Press.

Jacobs, K. M., & Donoghue, J. P. (1991). Reshaping the cortical motor map by unmasking latent intracortical connections. *Science, 251*, 944–947.

Jefferson, G. (1935). Jacksonisan epilepsy: A background and postscript. *Postgraduate Medical Journal, 11*, 150–162.

Jenkins, W. M., & Merzenich, M. M. (1987). Reorganization of neocortical representations after brain injury: A neurophysiological model of the bases of recovery from stroke. *Progress in Brain Research, 71*, 249–266.

Kass, J. H. (1987). The organization of the neocortex in mammals: Implications for theories of brain function. *Annual Review of Psychology, 38*, 129–151.

Katz, R. B., & Sevush, S. (1987). Accurate reading by nonlexical means: A case study. *Brain and Language, 31*, 252–266.

Kluin, K., Abou-Khalil, B., & Hood, T. (1988). Inferior speech area in patients with temporal lobe epilepsy. *Neurology, (Suppl. 1), 38*, 277.

Landau, W. M., Bishop, G. H., & Clare, M. H. (1965). Site of excitation in stimulation of the motor cortex. *Journal of Neurophysiology, 28*, 1206–1222.

Laxer, K. D., Needleman, R., & Rosenbaum, T. J. (1984). Subdural electrodes for seizure focus localization. *Epilepsia, 25*, 651.

Lesser, R. P., Hahn, J. F., Lüders, H., Rothner, A. D., & Erenberg, G. (1981). The use of chronic subdural electrodes for cortical mapping of speech. *Epilepsia, 22*, 240.

Lesser, R. P., Lüders, H., Dinner, D. S., Hahn, J. F., & Cohen, L. (1984). The location of speech and writing functions in the frontal language area. Results of extraoperative cortical stimulation. *Brain, 107*, 275–291.

Lesser, R. P., Lüders, H., Dinner, D. S., Klem, G., Hahn, J. F., & Harrison, M. (1986). Electrical stimulation of Wernicke's area interferes with comprehension. *Neurology, 36*, 658–663.

Lesser, R. P., Lüders, H., Klem, G., Dinner, D. S., Morris, H. H., & Hahn, J. F. (1984). Cortical afterdischarge and functional response thresholds: Results of extraoperative testing. *Epilepsia, 25*, 615–621.

Lesser, R. P., Lüders, H., Klem, G., Dinner, D. S., Morris, H. H., & Hahn, J. F. (1985). Ipsilateral trigeminal sensory responses to cortical stimulation by subdural electrodes. *Neurology, 35*, 1760–1763.

Lesser, R. P., Lüders, H., Klem, G., Dinner, D. S., Morris, H. H., Hahn, J. F., & Wyllie, E. (1987a). Extraoperative cortical localization in patients with epilepsy. *Journal of Clinical Neurophysiology, 4*, 27–53.

Lesser, R. P., Lüders, H., Klem, G., Dinner, D. S., Morris, H. H., Hahn, J. F., & Wyllie, E. (1987b). Extraoperative cortical functional localization in patients with epilepsy. *Journal of Clinical Neurophysiology, 4*, 27–53.

Lilly, J. C. (1961). Injury and excitation by electric currents. A. The balanced pulse-pair waveform. In D. E. Sheer (Ed.), *Electrical stimulation of the brain* (pp. 60–64). Austin: University of Texas Press.

Lüders, H., Hahn, J. F., Lesser, R. P., Dinner, D. S., Morris, H. H., Wyllie, E., Friedman, L., Friedman, D., & Skipper, G. (1989). Basal temporal subdural electrodes in the evaluation of patients with intractable epilepsy. *Epilepsia, 30*, 131–142.

Lüders, H., Hahn, J. F., Lesser, R. P., Dinner, D. S., Rothner, A. D., & Erenberg, G. (1982). Localization of epileptogenic spike foci: Comparative study of closely spaced scalp electrodes, nasopharyngeal, sphenoidal, subdural, and depth electrodes. In H. Akimoto, H. Kazamatsuri, M. Seino, & A. Ward (Eds.), *Advances in Epileptology: XII Epilepsy International Symposium* (pp. 185–189). New York: Raven Press.

Lüders, H., Lesser, R. P., Dinner, D. S., Morris, H. H., Wyllie, E., & Klem, G. (1986a). Comprehension deficits elicited by electrical stimulation of Broca's area. *Epilepsia, 27*, 598.

Lüders, H., Lesser, R. P., Hahn, J. F., Dinner, D. S., Morris, H. H., Resor, S., & Harrison, M. (1986b). Basal temporal language area demonstrated by electrical stimulation. *Neurology, 36*, 505–510.

Lüders, H., Lesser, R. P., Hahn, J. F., Dinner, D. S., Morris, H. H., Wyllie, E., & Godoy, J. (1991). Basal temporal language area. *Brain, 114*, 743–754.

Marshall, J. C. (1986). The description and interpretation of aphasic language disorder. *Neuropsychologia, 24*, 5–24.

McClelland, J. L., & Rumelhart, D. E. (1981). An interactive activation model of context effects in letter perception. Part 1. An account of basic findings. *Psychological Review, 88*, 375–407.

McCreery, D. B., & Agnew, W. F. (1983). Changes in extracellular potassium and calcium concentration and neural activity during prolonged electrical stimulation of the cat cerebral cortex at defined charge densities. *Experimental Neurology, 79*, 371–396.

Morris, H. H., Lüders, H., Hahn, J. F., Lesser, R. P., Dinner, D. S., & Estes, M. L. (1986). Neurophysiological techniques as an aid to surgical treatment of primary brain tumors. *Annals of Neurology, 19*, 559–567.

Morris, H. H., Lüders, H., Lesser, R. P., Dinner, D. S., & Hahn, J. F. (1984). Transient neuropsychological abnormalities (including Gerstmann's syndrome) during cortical stimulation. *Neurology, 34*, 877–883.

Morris, J. C., Cole, M., Banker, B. Q., & Wright, D. (1984). Hereditary dysphasic dementia and the Pick-Alzheimer spectrum. *Annals of Neurology, 16*, 455–466.

Mortimer, J. T., Shealy, C. N., & Wheeler, C. (1970). Experimental nondestructive electrical stimulation of the brain and spinal cord. *Journal of Neurosurgery, 32*, 553–559.

Ojemann, G. A. (1979). Individual variability in cortical localization of language. *Journal of Neurosurgery, 50*, 164–169.

Ojemann, G. A. (1983a). Brain organization for language from the prespective of electrical stimulation mapping. *The Behavioral and Brain Sciences, 2*, 189–230.

Ojemann, G. A. (1983b). Electrical stimulation and the neurobiology of language. *The Behavioral and Brain Sciences, 6*, 221–226.

Ojemann, G. A., & Mateer, C. (1979). Human language cortex: Localization of memory, syntax, and sequential motor–phoneme identification systems. *Science, 205*, 1401–1403.

Ojemann, G. A., & Whitaker, H. (1978). Language localization and variability. *Brain and Language, 6*, 239–260.

Patterson, K. (1981). Neuropsychological approaches to the study of reading. *British Journal of Psychology, 72*, 151–174.

Patterson, K., Marshall, J. C., & Coltheart, M. (1985). *Surface dyslexia*. Hillsdale, New Jersey: Erlbaum Associates.

Penfield, W., & Jasper, H. (1954). *Epilepsy and the functional anatomy of the human brain*. Boston: Little Brown.

Penfield, W., & Perot, P. (1963). The brain's record of auditory and visual experience—A final summary and discussion. *Brain, 86*, 595–696.

Penfield, W., & Roberts, L. (1959). *Speech and brain mechanisms*. Princeton, New Jersey: Princeton University Press.

Racine, R. J. (1978). Kindling: The first decade. *Neurosurgery, 3*, 234–252.

Ranck, J. B., Jr. (1975). Which elements are excited in electrical stimulation of mammalian central nervous system: A review. *Brain Research, 98*, 417–440.

Rapport, R. L., Tan, C. T., & Whitaker, H. A. (1983). Language function and dysfunction among Chinese-and-English-speaking polyglots: Cortical stimulation, Wada testing, and clinical studies. *Brain and Language, 18*, 342–366.

Rasmussen, T., & Milner, B. (1975). Clinical and surgical studies of the cerebral speech areas in man. In K. J. Zulch, O. Creutzfeldt, & G. C. Galbraith (Eds.), *Otfried Foerster Symposium on cerebral localization* (pp. 238–257). New York: Springer-Verlag.

Rasmussen, T., & Milner, B. (1977). The role of early left brain injury in determining lateralization of cerebral speech functions. *Annals of the New York Academy of Sciences, 299*, 355–369.

Safran, E. M. (1982). Neuropsychological approaches to the study of language. *British Journal of Psychology, 73*, 317–337.

Sato, S., Racine, R. J., & McIntyre, W. J. (1990). Kindling: Basic mechanisms and clinical validity. *Electroencephalography and Clinical Neurophysiology, 76*, 459–472.

Schomer, D. L., Erba, G., Blume, H., Spiers, P., & Ives, J. (1984). The utility of subdural strip recordings for the localization of epileptic activity. A case report. *Electroencephalography and Clinical Neurophysiology, 58*, 125P.

Schwartz, M. F. (1984). What the classical aphasic categories can't do for us, and why. *Brain and Language, 21*, 3–8.

Seidenberg, M. S. (1985). Constraining models of word recognition. *Cognition, 20*, 169–190.

Seidenberg, M. S., Bruck, M., Fornarolo, G., & Backman, J. (1985). Word recognition processes of poor disabled readers: Do they necessarily differ? *Applied Psycholinguistics, 6*, 161–180.

Serafetinides, E. A. (1966). Speech findings in epilepsy and electro-cortical stimulation: An overview. *Cortex, 2*, 463–473.

Shallice, T. (1979). Case study approach in neuropsychological research. *Journal of Clinical Neuropsychology, 1*, 183–211.

Shallice, T. (1981). Phonological agraphia and the lexical route in writing. *Brain, 104*, 413–429.

Smits, E., Gordon, D. C., Witte, S., Rasmusson, D. D., & Zarzecki, P. (1991). Synaptic potentials evoked by convergent somatosensory and corticocortical inputs in raccoon motor cortex: Substrates of plasticity. *Journal of Neurophysiology, 66*, 688–695.

Spencer, D. D., Spencer, S. S., Mattson, R., & Williamson, P. D. (1984). Intracerebral masses in patients with intractable partial epilepsy. *Neurology, 34*, 432–436.

Spencer, S. S. (1989). Depth versus subdural electrode studies for unlocalized epilepsy. *Journal of Epilepsy, 2*, 123–127.

Spencer, S. S., Spencer, D. D., Williamson, P. D., & Mattson, R. (1990). Combined depth and subdural electrode investigation in uncontrolled epilepsy. *Neurology, 40*, 74–79.

Uematsu, S., Lesser, R. P., Fisher, R. S., Gordon, B., Hara, K., Krauss, G. L., Vining, E. P., & Webber, R. W. (1992a). Motor and sensory cortex in humans: Topography studied with chronic subdural stimulation. *Neurosurgery, 31*, 59–72.

Uematsu, S., Lesser, R. P., & Gordon, B. (1992b). Localization of sensorimotor cortex: The influence of Sherrington and Cushing on the modern concept. *Neurosurgery, 30,* 904–913.

Van Buren, J., Ajmone-Marsan, C., & Mutsuga, N. (1975). Temporal lobe seizures with additional foci treated by resection. *Journal of Neurosurgery, 43,* 596–607.

Van Buren, J., Fedio, P., & Frederick, G. C. (1978). Mechanism and localization of speech in the parietotemporal cortex. *Journal of Neurosurgery, 2,* 233–238.

Wada, J. A., Mizuguchi, T., & Osawa, T. (1978). Secondarily generalized convulsive seizures by daily amygdaloid stimulation in the rhesus monkeys. *Neurology, 28,* 1026–1036.

White, R. L., & Gross, T. J. (1974). An evaluation of the resistance to electrolysis of metals for use in biostimulation microprobes. *IEEE Transactions of Biomedical Engineering, 21,* 487–490.

Wyler, A. R., Ojemann, G. A., Lettich, E., & Ward, A. A. (1984a). Subdural strip electrodes for localizing epileptogenic foci. *Journal of Neurosurgery, 60,* 1195–1200.

Wyler, A. R., Ojemann, G. A., Lettich, E., & Ward, A. A. (1984b). Subdural strip electrodes for localizing epileptogenic foci. *Journal of Neurosurgery, 60,* 1195–1200.

Wyler, A. R., Richey, E. T., & Hermann, B. P. (1989). Comparison of scalp to subdural recordings for localizing epileptogenic foci. *Journal of Epilepsy, 2,* 91–96.

Yuen, T. G. H., Agnew, W. F., Bullara, L. A., Jacques, S., & McCreery, D. B. (1981). Histological evaluation of neural damage from electrical stimulation: Considerations for the selection of parameters for clinical application. *Neurosurgery, 9,* 292–299.

Chapter 4

Localizing the Neural Generators of Event-Related Brain Potentials

Diane Swick, Marta Kutas, and Helen J. Neville

I. INTRODUCTION

A major approach to neural localization has been inferring the location of cognitive operations within "normal" brains from the behavioral deficits and brain damage observed in neurological patient populations (Caramazza, 1992; Kosslyn & Intriligator, 1992). Years of research attest to the difficulty of this enterprise and underscore the need for converging evidence from multiple techniques. This chapter focuses specifically on describing how recordings of the electrical activity of the brain (in particular, transient responses to various events) in patients and in control subjects can enrich our understanding of neuropsychological issues. Although the imaging methods currently available for investigating neural and psychological function in awake humans have improved markedly over the last decade, the nature of the inferences that can be drawn is constrained by the limits inherent in each technique. Generally, the trade off is between precision in space and time. At present, no technique provides both very high spatial and very high temporal resolution; however, within 5–10 years we are likely to have developed a technique that does.

The measurement of brain electrical activity from the scalp is noninvasive, has very high temporal resolution (on the order of tenths of milliseconds), and can provide an on-line record of brain function at the level of large neuronal populations. The electroencephalogram (EEG) consists of continuous voltage fluctuations caused by the summation of graded postsynaptic potentials from thousands of neurons and has long been used by clinicians to monitor behavioral state. (For discussions of neuronal electrogenesis, see Freeman, 1975; Nunez, 1981; Wood & Allison, 1981; for clinical use of the EEG, see Niedermeyer & Lopes da Silva, 1987.) In contrast to its fine temporal resolution, the spatial resolution of scalp-recorded electrical

Localization and Neuroimaging
in Neuropsychology

73

activity is relatively poor. However, recently implemented techniques (discussed subsequently) have improved localization accuracy and have increased our knowledge of the underlying physiological mechanisms and anatomical substrates.

Event-related potentials (ERPs) are brain potentials that are time-locked to the occurrence of sensory, motor, or cognitive events and extracted from the ongoing EEG by signal averaging techniques (Fig. 1; see Coles, Gratton, Kramer, & Miller, 1986; Hillyard & Kutas, 1983; Hillyard & Picton, 1987, for reviews). The resultant waveform consists of a series of overlapping peaks and troughs that have been separated into relatively distinct components on the basis of polarity, latency, scalp distribution, and experimental manipulation. Although labeled peaks offer a convenient shorthand, it is important to note that component effects need not

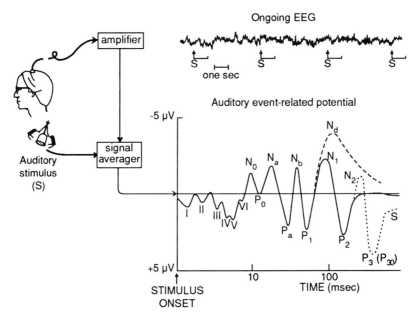

Figure 1 Idealized waveform of the computer-averaged auditory event-related potential (ERP) to a brief sound. The ERP is generally too small to be detected in the ongoing EEG (*top*) and requires computer averaging over many stimulus presentations to achieve adequate signal:noise ratios. The logarithmic time display allows visualization of the early brainstem responses (Waves I–IV), the midlatency components (N_o, P_o, N_a, P_a, N_b), the "vertex potential" waves (P_1, N_1, P_2), and task-related endogenous components (N_d, N_2, P300, and slow wave). Reproduced with permission from Hillyard & Kutas (1983).

map directly onto peaks and troughs, especially if the component is defined in terms of experimental manipulation. A direct mapping is most evident in the case of short latency "exogenous" components, which are primarily responsive to the physical parameters of the evoking stimulus (although attention effects have been noted as early as 20 msec[1]). In contrast, longer latency "endogenous" components, which are most sensitive to the psychological variables surrounding an event, often span multiple peaks and troughs. ERPs can provide a useful index of the timing of covert sensory, cognitive, and linguistic processing in humans to complement the traditional behavioral measures of cognitive psychology. By also providing a measure of the activity of neuronal ensembles, ERPs can help narrow the huge conceptual gap between psychological theories and cellular neurophysiology.

One of the most intensely investigated endogenous ERP components is the P300 or P3, a positive potential that is maximal at centro-parietal scalp sites, peaking 300–600 msec after infrequent target stimuli that are detected within a repetitive series of background events (the "oddball" paradigm; see Squires, Squires, & Hillyard, 1975). This P300, or P3b, can be elicited by stimuli of different modalities (Snyder, Hillyard, & Galambos, 1980) and even by missing or omitted stimuli (Simson, Vaughan, & Ritter, 1976). Its amplitude is responsive to stimulus probability (Fig. 2),

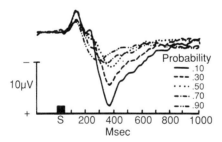

Figure 2 Grand-mean waveforms from Pz for auditory stimuli in an oddball paradigm under RT instructions at five levels of a priori probability. In this and subsequent figures, negative voltages are plotted as upward deflections. Stimulus presentation is indicated by the filled rectangle on the time scale. Reproduced with permission from Johnson (1986).

[1]In a dichotic listening experiment, Woldorff and Hillyard (1991) demonstrated that an early auditory component peaking just after the Pa wave, the P20–50, was enhanced when subjects attended to rapidly presented tones in one ear relative to when the same tones were ignored. However, another laboratory (Woods & Alain, 1993) has failed to replicate this finding.

subjective probability, temporal probability, stimulus meaning, and task relevance (see Donchin & Coles, 1988; Fabiani, Gratton, Karis, & Donchin, 1987; Johnson, 1988a; Pritchard, 1981, for reviews). The cognitive processes associated with P300 have been described by a number of psychological constructs including context updating, information delivery, stimulus categorization, and cognitive closure, although no consensus has been reached (for discussion, see Donchin & Coles, 1988; Verleger, 1988). In part, this disagreement may reflect the fact that there exists a family of late positivities with similar, albeit not identical, characteristics. For example, the P3a subcomponent, largest at fronto-central sites, is elicited by rare task-irrelevant stimuli and has been associated with orienting, arousal, and response to novelty (Courchesne, Hillyard, & Galambos, 1975; Ritter, Vaughan, & Costa, 1968; Squires et al., 1975). Finally, late positivities have been obtained under conditions other than those of the oddball task, including signal detection paradigms (Paul & Sutton, 1972) and the repetition of words in lists, sentences, and text (Besson, Kutas, & Van Petten, 1992; Rugg & Nagy, 1989; Van Petten, Kutas, Kluender, Mitchiner, & McIsaac, 1991). Whether all these positive components can be considered P300s per se is an unresolved question.

The context-updating hypothesis (Donchin, 1981; Donchin & Coles, 1988) is currently one of the most prevalent explanations for P3b and suggests that surprising or unexpected events interrupt ongoing cognitive processing and cause the subject to revise the current model of the environment. The relationship between late positive potentials and memory has been demonstrated in incidental learning paradigms. Words that were later remembered by the subjects elicited larger late positivities than words that were not remembered (Fabiani, Karis, & Donchin, 1986; Neville, Kutas, Chesney, & Schmidt, 1986; Paller, Kutas, & Mayes, 1987). Some groups have proposed a relationship between P300 and subsequent recall/recognition of words (Fabiani et al., 1986; Karis, Fabiani, & Donchin, 1984); whereas others have suggested that, although a memory-related positivity exists, it can be distinguished from P300 on the basis of its scalp distribution (Paller et al., 1987). Localizing the neural substrates of P300 and other ERPs (such as Nd, MMN, and N400) would clarify the distinctions between related subcomponents and would contribute to a better understanding of which brain regions are performing various information processing functions.

In principle, the problem of locating the generator(s) of scalp-recorded electrical activity from that activity alone is insoluble, since there is no axiomatic method for choosing among the many possible solutions that account for equal amounts of the variance (Balish & Muratore, 1990; van Oosteram, 1991; Wong, 1991; Wood, 1982). Unless additional constraints are applied, the "inverse problem" (i.e., obtaining the number and configuration of sources from the scalp potential field) has an infinite number of solutions. Hence, researchers have attempted to winnow the list of candidate sources for various ERP components by adding further restrictions to the localization problem (e.g., Vaughan, 1982). Among the various approaches used to localize ERP components are (1) developing animal models of a component that then can be evaluated using invasive techniques (lesioning, pharmacological intervention, intracranial recording); (2) analyzing recordings from various patient populations with naturally occurring lesions (strokes), surgical removals of brain tissue, or intracranial electrodes implanted for clinical purposes; (3) combining electrical with magnetic recording; (4) topographically mapping gradients of current, as in current source density analyses; and (5) dipole modeling methods such as brain electric source analysis. The remainder of this chapter provides illustrations of how these procedures have been employed in the localization of various ERP components. P300 is emphasized as a case study because it has been investigated for a longer period of time than most other cognitive components and therefore exemplifies the different approaches to localization. Additional components will be reviewed briefly as they arise in the text.

II. ANIMAL MODELS

Animal models have illuminated a range of issues relevant to the localization of ERPs by utilizing the techniques of lesioning, pharmacology, and intracranial recording (Galambos & Hillyard, 1981). One limitation of animal models is the difficulty of proving the equivalence of an ERP component across species. Similarities in the anatomical, physiological, and behavioral correlates of the component must be demonstrated. The assumption of phylogenetic continuity is problematic, particularly when comparing ERP recordings from species such as the rat with those from humans. Differences in brain organization due to, for example, the rat's greater reliance on olfactory and tactile cues and the expanded number of visual cortical areas in humans raise questions of homology. Moreover, animal models cannot reveal information about the generators of the several ERP components that have been hypothesized to reflect syntactic processing and other linguistic dimensions that are not applicable to animal communication (Kluender, 1991; Neville, Nicol, Barss, Forster, & Garrett, 1991; Osterhout & Holcomb, 1993).

Animal models of P300-like activity have been developed in rats, cats, rabbits, and monkeys to examine possible neural substrates more systematically (Arthur & Starr, 1984; Buchwald & Squires, 1982; Ehlers, Wall, & Chapin, 1991; Gabriel, Sparenborg, & Donchin, 1983; Glover, Ghilardi, Bodis-Wollner, Onofrj, & Mylin, 1991; Glover, Onofrj, Ghilardin, & Bodis-Wollner, 1986; Paller, Zola-Morgan, Squire, & Hillyard, 1988; Pineda, Foote, & Neville, 1987; Pineda, Foote, Neville, & Holmes, 1988; Wilder, Farley, & Starr, 1981). Some of these studies, however, have used paradigms that bear little resemblance to paradigms used to elicit P300 in humans. For example, Wilder et al. (1981) demonstrated late positive potentials sensitive to stimulus probability and signal meaning in cats, but the subjects were paralyzed, artificially respired, and exposed to an aversive conditioning paradigm. On the other hand, awake cats exposed to an auditory oddball paradigm showed a P300-like potential in response to rare loud clicks or to omitted stimuli (Buchwald & Squires, 1982). This "cat P300" has a smaller amplitude and longer latency in old cats, similar to the P300 changes seen in aged human subjects (Harrison & Buchwald, 1985).

Neville and Foote (1984) recorded a broad positive complex in squirrel monkeys that began ~300 msec after infrequent tones and decreased in amplitude with repeated presentations. This component was largest over frontal and temporal electrodes of the left hemisphere. In a second block of trials, an occasional "dog bark" also elicited a large P300-like component that was largest parietally and at lateral temporal electrodes. Pineda and associates (1987) further demonstrated that monkey P300 was sensitive to stimulus probability and trial-to-trial changes in stimulus sequence and largest over lateral parietal sites. A subsequent experiment examined the role of task relevance and behavioral response in a group of squirrel monkeys trained to press a lever after target tones were presented in an oddball paradigm (Pineda et al., 1988). A long-latency positive component (LPC), also inversely related to stimulus probability, was elicited by the targets (Fig. 3). This LPC showed a broader scalp distribution and had a greater amplitude when response rates were high. Arthur and Star (1984) also reported a P300-like component in macaque monkeys trained to release a lever within 600 msec of target tone offset. This potential, maximal over sensorimotor and parietal cortices and negligible over frontal and temporal areas, was not present when the lever was removed in the no-task condition.

Glover et al. (1991) recorded ERPs from monkeys who were being trained on a visual oddball task. Target probability was 0.5 in early sessions; the monkeys correctly discriminated on 55–60% of the trials. The ERP responses to targets and standards were not different at this stage of training but, as the monkeys improved their performance to 75–80%, a P300-like potential emerged over the following 2–4 wk. As the monkeys

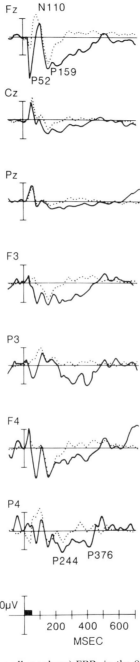

Figure 3 Grand average (across all monkeys) ERPs in the 90–10 oddball paradigm at all electrode sites in response to target (———) and background (•••••) tones. Black rectangle on the time scale represents the onset and duration of tone stimuli. Note the enhanced positivity occurring in the 200–600 msec latency interval following target presentations. Reproduced with permission from Pineda, Foote, Neville, & Holmes (1988).

continued to show behavioral progress, the amplitude of the P300-like component increased and its latency decreased. P300 amplitude also increased when target probability was decreased to 0.3.

Hence, there are several instances in nonhuman animals of P300-like activity that behaves like the human component in response to manipulations of probability, task relevance, and so on. A reasonable premise, therefore, is that these examples are valid animal models of P300, and that hypotheses about its loci can be tested by lesioning, pharmacological manipulation, and intracranial recording.

A. Lesions

One test of the postulated relationship between memory and P3a/P3b is damaging the brain regions involved in memory function in monkeys. Paller and colleagues (1988) presented both improbable pure tones and rare complex tones to untrained and trained macaques. Bilateral ablations of the medial temporal lobe (MTL) which included hippocampus, amygdala, and adjacent cortical areas did not abolish either passive (Fig. 4) or active P300-like waves. These results coincide with findings in human temporal lobectomy patients (reviewed subsequently) and suggest that MTL structures are not essential in generating P300 (at least P300 recorded at midline electrode sites).

Lesions of other brain regions have been performed in the cat model. Bilateral ablations of primary auditory cortex (Harrison, Buchwald, & Kaga, 1986) or polysensory association cortex (Harrison, Dickerson, Song, & Buchwald, 1990) did not significantly reduce the amplitude of the epidurally recorded cat P300. The association cortex lesions—which included either pericruciate cortex, anterior lateral and medial suprasylvian gyri, or all three areas—substantially diminished two earlier components: a 30- to 35-msec negativity (wave N_A) and a 50- to 75-msec positivity (wave C). Although polarity-reversing potentials have been recorded from these regions at latencies of 200–300 msec, they are apparently unnecessary for the generation of vertex P300-like potentials in the cat.

Another line of work has examined how neurotransmitter systems might modulate the neuronal activity that generates ERPs, particularly cortical slow potentials (reviewed by Marczynski, 1978; Pineda, Swick, & Foote, 1991). Cell groups containing norepinephrine, acetylcholine, dopamine, and serotonin comprise four major subcortical transmitter systems that modulate neocortical function and behavioral state (Foote & Morrison, 1987). These extrathalamic nuclei innervate widespread areas of primate neocortex with regional and laminar specificities. Their neurons have slow spontaneous firing rates, their action potentials can take tens or

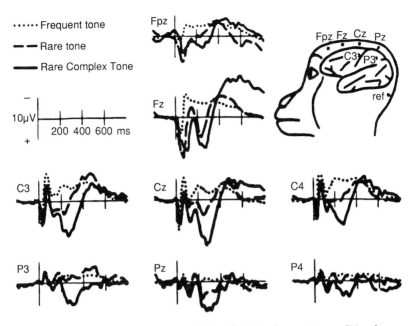

•••••• Frequent tone

— — Rare tone

——— Rare Complex Tone

$10\mu V$ 200 400 600 ms

Figure 4 Event-related brain potentials elicited during the passive condition from a group of five monkeys with bilateral medial temporal lobectomies. Ref, Reference electrode. Reproduced with permission from Paller, Zola-Morgan, Squire, & Hillyard (1988).

even hundreds of milliseconds to reach target areas, and their neurotransmitters have long-duration effects on postsynaptic target cells. Halgren and Smith (1987) suggest that the EEG is largely dominated by the effects of these systems; some ERP components might not be "generated by the synapses active in specifically processing the stimulus, but rather are generated by a diffuse synaptic network that modulates the specific information processing" (p. 131).

In fact, evidence exists for specific modulation of various ERP components following disruption of neurotransmitter function by lesions or neuropharmacological interventions. Bilateral lesions of the medial septal nucleus and the vertical limb of the diagonal band, the major cholinergic projections to hippocampus, resulted in a transient increase followed by a progressive decrease and disappearance of P300-like activity in cats (Harrison, Buchwald, Kaga, Woolf, & Butcher, 1988). Lesions of the noradrenergic (NA) nucleus locus coeruleus (LC) in squirrel monkeys produced a significant reduction of P300-like potentials recorded in a passive auditory oddball paradigm (Pineda, Foote, & Neville, 1989). Extensive damage to cell bodies in the nucleus and knife cuts disrupting ascending axons in the dorsal bundle (DB) were both necessary since DB damage alone produced no changes. There was a significant correlation between the size of lesion and percent reduction in P300 area.

B. Pharmacology

In addition to eliminating the intended neurons, electrolytic lesions can damage areas along the microelectrode track, fibers of passage, and areas adjacent to the nucleus. Pharmacological manipulations have the advantage of being reversible and more selective for a particular transmitter system. The alpha-2 adrenergic agonist clonidine, for example, suppresses LC unit activity by binding to autoreceptors on these neurons (Cedarbaum & Aghajanian, 1977). A preliminary study in squirrel monkeys demonstrated that systemic administration of clonidine produced dose-related decreases in the area of P300-like potentials, with recovery to control levels in post-drug sessions (Swick, Pineda, Holmes, & Foote, 1988). The highest dose was the most effective in reducing P300-like waves and was chosen for a subsequent investigation with a larger number of subjects (Swick, Pineda, & Foote, 1993). In this experiment, clonidine specifically increased the latency and decreased the area of a P300-like potential elicited in response to rare tones (Fig. 5).

In contrast to these alterations in the auditory modality, the same dose of clonidine did not affect the latency, amplitude, or area of a visual P300-like potential in squirrel monkeys (Pineda & Swick, 1992). This finding suggests that the LC–NA system makes a modality-dependent contribution to the generation of P300-like potentials. Variations in the distribution of NA fibers across different neocortical regions may result in distinct influences on the processing of signals from different sensory modalities. Lesion data from human patients suggests that P300 has modality-specific generators (Johnson, 1989a; Knight, Scabini, Woods, & Clayworth, 1989; see subsequent discussion). Differential NA innervation patterns could contribute to such effects.

The neurotoxin 1-methyl-4-phenyl-1,2,5,6-tetrahydropyridine (MPTP), which produces parkinsonian symptoms by depleting dopamine and norepinephrine, initially abolished P300-like potentials in a group of 5 monkeys (Glover, Ghilardi, Bodis-Wollner, & Onofrj, 1988). Acute administration of a dopamine precursor did not restore P300, although tremor and rigidity were improved temporarily. P300 returned 30–40 days later in 2 monkeys that showed partial behavioral recovery. Whereas dopamine systems remain chronically suppressed after MPTP treatment,

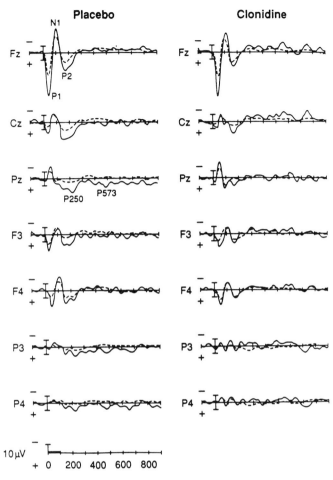

Figure 5 (*Left*) Grand average ERPs across 6 squirrel monkeys follow-ing placebo administration. ERPs were recorded in response to back-ground (dashed lines) and target (solid lines) tones presented in the 90–10 oddball paradigm. (*Right*) Grand average ERPs for the same 6 mon-keys following administration of clonidine (0.1 mg/kg IM). ERPs were recorded 15 min post-drug; placebo and drug sessions were separated by at least 3 weeks. Note the dramatic decrease in the magnitude of P250 and P573, whereas the earlier potentials are unchanged. Reproduced with permission from Swick, Pineda, & Foote (1993).

norepinephrine (NE) systems recover, perhaps accounting for the return of P300 in some animals. Partial depletions of dopamine and serotonin in rats did not affect a late positive potential recorded from dorsal hippocampus and amygdala (Ehlers et al., 1991), but these lesions may not have been substantial enough to produce alterations. Ethanol, which may produce some of its effects through endogenous opioid systems and the benzodiazepine/gamma-aminobutyric acid (GABA) receptor complex, significantly decreased N1 and P3 amplitudes in squirrel monkeys (Ehlers, 1988). Diazepam, which enhances activity of the $GABA_A$ receptor with no known actions on opioid systems, reduced only N1 amplitude (Ehlers, 1988). Utilizing the same passive auditory oddball paradigm, the opiate antagonist naloxone was found to decrease P300 latency in squirrel monkeys (Ehlers, 1989).

Neurons in the nucleus basalis of Meynert (NBM) comprise the major cholinergic projection to amygdala and neocortex in primates (Mesulam, Mufson, Levey, & Wainer, 1983). Pirch, Corbus, Rigdon, and Lyness (1986) demonstrated that NBM lesions, pharmacological depression of NBM neurons, and blockade of muscarinic cholinergic receptors in cortex reduce the low frequency negative potentials recorded from rat frontal cortex during an associative conditioning paradigm. These potentials were elicited by a 2-sec light cue that preceded rewarding stimulation of the medial forebrain bundle. This general type of paradigm, in which a warning stimulus (S_1) precedes an imperative stimulus (S_2) that requires a motor response, elicits the "contingent negative variation" (CNV), slow negative potentials that are largest over frontal–central sites, during the interval between S_1 and S_2 (Borda, 1970; Rohrbaugh, Syndulko, & Lindsey, 1976; Walter, Cooper, Aldridge, McCallum, & Winter, 1964).

A summary of drug effects on P300 and other potentials is presented in Table 1. These studies suggest that neurotransmitter systems, particularly the monoaminergic and cholinergic systems, are critical in modulating the synaptic events that give rise to ERP components. For instance, these systems alter the signal-to-noise processing characteristics of postsynaptic neurons, either enhancing or diminishing a cell's response to other inputs. Additionally, these systems are capable of providing a synchronizing input to anatomically distinct generators, the activity of which sums to produce an ERP component that is recorded as a single entity from the scalp.

C. Intracranial Recordings

A major advantage of recording within the brain rather than from the scalp or even the brain surface is, of course, the ability to get closer to the neural sources generating ERPs. Field potentials and single or multi-unit activity can be recorded from the same electrode in some cases. Additionally, it is possible to observe inversions of polarity, which are considered signs of local generation of the component in question. The majority of intracranial recordings has focused on somatosensory, auditory, and visual evoked potential responses. Locating the generators of these potentials, particularly those in brainstem and thalamic relay centers, is of clinical interest and can assist in diagnosing patients with sensory deficits.

Arezzo and co-workers (Arezzo, Legatt, & Vaughan, 1979; Arezzo et al., 1981) mapped the surface and depth components of the somatosensory evoked potential (SEP) in rhesus monkeys exposed to median nerve stimulation. Field potentials and multi-unit activity were recorded from the same intracranial electrodes. The earliest surface waves were thought to reflect activity of primary somatosensory neurons ascending in the dorsal columns, the summation of synchronized action potentials traveling along the medial lemniscus to the thalamus, and the thalamocortical radiations. The first cortical components recorded at precentral surface sites were the P10–N20 complex, analogous to the human P20–N30 which inverts across the central sulcus. The simian P10–N20 was generated in the posterior bank of the central sulcus, inverting in polarity across the deep layers of areas 3a and 3b of primary somatosensory cortex. The P12–N25 complex, recorded at surface sites posterior to the central sulcus and analogous to the human P25–N35, displayed a transcortical polarity inversion within areas 1 and 2. The monkey P20 was generated in area 5 and is probably analogous to a human peak recorded at 45–50 msec over parietal areas contralateral to stimulation. The monkey P40 had a bilateral source in area 7b and may be comparable to the human P80.

Similarly, the sources of auditory evoked potentials (AEPs) were mapped with depth recordings in monkeys (Arezzo, Pickoff, & Vaughan, 1975). The earliest cortical waves, P12 and P22, were generated in the posteromedial region of the supratemporal plane (STP) within primary auditory cortex. The sources of N38, N60, and N100 were more anterior in the STP, whereas P73 and N140 were generated in a broader posterior region of the STP. Some potentials also showed polarity inversions in motor cortex, summating with STP potentials to produce the surface-recorded peaks.

More recently, a few researchers have searched for the intracranial correlates of long-latency surface-recorded ERPs such as P300. Depth recordings in cats demonstrated polarity-reversing potentials in the marginal and suprasylvian gyri and hippocampus at 200–350 msec in response to a rare stimulus (O'Connor & Starr, 1985). Other investigators have found polarity-reversing potentials at 200–500 msec in the medial septal area, hippocampus (primarily the pyramidal cell layer), entorhinal cortex, and amygdala but not in auditory cortex (Harrison & Buchwald, 1987; Kaga,

Table 1 Summary of the Effects of Pharmacological Manipulations on P300 Potentials and Other Selected Components in Animals and Humans[a]

Drug	Actions	Species	Effects on P300[b]	Reference
Clon	Suppresses LC firing and NE release	Squirrel monkey	**aud:** eliminated in passive and active oddball	Swick et al. (1988, 1993) Pineda & Swick (1992)
		Human	**vis:** no effect in passive oddball **aud:** reduced in oddball task	Duncan & Kaye (1987); Joseph & Sitaram (1989)
MP	Increases release and blocks reuptake of CA	Human	**vis:** no effect on latency in detection task (amplitude not reported); RT decreased	Naylor et al. (1985)
Amph	Increases release and blocks reuptake of CA	Human	**vis:** no effect on latency in detection task (amplitude not reported); RT decreased	Halliday et al. (1987)
Coke	Blocks reuptake of CA	Human	**aud:** reduced amplitude in oddball task, N1 and P2 amplitude reduced **aud target detection following cue:** no effect on amplitude or latency to target, enhanced N1 and CNV amplitude to cue	Herning et al. (1985) Herning, Hooker, & Jones (1987)
MPTP	Depletes DA, NE	Macaque monkey	**aud classical conditioning:** initially abolished, returned 30–40 days later	Glover et al. (1988)
6-OHDA in VTA	Depletes DA	Rat	**aud:** no effect in passive oddball (depth potential in dorsal hippocampus & amygdala)	Ehlers et al. (1991)
PCPA	Depletes 5-HT	Rat	**aud:** no effect in passive oddball (depth potential in dorsal hippocampus and amygdala)	Ehlers et al. (1991)
Mths	5-HT antagonist	Human	**aud:** no effect on latency or amplitude in oddball task	Meador et al. (1989)

continues

Table 1 (*Continued*)

Scop	Cholinergic antagonist	Human	**aud:** increases latency and decreases amplitude in oddball task; reversed by Phys, an AChE inhibitor Altered hippocampal P3s: increases amplitude in epileptic hemisphere, increases latency in normal hemisphere **vis:** increases latency and RT in detection task (amplitude not reported)	Hammond et al. (1987); Meador et al. (1987) Meador et al. (1988) Callaway et al. (1985)
Atrp	Cholinergic antagonist	Rat	Reduced CNV recorded from frontal cortex	Pirch et al. (1986)
Etoh	Possibly at BDZ/GABA, opioid, other sites	Squirrel monkey	**aud:** reduced in passive oddball, N1 amplitude also reduced	Ehlers (1988)
DZP	Enhances activity of GABA$_A$ receptor	Squirrel monkey	**aud:** no effect in passive oddball, but N1 amplitude reduced	Ehlers (1988)
Nalox	opiate antagonist	Squirrel monkey	**aud:** decreases latency in passive oddball	Ehlers (1989)
		Human	**selective listening:** increases Nd amplitude in selective attention but not in single channel or divided attention conditions	Arnsten et al. (1984)

[a]Abbreviations: AChE, acetylcholinesterase; Amph, amphetamine; Atrp, atropine; BDZ, benzodiazepine; CA, catecholamine; Clon, clonidine; Coke, cocaine; DA, dopamine; DZP, diazepam; Etoh, ethanol; 5-HT, 5-hydroxytryptamine (serotonin); GABA, γ-aminobutyric acid; LC, locus coeruleus; MP, methylphenidate; MPTP, 1-methyl-4-phenyl-1,2,5,6-tetrahydropyridine; Mths, methysergide; Nalox, naloxone; NE, norepinephrine; PCPA, parachlorophenylalanine; Phys, physostigmine; Scop, scopolamine; 6-OHDA, 6-hydroxydopamine; VTA, ventral tegmental area.

[b]aud, auditory; vis, visual; RT, reaction time; CNV, contingent negative variation.

Harrison, Butcher, Woolf, & Buchwald, 1992). Rabbits previously trained in a discriminative avoidance task showed large positive potentials in cingulate cortex and anterior thalamus 150–300 msec after the infrequent CS⁺ (conditioned stimulus) (Gabriel et al., 1983). No polarity inversions were reported.

Negative potentials with 210-msec latency (50 msec later than a positive-going epidural potential recorded in response to rare tones[2]) were recorded in medial MTL structures of two macaque monkeys during a passive auditory oddball paradigm (Paller, McCarthy, Roessler, Allison, & Wood, 1992). A steep potential gradient was located in the hippocampal region. In a visual oddball task, negative MTL potentials that followed correct responses peaked 50–100 msec after an epidural positivity at about 250 msec. Recordings from human patients were similar in that negative potentials in the MTL peaked 50–100 msec later than scalp P300s (see also Halgren et al., 1980; McCarthy, Wood, Williamson, & Spencer, 1989). No polarity reversals were noted in either paradigm. The depth electrodes, located medial to the hippocampus in most cases, had 8 contacts spaced 1 mm apart (from thalamus dorsally to subiculum and entorhinal cortex ventrally). However, a transcortical polarity inversion was observed in the posterior parietal cortex following exposure to visual targets.

Another strategy has been to record from single cells in regions hypothesized to generate or modulate a component. Since one hypothesized function of the LC–NA system is a role in the control of attention, arousal, and response to novel events (Aston-Jones, Chiang, & Alexinski, 1991; Foote, Berridge, Adams, & Pineda, 1991), this system might also modulate P300 activity by providing a synchronizing input to the generators, thereby producing signal-to-noise enhancements in the neuronal ensembles that generate these potentials. An initial attempt to record LC unit activity in monkeys exposed to a passive auditory oddball paradigm failed to find evoked responses following presentation of either oddball or frequent stimuli (Grand, Aston-Jones, & Redmond, 1988). Louder, more alerting tones, all of the same frequency, did produce a phasic enhancement of firing. The work of Aston-Jones and colleagues (1991) suggested that tentatively identified LC cells respond more vigorously to a rare visual stimulus than to a highly probable one. Macaques were trained to discriminate two colors in a visual oddball paradigm. Cell firing was significantly higher in response to the targets. When the color of the infrequent stimulus was switched, the cells also showed a reversal in firing patterns. Although LC cells respond to alerting stimuli of all modalities,

[2]The epidural peak at 160 msec is probably too early to be a P300 analog; a smaller positivity between 200 and 300 msec is a more likely candidate.

some differences between modalities and between active and passive paradigms may exist.

Another group recorded LC unit activity and ERPs concurrently in monkeys exposed to an auditory oddball paradigm to determine whether the activity of individual LC neurons is enhanced during the occurrence of P300-like potentials (Swick, 1991). Some LC cells (25%) showed a phasic activation after presentation of infrequent tones. Like the P3a, cellular responses were heterogeneous, related to stimulis sequence, and influenced by the subjects' behavioral state. The occurrence of a P3a-like component was not necessarily correlated with a phasic LC response to infrequent tones in the passive condition. In one trained monkey, LC cells tended to show a tonic elevation in firing after presentation of target tones. This tonic activation ws enhanced when the monkey performed the task; a phasic activation was related to behavioral response rather than to stimulus presentation. Collectively, the lesion data, the pharmacological evidence, and the single unit data suggest a link between the LC–NA system and the generation of P300.

III. HUMAN PATIENTS

Determining the neural structures involved in generating ERP components in human patients with focal brain lesions has been suggestive of specific generator sites, although problems with localization and interpretation can make these results somewhat inconclusive. If a lesion abolishes an ERP component, it may not necessarily be because the neural substrate itself has been destroyed, but rather because of damaged input to the generators. Lesions also may act indirectly by altering behavioral state or through nonneural mechanisms (see Wood et al., 1984, for discussion). For instance, the removal of skull in temporal lobectomy patients presumably alters intracranial conductivity and surface ERP topography (Vaughan, 1987). Likewise, it is difficult to determine whether a particular intracranial potential contributes to a potential recorded concurrently on the scalp. Late positive potentials resembling P300 have been recorded in the MTL, for example (Halgren et al., 1980; McCarthy et al., 1989), but MTL lesions do not alter scalp-recorded P300 in a manner consistent with an MTL primary generator (Johnson, 1988b; Stapleton, Halgren, & Moreno, 1987). A final complication is the fact that most depth electrodes are placed in the abnormal compromised tissue of neurological patients; brain pathology may affect these intracranial potentials. Nevertheless, studies in brain-damaged patients have limited the range of possible sources for a given ERP component. Particular attention has been devoted to P300, which is discussed in detail in the following sections.

Table 2 Summary of Lesion Effects on P300 in Human Patients

Lesion	Modality	Effects on P300	Reference
Prefrontal cortex	Auditory	P3a to novels reduced bilaterally, P3b to targets normal	Knight (1984)
		P3a to novels decreased over lesioned hemisphere, with greater reduction for right lesions	Scabini et al. (1989)
	Visual	P3a to novels decreased over lesioned hemisphere, P3b to targets normal	Knight (1990)
	Somatosensory	P3a to novels decreased bilaterally, slight decrease in P3b to targets (only at frontal site ipsilateral to lesion)	Yamaguchi & Knight (1991)
Lateral parietal	Auditory	No change in P3a or P3b	Knight et al. (1989)
	Visual	No change in P3a or P3b	Knight (1990)
	Somatosensory	P3a decreased to contralateral shock novels, normal P3b to targets	Yamaguchi & Knight (1991)
Temporal–parietal junction	Auditory	P3a to novels and P3b to targets abolished posteriorly	Knight et al. (1989)
	Visual	P3a to novels reduced, lesser reductions for P3b to targets	Knight (1990)
	Somatosensory	P3a to novels & P3b to targets reduced bilaterally, both contralateral and ipsilateral to lesion	Yamaguchi & Knight (1991)
Anterior temporal lobe (unilateral)	Auditory	P3b to targets normal, P3a to novels reduced in left ATL patients	Stapleton et al. (1987)
		P3b to targets normal in right ATL, slightly reduced in left ATL; no left–right asymmetries in either group	Johnson (1988b)
		P3b to targets normal in patient with left MTL damage	Rugg et al. (1991)
		P3b reduced at parietal site ipsilateral to right temporal lesion	Daruna et al. (1989)
		P3b to targets normal at parietal sites, decreased at frontal sites for left ATL	Johnson (1989a)

continues

Table 2 (*Continued*)

	Visual	P3b to targets normal	Stapleton et al. (1987); Scheffers et al. (1991)
		P3b to targets normal in patient with left MTL damage	Rugg et al. (1991)
		P3b to targets normal at parietal sites, decreased at frontal sites for right ATL	Johnson (1989a)
Posterior hippocampus and inferior temporal cortex	Auditory	P3a to novels decreases frontally but not parietally, P3b to targets normal	Knight (1991)
	Visual	Same	Knight (1991)
	Somatosensory	Same	Knight (1991)
Anterior and midtemporal lobes (bilateral)	Auditory	P3b to targets normal at F3/4, C3/4, T5/6, P3/4, and O1/2 and at midline sites, but reduced at Fp1/2, F7/8, and T3/4	Onofrj et al. (1992)
Commissurotomy	Auditory	Binaural targets: P3 amplitude greater over right hemisphere	Kutas et al. (1990)
	Visual	Bilateral targets: P3 amplitude greater over right hemisphere	
		LVF targets: greater over right hemisphere	
		RVF targets: equal over both hemispheres	

A. Lesions

Substantial lesion and intracranial evidence has implicated a number of brain regions in P300 electrogenesis, including portions of frontal, temporal, and parietal cortices, as well as several subcortical areas (Halgren et al., 1980; Knight, 1984; Knight, Scabini, Woods, & Clayworth, 1989; McCarthy et al., 1989; Smith et al., 1990; Yingling & Hosobuichi, 1984). Table 2 summarizes the effects of different brain lesions on P3a and P3b. Patients with unilateral lesions of prefrontal cortex showed normal, parietally distributed P300s in response to targets in an auditory discrimination task (Knight, 1984). Responses to unexpected "novel" stimuli (rare nontarget stimuli) at fronto-central sites were reduced and instead exhibited a more parietal distribution, resembling the P300 elicited by targets. A lateralized decrease over lesioned cortex was not observed, however, leading the author to conclude that the prefrontal cortex, although not the primary generator of P3a, modulates its activity. A more recent preliminary study by Scabini, Knight, & Woods (1989), however, indicated that P3a is reduced over lesioned prefrontal cortex, maximally for right hemisphere lesions. Somatosensory P3as to "novel" tactile stimuli and shocks were decreased by lesions of dorsolateral prefrontal cortex, particularly at frontal sites (Yamaguchi & Knight, 1991); P3b to tactile targets showed only slight reductions.

In the auditory modality, the P3b to targets was abolished and P3a to novels was reduced substantially at central and parietal sites in patients with unilateral lesions that included both caudal inferior parietal cortex and superior temporal gyrus (Fig. 6; Knight et al., 1989). Patients with lesions including only lateral parietal cortex exhibited normal P3a and P3b potentials. Similar results were seen in the somatosensory modality (Yamaguchi & Knight, 1991). Lesions of the temporal–parietal junction eliminated the visual P3a but had lesser effects on the amplitude of the visual P3b (Knight, 1990, personal communication). In contrast, auditory and somatosensory P3bs were abolished by temporal–parietal lesions (Knight et al., 1989; Yamaguchi & Knight, 1991). One likely explanation is that P300 has multiple generators, some of which are modality specific. For instance,

Figure 6 Group-averaged ERPs recorded to target (*above*) and novel (*below*) stimuli in the monaural tone detection task. The arrows (S) denote stimulus onset. Solid lines show ERPs from controls, dotted lines from temporal patients, and dashed lines from patients with parietal lesions. Data are shown from the midline and parasagittal scalp sites. Scalp sites are shown ipsilateral (i) and contralateral (c) to the lesioned hemisphere for patients, or on the left and right for controls. Lesions in the temporal–parietal junction abolished the P3a and P3b responses at all posterior scalp sites. ERPs are grand averages over 6 patients in each group. Reproduced with permission from Knight, Scabini, Woods, & Clayworth (1989).

Targets

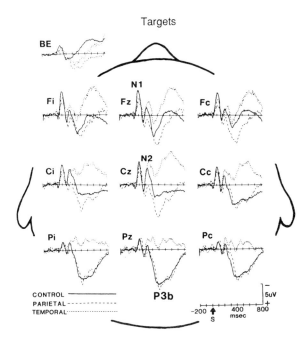

Novels

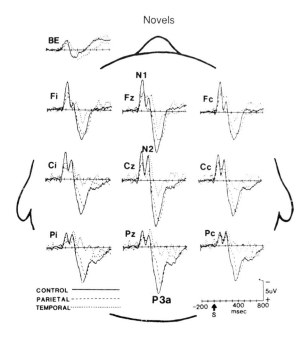

the temporal–parietal junction is critical for P3a in all modalities and for P3b in the auditory and somatosensory modalities; this region makes a significant but smaller contribution to visual P3b. These results also support previous suggestions that P3a and p3b are distinct subcomponents with different neural sources. Prefrontal cortex, for example, is necessary for the generation of P3a but not P3b.

In oddball-type paradigms, the modality-specific N200 component precedes P3a and P3b (Simson, Vaughan, & Ritter, 1977; Squires et al., 1975). Naatanen (1988, 1990, 1991) has divided N200 into two subcomponents: an earlier N2a or mismatch negativity (MMN) that reflects an "automatic" process and a later N2b that precedes P3b and is related to the shifting of attention toward a target. Woldorff, Hackley, and Hillyard (1991), however, reported that the MMN in unattended channels was suppressed under conditions of highly focused attention and suggested that it may be only "weakly automatic." These authors interpreted this effect as an attentuation of early sensory processing in the unattended channels. An alternative explanation contends that overlap from the N2b contributed to the larger negativity observed in attended channels, accounting for most of the MMN attention effect (Naatanen, 1991).

In the same groups of temporal–parietal patients discussed earlier, lesions of lateral parietal cortex significantly reduced N200 in response to target and novel stimuli in both auditory (Knight et al., 1989) and somatosensory (Yamaguchi & Knight, 1991) modalities. Lesions of the temporal–parietal junction, conversely, did not affect N200 amplitude in either modality. The dissociation between N200 and P300 observed in these studies suggests that the two potentials have different neural sources. Additionally, lesions of dorsolateral frontal cortex reduced the amplitude of somatosensory N200 to novels and targets (Yamaguchi & Knight, 1991) and auditory N200 to novels only (Knight, 1984).

Since P300 has been proposed to be a reflection of memory encoding (Donchin & Coles, 1988), temporal lobe structures have been considered primary sources of this potential. Contrary to the prediction of this hypothesis, unilateral temporal lobectomies that included removal of hippocampus, amygdala, and anterior temporal lobe did not significantly affect scalp-recorded P3b in auditory or visual oddball tasks (Johnson, 1988b; Stapleton et al., 1987). No hemispheric asymmetries were observed in these patients, as would be expected if MTL structures were major contributors to the P3b recorded on the scalp. Additionally, P300s from anterior temporal lobectomy (ATL) patients did not differ significantly from those of normal controls when stimulus quality in a visual discrimination tasks was reduced, although the right-lesioned group made significantly more errors (Scheffers, Johnson, & Ruchkin, 1991). A patient with a glioma affecting the entire left MTL exhibited unimpaired performance and symmetrical P300s of normal amplitude in response to both auditory and visual oddball stimuli, but a large negativity was evoked by visual targets at frontal sites (Rugg, Pickles, Potter, & Roberts, 1991). Slightly dissimilar results were obtained in another experiment that used an auditory oddball paradigm with longer and variable interstimulus intervals. Patients with right temporal lobe lesions exhibited smaller P300 amplitudes at the parietal site ipsilateral to the lesion in this paradigm (Daruna, Nelson, & Green, 1989).

Although ATL did not affect the P3b to auditory targets in the study of Stapleton et al. (1987), these investigators did observe a reduction in the P3a to novel auditory stimuli in the group with left ATLs. A similar report indicated that posterior hippocampus and adjacent inferior temporal cortex may make a modality-independent contribution to the frontal P3a (Knight, 1991). Lesions of these areas had no effect on the parietal P3b to targets or the parietal P3a to novel stimuli in auditory, visual, and somatosensory modalities. Patients with bilateral damage of anterior and medial temporal lobes and severe anterograde amnesia showed reductions in P3b amplitude recorded in an auditory oddball task (Onofrj et al., 1992). These decrements were significant at lateral frontal and midtemporal sites but not at midline, posterior temporal, lateral parietal, or occipital electrodes. In addition to demonstrating once again that intact MTLs are not necessary for generating the P300 that is maximal at Cz and Pz, this finding suggests that MTL contributes to the positivity recorded at other scalp sites and strongly implies multiple P300 sources.

Johnson (1989a) provides additional evidence for modality-dependent generators. Auditory P300 amplitude was reduced at frontal sites in patients with left temporal lobectomies, whereas visual P300 was normal at all electrodes. Conversely, patients with right temporal lobectomies showed normal auditory P300s but reduced visual P300s at frontal electrodes. Additionally, studies of the developmental changes in P300 recorded from children have demonstrated differences between auditory and visual P300s (Johnson, 1989b).

Unilateral stimuli elicit a bilaterally symmetric P300 in control subjects, whereas split-brain patients show an asymmetric distribution for both the auditory and visual modalities (Kutas, Hillyard, Volpe, & Gazzaniga, 1990). Targets in the left visual field and bilateral visual targets elicited an LPC that was larger over the right than over the left hemisphere. Targets in the right visual field produced an LPC of approximately equal amplitude over both hemispheres. Binaurally presented tones also produced a P300 that was greater in amplitude over the right hemisphere. This interesting pattern of results suggests that P300 arises from neither a diffuse bilateral source (since laterality in the visual task depended on which hemisphere was stimulated) nor from totally independent cortical generators (since separate lateralized generators would yield predictable, reversed asymmetries). Intact subcortical systems, however, are probably important in generating or modulating P300 activity, as discussed earlier.

B. Pharmacology

Pharmacological studies in humans also have contributed to an increasing awareness of the importance of neurotransmitters in various aspects of information processing and their ERP indices (Table 1). The stimulant drugs methylphenidate and amphetamine, which increase release and block reuptake of catecholamines, decreased reaction time (RT) but did not affect P300 latency, suggesting that stimulants act on response selection processes but not on stimulus evaluation (Halliday, Naylor, Callaway, Yano, & Walton, 1987; Naylor, Halliday, & Calloway, 1985). Herning, Jones, Hooker, and Tulunay (1985), conversely, reported that cocaine reduced not only P300 amplitude but also the amplitudes of N100 and P200 in an auditory oddball task, indicating that stimulants may influence both earlier processes such as those related to selective attention as well as later stimulus evaluation processes.

Scopolamine, a cholinergic antagonist acting at muscarinic receptors, increased both P300 latency and RT in a visual discrimination task (Callaway, Halliday, Naylor, & Schechter, 1985). Scopolamine also interacted with stimulus complexity, increasing P300 latency and RT for more easy "pop out" visual stimuli than for complex stimuli that required a serial search (Naylor, Brandeis, Halliday, Yano, & Callaway, 1988). Another group reported that scopolamine not only increases P300 latency but also decreases P300 amplitude and impairs recent memory (Hammond, Meddor, Aung-Din, & Wilder, 1987; Meador et al., 1987). These effects were partially reversed by physostigmine, an anticholinesterase. Depth electrodes placed in the hippocampi of epileptic patients showed polarity-inverting potentials that were altered by scopolamine (Meador et al., 1988). In another experiment, scopolamine was compared with methysergide, an antiserotonergic drug (Meador et al., 1989). Both drugs impaired recent memory, but only scopolamine had effects on P300, implying that serotonergic systems are not essential in generating P300.

A preliminary report by Duncan and Kaye (1987) showed that clonidine decreased P300 amplitude in an auditory discrimination task, to the greatest extent for the easiest discrimination. After placebo administration, P300 amplitude increased as discriminability increased, but the opposite effect was observed with clonidine. Joseph and Sitaram (1989) also reported that clonidine reduced P300 amplitude in an easily discriminable auditory paradigm.

The opiate antagonist naloxone was administered to subjects performing an auditory selective attention task (Arnsten, Neville, Hillyard, Janowsky, & Segal, 1984). Tones were presented in three different spatial locations, and subjects had to detect longer-duration targets in one of these channels. Typically, an enhanced negativity is elicited by tones in the attended channels compared with those in unattended channels (Hillyard, Hink, Schwent, & Picton, 1973). Naloxone increased this attention effect in the selective attention condition but not in undistracted or divided attention conditions, suggesting that endogenous opiates can influence ERP measures of selective attention. Pharmacological manipulations in humans, as well as in animals, are thus capable of altering the output of neural groups generating ERPs recorded from the scalp.

C. Intracranial Recordings

Intracranial recordings in human patients have demonstrated steep potential gradients and polarity inversions in P300-like activity recorded from a number of different cortical and subcortical regions, including hippocampus and medial temporal lobe structures (Halgren et al., 1980; McCarthy et al., 1989; Stapleton & Halgren, 1987), frontal lobe (Smith et al., 1990; Wood & McCarthy, 1986), parieto-occipital junction (Kiss, Dashieff, & Lordeon, 1989), inferior parietal lobe (Smith et al., 1990), and thalamus (Kropotov & Ponomarev, 1991; Velasco, Velasco, Velasco, Almonza & Olivera, 1986; Yingling & Hosobuichi, 1984). Very large positive potentials were elicited by auditory (Fig. 7), visual, or somatosensory targets in regions anterior and posterior to the hippocampus, inverting in polarity within the hippocampus (McCarthy et al., 1989). These potentials were task relevant, sequentially dependent, and could be elicited by omitted stimuli, demonstrating similarities to scalp-recorded P300 although they may not be a major source of its activity. In another study, stimulus omissions evoked typical late potentials from temporal but not from frontal or parietal depth electrodes, suggesting that P300 has both endogenous and exogenous sources (Alain, Richer, Achim, & Saint Hilaire, 1989).

A negative component with the same latency as the vertex P300 was recorded in the somatosensory thalamus and periaqueductal gray of a chronic pain patient, implying that the hippocampal activity is not volume conducted to the scalp (since positive potentials should be recorded dorsal to the hippocampus; Yingling & Hosobuichi, 1984). Polarity reversals were not observed in this experiment, but Velasco et al. (1986) found polarity inversions between subthalamus and dorsal thalamus. Further, they

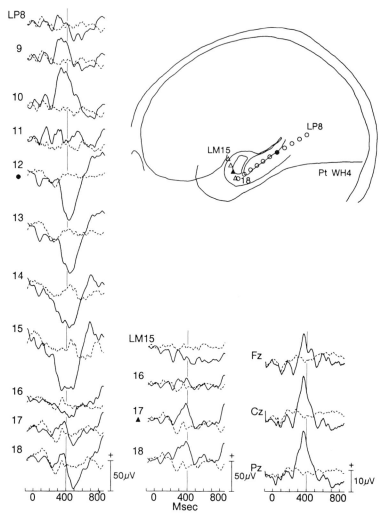

Figure 7   ERPs elicited by auditory count (solid) and ignore (dashed) targets for LP (circle) and LM (triangle) contacts in patient WH4. LP 12 and LM17 are depicted as filled symbols. Scalp-recorded ERPs elicited by auditory count targets (solid) and standards (dashed) prior to implant surgery are shown in the right-most column for comparison. Reproduced with permission from McCarthy, Wood, Williamson, & Spencer (1989).

named the region of the medial geniculate as a possible generator, since the "P300" component had the largest amplitude and shortest latency there. Kropotov and Ponomarev (1991) recorded ERPs and multiple unit activity from the globus pallidus and ventro-lateral thalamus of Parkinson's patients performing an oddball task. These investigators reported a P300-like component in these structures in response to task-relevant, rare visual stimuli. No phase reversals were observed, but the authors speculated that the generators might be found in adjacent regions. Approximately 20% of all multi-unit populations showed robust, long-latency firing only after the rare, relevant stimuli. Another 22% had short-latency responses to both targets and standards but showed enhanced activity in response to targets that started at 200–400 msec.

Although P300-like activity has been recorded from a number of intracranial sites in humans, the findings of Smith and colleagues (1990), combined with lesion data implicating temporal–parietal junction (Knight et al., 1989; Yamaguchi & Knight, 1991), suggest that a major generator of the scalp component may be in the inferior parietal lobe (IPL) and the posterior superior temporal plane. Smith et al. (1990) claimed that the diencephalon makes a minimal contribution to scalp P300; the small potentials recorded there may be volume conducted from other sources. In the frontal lobe, polarity inversions were found in the premotor area but not dorsolateral prefrontal cortex, anterior cingulate gyrus, or supplementary motor area. The frontal potentials may be too small in amplitude to make a substantial contribution to scalp P300. Conversely, very large positive potentials were observed in the IPL. No polarity reversals were seen, but steep voltage gradients were observed at sites anterior, posterior, superior, inferior, and medial to the IPL contacts.

The studies reviewed here illustrate the utility of examining ERPs in patients with brain lesions and intracranial electrodes. Contributions from particular neural sources can be inferred indirectly through detailed analysis of magnetic resonance images and computed tomographs in lesioned patients and more directly by localizing polarity inversions and voltage gradients with depth recordings. In turn, locating the neural generators with greater anatomical precision allows inferences to be made regarding the brain areas that are critical to the aspects of attention, memory, language processing, and so on reflected by the ERP in question.

IV. ELECTRICAL AND MAGNETIC RECORDINGS

The combined use of electrical and magnetic recordings has proven to be a valuable approach in the study of the neural generators of ERPs. Both

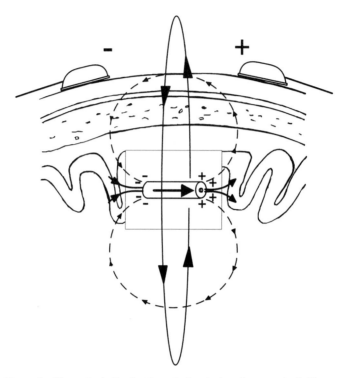

Figure 8 Theoretical distribution of electrical and magnetic fields produced by a current dipole within the cerebral cortex, schematically localized
beneath the scalp and skull. Dotted lines represent the electrical volume
currents that constitute the EEG. The solid line represents the magnetic field
produced by the current dipole. Adapted from Fig. 1 of Kaufman, Okada,
Brenner, & Williamson, (1981). On the relation between somatic evoked
potentials and fields. *International Journal of Neuroscience, 15,* 223–239.

magnetoencephalography (MEG) and magnetic resonance imaging (MRI)
have improved the accuracy of localizing current dipoles in the brain.
MEG measures the neuromagnetic field that results from intracellular current flux in populations of activated neurons (Fig. 8; for reviews of MEG,
see Beatty, 1990; Pizzella & Romani, 1990; Williamson & Kaufman, 1991).
The skull and scalp are essentially transparent to these magnetic fields, a
major advantage over EEG recordings which must contend with attenuation and smearing. A major disadvantage of MEG is its insensitivity to
radially oriented fields.

A great deal of controversy surrounds discussions of the relative utility
of MEG versus EEG in the localization of electrical sources in the brain
(see Crease, 1991). Some researchers claim that MEG can achieve more

accurate localization than ERPs, but the two methodologies can be seen as complementary. MEG is reflective of sulcal sources tangential to the brain surface, whereas EEG is composed of both radial and tangential sources. A typical figure cited for MEG accuracy is within approximately 3 mm, based on measurements of magnetic fields from current dipoles placed in a saline-filled lucite sphere and a plastic skull (Yamamoto, Williamson, Kaufman, Nicholson, & Llinas, 1988). An experiment performed by Cohen and colleagues (1990), however, did not support this contention. Weak current pulses were passed through depth electrodes of known locations implanted in epilepsy patients for clinical purposes. MEG and EEG were recorded separately, and the inverse solutions were calculated. The average error of localization was 8 mm for MEG and 10 mm for EEG. The methods of Cohen and colleagues have been criticized by other groups as inadequate and not up to current standards (Hari, Hamalainen, Ilmoniemi, & Lounasmaa, 1991; Williamson, 1991). For example, only one sensor was used to make sequential measurements, rather than using a 7- or 37-channel magnetometer. Since the position of the single sensor may not have been specified properly, a possible source of error was introduced. Despite these differences of opinion, the general conclusion is that, when combined, MEG and EEG can provide a wealth of information.

For instance, both magnetic fields and electrical potentials were recorded after stimulation of the median nerve in human subjects (Wood, Cohen, Cuffin, Yarita, & Allison, 1985). The SEP recorded from the scalp and cortical surface included the parietal N20–P30, largest over the hand area of somatosensory cortex, and the frontal P20–N30, maximal over the hand area of motor cortex (Allison, Goff, Williamson, & Vanglider, 1980). Three alternative sources for N20 and P20 have been proposed: thalamus and thalamocortical afferents, two radial sources in somatosensory and motor cortices, and one tangential source in somatosensory cortex; Wood and colleagues (1985) examined these possibilities. The resultant magnetic and potential distributions were dipolar, centered over sensorimotor cortex, and diverged in orientation by about 90°. The thalamus and thalamocortical afferents were too deep to account for the observed magnetic and potential extrema, nor could the combined data accommodate two radially oriented dipoles. One equivalent dipole with tangential current flow in somatosensory area 3b best explained the data. The magnetic and potential waveforms were similar but not identical, however, implying a smaller contribution from radial source(s) in somatosensory and/or motor cortices. Lesion and intracranial results from monkeys and humans have further indicated that these early SEPs are generated entirely within primary somatosensory cortex (reviewed by Allison, McCarthy, Wood, & Jones, 1991).

Magnetic and electrical recordings were also obtained from subjects performing a visual oddball task to determine the source locations of the magnetic equivalents of N200 and P300 (Okada, Kaufman, & Williamson, 1983). Isofield contour maps were plotted from magnetic field measurements for the two components, the sources of which were assumed to be a single equivalent dipole in each hemisphere (see subsequent discussion for descriptions of contour maps and equivalent dipoles). Calculations using the locations of the field extrema and a spherical head model estimated the sources of both components to be in or near the hippocampal formation. The spherical head model, however, may be an oversimplification that could lead to errors of 5–10 mm, particularly for deep brain structures (Barth, Sutherling, Broffman, & Beatty, 1986). Additionally, simultaneously active generators, hence contributions from other areas, must be considered as well.

Using a 7-channel magnetometer, another group recorded magnetic fields over the right hemisphere of subject counting auditory targets presented to the left ear in an oddball task (Rogers et al., 1991). Sources were estimated for successive 5-msec intervals using a single equivalent dipole model and projections onto MRIs. The P3m sources (300–450 msec) moved from medial (thalamus) to lateral (near superior temporal lobe) in some subjects, although a great deal of variability was noted. The mean locus of activity was near auditory cortex. Again, the single dipole assumption may discount multiple generators overlapping in time.

Neuromagnetic recordings, dipole localization methods, and reference to individual subject MRIs were combined to localize the source of the M100, the magnetic counterpart to the N1 component of the auditory evoked potential, within the transverse temporal (Heschl's) gyrus (Pantev et al., 1990). Primary auditory cortex is located on Heschl's gyrus and is thought to be the generator of M100 (see also Yamamoto et al., 1988). Woldorff and colleagues (1993) simultaneously recorded 37 channels of magnetic fields and 3 channels of ERPs over the left hemisphere of subjects performing a dichotic listening task. Stimuli were presented rapidly to maximize the selective focusing of attention to tones in one ear while ignoring tones of a different pitch in the other ear. Preliminary results from this experiment suggested that, as in previous ERP studies, tones in the attended ear elicited an early enhanced positivity between 20 and 50 msec (the P20–50 described by Woldorff & Hillyard, 1991), followed by an attention-related negativity, partially composed of an enhancement of the N1 component between 50 and 150 msec (early Nd or N1 attention effect; Hansen & Hillyard, 1980; Giard, Perrin, Pernier, & Peronnet, 1988; Hillyard et al., 1973; Woldorff & Hillyard, 1991; Woods, Alho, & Algazi, 1993). Magnetic counterparts to these electrical components showed highly dipolar field distributions, inverting in polarity from anterior to posterior sites. Dipole source localization techniques and comparisons with MRIs placed the generators of the M50, the M1, and the M1 attention effect within Heschl's gyrus.

In an attempt to dissociate the magnetic N1 (N1m) and Nd (Ndm) components, Arthur and co-workers used constant 800-msec interstimulus interval (ISIs) in subjects performing an auditory selective attention task (Arthur, Lewis, Medvick, & Flynn, 1991). Sources were modeled as single equivalent current dipoles in a conducting sphere (since the field distributions were dipolar), and their coordinates were related to MRI sections. Based on field distributions and dipole models, the generator of Ndm was near but anterior to that of N1m; both were located in auditory cortex in the region of the posterior superior temporal plane (Fig. 9). These findings and those of Woldorff et al. (1993) address the hotly debated question of whether the Nd reflects an "early selection/sensory gating" theory of attention, in which irrelevant stimuli are filtered out before full analysis, or a "late selection/attentional trace" model, which predicts complete analysis of stimuli before matching to a representation maintained in sensory memory (see Hansen & Woldorff, 1991; Naatanen, 1988, 1990; Woods, 1990). For instance, if Ndm represents modulation of the exogenous N1m component, the sources of both would likely be in the same location. This example illustrates how localization of an ERP component can illuminate theories in cognitive psychology and how MEG can assist in this endeavor.

V. CURRENT SOURCE DENSITY ANALYSES

Important information on the generators of the brain processes that give rise to ERPs has come from the topographic analysis of scalp-recorded electrical activity. Topographic displays of scalp potentials permit informative visual analyses useful in generating hypotheses about the localization of ERP generators. However, further analysis is required to specify ERP generators more accurately. A simple and direct way to improve the spatial resolution of scalp potential topography is to apply the current source density (CSD) technique, that is, to calculate one of several surface Laplacians (see Nunez, 1990, for a comparison of the different Laplacians). Currently most widely applied, the spline Laplacian provides estimates of local radial current density through the skull into the scalp. Since the Laplacian method estimates the second spatial derivative of the surface potential distribution, it eliminates the effects of tangential current flow. Thus, this method acts as a spatial filter that amplifies the contribution of local sources and diminishes the contribution of distant sources. An

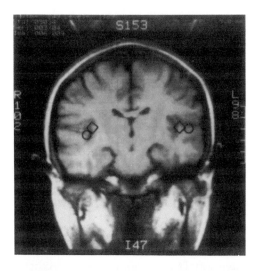

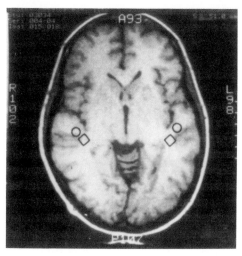

Figure 9 Coronal (Y-Z) (*top*) and horizontal (X-Y) (*bottom*) MRI sections from subject DA with N1m (diamond) and Ndm (circle) centroid locations superimposed. Reproduced with permission from Arthur, Lewis, Medvick, & Flynn (1991).

additional advantage provided by this approach is that the density of current flow is independent of the recording reference, thus mitigating concerns about estimates of the relative degree of activation of reference sites. Most published CSD data have been based on a three-concentric-sphere model of the head. In the future, extending the spline Laplacian approach to realistic ("finite element") head models including estimates of local skull resistance (which is highly variable over the surface), and the use of very high density electrode placements, will markedly improve the accuracy of this approach. Nonetheless, even as currently employed, a comparison of potential maps and surface Laplacians of modeled cortical dipoles, of epileptic spikes, and of ERPs clearly indicates the greater power of the Laplacian in localizing dipole sources (Giard et al., 1988; Mangun, Hillyard, & Luck, 1993; Nunez, 1990; see Fig. 10). According to the modeled results in Fig. 10, the spatial resolution of the CSD is on the order of a few centimeters.

Analysis of the CSD has been employed to explore the question of whether visual–spatial attention acts by modulating the flow of sensory information or by the activation of cortical areas additional to those along the afferent sensory pathway. Several studies have demonstrated that, during visual–spatial selective attention, ERP components P1 (at 75–100 msec) and N1 (at 100–200 msec) are significantly enhanced over visual cortical areas (Eason, Harter, & White, 1969; Mangun & Hillyard, 1987; Van Voorhis & Hillyard, 1977). Moreover, the amplitude increases of these

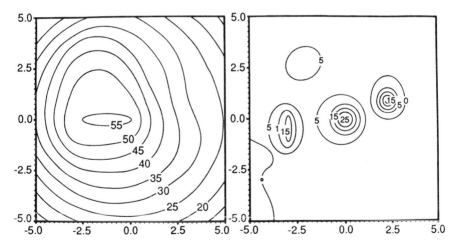

Figure 10 Theoretical potential (*left*) and surface Laplacian (*right*) for five dipoles in the three-concentric-spheres model. Four dipoles with various orientations are located within 2 cm of the outer spherical surface, and one stronger dipole is located near the center of the spheres. Potential distribution is maximum along the axis of the deep dipole and shows no evidence of four superficial dipoles. In contrast, surface Laplacian indicates the presence of all four superficial dipoles. Reproduced with permission from Nunez (1990).

ERP components occur without attendant changes in morphology or distribution. These results are consistent with the hypothesis that early sensory gating may play an important role in attentional processes. To localize the source of the attention effects and to determine further whether these generators are identical for attended and unattended stimuli, Hillyard, Mangun, and colleagues have related CSD distributions of these effects to the underlying cortical anatomy as revealed by MRI scans of the same subjects' brains (Hillyard, Mangun, Luck, & Heinze, 1990; Mangun et al., 1993). The CSD maps revealed that the earliest attention effect (on P1) displayed a current source identical in location to that of the sensory evoked response (see Fig. 11). This similarity of location supported the hypothesis that visual–spatial attention acts by modulating the

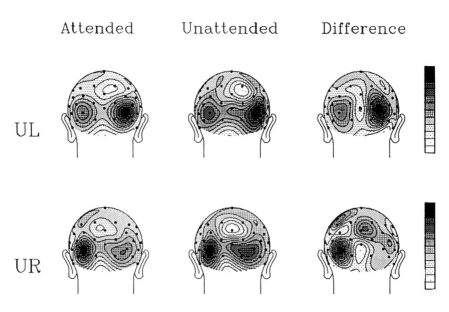

Figure 11 Scalp topography of grand average current source densities (CSD) calculated for the P1 component (at 108 msec) in response to upper left (UL) and upper right (UR) stimuli. Separate CSD maps are shown for the P1 component elicited by those flashes when attended and when unattended (averaged over the 3 other attention conditions). (*Far right*) The CSD distribution of the attention-related P1 difference formed by subtracting the ERPs in the unattended condition form those in the attended. The darkest zones represent current sources (current flowing out of the head), whereas the lightest zones represent current sinks. Each map is scaled individually to indicate 10 levels of CSD between the minimum and maximum values observed for that map. Reproduced with permission from Mangun, Hillyard, & Luck (1993).

amplitude of sensory activity along the afferent pathway but does not, at this latency, activate additional neural generators.

Additional analyses determined that the P1 attention effect was not generated in striate cortex. The CSD maps were similar in polarity and location in response to upper and lower visual field stimuli (see Fig. 11). A striate generator would have produced different CSD distributions because of the opposed orientation of cortical neurons for upper and lower field stimuli. In fact the CSD maps, when brought into register with the MRIs, indicated that the likely generator of P1 and the effect of attention on it (i.e., the locus of maximum current density) occurs over ventrolateral prestriate cortex, on the border of areas 18 and 19. This result is consistent with studies of monkeys that report no effects of attention on single neurons in striate cortex but significant effects in prestriate areas (Moran & Desimone, 1985; Wurtz, Goldberg, & Robinson, 1980).

Another study also utilized CSD analyses to explore the idea that nonidentical brain systems mediate different aspects of language (Neville, Mills, & Lawson, 1992). Words that convey primarily grammatical or semantic information were associated with different distributions of current at specific times. Additionally, a comparison of CSD maps from normal hearing subjects and congenitally deaf adults who learned English late and imperfectly raised the hypothesis that some of the brain systems important in grammatical processing are more dependent on early exposure to language for their normal development than are systems important in semantic processing. The results raise several hypotheses about the role of early experience in the development of different brain systems within and between the two hemispheres.

Clearly, the CSD approach will continue to be useful in generating and exploring hypotheses about the location of brain systems important to specific sensory and cognitive functions. In addition to being useful on their own, CSD analyses will be powerful in generating hypotheses that can be tested using dipole modeling approaches.

VI. DIPOLE MODELING TECHNIQUES

Another approach used by researchers to assist in the identification and localization of neural sources underlying ERP components is dipole modeling. The flow of ionic currents inside a volume conductor generates electrical potentials that can be recorded on the scalp. Electrical field theory can be employed to calculate the sources within the volume conductor from the potential field distribution measured at the surface, a query known as the inverse problem (Balish & Muratore, 1990; van Oosteram,

1991; Wood, 1982). Additional constraints and a basis in physiological reality are essential to accurately solve the "mathematically ill-posed" inverse problem, since myriad source configurations can yield the same potential distribution due to the principle of superposition. This principle states that the fields generated by any number of sources will sum linearly, rendering inverse solutions meaningless without some prior knowledge about source locations. Wood (1982) provides an excellent overview of dipole localization methods (DLMs), particularly those of Darcy, Ary, and Fender (1980) and Sidman, Giambalvo, Allison, and Bergy (1978), and their application to identifying the sources of ERPs.

A number of assumptions about the sources and the volume conductor are made in these models. Neuronal sources are modeled as dipolar to more easily approximate the generators of a resultant potential field, although the sources are not actually thought to be physical dipoles. The potential field generated by an "equivalent" dipole can be used as a "best estimate" of the fields generated by several neighboring sources, for example. With these assumptions in place, the inverse problem can be addressed by least squares parameter estimation. Each generator, or dipole, can be described by six parameters: three of these specify location on the x, y, and z axes and three specify dipole moment (strength) and orientation (for details, see Snyder, 1991; van Oosteram, 1991; Wood, 1982). DLM uses numerical minimization algorithms to estimate the best-fitting dipole for a particular scalp potential field. A single equivalent dipole (ED) is anatomically fictitious in most cases, however (Vaughan, 1987). More authentic are source models incorporating constructs such as dipole layers (de Munck, 1988) and distributed sources (Balish & Muratore, 1990).

Additional assumptions must be made in models of the volume conductor. Differences in conductivity between brain, cerebrospinal fluid (CSF), skull, and scalp necessitate that the simplest model consist of 3 or 4 concentric spheres (for EEG). A major advantage of MEG, of course, is that magnetic fields are not affected by skull and scalp, so a single sphere model can be used. A sphere may be a fair approximation for superficial sources, but more realistic head models are needed for deeper sources (van Oosteram, 1991). These newer, more complicated models based on MRIs from individual subjects may allow localization accuracy of a few millimeters (Dale & Sereno, 1993; Meijs & Peters, 1987; Stok, Meijs, & Peters, 1987). Other mathematical methods simulate the potentials that would be recorded directly from the brain surface. The cortical imaging technique of Sidman and colleagues (summarized by Sidman, 1991) and the finite element model deblurring technique of Gevins and associates (Gevins, Brickett, Costales, Le, & Reutter, 1990; Gevins, Le, Brickett,

Reutter, & Desmond, 1991) aim to reduce the skull's smearing and attenuating effects on scalp potential fields.

Snyder (1991) has pointed out that inverse DLMs work best for early sensory evoked responses, which can be modeled by one or two EDs. The number of recording electrodes must exceed the number of parameters to be estimated; these parameters or constraints increase with each additional generator. The technique has rarely been applied to later "cognitive" components such as P300 because of possible intrinsic limitations (however, see Sidman, Ford, Ramsey, & Schlichting, 1990; Turetsky, Raz, & Fein, 1990). For instance, the ability to distinguish correct from incorrect models drastically deteriorates as the number of generator loci increases to the theoretical limit allowed by the number of electrodes.

Furthermore, assumptions about multiple generators should be biologically plausible. Wood (1982) reminds us that, for components generated not by a single focal source that is stable over time but by many asynchronous sources, "it is particularly important not to confuse the numerical adequacy of DLM solutions with their physiological validity" (p. 148). This cautionary note holds even for scalp fields generated by a single focal source. For example, the DLM solution for the human SEP N20–P20, although mathematically reasonable, placed the equivalent dipole outside of the head (Wood, 1982). Vaughan (1987), however, is more optimistic about solving the inverse problem. Although an infinite number of source configurations exists in theory, this number is much smaller in reality, restricted by the brain's actual anatomy and physiology.

The benefits and drawbacks of models utilizing a single moving dipole versus multiple fixed dipoles are discussed by Lopes da Silva and Spekreijse (1991). These investigators applied these two strategies to visual evoked potential (VEP) data elicited by the appearance of a checkerboard pattern. For the single dipole model, a moving ED was calculated for consecutive 5-msec intervals starting at 35 msec and ending at 160 msec after pattern onset. Two clusters of EDs were obtained at 80–120 and 125–150 msec. Principle component analysis (PCA), which decomposes scalp-recorded ERPs into factors representing independent sources of variance (for discussion of PCA, see Turetsky et al., 1990), was used to describe two major components in the multiple fixed dipole model (see also Maier, Dagnelie, Spekreijse, & van Dijk, 1987). Both models obtained similar solutions: the earlier VEP component CI was accounted for predominantly by a radially oriented ED located in area 18/19; the second component CII was accounted for by a tangentially oriented ED located in area 17. Other models, however, have obtained different solutions. For example, the iterative minimization technique used by Butler et al. (1987) placed the ED corresponding to the CI generator in area 17. The sources of

CI and CII have been disputed since the initial studies of Jeffries and Axford (1972a,b) proposed them to be in areas 17 and 18, respectively.

Scherg and co-workers have developed a method, called brain electric source analysis (BESA), for decomposing scalp-recorded ERPs into their constituent or source waveforms (reviewed in Scherg & Picton, 1991). In their spatio-temporal dipole model, sources must have fixed positions and orientations. This multiple stationary dipole assumption allows more sources to be active at once. The number of sources (m) must merely be less than the number of recording sites (n), instead of less than the number of parameters (6) times n, as in other methods of calculating the inverse solution. For example, Scherg and Von Cramen (1986) decomposed middle- and long-latency AEPs into "dipole source potentials." To restrict the number of possible solutions, two dipoles in each hemisphere were assumed to represent the AEP sources. These dipoles were allowed to vary in strength over time. After solving the requisite equations, bilateral tangential and radial dipoles in auditory cortex accounted for the empirical AEP data recorded from normal subjects.

BESA is most effective when used to test hypotheses about the neural generators of specific ERPs, given that the sources are tentatively identified through other means (e.g., imaging, lesion, and depth data). Posing initial constraints on the possible solutions can overcome the difficulty of an arbitrary starting point. For instance, BESA and comparison with individual subject MRIs were used to localize component sources of the VEP to circular checkerboard stimuli (Clark, Fan, & Hillyard, 1991). The effects of different stimulus positions on scalp topography and component amplitude were also examined. The C1 (i.e., CI) showed a polarity inversion about 20–40° below the horizontal meridian instead of at the horizontal meridian as previously believed. The C1 component source was placed just lateral to the medial surface of the contralateral occipital lobe near the calcarine fissure, implicating striate cortex as the most likely generator. The tentative locations of N150 and P220 sources were found to be in several extrastriate regions.

Dale and Sereno (1993) criticize the nonlinear optimization technique used by Scherg and co-workers as computationally difficult. These researchers also question the a priori determination of the number of equivalent dipoles and the lack of confidence measures with which to evaluate the solutions. Their alternative approach, based on work in progress (Dale & Sereno, 1993), takes MRI data from individual subjects and shrinks a deformable template onto the images to determine the shape of the cortical sheet. The position and orientation of possible current sources is established from these transformed MRIs, with the assumption that current flow is perpendicular to the cortical sheet. Linear techniques then can be used to solve the inverse problem, which becomes well posed and constrained with the addition of regularization terms based on empirical EEG, MEG, and positron emission tomography (PET) data.

Other critiques of BESA include those of Turetsky et al. (1990) and Achim, Richer, and Saint-Hilaire (1991). Although the refined nonparametric model (Scherg & Von Cramen, 1986) avoids the arbitrary constraints of the earlier parametric model, it is more likely to be affected by noise (Turetsky et al., 1990). Overmodeling is a chronic problem; the model assumes a very high signal-to-noise ratio and fails to adequately account for the signal in the presence of physiological noise (Achim et al., 1991). Furthermore, there are no summary measures of dipole activity and therefore no clear procedures for analyzing results, as others have recognized. Turetsky and colleagues (1990) offered an alternative dipole component model, which they applied to auditory P300. A major advantage of their model is the incorporation of latency variations across subjects and conditions, although one problem is that the head was modeled as a homogeneous sphere instead of three concentric spheres. Fitting the model to the data proved to be a very computationally intensive procedure, perhaps prohibiting these additional constraints.

VII. SUMMARY

This chapter has reviewed several different approaches currently being implemented to improve the spatial resolution of (i.e., localize the neural generators of) ERPs. To the extent that specific ERP components index particular sensory and cognitive processes (an issue not addressed here but reviewed by Brunia, Gaillard, & Kok, 1990; Hillyard & Picton, 1987; Kutas & Van Petten, 1990; Regan, 1989), knowledge about their neural origins will contribute to several different issues in neuropsychology and cognitive neuroscience. Of course knowledge about the neural structure(s) that supports different aspects of perception and cognition is of intrinsic interest and will lay the groundwork for determining the inputs and outputs of these systems. Additionally, this information will provide information about the biological validity of different conceptions of the organization of specific sensory and cognitive systems. To the extent that different putative subsystems within a domain rely on nonidentical neural structures, this information may be considered evidence for their isolability and independence. The techniques for localizing these cognitive systems have improved considerably over the past 10 years. The next 10 years are likely to see improvements an order of magnitude greater than the last.

ACKNOWLEDGMENTS

We thank Vince Clark, Anders Dale, and Bob Knight for illuminating discussions and David Woods for critical comments on the chapter. We appreciate Margaret Mitchell's assistance in preparing the manuscript. This work was supported in part by grants from the McDonnell-Pew Foundation to Diane Swick, HD22614 and AG08313 to Marta Kutas, and DC00128 to Helen J. Neville.

REFERENCES

Achim, A., Richer, F., & Saint-Hilaire, J. M. (1991). Methodological considerations for the evaluation of spatio-temporal source models. *Electroencephalography and Clinical Neurophysiology, 79*, 227–240.

Alain, C., Richer, F., Achim, A., & Saint Hilaire, J.-M. (1989). Human intracerebral potentials associated with target, novel, and omitted auditory stimuli. *Brain Topography, 4*, 237–245.

Allison, T., Goff, W. R., Williamson, P. D., & Vangilder, J. C. (1980). On the neural origin of early components of the human somatosensory evoked potential. *Progress in Clinical Neurophysiology, 7*, 51–68.

Allison, T., McCarthy, G., Wood, C. C., & Jones, S. J. (1991). Potentials evoked in human and monkey cerebral cortex by stimulation of the median nerve. *Brain, 114*, 2465–2503.

Arezzo, J., Legatt, D., & Vaughan, H. G. (1979). Topography and intracranial sources of somatosensory evoked potentials in the monkey. I. Early components. *Electroencephalography and Clinical Neurophysiology, 46*, 155–172.

Arezzo, J. C., Vaughan, H. G., & Legatt, A. D. (1981). Topography and intracranial sources of somatosensory evoked potentials in the monkey. II. Cortical components. *Electroencephalography and Clinical Neurophysiology, 51*, 1–18.

Arezzo, J., Pickoff, A., & Vaughan, H. G. (1975). The sources and intracerebral distribution of auditory evoked potentials in the alert rhesus monkey. *Brain Research, 90*, 57–73.

Arnsten, A. F. T., Neville, H. J., Hillyard, S. A., Janowsky, D. S., & Segal, D. S. (1984). Naloxone increases electrophysiological measures of selective information processing in humans. *Journal of Neuroscience, 4*, 2912–2919.

Arthur, D. L., Lewis, P. S., Medvick, P. A., & Flynn, E. R. (1991). A neuromagnetic study of selective auditory attention. *Electroencephalography and Clinical Neurophysiology, 78*, 348–360.

Arthur, D. L., & Starr, A. (1984). Task-relevant late positive component of the auditory event-related potential in monkeys resembles P300 in humans. *Science, 223*, 186–188.

Aston-Jones, G., Chiang, C., & Alexinsky, T. (1991). Discharge of noradrenergic locus coeruleus neurons in behaving rats and monkeys suggests a role in vigilance. *Progress in Brain Research, 88*, 501–520.

Balish M., & Muratore, R. (1990). The inverse problem in electroencephalography and magnetoencephalography. *Advances in Neurology, 54*, 79–88.

Barth, D. S., Sutherling, W., Broffman, J., & Beatty, J. (1986). Magnetic localization of a dipolar current source implanted in a sphere and a human cranium. *Electroencephalography and Clinical Neurophysiology, 63*, 260–273.

Beatty, J. (1990). Magnetoencephalographic analysis of human cognitive processes. In A. B. Scheibel & A. F. Wechsler (Eds.), *Neurobiology of higher cognitive function* (pp. 151–166). New York: Guilford Press.

Beatty, J., Barth, D. S., Richer, F., & Johnson, R. A. (1986). Neuromagnetometry. In M. G. H. Coles, E. Donchin, & S. W. Porges (Eds.), *Psychophysiology: Systems, processes, and applications* (pp. 26–40). New York: Guilford Press.

Besson, M., Kutas, M., & Van Petten, C. (1992). An event-related potential (ERP) analysis of semantic congruity and repetition effects in sentences. *Journal of Cognitive Neuroscience, 4,* 132–149.

Borda, R. P. (1970). The effect of altered drive states on the contingent negative variation (CNV) in rhesus monkeys. *Electroencephalography and Clinical Neurophysiology, 29,* 173–180.

Brunia, C. H. M., Gaillard, A. W. K., and Kok, A. (Eds.), (1990). *Psychophysiological brain research.* Le Tilburg, The Netherlands: Tilburg University Press.

Buchwald, J. S., & Squires, N. K. (1982). Endogenous auditory potentials in the cat: A P300 model. In C. Woody (Ed.), *Conditioning: Representation of involved neural function* (pp. 503–515). New York: Plenum Press.

Butler, S. R., Georgiou, G. A., Glass, A., Hancox, R. J., Hopper, J. M., & Smith, K. R. H. (1987). Cortical generators of the CI component of the pattern-onset visual evoked potential. *Electroencephalography and Clinical Neurophysiology, 68,* 256–267.

Callaway, E., Halliday, R., Naylor, H., & Schechter, G. (1985). Effects of oral scopolamine on human stimulus evaluation. *Psychopharmacology, 85,* 133–138.

Caramazza, A. (1992). Is cognitive neuropsychology possible? *Journal of Cognitive Neuroscience, 4,* 80–95.

Cedarbaum, J. M., & Aghajanian, G. K. (1977). Catecholamine receptors on locus coeruleus neurons: Pharmacological characterization. *European Journal of Pharmacology, 44,* 375–385.

Clark, V. P., Fan, S., & Hillyard, S. A. (1991). The effects of stimulus position in the visually evoked potential: Analysis and localization with MRI. *Society for Neuroscience Abstracts, 17,* 656.

Cohen, D., Cuffin, B. N., Yunokuchi, M. S., Maniewski, R., Purcell, C., Cosgrove, G. R., Ives, J., Kennedy, J. G., & Schomer, D. L. (1990). MEG versus EEG localization test using implanted sources in the human brain. *Annals of Neurology, 28,* 811–817.

Coles, M. G. H., Gratton, G., Kramer, A. F., & Miller, G. A. (1986). Principles of signal acquisition and analysis. In M. G. H. Coles, E. Donchin, & S. W. Porges (Eds.), *Psychophysiology: Systems, processes, and applications* (pp. 183–221). New York: Guilford Press.

Courchesne, E., Hillyard, S. A., & Galambos, R. (1975). Stimulus novelty, task relevance, and the visual evoked potential. *Electroencephalography and Clinical Neurophysiology, 39,* 131–143.

Crease, R. P. (1991). Images of conflict: MEG vs. EEG. *Science, 253,* 374–375.

Dale, A. M., & Sereno, M. I. (1993). Improved localization of cortical activity by combining EEG and MEG with MRI cortical surface reconstruction: a linear approach. *Journal of Cognitive Neuroscience, 5,* 162–176.

Darcy, T. M., Ary, J. P., & Fender, D. H. (1980). Methods for the localization of electrical sources in the human brain. *Progress in Brain Research, 54,* 128–134.

Daruna, J. H., Nelson, A. V., & Green, J. B. (1989). Unilateral temporal lobe lesions after P300 scalp topography. *International Journal of Neuroscience, 46,* 243–247.

de Munck, J. (1988). The potential distribution in a layered anisotropic spherical volume conductor. *Journal of Applied Physics, 64,* 464–470.

Donchin, E. (1981). Surprise! . . . Surprise? *Psychophysiology, 18,* 493–513.

Donchin, E., & Coles, M. G. H. (1988). Is the P300 component a manifestation of context updating? *The Behavioral and Brain Sciences, 11,* 357–374.

Duncan, C. C., & Kaye, W. H. (1987). Effects of clonidine on event-related potential measures of information processing. *Current Trends in Event-Related Potential Research (EEG Suppl.), 40,* 527–531.

Eason, R., Harter, M., & White, C. (1969). Effects of attention and arousal on visually evoked cortical potentials and reaction time in man. *Physiology and Behavior, 4,* 283–289.

Ehlers, C. L. (1988). ERP responses to ethanol and diazepam administration in squirrel monkeys. *Alcohol, 5,* 315–320.

Ehlers, C. L. (1989). EEG and ERP responses to naloxone and ethanol in monkeys. *Progress in Neuro-Psychopharmacology and Biological Psychiatry, 13,* 217–228.

Ehlers, C. L., Wall, T. L., & Chapin, R. I. (1991). Long latency event-related potentials in rats: Effects of dopaminergic and serotonergic depletions. *Pharmacology, Biochemistry, and Behavior, 38,* 789–793.

Fabiani, M., Gratton, G., Karis, D., & Donchin, E. (1987). P300: Methodological and theoretical issues. In P.K. Ackles, J. R. Jennings, & M. G. H. Coles (Eds.), *Advances in psychophysiology* (Vol. 2, pp. 1–78). Greenwich, Connecticut: JAI Press.

Fabiani, M., Karis, D., & Donchin, E. (1986). P300 and recall in an incidental memory paradigm. *Psychophysiology, 23,* 298–308.

Foote, S. L., Berridge, C. W., Adams, L. M., & Pineda, J. A. (1991). Electrophysiological evidence for the involvement of the locus coeruleus in alerting, orienting, and attending. *Progress in Brain Research, 88,* 521–532.

Foote, S. L., & Morrison, J. H. (1987). Extrathalamic modulation of cortical function. *Annual Review of Neuroscience, 10,* 67–95.

Freeman, W. J. (1975). *Mass action in the nervous system.* New York: Academic Press.

Gabriel, M., Sparenborg, S. P., & Donchin, E. (1983). Macropotentials recorded from the cingulate cortex and anterior thalamus in rabbits during the "oddball" paradigm used to elicit P300 in normal human subjects. *Society for Neuroscience Abstracts, 9,* 1200.

Galambos, R., & Hillyard, S. A. (1981). Electrophysiological approaches to human cognitive processing. *Neurosciences Research Program Bulletin, 20,* 141–265.

Gevins, A., Brickett, P., Costales, B., Le, J., & Reutter, B. (1990). Beyond topographic mapping: Towards functional-anatomical imaging with 124-channel EEGs and 3-D MRIs. *Brain Topography, 3,* 53–64.

Gevins, A., Le, J., Brickett, P., Reutter, B., & Desmond, J. (1991). Seeing through the skull: Advanced EEGs use MRIs to accurately measure cortical activity from the scalp. *Brain Topography, 4,* 125–131.

Giard, M. H., Perrin, F., Pernier, J., & Peronnet, F. (1988). Several attention-related wave forms in auditory areas: A topographic study. *Electroencephalography and Clinical Neurophysiology, 69,* 371–384.

Glover, A., Ghilardi, M. F., Bodis-Wollner, I., & Onofrj, M. (1988). Alterations in event-related potentials (ERPs) of MPTP-treated monkeys. *Electroencephalography and Clinical Neurophysiology, 71,* 461–468.

Glover, A., Ghilardi, M. F., Bodis-Wollner, I., Onofrj, M., & Mylin, L. H. (1991). Visual "cognitive" evoked potentials in the behaving monkey. *Electroencephalography and Clinical Neurophysiology, 90,* 65–72.

Glover, A. A., Onofrj, M. J., Ghilardi, M. F., & Bodis-Wollner, I. (1986). P300-like potentials in the normal monkey using classical conditioning and an auditory "oddball" paradigm. *Electroencephalography and Clinical Neurophysiology, 65,* 231–235.

Grant, S. J., Aston-Jones, G., & Redmond, D. E. (1988). Responses of primate locus coeruleus neurons to simple and complex sensory stimuli. *Brain Research Bulletin, 21,* 401–410.

Halgren, E., and Smith, M. E. (1987). Cognitive evoked potentials as modulatory processes in human memory formation and retrieval. *Human Neurobiology, 6,* 129–139.

Halgren, E., Squires, N. K., Wilson, C. L., Rohrbaugh, J. W., Babb, T. L., & Crandall, P. H. (1980). Endogenous potentials generated in human hippocampal formation and amygdala by infrequent events. *Science, 210,* 803–805.

Halliday, R., Naylor, H., Callaway, E., Yano, L., & Walton, P. (1987). What's done can't always be undone: The effects of stimulant drugs and dopamine blockers on information processing. *Current Trends in Event-Related Potential Research (EEG Suppl.)*, 40, 532–537.

Hammond, E. J., Meador, K. J., Aung-Din, R., & Wilder, B. J. (1987). Cholinergic modulation of human P3 event-related potentials. *Neurology*, 37, 346–350.

Hansen, J. C., & Hillyard, S. A. (1980). Endogenous brain potentials associated with selective auditory attention. *Electroencephalography and Clinical Neurophysiology*, 49, 277–290.

Hansen, J. C., & Woldorff, M. (1991). Mechanisms of auditory selective attention as revealed by event-related potentials. *Event-Related Brain Research (EEG Suppl.)*, 42, 195–209.

Hari, R., Hamalainen, M., Ilmoniemi, R., & Lounasmaa, O. V. (1991). MEG versus EEG localization test. *Annals of Neurology*, 30, 222–223.

Harrison, J., & Buchwald, J. (1985). Aging changes in the cat P300 mimic the human. *Electroencephalography and Clinical Neurophysiology*, 62, 227–234.

Harrison, J. B., & Buchwald, J. S. (1987). A cat model of the P300: Searching for generator substrates in the auditory cortex and medial septal area. *Current Trends in Event-Related Potential Research (EEG Suppl.)*, 40, 473–480.

Harrison, J., Buchwald, J., & Kaga, K. (1986). Cat P300 present after primary auditory cortex ablation. *Electroencephalography and Clinical Neurophysiology*, 63, 180–187.

Harrison, J., Buchwald, J., Kaga, K., Woolf, N. J., & Butcher, L. L. (1988). "Cat P300" disappears after septal lesions. *Electroencephalography and Clinical Neurophysiology*, 69, 55–64.

Harrison, J. B., Dickerson, L. W., Song, S., & Buchwald, J. B. (1990). Cat-P300 present after association cortex ablation. *Brain Research Bulletin*, 24, 551–560.

Herning, R. I., Hooker, W. D., & Jones, R. T. (1987). Cocaine effects on electroencephalographic cognitive event-related potentials and performance. *Electroencephalography and Clinical Neurophysiology*, 66, 34–42.

Herning, R. I., Jones, R. T., Hooker, W. D., & Tulunay, F. C. (1985). Information processing components of the auditory event related potential are reduced by cocaine. *Psychopharmacology*, 87, 178–185.

Hillyard, S. A., Hink, R. F., Schwent, V. L., & Picton, T. W. (1973). Electrical signs of selective attention in the human brain. *Science*, 182, 177–180.

Hillyard, S. A., & Kutas, M. (1983). Electrophysiology of cognitive processing. *Annual Review of Psychology*, 34, 33–61.

Hillyard, S. A., Mangun, G. R., Luck, S. J., and Heinze, H. (1990). Electrophysiology of visual attention. In E. R. John, T. Harmony, L. Prichep, M. Valdez, & P. Valdez (Eds.), *Machinery of mind* (pp. 186–205). Boston: Birkhausen.

Hillyard, S. A., & Picton, T. W. (1987). Electrophysiology of cognition. In F. Plum (Ed.), *Handbook of physiology: The nervous system* (Vol. V, pp. 519–584). Bethesda, Maryland: American Physiological Society.

Jeffries, D. A., & Axford, J. G. (1972a). Source locations of pattern-specific components of human visual evoked potentials. I. Component of striate cortical origin. *Experimental Brain Research*, 16, 1–21.

Jeffries, D. A., & Axford, J. G. (1972b). Source locations of pattern-specific components of human visual evoked potentials. II. Component of extrastriate cortical origin. *Experimental Brain Research*, 16, 22–40.

Johnson, R., Jr. (1986). A triarchic model of P300 amplitude. *Psychophysiology*, 23, 367–384.

Johnson, R., Jr. (1988a). The amplitude of the P300 component of the event-related potential: Review and synthesis. In P. K. Ackles, J. R. Jennings, & M. G. H. Coles (Eds.), *Advances in psychophysiology* (Vol. 3, pp. 69–138). Greenwich, Connecticut: JAI Press.

Johnson, R., Jr. (1988b). Scalp-recorded P300 activity in patients following unilateral temporal lobectomy. *Brain*, 111, 1517–1529.

Johnson, R., Jr. (1989a). Auditory and visual P300s in temporal lobectomy patients: Evidence for modality-dependent generators. *Psychophysiology, 26,* 633–650.

Johnson, R., Jr. (1989b). Developmental evidence for modality-dependent P300 generators: A normative study. *Psychophysiology, 26,* 651–667.

Joseph, K. C., & Sitaram, N. (1989). The effect of clonidine on auditory P300. *Psychiatry Research, 28,* 255–262.

Kaga, K., Harrison, J. B., Butcher, L. L., Woolf, N. J., & Buchwald, J. S. (1992). Cat "P300" and cholinergic septohippocampal neurons: Depth recordings, lesions, and choline acetyl-transferase immunohistochemistry. *Neuroscience Research, 13,* 53–71.

Karis, D., Fabiani, M., & Donchin, E. (1984). "P300" and memory: Individual differences in the von Restorff effect. *Cognitive Psychology, 16,* 177–216.

Kiss, I., Dashieff, R. M., & Lordeon, P. (1989). A parieto–occipital generator for P300: Evidence from human intracranial recordings. *International Nournal of Neuroscience, 49,* 133–139.

Kluender, R. (1991). *Cognitive constraints on variables in syntax.* Ph.D. Thesis. San Diego: University of California.

Knight, R. T. (1984). Decreased response to novel stimuli after prefrontal lesions in man. *Electroencephalography and Clinical Neurophysiology, 59,* 9–20.

Knight, R. T. (1990). ERPs in patients with focal brain lesions. *Electroencephalography and Clinical Neurophysiology, 75,* S72.

Knight, R. T. (1991). Effects of hippocampal lesions on the human P300. *Society for Neuroscience Abstracts, 17,* 657.

Knight, R. T., Scabini, D., Woods, D. L., & Clayworth, C. C. (1989). Contributions of temporal–parietal junction to the human auditory P3. *Brain Research, 502,* 109–116.

Kosslyn, S. M., & Intriligator, J. M. (1992). Is cognitive neuropsychology plausible? The perils of sitting on a one-legged stool. *Journal of Cognitive Neuroscience, 4,* 96–106.

Kropotov, J. D., & Ponomarev, V. A. (1991). Subcortical correlates of component P300 in man. *Electroencephalography and Clinical Neurophysiology, 78,* 40–49.

Kutas, M., Hillyard, S. A., Volpe, B. T., & Gazzaniga, M. S. (1990). Late positive event-related potentials after commissural section in humans. *Journal of Cognitive Neuroscience, 2,* 258–271.

Kutas, M., & Van Petten, C. (1990). Electrophysiological perspectives on comprehending written language. *New Trends and Advanced Techniques in Clinical Neurophysiology (EEG Suppl.), 41,* 155–167.

Lopes da Silva, F. H., & Spekreijse, H. (1991). Localization of brain sources of visually evoked responses: Using single and multiple dipoles. An overview of different approaches. *Electroencephalography and Clinical Neurophysiology. (Suppl.), 42,* 38–46.

Maier, J., Dagnelie, G., Spekreijse, H., & van Dijk, B. W. (1987). Principal components analysis for source localization of VEPs in man. *Vision Research, 27,* 165–177.

Mangun, G. R., & Hillyard, S. A. (1987). The spatial allocation of visual attention as indexed by event-related brain potentials. *Human Factors, 29,* 195–211.

Mangun, G. R., Hillyard, S. A., & Luck, S. J. (1993). Electrocortical substrates of visual selective attention. In D. Meyer & S. Kornblum (Eds.). *Attention and performance XIV.* Cambridge, Massachusetts: MIT Press.

Marczynski, T. J. (1978). Neurochemical mechanisms in the genesis of slow potentials: A review and some clinical implications. In D. A. Otto (Ed.), *Multidisciplinary perspectives in event-related brain potential research* (pp. 25–35). Washington, D.C.: U.S. Government Printing Office.

McCarthy, G., Wood, C. C., & Williamson, P. D., & Spencer, D. D. (1989). Task-dependent field potentials in human hippocampal formation. *Journal of the Neurosciences, 9,* 4253–4268.

Meador, K. J., Loring, D. W., Adams, R. J., Patel, B. R., Davis, H. C., & Hammond, E. J. (1987). Central cholinergic systems and the P3 evoked potential. *International Journal of Neuroscience, 33,* 199–205.

Meador, K. J., Loring, D. W., Davis, H. C., Sethi, K. D., Patel, B. R., Adams, R. J., & Hammond, E. J. (1989). Cholinergic and serotonergic effects on the P3 potential and recent memory. *Journal of Clinical and Experimental Neuropsychology, 11,* 252–260.

Meador, K. J., Loring, D. W., King, D. W., Gallagher, B. B., Gould, M. J., Smith, J. R., & Flanigin, H. F. (1988). Cholinergic modulation of human limbic evoked potentials. *International Journal of Neuroscience, 38,* 407–414.

Meijs, J. W. H., & Peters, M. J. (1987). The EEG and MEG, using a model of eccentric spheres to describe the head. *IEEE Transactions of Biomedical Engineering,* BME-*34,* 913–920.

Mesulam, M.-M., Mufson, E. J., Levey, A. I., & Wainer, B. H. (1983). Cholinergic innervation of cortex by the basal forebrain: Cytochemistry and cortical connections of the septal area, diagonal band nuclei, nucleus basalis (substantia innominata) and hypothalamus in the rhesus monkey. *Journal of Comparative Neurology, 214,* 170–197.

Moran, J., & Desimone, R. (1985). Selective attention gates visual processing in the extrastriate cortex. *Science, 229,* 782–784.

Naatanen, R. (1988). Implications of ERP data for psychological theories of attention. *Biological Psychology, 26,* 117–163.

Naatanen, R. (1990). The role of attention in auditory information processing as revealed by event-related potentials and other brain measures of cognitive function. *The Behavioral and Brain Sciences, 13,* 201–288.

Naatanen, R. (1991). Mismatch negativity outside strong attentional focus: A commentary on Woldorff et al. (1991). *Psychophysiology, 28,* 478–483.

Naylor, H., Brandeis, D., Halliday, R., Yano, L., & Callaway, E. (1988). Why does scopolamine make "easy" difficult? *Psychophysiology, 25,* 472.

Naylor, H., Halliday, R., & Callaway, E. (1985). The effect of methylphenidate on information processing. *Psychopharmacology, 86,* 90–95.

Neville, H. J., & Foote, S. L. (1984). Auditory event-related potentials in the squirrel monkey: Parallels to human late wave responses. *Brain Research, 298,* 107–116.

Neville, H. J., Kutas, M., Chesney, G., & Schmidt, A. (1986). Event-related brain potentials during initial encoding and recognition memory of congruous and incongruous words. *Journal of Memory and Language, 25,* 75–92.

Neville, H. J., Mills, D. M., and Lawson, D. S. (1992). Fractionating Language: Different neural subsystems with different sensitive periods. *Cerebral Cortex, 2,* 244–258.

Neville, H. J., Nicol, J. L., Barss, A., Forster, K. I., & Garrett, M. F. (1991). Syntactically based sentence processing classes: Evidence from event-related brain potentials. *Journal of Cognitive Neuroscience, 3,* 151–165.

Niedermeyer, E., & Lopes da Silva, F. H. (1987). *Electroencephalography.* Baltimore: Urban and Schwarzenberg.

Nunez, P. L. (1981). *Electric fields of the brain.* New York: Oxford University Press.

Nunez, P. L. (1990). Localization of brain activity with electroencephalography. *Advances in Neurology, 54,* 39–65.

O'Connor, T., & Starr, A. (1985). Intracranial potentials correlated with an event-related potential, P300, in the cat. *Brain Research, 339,* 27–38.

Okada, Y. C., Kaufman, L., & Williamson, S. J. (1983). The hippocampal formation as a source of the slow endogenous potentials. *Electroencephalography and Clinical Neurophysiology, 55,* 417–426.

Onofrj, M., Fulgente, T., Nobilio, D., Malatesta, G., Bazzano, S., Colamartino, P., & Gambi, D. (1992). P3 recordings in patients with bilateral temporal lobe lesions. *Neurology, 42,* 1762–1767.

Osterhout, L., & Holcomb, P. (1993). Event-related brain potentials elicited by syntactic anomaly. *Journal of Memory and Language, 31,* 785–806.

Paller, K. A., Kutas, M., & Mayes, A. R. (1987). Neural correlates of encoding in an incidental learning paradigm. *Electroencephalography and Clinical Neurophysiology, 67,* 360–371.

Paller, K. A., McCarthy, G., Roessler, E., Allison, T., & Wood, C. C. (1992). Potentials evoked in human and monkey medial temporal lobe during auditory and visual oddball paradigms. *Electroencephalography and Clinical Neurophysiology, 84,* 269–279.

Paller, K. A., Zola-Morgan, S., Squire, L. R., & Hillyard, S. A. (1988). P3-like brain waves in normal monkeys and monkeys with medial temporal lesions. *Bevhavioral Neuroscience, 102,* 714–725.

Pantev, C., Hoke, M., Lehnertz, K., Lutzkenhoner, B., Fahrendorf, G., & Stober, U. (1990). Identification of sources of brain neuronal activity with high spatiotemporal resolution through combination of neuromagnetic source localization (NMSL) and magnetic resonance imaging (MRI). *Electroencephalography and Clinical Neurophysiology, 75,* 173–184.

Paul, D. D., & Sutton, S. (1972). Evoked potential correlates of response criterion in auditory signal detection. *Science, 177,* 362–364.

Pineda, J. A., Foote, S. L., & Neville, H. J. (1987). Long-latency event-related potentials in squirrel monkeys: Further characterization of wave form morphology, topography, and functional properties. *Electroencephalography and Clinical Neurophysiology, 67,* 77–90.

Pineda, J. A., Foote, S. L., & Neville, H. J. (1989). Effects of locus coeruleus lesions on auditory, long-latency, event-related potentials in monkey. *Journal of Neuroscience, 9,* 81–93.

Pineda, J. A., Foote, S. L., Neville, H. J., & Holmes, T. C. (1988). Endogenous event-related potentials in monkey: the role of task relevance, stimulus probability, and behavioral response. *Electroencephalography and Clinical Neurophysiology, 70,* 155–171.

Pineda, J. A., & Swick, D. (1992). Visual P3-like potentials in squirrel monkey: Effects of a noradrenergic agonist. *Brain Research Bulletin, 28,* 485–491.

Pineda, J. A., Swick, D., & Foote, S. L. (1991). Noradrenergic and cholinergic influences on the genesis of P3-like potentials. *Event-Related Brain Research (EEG Suppl.), 42,* 165–172.

Pirch, J. H., Corbus, M. J., Rigdon, G. C., & Lyness, W. H. (1986). Generation of cortical event-related slow potentials in the rat involves nucleus basalis cholinergic innervation. *Electroencephalography and Clinical Neurophysiology, 63,* 464–475.

Pizzella, V., & Romani, G. L. (1990). Principles of magnetoencephalography. *Advances in Neurology, 54,* 1–9.

Pritchard, W. S. (1981). Psychophysiology of P300. *Psychological Bulletin, 89,* 506–540.

Regan, D. (1989). *Human brain electrophysiology.* New York: Elsevier Science.

Ritter, W., Vaughan, H. G., & Costa, L. D. (1968). Orienting and habituation to auditory stimuli: A study of short-term changes in average evoked responses. *Electroencephalography and Clinical Neurophysiology, 25,* 550–556.

Rogers, R. L., Baumann, S. B., Papanicolaou, A. C., Bourbon, T. W., Alagarsamy, S., & Eisenberg, H. M. (1991). Localization of the P3 sources using magnetoencephalography and magnetic resonance imaging. *Electroencephalography and Clinical Neurophysiology, 79,* 308–321.

Rohrbaugh, J. W., Syndulko, K., & Lindsley, D. B. (1976). Brain wave components of the contingent negative variation in humans. *Science, 191,* 1055–1057.

Rugg, M. D. & Nagy, M. E. (1989). Event-related potentials and recognition memory for words. *Electroencephalography and Clinical Neurophysiology, 72,* 395–406.

Rugg, M. D., Pickles, C. D., Potter, D. D., & Roberts, R. C. (1991). Normal P300 following extensive damage to the left medial temporal lobe. *Journal of Neurology, Neurosurgery, and Psychiatry, 54,* 217–222.

Scabini, D., Knight, R. T., & Woods, D. L. (1989). Frontal lobe contributions to the human auditory P3a. *Society for Neuroscience Abstracts, 15,* 477.

Scheffers, M. K., Johnson, R., Jr., & Ruchkin, D. S. (1991). P300 in patients with unilateral temporal lobectomies: The effects of reduced stimulus quality. *Psychophysiology, 28,* 274–284.

Scherg, M., & Picton, T. W. (1991). Separation and identification of event-related potential components by brain electric source analysis. *Event-Related Brain Research (EEG Suppl.), 42,* 24–37.

Scherg, M., & Von Cramon, D. (1986). Evoked dipole source potentials of the human auditory cortex. *Electroencephalography and Clinical Neurophysiology, 65,* 344–360.

Sidman, R. D. (1991). A method for simulating intracerebral potential fields: The cortical imaging technique. *Journal of Clinical Neurophysiology, 8,* 432–441.

Sidman, R. D., Ford, M. R., Ramsey, G., & Schlichting, C. (1990). Age-related features of the resting and P300 auditory evoked responses using the dipole localization method and cortical imaging technique. *Journal of Neuroscience Methods, 33,* 23–32.

Sidman, R. D., Giambalvo, V., Allison, T., & Bergy, P. (1978). A method for localization of sources of human cerebral potentials evoked by sensory stimuli. *Sensor Processes, 2,* 116–129.

Simson, R., Vaughan, H. G., & Ritter, W. (1976). The scalp topography of potentials associated with missing visual or auditory stimuli. *Electroencephalography and Clinical Neurophysiology, 40,* 33–42.

Simson, R., Vaughan, H. G., & Ritter, W. (1977). The scalp topography of potentials in auditory and visual discrimination tasks. *Electroencephalography and Clinical Neurophysiology, 42,* 528–535.

Smith, M. E., Halgren, E., Sokolik, M., Baudena, P., Musolino, A., Liegeois-Chauvel, C., & Chauvel, P. (1990). The intracranial topography of the P3 event-related potential elicited during auditory oddball. *Electroencephalography and Clinical Neurophysiology, 76,* 235–248.

Snyder, A. Z. (1991). Dipole source localization in the study of EP generators: a critique. *Electroencephalography and Clinical Neurophysiology, 80,* 321–325.

Snyder, E., Hillyard, S.A., & Galambos, R. (1980). Similarities and differences among the P3 waves to detected signals in three modalities. *Psychophysiology, 17,* 112–122.

Squires, N. K., Squires, K. C., & Hillyard, S. A. (1975). Two varieties of long-latency positive waves evoked by unpredictable auditory stimuli in man. *Electroencephalography and Clinical Neurophysiology, 38,* 387–401.

Stapleton, J. M., & Halgren, E. (1987). Endogenous potentials evoked in simple cognitive tasks: Depth components and task correlates. *Electroencephalography and Clinical Neurophysiology, 67,* 44–52.

Stapleton, J. M., Halgren, E., & Moreno, K. A. (1987). Endogenous potentials after anterior temporal lobectomy. *Neuropsychology, 25,* 549–557.

Stok, C. J., Meijs, J. W. H., & Peters, M. J. (1987). Inverse solutions based on EEG and MEG applied to volume conductor analysis. *Physics in Medicine and Biology, 32,* 99–104.

Swick, D., Pineda, J. A., & Foote, S. L. (1993). Effects of systemic clonidine on auditory event-related potentials in squirrel monkeys. *Brain Research Bulletin* (in press).

Swick, D., Pineda, J. A., Holmes, T. C., & Foote, S. L. (1988). Effects of clonidine on P300-like potentials in squirrel monkeys. *Society for Neuroscience Abstracts, 14,* 1014.

Swick, D. (1991). The role of the noradrenergic nucleus locus coeruleus in modulating P300-like potentials. Ph.D. Thesis, University of California, San Diego.

Turetsky, B., Raz, J., & Fein, G. (1990). Representation of multi-channel evoked potential data using a dipole component model of intracranial generators: Application to the auditory P300. *Electroencephalography and Clinical Neurophysiology, 76,* 540–556.

van Oosterom, A. (1991). History and evolution of methods for solving the inverse problem. *Journal of Clinical Neurophysiology, 8,* 371–380.

Van Petten, C., Kutas, M., Kluender, R., Mitchiner, M., & McIsaac, H. (1991). Fractionating the word repetition effect with event-related potentials. *Journal of Cognitive Neuroscience, 3,* 131–150.

Van Voorhis, S. T., & Hillyard, S. A. (1977). Visual evoked potentials and selective attention to points in space. *Perception and Psychophysics, 22,* 54–62.

Vaughan, H. G. (1982). The neural origins of human event-related potentials. *Annals of the New York Academy of Sciences, 388,* 125–137.

Vaughan, H. G. (1987). Topographic analysis of brain electrical activity. *The London Symposia (EEG Suppl.), 39,* 137–142.

Velasco, M., Velasco, F., Velasco, A. L., Almanza, X., & Olivera, A. (1986). Subcortical correlates of the P300 potential complex in man to auditory stimuli. *Electroencephalography and Clinical Neurophysiology, 64,* 199–210.

Verleger, R. (1988). Event-related potentials and cognition: a critique of the context updating hypothesis and an alternative interpretation of P3. *The Behavioral and Brain Sciences, 11,* 343–427.

Walter, W. G., Cooper, R., Aldridge, V. J., McCallum, W. C., & Winter, A. L. (1964). Contingent negative variation: An electric sign of sensorimotor association and expectancy in the human brain. *Nature (London), 203,* 380–384.

Wilder, M. B., Farley, G. R., & Starr, A. (1981). Endogenous late positive component of the evoked potential in cats corresponding to P300 in humans. *Science, 211,* 605–607.

Williamson, S. J. (1991). MEG versus EEG localization test. *Annals of Neurology, 30,* 222.

Williamson, S. J., & Kaufman, L. (1991). Evolution of neuromagnetic topographic mapping. *Brain Topography, 3,* 113–127.

Woldorff, M. G., Gallen, C. C., Hampson, S. R., Hillyard, S. A., Pantev, C., Sobel, D., & Bloom, F. E. (1993). Modulation of early sensory processing in human auditory cortex during auditory selective attention. *Proceedings of the National Academy of Science, 90,* 8722–8726.

Woldorff, M. G., Hackley, S. A., & Hillyard, S. A. (1991). The effects of channel-selective attention on the mismatch negativity wave elicited by deviant tones. *Psychophysiology, 28,* 30–42.

Woldorff, M., & Hillyard, S. A. (1991). Modulation of early auditory processing during selective listening to rapidly presented tones. *Electroencephalography and Clinical Neurophysiology, 79,* 170–191.

Wong, P. K. H. (1991). Topographic representation of event-related potentials. *Event-Related Brain Research (EEG Suppl.), 42,* 5–12.

Wood, C. C. (1982). Application of dipole localization methods to source identification of human evoked potentials. *Annals of the New York Academy of Sciences, 388,* 139–155.

Wood, C. C., & Allison, T. (1981). Interpretation of evoked potentials: A neurophysiological perspective. *Canadian Journal of Psychology, 35,* 113–135.

Wood, C. C., Cohen, D., Cuffin, B. N., Yarita, M., & Allison, T. (1985). Electrical sources in human somatosensory cortex: Identification by combined magnetic and potential recordings. *Science, 227,* 1051–1053.

Wood, C. C., & McCarthy, G. (1986). A possible frontal lobe contribution to scalp P300. In J. W. Rohrbaugh, R. Johnson, Jr., & R. Parasuraman (Eds.), *Eighth international conference on event-related potentials of the brain (EPIC VIII): Research reports.* Stanford, California. pp. 164–166.

Wood, C. C., McCarthy, G., Squires, N. K., Vaughan, H. G., Woods, D. L., & McCallum, W. C. (1984). Anatomical & physiological substrates of event-related potentials: Two case studies. *Annals of the New York Academy of Sciences, 425,* 681–721.

Woods, D. L. (1990). The physiological basis of selective attention: Implications of event-related potential studies. In J. W. Rohrbaugh, R. Johnson, and R. Parasuraman (Eds.), *Event-related brain potentials: Issues and interdisciplinary vantages* (pp. 178–209). New York: Oxford University Press.

Woods, D. L., & Alain, C. (1993). Feature processing during high-rate auditory selective attention. *Perception and Psychophysics, 53,* 391–402.

Woods, D. L., Alho, K., & Algazi, A. (1993). Stages of auditory feature conjunction: An event-related brain potential study. *J. Exp. Psychology: Human Perception and Performance* (in press).

Wurtz, R. H., Goldberg, M. E., & Robinson, D. L. (1980). Behavioral modulation of visual responses in the monkey. *Progress in Psychobiology and Physiological Psychology, 9,* 43–83.

Yamaguchi, S., & Knight, R. T. (1991). Anterior and posterior association cortex contributions to the somatosensory P300. *Journal of Neuroscience, 11,* 2039–2054.

Yamamoto, T., Williamson, S. J., Kaufman, L., Nicholson, C., & Llinas, R. (1988). Magnetic localization of neuronal activity in the human brain. *Proceedings of the National Academy of Sciences of the United States of America, 85,* 8732–8736.

Yingling, C. D., & Hosobuchi, Y. A. (1984). Subcortical correlate of P300 in man. *Electroencephalography and Clinical Neurophysiology, 59,* 72–76.

Chapter 5

Use of Positron Emission Tomography to Study Aphasia

E. Jeffrey Metter and Wayne R. Hanson

I. INTRODUCTION

This chapter focuses on our results with aphasic patients that were studied at least 3 months poststroke, and serves as an example of how positron emission tomography (PET) has been useful in expanding our understanding of brain–behavior relationships. Observations during the acute period poststroke may be quite different among patients. Researchers long have found evidence that many regions of the left hemisphere play an important role in language processing, so much so that damage to these areas may bring about different forms of aphasia. Consequently, a uniform model of the pathoanatomy of aphasia clearly must consider the role of multiple brain regions, including cortical and subcortical structures.

Current approaches to studying brain–behavior relationships are dependent on modern imaging techniques including X-ray computerized tomography (CT), magnetic resonance imaging (MRI), PET, and single photon emission tomography (SPECT). The first two techniques examine structural anatomy, whereas the last two focus on physiological measures. Each technique takes advantage of the power of computers to create detailed tomographic images. Brain–behavior relationships are typically studied through experiments of nature, by examining subjects who develop focal or diffuse brain damage. Physiological imaging techniques not only help identify the size and location of the structural damage but also describe the nature of the physiological consequences of the injury for other brain regions. Further, by controlling the behavioral activity during imaging, brain function while completing a specific task can be analyzed under normal and pathological conditions.

Cerebral blood flow, glucose and oxygen metabolism, and evoked potential studies have demonstrated biochemical and physiological alterations

Localization and Neuroimaging
in Neuropsychology

in regions remote from hemispheric structural lesions shown on CT (Baron, Bousser, Comar, & Castiagne, 1981; Kanaya, Endo, Sugiyama, & Kuroda, 1983; Kuhl, Phelps, Kowell, Metter, Selin, & Winter, 1980; Lenzi, Frackowiak, & Jones, 1981; Metter, Wasterlain, Kuhl, Hanson, & Phelps, 1981; Metter, et al., 1985a; Phelps, Mazziotta, & Huang, 1982). However, little emphasis has been placed on understanding the meaning of remote effects, and on identifying the presence and types of patterns of structural and remote metabolic changes that follow focal brain damage. An awareness of such patterns could improve our understanding of the correlation of changes in behavior to brain pathophysiology.

In this chapter, we identify and describe patterns of metabolic abnormality that we have observed in our studies of aphasic subjects using PET.

II. METHODOLOGY

PET takes advantage of the unique characteristics of positron-emitting radioisotopes. When positrons are emitted and collide with electrons, both are annihilated, creating two high-energy photons that travel 180° in opposite directions. If both photons are sensed by detectors, then a line of origin for the annihilation can be established, allowing mapping of the distribution of annihilations by computer, as is done for the amount of X-ray transmission using CT.

Several positron radionuclides are available that can be used to create isotopes that model physiological and biochemical processes. Currently, [18F]fluoro-deoxyglucose (FDG) is used to examine glucose metabolism, [15O]oxygen in combination with [15O]water is used to measure oxygen metabolism, [15O]water is used to measure cerebral blood flow, and [18F]-labeled DOPA is used to measure L-DOPA uptake and presumably dopamine metabolism. The discussions focus on the use of FDG to study cerebral glucose metabolism. FDG is transported into the cell using the same mechanisms as glucose. Once inside the cell, FDG is phosphorylated as glucose but cannot be processed further, so the isotope accumulates within the cell. Thus, the uptake of FDG parallels cell utilization of glucose, making FDG a marker for glucose uptake and metabolism.

When used clinically, FDG is injected intravenously into the subject. The subject lies quietly for a period of about 40 min, during which repeated arterialized venous blood samples are taken to describe the blood curve for the isotope. During the 40 min, the FDG is taken up by cells and will reach equilibrium. The brain is then scanned to measure the distribution of 18F in multiple tomographic sections of the brain. The distribution of the isotope will be based on cerebral glucose utilization under the specific pathological or mental condition of the brain. Models have been developed that permit calculation of glucose metabolism based on the blood curve and the distribution of ^{18}F uptake (Phelps, Huang, Hoffman, Selin, Sokoloff, & Kuhl, 1979). Interactive video monitoring programs allow local metabolic rates of glucose (LCMRGlc) to be calculated for regions of interest. The accuracy of this method is dependent on the stability of the model under pathological conditions. The major advantage of FDG over other currently available PET isotopes is that multiple high resolution images can be obtained.

Two basic types of imaging studies can be done with FDG. First is the resting study, as described. The second approach involves controlling the mental activity of the brain by having the subject perform a specific task. FDG is not ideal for this latter approach because of the long uptake time before equilibrium is established. Activation has been done with FDG (Mazziotta, Phelps, Carson, & Kuhl, 1982), but interpretation is not always direct because of the long time course. Activation studies are more appropriate with cerebral blood flow studies, in which the scan is completed in less than 10 min.

The data reported in this chapter are derived from aphasic subjects who were studied on the NeuroECAT scanner (CTI, Knoxville, Tennessee), which has a resolution of about 10 mm full-width-half-max (FWHM) in a resting state. Regions of interest were determined using interactive computer programs. Regional metabolic rates and PET tomographs then were compared with CT scans obtained at the time of PET imaging. Metabolic rates typically are expressed as a local metabolic rate of glucose expressed as mg/100 cc tissue/min. Because of large interindividual variability, data typically are examined intrasubject by comparing a region in one hemisphere to the same region in the contralateral hemisphere, to a global measure, or to a specific region. In this chapter, data are presented as left-to-right regional ratios or a regional measure to the average right hemisphere metabolic ratio (referred to as a reference ratio). Major efforts have been under way to develop more automated and repeatable methods by which to determine regional metabolic rates of glucose that can be compared across laboratories. Further, techniques are now available that allow overlaying CT or MRI with PET images to improve the anatomical localization of the PET images.

III. REMOTE EFFECTS: GENERAL CONSIDERATIONS

Comparing CT and FDG PET has allowed us to explore the physiological effects of a structural lesion in aphasia. Metabolic abnormalities occur in regions that are damaged structurally and presumably reflect the degree and extent of that damage. Metabolic abnormalities also occur in brain regions with no apparent structural damage. These remote metabolic changes appear to reflect the long-range consequences of focal damage. Researchers assume that changes in regional glucose metabolism represent alterations in the functional capability of the region.

Reduced glucose metabolism in nondamaged brain regions can be understood by considering factors that contribute to regional metabolism, including the metabolic activity of neurons, glia, and dendrites. Dendrites derive from regional cell–cell interactions, connections from adjacent brain regions, and long fibers arising from remote regions. Dendritic synaptic activity is a heavy utilizer of energy derived from glucose. The total regional metabolic activity depends on the summation of the contributions from all regional cellular components. A reduction in regional metabolism can occur as a result of damage to one or more of these components or as a result of changes in the firing patterns of neurons within the region. Firing changes would result from increases or decreases in either excitatory or inhibitory inputs to the region. Structural damage in one region can result in metabolic changes in a distant region based on changes in or loss of the message sent to the second region. The degree of contribution from each factor is currently difficult or impossible to determine.

Lesions resulting in remote effects can damage white matter tracts, resulting in a disconnection (Metter et al., 1985a). A subject was studied who had a focal lesion that destroyed the genu and anterior limb of the left internal capsule (Fig. 1), producing a 20–25% reduction in metabolism in the overlying inferior frontal cortex. The subject died 10 days after his PET study. Histological sections of the hypometabolic left frontal area showed normal anatomical structure, with no evidence of ischemic changes, and a neuronal cell count similar to the corresponding region in the right hemisphere. This subject demonstrated the metabolic consequences of a focal lesion that disrupted major white matter pathways. The observation suggests that disconnections could be associated with distant cortical metabolic changes and may have specific metabolic signatures.

In the disconnection, proximal and distal regions presumably function normally; the communication between the two is destroyed. A gray matter lesion can produce a similar remote metabolic abnormality. In this case one region is damaged, so communications between regions may remain but the signal has been disrupted or altered. This event suggests that a remote metabolic abnormality will have different significance based on the location and extent of structural damage elsewhere.

The contribution that remote metabolic changes may make to behavior may be distinctly different from that made by direct structural damage.

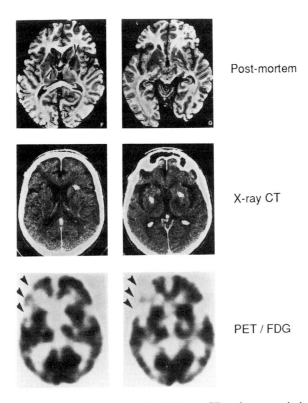

Post-mortem

X-ray CT

PET / FDG

Figure 1 Comparison of FDG PET, X-ray CT, and gross pathology at postmortem of a 69-year-old man with multiple cerebral infarctions. A metabolic abnormality is present in the left lateral frontal cortex (black arrows on FDG PET study). No gross structural lesions were found on CT or at postmortem in this frontal region. A lacunar infarct was present in the genu of the left internal capsule that was associated with degeneration of the anterior limb of the internal capsule. Destruction of the anterior limb of the internal capsule appears to be responsible for the frontal hypometabolism. Multiple lacunar infarctions were present in the right corpus striatum, whereas the right internal capsule appeared to be intact. No apparent cortical hypometabolism was associated with these lesions. Thus, slight differences in location of the lacunes in the left and right subcortical regions were associated with very different metabolic consequences. Reproduced with permission from Metter, Mazziotta, Itabashi, Mankovich, Phelps, & Kuhl (1985b).

Structural lesions can cause specific behavior changes by disrupting the structural integrity within the region. Remote metabolic effects may cause functional loss similar to that of direct damage, or may modify the functional integrity in its ability to carry out its goals, that is, the distant region may operate in a normal way but without information normally received from the structurally damaged area. Thus, a 50% loss of metabolic activity at the site of a structural lesion may result in different behavioral changes than a 50% metabolic reduction caused by a lesion at a remote site. In some situations, the metabolic changes may be caused by alterations in regional blood flow that result from stenosis or acute physiological effects of cerebral infarction. New concepts are needed to understand how remote metabolic effects are similar to and different from the direct effects of structural damage.

Another issue in understanding brain organization is that not all behaviors may be strictly localizable. Language is clearly localizable to the left perisylvian area, particularly temporal and parietal regions. Some aspects of linguistic organization, however, may rely on circuitry that is common to many or all brain regions, and therefore cannot be strictly localized.

IV. OBSERVATIONS IN APHASIC PATIENTS: TWO-COMPARTMENT MODEL

Based on our observations, we have argued for a two compartment brain model of the pathophysiology of aphasia. Each compartment may contain one or more functional systems. The temporo-parietal compartment is proposed to be central to the aphasic deficit and is related to language processing. The frontal–subcortical compartment is seen as more peripheral, differentiating aphasic patients on the basis of various behavioral parameters (e.g., fluency). Parietal cortex and, to some degree, Broca's region tend to occupy an overlapping area in both compartments.

This model resulted from the following findings concerning the neuropathophysiology of aphasia. Essentially, all aphasic patients that we have studied have metabolic abnormalities in the left temporo-parietal region as measured by FDG PET (Metter, Hanson, Jackson, Kempler, & van Lancker, 1989). The metabolic abnormalities were associated with structural damage in the parietal lobe in 67%, Wernicke's area in 67%, and posterior middle temporal region in 58% of the subjects. Of these subjects, 97% showed metabolic abnormalities in the angular gyrus, 89% in supramarginal gyrus, and 87% in the lateral and transverse superior temporal gyrus.

These observations suggested that common behavioral features were caused by left temporo-parietal dysfunction, that is, the presence of an

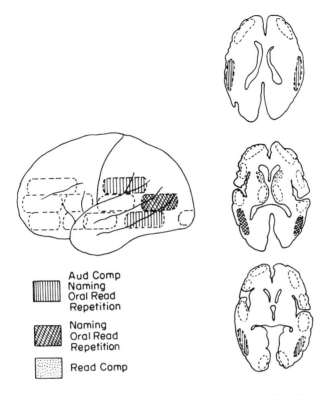

Aud Comp
Naming
Oral Read
Repetition

Naming
Oral Read
Repetition

Read Comp

Figure 2 Correlations of regional glucose metabolism with subtests of the Boston Diagnostic Aphasia Evalaution from 11 aphasic subjects. Regional glucose measures are presented on a lateral view of the left hemisphere (*left*) and on three diagrams of the tomographic sections (*right*). Cross-hatched areas correspond to a correlation of $p < .01$. The figure demonstrates the correlations to comprehension, naming, repetition, oral reading, and reading comprehension. Reproduced with permission from Metter, Riege, Hanson, Camras, Kuhl, & Phelps (1984).

aphasia resulted from the extent of temporo-parietal changes. Further, a previous report showed that comprehension deficits are correlated primarily to left temporo-parietal glucose metabolism (Fig. 2) (Metter, Riege, Hanson, Camras, Kuhl, & Phelps, 1984). Karbe, Hercholz, Szelies, Pawlik, Wienhard, and Heiss (1989) noted similar high correlations of the Token Test to temporo-parietal metabolism in 24 aphasic subjects.

A fundamental issue in aphasiology is whether the disorder consists of a single syndrome or multiple syndromes. The observations just presented support a unitary locus if the temporo-parietal region is considered a single functional system. The data demonstrate the sensitivity of temporo-parietal metabolism to aphasia, but do not give any information about the specificity. Features that distinguish aphasic and nonaphasic subjects with temporo-parietal hypometabolism have not been revealed by these studies. Temporo-parietal cortex appears to be critical for the presence of aphasia, independent of the localization of structural damage. The combined action of the lesion and the functional consequences of the lesion on the temporo-parietal region is most likely to dictate the severity and many of the characteristics of the aphasia.

Approximately 57% of our subjects have demonstrated hypometabolism in left prefrontal regions, that is, regions anterior and anterior–superior to Broca's area, whereas only 22% have structural damage there. Subjects with left prefrontal hypometabolism had greater structural damage to subcortical regions—including the internal capsule, caudate, lenticular nuclei, thalamus, and insula—than did subjects without prefrontal hypometabolism (Metter et al., 1987). With middle cerebral artery distribution lesions, the deeper the lesion extends into basal ganglia, internal capsule, and thalamus, the greater is the likelihood of having prefrontal hypometabolism. Except in cases with prefrontal structural damage, prefrontal hypometabolism seems to depend on the presence of subcortical structural damage.

Prefrontal hypometabolism corresponded in some measure to aphasia type. Wernicke's, Broca's, and conduction aphasias differed in the extent of subcortical structural damage and of left prefrontal hypometabolism (Metter, Kempler, Jackson, Hanson, Mazziotta, & Phelps, 1989). In Fig. 3, the regional metabolic differences are displayed when comparing left and right hemisphere metabolism for each region. Broca's aphasia showed the greatest global hypometabolism with the lowest values in the head of the caudate nucleus. The frontal regions are the most markedly depressed in Broca's aphasia, relative to Wernicke's and conduction aphasias. When comparing aphasia severity as judged by the Western Aphasia Battery (Kertesz, 1982), Wernicke's aphasia showed the most severe aphasia, a result that agreed with the greater temporo-parietal hypometabolism in this group of aphasic subjects.

The studies just described argue that the temporo-parietal cortex is primarily responsible for the language abnormalities. The subcortical–frontal system appears to be associated with factors that tend to modulate and modify the underlying language problems. In several studies, we have found that the frontal lobe hypometabolism is associated with the presence of hemiplegia and with problems with the expressive aspects of language.

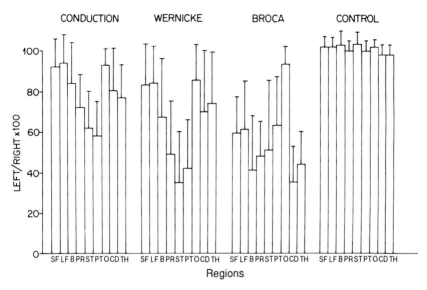

Figure 3   Comparison of regional left hemisphere to regional right hemisphere ratios for conduction, Wernicke's, and Broca's aphasias. SF, superofrontal (prefrontal); LF, inferofrontal (prefrontal); B, Broca's area; P, parietal; ST, Wernicke's area; PT, posterotemporal; O, occipital; CD, head of caudate; and TH, thalamus. Reproduced with permission from Metter, Kempler, Jackson, Hanson, Mazziotta, & Phelps (1989).

V. ROLE OF SUBCORTICAL STRUCTURES

Aphasia is known to occur with lesions in subcortical structures including caudate, putamen, thalamus, and internal capsule (Damasio, Damasio, Rizzo, Varney, & Gersch, 1982; Naeser, Alexander, Helm-Estabrooks, Levine, Laughlin, & Geschwind, 1982). In our studies, we have examined subjects with either lacunes or hemorrhages in the subcortical regions (Metter et al., 1981, 1986). Baron, D'Antona, Pantano, Serdaru, Samson, and Bousser (1986) have made similar observations with a major emphasis on thalamic lesions. We have observed differing patterns of hypometabolism associated with subcortical lesions. A critical question is whether disruption of cortical–subcortical feedback loops are important in subcortical aphasia or whether (or to what extent) intrinsic characteristics of subcortical structures have a role. We have used path analysis to build a model, from observation of glucose metabolism in aphasia, to study this question (Fig. 4) (Metter, Riege, Hanson, Jackson, Kempler, & Van Lancker, 1988a). Specifically, the model was intended to determine whether subcortical

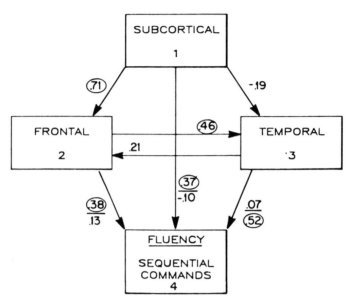

Figure 4 Path analysis model of the effect of subcortical structural damage on fluency and sequential command scores from the Western Aphasia
Battery. Arrows represent relationships between boxes, with direction
implying the causal relationships. The number next to each arrow is an
estimate of the association between the two boxes. Numbers that are
circled are considered to represent important connections. The analysis
showed that subcortical structural damage has a direct effect on fluency
but not on sequential commands. In addition, such damage has an indirect effect on fluency through its action on the frontal lobe. Reproduced
with permission from Metter, Riege, Hanson, Jackson, Kempler, & Van
Lancker (1988a).

structural damage had an effect on behavior that could be separated from
remote effects on cortex. The data clearly showed both a direct and an
indirect effect of subcortical damage on word fluency. The results indicated that a primary effect of subcortical damage on some aspects of linguistic behavior can be separated from secondary effects associated with
a depression of frontal cortical metabolism. In our studies, the internal
capsule appeared to be the most critical of the subcortical structures studied. Damage to the internal capsule may disrupt motor pathways involved in the control and management of speech production. These
pathways presumably include the frontal–basal ganglia feedback loop.
When restricting the analysis to the head of the caudate, correlations of

glucose metabolism from this nucleus with behavioral measures seem to be at a more basic level that is related to recognition or motor planning of simple and overlearned materials including simple syntax, low levels of abstraction, and identification of sequencing of phonetic and semantic material (Metter, Riege, Hanson, Phelps, & Kuhl, 1988b).

VI. COMMON METABOLIC PATTERNS

Descriptions of the patterns of hypometabolism that we have identified in association with cortical and subcortical structural lesions follow.

Pattern 1 is metabolic abnormality involving the temporo-parietal region but not the prefrontal regions. Of 45 aphasic patients, 17 had this pattern in the left hemisphere. Figure 5 shows a typical patient with this pattern. Table 1 presents the mean regional metabolic reference ratios and CT ratings; Table 2 gives descriptive information for Patterns 1 and 2. The CT data (Fig. 6) showed temporal/parietal lesions without major extension into the basal ganglia or internal capsule, although in four patients the lesion extended into the insula, in one patient into lenticular nuclei, and in one patient into the posterior internal capsule. Small lesions in the left posterior putamen, the posterior part of the posterior limb of the internal capsule, or the thalamus also may result in glucose hypometabolism in the left temporo-parietal cortex (Fig. 7). The findings argue that these subcortical lesions interact with temporo-parietal cortex in producing the aphasia. As shown in Table 2, Pattern 1 patients tend to display the less severe, fluent forms of aphasic disturbance.

Pattern 2 is prominent metabolic asymmetry involving perisylvian and both prefrontal and posterior temporal regions. Of 45 aphasic patients, 24 demonstrated this pattern. Figure 8 shows a typical patient with this pattern. Table 1 presents the means for regional reference ratios and CT ratings. Figure 9 shows the CT templates. The lesions tended to be larger than those of Pattern 1, although not in all patients. All subjects had lesions extending deeply, with damage to the insula, lenticular nuclei, or internal capsules. The prefrontal metabolic changes were not associated with direct structural damage. Clinical aphasia diagnoses are presented in Table 2; both fluent and nonfluent aphasic patients are in this group.

Patterns 1 and 2 differed because Pattern 2 had lower reference ratios by univariate analysis ($p < 0.005$) in the prefrontal, posterior inferior frontal (Broca), parietal, caudate, and thalamic regions. The structural damage differentiating Patterns 1 and 2 was the presence of structural damage in Pattern 2 in the anterior internal capsule, posterior internal capsule, caudate, thalamus, lenticular nuclei, and insula.

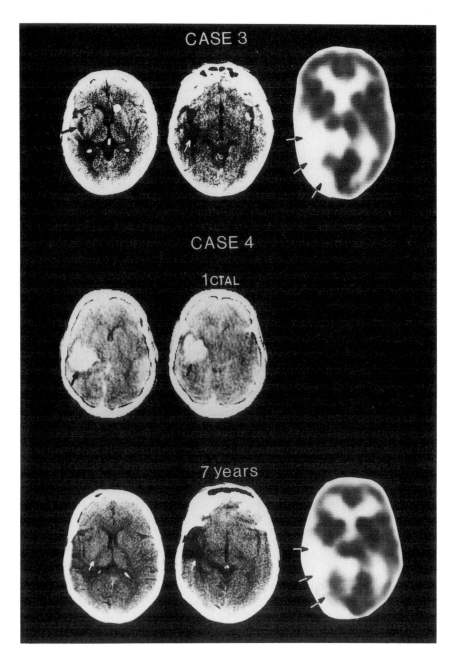

Figure 5 Pattern 1, with posterior temporal metabolic asymmetry, shows two patients who demonstrate prominent hypometabolism in the temporal and parietal regions but no asymmetry in the prefrontal regions. Reproduced with permission from Metter, Jackson, Kempler, Riege, Hanson, Mazziotta, & Phelps (1986).

Table 1 Means and Standard Deviations for Metabolic Reference Ratios and CT Ratings for Patterns 1 and 2[a]

	SF[b]		IF		B		PR		ST		PT		O		CD		TH		AI	PI	LN	IN
	1[c]	2	1	2	1	2	1	2	1	2	1	2	1	2	1	2	1	2	2	2	2	2
Pattern 1																						
$\bar{x}$	87	2	95	1	90	1	66	16	70	15	58	21	134	1	95	1	95	1	0	4	7	12
SD	7	3	6	1	15	2	16	13	23	12	19	14	13	3	20	2	19	2	0	7	10	13
Pattern 2																						
$\bar{x}$	61	7	64	5	47	16	47	25	55	23	57	13	126	3	41	17	52	9	26	16	34	28
SD	16	10	19	10	25	16	19	12	32	11	22	12	25	9	20	15	20	10	15	11	8	13
Normal																						
Left hemisphere																						
$\bar{x}$	97		96		108		92		107		93		136		99		108					
SD	8		7		7		5		7		7		11		11		9					
Right hemisphere																						
$\bar{x}$	95		94		98		92		105		93		134		102		111					
SD	5		6		6		5		5		5		10		10		10					

[a] The CT rating is based on a clinical reading of the CT scan with 0 = normal, 1 = atrophy, 2 = mild regional damage, 3 = moderate damage, 4 = severe damage (Metter et al., 1987). Numbers for normals indicate regional means and standard deviations for reference ratios in each hemisphere for normal control subjects.

[b] SF, Superior frontal (regions 1 and 2); IF, inferior frontal (regions 3 and 4); B, posterior inferior frontal (Broca's area, regions 5 and 6); PR, parietal (regions 7 and 8); ST, posterior superior temporal (Wernicke's area, regions 9 and 10); PT, posterior temporal (regions 11 and 12); O, Occipital (region 13); CD, Caudate (region 14); TH, Thalamus (region 15); AI, anterior internal capsule; PI, posterior internal capsule; LN, lenticular nuclei; IN, insula.

[c] 1 is the metabolic reference ratio ×100; 2 is the CT rating ×10.

Table 2 Descriptive Information for Pattern 1 and 2 Subjects[a]

		Age	MPO[c]	Western Aphasia Battery[b]							
				AQ	IC	F	AC	Rp	N	RD	W
Pattern 1 (N = 17)											
	$\underline{x}$	61.4	18.9	81.4	8	8	17	76	8	82	80
	sd	9	25	15	2	1	3	15	2	20	20
Pattern 2 (N = 24)											
	$\underline{x}$	61.7	23.6	55.4	5	5	15	47	4	59	41
	sd	6.3	24	25	3	3	4	34	3	26	33

[a] Types of aphasia (WAB classifications) present in each pattern are, for Pattern 1, anomic (9), conduction (5), transcortical sensory (1), and Wernicke (2) and, for Pattern 2, anomic (7), Broca (10), conduction (2), global (1), and Wernicke (4).

[b] AQ, WAB aphasia quotient; IC, information content; F, fluency; AC, auditory comprehension; Rp, repetition; N, naming; RD, reading; W, writing.

[c] MPO, months postonset.

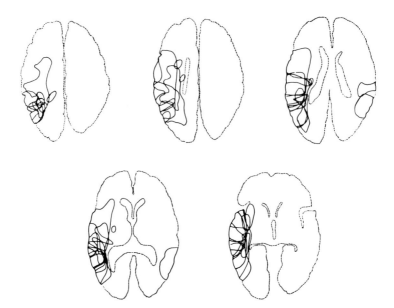

Figure 6 Distribution of CT lesions of Pattern 1 with posterior temporal metabolic asymmetry. CT scans were drawn onto templates that were overlaid onto a single set of templates to show the regions of structural damage for all 17 subjects.

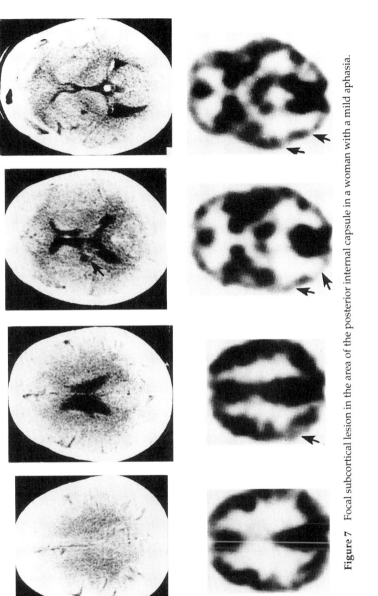

Figure 7 Focal subcortical lesion in the area of the posterior internal capsule in a woman with a mild aphasia.

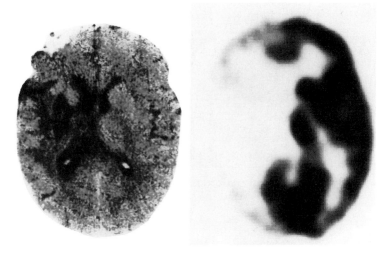

Figure 8 Pattern 2 with both prefrontal and posterior temporal asymmetry. This patient demonstrates the typical localization of structural lesions that cause this pattern of metabolic asymmetry. Reproduced with permission from Metter, Kempler, Jackson, Hanson, Mazziotta, & Phelps (1989).

Since all the subjects showing either Pattern 1 or 2 were aphasic, describing the nature of the aphasia (Table 2) seemed important. Univariate analysis indicated group differences for aphasia quotient (AQ), information content, fluency, repetition, naming, reading, and writing; Pattern 2 subjects showed lower scores on all measures. No difference was found on a separate univariate analysis for age. From these results, patients with Pattern 2 apparently had more severe aphasia than those with Pattern 1. Note that Pattern 2 included many patients with Broca's aphasia whereas Pattern 1 had no patients with those symptoms.

The presence of frontal hypometabolism from deep lesions has several explanations including (1) disconnection or disruption of structures that connect frontal regions with other regions causing hypometabolism in disconnected regions, (2) structural damage to the basal ganglia by interrupting their role in regulating corticocortical communication, and (3) large lesion size that causes enough damage for function in large areas of cortex to be disrupted, resulting in overall decrease in brain activity. The latter suggestion is supported by the fact that Pattern 2 subjects tend to show lower parietal and temporal metabolism than Pattern 1 patients. Several different mechanisms are likely to be responsible for Pattern 2. The smallest lesions may yield this pattern by causing a disconnection via

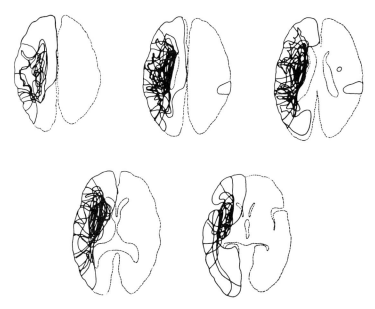

Figure 9 Distribution of CT lesions for Pattern 2 with prefrontal and posterior temporal metabolic asymmetry. The CT lesions for all patients with this pattern were overlaid onto one set of templates.

white matter destruction (as in Fig. 1), whereas large lesions may be explained by the size and extent of the lesion without invoking a disconnection.

Data presented early in this chapter suggest that differences between Patterns 1 and 2 that primarily reflect deep structural damage and frontal hypometabolism may be critical in the differential features of aphasic subjects. Wernicke's, Broca's, and conduction aphasias were shown to be distinguished by comparing the extent of frontal and temporo-parietal hypometabolism. Most conduction aphasia subjects show Pattern 1 with frontal sparing, whereas all Broca's subjects show Pattern 2; Wernicke's aphasia patterns were represented in both patterns (Metter et al., 1989).

Pattern 3 shows mild frontal, parietal, and temporal metabolic asymmetry. The most common structural lesion causing Pattern 3 is restricted to the subcortical regions, particularly the thalamus. The metabolic abnormality consists of a mild hypometabolism that involves most of the hemisphere including frontal, parietal, and temporal lobes (Metter et al., 1983, 1986). The hypometabolism shows a 5–20% reduction relative to the same region in the contralateral hemisphere (Table 3). This mild uniform

Table 3 Reference Metabolic Ratios for Pattern 3 Subjects[a]

Subject[b]	SF	IF	B	PR	ST	PT	O	CD	TH
3a									
L	93	96	102	86	101	79	105	90	81
R	100	102	111	94	111	84	103	99	90
3b									
L	100	95	96	84	98	85	129	95	113
R	102	100	100	97	103	89	133	99	113
NA									
L	112	83	78	—	123	85	98	48	49
R	113	96	97	—	127	98	103	60	75
Normal									
L	(97)[c]	(96)	(108)	(92)	(107)	(93)	(136)	(99)	(108)
	(8)	(7)	(7)	(5)	(7)	(7)	(11)	(11)	(9)
R	(95)	(94)	(98)	(92)	(105)	(93)	(134)	(102)	(111)
	(5)	(6)	(6)	(5)	(5)	(5)	(10)	(10)	(10)

[a] See Table 1 for abbreviations.

[b] L and R refer to the reference ratio for left and right hemisphere for each region.

[c] Numbers in parentheses indicate regional means and standard deviations for reference ratios in each hemisphere for normal control subjects.

reduction has not been observed in normal subjects. In normals, for any specific region a 5–15% reduction may be found, but not for essentially all regions in the hemisphere. In this situation, the overall picture, not the observation in a single region, leads to this pattern.

Three examples of this distribution are presented in Fig. 10. All patients have been reported on previously (Metter et al., 1983, 1986). One subject was a 32-year-old man who had an intracerebral hemorrhage while running that left him with word-finding problems and a right hemiparesis. When studied 5 years later, he had some mild difficulty with writing because of sensory loss with dysesthesias in the right hand. The second subject was a 23-year-old who had a left intracerebral hemorrhage that caused right-sided numbness, weakness, and incoordination. A third subject, N. A., had a traumatic injury to the dorsal medial nucleus of the left thalamus that left him with a permanent amnesia. FDG studies demonstrated a mild (5–20%) uniform reduction from normal mean values of left hemisphere regions except visual cortex (Table 3). None of 22 normal control subjects showed reference ratio asymmetry of this magnitude for all cortical regions in a single hemisphere. Thus, although no single region was severely or even moderately abnormal in patients with Pattern 3, all regions examined together presented an abnormal profile. Pattern 3 dif-

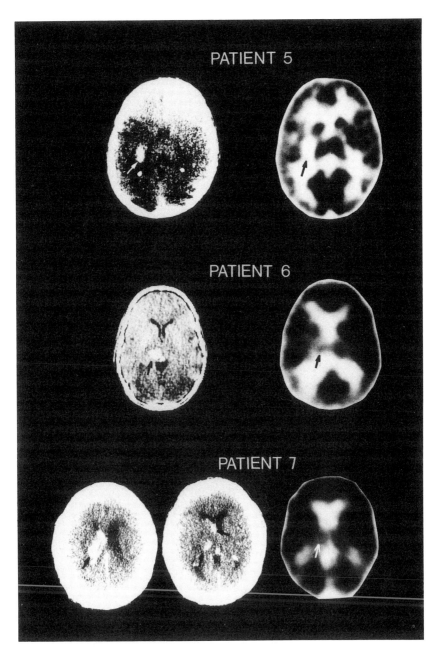

Figure 10 Pattern 3 characteristically shows mild diffuse metabolic asymmetry. These three subjects show thalamic lesions and mild metabolic asymmetry throughout the left hemisphere. Reproduced with permission from Metter, Jackson, Kempler, Riege, Hanson, Mazziotta, & Phelps (1986).

fered from Pattern 2 in the relative uniformity of the cortical metabolic change from region to region, and in the mild degree of asymmetry. Pattern 2 showed a gradient effect in which the greatest metabolic asymmetry was at the lesion and then decreased with distance from the lesion. Pattern 3 showed less metabolic asymmetry, but the asymmetry was more consistent from region to region, suggesting that a mechanism had been altered that acts on the overall level of cortical metabolic activity. None of the Pattern 3 patients were aphasic, although N. A. had been amnesic for 20 years, the amnesia being attributed to damage directly to the dorsal medial nucleus of the thalamus (Squire & Moore, 1979). We hypothesized that this pattern resulted from damage to the nonspecific thalamic activating system (Metter et al., 1983). Baron and colleagues (1986) have made similar observations in patients with thalamic lesions, and have suggested that this effect represents "either loss of nonspecific activating afferences or a degenerative deafferentation–deefferentation process, or both." Whatever the specific process, the mechanism may act as a physiological gain control set by deep gray structures and mediated through the thalamus. The overall function of the cortex can be adjusted downward without persistent symptomatology. Thus, the hypometabolism may be related to an abnormality in the maintenance of overall cortical metabolic environment. Animal studies have shown a generalized cortical innervation of layer IV of cortex by various thalamic nuclei (Hersch & White, 1982; White, 1979). Anterior and medial thalamus project to the frontal cortex, whereas posterior and lateral thalamus project to posterior hemisphere regions. Most thalamic lesions will disrupt both sets of projections. The mild hypometabolism may reflect innervation of layer IV by this activating system.

Pattern 4 shows hemispheric metabolic abnormality equal in size to the structural lesion. Six subjects were found who had this pattern with focal structural abnormalities in the right hemisphere (although we do not believe that this pattern is restricted to the right hemisphere). Five of these subjects had small unexpected structural abnormalities in the right hemisphere along with an infarct or hemorrhage in the left hemisphere, resulting in an aphasia. The sixth had only a right hemisphere abnormality. Three of these patients were found to have small deep right hemisphere lesions (Fig. 11A), and three had right hemisphere cortical abnormalities (Fig. 11B). The right cortical abnormalities may represent focal infarctions limited to cortex or focal cortical atrophy. In the left hemisphere, three of these patients showed Pattern 1 and two had Pattern 2. All six subjects showed right hemisphere metabolic abnormalities only at the site of structural abnormality. Right hemisphere reference ratios are given in Table 4. The reference ratios were similar to those of controls and of aphasic sub-

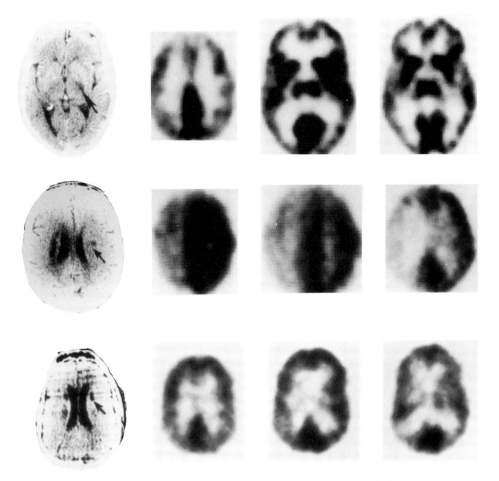

Figure 11 Pattern 4. (A) Three subjects (1, 107, 3) who had small deep right hemisphere lesions that showed no associated metabolic abnormalities and were clinically asymptomatic.

jects (without right hemisphere lesions) except at the lesion site (when the lesion was in a measured region).

Patient 1a (*top*, Fig. 11A) was of particular interest because she had volunteered as a "normal" control and was found to have a lesion in the right hemisphere bordering on the right posterior putamen and posterior limb of the internal capsule. The lesion did not destroy either structure. As with all subjects with this pattern, right hemisphere regional ratios were within the normal range (Table 4). Detailed neuropsychological and

B

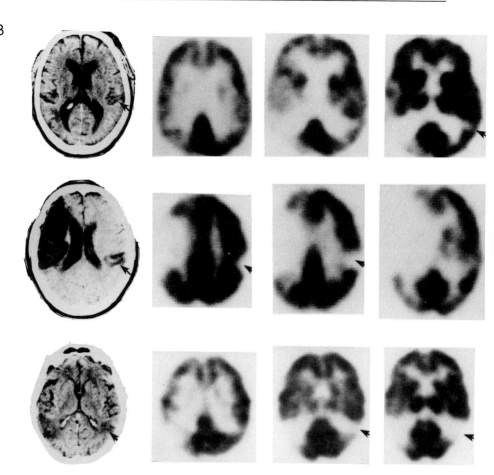

Figure 11 (Continued) (B) Three subjects (121, 128, 132) who had focal cortical structural and metabolic abnormalities associated with no clinical symptomatology.

memory evaluation were normal (Riege, Tomaszewski, Lanto, & Metter, 1984).

In all six subjects, the right hemisphere structural abnormality was an incidental finding; the patient had no previous history of a neurological event that would be explained by the lesion. No metabolic abnormalities were found other than at the site of structural damage. The observation raises the hypothesis that, for a lacunar lesion to be symptomatic, it must have a functional effect on other brain areas outside the locus of structural damage.

Table 4   Right Hemisphere Reference Ratios for Pattern 4 Subjects [a,b]

Patient	Sex	Age	SF	IF	B	PR	ST	PT	O	CD	TH
1	F	72	85	93	104	90	110	94	125	113	116
107	M	61	85	97	95	100	105	98	138	100	105
3	M	48	102	94	110	86	119	95	—	97	106
121	M	69	90	91	100	94	104	94[c]	128	121	110
128	M	57	82	98	112	67[c]	107	95	164	11	107
132	M	73	89	95	108	96	116	59[c]	131	104	137

[a] Subjects are listed in the order they appear in Fig. 11A, 11B.
[b] See Table 1 for abbreviations.
[c] Implies that the region was structurally damaged. Note that, for 121, damage was found but no measurable change in metabolism was seen. As can be seen in Fig. 11B, only a small structural and metabolic defect existed that was lost in the larger region that was measured.

In summary, four distinct patterns of metabolic abnormalities were observed in patients who have had either cerebral infarctions or hemorrhages. These patterns reflect the location, size, and severity of the structural damage. Patterns 1 and 2 develop early after the insult and appear to persist indefinitely. We have studied patients as early as 1 mo and as long as 15 yr poststroke that show such patterns.

VII. COMMENTS

The four patterns identified appear to be influenced substantially by the location of lesions to both cortical and subcortical structures. The study of subcortical lesions with PET appears to be of value, since the resulting metabolic scans facilitate determining the consequences of the structural damage on the cortex. In our experience, these effects can vary from no metabolic consequence with no apparent resulting symptomatology, that is, the silent structural lesion, to mild generalized hypometabolism associated with an acute syndrome and perhaps minimal residual consequences. Symptomatic subcortical lesions appear to be associated with cortical hypometabolism that argues for disruption of either the cortical function or the communication processes between regions. The metabolic pattern for either of these functional consequences may appear to be the same. Strategies must be developed to separate these possibilities.

Small infarcts found surreptitiously on CT in subjects with no clinical disability have become a common occurrence (Kinkel, 1984). These lesions are typically lacunes in deep white or gray matter, presumably associated with noncritical axonal or deep gray damage. The small cortical lesions that were observed may be infarcts situated in "noncritical" regions or areas of focal cortical atrophy. Although we cannot resolve this issue fully in this discussion, we do have some evidence that small deviations in lesion sites can affect function greatly. In the postmortem study (Fig. 1), we demonstrated a subject who was found to have hypometabolism in his left frontal region associated with a lacunar infarct destroying the anterior internal capsule. However, on the contralateral side, a similar lacune that was slightly lateral in the striatum, sparing the internal capsule, was found. On this side, the overlying cortex had normal glucose metabolism (Metter, Mazziotta, Itabashi, Mankovich, Phelps, & Kuhl, 1985b). The slightest difference in lesion location spared the internal capsule on the contralateral side as well as its effects on frontal cortex.

Other factors might influence the patterns that occurred in our patients, including (1) technical issues involving the deoxyglucose model, (2) persisting ischemia in the remote regions, and (3) subtle cellular loss. Each of these possibilities is considered briefly.

Methodological difficulties with the FDG model can affect the accuracy of the measurements. In ischemia and infarction, changes in rate and lump constants may occur that result in inaccuracies in the estimate of LCMRG1c. LCMRG1c may be underestimated by about 20% in regions with severe ischemia when using average kinetic constants derived from normal subjects, as in this study (Hawkins, Phelps, Huang, & Kuhl, 1981). The remote effects that we have observed are in noninfarcted regions that are not severely ischemic. Thus, in the remote regions, the FDG model can be assumed to be reliable. The asymmetries, therefore, cannot be explained purely by methodological factors.

Another explanation for the remote effects is subtle cellular loss caused by mild persistent ischemia. Mies, Auer, Ebhardt, Traupe, and Heiss (1983) found, in cats, that cortical bloodflow correlated significantly with neuronal cell count for the involved region. These researchers also noted that regions in the infarcted hemisphere with normal cell counts had a 20% reduction in regional blood flow that was not explainable by cell loss. Further, Nedergaard, Astrup, and Klinken (1984) noted in human infarcts that, on histological examination, the border between infarcted and normal tissue was abrupt, thus arguing that distant blood flow and metabolic changes are not caused by cellular loss but are related to neuronal disconnection or deactivation. Ischemia and cell loss do not appear to be likely explanations for the remote phenomena.

From these structure–metabolism patterns, we can begin to generate hypotheses about the ways in which brain function, brain structure, and behavior are related. For instance, we hypothesize that, for a structural lesion to have lasting behavioral effects, it must affect regions beyond its boundaries. Second, mild (5–15% metabolic asymmetry, i.e., from Pattern 3) metabolic disturbance need not interfere with behavior. These findings collectively suggest that metabolic abnormalities are only behaviorally significant if they (1) are beyond a certain critical level and (2) affect regions beyond the zone of structural damage. A better understanding of the patterning of brain metabolism in aphasic patients should improve our understanding of pathoanatomy and its relationship to behavior.

As a whole, the data argue for two major functional components in the left hemisphere that are involved with language function: a temporo-parietal component that traditionally is involved with primary language processing and a frontal component that appears to be involved with modifying the effects of the language abnormalities associated with the aphasia. Of particular interest is the role of prefrontal regions that are not generally structurally damaged in aphasia. From our studies, prefrontal regions appear to be important in differentiating among aphasic individuals. Presumably their role is not directly language related, but reflects how the brain adapts or responds to disruption of the language system. Subcortical structures are critical here, because damage to these deeper structures appears to disrupt the prefrontal metabolic processes that can be measured by PET. For most aphasic subjects subcortical damage is associated more directly with frontal lobe than with temporo-parietal function.

REFERENCES

Baron, J. C., Bousser, M. G., Comar, D., & Castiagne, P. (1981). Crossed cerebellar diaschisis in human brain infarction. *Transactions of American Neurologic Association, 105,* 459–461.

Baron, J. C., D'Antona, R., Pantano, P., Serdaru, M., Samson, Y., & Bousser, M. G. (1986). Effects of thalamic stroke on energy metabolism of the cerebral cortex: A positron tomography study in man. *Brain, 109,* 1243–1259.

Damasio, A. R., Damasio, H., Rizzo, M., Varney, N., & Gersch, F. (1982). Aphasia with non-hemorrhagic lesions in the basal ganglia and internal capsule. *Archives of Neurology, 39,* 15–20.

Hawkins, R. A., Phelps, M. E., Huang, S. C., & Kuhl, D. E. (1981). Effect of ischemia on quantification of local cerebral glucose metabolic rate in man. *Journal of Cerebral Blood Flow and Metabolism, 1,* 37–51.

Hersch, S. M., & White, E. L. (1982). A quantitative study of the thalamocortical and other synapses in layer IV of pyramidal cells projecting from mouse SmI cortex to the caudate-putamen nucleus. *Journal of Comparative Neurology, 211,* 217–225.

Kanaya, H., Endo, H., Sugiyama, T., & Kuroda, K. (1983). "Crossed cerebellar diaschisis" in patients with putamenal hemorrhage. *Journal of Cerebral Blood Flow and Metabolism, 3 (Suppl. 1),* S27–S28.

Karbe, H., Hercholz, K., Szelies, B., Pawlik, G., Wienhard, K., & Heiss, W. D. (1989). Regional metabolic correlates of token test results in cortical and subcortical left hemispheric infarction. *Neurology, 39,* 1083–1088.

Kertesz, A. (1982). *Western Aphasia Battery.* New York: Grune & Stratton.

Kinkel, W. (1984). Computerized tomography in clinical neurology. In A. B. Baker & B. H. Baker (Eds.), *Clinical neurology.* Philadelphia: Harper and Row.

Kuhl, D. E., Phelps, M. E., Kowell, A. P., Metter, E. J., Selin, C., & Winter, J. (1980). Effects of stroke on local cerebral metabolism and perfusion: Mapping by emission computed tomography of 18FDG and 13NH3. *Annals of Neurology, 8,* 47–60.

Lenzi, G. L., Frackowiak, R. S., & Jones, T. (1981). Regional cerebral blood flow (CBF), oxygen utilization (CMRO2) and oxygen extraction ratio (OER) in acute hemispheric stroke. *Journal of Cerebral Blood Flow and Metabolism (Suppl.), 1,* S504–S505.

Mazziotta, J. C., Phelps, M. E., Carson, R. E., & Kuhl, D. E. (1982). Tomographic mapping of human cerebral metabolism: Auditory stimulation. *Neurology, 32,* 921–937.

Metter, E. J., Hanson, W. R., Jackson, C. A., Kempler, D., & Van Lancker, D. (1989). Temporoparietal cortex: The common substrate for aphasia. T. E. Prescott, (Ed.), *Clinical aphasiology conference* (Vol. 18, pp. 31–40). College Hill: Little, Brown & Co.

Metter, E. J., Jackson, C. A., Kempler, D., Riege, W. H., Hanson, W. R., Mazziotta, J. C., & Phelps, M. E. (1986). Left hemisphere intracerebral hemorrhages studied by (F-18)-fluorodeoxyglucose PET. *Neurology, 36,* 1155–1162.

Metter, E. J., Jackson, C. A., Mazziotta, J. C., Benson, D. F., Hanson, W. R., Riege, W. H., & Phelps, M. E. (1985a). Relationship of temporo-parietal lesions and distal glucose metabolic changes in the head of the caudate nucleus in aphasic patients. *Journal of Cerebral Blood Flow and Metabolism, 5 (Suppl. 1),* S43–S44.

Metter, E. J., Kempler, D., Jackson, C. A., Hanson, W. R., Mazziotta, J. C., & Phelps, M. E. (1989). Cerebral glucose metabolism in Wernicke's, Broca's and conduction aphasias. *Archives of Neurology, 46,* 27–34.

Metter, E. J., Kempler, D., Jackson, C. A., Hanson, W. R., Riege, W. H., Camras, L. R., Mazziotta, J. C., & Phelps, M. E. (1987). Cerebellar glucose metabolism in chronic aphasia. *Neurology, 37,* 1599–1606.

Metter, E. J., Mazziotta, J. C., Itabashi, H. H., Mankovich, N. J., Phelps, M. E., & Kuhl, D. E. (1985b). Comparison of x-ray CT, glucose metabolism and post-mortem data in a patient with multiple infarctions. *Neurology, 35,* 1695–1701.

Metter, E. J., Riege, W. R., Hanson, W. R., Camras, L., Phelps, M. E., & Kuhl, D. E. (1984). Correlations of cerebral glucose metabolism and structural damage to language function in aphasia. *Brain and Language, 21,* 187–207.

Metter, E. J., Riege, W. H., Hanson, W. R., Jackson, C. A., Kempler, D., & Van Lancker, D. (1988a). Subcortical structures in aphasia: An analysis based on FDG PET and CT. *Archives of Neurology, 45,* 1229–1234.

Metter, E. J., Riege, W. H., Hanson, W. R., Kuhl, D. E., Phelps, M. E., Squire, L. R., Wasterlain, C. G., & Benson, D. F. (1983). Comparisons of metabolic rates, language and memory in subcortical aphasias. *Brain and Language, 19,* 33–47.

Metter, E. J., Riege, W. H., Hanson, W. R., Phelps, M. E., & Kuhl, D. E. (1988b). Evidence for a caudate role in aphasia from FDG positron emission tomography. *Aphasiology, 2,* 33–44.

Metter, E. J., Wasterlain, C. G., Kuhl, D. E., Hanson, W. R., & Phelps, M. E. (1981). FDG positron emission computed tomography in a study of aphasia. *Annals of Neurology, 10,* 173–183.

Mies, G., Auer, L. M., Ebhardt, G., Traupe, H., & Heiss, W. D. (1983). Flow and neuronal density in tissue surrounding chronic infarction. *Stroke, 14,* 22–27.

Naeser, M. A., Alexander, M. P., Helm-Estabrooks, N., Levine, H. L., Laughlin, S. A., & Geschwind, N. (1982). Aphasia with predominantly subcortical lesion sites. *Archives of Neurology, 39,* 15–20.

Nedergaard, M., Astrup, J., & Klinken, L. (1984). Cell density and cortex thickness in the border zone surrounding old infarcts in the human brain. *Stroke, 15,* 1033–1039.

Phelps, M. E., Huang, S. C., Hoffman, E. J., Selin, C. S., Sokoloff, L., & Kuhl, D. E. (1979). Tomographic measurement of local cerebral metabolic rate in humans with (F-189) 2-Fluoro-2-Deoxyglucose: Validation of method. *Annals of Neurology, 6,* 371–388.

Phelps, M. E., Mazziotta, J. C., & Huang, S. C. (1982). Study of cerebral function with positron computed tomography. *Journal of Cerebral Blood Flow and Metabolism, 2,* 113–162.

Riege, W. H., Tomaszewski, R., Lanto, A., & Metter, E. J. (1984). Age and alcoholism: Independent memory decrements. *Alcoholism, Clinical and Experimental Research, 8,* 42–47.

Squire, L. R., & Moore, R. Y. (1979). Dorsal thalamic lesion in a noted case of human memory dysfunction. *Annals of Neurology, 6,* 503–506.

White, E. L. (1979). Thalamocortical synaptic relations: A review with emphasis on the projections of specific thalamic nuclei to the primary sensory areas of the neocortex. *Brain Research Review, 1,* 275–311.

Chapter 6

Functional Activation and Cognition: The ^{15}O PET Subtraction Method

Howard Chertkow and Daniel Bub

A researcher in this field can hardly imagine publishing a book on localization of cognitive function in the 1990s without a chapter including the dramatic results of recent positron emission tomographic (PET) activation studies using ^{15}O. On the other hand, the status of the data generated by such PET activation studies remains unclear. Although certain findings (for example, dorsolateral frontal lobe activation in the Wisconsin Card-Sorting Task and Self-Ordered Pointing Task; anterior cingulate activation in attentional tasks) seem consistent with previous neuropsychological literature, other findings (such as activation of the left inferior frontal region during semantic processing) are surprising and represent a divergence from previous lesion evidence of localization of function. In addition, enough contradictory data have been produced over the past 5 years (for example, divergent evidence for localization of the orthographic lexicon) to lead to a certain degree of uncertainty in the neuropsychological and cognitive community. The skeptical neuropsychologist must be forgiven for any unwillingness to abandon the results of many years of clinical and pathological correlation studies using focal lesions for the results of 1 or 2 controversial studies using ^{15}O PET activation. Currently, which of the findings of the past 5 years of PET activation studies will prove durable, replicable, and robust enough to last into the coming century is simply unclear.

This chapter therefore has been arranged in the following manner. We first consider the methodological advantages and limitations of the technique. The bulk of the chapter then involves presentation of selected results of ^{15}O activation studies carried out in the past 5 years.

Localization and Neuroimaging
in Neuropsychology

151

Thereafter, we present additional discussion of methodological issues and problems with the technique, problems that may limit its applicability or replicability, or the validity of the results to date. Although PET activation studies clearly have enormous potential for rewriting our knowledge of functional neuroanatomy, leaving the reader with the impression that we have "got all the bugs out" of this novel technique would be unjust.

I. METHODOLOGY: MEASURING FUNCTIONAL ACTIVATION WITH PET

PET scanning utilizes emission tomography. The tomograph is an external detection system that maps the distribution of an injected radiotracer in three-dimensional space. The radiotracer in the studies to be considered in this chapter is ^{15}O, a radioactive atom with a half-life of 2 min. This metabolic substrate either is injected intravenously in the form of $H_2^{15}O$ or is inhaled as $C^{15}O_2$. In either case, the oxygen radionuclide enters the bloodstream; therefore, the radioactivity that enters the brain is proportional to cerebral blood flow. As ^{15}O decays, it emits positrons. Since PET technology has been addressed in previous chapters, we will not review the details at this point.

The important theoretical points regarding the use of ^{15}O injections to study cognitive function include the following. First, after a bolus injection of ^{15}O, either as H_2O or as CO_2, the radioactivity detected in the brain occurs as a linear function of regional cerebral blood flow (rCBF) (Collins, 1991; Frackowiak, 1989; Sokoloff, 1984). When tissues of the brain are activated or are involved in increased neuronal function, a local increase in blood flow (rCBF) and neuronal glucose metabolism (rCMRGlu) is demonstrable within seconds, although the exact relationship between the two is far from clear (Collins, 1991). Increased rCBF is thus a reliable indicator of increased neuronal activity in response to increased local demands from a given stimulus (Fox & Raichle, 1986). Measurement of rCBF therefore provides an exciting tool with which to assess sites of increased or decreased cerebral activity that occurs with cognitive activation. These blood flow changes correlate with the activity of energy dependent sodium pump mechanisms (Mata et al., 1980), suggesting that blood flow changes are localized largely to synaptic junctions and the dendritic arbor. Alterations in rCBF therefore can be interpreted to reflect local synaptic activity. Note that activity of both inhibitory and excitatory synapses produces an increase in rCBF, although in theory the two might have opposing behavioral or processing effects. In addition, the overall global cerebral blood flow may be altered by many physiological controlling factors and also varies among individuals. Therefore, the usual practice is to normalize the rCBF by dividing by the mean CBF for that individual to reduce variability across individuals.

From the perspective of functional localization of cognition, the critical attraction of ¹⁵O studies is the short half-life of the radionuclide, which allows scanning to be carried out practically in cognitive "real time." In addition, $H_2^{15}O$ is a near ideal flow tracer, since it is extracted almost completely by brain tissue during normal cerebral blood flow. The relationship between rCBF and tissue activity is nearly linear after bolus injections; therefore, detection of rCBF changes can be accomplished by measurement of radioactivity counts in the brain without requiring measurements of arterial blood concentrations. In addition, the low radioactivity dose achieved with each bolus injection allows a sequence of scans to be carried out in the same subject without exceeding radiation safety limits. Repeat measurements of rCBF in the same subject provide a powerful tool for comparing blood flow during cognitive activation tasks with blood flow in a baseline state. This subtraction methodology, introduced first by Raichle, Fox, and others (Fox & Mintun, 1989; Raichle, Martin, Herscovitch, Mintun, & Markham, 1983), has provided the basis of the PET activation studies that are reviewed in this chapter.

The attraction of studying cognition with ¹⁵O PET activation is immediately obvious to the cognitive neuroscientist. Most previous correlation studies have investigated the neuroanatomy of cognition using patients with brain lesions measured pathologically or with computerized tomography (CT) and magnetic resonance imaging (MRI), which delineate the anatomical (but not necessarily the functional) extent of damaged tissue. As Steinmetz and Seitz (1991) point out, lesions are likely to reveal only the anatomy of areas essential to the carrying out of a particular function. Given the modern models of cognitive processing that emphasize distributed function, such *essential* areas might be limited to the input or output nodes of the system. ¹⁵O PET activation studies would reveal the "brain areas *involved* in a tested behavior" (Steinmetz & Seitz, 1991; p. 1150), including essential as well as nonessential focal brain areas. Clearly the goal would be for data from functional activation studies of normal individuals to converge with the previous lesion data and extend our previous knowledge to provide a complete picture of the functional neuroanatomy of cognition. In addition, although the great majority of the neuropsychological literature deals with correlation of general cognitive functions or syndromes (for instance, aphasic syndromes) with neuroanatomy, ¹⁵O PET activation provides a method for localizing specific components of a given cognitive function. Accumulating sufficient brain-damaged patients with functional impairment limited to that particular component might prove to be a virtually impossible task.

II. METHODOLOGICAL PROBLEMS WITH PET ACTIVATION

All investigators in the PET field have acknowledged a number of obvious limitations of this promising technique from the outset. First, the changes in rCBF associated with activation of higher order cognitive functions are small, on the order of 5% change in blood flow compared with a 25% change in blood flow to primary motor and sensory cortex in response to functional activation (Fox & Mintun, 1989; Raichle et al., 1983). This fact has necessitated attempts to increase the signal-to-noise ratio. The major solution to this problem has been the use of image pair subtractions (subtracting PET activity from a baseline scan from the activity seen in an activation task) followed by averaging results across a group of normal individuals (Fox, Mintun, & Reiman, 1988). In this approach, the stimuli are so organized that the scan task acts as a subtraction baseline for the next task in the form of a hierarchy of additive processing components. The subtraction images produced may be analyzed using an analysis of change distribution across the difference maps (referred to as the change distribution analysis or CDA technique) (Fox & Mintun, 1989). Another approach utilized by the Hammersmith group involves the averaging of rCBF maps across individuals for each of the series of stimulation tasks associated with a given behavior or process. These maps can be subjected to formal statistical comparisons of stimuli (i.e., control vs any cognitive task) or of any set of activation tasks. This technique generates statistical maps of rCBF change that are referred to as statistical parametric maps (SPM technique; Frith, Friston, Liddle, & Frackowiak, 1991), which depend less on the assumption of a hierarchy of additive processing components. These differences in statistical image analysis techniques ultimately may prove important, given certain methodological issues that are addressed later in this chapter.

A second problem limiting use of ¹⁵O activation studies is their reliance on cerebral blood flow changes in a focal area of the brain. Certain cognitive processes or stages in cognitive processing are represented as neural networks in the brain (Goldman-Rakic, 1987; Mesulam, 1990). Such neural networks will be essentially invisible to the ¹⁵O PET activation technique. Unfortunately, for any particular cognitive domain, which (if any) components are focal or distributed within an anatomical region (e.g., the "Language Zone"; Caplan, 1988) is generally unknown.

Another limitation of the technique is the limited spatial resolution of many PET images. Although adequate for visualizing cortex, the functional neuroanatomy of smaller structures such as the thalamus or particular areas of the hippocampus will be impossible to resolve using PET techniques. Perhaps a more important issue that is addressed later in this chapter is the temporal resolution of ^{15}O scanning. Although 60–120 sec is a relatively short period of time in terms of neuropsychology, it is immense in terms of cognitive processing. Postulated stages of representation in such cognitive processes as object recognition and reading occur over a time span under 100 msec (Sergent, Zuck, Terriah, & MacDonald, 1992). In ^{15}O PET activation studies, a series of stimuli usually is administered every few seconds, with repetition of the same cognitive processes multiple times during the course of 1 min. Therefore, the resultant blood flow changes represent the integration and accumulation of multiple stages of processing of multiple stimuli. The ability to extract information regarding the neural substrate of particular stages of processing thus relies on the clever use of paired activation and baseline states that can be subtracted. As discussed later in this chapter, the determination of the appropriate activation and baseline states remains a problem for cognitive neuropsychology. Thus, the current state of cognitive modeling and techniques available for measuring cognitive functions impose an important restriction on the use of PET activation for analyzing stages of cognitive processing (Ober, Reid, & Jagust, 1991).

Finally, mention should be made of the technical issues involved in matching PET images of radioactivity with actual neuroanatomy. PET scanning provides an image of the whole brain, including 10–15 cm in the z axis, which provides complete sampling of the cranial contents. A number of solutions have been developed to the problem of accurately correlating activation, which occurs in three-dimensional space, with the precise anatomical localization in each individual's brain. The most satisfactory solution appears to be correlation of each individual's PET scan to the same individual's MRI, which supplies accurate localization for that individual. The PET and MRI images are deformed computationally to the size and shape of a standard brain prior to intersubject averaging (Evans et al., 1988). Failing that, as a technically simpler but less accurate alternative, the activation peaks of one individual or a series of individuals can be matched to common stereotactic conventions, usually based on the reference atlas described by Talairach and Tournoux (1988).

III. SENSORY PROCESSING AND VISUAL COGNITION

Sensory processing has been studied with ^{15}O activation studies, generally confirming our standard "classical" knowledge of the localization of sensory cortex. Auditory stimulation with white noise or words produces increased cerebral blood flow within 15 sec bilaterally in superior temporal gyri (Zatorre, Evans, Meyer, & Gjedde, 1992a; see Color Plate 2). Vibrotactile stimulation of the hand has been shown to produce increased cerebral blood flow in the contralateral somaesthetic cortex in the primary sensory representation area for the hand (Meyer et al., 1991). When attention was directed toward the stimulus, this sensory activation was accompanied by increased blood flow in the anterior cingulate cortex as well. Olfactory stimulation produces significant increase in cerebral blood flow at the junction of the inferior frontal and temporal lobes bilaterally, corresponding to the pyriform cortex. Also, unilateral increase in cerebral blood flow occurs in the right orbito-frontal cortex (Zatorre, Jones-Gottman, Evans & Meyer, 1992b), corresponding well with the anatomy of the primary olfactory cortex in monkeys, and suggesting a functional specialization of the right orbital frontal cortex for higher order processing of olfactory stimuli. Even at the level of sensory processing, ^{15}O PET studies have demonstrated a cortical localization previously considered speculative. Talbot et al. (1991) carried out an investigation of sensory stimulation of the hand using warm and hot (mildly painful) stimuli applied to the right forearm. The subtraction of the two conditions produced significant residual increased blood flow in the anterior cingulate gyrus (Brodman area 24) in addition to activation in the parietal lobe in the S2 and S1 sensory areas. These investigators concluded that a cortical representation for pain perception existed in the cingulate gyrus.

Extensive investigations have been carried out to assess the human visual system at all levels, from primary visual stimulation to higher visual processing. This area has been among the richest and most exciting ^{15}O activation studies to date, largely because it has provided the possibility of investigating human analogs of the extensively studied visual system of primates. The posterior localization of macular retinal representation in the striate cortex vs more anterior representation of peripheral fields has been demonstrated in the human visual cortex using ^{15}O PET activation (Fox, Mintun, & Raichle, 1986).

Using ^{15}O PET, two higher order extrastriate cortical areas, known to exist in the monkey, have been demonstrated successfully in groups of human subjects. Zeki, Lueck, & Watson, (1991) presented subjects with a colored display in the active task and compared the regional cerebral blood flow in the brains of subjects viewing such displays with that produced when they viewed an identical pattern composed of equiluminant shades of gray. The comparison revealed that area V1 and the adjoining area V2 (primary visual and visual association cortex) were equally active under both conditions. In the cortical area outside V1 and V2, viewing the colored collage produced significant activation in the territory of the left lingual and fusiform gyri. The fusiform gyrus was thought to represent the human correlate of visual area V4, the color center in the macaque monkey (Zeki et al., 1991). The active zone in the lingual gyrus was thought to correspond to area V2 in the monkey. At the same time, a center for the perception of visual motion (human homolog of area V5 in the monkey) was demonstrated, using a moving collage, in Brodman area 39 at the junction of the temporal and occipital cortices (Lueck et al., 1989; Zeki et al., 1991) (see Color Plate 1). This dissociation of the locations of V4 and V5 human equivalents provides a convergence with animal experimentation that enhances our understanding of visual processing in ways that lesion studies have been unable to achieve, since human brain lesions simply do not respect the small areas of tissue involved in components of visual processing.

Interestingly, the degree of attention being paid to a particular visual attribute has a significant effect on the blood flow in the particular area of visual association cortex involved in processing that attribute. When subjects are attending to the color of squares of different sizes and velocities of motion, a significant response is seen in the lingual and fusiform gyri (collateral sulcus) at a location similar to the color center described by Lueck, Zeki, and colleagues. Similarly, attention to shape increases blood flow in the fusiform and parahippocampal gyri (collateral sulcus), parieto-occipital sulcus, and superior temporal gyrus. Attention to speed activates the left inferior parietal lobule (Corbetta, Miezen, Dobmeyer, Shulman, & Petersen, 1990,1991a) (Fig. 1). Overlap exists here, but no complete correspondence, with the focal areas demonstrated by Zeki et al. (1991). Thus, in ways that we do not understand, attention interacts with blood flow to the visual perceptual and visual cognitive cortical areas of the brain through a clearly "top-down processing" mechanism. Physiological demonstration of such interactions between attentional and sensory systems represents an important contribution of ^{15}O activation studies to cognitive neuroscience.

Similarly, the two visual pathways—one for "where" and one for "what"—delineated by Mishkin in monkeys (Mishkin, 1982; Mishkin, Ungerleider, & Macko, 1983) have been sought in human subjects using ^{15}O PET activation studies. The results of Corbetta's group are certainly consistent with temporal lobe preponderance for object identification and parietal lobe predominance for visuo-spatial and location processing of visual stimuli. Grady, Haxby, and colleagues (Grady et al., 1989,1990; Haxby et al., 1991) have used PET and the [^{15}O]water subtraction method to measure rCBF changes in normal subjects who are engaged in object discrimination vs spatial localization tasks. Both tasks involved a two-

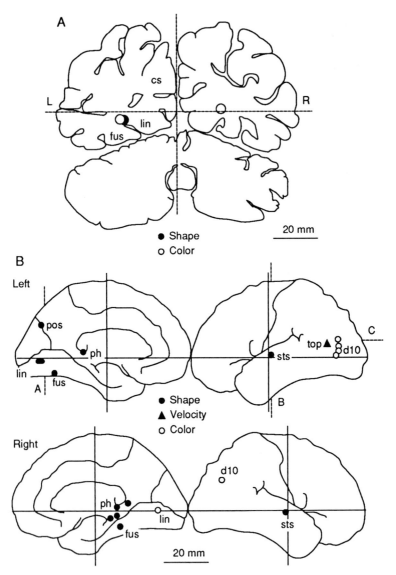

Figure 1 Visual association cortex activated by directed visual attention. (A) Coronal slice through occiput and cerebellum demonstrating PET activation of the left collateral sulcus for attention to color and shape, but not for attention to speed. (B) Sagittal views of lateral and medial aspects of both hemispheres. Focal areas activated by attention to visual shape, color, or motion are demonstrated. See text for details. CS, calcarine sulcus; LIN, lingual gyrus; FUS, fusiform gyrus; PH, parahippocampal gyrus. Reproduced with permission from Corbetta et al. (1990).

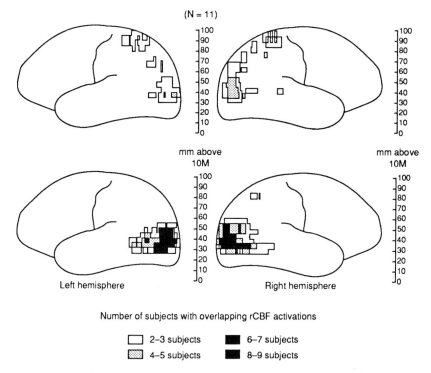

Figure 2 Mapping of rCBF in posterior brain regions during dot location matching (*top*) and face matching (*bottom*) experiments. Only regions showing a 30% or greater increase in rCBF on each of two experimental perception tasks relative to a sensorimotor control task are shown. See text for details. Reproduced with permission from Haxby et al. (1991).

choice match-to-sample discrimination. In the object discrimination task, subjects were asked to identify which of two lower squares contained a face that matched the one shown in the single top square. In the case of spatial localization, the task was for subjects to identify which of two lower squares contained a dot in the same location as in the single upper square. Subjects responded by pressing a left- or right-hand button. In the control, the squares were empty and subjects merely alternated left- and right-hand responses. After rCBF values were normalized, the areas for each subject that showed rCBF increases greater than 30% across 12 contiguous pixels (48 mm²) were represented as significant increases. A double dissociation was demonstrated for young and old subjects. In the face task, occipito-temporal regions and ventro-lateral occipital association cortex were more activated. In the spatial recognition task, superior parietal areas were activated. This dissociation seemed unaffected by normal aging (see Fig. 2). The authors interpreted these results as evidence for the existence of two distinct pathways in visual processing in humans, one for object discrimination and another for determination of spatial location. Interestingly, no clear relationship was seen between perceptual performance and ability and the degree of cortical blood flow change demonstrated.

IV. STUDIES OF ^{15}O PET ACTIVATION AND ATTENTION

Attention is mediated by a family of neurological processes for focusing awareness on relevant external stimuli while inhibiting intrusion from distracting stimuli. No satisfactory unitary model of all attentional processes exists; however, psychologists have focused on three basic attentional components—vigilance (tonic attention), directed or selective attention, and attentional processing capacity. Tonic attention designates a class of mechanisms that work to regulate the span or efficiency of vigilance and concentration. Another mechanism, often designated selective attention, is involved in determining the direction of attentional resources. Posner, Rafal, and colleagues (Posner, Walker, Friedrich, & Rafal, 1984; Rafal and Posner, 1987; Rafal, Posner, Friedman, Inhoff, & Bernstein, 1988) have delineated different characteristics of focal lesions in different parts of the brain, on an experimental paradigm measuring elements of selective visual attention. Lesions of the right parietal lobe, midbrain, and thalamus appear to impair particular elements in selective visual attention. Thalamic-lesioned patients are postulated to have a deficit in the ability to engage attention on the contralateral side, and it is thought that that the thalamus has a role in controlling the "attentional spotlight" for directed visual attention (Crick, 1984; LaBerge, 1990). This focal role postulated for the thalamus (probably the pulvinar) in directed visual attention has not received adequate study or clear demonstration with ^{15}O activation studies to date. Indeed, the major studies relevant to directed visual attention were done by Corbetta and colleagues (Corbetta, Miezen, Shulman, & Petersen, 1991c; Corbetta et al., 1990,1991a; Petersen, Fox, Posner, Mintun, & Raichle, 1989). As discussed earlier, attention to particular aspects of the stimulus induces selective increase in blood flow in focal areas of occipital, parietal, and temporal lobe that are implicated in visual processing of those features. LaBerge (1990) suggests that this top-down processing is mediated through the pulvinar. This top-down attentional effect on blood flow in lower level "perceptual" brain areas has been demonstrated even for simple vibrotactile stimulation of the hand (Meyer et al., 1991).

As indicated, a fundamental aspect of attentional function is the ability to sustain alertness to process high priority signals. The alert state has been studied behaviorally (Posner, 1978). This vigilance appears to depend heavily on the integrity of the right cerebral hemisphere (Coslett, Bowers, & Heilman, 1987; Cohen, Semple, Gross, Itolcomb, Dowling & Nordahl, 1988; Heilman & Valenstein, 1985). Therefore, interestingly, in a task requiring sustained somatosensory vigilance, Pardo, Fox, and Raichle (1991) produced increased blood flow selectively in the right frontal cortex.

Corbetta, Miezen, Shulman, and Petersen (1991b) also have obtained evidence from ¹⁵O PET activation for the existence of a posterior attentional system for directed visual attention in the parietal lobe. When subjects shifted attention toward the contralateral visual fields on a visual detection task, a region of the superior parietal lobule was found to be activated (see Fig. 4). This region of activation was not present when attention was maintained in the center of the visual field, suggesting the existence of a parietal mechanism for shifting attention in space regardless of direction of shift.

¹⁵O PET activation studies have demonstrated another cortical area that is implicated in attentional processes: the anterior cingulate gyrus. This area of the brain may exist as a major focus of the "anterior attentional system" postulated by Posner and Petersen (1990). This anterior attentional system (see Fig. 4) is related more closely to issues of attentional capacity and working memory capacity than to those of selective visual attention directed in space. The system is postulated to be a more general attentional system that also is implicated in the processing of language stimuli and in "attention for action" (Posner, Petersen, Fox, & Raichle, 1988).

Numerous ¹⁵O PET activation studies, almost all involving language processing or active responses, have produced significant increases in blood flow in regions of the anterior cingulate gyrus, located medially in the brain. Blood flow increases have been noted during semantic processing of single words, when subjects are asked to generate a verb from a noun or to make a foreign–local decision regarding words that are names of animals (Petersen, Fox, Posner, Mintun, & Raichle, 1988). Further, Posner et al. (1988) reported that anterior cingulate blood flow increases as the number of targets to be detected goes up. The requirement of action or decision making or of semantic selection appears to be critical for involvement of the anterior cingulate gyrus.

This finding of increased blood flow in the anterior cingulate gyrus for attentionally demanding tasks has shed some light on the question of whether access to meaning from visual words requires attention. In the experiments of Posner and Petersen, passive viewing of words did not activate the anterior cingulate gyrus whereas carrying out a semantic association task did. We have replicated this finding (Marrett et al., 1990). Interestingly, passive viewing of words by elderly subjects does produce increased activation of the anterior cingulate gyrus; this activation also is found in patients with Alzheimer's disease who are passively viewing words or pictures (Bub, Decter, & Chertkow, 1990; Chertkow et al., 1992), suggesting that simple presentation of words or pictures becomes more attentionally demanding with age and dementia, consistent with the behavioral literature (e.g., Herdman, 1992). Clearly, attention interacts strongly with semantic activation, and the act of attending to one meaning of a word or of activating associations involves the anterior attentional system (Posner & Petersen, 1990; Posner et al., 1988).

Other attentionally demanding tasks also appear to activate the anterior attentional system that includes the anterior cingulate cortex. For example, Pardo, Pardo, Janer, and Raichle (1990) demonstrated robust activation of blood flow in the anterior cingulate gyrus when subjects performed a Stroop task during scanning. In this task, subjects must name the color of words displayed on the screen when the words are color names incongruent with the actual color of the display. Left premotor cortex, left postcentral cortex, supplementary motor area, right superior temporal gyrus, and bilateral peristriate cortices also were activated during this task, demonstrating an extensive distributed network of activated regions involved in this complex task.

Other authors have presented an alternative interpretation of these data, stressing the known role of the cingulate cortex in preparation and initiation of movement. Paus, Petrides, Evans, and Meyer (1993) carried out ^{15}O PET scanning with normal subjects who were instructed to respond to visual stimuli either by moving their eyes, manipulating the response key, or saying a word. In the "overpracticed" version of these tasks, the association between stimuli and responses had been established in a training session prior to the scanning. This training was followed by a reversal version of the same tasks, when subjects were assigned a new combination of stimulus–response associations. Subtracting the overpracticed from the reversal condition produced activation in the anterior cingulate cortex for each condition, consistent with the novelty or increased attentional resources required for the new associations. Interestingly, however, a clear somatotopic representation of the activated anterior cingulate area appeared. The most rostral foci dorsal to the genu of the corpus callosum (Brodman area 24) were activated when the oculomotor responses were required. Vocal responses produced activation in the intermediate (Brodman's area 32) and rostral (face representation of area 24) part of the cingulate cortex, including the region involved in the vocalization of the monkey. The manual response condition produced a focus of activation more caudally in the vicinity of the callosal marginal sulcus. Paus and co-workers conclude that the anterior cingulate region is involved in facilitating the execution of the appropriate responses and/or the suppression of the inappropriate ones, and should be considered a motor rather than purely an attentional region of functional activation.

V. ¹⁵O PET ACTIVATION OF LANGUAGE PROCESSING

One of the most exciting and controversial areas of ¹⁵O activation studies has been their use in investigating the functional neuroanatomy of components of language processing. Although broad agreement exists among cognitive neuropsychologists on the constituents of these basic components, their functional neuroanatomy and their interrelationships remain unclear. A quick review of postulated components of simple word processing may be in order. Howard et al. (1992) discussed single word processing in terms of the following stages:

1. basic perceptual analysis for seen (orthographic material) or heard (acoustic material) words
2. a stored representation of the word's form, termed the input lexicon; there would be a visual or orthographic lexicon for visually presented words and an auditory lexicon for auditorily presented words
3. an amodal concept store (Patterson's "cognitive system") accessed by both visual and auditory words; this semantic information would come into play when such tasks such as semantic association or categorization are carried out; substantial evidence exists (e.g., Neely, 1991) that semantic information is activated automatically when words are seen or heard, even if no response is made and no task of semantic search is carried out
4. an output lexicon for the phonological form of words that will be spoken and for the orthographic form of words that will be written
5. the motor component of articulation involving a response buffer that accesses the bilateral motor cortices and their connections to the speech apparatus (Fig. 3)

Connections between these components allow recognition and pronunciation of known regular and irregular words, and also allow repetition and reading of new or unknown words (through direct input-to-output conversion systems).

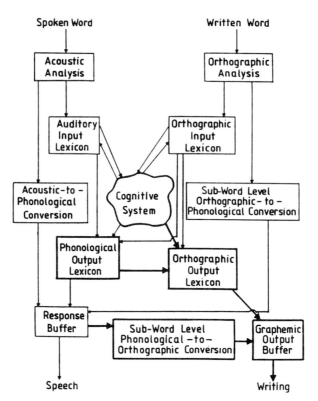

Figure 3   Processing diagram of basic components involved in the
processing of single words. The cognitive system indicated in the
center of the diagram is roughly equivalent to a semantic memory
store accessed from auditory and visual material. Although a full
discussion of the individual components of such a functional architecture is beyond the scope of this chapter, note that each individual
component might be focally or diffusely represented in the brain.
Reproduced with permission from Patterson & Shewell (1987).

Although broad agreement exists on these separable components of language processing (Posner et al., 1988), which tasks can best be used to
activate the various stages of representation is not entirely clear, nor are
the subtractions that will provide the most specific residual activation of a
desired component. Evidence from the behavioral and experimental psychology literature regarding localization of components is less than clear.
For example, Posner has suggested that evidence from pure alexia points
to a localization for the visual word form in the occipital lobe (Posner et al.,

1988). Other investigators (Bub et al., 1993; Howard et al., 1992; Vanier & Caplan, 1985) suggest that the visual word form is more likely to be localized in the left temporal lobe. Regarding semantic memory and conceptual storage, the usual or classical view is that conceptual knowledge is stored diffusely (Schwartz, Marchok, Kreinick, & Flynn, 1979). Growing evidence has accumulated that at least some component of semantic memory may be represented focally, most likely in the posterior left temporal lobe (Coughlan & Warrington, 1978; Hart, Lesser, & Gordon, 1992; Warrington & Shallice, 1984). The prime candidate for the auditory input lexicon is the left temporo-parietal cortex in the broad region termed Wernicke's area. The location of a phonological output lexicon is unclear. Certainly Broca's area in the posterior third of the left inferior frontal gyrus might be a good candidate. Impairment of articulatory rehearsal in patients with speech apraxia (Waters, Rochon, & Caplan, 1992) is also consistent with localization of the phonological output form of spoken words to the left frontal lobe.

With this sketch of the best "behavioral" candidates for components of language processing of single words as a backdrop, we turn to the data accumulated over the past 5 years from PET experimentation with ^{15}O subtractions. The initial set of experiments reported by Petersen et al. (1988; Peterson, Fox, Snyder, & Raichle, 1990) involved passive presentation of concrete words on a screen, subtracting a baseline condition in which plus signs were viewed flashing on the screen. Significant activation was produced in bilateral striate cortex and bilateral extrastriate cortex in the occipital lobes. The authors argued that this result was consistent with an occipital localization of the visual word form (orthographic lexicon) (Fig. 4).

In a follow-up study, similar presentation of visual words produced activation again in the occipital lobes, this time more restricted to the left side. In addition, activation was produced in the left inferior frontal area anterior to Broca's area. Passive presentation of false fonts and letters did not produce activation in the frontal zone. In the initial study, activation of the left inferior frontal area (see Fig. 4) was produced only when the active condition was a semantic task, subtracting away passive presentation of visual or auditory words. Thus, when subjects were asked to generate a verb associated with the noun or to decide whether the animal whose name was presented on the screen was safe or dangerous, increased blood flow was noted in the left inferior frontal region. Posner, Petersen, and colleagues argued that this left inferior frontal region is involved in semantic processing; this argument is strengthened by the fact that carrying out the same task for auditory word material produced activation in the identical left frontal region. In contrast, passive presentation of

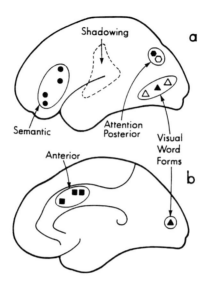

Figure 4   Summary of data from PET studies of visual and auditory words by Posner, Petersen, and colleagues. (A) Activation in the left lateral cerebral cortex. (B) Activation roughly in the medial portion of the brain. Solid symbols indicate left hemispheric and open symbols indicate right hemispheric activation. Triangles represent areas involved in visual word forms, circles indicate semantic analysis areas, squares indicate anterior attentional system, and hexagons indicate posterior attentional system. The areas thought to be activated by repeating auditory words (shadowing) are surrounded by a dotted area. Posner, Sandson, Dhawan, & Shulman (1989). Copyright © 1989 by The Massachusetts Institute of Technology Press, Cambridge, Massachusetts.

auditory words produced increased blood flow restricted to the posterior superior temporal cortex on the left and to the bilateral anterior superior temporal cortices. An output response (repeating the words that were heard or seen, subtracting passively hearing or seeing those words) produced increased blood flow in the bilateral motor strip area for mouth. This result is consistent with motor programming and articulation occurring in the motor strip. As noted earlier, carrying out the semantic association tasks also produced increased blood flow to the anterior cingulate gyrus.

Attempts at replicating these important initial findings of Petersen, Posner, and colleagues have met with variable success. In our laboratory, passive presentation of concrete words (animal names) also produced increased blood flow in left occipital visual association cortex, as well as in a left inferior region virtually identical to that documented by Petersen et al. (Bub et al., 1990; Marrett et al., 1990). On the other hand, Frackowiak and colleagues at Hammersmith Hospital have produced quite a different set of cortical activation areas in language tasks. In an auditory word–nonword decision task, Frith et al. (1991) demonstrated increased activity bilaterally in the superior temporal gyrus. This area was defined in terms of the statistical parametric map, comparing blood flow during this task with that in a set of contrasting tasks. The authors suggested that this result was compatible with localization of a bilateral auditory lexicon in the superior temporal gyrus. Howard et al. (1992) focused on demonstrating activation of auditory lexicon, visual lexicon, and semantics for single word presentation. Their goal was testing whether evidence from ¹⁵O PET studies existed for separate activation of auditory and visual lexicons from their respective presentation materials, or whether (as Dejerine suggested) visual stimuli had to be recoded into a phonological form before accessing conceptual knowledge in semantic memory. These researchers hypothesized that testing these contrasting theories of visual word processing would be possible using task comparisons that would variably activate the respective lexicons, in comparison with visual and auditory tasks that would not activate the lexicon. Howard and co-workers speculated that both visual and auditory word presentation would activate a shared area for semantic processing and storage. Clearly, the aim was to show that presentation of visual single words did not, in fact, activate the auditory lexicon as Dejerine suggested.

For visual word presentation, Howard and colleagues used an active task in which words were presented on the screen and read aloud by the subject. A baseline condition for comparison consisted of presentation of false fonts, at which time the subject said the word "crime" aloud. This control condition was hoped to balance brain activity for input and output procedures without activating the visual lexicon or semantic storage. When assessed via statistical parametric mapping, the comparison of the two conditions showed that presentation of visual words produced increased blood flow in the left posterior middle temporal gyrus as well as in a dorsal portion of the left anterior cingulate gyrus. Decreased blood flow was produced in the right inferior and middle frontal gyri, as well as in the right temporal lobe and middle occiput. A similar set of conditions was presented using auditory material. In the active condition, subjects heard a recorded word and repeated it aloud. The auditory control consisted of hearing unintelligible recordings and the subject saying the word "crime" aloud when each word was heard. The word repetition condition, when contrasted with its baseline, produced a significant peak of

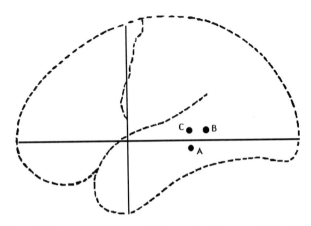

Figure 5 Posterior left temporal lobe peaks of increased ^{15}O activity. (A) Peak demonstrated by Howard et al. (1992), postulated to be the localization of the auditory input lexicon. (B) Peak demonstrated by Howard et al. (1992) postulated to be localization of the visual orthographic lexicon. (C) Peak within a statistical parametric map of Wise et al. (1991). This area was activated on each of four language tasks and is postulated to represent activation of a central semantic system.

increased blood flow in the left superior temporal gyrus and the middle temporal gyrus. Activity also was seen in the right striate cortex of the occipital lobe. Decreased blood flow was noted in the right middle frontal gyrus and the parietal lobe. The areas of blood flow increase in the posterior left temporal lobe for auditory and visual words were separated by about 16 mm (Fig. 5).

These results, although clearly important, are somewhat difficult to interpret. Howard argues for the existence of an orthographic lexicon in the posterior temporal lobe, in contrast to the results of Petersen and Posner. He claims to have demonstrated that visual stimuli do not activate the auditory lexicon. On the other hand, the close proximity of the two lexicons and their proximity to Wernicke's area might be interpreted as support for the classical model of Dejerine, which was revived by Geschwind (1971; Benson, 1985). Howard and colleagues remain uncommitted about whether semantic representations also were being activated in the posterior left temporal lobe, or whether they were not visible in the comparison conditions examined.

Additional data regarding phonological processing derive from two other studies. Zatorre et al. (1992a) examined auditory processing of speech. Passive presentation of pairs of speech syllables (for instance, "fat"–"tid") minus baseline presentation of noise bursts produced activation in right and left superior temporal gyri as well as in the left posterior temporal area (Color Plate 2). This subtraction also produced activation in the left inferior frontal gyrus, suggesting that this area is not specific to semantic processing. Alternatively, the left inferior frontal region may be a nonspecific mechanism for searching orthographic and semantic representations, regardless of stimulus. In a second condition, subjects were asked to make a phonetic decision that consisted of responding normally when the second syllable of the pair ended with the same consonant sound as the first. Subtracting away passive presentation of speech sounds, activation was noted in the posterior zone of Broca's area in the left inferior frontal gyrus, as well as in the cingulate gyrus anteriorly and posteriorly. The left superior parietal lobe was activated also. Zatorre proposes that, in making a phonetic judgment, subjects must access an articulatory representation involving neural circuits that include Broca's area. Alternatively, carrying out such a phonetic judgment might involve activation of a verbal short-term memory system (the phonological loop; Baddeley, Lewis, Elderidge, & Thompson, 1984). Activation of the same area during subvocalization for picture stimuli (Bub et al., 1990) is consistent with such a formulation.

In another condition, Zatorre's subjects carried out a pitch discrimination task, responding only when the second sound had a higher pitch than the first. This condition, in contrast to the phonetic decision, produced activation in the right hemisphere and in the right inferior frontal and right middle frontal gyri. Thus, with similar input stimuli and output responses, phonetic decisions produced blood flow changes lateralized to the left hemisphere whereas pitch discrimination produced increased blood flow that lateralized to the right hemisphere.

In another study of phonological processing, Démonet et al. (1992) carried out a series of tasks in which subjects monitored pairs of stimuli for particular properties. In a tone task, the subjects monitored a series of pure tones to detect a rising pitch. In a phonemic task, they monitored a series of syllables for a particular combination of phonemes. In a third task, subjects monitored a series of adjective–noun pairs, in which the nouns were all animal names, for a particular combination of semantic properties. Comparisons of the tasks were made using the SPM approach. The phonemic task, compared with the others, was found to produce increased blood flow in the left superior temporal gyrus. Some activation in the middle temporal gyrus on the left was seen also, as well as in the anterior superior temporal gyrus. Activation also was seen in the left inferior frontal lobe, somewhat dorsal to Broca's area. This region is similar in location to that activated by Zatorre and colleagues for phonetic discrimination. Thus, some definite convergence of the work of Démonet and Zatorre is seen, although quite different interpretations are attached to these areas of blood flow activation.

Posner and Petersen stressed the left inferior frontal lobe as a possible repository of semantic information. Other investigators have suggested that this area plays a more nonspecific or generalized role in search or access mechanisms. Suggestions have been made, however, that semantic processing does, in fact, activate semantic representations that are located in the left temporal lobe (rather than the left frontal lobe). Wise et al. (1991) carried out a series of tasks of semantic category judgment, semantic association judgment, and verb generation in response to presentation of nouns. These three conditions all produced increased blood flow in the left posterior superior temporal gyrus (see Fig. 5) that appeared to be independent of the rate of presentation of stimuli. Wise concluded that this result is likely to represent a cortical localization for semantic memory. As noted earlier, Howard and associates also produced activation in posterior left temporal lobe, but were led to conclude that this represented a cortical localization of the auditory and visual lexicons rather than of semantics. Démonet et al. (1992) assessed blood flow changes during their semantic judgment task compared with the phoneme and tone decision tasks. These researchers demonstrated increased blood flow for the auditory word task in the left inferior temporal gyrus quite posteriorly, as well as in the left supermarginal and angular gyri (Brodman area 40 and 39) and the left prefrontal cortex (Brodman area 8 and 9). This frontal area lies quite deep and, in fact, occupies white matter between cingulate gyrus and lateral frontal cortex. Démonet and colleagues argued that these three areas indicated a lexical semantic network. Here, again, is both correspondence and divergence from interpretations by other investigators. For instance, Démonet's is the first paper to suggest an inferior posterior left temporal lobe activation for semantic memory. The region activated in the angular gyrus is about 22 mm from the semantic region activated by Wise and co-workers. Whether these represent the same region or not is unclear. The frontal region activated by Démonet and colleagues is not clearly identical to the inferior left frontal region of Petersen and Posner, and may, in fact, represent an extension from the anterior cingulate cortex (Brodman area 47).

We also have carried out investigations into possible localization of semantic processing. When subjects hear an animal name or see its written word form and are asked to decide what protuberances (i.e., horns, tusks, or antlers) come out of its head, we find activation to occur similarly in the posterior left temporal region (Bub et al., 1990; Chertkow et al.,

1993). Visual imagery tasks without semantic content, however, produce blood flow activation that is more posterior and inferior at the occipital–temporal junction on the left. In summary, then, visual imagery, semantic processing, and activation of the visual orthographic and auditory lexicons all might be responsible for increased blood flow in the posterior left temporal lobe in any particular condition or task. At the same time, certain experimental conditions involving activation of these levels of representation have failed to increase blood flow in this area (Petersen et al., 1988). In the last part of this chapter, we comment on certain methodological problems that have been raised by results from these language studies, results that certainly do not present an entirely consistent set of data at this time.

VI. FRONTAL LOBE PROCESSES STUDIED WITH ¹⁵O PET ACTIVATION

¹⁵O activation studies have demonstrated successfully that various putative frontal tasks do, in fact, involve activation of frontal lobe cortex in terms of cerebral blood flow. In the monkey, behavioral dissociation has been demonstrated between the mid-dorsolateral frontal cortex (Brodman's area 46 and 9), which plays a critical role in the performance of self-ordered working memory tasks, and the immediately adjacent posterior dorsolateral frontal cortex (Brodman area 8). The latter is critical for the learning and performance of visual conditional association tasks. Petrides, Alivisatos, Evans, and Meyer (1993) have reported a study in which subjects carried out self-ordered pointing tasks during ¹⁵O injections. Patients with frontal lobe lesions are known to show impairment in performance of this task. In a control task, subjects merely pointed to a particular design on a card in response to a specific cue. When the control task was subtracted from the self-ordered pointing task, residual activation was noted in the left mid-dorsolateral frontal cortex (Brodman area 9 and area 46) as well as in the anterior cingulate cortex (Brodman area 32). The dorsolateral frontal activation was visible in the right hemisphere as well (Color Plate 3). Posterior parietal cortex was also activated bilaterally. When the activation condition involved a conditional association task (subjects were required to point to a different design, conditional on the color of the stripe at the top of the card), activation again was seen in the frontal lobe, but a different pattern of activation was noted. This time, unilateral activation deep in the left caudal superior frontal sulcus (Brodman area 8) was seen in conjunction with posterior parietal cortex activation bilaterally.

Weinberger and colleagues (Berman et al., 1991) carried out [15]O PET rCBF studies using a series of tasks known to implicate frontal lobe in working memory. During such tasks as the Wisconsin Card-Sorting Test and a version of the delayed alternation task, a significant bilateral elevation of rCBF in the dorsolateral prefrontal cortex was seen in young normal subjects that was localized largely to the inferior frontal gyrus, bilaterally, for both frontal lobe tasks. The frontal lobe activation was accompanied by increased blood flow in the anterior cingulate gyrus as well. When the tasks were explained fully to the subjects and then repeated, eliminating the need for creation of novel conceptual sets, interesting changes in the blood flow emerged. These subjects continued to show activation of the dorsolateral prefrontal cortex, but increased blood flow was no longer seen in the anterior cingulate cortex, suggesting that the physiological response in the anterior cingulate gyrus was more related to attentional factors than to working memory per se.

Another behavioral task with a putative functional localization to the frontal lobe is the verbal fluency task. Milner and Petrides (1984) showed that left dorsolateral prefrontal cortex lesions produce an impairment in the patient's ability to produce examplars on a verbal fluency task. In the study by Wise et al. (1991), subjects carried out one activation condition in which they were asked to generate verbs that were associated with a presented noun. This verb generation task produced increased blood flow in the left posterior middle frontal gyrus and in left Broca's area, as well as in the midline supplementary motor area. In the study by Frith et al. (1991), two verbal fluency conditions were carried out. In one task, subjects generated a list of "jobs" and, in another, a list of words beginning with the letter "a". No significant difference was found between these two conditions, both of which demonstrated increased blood flow in the left dorsolateral prefrontal cortex, predominantly in the middle frontal gyrus. In addition, increased blood flow was seen in the left parahippocampal area and the left and right anterior cingulate gyrus. Interestingly, this increased blood flow correlated with decreased blood flow to the superior temporal gyrus on the left side. Frith postulated the presence of reciprocal inhibition between dorsolateral prefrontal cortex and superior temporal gyrus during the intrinsic generation of words. The authors alluded to evidence from the animal literature of anatomical connections between left dorsolateral prefrontal cortex and left parahippocampal gyrus, as well as for connections between frontal lobe and superior temporal gyrus. They speculated that the auditory lexicon might be represented as a distributed neural network, bilaterally in superior temporal gyri. During performance of a verbal fluency task, they speculated, areas including the lexicon and semantic associations might need to be inhibited to prevent their interference with the task. Although this speculation may or may not prove correct, it represents an important attempt to use the data generated from ^{15}O PET activation studies to develop behavioral hypotheses that subsequently may be testable by cognitive neuropsychologists.

VII. PRACTICE, LEARNING, AND CEREBELLAR ACTIVATION

One surprising result of the series of ^{15}O PET activation studies noted earlier is the growing realization that activity in the cerebellum is implicated in many cognitive processes. Although isolated reports have been made stressing the importance of the cerebellum in higher cognitive processes (e.g., Leiner, Leiner, & Dow, 1987), this region of the brain commonly has been viewed as having a role restricted to coordination and motor programming. ^{15}O PET studies have demonstrated that the cerebellum certainly is involved in motor sequence learning (Jenkins, Brooks, Nixon, Frackowiak, & Passingham, 1993), with increased bilateral activation of the cerebellum in ^{15}O PET studies during learning of new sequences compared with automatic performance of overlearned sequences. The cerebellum shows increased cerebral blood flow during tasks in which subjects prepare to reach out to objects (Decety, Roland, & Eulyás, 1992). However, the cerebellum also is activated during the generation of word associations (Raichle, 1990), as well as during repeating of words, spontaneous speech, and semantic decisions (Bub et al., 1990; Raichle, 1990). Particularly, the posterior lobule of the cerebellum seems to participate in cognitive activity such as decision making and verbal encoding (Decety, Sjöhdm, Ryding, Stenberg, & Inguar, 1990). These PET data have been supported by reports of subtle impairments of higher cognitive processing in patients with focal cerebellar lesions (Leiner, Leiner, & Dow, 1986).

A growing interest in the ^{15}O PET literature has been the use of PET to assess the impact of subject variability, learning, and practice on the pattern of CBF change in individuals. The importance of stimulus novelty in the degree of increased CBF seen in anterior cingulate (Berman et al., 1991) and cerebellar (Jenkins et al., 1993) cortex has been alluded to. (Raichle, 1991; Raichle, Fiez, Videen, Fox, Pardo, & Petersen, 1991) has noted, in a conditional associative task ("say a verb" for a seen noun), that practice over six trials was accompanied by progressive reduction in CBF in left prefrontal cingulate, cerebellar, and posterior temporal cortex. At the same time, CBF increased over practice trials in bilateral insular cortex. When subjects are overpracticed on a sensorimotor association task with speech response, most subjects cease to show anterior cingulate cortex activation. Subjects who did poorly on the task, in terms of accuracy and latency measures, continued to demonstrate anterior cingulate cortex activation (Paus et al., 1992).

These data demonstrate the possibilities of ^{15}O PET for future investigations of the neurobiology of learning and memory in the human. This domain undoubtedly will prove to be one in which psychologists and neuroimaging teams can interact fruitfully.

VIII. OTHER AREAS OF COGNITION

This chapter is by no means all inclusive in its survey of the literature. Of necessity, several cognitive domains have been neglected in the interest of space, and we attempted to concentrate on areas for which the results of PET experimentation have been most rewarding using the ^{15}O activation method. Developments in imaging of memory systems have not been addressed here (Raife, Fiez, Raichle, Balota, & Petersen, 1992; Squire et al., 1992), nor have we discussed studies investigating imagery. At least one study has been done investigating face and object recognition (Sergent et al., 1992); additional studies in this domain are likely to be forthcoming.

IX. ADDITIONAL METHODOLOGICAL PROBLEMS AND ISSUES REGARDING ^{15}O PET ACTIVATION STUDIES

In the remainder of this chapter, we address four of the problems and limitations of the ^{15}O subtraction technique of PET. These comments are in the form of a general critique of the methodology and, at the same time, an explanation of the divergence of results from different laboratory, and of results from each laboratory from those presented in previous neuropsychological literature.

1. In the preceding discussion, particularly concerning language, little effort has been made to derive clear and unambigious conclusions from the data generated, reflecting the fact that the cognitive models used in these studies remain fairly basic and are riddled with potential pitfalls in terms of interpretation. In most cases, the cognitive components implicated in any particular task are not entirely clear. For example, the baseline conditions used by Howard et al. (1992), in which subjects view an abstract design and repeat the word "crime", were designed to avoid activation of the lexicon or semantics. However, repeating a meaningful word may in fact activate semantics to a considerable degree. If so, this activa-

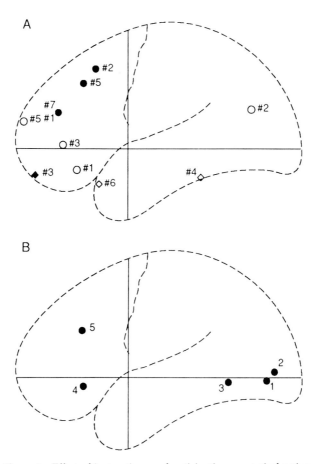

Figure 6 Effect of instructions and anticipation on cortical activation (Chertkow et al., 1993). Solid circles are peaks of increased CBF, open circles are peaks of decreased CBF. Circles are cortical locations in left hemisphere, diamonds are cortical locations in right hemisphere. (A) Activation of brain regions during performance of a perceptual semantic decision (judging whether animals whose names were seen on a screen had horns or tusks protruding from their heads). A baseline presentation of "+" signs was subtracted. (B) Carrying out of the identical task when the subtraction baseline consisted of anticipating the semantic task while subjects viewed the same "+" signs. Image (A) demonstrates lack of apparent temporal lobe activation during semantic processing; however, (B) shows clear evidence of posterior and inferior left temporal lobe activation during the semantic processing task, when baseline anticipation state is subtracted.

tion might explain why no residual semantic activation was visible when the baseline and active tasks were compared. Similarly, the behavioral evidence (mainly semantic priming experiments) points to the automatic activation of semantic knowledge on passive presentation of words, through either visual or auditory material (Neely, 1991). If such activation is occurring, one would expect to see activation of semantic areas during passive presentation of stimuli, even without active semantic processing. Such anticipated activation was not produced in the initial study by Petersen et al. (1988), and the behavioral expectations were rejected. When left inferior frontal lobe activation did appear in the subsequent study by Petersen et al. (1989), it was assumed to represent activation of semantics on passive presentation of words, consistent with behavioral data. Thus, lacking strong models of the components activated in any particular task, researchers have tended to fall back on post-hoc explanations of the PET activation peaks produced in any subtraction comparison.

2. As noted by Frith et al. (1991), many tasks produce both increase and decrease in blood flow in different parts of the brain. These changes may be due to linked and reciprocal activation networks, or might represent the activity of facilitatory and inhibitory neurons. Activation of inhibitory neurons would produce metabolic activity with subsequent increase in blood flow (Collins, 1991), but focal cortical areas that receive their inputs from such neurons will show decreased neuronal activity and, therefore, decreased cerebral blood flow. Alternatively, since many cortical neurons are themselves inhibitory, cognitive processing may, in fact, require such neurons to be "turned off" before processing can occur. Thus, a primary component of any particular cognitive process might include deactivation of significant inhibitory neurons. Thus, any particular cognitive processing activity might be accompanied by increased blood flow, decreased blood flow, or a combination of the two, in effect canceling each other out. The manner in which these positive and negative changes in blood flow relate to cognitive processing remains unclear, but the effect might be to invalidate experimental paradigms that rely on an "additive components" subtraction hierarchy.

3. The possibility has been raised that researchers have neglected certain important elements of the cognitive tasks and comparison states used in their studies. In a typical task used in ^{15}O PET studies, a subject lying in a PET scanner is given certain instructions. The subject is told that he or she will view certain stimuli on a screen and will be asked to carry out a certain task. Cognitive psychologists would argue that mere presentation of such instructions will alter the mental state of the subject and potentially alter the cerebral blood flow (Deutsch et al., 1986). In addition, in

many cases such as the language experiments alluded to, the actual cognitive processing required by the individual occupies only a fraction of the total time in the machine. For example, in the experiment by Howard et al. (1992), concrete words were presented every 1.5 sec for repetition. In other tasks, for example those of Frith et al. (1991), auditory lexical decisions were carried out every 1.5 sec. However, completing such a task actually requires about 300 or 400 msec. Thus, for about half the time period being evaluated, the subject is in a state of anticipation or preparation for the next stimulus. The PET results obtained may, then, represent the amalgamation and synthesis of the effects of instructions, the effects of anticipation and preparation, and the effects of the actual cognitive processing that is the object of study. Unless the physiological effects of instructions and preparation are identical in each condition, the subtraction images may reflect the effects of anticipation and instructions, rather than those of cognitive processing alone. Consistent with this speculation, Raichle (1991) has found that increasing the interstimulus interval from 1000 msec to 1500 msec (thus significantly increasing the fraction of time spent preparing for stimuli) produced significantly different results on a semantic activation task.

In an attempt to untangle these possible multiple effects, we have carried out the following PET experiment. In the baseline condition, subjects viewed "+" signs flashing on a screen. In a subsequent condition, they passively viewed words on the screen that were names of animals. In a second condition, they viewed the same words but now were instructed to decide whether the animal had horns, tusks, or antlers protruding from its head. In a final condition, they viewed the same animal words but were asked to decide whether the animal was a foreign or local creature. For each of the conditions described, the subject was scanned twice. In one instance, the scanning was undertaken while the subject viewed the words and carried out the required task. In the second instance, the subject was scanned after the instructions had been given, but during a 60-sec interval during which he or she viewed flashing "+" signs on the screen and awaited presentation of the relevant word stimuli. In other words, this interval represented a state of anticipation or preparation for the task at hand. We (Chertkow et al., 1993) found that presentation of instructions with anticipation caused focal decrease in cerebral blood flow in the posterior temporal regions bilaterally. The effects of instructions and anticipation were not identical for every condition studied, suggesting that they cannot be assumed to cancel each other out under baseline and activation conditions. Further, the significant peaks seen in the "horns, tusks, and antlers minus passive word" condition (Fig. 6) were drastically altered when the effects of anticipation were taken into consideration. When the effect of anticipation was subtracted from both activation condition and baseline, the peaks were no longer in the frontal lobe, but were seen predominantly in the left posterior temporal lobe. This result implies that the visible effects in the frontal lobe were related largely to preparation and anticipation, whereas the actual semantic processing was occurring in the posterior temporal region. If this result is found to extend beyond the particular paradigm utilized, there may be serious consequences for our interpretation of the many studies that suggest frontal lobe localization of various cognitive functions.

The effects of preparation and anticipation represent only one of a large number of variables that may have to be controlled in and across PET experiments. Currently, no uniformity exists across laboratories in tasks administered, the nature of the stimuli used, the technical aspects of the PET imaging, or the approach to statistical interpretation of the data. Therefore, that considerable disagreement exists in PET results for cognitive processing tasks should not be surprising.

4. A final point to be made with respect to interpretation of PET studies to date involves the ever present problem of variability in a group of individuals. Steinmetz and Seitz (1991) have noted that normal variation in the location of cytoarchitectonic borders, gyri, and sulci can produce variability of up to 10 mm in points represented in the same position of the motor cortex across a group of individuals. Although not excessive, such normal anatomical variability might be sufficient to interfere with summation of active peaks across a group of normal subjects. Far more important is the question of functional variability, particularly with respect to components of language processing. Good evidence exists that many of the focal components of the language system, such as the postulated module for syntactic comprehension and phonological processing, are localized variably within the language zone of the left hemisphere across a group of individuals (Caplan, 1988). This variability has been demonstrated in studies of brain-damaged subjects as well as in electrocorticography (Ojemann, 1983; Ojemann, Fried, & Lettich, 1989). If different brain areas are involved in cognitive processing of the same task in different individuals, a substantial portion of the neurophysiological effect will be lost during intersubject averaging. According to Steinmetz and Seitz (1991), this effect probably accounts for the relative absence of significant activation peaks in ^{15}O PET studies of language processing since, in theory, the task-induced activation in PET should reveal a greater number of cortical sites than identified in studies of brain-damaged individuals. This problem of functional variability represents a major challenge for the intersubject averaging approach to PET activation studies.

X. CONCLUSION

PET imaging to date has confirmed previous classical neuropsychological findings with respect to localization of sensory processing, and has defined in more detail areas that previously had only been suspected in the brain lesion literature. These results include demonstration of components of visual processing equivalent to area V4 and V5 in the primates, as well as a suggested demonstration of an analog of the two visual systems—one for "what" and one for "where". Additionally, the top-down effect of attention on cerebral blood flow to visual association cortical areas has been demonstrated. In terms of attention, PET has provided dramatic demonstration of the existence of an anterior attentional system, particularly involving the anterior cingulate gyrus, and of a localized area in the right frontal lobe activated during vigilance. Although studies of language processing remain somewhat controversial and incompatible, evidence is growing for involvement of the posterior left temporal lobe in semantic processing, with activation of the left inferior frontal lobe as a more general mechanism for semantic search. Demonstration of involvement of the frontal lobe in tasks generally believed to require frontal lobe involvement has increased our understanding of the functional neuroanatomy of frontal lobe regions. Finally, hitherto unsuspected or underappreciated involvement of the cerebellum in higher cognitive processing has become evident using ¹⁵O PET studies.

In the coming years, some of the obstacles and problems presently encountered in the ¹⁵O PET methodology doubtless will be overcome. On the other hand, some of the issues, such as intersubject averaging, may continue to remain a problem for this methodology. The advent of functional neuroimaging with ¹⁵O PET has demonstrated dramatically the gaps in our knowledge of the neurophysiology of cognition, and has provided a major stimulus for cross-discipline research involving cognitive psychologists and basic neuroscientists. We continue to hope that a day will come shortly when results of functional imaging with PET will indeed converge with the long history of brain–behavior studies in neuropsychology.

ACKNOWLEDGMENTS

This work was supported by an MRC Scholarship award to H. Chertkow and a chercheur-boursier personal award to D. Bub from the Fonds de la Recherche en Santé du Québec, as well as by operating grants from the Medical Research Council of Canada to H. Chertkow (MA-10355) and D. Bub (MA-10506). The collaboration of Alan Evans, Ernst Meyer, and

Albert Gjedde of the Positron Imaging Laboratory of the McConnell Brain Imaging Centre of the Montreal Neurological Institute is acknowledged also. Helpful discussions of various aspects of this chapter with Tomas Paus, Robert Zatorre, and Michael Petrides also are greatly appreciated.

REFERENCES

Baddeley, A., Lewis, V., Elderidge, M., & Thompson, N. (1984). Attention and retrieval from long-term memory. *Journal of Experimental Psychology: General, 113*, 518–540.

Benson, D. F. (1985) Aphasia. In K. M. Heilman & E. Valenstein (Eds.), *Clinical neuropsychology* (pp. 17–48). New York: Oxford University Press.

Berman, K. F., Randolph, C., Gold J., Holt, D., Jones, D. W., Goldberg, T. E., Carson, R. E., Herscovitch, P., & Weinberger, D. R. (1991). Physiological activation of frontal lobe studies with positron emission tomography and oxygen −15 water during working memory tasks. *Journal of Cerebral Blood Flow and Metabolism (Suppl. 2), 11*, S851.

Bub, D., Decter, M., and Chertkow, H. (1990). The functional neuroanatomy of word and picture processing studied by positron emission tomography. *Canadian Journal of Psychology, 31*, 427.

Bub, D., Decter, M., and Chertkow, H. (1993). Multiple representation of object concepts: Evidence form category-specific visual agnosia. *Cognitive Neuropsychology*, (submitted for publication).

Caplan, D. (1988). The language zone. In E. Newmeyer (Ed.), *Linguistics: The Cambridge Survey*. Cambridge: Cambridge University Press.

Chertkow, H., Bub, D., Waters, G., Evans, A., Whitehead, V., & Hosein, C. (1993). Separate effects of instructions and stimuli on cerebral blood flow: A ^{15}O positron emission tomographic study. *Neurology, 43*, A189.

Chertkow, H., Hamel, E., Bub, D., Meyer, E., Wisbord, S., Evans, A., and D'Antono, B. (1992). Increased activation of anterior cingulate region in dementia of the Alzheimer's type. *Neurology (Suppl.3), 42*, 316.

Cohen, R. M., Semple, W. E., Gross, M., Holcomb, H. J., Dowling, S. M., & Nordahl, T. E. (1988). Functional localization of sustained attention. *Neuropsychiatry, Neuropsychology, and Behavioral Neurology, 1*, 3–20.

Collins, R. C. (1991). Basic aspects of functional brain metabolism. In R. Collins (Ed.), *Exploring brain functional anatomy with positron tomography* (pp. 6–22). New York: John Wiley & Sons.

Corbetta, M., Miezen, F. M., Dobmeyer, S., Shulman, G. L., & Petersen, S. E. (1990). Attentional modulation of neural processing of shape, color, and velocity in humans. *Science, 248*, 1556–1559.

Corbetta, M., Miezen, F. M., Dobmeyer, S., Shulman, G. L., & Petersen, S. E. (1991a). Selective and divided attention during visual discrimination of shape, color, and speed: Functional anatomy by positron emission tomography. *Journal of Neuroscience, 11*, 2383–2402.

Corbetta, M., Miezin, F. M., Shulman, G. L., & Petersen, S. E. (1991b). PET studies of spatial attention: direction vs. visual hemifield. *Journal of Cerebral Blood Flow and Metabolism, 11*, S435.

Corbetta, M., Miezin, F. M., Shulman, G. L., & Petersen, S. E. (1991c). Selective attention modulates extrastriate visual regions in humans during visual feature discrimination and recognition. In R. Collins (Ed.), *Exploring brain functional anatomy with positron tomography* (pp. 165–180). New York: John Wiley & Sons.

Coslett, H. B., Bowers, D., & Heilman, K. M. (1987). Reduction in cerebral activation after right hemisphere stroke. *Neurology, 37,* 957–962.

Coughlan, A., & Warrington, E. (1978). Word comprehension and word retrieval in patients with localized cerebral lesions. *Brain, 101,* 163–185.

Crick, F. (1984). Function of the thalamic reticular complex: The searchlight hypothesis. *Proceedings of the National Academy of Sciences of the United States of America, 81,* 4586–4590.

Decety, J., Roland, P. E., & Gulyás, B. (1992). A PET study of the structures in the human brain engaged in the preparation phase of reaching. *Neuro Report, 3,* 761–764.

Decety, J., Sjöholm, H., Ryding, E., Stenberg, G., & Ingvar, D. H. (1990). The cerebellum participates in mental activity: Tomographic measurements of regional cerebral blood flow. *Brain Research, 535,* 313–317.

Démonet, J.-F., Chollet, F., Ramsay, S., Cardebat, D., Nespoulous, J. L., Wise, R., Rascol, A., & Frackowiak, R. (1992). The anatomy of phonological and semantic processing in normal subjects. *Brain, 115,* 1753–1768.

Deutsch, G., Papanicolaou, A. C., Bourbon, T., & Eisenberg, H. M. (1986). Cerebral blood flow evidence of right cerebral activation in attention demanding tasks. *International Journal of Neuroscience, 36,* 23–28.

Evans, A. C., Beil, C., Marrett, S., et al. (1988). Anatomical-functional correlation using an adjustable MRI based region-of-interest with positron emission tomography. *Journal of Cerebral Blood Flow and Metabolism, 8,* 513–530.

Fox, P. T., & Mintun, M. (1989). Noninvasive functional brain mapping by change distribution analysis of averaged PET images of H₂¹⁵O tissue activity. *Journal of Nuclear Medicine, 30,* 141–149.

Fox, P. T., Mintun, M., & Raichle, M. E. (1986). Mapping human visual cortex with positron emission tomography. *Nature (London), 323,* 806–809.

Fox, P. T., Mintun, M. A., & Reiman, E. M. (1988). Enhanced detection of focal brain responses using intersubject averaging and change distribution analysis of subtracted PET images. *Journal of Cerebral Blood Flow and Metabolism, 8,* 642–653.

Fox, P. T., & Raichle, M. E. (1986). Focal physiological uncoupling of cerebral blood flow and oxidative metabolism during somatosensory stimulation in human subjects. *Proceedings of the National Academy of Sciences of the United States of America, 83,* 1140–1144.

Frackowiak, R. S. J. (1989). Positron emission tomography. *Seminars in Neurology, 9,* 275–407.

Frith, C. D., Friston, K. J., Liddle, P. F., & Frackowiak, R. S. J. (1991). A PET study of word finding. *Neuropsychologia, 29,* 1137–1148.

Geschwind, N. (1971). Aphasia. *The New England Journal of Medicine, 248 (12),* 654–656.

Goldman-Rakic, P. S. (1987). Circuitry of primate prefrontal cortex and regulation of a behavior by representational memory. In F. Plum (Ed.), *Handbook of physiology: The nervous system* (Vol. V, pp. 373–417). Bethesda, Maryland: American Physiological Society.

Grady, C. L., Haxby, J. V., Horwitz, B., Schapiro, M., Ungerleider, L. G., Mishkin, M., Carson, R. E., Herscovitch, P., & Rapoport, S. I. (1990). Changes in regional cerebral blood flow (rCBF) demonstrate separate visual pathways for object discrimination and spatial location. *Journal of Clinical and Experimental Neuropsychology, 12,* 93 (Abstract).

Grady, C. L., Haxby, J. V., Horwitz, B., Schapiro, M., Ungerleider, L. G., Mishkin, M., Friedland, R. P., & Rapoport, S. I. (1989). Mapping human visual systems for object recognition and spatial localization by measurement of regional cerebral blood flow. *Journal of Clinical and Experimental Neuropsychology, 12,* 93 (Abstract).

Hart, J., Jr., Lesser, R. P., & Gordon, B. (1992). Selective interference with the representation of size in the human by direct cortical electrical stimulation. *Journal of Cognitive Neuroscience, 4,* 337–344.

Haxby, C. L., Grady, J. V., Horwitz, B., Ungerleider, L. G., Mishkin, M. M., Carson, R. E., Herscovitich, P., Schapiro, M. B., & Rapoport, S. I. (1991). Dissociation of object and spatial visual processing pathways in human extrastriate cortex. *Proceedings of the National Academy of Sciences of the United States of America, 88*, 1621–1625.

Heilman, K. M., & Valenstein, E. (Eds.) (1985). *Clinical neuropsychology* (2d Ed.). New York: Oxford University Press.

Herdman, C. M. (1992). Attentional resource demands of visual word recognition in naming and lexical decisions. *Journal of Experimental Psychology, 18*, 460–470.

Howard, D., Patterson, K., Wise, R., Brown, W. D., Friston, K., Weiller, C., & Frackowiak, R. (1992). The cortical localization of the lexicons. Positron emission tomography evidence. *Brain, 115 (pt.6)*, 1769–1782.

Jenkins, I. H., Brooks, D. J., Nixon, P. D., Frackowiak, R. S. J., & Passingham, R. E. (1993). A PET study of the functional anatomy of motor sequence learning. *Neurology, 43*, A188.

LaBerge, D. (1990). Thalamic and cortical mechanisms of attention suggested by recent positron emission tomographic experiments. *Journal of Cognitive Neuroscience, 2 (4)*, 358–372.

Leiner, H. C., Leiner, A., & Dow, R. S. (1986). Does the cerebellum contribute to mental skills? *Behavioral Neuroscience, 100*, 443–454.

Leiner, H. C., Leiner, A., & Dow, R. S. (1987). Cerebro-cerebellar learning loops in apes and humans. *Italian Journal of Neurological Science, 8*, 425–436.

Lueck, C., Zeki, S., Friston, K. J., et al. (1989). A colour centre in the cerebral cortex of man. *Nature (London), 340*, 386–389.

Marrett, S., Bub, D., Chertkow, H., Meyer, E., Gum, T., & Evans, A. (1990). Functional neuroanatomy of visual single word processing studied with PET/MRI. Presented at Society for Neuroscience, October, 1990. St. Louis, Missouri.

Mata, M., Fink, D. G., Gainer, N., et al. (1980). Activity dependent energy metabolism in rat posterior pituitary primarily reflects sodium pump activity. *Journal of Neurochemistry, 34*, 213–215.

Mesulam, M. M. (1990). Large scale neurocognitive networks and distributed processing for attention, language and memory. *Annals of Neurology, 28*, 597–613.

Meyer, E., Ferguson, S. S. G., Zatorre, R. J., Alivisatos, B., Marrett, S., & Evans, A. (1991). Attention modulates somatosensory cerebral blood flow response to vibrotactile stimulation as measured by positron emission tomography. *Annals of Neurology, 29*, 440–443.

Milner, B., & Petrides, M. (1984). Behavioural effects of frontal lobe lesions in man. *Trends in Neuropsychology, 7*, 403–407.

Mishkin, M. (1982). A memory system in monkeys. *Philosophical Transactions of the Royal Society, London, B298*, 85–95.

Mishkin, M., Ungerleider, L. G., & Macko, K. A. (1983). Object vision and spatial vision: two cortical pathways. *Trends in Neuroscience, 6*, 414–417.

Neely, J. (1991). Semantic priming effects in visual word recognition: A selective review of current findings and theories. In D. Besner and G. Humphreys (Eds.), *Basic processes in reading: Visual word recognition* (pp. 264–336). Hillsdale, New Jersey: Erlbaum Associates.

Ober, B. A., Reed, B. R., & Jagust, W. J. (1991). Neuroimaging and Cognitive Function. In D. I. Margolin (Ed.), *Cognitive neuropsychology in clinical research* (pp. 495–531). New York: Oxford University Press.

Ojemann, G. A. (1983). Brain organization for language from the perspective of electrical stimulation mapping. *The Behavioral and Brain Sciences, 6*, 189–206.

Ojemann, G. A., Fried, I., & Lettich, E. (1989a). Electrocorticographic (ECoG) correlates of language. I. Desynchronization in temporal language cortex during object naming. *Electroencephalography and Clinical Neurophysiology, 73*, 453–463.

Pardo, J. V., Fox, P. T., & Raichle, M. E. (1991). Localization of a human system for sustained attention by positron emission tomography. *Nature (London), 249*, 61–64.

Pardo, J. V., Pardo, P. J., Janer, K. W., & Raichle, M. E. (1990). The anterior cingulate cortex mediates processing selection in the Stroop attentional conflict paradigm. *Proceedings of the National Academy of Sciences of the United States of America, 87*, 256–259.

Patterson, K., & Shewell, C. (1987). Speak and spell: Dissociation and word-class effects. In M. Coltheart, R. Job, & G. Sartori (Eds.), *The cognitive neuropsychology of language* (pp. 273–294). London: Erlbaum.

Paus, T., Petrides, M., Evans, A. C., & Meyer, E. (1993). Role of the human anterior cingulate cortex in the control of oculomotor, manual, and speech responses: A positron emission tomography study. *Journal of Neurophysiology, 70*, 453–469.

Petersen, S. E., Fox, P. T., Posner, M. I., Mintun, M., & Raichle, M. E. (1988). Positron emission tomographic studies of the cortical anatomy of single word processing. *Nature (London), 331*, 585–589.

Petersen, S. E., Fox, P. T., Posner, M. I., Mintun, M., & Raichle, M. E. (1989). Positron emission tomographic studies of the processing of single words. *Journal of Cognitive Neuroscience, 1*, 153–170.

Petersen, S. E., Fox, P. T., Snyder, A. Z., & Raichle, M. E. (1990). Activation of extrastriate and frontal cortical areas by visual words and word-like stimuli. *Science, 249*, 1041–1044.

Petrides, M., Alivisatos, B., Evans, A. C., & Meyer, E. (1993). Dissociation of human mid-dorsolateral from posterior dorsolateral frontal cortex in memory processing. *Proceedings of the National Academy of Sciences of the United States of America, 90*, 873–877.

Posner, M. I. (1978). *Chronometric explorations of mind*. Englewood Cliffs, New Jersey: Erlbaum.

Posner, M. I., & Petersen, S. E. (1990). The attention system of the human brain. *Annual Review of Neuroscience, 13*, 25–42.

Posner, M. I., Petersen, S. E., Fox, P. T., & Raichle, M. E. (1988). Localization of cognitive operations in the human brain. *Science, 240*, 1627–1631.

Posner, M. I., Sandson, J., Dhawan, M., & Shulman, G. L. (1989). Is word cognitionautomatic? A cognitive-anatomical approach. *Journal of Cognitive Neuroscience, 1*, 50–60.

Posner, M. I., Walker, J. A., Friedrich, F. J., & Rafal, R. D. (1984). Effects of parietal injury on covert orienting of visual attention. *Journal of Neuroscience, 4*, 1863–1874.

Rafal, R. D., & Posner, M. I. (1987). Deficits in human visual spatial attention following thalamic lesions. *Proceedings of the National Academy of Sciences of the United States of America, 84*, 7349–7353.

Rafal, R. D., Posner, M. I., Friedman, J. H., Inhoff, A. W., & Bernstein, E. (1988). Orienting of visual attention in progressive supranuclear palsy. *Brain, 111*, 267–280.

Raichle, M. E. (1990). Exploring the mind with dynamic imaging. *Seminars in Neuroscience, 2*, 307–314.

Raichle, M. E. (1991). Memory mechanisms in the processing of words and word-like symbols. In R. Collins (Ed.), *Exploring brain functional anatomy with positron tomography* (pp. 198–217). New York: John Wiley & Sons.

Raichle, M. E., Fiez, J., Videen, T. O., Fox, P. T., Pardo, J. V., & Petersen, S. E. (1991). Practice related changes in human brain functional anatomy. *Society for Neuroscience Abstracts, 17*, 21.

Raichle, M. E., Martin, W. R. W., Herscovitch, P., Mintun, M. A., & Markham, J. (1983). Brain blood flow measured with intravenous H2 150. II. Implementation and validation. *Journal of Nuclear Medicine, 24*, 790–798.

Raife, E. A., Fiez, J. A., Raichle, M. E., Balota, D. A., & Petersen, S. E. (1992). A PET study of verbal working memory. *Society for Neuroscience Abstracts, 18*, 932.

Roland, P. E. (1993). Partition of the human cerebellum in sensory-motor activities, learning and cognition. *Canadian Journal of Neurological Science, 20*, S75–S77.

Schwartz, A. S., Marchok, P. L., Kreinick, C. J., & Flynn, R. E. (1979). The asymmetric lateralization of tactile extinction in patients with unilateral cerebral dysfunction. *Brain, 102*, 669–684.

Sergent, J., Zuck, E., Terriah, S., & MacDonald, B. (1992). Distribution neural network underlying musical sight-reading and keyboard. *Science, 257*, 61–63.

Sokoloff, L. (1984). Metabolic probes of central nervous system activity in experimental animals and man. In *Magnes lecture series 1*. (pp. 1–97). Sunderland, Massachusetts: Sinauer Association.

Squire, L. R., Ojemann, J. G., Miezen, F. M., Petersen, S. E., Videen, T. O., & Raichle, M. E. (1992). Activation of the hippocampus in normal humans: A functional anatomical study of memory. *Proceedings of the National Academy of Sciences of the United States of America, 89*, 1837–1841.

Steinmetz, H., & Seitz, R. J. (1991). Functional anatomy of language processing neuroimaging and the problem of individual variability. *Neuropsychologia, 29*, 1149–1161.

Tailarach, J., & Tournoux, P. (1988). *Co-planar stereotaxic atlas of the human brain: 3-Dimensional proportional system; An approach to cerebral imaging*. Stuttgart: George Thieme Verlag.

Talbot, J. D., Marrett, S., Evans, M. C., Meyer, E., Bushnell, M. C., & Duncan, G. H., (1991). Multiple representations of pain in human cerebral cortex. *Science, 251*, 1355–1358.

Vanier, M., & Caplan, D. (1985). CT scan correlates of surface dyslexia. In K. E. Patterson, J. C. Marshall, & M. Coltheart (Eds.), *Surface dyslexia: Neuropsychological and cognitive studies of phonological reading*. (pp. 511–525). London: Lawrence Erlbaum.

Warrington, E. K., & Shallice, T., (1984). Category specific semantic impairments. *Brain, 107*, 829–854.

Waters, G. S., Rochon, E., & Caplan, D. (1992). The role of high level speech planning in rehearsal: Evidence from patients with apraxia of speech. *Journal of Memory and Language, 31*, p. 54–73.

Wise, R., Chollet, F., Hadar, U., Friston, K., Hoffner, E., & Frackowiak, R. (1991). Distribution of cortical neural networks involved in word comprehension and word retrieval. *Brain, 114*, 1803–1817.

Zatorre, R. J., Evans, A. C., Meyer, E., & Gjedde, A. (1992a). Lateralization of phonetic and pitch discrimination in speech processing. *Science, 256*, 846–849.

Zatorre, R. J., Jones-Gotman, M., Evans, A. C., & Meyer, E. (1992b). Functional localization and lateralization of human olfactory cortex. *Nature (London), 360*, 339–340.

Zeki, S., Lueck, C. J., & Watson, J. (1991). A direct demonstration of functional specialisation in human visual cortex. *Journal of Neuroscience, 11*, 641–649.

Chapter 7

Human Brain Mapping with Functional Magnetic Resonance Imaging

Jeffrey R. Binder and Stephen M. Rao

I. INTRODUCTION

Functional magnetic resonance imaging (FMRI) is a new noninvasive tool for mapping human brain functions (Shulman, Blamire, Rothman, & McCarthy, 1993). Over the past few years, numerous published studies (Bandettini, Wong, Hinks, Tikofsky, & Hyde, 1992; Belliveau et al., 1991, 1992; Cao, Towle, Levin, & Balter, 1993; Frahm, Bruhn, Merboldt, & Hanicke, 1992a; Frahm, Merboldt, & Hanicke, 1993; Kim et al., 1993; Kwong et al., 1992; Menon et al., 1992; Ogawa et al., 1992) have shown that FMRI can detect regional signal intensity changes within human primary visual and motor cortex in response to simple task activations (viewing checkerboard patterns, repetitive finger tapping). Other studies conducted at our medical center (Binder et al., 1993c; Rao et al., 1992, 1993) and elsewhere (Le Bihan, Turner, Jezzard, Cuenod, & Zeffiro, 1992; McCarthy, Blamire, Rothman, Gruetter, & Shulman, 1993) have demonstrated application of this technique to mapping brain structures involved in complex mental operations. This chapter provides an overview of the technological, physiological, and analytical principles associated with this new imaging technique and summarizes the results of recently completed FMRI experiments conducted in our laboratory.

A. Physiological Principles of FMRI

FMRI is capable of measuring endogenous brain state changes without the use of injected contrast agents. The physiological mechanisms

Localization and Neuroimaging
in Neuropsychology

185

currently thought to underlie this endogenous signal change have been described in detail elsewhere (Bandettini et al., 1992; Brooks & Di Chiro, 1987). Studies have shown that deoxyhemoglobin is paramagnetic relative to oxyhemoglobin and surrounding brain tissue (Ogawa, Lee, Nayak, & Glynn, 1990; Thulborn, Waterton, Matthews, & Radda, 1982). Cerebral blood flow studies have suggested that a local increase in oxygen delivery beyond metabolic demand occurs in active cerebral tissue (Fox & Raichle, 1986; Frostig, Lieke, Tso, & Grinvald, 1990), resulting in a higher concentration of oxygenated blood and a decrease in deoxyhemoglobin concentration within the microvasculature of metabolically active brain areas. This decrement in paramagnetic substance results in a local increase in magnetic field homogeneity, less dephasing of spins, and an increased MR signal relative to the resting state (Ogawa & Lee, 1991; Ogawa et al., 1990). Physiological contributors to the increase in local oxygenation that underlies these changes may include increases in regional cerebral blood flow (rCBF), regional cerebral blood volume (rCBV), or arterial oxygenation.

B. FMRI Techniques

FMRI studies are based on fast image acquisition techniques, the two most common of which are echo-planar imaging (EPI) and fast low angle shot (FLASH). Both techniques measure endogenous changes in the vascular concentration of deoxyhemoglobin by following the MR signal over time using a series of rapidly repeated images. Single-shot gradient-echo EPI requires the application of only one radiofrequency (RF) pulse followed by multiple reversals of the frequency-encoding gradient to encode all data in a given brain slice. The total acquisition time can be as brief as 40 msec to obtain a 64×64 voxel image. EPI pulse sequences have demonstrated the highest signal-to-noise ratios (SNR) in FMRI experiments of functional brain activity (Shulman et al., 1993). The disadvantages of EPI are (1) the relatively low spatial resolution, which is limited by the number of phase-encoding steps and by the speed of gradient reversals; (2) a tendency for the images to distort (although several techniques have been developed to correct this problem; Jesmanowicz, Wong, DeYoe, & Hyde, 1992); (3) the considerable effort required to maintain a homogeneous magnetic field within the imaging space (referred to as "shimming"); and (4) a "drop-out" of signal in brain regions (e.g., olfactory bulbs and inferior temporal lobe) near sinuses and air passages caused by susceptibility differences between air and brain tissue. On the other hand, FLASH pulse sequences offer trade offs between temporal and spatial resolution. Spatial resolutions of 64×128 and 128×256 can be achieved; however, this information is acquired over 3- and 6-sec intervals, respectively (Frahm,

Gyngell, & Hanicke, 1992b). In addition, the SNR of FLASH functional experiments appears to be somewhat smaller than that of EPI studies.

FMRI requires additional apparatus to optimize SNR. Although some researchers have been able to extract functionally related signal changes using the gradient and RF coils provided with conventional clinical scanners (Cao et al., 1993), SNR appears to be greatest in studies that use specially designed surface or head coils that are positioned closer to the brain surface. Two disadvantages to using surface coils are that (1) the MR signal is greatest in brain tissue underlying the coil and decreases with distance from the coil, thereby preventing a direct comparison of relative signal changes across a brain slice, and (2) without whole-brain imaging capacity, the investigator must select a priori the brain regions of interest. Gradient head coils, on the other hand, can provide whole-brain imaging as well as a uniform MR signal across the entire imaging plane (Bandettini et al., 1993b; Wong, Bandettini, & Hyde, 1992).

Finally, note that FMRI can be accomplished with the use of exogenous contrast agents. In the first published FMRI study, investigators at Massachusetts General Hospital (Belliveau et al., 1991) injected a diethylenetriaminepentaacetic gadolinium (Gd-DTPA) bolus into the veins of subjects during a resting state and again during visual stimulation. The CBV maps generated during the two conditions were subtracted, revealing an increase in CBV in the primary visual cortex as a result of photic stimulation. Like positron emission tomography (PET), the bolus technique can provide an absolute measurement of CBV. As in PET, imaging requires 30–40 sec of activity, and the CBV maps of activation and resting states are obtained at different times, usually separated by several minutes. The number of functional tasks that can be studied is limited by restrictions on total Gd-DTPA dose.

In this chapter, we focus our discussion on FMRI investigations of endogenous signal changes, which have received the greatest attention among investigators interested in mapping brain functions.

II. CHARACTERISTICS OF FMRI TIME SERIES DATA

FMRI techniques track signal change over time by acquiring a series of consecutive images as subjects are presented with test stimuli or perform experimental tasks. In a typical procedure, as many as several hundred images are acquired at regular intervals of 1–3 sec. Signal changes can be examined in single voxels of the image or in groups of voxels constituting a general region of interest. Examples of this signal change over time have been provided in previously published studies of motor and visual cortex using fast EPI acquisitions (Bandettini et al., 1992; Kwong et al., 1992). Figure 1 shows similar data obtained from auditory cortex during performance of a pure tone discrimination task, in which multiple activation cycles have been averaged to remove random noise (Binder et al., 1993b). As has been typical of other brain regions, signal begins to rise within the first 3 sec of activity, reaching a maximum in this subject within 6 sec. Other subjects have shown slightly longer rise times, although in almost all cases this interval has been less than 10 sec (Bandettini et al., 1992; DeYoe, Neitz, Bandettini, Wong, & Hyde, 1992; Kwong et al., 1992). Signal begins to decrease within the first 3 sec of cessation of activity, returning to baseline values within 10–12 sec. A final feature of these data is an "undershoot" after activation consisting of a modest decrement below baseline values, occurring in this subject between 12 and 24 sec after activity cessation. Although noted in previous studies (Stern, Kwong, Belliveau, Baker, & Rosen, 1992; Turner, Jezzard, Wen, Kwong, Le Bihan, & Balaban, 1992), the meaning of this undershoot is unclear. The phenomenon is more apparent with gradient-echo than with spin-echo sequences (Stern et al., 1992), suggesting a relative deficit of capillary oxygenated hemoglobin concentration after activation.

The magnitude of signal change may be expressed in terms of the simple deviation in arbitrary signal units away from baseline (Frahm et al.,

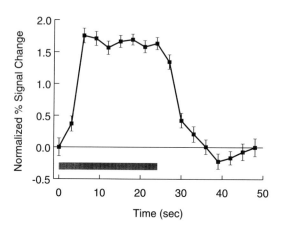

Figure 1 Time series data illustrating MR signal increase during a 24-sec activation task (shaded bar) in which the subject discriminated pure tone sequences. Data are normalized signal values from a region of interest in the left superior temporal gyrus. The interval between data points (TR) is 3 sec. Error bars indicate 95% confidence intervals.

1993), as a percentage change relative to a baseline or mean value (McCarthy et al., 1993), or as a more complex index incorporating measures of the correlation between signal response and an ideal response function (Bandettini, Jesmanowicz, Wong, & Hyde, 1993a). Regardless of the unit used, clearly response magnitudes vary considerably from voxel to voxel in the same individual during the same image series (Fig. 2), presumably

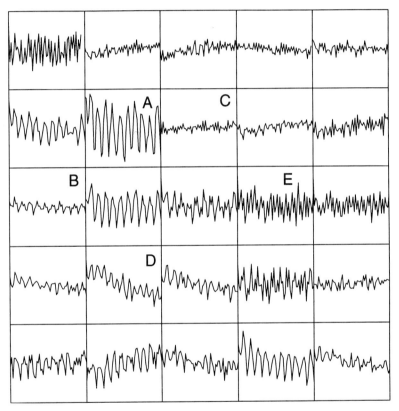

Figure 2 Temporal–spatial array of time series data from 25 contiguous voxels of the left perisylvian region during passive listening to speech sounds. Frontoparietal operculum occupies the upper right third of the array, superior temporal gyrus (slanting downward) the lower left half. (A) Signal increases of 6% coincident with 10 intermittent periods of word presentation. (B) Much smaller changes (approximately 1%) that are still readily distinguishable from background noise, such as that in C. (D) Baseline drift, which is opposite in direction from similar drift in a neighboring inferior voxel. (E) Large amplitude, irregularly oscillating changes that probably represent vessel artifact.

indicating differences in the amount of active tissue in different voxels, differences in neuronal activity in different brain regions, or combinations of these factors. The range of signal change obtained depends on a number of task and imaging acquisition variables, some of which are discussed in section III. Signal changes of 1.5–5% have been observed in motor, visual, and auditory cortex using a field strength of 1.5 Tesla (Bandettini et al., 1992; DeYoe et al., 1992; Kwong et al., 1992; Rao et al., 1992). In our experience, and using the analysis methods described subsequently, changes as small as 1% can be discriminated reliably from background noise (Fig. 2).

A useful technique for inspecting and presenting FMRI data that is employed in our laboratory is the "temporal–spatial array," in which time series plots from contiguous pixels of an image are juxtaposed (Fig. 2) (Bandettini et al., 1993a). As in individual plots, as in Fig. 1, the horizontal axis of each box in the array represents a time scale covering the duration of the image series, and the vertical axis represents the magnitude of signal change. Signal values can be represented with minima at the lower edge (Bandettini et al., 1993a) or can be centered about a mean value, as in Fig. 2. Such arrays allow detailed inspection of the activity in a selected region of interest, and emphasize the typically "patchy" distribution of signal changes that results from undulation of the cortical ribbon in and out of the slice plane.

Temporal–spatial arrays are invaluable for the detection of various signal alterations that are not caused by cerebral activity, such as those seen with subject movement, baseline drift, and the presence of pulsatile vessels. Figure 2 illustrates an example of baseline drift, which typically appears as a monotonic increase or decrease in mean signal over time. The effect, which may be due to subtle shifts in head position or artifacts arising from instability of the magnetic field, is not infrequently opposite in direction in neighboring pixels, and may affect some pixels prominently while sparing others nearby. The effects of such drift can be removed using vector mathematical techniques that orthogonalize the drift and true data functions according to the Gram–Schmidt process (Bandettini et al., 1993a). Figure 2 also illustrates the typical appearance of large vessels, characterized by large-amplitude rapid and irregular signal changes that may result from temporal "beating" of the systolic–diastolic rhythm with periodic image acquisitions. In this example, the large irregular signals are located at the dorsal surface of the superior temporal gyrus, where closely spaced proximal branches of the middle cerebral artery course laterally through the Sylvian fissure.

Temporal characteristics of the FMRI signal response must be understood to optimize experimental procedures and results. Because peak signals are reached in some individuals as late as 9 sec after task initiation, activation periods shorter than this may result in smaller response amplitudes (Bandettini et al., 1993c). This requirement, however, still allows for many repetitions of a task within a single image series that is obtained in minutes (Bandettini et al., 1993a; Rao et al., 1993). Combinations of different alternating or irregularly interleaved tasks can be incorporated in a single image series, thus controlling for the effects of test order on the activity measures obtained. Figure 3 illustrates data from two regions of the left temporal lobe in a subject who was presented alternately with white noise and speech sounds during a single image series. The two types of stimuli were presented at regularly spaced intervals; the interval between noise presentations was slightly shorter than that between speech presentations. Data from the dorsal superior temporal gyrus show two distinct series of peaks representing responses to both noise and speech sounds, whereas data obtained simultaneously from the superior temporal

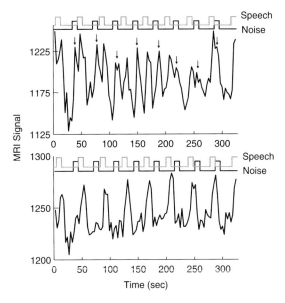

Figure 3 Time series data collected simultaneously from left primary (*top*) and association (*bottom*) auditory cortices during passive stimulation with white noise and speech sounds. Stimulation periods with each stimulus are indicated as upward square-wave deflections in the tracings above each graph. Primary cortex demonstrates signal responses of 4–5% to noise (arrows) as well as to speech. Association cortex, in contrast, shows no response to the noise stimulus.

sulcus show a single series of peaks representing responses to speech sounds only. Differences in the frequency of presentation of the two stimulus types also allows for alternative methods of data analysis using the Fourier transform (see subsequent discussion).

III. VARIABLES AFFECTING ACTIVITY-RELATED SIGNAL CHANGE

Several physical parameters defining the data acquisition may affect activity-related signal changes and may dictate features of the imaging protocol. In theory, increasing magnet strength should have the effect of increasing all dynamic aspects of the MR signal, potentially resulting in greater activity-related percentage signal change (Kim et al., 1993; Turner et al., 1993). The incremental effect on activity-related signal is expected to exceed effects on background noise, which should improve overall SNR in functional studies using higher field strengths. The impact of the larger artifacts associated with stronger fields remains to be determined in careful studies.

Size of the image voxels also may affect signal strength. Voxel dimensions are determined by the slice thickness, the in-plane resolution of the slice image, and the area of the imaged field ("field of view," FOV). For example, an 8-mm slice with 240-mm FOV and 64×64 in-plane resolution yields voxels measuring $3.75 \times 3.75 \times 8$ mm or 112.5 μl. Although overall signal magnitude is proportional to voxel volume, tissue showing activity-related signal enhancement may fill only a portion of larger voxels; the effect of this dynamic signal increment therefore may be lost by averaging with inactive tissue in the same voxel. A tentative guiding principle is planning voxel size to match the size of the "active" structures to be imaged. Because the width of cerebral cortex is typically less than 6 mm, voxels with dimensions smaller than 6 mm should be less susceptible to partial averaging with "inactive" structures such as white matter, meninges, or cerebrospinal fluid (CSF). In support of these predictions, several studies have demonstrated greater activity-related signal enhancement with slices 4 mm to 7 mm thick than with thicker slices (Baker, Cohen, Stern, Kwong, Belliveau, & Rosen, 1992; Frahm et al., 1993).

FMRI requires acquisition of images that is sufficiently fast that the dynamic event-related features of the MR signal can be followed over time. Of the several methods for rapid acquisition, EPI using either gradient-recalled or spin-echo sequences and variations of the more conventional low-angle multishot sequences (FLASH, GRASS, FISP, etc.) have been the most widely used. EPI techniques have the principal advantage of faster acquisitions and, therefore, greater inherent temporal resolution. Images can be acquired readily with this method as frequently as every 100 msec. Although decreasing the interscan interval (TR) of echo-planar images increases the temporal resolution at which signal changes can be studied, this has an adverse effect on overall signal strength because of incomplete relaxation of spins between repetitions. This effect is negligible when TR is 750 msec or longer, whereas progressively shorter TRs result in increasingly larger signal reductions. At TR of 100 msec, signal may be reduced by as much as 90%.

IV. IMAGE ANALYSIS METHODS

In this section, we discuss procedures for creating functional anatomic images from the signal changes observed in FMRI experiments. FMRI has the advantage of acquiring functional and anatomical information in the same scanning session, facilitating a direct registration of regional functional activity with standard high resolution MR images. Because FMRI does not yield an absolute measure of blood flow or volume, but a relative change in signal intensity, methods for extracting functional information are required that are objective, reproducible, and meet acceptable standards of statistical significance. The following sections describe various approaches to generating functional images from fast MRI data (see Bandettini et al., 1993a, for a more detailed discussion).

A. Subtraction Techniques

The most common technique for creating functional images from FMRI data is subtraction of images obtained during a resting (baseline) state from images of the same tissue during an activation state (Bandettini et al., 1992; Belliveau et al., 1991; Cao et al., 1993; Frahm et al., 1992a). In a typical experiment of this kind, only one or two activation periods may occur within a series of sequential images. Figure 4A illustrates a "baseline image" derived by averaging 10 echo-planar images while the subject is at rest; Fig. 4B provides an "activated image" derived by averaging 10 images during peak signal changes accompanying the presentation of speech stimuli. Although no obvious functional activity can be observed on the averaged echo-planar images, the "subtraction image" in Fig. 4C demonstrates the main focus of signal change in the superior temporal gyrus despite considerable residual noise in the image. A commonly used method for testing the statistical significance of such subtraction images is to convert the difference scores to standard normal deviates (z scores),

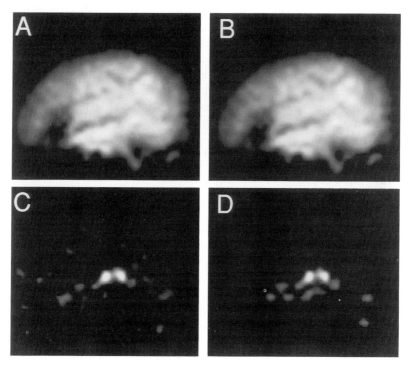

Figure 4 Echo-planar image data showing two methods of functional image formation. The condition consists of passive listening to pseudowords (e.g., "narb"). The
sagittal slice is over the left hemisphere. (A) "Baseline image" obtained by averaging
10 images taken during baseline periods. (B) "Activation image" obtained by averaging 10 images taken during peak activation periods. Signal increases of 1–5% in the
superior temporal gyrus during activation are virtually imperceptible. (C) "Subtraction image" obtained by subtracting A and B and excluding differences less than 1%.
(D) Functional image created by the cross-correlation technique.

calculated by dividing the difference scores by the standard deviation of
the resting images.

As implied earlier, we have become dissatisfied with the "noisy" functional images generated by the subtraction method. FMRI signal changes
are susceptible to gross motion, pulsatile blood and CSF flow, and pulsatile brain motion. These processes produce rather dramatic "spikes" or
"drifts" in signal, as visualized on temporal–spatial arrays such as Fig. 2.
When only one or two activation époques are used in an imaging series,
these artifactual signal changes can result in erroneous areas of signal
enhancement on functional images obtained by simple subtraction.

As an alternative to the subtraction method, we currently employ paradigms using multiple cycles (typically 5–15) of alternating baseline and activation states. This method enables us to take advantage of more advanced signal processing strategies to separate artifactual from activation-induced signal changes. These methods are described in the next two sections.

B. Fourier Analysis

Fast Fourier analysis techniques can be used to extract functional information from FMRI data. To illustrate this method, we examine an experiment conducted by Bandettini et al. (1993a) at our institution. A subject was cued to move the fingers of his right hand to an on–off switching rate of 0.08 Hz (15 12-sec cycles of alternating activation and rest periods of 6 sec each). With the left hand, the subject simultaneously moved his fingers to an on–off switching rate of 0.05 Hz (9 20-sec cycles with activation and rest periods of 10 sec each). Figure 5A presents two time course plots from the right (A) and left (B) primary motor cortex during this imaging sequence. Figure 5B demonstrates the Fourier transformations of the two data sets, plotted as spectral density against frequency. The peaks in the spectral plots (Fig. 5B) closely correspond to the separate activation frequencies used with each hand. By identifying only those

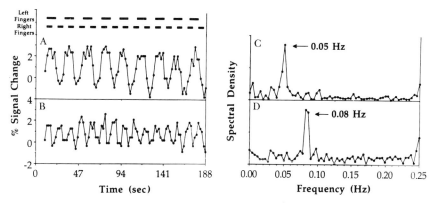

Figure 5   (*Left*) Dark bars across the top indicate the on/off frequencies for finger movements of the left (0.05 Hz) and right (0.08 Hz) hands. Below are time course plots from pixels of the right (A) and left (B) motor cortex. (*Right*) Plots of the spectral density vs. frequency from the same pixels shown in A. Peaks in C and D are observed at the frequencies corresponding to the left and right hand activation frequencies, respectively. Reproduced with permission from Bandettini, Jesmanowicz, Wong, & Hyde (1993a).

pixels responding at a particular frequency, functional images may be created from the Fourier transform data. Figure 6A demonstrates areas of signal enhancement confined to the left motor cortex by highlighting only those pixels responding at 0.08 Hz; conversely, by identifying pixels responding at 0.05 Hz, signal enhancement is observed in the right motor cortex (Fig. 6B).

Although Fourier analysis can provide a powerful technique for extracting functional information from FMRI data sets, it also has some drawbacks. This technique does not eliminate noise entirely in areas in which large magnitude signal changes create multiple peaks in the spectral density profile (note the areas of enhancement in the sagittal sinus of Fig. 6). Further, an acceptable method for testing the statistical significance of the areas of signal enhancement is not readily apparent.

C. Cross-Correlation Thresholding

Investigators at our medical center (Bandettini et al., 1993a) have developed a signal processing method for generating functional images that emphasizes the shape of the time course waveform. The first stage of this

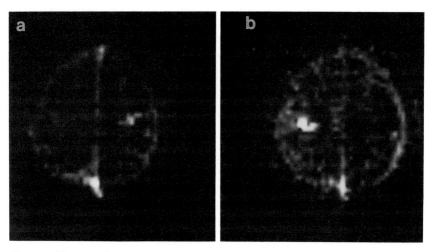

Figure 6 Functional images obtained from the Fourier transform data set in which finger movements of the left and right hands were cycled at on/off frequencies of 0.05 Hz and 0.08 Hz, respectively. The spectral density images based on either 0.08 Hz or 0.05 Hz resulted in high signal intensity changes in the left (a) and right (b) motor cortices, respectively. The subject's right is on the reader's left. Reproduced with permission from Bandettini, Jesmanowicz, Wong, & Hyde (1993a).

method involves a thresholding procedure that identifies only those pixels that display signal changes that correspond to the temporal pattern of the activation task. This identification is accomplished by correlating the normalized imaging data from each pixel with a reference waveform. Figure 7A presents a time course plot of MR signal intensity values (converted to percentage change from the mean) from the right motor cortex during repetitive movements of the left hand. The thick bars along the top of the graph indicate ten 10-sec intervals in which motor activity occurred. In this study, we have assumed that the functional activity (as shown in Fig. 7A) resembles a sinusoidal waveform (Fig. 7B). This assumption appears reasonable since the correlation between the data presented in Fig. 7A and the sinusoidal reference waveform in Fig. 7B was quite high ($r = 0.96$). Alternatively, we could have used another synthesized waveform (e.g., squarewave) or a waveform created from actual signal data. After selecting the most appropriate waveform, we establish a threshold correlational value. For the data set in Fig. 7A, the cutoff value was set to $| r | \geq 0.50$. This cutoff was selected to correspond with the Bonferroni-adjusted alpha level ($p = 2.4 \times 10^{-5}$) required when performing multiple statistical comparisons, and was dependent on the total number of pixels in the brain slice of interest and the total number of images in the series.

Because the normalized correlation coefficient contains no information concerning the magnitude of the signal change, a second process reintroduces amplitude information into those pixels surviving the threshold analysis by multiplying the correlation coefficient for each pixel by the standard deviation of the MR signal values constituting the time course plot. The derived index of activity thus depends both on the temporal relationship between signal change and activity and on the magnitude of the signal change. This value is multiplied by a constant to adjust the brightness of the pixels in the functional image (the same constant is used for all images). The final stage of creating the functional images involves the superimposition of the thresholded image on high resolution anatomical images of the same brain slice. Both functional and anatomical images typically are interpolated to 256×256 pixels. The functional images are colorized: positive values (i.e., pixels in phase with the reference waveform) are displayed on a red (minimum) to yellow (maximum) scale; negative values (i.e., pixels 180° out of phase with the reference waveform) are displayed on a blue (minimum) to cyan (maximum) scale; pixels not surviving the cutoff are made transparent.

This method has several advantages over the subtraction technique. The first is the virtual elimination of "noise" from the functional images. Figure 4D compares a functional image created using the correlation thresholding method with the image generated from the subtraction technique

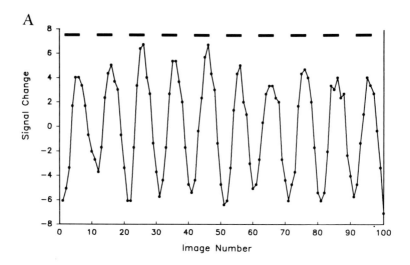

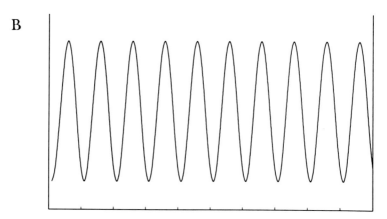

Figure 7   (A) MR signal change as a function of image number obtained from a pixel located within the right motor cortex during repetitive finger movements performed with left hand. The dark lines across the top of the graph indicate the beginning and end of motor activity. (B) Sinusoidal reference waveform matched to the phase of the plot in A. The correlation between the experimental (A) and reference (B) plots is 0.96. Reproduced with permission from Rao et al. (1993).

mentioned earlier in this chapter (Fig. 4C). The correlation image more clearly delineates areas of stimulus-related signal change. Second, the cross-correlation method can extract valuable information pertaining to the rise and fall of the signal; such information typically is eliminated when creating subtraction images. For example, by creating a set of functional images from multiple reference waveforms generated by shifting the phase in small increments of time, one can compare the temporal latency of signal change in different brain regions (Binder, Jesmanowicz, Rao, Bandettini, Hammeke, & Hyde, 1993a). Finally, visual inspection of temporal–spatial arrays suggest that functional activity can occur in pixels that also exhibit a linear drift in signal across time (Fig. 2). Such drift may diminish artificially the correlation coefficient between actual and reference waveforms using the correlation technique and may exaggerate or diminish changes observed in simple subtraction images. Our cross-correlation method allows the incorporation of a second reference waveform (typically a linear ramp function) that can serve as a covariate in the computation of a partial correlation, which in effect makes the drift "level." This computation also has been described in the terminology of vector orthogonalization (Bandettini et al., 1993a).

The success of the cross-correlation technique depends on the valid selection of reference waveform and threshold criteria. The selection of valid reference waveforms, in turn, will require the careful analysis of activation data from a variety of brain areas using a range of tasks (Binder et al., 1993b; DeYoe et al., 1992). In the next section, we apply the correlation method to data from recently completed brain mapping experiments conducted in our laboratory.

V. APPLICATIONS OF FMRI TO MAPPING HUMAN BRAIN FUNCTIONS

As noted at the beginning of this chapter, the first wave of FMRI studies focused on signal changes emanating from the primary motor and visual cortices. PET studies, however, have indicated that blood flow changes in these two regions are considerably greater than in secondary or tertiary cortex (Raichle, 1989). Thus, whether FMRI is sensitive to hemodynamic changes in the brain regions that are of greatest interest to the study of complex mental processes is unknown. Recently completed FMRI studies at our institution have addressed this issue by examining both primary and secondary brain regions activated by complex movements and language.

All our FMRI studies were conducted on a commercial 1.5-T scanner (Signa, General Electric Medical Systems, Milwaukee) equipped with prototype three-axis local gradient head and quadrature radiofrequency coils. Scanning began with the acquisition of images obtained with standard GRASS pulse sequences using the following imaging variables: 24-cm FOV, 600-msec TR, 10 msec TE, 90° flip angle, 5- to 12-mm slice thickness, and a 256×128 matrix. These standard images served as the high resolution anatomical scans on which functional data were superimposed. Functional imaging used a single-shot, blipped, gradient-echo EPI pulse sequence. Data acquisition time was 40 msec to acquire a 64×64 image (in-plane voxel dimensions: 3.75×3.75 mm).

A. FMRI of Complex Movements

During the past decade, several functional imaging studies have demonstrated increases in rCBF in the primary (Brodmann area 4) and nonprimary (Brodmann area 6) motor cortex in response to voluntary movements of varying complexity (Colebatch, Deiber, Passingham, Friston, & Frackowiak, 1991; Deiber, Passingham, Colebatch, Friston, Nixon, & Frackowiak, 1991; Orgogozo & Larsen, 1979; Roland, Larsen, Lassen, & Skinhoj, 1980a; Roland, Meyer, Shibaski, Yamamoto, & Thompson, 1982; Roland, Skinhoj, Lassen, & Larsen, 1980b; Seitz, Roland, Bohm, Greitz, & Stone-Elander, 1990). In an influential study, Roland et al. (1980a) showed that simple finger movements resulted in rCBF increases confined to the contralateral sensorimotor hand region. In contrast, performing a complicated finger sequencing task resulted in rCBF increases within the supplementary motor area (SMA) and bilateral premotor cortex of Brodmann area 6, in addition to the contralateral sensorimotor hand area. Imagining the complex finger task produced rCBF changes within the SMA, but not within primary sensorimotor cortex. On the basis of these results, Roland and co-workers proposed that the SMA is a higher order "supramotor" center involved in the generation and programming of complex movements. This view is supported by human lesion studies showing that patients with SMA lesions experience a severe decrease in spontaneous movement (although eliciting a motor act in response to command is possible) (LaPlane, Talairach, Meininger, Bancaud, & Orgogozo, 1977), bilateral ideomotor apraxia for transitive limb movements (Watson, Fleet, Rothi, & Heilman, 1986), and disturbances in performing alternating hand movements (LaPlane et al., 1977). These findings also converge with results from electrical stimulation and lesion studies in animals suggesting that cortical motor control is organized hierarchically (Goldberg, 1985; Rizzolatti, 1987; Wiesendanger, 1981; Wise, 1984, 1985).

Findings of PET studies, however, have challenged the hierarchical control hypothesis. Fox, Fox, Raichle, and Burde (1985), for example, have suggested that the results of Roland et al. (1980a) could have been due to differences in task performance rate, since the simple task in the study by Roland et al was performed at a slower rate than the complex task (1.0 vs 3.2 Hz). Thus, activation of the SMA and premotor regions may not have been detected during the simple motor task because of smaller overall increases in rCBF. In the Fox et al. study, simple finger movements at a faster rate (2.0 Hz) resulted in SMA activation (complex finger movements were not examined). In a similar study, Colebatch et al. (1991) had subjects perform simple and complex finger movements at a fixed rate of 1.5 Hz and found equivalent rCBF increases in the contralateral sensorimotor, SMA, and premotor cortical areas for both types of movement. As an alternative to the view that secondary motor cortex is involved in the programming of complex motor acts, Fox and associates proposed that the SMA and premotor cortex are involved in the initiation ("readiness to move") component of all motor tasks, regardless of their complexity. Differences in the results of these imaging studies also could be explained by dissimilarities in the motor tasks. The Colebatch et al. study, for example, used a "simpler" complex task than that did Roland and colleagues. In addition, Roland's group asked subjects to pace their own movements, whereas Fox et al. and Colebatch et al. asked subjects to pace their movements in response to a metronome. Pacing even simple movements in response to an external cue may require a higher level of programming than self-paced movements.

Our study, therefore, was designed to address the following questions. (1) Can FMRI detect activation within secondary motor cortex (SMA, premotor cortex) in response to motor tasks? (2) Are there differences in rCBF patterns of simple and complex self-paced movements? (3) Does pacing of motor functions change the pattern of cerebral activation in comparison to self-paced movements? (4) Can imagined complex movements illicit changes detectable by FMRI?

Six healthy, strongly right-handed volunteers performed self-paced simple or complex finger movements with the right or left hand. The simple movement consisted of a finger tapping task in which subjects flexed and extended all fingers (except the thumb) repeatedly in unison, as quickly as possible, without also moving the wrist. For the complex movement, subjects tapped the four fingers (excluding the thumb) in a repeating fixed sequence (e.g., a 2431 sequence required tapping the middle finger, followed by the little finger, the ring finger, and finally the index finger). Subjects repeated each sequence as quickly as possible. The sequence did not change during the activation condition; different sequences were used for each hand to minimize learning effects. Subjects tapped their fingers on a flat surface for all conditions. Each subject underwent four activation conditions: simple right hand, simple left, complex right, and complex left. The order of the activation conditions was counterbalanced across subjects. Two subjects underwent additional activation tasks to assess the effects of equating the rates of simple and complex movements; they were asked to pace their finger movements in response to a metronome set at 2 Hz. Finally, two subjects completed additional activation conditions involving imagined complex movements. For these conditions, subjects were provided four digit sequences, as in the complex condition, and were asked to imagine performing the complex motor task with either their right or their left hand. Explicit instructions were given to suppress all motor activity.

The results of this study can be summarized as follows. During the simple motor task, functional activity was observed primarily in the contralateral primary motor cortex (see Color Plate 4). In contrast, the complex motor task was associated with activity not only in the contralateral primary motor cortex but also in the SMA, premotor cortex bilaterally, and the somatosensory cortex. A small degree of activation also was observed in the ipsilateral primary motor cortex of both subjects during the complex movement conditions. The more widespread activation observed during performance of the complex motor task is particularly striking, since the complex task was performed at a slower rate than the simple task (mean of 2.2 Hz vs 2.9 Hz, respectively). This general finding, that is, more widespread activation during complex than during simple movements, was similar across all six subjects. In contrast, the paced simple condition resulted in no functional activity in either brain slice, including the contralateral primary motor cortex (see Color Plate 5A). Activity during the paced complex condition generally was reduced relative to the self-paced complex condition, with no functional activity observed in the somatosensory cortex and less in premotor and motor areas. Note that the performance rates for the self-paced and paced complex motor tasks were similar (self-paced, 2.2 Hz; paced, 2.0 Hz), whereas self-paced and paced simple motor tasks differed by almost 1 Hz (mean of 2.9 vs 2.0 Hz). During the imagined complex movement task (see Color Plate 5B), the largest signal changes were observed within the SMA. Less intense and less consistent activations were observed in the premotor cortex. No activation was observed in the primary motor or somatosensory cortex.

This preliminary study indicates that (1) FMRI is capable of detecting regional changes related to brain activity within nonprimary association motor cortex, (2) the overall findings are consistent with a hierarchical model of voluntary motor control, in which the SMA and premotor cortex participate to a greater extent during complex sequential motor acts, and (3) the signal changes detected in the SMA and premotor cortex during motor imagery suggest that FMRI is sensitive to "higher order" mental processing in the absence of motor activation.

B. FMRI of Language Cortex

Since our initial report describing superior temporal gyrus activity during passive word presentation (Rao et al., 1992), our laboratory has initiated a number of studies of auditory language processing. Passive sound presentation is of theoretical interest, since it presumably emphasizes stimulus-dependent "automatic" processing of auditory stimuli and minimizes speech, motor, attentional, and memory activity. Stimuli used to date have included pure tones, broad band white noise, consonant–vowel syllables and other "nonsense" speech sounds, English nouns, and narrative text. Pure tones and white noise differ inherently from speech because they lack many important acoustic cues that underlie speech perception. Chief among these is the absence of rapid frequency modulations that define component phonemic sounds in speech (Liberman, 1967). The existence of these complex and rapidly changing acoustic elements in speech suggests that some cortical areas, presumably in or near the classical auditory association areas, may be primarily responsible for their analysis and identification (Schwartz & Tallal, 1980; Seldon, 1985). In one study conducted in our laboratory to test this hypothesis, white noise presentation was compared with presentation of "pseudowords" (e.g., narb) and nouns, with stimuli matched for duration, presentation rate, average sound pressure level, and spectral range (Binder et al., 1994).

Color Plate 6 illustrates typical results from two subjects studied with white noise and speech sounds. In these and all other subjects studied, the area activated by white noise was considerably smaller than that activated by speech sounds, and was restricted to the dorsal aspect of the superior temporal gyrus. In most instances, this region coincided with or included the transverse temporal (Heschl's) gyrus, a structure containing primary auditory koniocortex (Galaburda & Sanides, 1980). Presentation of speech sounds, in contrast, activated a larger region including more anterior and posterior areas of the dorsal superior temporal gyrus, as well as cortex in or near the superior temporal sulcus bilaterally. Activity occurred fairly symmetrically in left and right temporal lobes. Unlike processing of white noise, therefore, processing of speech sounds appears to elicit participation of extensive auditory association areas even when the subject is not engaged in any "active" task. Although activation by speech was more extensive than activation produced by noise, activated areas did not differ when comparing words and pseudowords, a finding that emphasizes the presemantic "sensory/perceptual" nature of the passive presentation paradigm.

More numerous or more widespread areas may become active when subjects engage in cognitive tasks pertaining to auditory input. Whereas activity during passive presentation of pure tones, for example, typically is restricted to a small region of the dorsal superior temporal gyrus, subjects show considerably more activity when actively engaged in a complex pure tone discrimination task (Color Plate 7). In this paradigm, subjects heard clusters of 3–7 sequential pure tones of 500 or 750 Hz frequency. A response, consisting of briefly lifting the left index finger, was required if tone sequences contained two occurrences of the 750-Hz tone (Binder et al., 1993b). This task, therefore, requires not only low-level sensory perception of tone frequencies, but also various attentional and linguistic functions such as working memory for a set of instructions, application of a verbal criterion to a sensory experience, monitoring of the number of positive targets over a variable time époque, and criteria-driven performance of a motor response. Activated regions shown in this plate include not only auditory sensory cortices of the superior temporal gyrus, but also bilateral lateral frontal cortices and the right supramarginal gyrus.

We have used the pure tone task just described as one type of "control task" for other studies of cognitive auditory processing. In such designs, the control task is performed during intervals that would otherwise be resting or "baseline" periods. This technique results, in effect, in alternating between the control task and another task of interest, referred to as the "probe" task. Any signal increase observed during the probe task, therefore, represents activity exceeding that which occurs in the control task. One such study involved monitoring lists of auditory animal words (e.g., squirrel) for occurrences of animals possessing specified verbal semantic features (Binder et al., 1993). In one such task, subjects were required to respond when they considered the named animal to be both (1) found commonly in the United States and (2) used commonly by people as pets, food, or clothing. This "semantic monitoring" task alternated regularly with the pure tone control task through 8 complete cycles. Color Plate 8 illustrates activity during this paradigm in a representative subject, whose responses to the pure tone control task alone were provided in Color Plate 7. In Color Plate 8, the red–yellow scale indicates relative signal increases occurring during the probe task (semantic monitoring), whereas the blue–cyan scale indicates relative increases during the control task (pure tone discrimination). Several features of these data deserve brief comment.

First, a comparison of the lateral (top row) slices of Color Plate 8 with Color Plate 7 demonstrates the "subtraction effect" of using a control task to generate Color Plate 8. The strongly responding areas in superior temporal gyri in Color Plate 7 are almost entirely absent from Color Plate 8, indicating an equivalent activation of these auditory sensory areas during each of the two tasks used for Color Plate 8. Second, multiple polymodal areas in lateral frontal (Brodmann areas 6, 45, and 46 near the inferior frontal sulcus), parieto-occipital (Brodmann's area 39), and temporo-occipital (Brodmann's area 19 and 37) cortex of the left hemisphere are active during the semantic task but not the control task. Third, small but significant foci in the right frontal, right supramarginal, and superior temporal cortices are relatively more active during the pure tone control task in Color Plate 8, suggesting a double dissociation between the cortical areas (and hemispheres) that are uniquely active during these contrasting tasks. These data are in general agreement with results of PET studies describing similar or related tasks (Démonet et al., 1992; Zatorre, Evans, Meyer, & Gjedde, 1992). Earlier PET research had suggested an exclusively left frontal localization for semantically based language behavior (Petersen, Fox, Posner, Mintun, & Raichle, 1988,1989), a proposal that was at odds with classical concepts of fluent aphasia (Wernicke, 1874) and with numerous modern lesion studies suggesting major semantic knowledge deficits after posterior dominant temporal or temporo-parietal lesions (Alexander, Hiltbrunner, & Fischer, 1989; Bub, Black, Hampson, & Kertesz, 1992; Coughlan & Warrington, 1978; Warrington & McCarthy, 1987; Warrington & Shallice, 1984). Our results confirm the importance of both left frontal and left posterior polymodal areas in semantic performance. One explanation for this widespread activity is that the two general regions may provide complementary functions in a distributed network that controls semantically mediated behavior. Lesion studies suggest that the left posterior components of this network constitute sites of storage for highly structured forms of semantic information (Hart & Gordon, 1990; Hillis & Caramazza, 1991; Warrington & McCarthy, 1987). The left frontal component, in contrast, may contain mechanisms that facilitate access to the posterior information store (Milberg, Blumstein, & Dworetzky, 1987; Swinney, Zurif, & Nicol, 1989). The effects of posterior lesions in this model would be to produce fixed gaps in available information, whereas frontal lesions would produce less severe disturbances characterized by global deficits of information manipulation and retrieval (Shallice, 1989). The frontal "executive" component of the system, however, might be more continuously and more diffusely active than the posterior information store and, therefore, be more readily apparent on functional imaging studies of intact brains.

McCarthy et al. (1993) have reported FMRI data obtained while subjects performed a word fluency task consisting of generating verbs associated with presented nouns (Petersen et al., 1988). Imaging was limited to a small region of the inferior frontal cortex, which showed bilateral activation during both the verb fluency task and a control task consisting of word repetition. No activation was obtained during "covert" verb generation in which subjects refrained from saying aloud the generated verb, suggesting that the signal changes observed during "overt" tasks resulted either from activity in motor response systems or from disturbances of the magnetic field induced by orofacial movements. Pilot studies with a similar task in our laboratory used longer response periods (7 sec) and permitted only silent responses (tapping the left index finger for each verb generated), potentially emphasizing the cognitive component and minimizing the oral response component of the task. As in our other language studies, full-field sagittal images were obtained simultaneously through the lateral aspect of both hemispheres. Representative data are illustrated in Color Plate 9, showing an extensive left frontal area of activation defined by signal increment during the verb generation task.

VI. FUTURE DIRECTIONS AND CHALLENGES

FMRI is an evolving brain imaging technology that promises to yield important new information regarding the functional organization of brain systems. The current state of development of FMRI is analogous to the period in the 1980s during which various methodologies were refined for acquiring and analyzing PET data. Although FMRI has the potential to become as powerful a brain mapping tool as PET, a number of technical issues will need to be addressed in coming years. Not surprisingly, considerable debate continues among MRI investigators over the optimal pulse sequences and hardware (gradient and RF coils) required to achieve functional images of the whole brain at the highest levels of spatial and temporal resolution and SNR, while minimizing anatomical distortion and susceptibility artifacts. Image analysis techniques gradually are moving away from the subtraction methods pioneered by PET researchers toward signal processing approaches that are better suited to extracting information from FMRI data. On a practical level, designing an FMRI experiment presents unique and challenging problems for the investigator. Traditional ferromagnetic instrumentation (cathode ray tubes, audio headphones, keypress devices to measure reaction time) cannot be used in the scanning environment. Several laboratories have developed creative solutions (e.g., pneumatic audio systems, LCD visual projection systems) to some of these problems. In addition, subjects must lie motionless on their back for 2–3 hr to insure exact spatial registration of images, suggesting a need for improved immobilization methods. Finally, most published FMRI investigations present representative functional images from a small subset of subjects. As investigators begin to use FMRI techniques to map complex mental processes, methods will be needed for averaging functional imaging data across subjects, as is the practice in PET studies.

Although many methodological issues confronting FMRI await resolution, the potential benefits of the technique to clinical neuroscience are becoming increasingly clear. With its high spatial resolution and noninvasive administration, FMRI promises to make relatively routine the precise *in vivo* mapping of human brain functions. In caring for persons with epilepsy or mass lesions, FMRI could provide functional information and hemisphere dominance measures before brain surgery, potentially supplanting current invasive methods such as intraoperative stimulation mapping and intracarotid amobarbital injection. The technique is well suited to longitudinal studies of individual neural reorganization after brain injury, as has been done on a limited basis using PET (Chollet, Di-Piero, Wise, Brooks, Dolan, & Frackowiak, 1991). Well-defined deviations from normal activation patterns may provide insights into the mechanisms underlying developmental, metabolic, and degenerative behavioral disorders, and may provide a means of diagnosing or monitoring the effects of therapy in such illnesses. In addition to its potential clinical uses, FMRI could contribute significantly to the field of cognitive neuroscience. In particular, the periodic activation methods described earlier appear to be sensitive to shifts in cognitive processing evoked by changes in stimulus features or in attentional processing strategies. Such considerations clearly mandate further exploration and development of this new technology.

ACKNOWLEDGMENTS

The studies reported in this chapter were supported in part by grants from the McDonnell-Pew Program in Cognitive Neuroscience and the National Institutes of Health (CA41464). The support of our colleagues at the Medical College of Wisconsin—Peter A. Bandettini, Edgar A. DeYoe, Thomas A. Hammeke, Michael D. Goldstein, Victor M. Haughton, James S. Hyde, Andrej Jesmanowicz, George L. Morris, Wade M. Mueller, Eric C. Wong, F. Zerrin Yetkin, and Jeffrey R. Zigun—is gratefully appreciated.

REFERENCES

Alexander, M. P., Hiltbrunner, B., & Fischer, R. S. (1989). Distributed anatomy of transcortical sensory aphasia. *Archives of Neurology, 46,* 885–892.

Baker, J. R., Cohen, M. S., Stern, C. E., Kwong, K. K., Belliveau, J. W., & Rosen, B. R. (1992). The effect of slice thickness and echo time on the detection of signal change during echo-planar functional neuroimaging. *Book of Abstracts, 11th Annual Meeting, Society for Magnetic Resonance in Medicine*, Berlin, 1822.

Bandettini, P. A., Jesmanowicz, A., Wong, E. C., & Hyde, J. S. (1993a). Processing strategies for time-course data sets in functional MRI of the human brain. *Magnetic Resonance in Medicine, 30*, 161–173.

Bandettini, P. A., Rao, S. M., Binder, J. R., Hammeke, T. A., Jesmanowicz, A., Yetkin, F. Z., Bates, S., Estkowski, L. D., Wong, E. C., Haughton, V. M., Hinks, R. S., & Hyde, J. S. (1993b). Magnetic resonance functional neuroimaging of the entire brain during performance and mental rehearsal of complex finger movement tasks. *Book of Abstracts, 12th Annual Meeting, Society for Magnetic Resonance in Medicine*, New York, 1396.

Bandettini, P. A., Wong, E. C., DeYoe, E. A., Binder, J. R., Rao, S. M., Birzer, D., Estkowski, L. D., Jesmanowicz, A., Hinks, R. S., & Hyde, J. S. (1993c). The functional dynamics of blood oxygen level dependent contrast in the motor cortex. *Book of Abstracts, 12th Annual Meeting, Society for Magnetic Resonance in Medicine*, New York, 1382.

Bandettini, P. A., Wong, E. C., Hinks, R. S., Tikofsky, R. S., & Hyde, J. S. (1992). Time course EPI of human brain function during task activation. *Magnetic Resonance in Medicine, 25*, 390–397.

Belliveau, J. W., Kennedy, D. N., McKinstry, R. C., Buchbinder, B. R., Weisskoff, R. M., Cohen, M. S., Vevea, J. M., Brady, T. J., & Rosen, B. R. (1991). Functional mapping of the human visual cortex by magnetic resonance imaging. *Science, 254*, 716–719.

Belliveau, J. W., Kwong, K. K., Kennedy, D. N., Baker, J. R., Stern, C. E., Benson, R., Chesler, D. A., Weisskoff, R. M., Cohen, M. S., Tootell, R. B. H., Fox, P. T., Brady, T. J., & Rosen, B. R. (1992). Magnetic resonance imaging mapping of brain function: Human visual cortex. *Investigative Radiology, 27 (Suppl. 2)*, S59–S65.

Binder, J. R., Jesmanowicz, A., Rao, S. M., Bandettini, P. A., Hammeke, T. A., & Hyde, J. S. (1993a). Analysis of phase differences in periodic functional MRI activation data. *Book of Abstracts, 12th Annual Meeting, Society for Magnetic Resonance in Medicine*, New York, 1383.

Binder, J. R., Rao, S. M., Hammeke, T. A., Bandettini, P. A., Jesmanowicz, A., Frost, J. A., Wong, E. C., Haughton, V. M., & Hyde, J. S. (1993b). Temporal characteristics of functional magnetic resonance signal change in lateral frontal and auditory cortex. *Book of Abstracts, 12th Annual Meeting, Society for Magnetic Resonance in Medicine*, New York, 5.

Binder, J. R., Rao, S. M., Hammeke, T. A., Yetkin, F. Z., Wong, E. C., Mueller, W. M., Morris, G. L., & Hyde, J. S. (1993c). Functional magnetic resonance imaging (FMRI) of auditory semantic processing. *Neurology (Suppl. 2), 43*, 189.

Binder, J. R., Rao, S. M., Hammeke, T. A., Yetkin, F. Z., Jesmanowicz, A., Bandettini, P. A., Wong, E. C., Estkowski, L. D., Goldstein, M. D., Haughton, V. M., & Hyde, J. S. (1994). Functional magnetic resonance imaging of human auditory cortex. *Annals of Neurology* (in press).

Brooks, R. A., & Di Chiro, G. (1987). Magnetic resonance imaging of stationary blood: A review. *Medical Physics, 14*, 903–913.

Bub, D. N., Black, S., Hampson, E., & Kertesz, A. (1992). Semantic encoding of picture and words: Some neuropsychological observations. *Cognitive Neuropsychology, 5*, 27–66.

Cao, Y., Towle, V. L., Levin, D. N., & Balter, J. M. (1993). Functional mapping of human motor cortical activation by conventional MRI at 1.5T. *Journal of Magnetic Resonance Imaging*.

Chollet, F., DiPiero, V., Wise, R. J., Brooks, D. J., Dolan, R. J., & Frackowiak, R. S. (1991). The functional anatomy of motor recovery after stroke in humans: A study with positron emission tomography. *Annals of Neurology, 29*, 63–71.

Colebatch, J. G., Deiber, M. P., Passingham, R. E., Friston, K. J., & Frackowiak, R. S. (1991). Regional cerebral blood flow during voluntary arm and hand movements in human subjects. *Journal of Neurophysiology, 65*, 1392–1401.

Coughlan, A. K., & Warrington, E. K. (1978). Word-comprehension and word-retrieval in patients with localized cerebral lesions. *Brain, 101*, 163–185.

Deiber, M. P., Passingham, R. E., Colebatch, J. G., Friston, K. J., Nixon, P. D., & Frackowiak, R. S. (1991). Cortical areas and the selection of movement: A study with positron emission tomography. *Experimental Brain Research, 84*, 393–402.

Démonet, J.-F., Chollet, F., Ramsay, S., Cardebat, D., Nespoulous, J.-L., Wise, R., Rascol, A., & Frackowiak, R. (1992). The anatomy of phonological and semantic processing in normal subjects. *Brain, 115*, 1753–1768.

DeYoe, E. A., Neitz, J., Bandettini, P. A., Wong, E. C., & Hyde, J. S. (1992). Time course of event-related MR signal enhancement in visual and motor cortex. *Book of Abstracts, 11th Annual Meeting, Society for Magnetic Resonance in Medicine*, Berlin, 1824.

Fox, P. T., Fox, J. M., Raichle, M. E., & Burde, R. M. (1985). The role of cerebral cortex in the generation of voluntary saccades: a positron emission tomographic study. *Journal of Neurophysiology, 54*, 348–369.

Fox, P. T., & Raichle, M. E. (1986). Focal physiological uncoupling of cerebral blood flow and oxidative metabolism during somatosensory stimulation in human subjects. *Proceedings of the National Academy of Sciences of the United States of America, 83*, 1140–1144.

Frahm, J., Bruhn, H., Merboldt, K. D., & Hanicke, W. (1992a). Dynamic MR imaging of human brain oxygenation during rest and photic stimulation. *Journal of Magnetic Resonance Imaging, 2*, 501–505.

Frahm, J., Gyngell, M. L., & Hanicke, W. (1992b). Rapid scan techniques. In D. D. Stark & W. G. Bradley (Eds.), *Magnetic resonance imaging* (pp. 165–204). St. Louis: Mosby.

Frahm, J., Merboldt, K.-D., & Hanicke, W. (1993). Functional MRI of human brain activation at high spatial resolution. *Magnetic Resonance in Medicine, 29*, 139–144.

Frostig, R. D., Lieke, E. E., Tso, D. Y., & Grinvald, A. (1990). Cortical functional architecture and local coupling between neuronal activity and the microcirculation revealed by in vivo high-resolution optical imaging of intrinsic signals. *Proceedings of the National Academy of Sciences of the United States of America, 87*, 6082–6086.

Galaburda, A., & Sanides, F. (1980). Cytoarchitectonic organization of the human auditory cortex. *Journal of Comparative Neurology, 190*, 597–610.

Goldberg, G. (1985). Supplementary motor area structure and function: Review and hypotheses. *The Behavioral and Brain Sciences, 8*, 567–588.

Hart, J., & Gordon, B. (1990). Delineation of single-word semantic comprehension deficits in aphasia, with anatomic correlation. *Annals of Neurology, 27*, 226–231.

Hillis, A. E., & Caramazza, A. (1991). Category-specific naming and comprehension impairment. *Brain, 114*, 2081–2094.

Jesmanowicz, A., Wong, E. C., DeYoe, E. A., & Hyde, J. S. (1992). Method to correct anatomic distortion in echo planar images. *Book of Abstracts, 11th Annual Meeting, Society for Magnetic Resonance in Medicine*, Berlin, 4260.

Kim, S.-G., Ashe, J., Georgopoulos, A. P., Merkle, H., Ellermann, J. M., Menon, R. S., Ogawa, S., & Ugurbil, K. (1993). Functional imaging of human motor cortex at high magnetic field. *Journal of Neurophysiology, 69*, 297–302.

Kwong, K. K., Belliveau, J. W., Chesler, D. A., Goldberg, I. E., Weisskoff, R. M., Poncelet, B. P., Kennedy, D. N., Hoppel, B. E., Cohen, M. S., Turner, R., Cheng, H., Brady, T. J., &

Rosen, B. R. (1992). Dynamic magnetic resonance imaging of human brain activity during primary sensory stimulation. *Proceedings of the National Academy of Sciences of the United States of America, 89,* 5675–5679.

LaPlane, D., Talairach, J., Meininger, V., Bancaud, J., & Orgogozo, J. M. (1977). Clinical consequences of corticectomies involving the supplementary motor area in man. *Journal of the Neurological Sciences, 34,* 310–314.

Le Bihan, D., Turner, R., Jezzard, P., Cuenod, C. A., & Zeffiro, T. (1992). Activation of human visual cortex by mental representation of visual patterns. *Book of Abstracts, 11th Annual Meeting, Society for Magnetic Resonance in Medicine, Berlin,* 311.

Liberman, A. M. (1967). Perception of the speech code. *Psychological Review, 74,* 431–461.

McCarthy, G., Blamire, A. M., Rothman, D. L., Gruetter, R., & Shulman, R. G. (1993). Echo-planar magnetic resonance imaging studies of frontal cortex activation during word generation in humans. *Proceedings of the National Academy of Sciences of the United States of America, 90,* 4952–4956.

Menon, R. S., Ogawa, S., Kim, S.-G., Ellermann, J. M., Merkle, H., Tank, D. W., & Ugurbil, K. (1992). Functional brain mapping using magnetic resonance imaging: Signal changes accompanying visual stimulation. *Investigative Radiology, 27 (Suppl. 2),* S47–S53.

Milberg, W., Blumstein, S. E., & Dworetzky, B. (1987). Processing of lexical ambiguities in aphasia. *Brain and Language, 31,* 138–150.

Ogawa, S., & Lee, T.-M. (1991). Magnetic resonance imaging of blood vessels at high fields: *In vivo* and *in vitro* measurements and image stimulation. *Magnetic Resonance in Medicine, 16,* 9–18.

Ogawa, S., Lee, T.-M., Nayak, A. S., & Glynn, P. (1990). Oxygenation-sensitive contrast in magnetic resonance image of rodent brain at high magnetic fields. *Magnetic Resonance in Medicine, 14,* 68–78.

Ogawa, S., Tank, D. W., Menon, R., Ellermann, J. M., Kim, S., Merkle, H., & Ugurbil, K. (1992). Intrinsic signal changes accompanying sensory stimulation: Functional brain mapping using MRI. *Proceedings of the National Academy of Sciences of the United States of America, 89,* 5951–5955.

Orgogozo, J. M., & Larsen, B. (1979). Activation of the supplementary motor area during voluntary movement in man suggests it works as a supramotor area. *Science, 206,* 847–850.

Petersen, S. E., Fox, P. T., Posner, M. I., Mintun, M., & Raichle, M. E. (1988). Positron emission tomographic studies of the cortical anatomy of single-word processing. *Nature (London), 331,* 585–589.

Petersen, S. E., Fox, P. T., Posner, M. I., Mintun, M., & Raichle, M. E. (1989). Positron emission tomographic studies of the processing of single-words. *Journal of Cognitive Neuroscience, 1,* 153–170.

Raichle, M. E. (1989). Developing a functional anatomy of the human brain with positron emission tomography. *Current Neurology, 9,* 161–178.

Rao, S. M., Bandettini, P. A., Wong, E. C., Yetkin, F. Z., Hammeke, T. A., Mueller, W. M., Goldman, R. S., Morris, G. L., Antuono, P. G., Estkowski, L. D., Haughton, V. M., & Hyde, J. S. (1992). Gradient-echo EPI demonstrates bilateral superior temporal gyrus activation during passive word presentation. *Book of Abstracts, 11th Annual Meeting, Society for Magnetic Resonance in Medicine, Berlin,* 1827.

Rao, S. M., Binder, J. R., Bandettini, P. A., Hammeke, T. A., Yetkin, F. Z., Jesmanowicz, A., Lisk, L., Morris, G. L., Mueller, W. M., Estkowski, L. D., Wong, E. C., Haughton, V. M., & Hyde, J. S. (1993). Functional magnetic resonance imaging of complex human movements. *Neurology.*

Rizzolatti, G. (1987). Functional organization of inferior area 6. In G. Bock, M. O'Connor, & J. Marsh (Eds.), *Motor areas of the cerebral cortex* (pp. 171–186). Chichester: Wiley.

Roland, P. E., Larsen, B., Lassen, N. A., & Skinhoj, E. (1980a). Supplementary motor area and other cortical areas in organization of voluntary movements in man. *Journal of Neurophysiology, 43,* 118–136.

Roland, P. E., Meyer, E., Shibaski, T., Yamamoto, Y. L., & Thompson, C. J. (1982). Regional cerebral blood flow changes in cortex and basal ganglia during voluntary movements in normal human volunteers. *Journal of Neurophysiology, 48,* 467–478.

Roland, P. E., Skinhoj, E., Lassen, N. A., & Larsen, B. (1980b). Different cortical areas in man in organization of voluntary movements in extrapersonal space. *Journal of Neurophysiology, 43,* 137–150.

Schwartz, J., & Tallal, P. (1980). Rate of acoustic change may underlie hemispheric specialization for speech perception. *Science, 207,* 1380–1381.

Seitz, R. J., Roland, E., Bohm, C., Greitz, T., & Stone-Elander, S. (1990). Motor learning in man: A positron emission tomographic study. *Neuroreport, 1,* 57–60.

Seldon, H. L. (1985). The anatomy of speech perception: Human auditory cortex. In A. Peters and E. G. Jones (Eds.), *Cerebral cortex. Vol. 4. Association and auditory cortices* (pp. 273–327). New York: Plenum.

Shallice, T. (1989). *From neuropsychology to mental structure.* Cambridge: Cambridge University Press.

Shulman, R. G., Blamire, A. M., Rothman, D. L., & McCarthy, G. (1993). Nuclear magnetic resonance imaging and spectroscopy of human brain function. *Proceedings of the National Academy of Sciences of the United States of America, 90,* 3127–3133.

Stern, C. E., Kwong, K. K., Belliveau, J. W., Baker, J. R., & Rosen, B. R. (1992). MR tracking of physiological mechanisms underlying brain activity. *Book of Abstracts, 11th Annual Meeting, Society for Magnetic Resonance in Medicine,* Berlin, 1821.

Swinney, D., Zurif, E., & Nicol, J. (1989). The effects of focal brain damage on sentence processing: An examination of the neurological organization of a mental module. *Journal of Cognitive Neuroscience, 1,* 25–37.

Thulborn, K. R., Waterton, J. C., Matthews, P. M., & Radda, G. K. (1982). Oxygenation dependence of the transverse relaxation time of water protons in whole blood at high field. *Biochimica et Biophysica Acta, 714,* 265–270.

Turner, R., Jezzard, P., Wen, H., Kwong, K., Le Bihan, D., & Balaban, R. (1992). Functional mapping of the human cortex at 4 tesla using deoxygenation contrast EPI. *Book of Abstracts, 11th Annual Meeting, Society for Magnetic Resonance in Medicine,* Berlin, 304.

Turner, R., Jezzard, P., Wen, H., Kwong, K. K., Le Bihan, D., Zeffiro, T., & Balaban, R. S. (1993). Functional mapping of the human visual cortex at 4 Tesla and 1.5 Tesla using deoxygenation contrast EPI. *Magnetic Resonance in Medicine, 29,* 277–279.

Warrington, E. K., & McCarthy, R. A. (1987). Categories of knowledge: Further fractionations and an attempted integration. *Brain, 110,* 1273–1296.

Warrington, E. K., & Shallice, T. (1984). Category specific semantic impairments. *Brain, 107,* 829–854.

Watson, R. T., Fleet, W. S., Rothi, L. G., & Heilman, K. M. (1986). Apraxia and the supplementary motor area. *Archives of Neurology, 43,* 787–792.

Wernicke, C. (1874). *Der aphasische Symptomenkomplex.* Breslau: Cohn & Weigert.

Wiesendanger, M. (1981). Organization of secondary motor areas of cerebral cortex. In J. M. Brookhart, V. B. Mountcastle, V. B. Brooks, & S. R. Geiger (Eds.), *Handbook of physiology. Vol. II. Motor control. Part 2* (pp. 1121–1147). Bethesda, Maryland: American Physiological Society.

Wise, S. P. (1984). The nonprimary motor cortex and its role in the cerebral control of move-ment. In G. M. Edelman, W. E. Gall, & W. M. Cowan (Eds.), *Dynamic aspects of neocortical function* (pp. 525–556). New York: John Wiley and Sons.

Wise, S. P. (1985). The primate premotor cortex: Past, present, and preparatory. *Annual Review of Neuroscience, 8,* 1–19.

Wong, E. C., Bandettini, P. A., & Hyde, J. S. (1992). Echo-planar imaging of the human brain using a three axis local gradient coil. *Book of Abstracts, 11th Annual Meeting, Society for Magnetic Resonance in Medicine,* Berlin, 105.

Zatorre, R. J., Evans, A. C., Meyer, E., & Gjedde, A. (1992). Lateralization of phonetic and pitch discrimination in speech processing. *Science, 256,* 846–849.

Chapter 8

Anatomical Asymmetries and Cerebral Lateralization

Andrew Kertesz and Margaret A. Naeser

I. EVOLUTION OF ANATOMICAL ASYMMETRIES

Anatomical asymmetries have been recognized as part of the evolution of living organisms. Goethe, in his book *Metamorphosis*, presented a pre-Darwinian theory proposing that evolution leads to a stricter definition of parts and toward a lack of symmetry. Asymmetry then was viewed as a progressive development, differentiating the complex from the simple and the advanced from the primitive. However, recent investigations have uncovered a great deal of asymmetry in the nervous systems of species at a lower scale of evolution, for example, in birds and rats (Glick, 1985; Nottebohm & Nottebohm, 1976). The phylogeny of cerebral asymmetry has been of great interest to anthropologists, linguists, neurologists, and pathologists.

The development of cerebral asymmetries was studied extensively in anthropoid apes, in addition to other species. Endocranial casts revealed fewer asymmetries in monkeys than in anthropoid apes (Holloway & De La Coste, 1982). The most constant asymmetry on endocasts is the deviation of the superior sagittal sinus to the right and a lower position of the left transverse sinus. Asymmetries were noted in the orangutan, in which the posterior end of the Sylvian fissure is higher on the right than on the left in the majority of cases. Differences in the occipital poles were noted in the chimpanzee; in 6 of 9, the left was wider than the right, as in humans. The right hemisphere of the gorilla was found to be longer in 5 of 7 brains (LeMay, 1976).

The gorilla manifests the closest approximation to human asymmetries; the pre-rolandic cortex favors the right side and the post-rolandic cortex the left hemisphere. The gorilla appears to be the most sexually dimorphic pongid in terms of body and brain size, but only minor sex differences were seen in brain asymmetries (Horvath, Woodward, & De La Coste-Lareymondie, 1985). Yeni-Komshian and Benson (1976) showed that in the chimpanzee the left Sylvian fissure is longer than the right, but in the rhesus monkey this difference does not reach statistical significance. Even cat brains appear to have asymmetry in their surface structure (Webster, 1977). Less than half the cat brains appeared to be asymmetrical, but included among the asymmetrical brains was those of three newborn kittens, indicating that the asymmetry may be inborn. Webster (1977) did not find that the asymmetry was related to paw preference. The asymmetry was more predominant in the posterior or visual area, indicating that this area might be a substrate of functional asymmetry in the visual processes in cats.

Asymmetries are postulated for the brains of ancient humans. The endocasts of the skull of Peking man, who lived more than a half million years ago, were believed to show the forward extension of the right hemisphere and the posterior extension of the left hemisphere as the most common pattern of asymmetry. The skull of *Pithecanthropus* I appeared to have a slight right occipital petalia, but that of *Pithecanthropus* II may have had a slight left occipital petalia (LeMay, 1976). Asymmetries have been shown for the Neanderthal man of 50,000 years ago, with higher Sylvian fissure on the right than on the left (Boule & Anthony, 1911; LeMay & Culebras, 1972). Most of the fossil skulls are damaged and fragmentary. One can only conjecture about the functional significance of the asymmetries in the ancient skulls. It is reasonable to assume, however, that they were precursors of development of the asymmetries found in modern man. Holloway and De La Coste (1982) summarized the complex issue of paleontology of cerebral dominance in 190 hominid endocasts. Only modern *Homo* and hominids such as the Neanderthals show a distinct left occipital–right frontal petalia.

II. FUNCTIONAL LATERALIZATION

The heuristic value of cerebral dominance is probably related to the advantage of increasing specialization of each hemisphere, allowing a greater repertoire of special functions to be performed. For most basic biological mechanisms, such as locomotion and orienting, bilateral symmetry appears to be an advantageous and necessary condition. However, higher cognitive function has less need for symmetry, which allows for the development of human asymmetries, according to Kinsbourne (1978). Motor asymmetry, such as handedness, then would follow the cognitive asymmetry as a result of linkage related to the complex motor requirements of speech (Kimura, 1979). On the other hand, many investigators believe that the opposite is true, that is, that the development of motor asymmetry, because of the manipulative function of the right hand while the left hand was engaged in holding or protecting, preceded the development of language in the left hemisphere.

Broca's (1863) notion of cerebral dominance for language developed from his anthropological interest, fortuitously combined with his function as a clinician. He and his anthropological colleagues in the middle of the 19th century in France continued to pursue Gall's observation of frontal lobe development as significant for language development. Broca's major contribution was the discovery that speech loss was associated with left-sided damage in patients who subsequently came to autopsy ("Nous parlons avec l'hémisphère gauche"). Ever since Broca, numerous lesion studies and other techniques confirmed the left perisylvian lateralization of language.

Functional asymmetries have been studied extensively with lateralized tachistoscopic presentation and dichotic listening techniques in neurologically intact subjects. The assumption is that the presentation of stimulus contralateral to the dominant hemisphere for that particular function will result in superior performance relative to projection to the nondominant or unspecialized hemisphere. A great deal of variability in the nature of the stimulus and the processing, as well as in the output, introduced a large methodological variance in these studies. Such studies reached their peak in the 1960s and 1970s and declined somewhat after cognitive psychology and functional activation became more popular, but they remain useful tools in investigating functional laterality in normal subjects.

Dichotically presented spoken digits and words resulted in a right ear preference, assuming a left hemisphere process (Kimura, 1967). Even meaningless speech sounds produce a right ear superiority. Tachistoscopically presented words and letters produced a right field preference, as in the earlier reaction time studies of Mishkin and Forgays (1952). Recognition of melodies and familiar and environmental noises had a left ear advantage and a presumed right hemisphere superiority (Curry, 1967; Kimura, 1964). The differences were quite small and only 70–90% of right-handed subjects would show this asymmetry. Recognition of faces and location of dots also have a left visual field/right hemisphere advantage (Geffen, Bradshaw, & Wallace, 1971; Kimura, 1969; Rizzolatti, Umilta, & Berlucchi, 1971). However, not all nonverbal stimuli have left field advantage. Paradoxical lateralization can be achieved if the processing strategies differ from the expectations related to the nature of the stimulus. In other words, some pictures could be processed verbally and some letters could be matched visually, resulting in preferences opposite those expected

(Klatzky & Atkinson, 1971; Seaman & Gazzaniga, 1973). The discrepancy between the results of Wada testing and those of behavioral measures of asymmetry could be explained by the fact that the Wada test assesses speech output and behavioral tests of lateralization measure perception.

Dichotic and visual measures of lateralization are not correlated highly with each other (Hines & Satz, 1971; Zurif & Bryden, 1969). Some studies also found that the tests did not always produce the same results in the same individuals. The reliability could be improved in the dichotic tasks if the subject was asked to pay attention to one ear for some of the stimuli and to the other for the rest (Bryden, 1978). Attentional shifts also explain some of the variability in lateralization studies. Each hemisphere may be activated or primed by the appropriate stimulus material that precedes the task presented to the hemisphere (Kinsbourne, 1974). Handedness, of course, is one of the most commonly used measures of functional laterality, but its correlation with other functional measures, especially with language dominance, is complex. The extensive research on this subject has been well reviewed (Porac & Coren, 1981). The preceding discussion is just a sample of the extensive literature on dichotic listening, visual field preference, and hand performance that was selected for relevance to our methodology. For more extensive reviews of the subject, Bryden (1982), Bradshaw (1989), and Hellige (1990) should be consulted.

III. ANATOMICAL STUDIES OF ASYMMETRY

Anatomical asymmetries also have been observed, but are less obvious than functional lateralization. On casual inspection, the hemispheres appear symmetrical, but studies of hemispheric weight indicate a slightly heavier right side (Braune, 1891; Broca, 1885; Wagner, 1864; Wilde, 1926) that is longer (Gundara & Zivanovic, 1968; Hoadley, 1929; Schwartz et al., 1985). However, longer left hemispheres were found by Cunningham (1892; Hadziselimovic & Cus, 1966; Hrdlicka, 1907). Connolly (1950) found a sex difference in hemispheric length; the left was longer in males and the right longer in a small number of females. Von Bonin (1962) was more impressed by the small magnitude of the differences and the relative symmetry of the two hemispheres. Lobar asymmetries have been observed on autopsied specimens, as well as on the endocasts of skulls. These protuberances are called petalias; the most frequently investigated and confirmed petalia is the left parieto-occipital one in the majority of human and hominid skulls (Hadziselimovic & Cus, 1966), corresponding to a larger left parieto-occipital area on cross section (Inglessis, 1925).

Temporal lobe asymmetries were noted by Heschl (1878), who noted the regularity of two transverse gyri in the temporal operculum on the right and only one on the left. Pfeiffer (1936) described a larger planum temporale, the area behind Heschl's gyrus, on the left side. Geschwind and Levitsky (1968) confirmed this finding quantitatively and determined that the planum was larger on the left side in 65%, on the right side in 11%, and equal in 24% of brains examined. That this asymmetry might be related to language dominance on the left side generated a great deal of interest. The difference has been confirmed to exist in newborns by Witelson and Pallie (1973) and by Wada, Clarke, and Hamm (1975), suggesting that our brain may be prewired at birth for language dominance on the left. Eberstaller (1884) and Cunningham (1892) observed that the left Sylvian fissure was longer and more horizontal and that the right curved up sharply. This result was confirmed quantitatively by Rubens, Mahowald, and Hutton (1976).

Asymmetries in Broca's area were reported less consistently. Wada et al. (1975) found the area on the right side to be larger than on the left, but Nikkuni and co-workers (1981) described a larger left than right Broca's area. In right-handed subjects, the area of the frontal operculum visible on the surface of the brain is larger on the right, and the surface area in the sulci of this region is greater on the left (Falzi, Perrone, & Vignolo, 1982). The right frontal lobe was found to be larger on the right side (Weinberger, Luchins, Morihisa, & Wyatt, 1982; Weis, Haug, Holoubek, & Orun, 1989). Albanese, Merlo, Albanese, and Gomez (1989) re-examined Broca's area asymmetries. In this quantitative study, 62.5% of the brains showed larger left and 12.5% showed larger right frontal opercular and posterior triangularis regions of Broca's area. No asymmetry was found in the more anterior portions of the inferior frontal gyrus that are not involved with language. Although their data of the planum temporale analysis was not quantified, these researchers noted that the larger side was not the same for the frontal and temporal language areas. Subcortical structures were examined extensively by Orthner and Seler (1975), who considered sex differences as well. The left pallidum was larger in both sexes. Prominence of the right frontal region anterior to the commissure also was found in both sexes. Another study confirmed a larger pallidum on the left side (Kooistra & Heilman, 1988).

IV. CYTOARCHITECTONIC ASYMMETRIES

Efforts were made to detect cytoarchitectonic asymmetries corresponding to functional laterality or anatomical landmarks. These studies are restricted primarily to the planum temporale (Galaburda, Sanides, &

Geschwind, 1978) and the primary auditory cortex (Seldon, 1985). Von Economo and Horn (1930) found a larger Ta1 posterior temporal lobe region in the left side, a result that was confirmed by a study by Blinkov (1940), reporting a larger Brodmann's area 22 on the left, and by Galaburda et al. (1978) using pigment architectonics in a region defined as tpt. The planum temporale was analyzed further for area. The results were interpreted to support the hypothesis that larger asymmetries represented a reduced area on the small side, since the total area seemed to be less in the asymmetrical cases (Galaburda, Corsiglia, Rosen, & Sherman, 1987). Galaburda interpreted these results, on the basis of histological studies in rats, to mean that the cytoarchitectonic asymmetries were related to cell numbers and not to cell packing density. Initially, Geschwind and Galaburda (1985) advanced the theory that testosterone in the embryo may retard the development of the left hemisphere and generate the compensatory growth of the right hemisphere. However, on subsequent studies, asymmetrical brains were shown to have a relatively smaller right side with a presumably lower complement of neurons (Galaburda et al., 1987).

Heschl's gyrus contains the cytoarchitectonic primary cortex area tc (in Von Economo and Horn's classification), which also appears to have hemispheric differences. The central portion of Heschl's gyrus is the most granular and appears to be larger in the left hemisphere. This area has been postulated to correspond to the representation of the frequency range of the human voice. The center-to-center distance between the cell clusters was found to be greater on the left side (Seldon, 1981). These studies also showed that the dendritic input region for each unit is greater on the left side than on the right, but these input regions on the left are farther apart. The results imply that, functionally, neurons in the left hemisphere can react more selectively than those on the right, which may explain the specificity of speech perception on the left side (Seldon, 1985). Dendritic branching was examined in the left and right hemispheres in right-handed and left-handed brains; the left-sided cortex had more neurons with higher order branching in right handers. (Scheibel et al., 1985). The patterns were reversed in the brains of the left-handed subjects. Parietal lobe cytoarchitectonic asymmetries indicated a larger language-related area (designated PG), architectonically, on the left side, located on the angular gyrus, and a larger area of PEG, a nonlanguage area of the superior parietal lobule on the right side (Eidelberg & Galaburda, 1984).

V. NEUROIMAGING AND FUNCTIONAL ASYMMETRY

Neuroimaging asymmetries provide in vivo measurements that may be correlated with functional lateralization. Di Chiro (1962), looking at the venous patterns on angiograms, found that in the dominant hemisphere for speech, as tested by sodium amytal, the vein of Labbé and in the nondominant hemisphere for speech the vein of Trolard appeared larger. The left ventricle has been shown to be larger than the right in several studies, using ventricular casts and pneumoencephalograms (Last & Thompsett, 1953). The study by McRae, Branch, and Milner (1968) found the left occipital horn of the lateral ventricle larger in right-handers, but found the horns equal in left-handers. The asymmetry did not correlate with speech dominance as tested by sodium amytal. LeMay and Culebras (1972) observed a higher Sylvian point or end of Sylvian fissure on the right side on angiograms and a narrower angle on the left because of a larger left parietal operculum in right-handers. In left-handers, the configuration of the carotid arteries at the posterior end of the Sylvian fissure was more similar on both sides. The difference in the Sylvian points also was confirmed in fetuses.

A few studies attempted to establish the relationship between anatomical asymmetries and functional laterality. Ratcliff, Dila, Taylor, and Milner (1980) correlated the asymmetry of the Sylvian point on angiograms with intracarotid amytal studies of cerebral dominance in an epileptic population. A larger left parietal operculum was found in 66% of left-positive Wada test subjects and in 35% of right-positive subjects. Strauss, LaPointe, Wada, Gaddes, and Kosaka (1985) found no correlation between Sylvian point measures and amytal testing of language dominance, but found some correlation between right ear advantage (REA) on dichotic tests and posterior temporal widths. The Sylvian asymmetries were distributed equally in both dominance groups. Witelson and Kigar (1988) reported pilot data of postmortem cerebral asymmetry in cancer patients tested with dichotic listening before they died. Four patients with REA had a larger left planum, whereas three "non-REA" subjects had an equal or smaller left planum.

VI. COMPUTERIZED TOMOGRAPHY SCAN ASYMMETRIES

Since 1976, cerebral hemispheric asymmetries have been measured on computerized tomography (CT) scans using a technique pioneered by LeMay and associates (LeMay, 1976,1977; LeMay & Geschwind, 1978; LeMay & Kido, 1978). LeMay observed a possible relationship between cerebral hemispheric asymmetries and cerebral dominance for handedness. In her 1977 study, she examined the CT scan hemispheric asymmetries of right- and left-handed males. Among the 100 right-handed males in her study, 78% had a greater left than right occipital length and only

13% had a greater right than left occipital length. The left-handed males in her study were distributed more evenly across asymmetry type: 37% had a greater left occipital length, 24% had equal left and right lengths, and 39% had a greater right occipital length.

Subsequent studies, however, did not confirm all these findings (Chui & Damasio, 1980; Koff, Naeser, Pieniadz, Foundas, & Levine, 1986; Naeser & Borod, 1986; Palumbo, Naeser, & York, 1990; Pieniadz, Naeser, Koff, & Levine, 1983). These studies found the most typical pattern of CT scan hemispheric asymmetries to be equal frontal asymmetries and left occipital asymmetries for left- and right-handers. The emphasis on this portion of the chapter is on occipital asymmetries, because no previous CT scan studies have observed any specific behavioral measures to correlate with frontal asymmetries.

The asymmetries illustrated here are determined with modifications of LeMay's techniques (LeMay, 1977; LeMay & Geschwind, 1978). These methods have been published by Pieniadz et al. (1983). CT scan slices, such as slice W and slice SM, are labeled according to criteria defined by Naeser and Hayward (1978). Occipital width and length asymmetry measurements are taken on slice W and slice SM. Slice W is at the level of the roof of the third ventricle. Slice SM is at the level of the bodies of the lateral ventricles, where the bodies of the lateral ventricles are close together. Slice W contains, in part, the posterior portion of the superior temporal gyrus, which is the posterior half of Wernicke's area. Slice SM contains, in part, the deep periventricular white matter. For more information on neuroanatomical structures present on CT scan slice W and slice SM, see Fig. 1 in Chapter 9.

Occipital lengths are measured by placing a clear plastic polar coordinate overlay on the CT slice to be measured (Fig. 1). The inner table of the skull is visible through the concentric circle overlay at the occipital pole. A concentric circle is placed on the shorter of the two hemispheres and the length of the longer hemisphere beyond that concentric circle is measured. Length asymmetries are quantified in the following manner. If the lengths of the two hemispheres differ by less than 0.25 mm, the occipital length is considered equal, or symmetrical, and assigned a value of zero. If the difference between the left and right hemisphere is 0.25 mm or greater, the difference is quantified in millimeters. This length difference (length asymmetry measurement) is assigned a positive or negative value. The asymmetry is positive if it is in the typical left direction and negative if it is in the atypical right direction. The millimeter difference in lengths is multiplied by 10 for ease of data manipulation. For example, in Fig. 1 on slice W, the occipital length difference between the left and right hemispheres is 3 mm, with the left longer than the right. The 3-mm difference

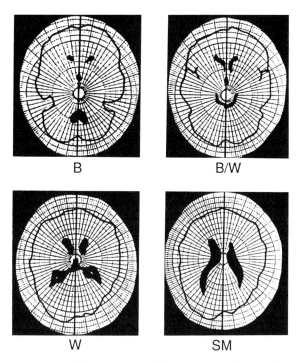

Figure 1 Method for measuring CT scan occipital lengths (*bottom*) and frontal lengths (*top*). In this case, occipital length asymmetry on slice W was 3-mm difference, with left greater than right. Asymmetry value was +30. Asymmetry value on slice SM was +40. Similar methods were employed in the measurement of frontal lengths on slices B and B/W. Reproduced with permission.

is multiplied by 10, thus producing an occipital length asymmetry value for slice W of +30. In this case, a plus value is assigned because a left occipital asymmetry is the typical occipital asymmetry. In addition to the occipital length asymmetry measurements on slice W and slice SM, the mean occipital length asymmetry is calculated across these two slices. In Fig. 1, the occipital length asymmetry value on slice W is +30 and the occipital length asymmetry value on slice SM is +40. Thus, the mean occipital length asymmetry is +35 [i.e., (30 + 40)/2 = 35]. The mean occipital length asymmetry value then is categorized as left (values equal to or beyond +2.5), equal (values between +2.5 and −2.5), or right (values equal to or beyond −2.5).

Occipital width asymmetries are measured by placing a clear millimeter ruler at the most posterior extension of the interhemispheric fissure at the

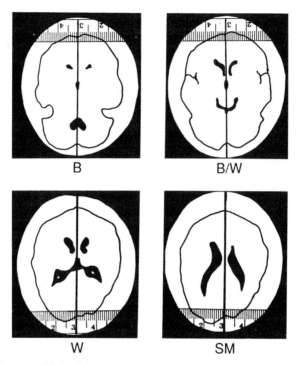

Figure 2 Method for measuring CT scan occipital widths (*bottom*) and frontal widths (*top*). In this case, occipital width asymmetry on slice W was 3-mm difference, with left greater than right. Asymmetry value was +30. Asymmetry value on slice SM was +10. Similar methods were employed in the measurement of frontal widths on slices B and B/W. Reproduced with permission.

posterior junction of the inner table of the skull (Fig. 2). The ruler is moved inward, toward the occipital horns, perpendicular to the longitudinal axis of the skull, until either the left or right inner table of the skull measures 10 mm from the midline interhemispheric fissure. The width measurements are taken for both hemispheres at this point. Width asymmetries are quantified in the following manner. If the widths of the two hemispheres differ by less than 0.5 mm, the occipital width is considered equal, or symmetrical, and assigned a value of zero. If the difference between the left and right hemispheres is 0.5 mm or greater, the difference is quantified in millimeters. This width difference (width asymmetry measurement) is assigned a positive or negative value. The asymmetry is positive if it is in the typical left direction and negative if it is in the atypical right direction.

In addition to the occipital width measurements on slices W and SM, a mean occipital width asymmetry is calculated across slice W and slice SM. In Fig. 2, the occipital width asymmetry value on slice W is $+30$ and the occipital width asymmetry on slice SM is $+10$. Thus, the mean occipital width asymmetry is $+20$ [i.e., $(30 + 10)/2 = 20$]. The mean width asymmetry values then are categorized as left (values equal to or beyond $+5$), equal (values between $+5$ and -5), or right (values equal to or beyond -5.).

A study in which the occipital asymmetries on 87 CT scans were measured by two independent raters using the measurement techniques just described showed good inter-rater reliability (Palumbo et al., 1990). The correlation coefficient for mean occipital length measurements was $+ .892$ ($p < .001$) and the correlation coefficient for mean occipital width measurements was $+ .886$ ($p < .001$). In addition, this study showed that occipital length and width asymmetry categories (left, equal, or right) were usually consistent across three separate scan times in 29 subjects. The angulation of the CT scan did not appear to affect the consistency of the CT scan asymmetry categories.

The category of CT scan occipital length asymmetry (left, equal, or right) cannot be used to predict handedness or hemispheric side of language dominance. For example, for two aphasia populations (right-handed and left-handed men) studied at the Boston University Aphasia Research Center (Boston V.A. Medical Center), no significant differences have been observed on CT scans in the distribution (left, equal, or right) of occipital (or frontal) asymmetries (Naeser & Borod, 1986; Pieniadz et al., 1983). These studies included 89 right-handed men who were aphasic from left hemisphere lesion and 26 left-handed men who also were aphasic from left hemisphere lesion. In the study with right-handed patients, 33% (29 of 89 cases) had right occipital length asymmetry, but all were aphasic from a left hemisphere lesion. In the study with left-handed patients, 15% (4 of 26 cases) had right occipital length asymmetry, but all were aphasic from a left hemisphere lesion. Thus, the results from these two aphasia population studies suggest that handedness cannot be predicted from CT scan occipital asymmetry, and that hemispheric side of language dominance also cannot be predicted from CT scan occipital asymmetry.

The notion that CT scan occipital asymmetry may not be used to predict hemispheric side of language dominance is strengthened also when CT scan asymmetries are examined in crossed aphasia cases (Henderson, Naeser, Weiner, Pieniadz, & Chui, 1984). In this study of crossed aphasia, 11 of 15 right-handed aphasia cases with right hemisphere lesion had left occipital length asymmetry on CT scan. Hence, the majority of crossed aphasia cases (right-handers with right hemisphere lesions) had left occipital length on CT scan, but this result would not have been accurate in predicting which hemisphere would produce aphasia, if a lesion were present. It is, of course, possible that some of these crossed aphasia cases or the left-handed aphasia cases discussed earlier could have become aphasic from a lesion in either hemisphere, but that is not known. Thus, although CT scan hemispheric asymmetries may not be useful in predicting handedness or hemispheric side of language dominance, Section VIII presents a discussion of how they may be useful in predicting potential for better recovery of some "posterior" language functions in a subset of global aphasia patients.

VII. RELATIONSHIP BETWEEN OCCIPITAL ASYMMETRIES ON CT SCANS AND PLANUM TEMPORALE ASYMMETRIES IN THE BRAINS OF THE SAME CASES POSTMORTEM

The relationship between occipital asymmetries on CT scans and planum temporale asymmetries in the brains of the same cases postmortem was investigated in 15 right-handed male patients, ages 41 to 76, who later came to postmortem examination (Pieniadz & Naeser, 1984). The CT scans had been performed prior to death. The CT scan films were obtained with two window settings: (1) the original routine window (w = 130–140) and center (c = 30–40) and (2) the bone window (w = approximately 400). The bone window setting improved visualization of the inner table of the skull and thus increased accuracy in measurement of the asymmetries. The occipital length and width asymmetry values were quantified as explained earlier. In addition, head size was adjusted for, in each case, by computing a ratio similar to that described by Eidelberg and Galaburda (1982). For example, for occipital width measurements $L - R/[(L + R)/2]$ was used, where L is left and R is right. For occipital length measurements, $L - R/(AP/2)$ was used, where AP is the greatest distance from the anterior pole to the posterior pole.

The planum temporale length measurements were taken at postmortem examination. The length of the outer border of the left planum temporale

Figure 3 *(Top)* Method for measuring planum temporale length (PTl). Distance between horizontal black lines on left and right represents length of outer border of left and right plana temporale. In this case of typical left PTl asymmetry, left PTl was 38.00 mm and right PTl was 13.15 mm. Measurements were from 20 × 25-cm photographs of brains. *(Bottom)* CT scan for this case, in which typical left occipital length asymmetry was observed on slice W (+10) and on slice SM (+10) (arrows). CT scan is dark because bone windows were used to better visualize the inner table of the skull. Reproduced with permission.

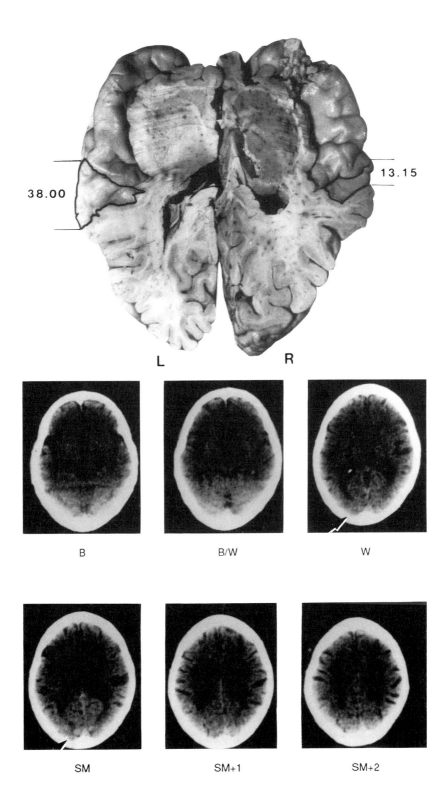

38.00

13.15

L   R

B   B/W   W

SM   SM+1   SM+2

was measured with a millimeter ruler from the posterior sulcus of Heschl (first transverse gyrus) to the posterior margin of the left planum temporale, in the manner described by Geschwind and Levitsky (1968). The length of the outer border of the right planum temporale was measured from the sulcus immediately posterior to the first transverse gyrus to the posterior margin of the right planum temporale. These straight line measurements were made on 8 × 10-inch photographs containing the two plana that were photographed together to control for equal magnification of the two hemispheres. The difference between the left and right raw measurements for the left and right plana lengths also was computed into an asymmetry ratio, as described earlier for the CT measurements.

Figure 3 shows the CT scan of a case with typical left occipital length asymmetry and longer left planum temporale length postmortem. Figure 4 shows the CT scan of a case with atypical right occipital length asymmetry and longer right planum temporale length postmortem. The correlation between CT scan occipital length at slice SM and postmortem planum temporale length was +0.674 ($p <.01$; $df = 13$). For 11 of the 15 cases, the category of occipital length asymmetry on CT scan (left, equal, or right) was consistent with the category of planum temporale length asymmetry postmortem (left, equal, or right). Errors in the four cases were between the equal and left asymmetry categories. Overall, 12 cases had left occipital length asymmetry at slice SM, two cases were equal, and one case had right asymmetry. The single case with right occipital length asymmetry at slice SM on CT scan was also the single case with greater right planum temporale length postmortem (Fig. 4). No other significant correlations were found between slice SM occipital width or slice W occipital width or length and planum temporale length, planum temporale area, Sylvian fissure length, or Sylvian fissure height.

In summary, the finding of a significant correlation on CT scan between slice SM occipital length asymmetry and postmortem planum temporale length asymmetry is important because it demonstrates that at least one gross CT scan asymmetry measure (occipital length at slice SM) may help predict an asymmetry of the brain in a language-related region (planum temporale length). The possibility that occipital asymmetries on CT scan may have relevance to some aspects of language recovery in aphasia is discussed in the next section.

Figure 4   (*Top*) Atypical right planum temporale length (PTl) asymmetry in which left PTl was 17.25 mm and right PTl was 28.00 mm. (*Bottom*) CT scan for this case, in which atypical right occipital length asymmetry was observed on slice W (−10) and on slice SM (−15) (arrows). CT scan is dark because bone windows were used. Reproduced with permission.

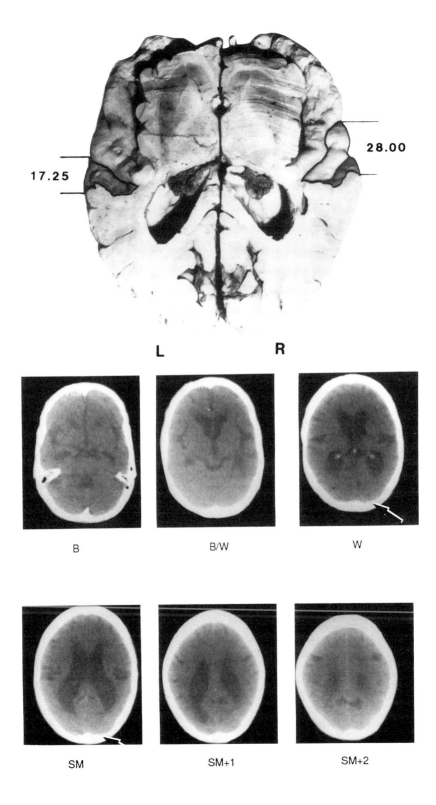

17.25

28.00

L   R

B

B/W

W

SM

SM+1

SM+2

VIII. RELATIONSHIP BETWEEN OCCIPITAL ASYMMETRIES
ON CT SCANS AND RECOVERY OF SOME "POSTERIOR"
LANGUAGE FUNCTIONS IN GLOBAL APHASIA PATIENTS

In a retrospective study, language recovery and CT scan asymmetries for 14 right-handed global aphasia patients with large left hemisphere frontal, parietal, and temporal lobe lesions were examined (Pieniadz et al., 1983). Handedness was assessed by hand used for writing and hand used for most activities. Scores from the Boston Diagnostic Aphasia Exam (Goodglass & Kaplan, 1972) were available in most cases. Test scores were reported in two phases when possible. Time 1 (T_1) was 1–7 mo after stroke (MPO) and time 2 (T_2) was 7–303 MPO. Each case had large left frontal, parietal, and temporal lobe lesion, both cortical and subcortical (including the internal capsule region), consistent with left middle cerebral artery occlusion. CT scan hemispheric asymmetries were measured and quantified as described earlier.

No significant correlations were found between scores on tests administered at either T_1 or T_2 (or values derived from subtracting T_1 and T_2 scores), and age at stroke onset, MPO at testing times, or mean lesion size. No significant correlations were found between T_1 or T_2 language test scores and any category or degree of CT scan asymmetry. Significant correlations were observed, however, between CT scan occipital asymmetries and "recovery," defined as "amount of change over time" (T_2 minus T_1 scores) in "posterior" language functions at the one-word level. Mean occipital width (slices W and SM combined) correlated -0.64 ($p < .05$) with word discrimination (single-word comprehension) and -0.68 ($p < .05$) with visual confrontation naming (single-word naming). Mean occipital length (slices W and SM combined) correlated $-.073$ ($p < .05$) with single-word repetition and -0.72 ($p < .01$) with visual confrontation naming.

The CT scan for a right-handed globally aphasic patient with typical left occipital asymmetry who had poor "recovery" is shown in Fig. 5. The CT scan for a right-handed globally aphasic patient with atypical right occipital asymmetry who had better "recovery" at the one-word level for these "posterior" language functions is shown in Fig. 6.

The "recovery" for the globally aphasic patients with atypical right or equal occipital asymmetry at the one-word level for comprehension, repetition, and naming was not complete and, although limited, was better than the "recovery" observed for those with typical left asymmetries. For example, when median scores were computed for T_2 data for word comprehension, word repetition, and naming, the following results were obtained:

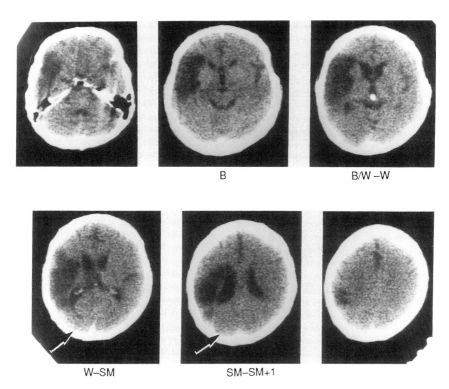

Figure 5   CT scan for 61-year-old man with global aphasia who had poor recovery. Typical left occipital asymmetry was present on CT scan (arrows). The mean occipital length across CT slices W and SM was +16.5. The mean occipital width was +10. The mean total occipital asymmetry was +13. Reproduced with permission.

1. The median on word comprehension was 45 of 72. Of 7 cases above the median, 6 exhibited atypical occipital asymmetry (TOA, mean of total occipital width plus length combined); the range for these 6 cases was 47–66 of 72. Of 7 cases below the median for word comprehension, 4 exhibited typical TOA; the range for these 4 cases was 23–40 of 72.

2. The median one-word repetition was 2.5 of 10. Of 7 cases above the median, 5 exhibited atpyical TOA; the range for these 5 cases was 4–8 of 10. Of 7 cases below the median, 3 exhibited typical TOA; the range for these 3 cases was 0–2 of 10.

3. The median score on the naming task was 7 of 105. Of 7 cases above the median, 6 exhibited atypical TOA; the range for these 6 cases was

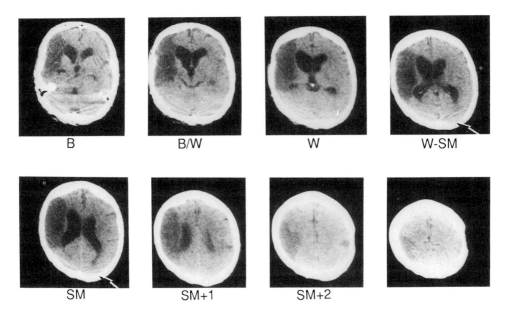

Figure 6 CT scan for a 24-year-old man with global aphasia who had good recovery of some single-word comprehension, repetition, and naming but no recovery of spontaneous speech output. Atypical right occipital asymmetry was present on CT scan (arrows). The mean occipital length across CT slices W and SM was −15. The mean occipital width was −50. The mean total occipital asymmetry was −32.5. Reproduced with permission.

10–46. Of 7 cases below the median, 4 exhibited typical TOA; the range for these 4 cases was 0–4.

Thus, although some recovery of posterior language functions at the one-word level occurred for the globally aphasic patients with atypical occipital asymmetry, the recovery was still limited. No significant correlations were observed between any CT scan asymmetry measurements (occipital or frontal) and recovery of spontaneous speech in this study.

The pattern of recovery in the right-handed global aphasia patients with atypical asymmetries parallels the pattern of right hemisphere linguistic capabilities observed in commissurotomy cases (Gazzaniga & Hillyard, 1971; Sidtis, Volpe, Wilson, Rayport, & Gazzaniga, 1981; Zaidel, 1976, 1977, 1978). For example, a greater ability to process at the one-word level was seen in relatively "posterior" language functions (comprehension, repetition, naming) than in spontaneous speech. The right hemisphere in patients who have atypical asymmetries and who have exhibited some "recovery" in global aphasia may aid in this recovery by complementing or cooperating with the few preserved areas of the left hemisphere in some unique manner that is not ordinarily seen in other globally aphasic patients who have typical asymmetries and who have less "recovery." In cases with atypical asymmetries, some type of anomalous language dominance may exist, and the degree or extent to which the left hemisphere is capable of inhibiting the right hemisphere may be reduced. Because the recovery is limited to posterior language functions at the one-word level, and did not extend to spontaneous speech in these right-handed globally aphasic patients with large left hemisphere lesion, CT scan occipital asymmetries may be more related to some aspect or degree of posterior language dominance only, and not to motor speech output dominance. Thus, two separate language dominances (motor speech output dominance and posterior language function dominance) may exist. Support for this notion was observed in our study of aphasia in left-handers, in whom lesion size, lesion site, and hemispheric asymmetries were studied (Borod, Carper, & Naeser, 1990; Naeser & Borod, 1986). Space limitations do not permit a detailed discussion of these left-handed aphasia patients here. However, in the 1986 study, three of the left-handed cases had aphasia as a result of massive right hemisphere lesion including frontal, temporal, and parietal lobes, both surface and deep. Two of these patients had limitations in speech output, but good auditory language comprehension. Each of these two cases had left occipital length asymmetry. Each of these two cases probably had speech output and handedness dominance in the right frontal lobe, but auditory language comprehension dominance in the left temporal lobe (or bilaterally). The occipital length asymmetry (left) in these two cases was contralateral to the side of the lesion causing the aphasia (right). Each case had good comprehension. This pattern is parallel (but opposite) to that observed in the right-handed global aphasia patients by Pieniadz et al. (1983). In that study, when the aphasia-producing lesion was in the left hemisphere but the occipital asymmetry was in the right contralateral hemisphere, some better recovery in "posterior" language functions (comprehension, repetition, naming) at the single word level was seen.

One of the three left-handed patients with massive right hemisphere lesion—including frontal, temporal, and parietal structures, both surface and deep—had right occipital length asymmetry. This patient had limitation in speech output and poor auditory language comprehension. This case probably had speech output and handedness dominance in the right frontal lobe, as well as auditory language comprehension dominance in the right temporal lobe. This patient had all three dominances—handedness, speech, and comprehension—in the right hemisphere. The asymmetry was ipsilateral to the aphasia-producing lesion, and poor recovery in auditory comprehension was noted.

In summary, CT scan occipital length asymmetry appears to be useful in predicting better recovery of some "posterior" language functions when the CT scan occipital length asymmetry is contralateral to the hemisphere with a lesion (including the temporal lobe) producing the aphasia. This appears to be the case in right-handed or left-handed aphasia patients. In addition, the hemispheric side for location of handedness dominance, language comprehension dominance, or speech output dominance may be totally left, totally right, or separated in various combinations between the two hemispheres. These latter cases are particularly challenging and interesting to study.

IX. MAGNETIC RESONANCE IMAGING STUDIES
OF ANATOMICAL ASYMMETRY

Magnetic resonance imaging (MRI) offers a unique opportunity to study the relationship of functional lateralization and in vivo anatomical asymmetry in normal subjects. MRI techniques show excellent distinction between gray and white matter, brain, and cerebrospinal fluid (CSF), producing anatomical images that approximate those obtained at autopsy (Doyle, Pennock, Orr, Gore, Bydder, & Steiner, 1981). Unlike X-ray computerized tomography, MRI does not use ionizing radiation so normal volunteers, in whom tests of functional lateralization can be carried out, can be imaged safely (Sweetland, Kertesz, Prato, & Nantau, 1987). We observed one of the most obvious asymmetries on inversion recovery MRI images at the posterior opercular regions (Kertesz, Black, & Howell, 1984; Kertesz, Black, Polk, & Howell, 1986). Axial sections across the insula showed a larger left planum and a sharper demarcation of the opercular region from the parietal sulci on the right (Fig. 7). In a pilot study, we found this asymmetry in 88% of all scans (Kertesz et al., 1984). In corresponding anatomical sections in the same plane, the acute demarcation on the right and a gradual slope associated with a larger planum temporale on the left were confirmed (Kertesz et al., 1986).

In our first study correlating functional and anatomical asymmetries, we found that the opercular parietal demarcation of the sulci had a sharper angle on the right side in 60% of right-handers and in only 10% of left-handers (Fig. 7). Occipital width measurements were larger on the left in 90% of the right-handers and in only 30% of the left-handers. Three of the anatomical features—the anterior frontal width, the parietal width, and the sulcal demarcation—collectively discriminated well between the left- and right-handed groups. From all the behavioral measures, hand

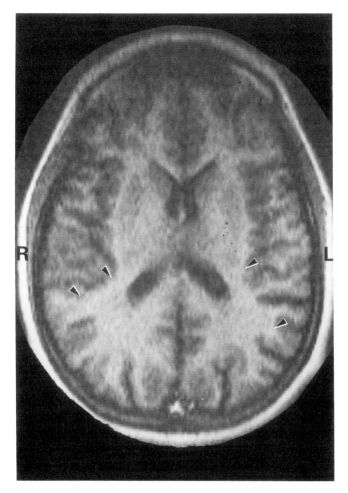

Figure 7   Axial (horizontal) MRI slice of the brain at the level of the parietal operculum in a normal volunteer. The arrows show the demarcation from the parietal cortex, at an acute angle on the right and a greater angle on the left, corresponding to the larger planum temporale on the left.

performance correlated best with the anatomical measures (Kertesz et al., 1986).

Other investigators have measured the planum temporale with different techniques. Steinmetz and co-workers (1989) described a method verified on 10 postmortem brains. The total surface area was obtained by adding the manually traced lines of planum temporale at each sagittal MRI slice. The total surface of the planum temporale showed significant asymmetry toward the left, but other measures did not. Similar linear measures were juxtaposed to obtain planimetric measures between Heschl's gyrus and the posterior end of the Sylvian fissure by Habib (1989). A correlation existed between handedness and asymmetry coefficients for planum areas in this study.

Temporal area measurements showed a slightly larger right side in right-handed subjects without any effect of age, sex, or handedness in a study by Jack, Twomey, Zinsmeister, Sharbrough, Petersen, and Cascino (1989). Three-dimensional MRI measures of central sulcus, Sylvian fissure, and pars triangularis were compared favorably with digitized length measures on cadaver brains (Vannier et al., 1991). Several parameters of the pars triangularis, including the length of the ascending rami and the area, appeared to be larger in the right hemisphere.

X. SEX, HANDEDNESS, AND ANATOMICAL ASYMMETRIES IN MRI

In a subsequent study (Kertesz, Polk, Black, & Howell, 1990), we studied 52 right-handed (25 males, 27 females) and 52 left-handed (26 males, 26 females) normal individuals, aged 18–49 years. All subjects were university students or were gainfully employed, and were free from neurological or psychiatric disease. Handedness was defined by the hand used for writing and other skilled activities, on the basis of self-report on a handedness questionnaire (Bryden, 1982). We used a 0.15-Tesla resistive magnet, spin echo technique with short time-to-echo interval to obtain good gray matter and CSF contrast for brain contours. The head was positioned in the midline with a laser centering device through the nose, philtrum, and chin. A midsagittal cut was obtained to visualize the corpus callosum and other central structures. The line between the splenium and the optic chiasm, which is parallel to the Sylvian fissures (usually −15° to the orbitomeatal line), was used as the plane for the axial (horizontal) slices. The coronal planes were perpendicular to the axial slices. The first axial slice above the third ventricle in which the bodies of the lateral ventricles appeared to touch in the midline was used for linear and area measures of the hemispheres. After measuring the maximum sagittal length of each hemisphere, width measurements were performed at 10 and 30% of the greater length from the frontal and occipital lobes and designated as anterior frontal, parietal, and occipital. To avoid bias, the labels on the

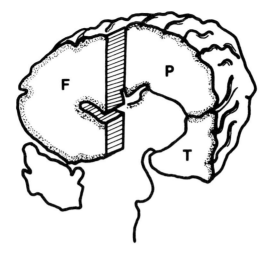

Figure 8 A diagrammatic representation of the coronal MRI cuts used for the lobar area measures. Note that the frontal slice (F) is more anterior than the temporal (T) and the parietal (P) measures, which are at the level of the slice through the pons. Reproduced with permission from Kertesz, Polk, Black, & Howell (1992).

images were masked in unmarked envelopes and the sides of the images were reversed randomly for tracing and measuring.

Areas were traced and digitized with a microcomputer from MRI scans after enlargement to life size on a photographic enlarger. Tracing and measuring was done twice by two independent observers blind to the side of the brain and to the behavioral data. Correlations between sets of measurements ranged from 0.88 to 0.97 ($x = 0.91$); therefore the two sets were averaged. For hemispheric areas, the same slice was used as for the linear measures. Coronal cuts exactly perpendicular to the axial image were used to trace frontal, temporal, and parietal lobe areas (Fig. 8). The frontal lobes were traced on the cut that was the first one to show the temporal lobes separate from the frontal lobes. The tracing included the frontal horns in the area. The corpus callosum was bisected by a straight line and the brain contours were traced following sulcal indentations if they were wider than 90°. Sharper infoldings or sulci were bridged rather than followed to their depth. The coronal cut for temporal lobe measurements was selected on the image in which both the pons and the medulla could be seen. In this image, the Sylvian fissure was usually the deepest, and the parietal and temporal operculum were well outlined. The inferior ramus of the Sylvian fissure on the coronal view was connected to the hippocampal sulcus by a straight line to outline the temporal lobe area; the rest of the surface contour was traced. For the parietal lobe, the superior ramus was connected with the tip of the anterior horn of the lateral ventricle and the inferior end of the interhemispheric fissure (Fig. 8).

The results showed that linear and area measurements were larger in the right hemisphere across all the sex and handedness groups. An exception to this rule occurred in left-handers for anterior and posterior frontal widths and for the frontal lobe area, which was larger on the left side. Also, a significant interaction was seen between sex and handedness in parietal asymmetries. Right-handed male and left-handed female parietal areas were larger on the left side and right-handed female and left-handed male parietal areas were larger on the right. A three-dimensional torque was postulated to explain the rostrocaudal difference of showing a larger right side above the third ventricle and a slightly more prominent left side below. This torque is similar to that of right frontal and left occipital petalia measures on CT. Contrary to the expectation of a larger contralateral frontal lobe to be associated with superior motor skills in the hand, left-handers had larger left and right-handers had equal or slightly larger right frontal lobes. A sex difference consisting of more symmetrical frontal length or petalia in females also was seen. Finally, the temporal lobe areas were relatively larger in females, as was evidenced by less sex difference in either temporal area. This difference may be related to the slightly higher language abilities of females, although this kind of relationship is only speculative.

XI. AUDITORY, MANUAL, AND VISUAL LATERALITY AND MRI ASYMMETRIES

In another study, we correlated behavioral measures with anatomical asymmetries (Kertesz, Polk, Black, & Howell, 1992). The following behavioral measures of cerebral dominance were used.

1. *Dichotic Listening*: Stop consonant–vowel pairs (e.g., Ba–Da) were presented over 30 practice trials and 4 blocks of 30 test trials. An attention-directed paradigm was used to measure laterality. The subjects were instructed to attend to items presented to one ear for 60 trials and to the other ear for the rest. A laterality index was derived from an odds ratio of the right ear responses exceeding left ear responses (Bryden & Sprott, 1981).
2. *Visual Half-Field*: A tachistoscopic presentation of a series of 3- and 6-letter monosyllabic high-frequency words was made unilaterally

once in each visual field for 150 msec, for a total of 24 practice and 80 test trials. A voice-activated timer recorded reading latencies in milliseconds. This laterality index was derived from a slope formula subtracting the differences between the reaction times to 6- and 3-letter words for left visual field from those to 6- and 3-letter words for the right visual field (Bub & Lewine, 1988).

3. *Hand Performance*: Subjects were instructed to mark a series of small circles with dots as quickly as possible in 20 sec. This task was performed twice with each hand. This laterality index was calculated using a ratio of the difference between number of dots completed by each hand divided by total number of dots (Tapley & Bryden, 1985).

The results of this study showed the greatest differences between groups separated for hand performance. Left hemisphere dominance subjects showed larger right frontal and left occipital widths, similar to the results of the previous CT studies. Dichotic listening and handedness interaction showed a uniquely consistent direction. Subjects in whom the same hemisphere is dominant for handedness and language have larger structures than those in whom dominance is crossed. The consistency throughout all the structures supports the validity of these findings and suggests that the coincidence of certain functional lateralities is associated with larger structures. Anatomical measures were found to be less significantly different among visual field laterality groups. However, left visual field dominance for reading, interacting with right handedness, was associated with wider left parietal width.

XII. CONCLUSION

The multidimensional associations highlight the complexity of the relationships between functional and anatomical laterality, and may provide a basis to explain some individual differences in cerebral organization, behavior, and recovery from brain damage. Sex and handedness are important to a certain extent, but a multitude of other factors contributes to the structural and functional diversity of the brain. Genetic factors in anatomical variability have been suggested (Galaburda, Rosen, & Sherman, 1990) and some experimental evidence supports environmental influences (Rakic & Williams, 1986). Some asymmetries have been postulated to be related to hormonal effects on intrauterine development, for example, testosterone retarding the growth of the left side of the brain (Geschwind & Galaburda, 1985).

The most obvious biological advantage of hemispheric specialization is the enhanced ability of a complex structure to handle complex tasks. A purely symmetrical structure is more likely to represent a duplication of functions. Asymmetrical representation in the brain allows for diversity of complementary function, permitting greater adaptability. Some investigators have contended that the further an organism advances on the phylogenetic scale, the more functional and anatomical asymmetries can be demonstrated in the nervous system. Although attempts have been made to associate left parietal petalia with the development of language, phylogenetically many studies caution against such an obviously straightforward relationship. A surprising amount of cerebral asymmetry has been shown in many species throughout the phylogenetic scale.

In addition to the theoretical issues, imaging certain asymmetries that may have predictable associations with language or with other cognitive functions has practical and clinical relevance. For instance, invasive tests of cerebral dominance for language, such as the Wada test during angiograms, could be supplanted by noninvasive imaging of anatomical symmetries if reliable associations could be demonstrated. To date, such associations have not been proven. Another potential clinical use of imageable cerebral asymmetry is the possible prediction of recovery from language deficit after stroke or other acute recoverable cerebral insult. If cerebral function is distributed more symmetrically, and this distribution is associated with anatomical symmetry, then recovery may be faster, more complete, or both because the homologous structures may have more compensatory capacity. Anatomical asymmetries that are systematic may be important in stereotactic surgery in which the difference of a few millimeters can influence functional outcomes. Metabolic imaging also requires increasing accuracy in anatomical localization for the interpretation of physiological data. Cerebral morphometric techniques will be essential to many areas of neurology and neuroimaging in the development of diagnostic and clinical tools and in the assessment of cerebral function.

ACKNOWLEDGMENTS

A. Kertesz acknowledges the assistance of S. Black, M. Polk, J. Howell, D. Drost, T. Carr, and L. Nicholson for their collaboration in the MRI studies and Bonita Caddel for her secretarial assistance. M. Naeser acknowledges the invaluable assistance of Carole Palumbo, Neva Frumkin, Jessica Lydon, Patricia Emery, and Sulochana Naidoo in CT scan analysis and data collection, and Claudia Cassano and Roger Ray for assistance with manuscript preparation. M. Naeser also thanks the Radiology Service of the Boston V.A. Medical Center, including A. Robbins and R. N. Samaraweera, and the Medical Media Service of the Boston, V.A. Medical Center for photography and illustrations (John Dyke and Mary Burke).

This research was supported in part by Medical Research Council Grant MA8907 to A. Kertesz, by the Medical Research Service of the Department of Veterans Affairs, and by USPHS Grant DC00081 to M. A. Naeser.

REFERENCES

Albanese, E., Merlo, A., Albanese, A., & Gomez, E. (1989). Anterior speech region: Asymmetry and weight-surface correlation. *Archives of Neurology, 46(3)*, 307–310.

Blinkov, S. M. (1940). *Transactions of the Brain Institute, Moscow, 5*, 159 (in Russian). Cited in O. S. Adrianov, Structural basis for functional interhemispheric brain asymmetry. *Human Physiology, 5*, 359–363, 1979.

Borod, J. C., Carper, J. M., & Naeser, M. A. (1990). Long-term language recovery in left-handed aphasic patients. *Aphasiology, 4*, 561–572.

Boule, M., & Anthony, R. (1911). L'encéphale de l'homme fosile de la Chapelle-aux-Saints. *Anthropologie, 4*, 65–78.

Bradshaw, J. L. (1989). *Hemispheric specialization and psychological function*. Chichester: John Wiley.

Braune, C. (1891). Das Gewichtsverhältniss der rechten zur linken Hirnhälfte beim Menschen. *Archiv für Anatomie und Entwickelungsgeschichte*, 253–270.

Broca, P. (1863). Localisation des fonctions cérébral siège de la faculté du langage articulé. *Bulletin Société Anthropologie, Paris, 4*, 200–204.

Broca, P. (1885). Data reported by P. Topinard. In A. Delabraye & E. Delabraye (Eds.), *Éléments d'anthropologie générale* (p. 591). Paris: Lecrosnier.

Bryden, M. P. (1978). Strategy effects in the assessment of hemispheric asymmetry. In G. Underwood (Ed.), *Strategies of information processing*. London: Academic Press.

Bryden, M. P. (1982). *Laterality: Functional asymmetry in the intact brain*. New York: Academic Press.

Bryden, M. P., & Sprott, D. A. (1981). Statistical determination of degree of laterality. *Neuropsychologia, 19*, 571–581.

Bub, D., & Lewine, J. (1988). Different modes of word recognition in left and right visual fields. *Brain and Language, 33*, 161–188.

Chui, H. C., & Damasio, A. R. (1980). Human cerebral asymmetries evaluated by computerized tomography. *Journal of Neurology, Neurosurgery, and Psychiatry, 43*, 873–878.

Connolly, C. J. (1950). *External morphology of the primate brain*. Springfield, Illinois: Thomas.

Cunningham, D. J. (1892). *Surface anatomy of the cerebral hemispheres*. Dublin: Academy House.

Curry, F. W. K. (1967). A comparison of left-handed and right-handed subjects on verbal and nonverbal dichotic listening tasks. *Cortex, 3*, 343–352.

Di Chiro, G. (1962). Angiographic patterns of cerebral convexity veins and superficial dural sinuses. *American Journal of Roentgenology, Radium Therapy, and Nuclear Medicine, 87*, 308–321.

Doyle, F. H., Pennock, J. M., Orr, J. S., Gore, J. C., Bydder, G. M., & Steiner, R. E. (1981). Imaging of the brain by nuclear magnetic resonance. *Lancet, 1*, 53–58.

Eberstaller, O. (1884). Zur Oberflächenanatomie der Grosshirnhemisphären. *Wiener Medizinische Blätter, 7*, 479–482.

Eidelberg, D., & Galaburda, A. (1982). Cytoarchitectonic asymmetries in the human parietal lobe. Read before the 34th Annual Meeting of the American Academy of Neurology, Washington, D.C., April 29.

Eidelberg, D., & Galaburda, A. M. (1984). Inferior parietal lobule. Divergent architectonic asymmetries in the human brain. *Archives of Neurology, 41*, 843–852.

Falzi, G., Perrone, P., & Vignolo, L. (1982). Right–left asymmetry in anterior speech region. *Archives of Neurology, 39*, 239–240.

Galaburda, A. M., Corsiglia, J., Rosen, D., Sherman, G. F. (1987). Planum temporale asymmetry: Reappraisal since Geshwind and Levitsky. *Neuropsychologia, 25*, 853–868.

Galaburda, A. M., Rosen, G. D., & Sherman, G. F. (1990). Individual variability in cortical organization: Its relationship to brain laterality and implications to function. *Neuropsychologia, 28(6)*, 529–546.

Galaburda, A. M., Sanides, F., & Geschwind, N. (1978). Human brain. Cytoarchitectonic left–right asymmetries in the temporal speech region. *Archives of Neurology, 35*, 812–817.

Gazzaniga, M. S., & Hillyard, S. A. (1971). Language and speech capacity of the right hemisphere. *Neuropsychologia, 9*, 272.

Geffen, G., Bradshaw, J. L., & Wallace, G. (1971). Interhemispheric effects on reaction time to verbal and nonverbal visual stimuli. *Journal of Experimental Psychology, 87*, 415–422.

Geschwind, N., & Galaburda, A. M. (1985). Cerebral lateralization: Biological mechanisms, associations, and pathology. *Archives of Neurology, 42*, 428–654.

Geschwind, N., & Levitsky, W. (1968). Human brain: Left–right asymmetries in temporal speech regions. *Science, 161*, 186–187.

Glick, S. D. (Ed.) (1985). *Cerebral lateralization in nonhuman species*. New York: Academic Press.

Goodglass, H., & Kaplan, E. (1972). *The assessment of aphasia and related disorders*. Philadelphia: Lea and Febiger.

Gundara, N., & Zivanovic, S. (1968). Asymmetry in East African skulls. *American Journal of Physiology and Anthropology, 28*, 331–338.

Habib, M. (1989). Anatomical asymmetries of the human cerebral cortex. *International Journal of Neuroscience, 47*, 67–79.

Hadziselimovic, H., & Cus, M. (1966). The appearance of the internal structures of the brain in relation to the configuration of the human skull. *Acta Anatomica, 63*, 289–299.

Hellige, J. B. (1990). Hemispheric asymmetry. *Annual Review of Psychology, 41*, 55–80.

Henderson, V. W., Naeser, M. A., Weiner, J. M., Pieniadz, J. M., & Chui, H. C. (1984). CT criteria of hemisphere asymmetry fails to predict language laterality. *Neurology, 34*, 1086–1089.

Heschl, R. L. (1878). *Uber die vordere quere Schlafenwindung des menschlichen Grosshirns*. Wien: Wilhelm Braumuller.

Hines, D., & Satz, P. (1971). Superiority of right visual half-fields in right-handers for recall of digits presented at varying rates. *Neuropsychologia, 9*, 21–25.

Hoadley, N. F. (1929). On measurement of the internal diameter of the skull in relation (i) to the prediction of its capacity (ii) to the 'pre-eminence' of the left hemisphere. *Biometrika, 21*, 85–123.

Holloway, R. L., & De La Coste, M. C. (1982). Brain endocast asymmetry in Pongids and Hominids: some preliminary findings on the paleontology of cerebral dominance. *American Journal of Physiology and Anthropology, 58*, 101–110.

Horvath, D. S., Woodward, D. J., & De La Coste-Lareymondie, M. C. (1985). Cerebral asymmetries and sex differences in hominoids. Abstract. *Proceedings of the Society for Neuroscience*.

Hrdlicka, A. (1907). Measurements of the cranial fossae. *Proceedings of the United States National Museum, 32*, 177–232.

Inglessis, M. (1925). Untersuchungen über Symmetrie und Asymmetrie der Menschlichen Grosshirnhemispharen. *Zeitschrift für des Gesamte Neurologie Psychiatrie, 95/96*, 464–474.

Jack, C. R., Twomey, C. K., Zinsmeister, A. R., Sharbrough, F. W., Petersen, R. C., & Cascino, G. D. (1989). Anterior temporal lobes and hippocampal formations: Normative volumetric measurements from MR images in young adults. *Radiology, 172,* 549–554.

Kertesz, A., Black, S. E., & Howell, J. (1984). Functional lateralization and anatomical asymmetries on nuclear magnetic resonance. Abstract. Presented at Academy of Aphasia, Los Angeles.

Kertesz, A., Black, S. E., Polk, M., & Howell, J. (1986). Cerebral asymmetries on magnetic resonance imaging. *Cortex, 22,* 117–127.

Kertesz, A., Polk, M., Black, S. E., & Howell, J. (1990). Sex, handedness, and the morphometry of cerebral asymmetries on magnetic resonance imaging. *Brain Research, 530,* 40–48.

Kertesz, A., Polk, M., Black, S. E., & Howell, J. (1992). Anatomical asymmetries and functional laterality. *Brain, 115,* 589–605.

Kimura, D. (1964). Left-right differences in the perception of melodies. *Quarterly Journal of Experimental Psychology, 16,* 355–358.

Kimura, D. (1967). Functional asymmetry of the brain in dichotic listening. *Cortex, 3,* 163–178.

Kimura, D. (1969). Spatial localization in left and right visual fields. *Canadian Journal of Psychology, 23,* 445–458.

Kimura, D. (1979). Neuromotor mechanisms in the evolution of human communications. In H. D. Steklis & M. J. Raleigh (Eds.), *Neurobiology of social communications in primates* (pp. 197–219). New York: Academic Press.

Kinsbourne, M. (1974). The mechanisms of hemisphere asymmetry in man. In M. Kinsbourne & A. Smith (Eds.), *Hemispheric disconnection and cerebral function.* Springfield, Illinois: Thomas.

Kinsbourne, M. (1978). Biological determinants of functional bisymmetry and asymmetry. In M. Kinsbourne (Ed.), *Asymmetrical function of the brain* (Vol. 1, pp. 3–13). Cambridge: Cambridge University Press.

Klatzky, R., & Atkinson, R. (1971). Specialization of the cerebral hemispheres in scanning for information in short-term memory. *Perception Psychophysics, 10,* 335–338.

Koff, E., Naeser, M. A., Pieniadz, J. M., Foundas, A. L., & Levine, H. (1986). CT scan hemispheric asymmetries in right- and left-handed males and females. *Archives of Neurology, 43,* 487–491.

Kooistra, C. A., & Heilman, K. M. (1988). Motor dominance and lateral asymmetry of the globus pallidus. *Neurology, 3,* 388–390.

Last, R. J., & Thompsett, D. H. (1953). Casts of cerebral ventricles. *British Journal of Surgery, 40,* 525–542.

LeMay, M. (1976). Morphological cerebral asymmetries of modern man, fossil man, and nonhuman primate. *Annals of the New York Academy of Sciences, 280,* 349–366.

LeMay, M. (1977). Asymmetries of the skull and handedness-Phrenology revisited. *Journal of Neurological Sciences, 32,* 243–253.

LeMay, M., & Culebras, A. (1972). Human brain: morphologic differences in the hemispheres demonstrable by carotid arteriography. *New England Journal of Medicine, 287,* 168–170.

LeMay, M., & Geschwind, N. (1978). Asymmetries of the human cerebral hemispheres. In A. Caramazza & E. B. Zurif (Eds.), *Language acquisition and language breakdown: Parallels and divergencies* (pp. 311–328). Baltimore: Johns Hopkins University Press.

LeMay, M., & Kido, D. K. (1978). Asymmetries of the cerebral hemispheres on computed tomographs. *Journal of Computed Assisted Tomography, 2,* 471–476.

McRae, D. L., Branch, C. L., & Milner, B. (1968). The occipital horns and cerebral dominance. *Neurology, 18,* 95–98.

Mishkin, M., & Forgays, D. G. (1952). Word recognition as a function of retinal locus. *Journal of Experimental Psychology, 43,* 43–48.

Naeser, M. A., & Borod, J. C. (1986). Aphasia in left-handers: CT scan lesion site, lesion side, and hemispheric asymmetries. *Neurology, 36,* 471–489.

Naeser, M. A., & Hayward, R. W. (1978). Lesion localization in aphasia with cranial computed tomography and the Boston Diagnostic Aphasia Exam. *Neurology, 28,* 545–551.

Nikkuni, S., Yashima, Y., Ishige, K., Suzuki, S., Ohno, E., Kumashiro, H., Kobayashi, E., Awa, H., Mihara, T., & Asakura, T. (1981). Left–right hemispheric asymmetry of critical speech zones in Japanese brains. *Brain and Nerve, 33,* 77–84.

Nottebohm, F., & Nottebohm, M. E. (1976). Left hypoglossal dominance in the control of canary and white-crowned sparrow song. *Journal of Comparative Physiology, 108,* 171–192.

Orthner, H., & Seler, W., (1975). Planimetrische volumetrie und menschlichen Gehirnen. *Fortschritte der Neurologie-Psychiatrie, 43,* 191–209.

Palumbo, C. L., Naeser, M. A., & York, A. D. (1990). Reliability of CT scan cerebral hemispheric asymmetry measurements across multiple CT scans and variability in CT scanning angulation. *Neuropsychology, 3,* 231–241.

Pfeifer, F. A. (1936). Pathologie der Hörstrahlung und der corticaler Hörsphare. In O. Bumke and O. Foerster (Eds.), *Handbuch der Neurologie* (pp. 523–626). Berlin: Springer-Verlag.

Pieniadz, J. M., & Naeser, M. A. (1984). Computed tomographic scan, cerebral asymmetries and morphologic brain asymmetries. *Archives of Neurology, 41,* 403–409.

Pieniadz, J. M., Naeser, M. A., Koff, E., & Levine, H. L. (1983). CT scan cerebral hemispheric asymmetry measurements in stroke cases with global aphasia: Atypical asymmetries associated with improved recovery. *Cortex, 19,* 371–391.

Porac, C., & Coren, S. (1981). *Lateral preferences and human behavior.* New York: Springer-Verlag.

Rakic, P., & Williams, R. W. (1986). Thalamic regulation of cortical parcellation: An experimental perturbation of the striate cortex in rhesus monkeys. *Society for Neuroscience Abstracts, 12,* 1499.

Ratcliff, G., Dila, C., Taylor, L., & Milner, B. (1980). The morphological asymmetry of the hemispheres and cerebral dominance for speech: A possible relationship. *Brain and Language, 11,* 87–98.

Rizzolatti, G., Umilta, C., & Berlucchi, G. (1971). Opposite superiorities of the right and left cerebral hemispheres in discriminative reaction time to physiognomical and alphabetical material. *Brain, 94,* 431–442.

Rubens, A. B., Mahowald, M. W., & Hutton, J. T. (1976). Asymmetry of the lateral (sylvian) fissures in man. *Neurology, 26,* 620–624.

Scheibel, A. B., Paul, L. A., Fried, I., Forsythe, A. B., Tomiyasu, U., Wechsler, A., Kao, A., & Slotnick, J. (1985). Dendritic organization of the anterior speech area. *Experimental Neurology, 87,* 109–117.

Schwartz, M., Creasey, H., Grady, C. L., DeLeo, J. M., Frederickson, H. A., Cutler, N. R., & Rapoport, S. I. (1985). Computed tomographic analysis of brain morphometrics in 30 healthy men, aged 21 to 81 years. *Annals of Neurology, 17,* 146–157.

Seaman, J. G., & Gazzaniga, M. S. (1973). Coding strategies and cerebral laterality effects. *Cognitive Psychology, 5,* 249–256.

Seldon, H. L. (1981). Structure of human auditory cortex. I. Cytoarchitectonics and dendritic distribution. *Brain Research, 229,* 277–294.

Seldon, H. L. (1985). The anatomy of speech perception—Human auditory cortex. In A. Peters & E. G. Jones (Eds.), *Cerebral cortex* (Vol. 4, pp. 273–327). New York: Plenum Press.

Sidtis, J. J., Volpe, B. T., Wilson, D. H., Rayport, M., & Gazzaniga, M. S. (1981). Variability in right hemisphere language function after callosal section: Evidence for a continuum of generative capacity. *Journal of Neuroscience, 1*, 323–331.

Steinmetz, H., Rademacher, J., Huang, Y., Hefter, H., Zilles, K., Thron, A., & Freund, H-J. (1989). Cerebral asymmetry: MRI planimetry of the human planum temporale. *Journal of Computer Assisted Tomography, 13*, 996–1005.

Strauss, E., LaPointe, J. S., Wada, J. A., Gaddes, W., & Kosaka, B. (1985). Language dominance: Correlation of radiological and functional data. *Neuropsychologia, 23*, 415–420.

Sweetland, J., Kertesz, A., Prato, F., & Nantau, K. (1987). The effect of magnetic resonance imaging on human cognition. *Annals of Neurology, 17*, 146–157.

Tapley, S. M., & Bryden, M. P. (1985). A group test for the assessment of performance between the hands. *Neuropsychologia, 23*, 215–222.

Vannier, M. W., Brunsden, B. S., Hildebolt, C. F., Falk, D., Cheverud, J. M., Figiel, G. S., Perman, W. H., Kohn, L. A., Robb, R. A., Yoffie, R. L., & Bresina, S. J., (1991). Brain surface cortical sulcal lengths: Quantification with three-dimensional MR imaging. *Radiology, 180*, 479–484.

Von Bonin, G. (1962). Anatomical asymmetries of the cerebral hemispheres. In V. B. Mountcastle (Ed.), *Interhemispheric relations and cerebral dominance* (pp. 1–6). Baltimore: Johns Hopkins Press.

Von Economo, C., & Horn, L. (1930). Über Windungsrelief, Masze, und Rindenarchitektonik der Supratemporalfläche, ihre individuellen und ihre Seitenunterschiede. *Zeitschrift der Gesamten Neurologie Psychiatrie, 130*, 678–757.

Wada, J. A., Clarke, R., & Hamm, A. (1975). Cerebral hemispheric asymmetry in humans: Cortical speech zones in 100 adult and 100 infant brains. *Archives of Neurology, 32*, 239–246.

Wagner, H. (1864). *Massbestimmungen der Oberfläche des grossen Gehirns.* Kassel: Wiland.

Webster, W. G. (1977). Hemispheric asymmetry in cats. In S. Harnard, R. W. Doty, L. Goldstein, J. Jaynes, & G. Krauthamer (Eds.), *Lateralization in the nervous system* (pp. 471–490). New York: Academic Press.

Weinberger, D. R., Luchins, D. J., Morihisa, J., & Wyatt, R. J. (1982). Asymmetrical volumes of the right and left frontal and occipital regions of the human brain. *Annals of Neurology, 11*, 97–100.

Weis, S., Haug, H., Holoubek, B., & Orun, H. (1989). The cerebral dominances: Quantitative morphology of the human cerebral cortex. *International Journal of Neuroscience, 47*, 165–168.

Wilde, J. (1926). Über das Gewichtsverhältnis der Hirnhälften beim Menschen. *Litvijas Universiti Raksti, 14*, 271–288.

Witelson, S. F., & Kigar, D. L. (1988). Asymmetry in brain function follows asymmetry in anatomical form: Gross, microscopic, postmortem and imaging studies. In F. Boller & J. Grafman (Eds.), *Handbook of neuropsychology* (Vol. 1, pp. 111–142). Amsterdam: Elsevier Science.

Witelson, S. F., & Pallie, W. (1973). Left hemisphere specialization for language in the newborn: Anatomical evidence of asymmetry. *Brain, 96*, 641–646.

Yeni-Komshian, G. H., & Benson, D. A. (1976). Anatomical study of cerebral asymmetry in the temporal lobe of humans, chimpanzees and rhesus monkeys. *Science, 192*, 387–389.

Zaidel, E. (1976). Auditory vocabulary of the right hemisphere following brain bisection or hemidecortication. *Cortex, 12*, 191–212.

Zaidel, E. (1977). Auditory language comprehension in the right hemisphere following cerebral commissurotomy and hemispherectomy: A comparison with child language and aphasia. In A. Caramazza & E. Zurif (Eds.), *Language acquisition and language breakdown.* Baltimore: Johns Hopkins University Press.

Zaidel, E. (1978). Lexical organization in the right hemisphere. In P. A. Buser and T. Rougeul-Buser (Eds.), *Cerebral correlates of conscious experience* (Institut National de la Santé et de la Recherche Médicale (INSERM) Symposium No. 6., pp. 277–299). Amsterdam: Elsevier/North-Holland.

Zurif, E. B., & Bryden, W. P. (1969). Familial handedness and left-right difference in auditory and visual perception. *Neuropsychologia, 7,* 179–187.

Chapter 9

Neuroimaging and Recovery of Auditory Comprehension and Spontaneous Speech in Aphasia with Some Implications for Treatment in Severe Aphasia

Margaret A. Naeser

I. INTRODUCTION

Since the advent of *in vivo* brain imaging with computerized tomography (CT) scans over 20 years ago (Hounsfield, 1973), the field of lesion localization and aphasia research has advanced markedly. Most CT scan studies have focused primarily on the relationship between lesion in major cortical language areas and resulting aphasia syndromes including Broca's aphasia (Alexander, Naeser, & Palumbo, 1990; Knopman, Selnes, Niccum, Rubens, Yock, & Larson, 1983; Mohr, Pessin, Finkelstein, Funkenstein, Duncan, & Davis, 1978), Wernicke's aphasia (Kertesz, 1983; Naeser, Helm-Estabrooks, Haas, Auerbach, & Srinivasan, 1987), conduction aphasia (Damasio & Damasio, 1980), transcortical motor aphasia (Freedman, Alexander, & Naeser, 1984), and others (Barat, Constant, Mazaux, Caille, & Arne, 1978; Damasio, 1981; Kertesz, Harlock, & Coates, 1979; Mazzocchi & Vignolo, 1980; Naeser & Hayward, 1978; Selnes, Knopman, Niccum, Rubens, & Larson, 1983; Selnes, Niccum, Knopman, & Rubens, 1984).

Some CT scan studies have focused on the relationship between lesion in primarily subcortical areas and resulting aphasia (Cappa, Cavallotti, Guidotti, Papagno, & Vignolo, 1983; Mazzocchi & Vignolo, 1979). Subcortical lesion sites usually include primarily the thalamus (Alexander &

LoVerme, 1980) or the basal ganglia (putamen/caudate) and surrounding white matter (Alexander & Naeser, 1988; Alexander, Naeser, & Palumbo, 1987; Damasio, Damasio, Rizzo, Varney, & Gersh, 1982; Naeser, Alexander, Helm-Estabrooks, Levine, Laughlin, & Geschwind, 1982).

Previous CT scan studies have taught us about the precise location of neuroanatomical structures and white matter pathways on CT scans associated with aphasia. This chapter focuses on the relationship between lesion in a few of these cortical and subcortical areas on CT scan and two specific areas of language recovery in aphasia: (1) recovery of auditory language comprehension in Wernicke's aphasia patients and global aphasia patients and (2) recovery of some spontaneous speech (nonfluent, agrammatic speech), versus recovery of no spontaneous speech, in patients with severe limitation in speech. The emphasis in this chapter is not on aphasia syndromes per se, but on recovery in auditory comprehension and in spontaneous speech. This chapter explains how to examine a chronic CT scan of any aphasia patient, and how to understand recovery potential for these two aspects of language behavior in the chronic phase, poststroke. The CT scans on which the predictions for recovery are based are performed at least 2–3 months after stroke onset (MPO). The recovery of auditory comprehension or spontaneous speech is based on expected language behavior at least 6 months or 1 year postonset. Most material in this chapter is based on our CT scan research at the Boston University Aphasia Research Center (Boston V.A. Medical Center).

The term "recovery" in this chapter usually refers to the actual late language scores, that is, the language scores obtained by a patient 6 MPO or even 1–3 years after stroke onset. The term "recovery" does not usually refer to "amount of change" or "rate of change" from early Time 1 scores (1–2 MPO) to late Time 2 scores (6, 12, or 36 MPO). These are different areas of investigation. The Time 1 and Time 2 scores, and therefore the "amount of change," are, however, listed in most of our published papers. All our CT scan recovery studies have been retrospective recovery studies.

The latest postonset language scores were always used for any patient, that is, scores from the longest time after onset, and, thus, the more stable later scores. The early language scores 1–2 MPO were examined also, but these individual scores usually were not useful for predictions of recovery in the chronic phase.

Lesion size analysis is not included in most of the material presented in this chapter, because research from our laboratory and others has not found total lesion size to be helpful in making predictions for recovery, except in very large or very small lesions (Kertesz et al., 1979; Naeser, Palumbo, Helm-Estabrooks, Stiassny-Eder, & Albert, 1989; Naeser et al., 1987; Selnes et al., 1983,1984; Vignolo, 1979). For example, in our previous lesion size studies, the mean lesion size for transcortical motor aphasics was 5.7% left hemisphere tissue damage (SD 3.3) and the mean lesion size for conduction aphasics was 3.6% (SD 2.4); no significant difference existed (Naeser, Hayward, Laughlin, Becker, Jernigan, & Zatz, 1981a; Naeser, Hayward, Laughlin, & Zatz, 1981b). The transcortical motor aphasics had primarily frontal lobe lesions, and the conduction aphasics had primarily parietal lobe lesions. The two groups had very different language behavior characteristics, including excellent sentence repetition for the transcortical motor aphasics and poor sentence repetition for the conduction aphasics. In our experience, knowing only the percentage lesion size without knowing the exact lesion sites is not useful in predicting language recovery.

The mean lesion size for global aphasia patients with cortical and subcortical lesion is 28.1% left hemisphere tissue damage (SD 11.2) (complete left middle cerebral artery territory ischemic infarction). The mean lesion size for global aphasia patients with primarily subcortical lesion is only 13.6% left hemisphere tissue damage (SD 3.2) (Naeser, 1983). The subcortical global aphasia patients have total lesion size (13.6%) that is less than half that observed in global aphasia patients with cortical and subcortical lesion (28.1%). The subcortical global aphasia patients have lesion primarily in the internal capsule and putaminal area, with white matter lesion extension in three directions: (1) anterior (across the anterior limb, internal capsule, and white matter near the frontal horn), (2) posterior (across white matter in the temporal isthmus), and (3) superior (into the white matter adjacent to the body of the lateral ventricle) (Naeser, et al., 1982). The subcortical global aphasia patients have lesions that undercut the critical white matter pathways for speech and comprehension. These small subcortical white matter lesions may extend only a few millimeters in one direction or another, yet produce profound deficits. Hence, over the last decade, our research has increasingly shifted away from overall lesion size analysis to precise CT scan lesion site analysis.

A. CT Scan Lesion Site Analysis

Most CT scans examined in our research are obtained at approximately 15–20° to the canthomeatal line, with 10 mm slice thickness, at 7-mm intervals, through the ventricles beginning at the level of the suprasellar cistern. CT scan lesion site analysis includes cortical language areas as well as subcortical areas. These areas are diagrammed on the CT scan slices shown in Fig. 1. Most of these neuroanatomical areas are listed in various CT scan atlases (DeArmond, Fusco, & Dewey, 1976; Hanaway, Scott, & Strother, 1977; Matsui & Hirano, 1978).

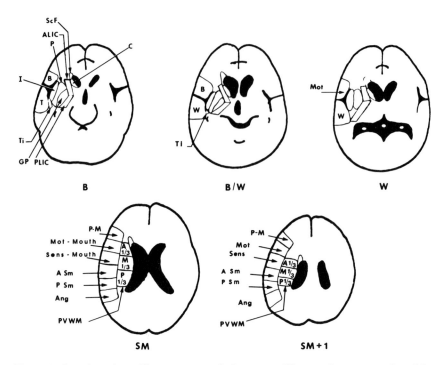

Figure 1 Location of specific neuroanatomical areas on CT scans that were analyzed for presence of lesion in the CT scan lesion site analysis. The CT scan slices B, B/W, W, SM, and SM + 1 are labeled according to the Naeser and Hayward (1978) slice-labeling system. Each neuroanatomical area was examined for extent of lesion using a 0–5 point scale (0, no lesion; 5, entire area has solid lesion; see text). B, Broca's area (45 on slice B; 44 on slice B/W); T, temporal lobe anterior–inferior to Wernicke's area on slice B; Ti, temporal isthmus; I, insular structures including insula, extreme capsule, claustrum, external capsule; P, putamen; GP, globus pallidus; ALIC, anterior limb, internal capsule; PLIC, posterior limb, internal capsule; Sc F, medial subcallosal fasciculus; C, caudate; W, Wernicke's area (22); Mot, motor cortex; PM, premotor cortex; Sens, sensory cortex; A Sm, anterior supramarginal gyrus; P Sm, posterior supramarginal gyrus; Ang, angular gyrus; PVWM, periventricular white matter area (A 1/3, anterior 1/3 PVWM; M 1/3, middle 1/3 PVWM; P 1/3, posterior 1/3 PVWM). Reproduced with permission from Naeser et al. (1989).

The extent of lesion within each neuroanatomical area is assessed visually using a 0-to-5 point scale, where 0 = no lesion; 1 = equivocal lesion; 2 = small, patchy, or partial lesion; 2.5 = patchy, less than half of area has lesion; 3 = half of area has lesion; 3.5 = patchy, more than half of area has lesion; 4 = more than half of area has solid lesion; and 5 = total area has solid lesion. Lesion extent values >3 are of special importance. Lesion extent values >3 (indicating lesion in more than half of a specific area) have been observed to correlate with increased severity of language deficit (Naeser et al., 1987,1989; Naeser, Gaddie, Palumbo, & Stiassny-Eder, 1990). All scans were rated by at least two experienced raters and conferenced data were used. In previous studies, we have observed an inter-rater reliability coefficient of .93 (Borod, Carper, Goodglass, & Naeser, 1984).

II. RECOVERY OF AUDITORY LANGUAGE COMPREHENSION

A. In Wernicke's Aphasia

In this retrospective study, CT scans and auditory comprehension scores were examined for 10 male Wernicke's aphasia patients classified after 6 MPO as mild, good recovery cases (GR) ($n = 5$) or moderate–severe, poor recovery cases (PR) ($n = 5$) (Naeser et al., 1987). Each patient was right-handed and had suffered single-episode left hemisphere occlusive vascular stroke between the ages of 47 and 71 years ($x = 58.4$; SD 6.9), with no significant group differences. The CT scans used for lesion localization were performed between 3 and 36 MPO.

The auditory comprehension test scores from the Boston Diagnostic Aphasia Exam (BDAE) (Goodglass & Kaplan, 1972) were examined from two time periods. Time 1 (T_1) scores were obtained 1–2 MPO; Time 2 (T_2) scores were obtained after 6 MPO.

Patients could not be differentiated on a case-by-case basis using only the early T_1 test scores. The second test to determine T_2 scores was administered 6–13 MPO for the mild group and 12–38 MPO for the moderate–severe group. The T_2 scores for the moderate–severe group were taken as long after onset as possible to extend the potential recovery period.

Patients were differentiated on the basis of T_2 scores as follows: (1) GR cases scored above 0 (above the 50th percentile) on the BDAE Overall Auditory Comprehension Z-Score; (2) PR cases scored below 0 (below the 50th percentile) (see Fig. 2). The reader is referred to the original paper for exact T_1 and T_2 test scores for all 10 Wernicke's aphasia patients (Naeser et al., 1987).

The CT scans were analyzed with two methods: (1) the CT scan lesion site analysis described earlier in which the 0–5 extent-of-lesion scale was used to rate visually the amount of infarction (extent of lesion) within specific cortical and subcortical areas and (2) a computer-based lesion size analysis that quantified the total percentage of left hemisphere temporo-parietal lesion size (Jernigan, Zatz, & Naeser, 1979; Naeser et al., 1981a).

Since the time of Wernicke, multiple interpretations have been made regarding the exact location and limits of the so-called "Wernicke's area"

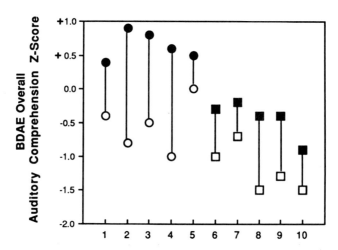

Figure 2 Time 1 (1–2 MPO) and time 2 (6 MPO or 1–3 yr) Overall Auditory Comprehension Z-Scores for 10 Wernicke's aphasia patients. Graph shows overlap in the T_1 scores among some of the good recovery (GR) cases (1–5) and poor recovery (PR) cases (6–10). Thus, the T_1 test scores could not be used on a case-by-case basis to predict GR or PR after 6 months post onset. ○; T_1, GR; ●, T_2, GR; □, T_1, PR; ■, T_2, PR.

(Bogen & Bogen, 1976). For the purposes of this study, Wernicke's area was defined as the posterior two-thirds of the left superior temporal gyrus area. On CT scans, the anterior half of Wernicke's area, that is, the middle third of the superior temporal gyrus area, was located lateral to the maximum width of the third ventricle on slice B/W (Fig. 1). In addition, the posterior half of Wernicke's area, that is, the posterior third of the superior temporal gyrus area, was located lateral to the roof of the third ventricle on slice W (Fig. 1; see area marked "W" on slices B/W and W). The supramarginal and angular gyrus areas in the parietal lobe also were analyzed on slices SM and SM+1 (Fig. 1).

All GR Wernicke's patients with T_2 Auditory Comprehension Z-Scores above 0 had lesion in only half or less than half of Wernicke's area. All PR Wernicke's patients with T_2 Auditory Comprehension Z-Scores below 0 had lesion in more than half of Wernicke's area (see Fig. 3, *top*). The correlation between T_2 BDAE Overall Auditory Comprehension Z-Scores and the extent of lesion within Wernicke's Area was $-.91$ ($p < .001$).

The total percentage of left temporo-parietal lesion size was not useful in distinguishing between cases with good recovery and poor recovery of

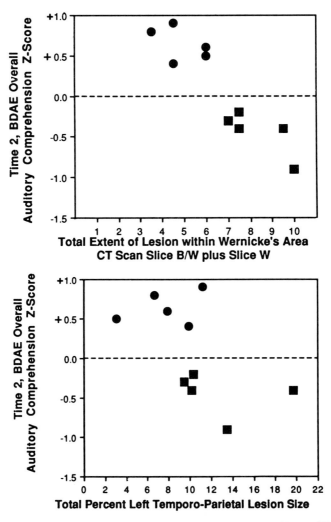

Figure 3 (Top) Graph showing a highly significant correlation ($r = -.91, p < .001$) between total extent of lesion within Wernicke's area on CT scan slices B/W and W and T_2 BDAE Overall Auditory Comprehension Z-Scores for 10 Wernicke's aphasia patients. A total extent of lesion value of 10 reflects a rating of 5 (complete solid lesion) in Wernicke's area on both CT slices B/W and W. Patients with total extent of lesion values ≤6 have lesion in only half or less than half of Wernicke's area; these patients (●) had good recovery after 6 MPO, and T_2 BDAE Z-Scores above 0 or above the 50th percentile. Patients with total extent of lesion values >6 have lesion in more than half of Wernicke's area; these patients (■) still had poor recovery 1–3 years after stroke onset. (Bottom) Graph showing no significant correlation ($r = -.56$, ns) between total percentage of left temporo-parietal lesion size on CT scan slices B, B/W, W, SM, and SM + 1 and T_2 BDAE Overall Auditory Comprehension Z-Scores for 10 Wernicke's aphasia patients. Overlap was seen between the GR cases (●) and the PR (■) cases around the 10% lesion size value. Total lesion size could not be used to separate all GR and PR cases at T_2 testing. Only total extent of lesion within Wernicke's area could be used to separate all GR from PR cases at T_2 testing (see *top*).

auditory comprehension at T_2 (see Fig. 3, *bottom*). The correlation between the T_2 Auditory Comprehension Z-Scores and total percentage of left temporo-parietal lesion size was $-.56$ (n.s.). Also no significant correlation existed between "amount of change" between T_1 and T_2, and extent of lesion within Wernicke's area ($r = -0.494$; n.s.) or total percentage of left hemisphere temporo-parietal lesion size ($r = -0.013$; n.s.). A significant correlation, however, existed between the total percentage of left temporo-parietal lesion size and the T_2 Visual Confrontation Naming scores from the BDAE ($-.88$; $p < .001$). This finding (naming) is in general agreement with those of Kertesz (1979), who found that the highest degree of correlation between total lesion size and severity of aphasia existed for anomic aphasia patients.

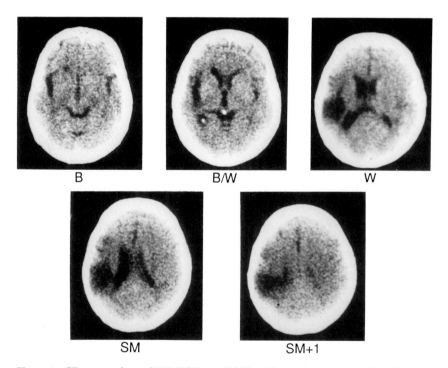

Figure 4 CT scan performed 24 MPO in a mild Wernicke's aphasia patient (Case 2) who had good recovery ($+0.9$ on BDAE Auditory Comprehension Z-Score) at T_2 testing (7 MPO). Lesion was present only in the posterior half of Wernicke's area on slice W (lesion extent value, 4.5). Additional parietal lobe lesion was found in the anterior and posterior supramarginal gyrus areas, surface and deep. Reproduced with permission from Naeser et al. (1987).

Case examples

Figure 4 shows the CT scan of a Wernicke's aphasia patient with lesion in Wernicke's cortical area, only on slice W, and good recovery of auditory comprehension 7–10 MPO (lesion in about half of Wernicke's total area). Figure 5 shows the CT scan of a Wernicke's aphasia patient with complete lesion in Wernicke's cortical area on slice B/W and slice W, and poor recovery of auditory comprehension 14 MPO (lesion in all of Wernicke's area).

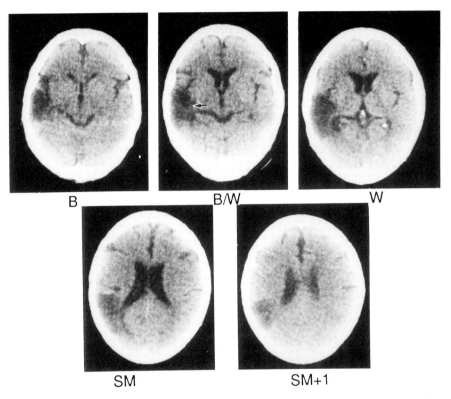

Figure 5 CT scan performed 7 MPO in a severe Wernicke's aphasia patient (Case 10) who had poor recovery (−0.9 on BDAE Auditory Comprehension Z-Score) at T_2 testing (14 MPO). Extensive lesion was present in Wernicke's area on both slices B/W and W (lesion extent value, 5 on each slice; total lesion extent value of 10). Large temporal lobe lesion also was present on slice B (lesion extent value, 4.5), anterior and inferior to Wernicke's area. Additional parietal lobe lesion was found in anterior and posterior supramarginal gyrus areas, surface and deep. There was some lesion in angular gyrus on slices SM and SM + 1. Reproduced with permission from Naeser et al. (1987).

Results from this study support the notion that careful examination of extent of lesion within Wernicke's area on a chronic CT scan (performed after 2–3 MPO) may be useful in predicting long-term recovery of auditory comprehension in Wernicke's aphasia patients. Those patients with lesion in only half or less than half of Wernicke's area have a better prognosis for recovery of auditory comprehension within the first year after stroke onset.

B. In Global Aphasia

In this retrospective study, CT scans and auditory comprehension scores were examined for 14 right-handed stroke patients with global aphasia (12 men and 2 women; age 50–66 years) who had unilateral left hemisphere ischemic infarcts (Naeser et al., 1990). All patients had been tested at least twice with the BDAE (Goodglass & Kaplan, 1972). Time 1 testing ranged from 1 to 4 MPO. All patients had been classified as globally aphasic at T_1 on the basis of the BDAE. All patients had BDAE Auditory Comprehension Z-scores at T_1 that were below -1.0, that is, indicative of severe auditory comprehension deficits. Time 2 testing was approximately 1 to 2 years after stroke onset.

All patients had CT scans that were obtained after 2 MPO (range, 2–110 MPO). CT scan lesion site analysis was performed. Most of the cortical and subcortical areas shown in Fig. 1 were assessed visually for extent of lesion, including major frontal, parietal, and temporal lobe areas as well as subcortical structures. Special emphasis was placed on analyzing lesion extent in Wernicke's cortical area (and immediate subjacent white matter) on CT scan slices B/W and W, and lesion extent in the subcortical temporal lobe structure, the temporal isthmus area, on CT scan slices B and B/W.

The temporal isthmus area contains auditory pathways from the medial geniculate body to Heschl's gyrus. Lesion in the temporal isthmus area has been associated with auditory language comprehension deficits since the time of Nielsen (1946). The location of this area was defined as the white matter that is inferior to the sylvian fissure/insular area and superior to the temporal horn (Naeser et al., 1982; Nielsen, 1946) (see Fig. 6).

Nielsen has described the following measurements of the small subcortical temporal isthmus area: "It measures from 10 to 15 mm across and is in height nearly equal to that of the thalamus.... The artery of supply of the isthmus is the anterior choroidal." In the current study, only the anterior half of the temporal isthmus was evaluated for extent of lesion in auditory pathways; the posterior half of the temporal isthmus contains visual pathways.

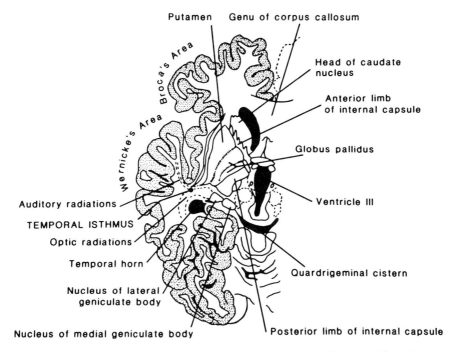

Figure 6 Schematic of CT scan slice B/W (left hemisphere) showing location of the auditory radiations within the anterior half of the temporal isthmus (Ti). The Ti is located in the white matter inferior to the sylvian fissure and superior to the temporal horn. Reproduced with permission from Naeser et al. (1990).

On the basis of CT scan lesion site analysis, the subjects were classified into two groups. Group 1 global aphasia cases had cortical/subcortical lesion in the frontal, parietal, and temporal lobes including Wernicke's cortical area. Each case in Group 1 ($n = 9$) had lesion in at least half of Wernicke's cortical area. Group 1 cases are labeled FPT cases to reflect cortical/subcortical lesion in the frontal, parietal, and temporal lobes.

Group 2 global aphasia cases ($n = 5$) also had cortical/subcortical lesion in the frontal and parietal lobes, but had only subcortical lesion in the temporal lobe including the subcortical temporal isthmus area (Ti). Group 2 cases are labeled FPTi cases to reflect cortical/subcortical lesion in the frontal and parietal lobes, but only subcortical temporal lobe lesion including the Ti. Both groups had similar mean lesion extent values in frontal, parietal, and subcortical areas, including the subcortical Ti area. All cases in the FPT group had lesion in more than half of Wernicke's cortical area; none of the cases in the FPTi group had cortical lesion in Wernicke's area.

No significant difference in age at stroke onset existed between the two groups (FPT group: mean, 58.2 years; SD 4.2; FPTi group: mean, 57.8 years; SD 5.0). Each group contained only one woman. No significant differences existed between the two groups in terms of MPO when T_1 or T_2 testing was performed.

In four of the five FPTi cases, the T_2 Auditory Comprehension Z-scores were above -0.5. In eight of the nine FPT cases, the T_2 Auditory Comprehension Z-scores were below -0.5 (see Fig. 7). A significantly greater increase ($p < .01$) was seen in the amount of recovery that had taken place from T_1 to T_2 for the FPTi group relative to the FPT group in the BDAE Overall Auditory Comprehension Z-score. The mean change from T_1 to T_2 for the FPTi group was $+1.58$. The mean change from T_1 to T_2 for the FPT group was only $+0.65$.

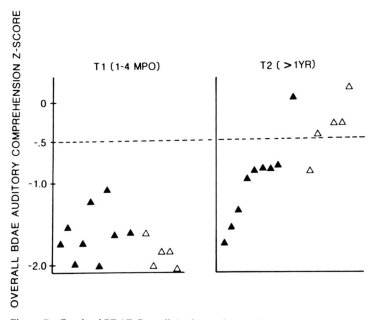

Figure 7 Graph of BDAE Overall Auditory Comprehension Z-scores for all cases at Time 1 (T_1) testing and Time 2 (T_2) testing. Note that, at T_1 testing, not one FPT case (▲; cortical/subcortical lesion in the frontal, parietal, and temporal lobe, including Wernicke's cortical area) or FPTi case (△; cortical/subcortical lesion in the frontal and parietal lobes, but only subcortical temporal lobe lesion including the temporal isthmus) achieved a Z-score that was better than -1.0. At T_2 testing, four of five FPTi cases achieved Z-scores better than -0.5. At T_2 testing, only one of nine FPT cases achieved a Z-score better than -0.5. Reproduced with permission from Naeser et al. (1990).

The FPTi cases had a significantly greater ($p < .01$) amount of recovery from T_1 to T_2 at the single-word level of comprehension (Word Discrimination and Body-Part Identification subtests) relative to the FPT cases. Patients in the FPTi group actually had significantly higher ($p < .01$) Body-Part Identification absolute scores at T_2 than did patients in the FPT group (T_2 FPTi: mean, 14.3; SD 3.6; T_2 FPT: mean, 5.7; SD 4.5).

Thus, most global aphasia cases with temporal lobe lesion that included at least half of Wernicke's cortical area had poor recovery of auditory comprehension 1–2 years after stroke onset, whereas most global aphasia cases with only subcortical temporal lobe lesion including the subcortical temporal isthmus had better recovery of auditory comprehension 1–2 years poststroke.

No significant differences were seen between the two groups in the amount of recovery that had taken place from T_1 to T_2 in the number of words per phrase length in spontaneous speech, single-word repetition, or naming. Most subjects in each group remained severely impaired in these three areas at T_2. The reader is referred to the original paper for all exact T_1 and T_2 scores (Naeser et al., 1990).

Case examples

Figure 8 shows the CT scan and BDAE Auditory Comprehension Z-scores for an FPTi case with relatively good recovery of auditory comprehension after 1 year. Figure 9 shows the CT scan and BDAE Auditory Comprehension Z-scores for an FPT case with poor recovery of auditory comprehension, even 8 years after stroke onset.

Results from this study suggest that careful examination of cortical and subcortical lesion in the temporal lobe on CT scan provides information regarding potential for recovery of some auditory language comprehension (especially single-word comprehension) after 1 year post onset in a subset of global aphasia patients. A majority of the patients (approximately 80%) with only subcortical temporal isthmus lesion in the temporal lobe (compared with cortical lesion in Wernicke's area in the temporal lobe) had increased recovery of single-word comprehension 1 year after onset.

The results from this study support the notion of Sarno and Levita (1979,1981) that global aphasia patients are not a homogeneous group. These results suggest that careful examination of cortical and subcortical lesion in the temporal lobe can result in information that may be useful in predicting a subset of global aphasia patients who have potential for increased recovery of auditory comprehension after 1–2 years poststroke onset. The CT scans used for this predictive information should be obtained 2–3 MPO because the exact borders of an infarct are not well visualized on CT scans that are performed earlier.

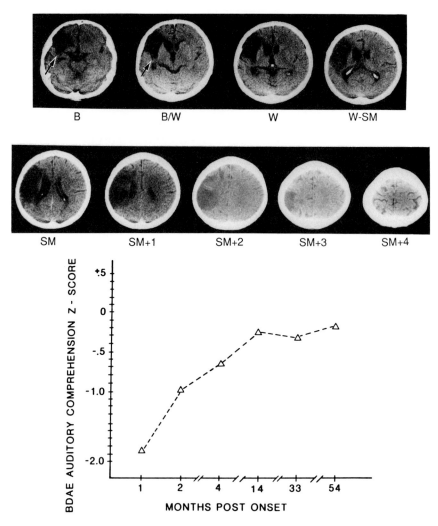

Figure 8   (Top) CT scan at 33 MPO of an FPTi case (age, 61 years) showing extensive corti-
cal/subcortical lesion in the frontal and parietal lobes, but only subcortical temporal lobe
lesion in the temporal isthmus at slices B and B/W (arrows). Note complete sparing of
Wernicke's cortical area on slices B/W and W. (Bottom) Graph showing this patient's BDAE
Overall Auditory Comprehension Z-scores over a period of several months after onset. Note
good recovery of auditory comprehension beginning at 2–4 MPO. His BDAE Auditory Comprehension Z-scores were −0.24, −0.33, and −0.18 at 14, 33, and 54 MPO, respectively.
Reproduced with permission from Naeser et al. (1990).

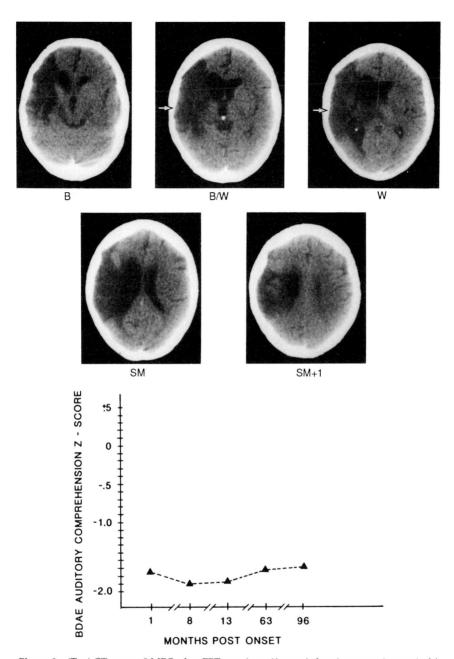

Figure 9 (Top) CT scan at 8 MPO of an FPT case (age, 61 years) showing extensive cortical/ subcortical lesion in the frontal, parietal, and temporal lobes, including Wernicke's cortical area, compatible with global aphasia. Complete lesion was seen in Wernicke's cortical area, including the immediately subjacent white matter on slices B/W and W (arrows). (Bottom) Graph showing this patient's BDAE Overall Auditory Comprehension Z-scores over a period of several months after onset. A severe auditory comprehension deficit was still present at 8 years after onset (Z-score = −1.7). Reproduced with permission from Naeser et al. (1990).

III. RECOVERY OF SPONTANEOUS SPEECH

Research in our laboratory and others has demonstrated that the presence of lesion in certain subcortical white matter areas, as detected by CT scan, can have a profound effect on limiting spontaneous speech (Alexander et al., 1987; Hier, Davis, Richardson, & Mohr, 1977; Naeser et al., 1982,1989). In our 1989 study, for example, we observed recovery of spontaneous speech to be related to amount of lesion in two specific white matter areas combined: (1) the medial subcallosal fasciculus area, which is located deep to Broca's area on slices B and B/W, plus (2) the middle third of the periventricular white matter area, which is located deep to the motor/sensory cortex area for mouth on slice SM. These two important areas are described in detail later in this section.

In this retrospective study, CT scans and number of words per phrase length in spontaneous speech were examined for 27 right-handed aphasia patients (24 men and 3 women) with single episode left hemisphere occlusive vascular strokes (thromboembolic infarcts) (Naeser et al., 1989). Their mean age at onset was 57.6 years (SD 7.6; range, 35–69 years). Each patient had a CT scan performed from 2 months to 9 years following stroke onset.

The number of words per phrase length for spontaneous speech was determined from the elicited spontaneous speech sample for description of the Cookie Theft Picture from the BDAE (Goodglass & Kaplan, 1972). These speech samples were obtained from the latest testing time available after stroke onset (6 MPO to 9 years). The samples were used to assign patients to one of four groups, based on severity of impairment of spontaneous speech. The classification of patients according to severity of spontaneous speech was carried out independently from the CT scan analysis.

Group 1: No speech or only a few irrelevant words. Group 1 consisted of 7 patients (6 men, 1 woman) who were able to provide either no speech or only a few irrelevant words in describing the Cookie Theft Picture (see Table 1 for speech samples). The speech samples for these patients were obtained at a variety of times after onset, ranging from 9 MPO to 8 years. For information on auditory comprehension, repetition, and naming for each patient, see Table 2. Not all cases were globally aphasic in all areas of language.

Group 2: Only stereotypies. Group 2 consisted of 10 patients (all men) who were able to provide only stereotypies in describing the Cookie Theft Picture (see Table 1 for speech samples). The speech samples for these patients were obtained ranging from 6 MPO to 9 years. This group, like Group 1, included cases who were not globally aphasic in all areas of language. For more information on each patient, see Table 2.

Table 1 Spontaneous Speech Samples for Patients Studied Regarding Recovery of Spontaneous Speech

Case	Time postonset	Spontaneous speech samples for description of the Cookie Theft Picture
Group 1		
1	27 months	"Yeah . . . yeah"
2	8 years	No speech
3	9 months	No speech
4	47 months	"Juh . . . ah . . . jou . . . juhjuh . . . uhpai . . . uhnouer"
5	2.5 years	No speech
6	15 months	"No . . . " (and grunts)
7	9 months	No speech
Group 2		
8	6 years	"Boom . . . boom"
9	18 months	"Ai . . . da tu . . . dididi"
10	13 months	"Senny fenny"
11	9 years	"I don't know . . . good good . . . yes, yes . . . tu, tu . . . no, no"
12	4 years	"Wa, wa . . . for Christ sake"
13	15 months	"Guhdi, guhdi . . . wazuh waz"
14	2 months	"Bee bee . . . bye bye"
	(2 years)	At this time there was almost no speech
15	6 months	"Yes, yes"
16	33 months	"Morning, morning . . . boy, boy"
17	13 months	"1, 2, 3, 4, 5 . . . boom, boom"
Group 3		
18	35 months	Unintelligible vowel sound; "Siuhl . . . yeah down . . . un . . . cookies wash um um wahs eeah no water here, fuhee, no good over here."
19	7 months	"Goddam . . . Chrissakes . . . I forgot it . . . well goddam"
20	15 months	"There, too . . . there, too . . . um . . . I don't know . . . that's all I guess gee whiz. I don't know, that's all . . . well . . . that, too and there and there."
21	52 months	"Well . . . uh . . . Duh um . . . glasses . . . run"
22[a]	8 months	"Nothing. The kid break'in an . . . on an that one. He gonna get gett'in, gah. It's running. He given one to give one. She's dissing."
Group 4		
23	7 months	"The wady is doing her dishes. Sink undis over uh . . . The window is open and the "w" won . . . a very funny day outside . . . ook children . . . a boy and a girl."
24	7 months	"A kids . . . a cookies . . . and uh, uh, fall down, . . . wash'in de dishes . . . un runn'in water . . . fish fash . . . uh foor . . . he was . . . girl a cookie."
25	24 months	"The girl . . . uh, sh-sh sheez, the boy fall down . . . the ch-ch chair . . . the boy . . . is . . . cookies . . . the boy, the lady . . . is . . . raiping the dishes."
26	17 months	"Well, wiss . . . watcheez, water, . . . uh, this kaitee jar . . . uh . . . do . . . eee . . . dee . . . deezeez . . . uh, ahniz . . . ahniz, uh, whoops . . . bay . . . birl . . . no . . . girl . . . boy . . . girl, I/ton/know."
27	6 years	"Dis iz . . . bee out a lawn built up . . . This kid . . . fall down . . . This kid waking up here."

[a] At this time the patient's speech output was almost compatible with nonfluent Broca's aphasia; however, her comprehension was still too impaired to be considered a Broca's aphasic.

Table 2 Patient Data and Boston Diagnostic Aphasia Examination Test Scores for Patients Studied Regarding Recovery of Spontaneous Speech, Groups 1–4

Case	Sex	Age at onset (yr)	Testing (time postonset)	BDAE auditory comprehension z score	Auditory comprehension Words (72)	Auditory comprehension Commands (15)	Word repetition (10)	Visual confrontation naming (105)
Group 1								
1 (S.F.)	F	66	27 months	-1.60	28.5	2	5	0
2 (L.P.)	M	61	8 years	-1.7	30	1	8	6
3 (H.J.)	M	35	7 months	-0.21	51.5	10	0	0
4 (T.F.)	M	65	4 years	-1.19	15.5	9	5	0
5 (M.W.)	M	68	2.5 years	-0.58	42.5	10	7	19
6 (L.N.)	M	52	15 months	-1.0	22.5	6	0	0
7 (H.L.)	M	53	9 months	-0.11	60	13	1	0
Group 2								
8 (W.C.)	M	53	3 months	-1.9	11	4	0	0
9 (D.A.)	M	54	18 months	-1.34	47	4	0	0
10 (D.E.)	M	56	13 months	-0.60	53	7	1	0
11 (G.P.)	M	55	9 years	+0.05	60	12	5	24
12 (H.M.)	M	58	8 years	-0.90	35	9	5	4
13 (A.G.)	M	64	15 months	-0.94	43	9	5	0
14 (J.N.)	M	53	2 years	-0.21	50	4	0	0
15 (E.H.)	M	55	6 months	-1.95	13.5	9	4	0
16 (K.M.)	M	59	33 months	-0.33	60	8	0	0
17 (G.J.)	M	59	13 months	+0.09	57	10	7	0
Group 3								
18 (H.D.)	M	63	35 months	+0.29	57	12	6	42
19 (C.A.)	M	69	7 months	-2.1	10.5	2	0	0
20 (K.W.)	M	61	15 months	-0.90	55.5	5	0	4
21 (A.A.)	M	58	52 months	-0.70	41	8	9	13
22 (Z.J.)	F	64	11 months	-0.44	57	5	6	0
Group 4								
23 (W.A.)	M	50	7 months	+0.75	66.5	14	9	83
24 (M.E.)	M	58	5 months	+0.55	60	13	6	62
25 (M.L.)	F	56	24 months	+0.93	71	15	9	85
26 (B.J.)	M	67	17 months	+0.84	70	15	DNT[a]	101
27 (T.H.)	M	42	6 years	+0.38	58.5	12	7	58

[a] DNT, Did not test.

Group 3: A few words and/or some overlearned phrases. Group 3 consisted of 5 patients (4 men and 1 woman) who were able to provide a few words and/or some overlearned phrases in describing the Cookie Theft Picture (see Table 1 for speech samples). Their spontaneous speech was more difficult to classify and was considered "borderline" between the most severe cases in Groups 1 and 2 and the least severe cases in Group 4. The speech samples for these patients were obtained ranging from 7 MPO to 4.5 years. This group was similar to Groups 1 and 2 because not all cases were globally aphasic in all areas of language. For more information on each patient, see Table 2.

Group 4: Nonfluent Broca's. Group 4 consisted of 5 patients (4 men and 1 woman) who were able to provide verbal information relevant to the Cookie Theft Picture with reduced, hesitant, poorly articulated, agrammatical speech (see Table 1 for speech samples). The speech samples for these patients were obtained ranging from 7 MPO to 6 years. This group was milder in all language modalities than the other three groups. For more information on each patient, see Table 2.

When t tests were used to compare the BDAE scores between the groups, the patients in Group 4 had significantly higher ($p < .005$) Auditory Comprehension Z-Scores and Visual Confrontation Naming scores than the patients in Groups 1, 2, and 3. In addition, the patients in Group 4 had significantly higher ($p < .005$) Word Repetition scores than the patients in Group 2. No other significant differences in auditory comprehension, word repetition, or naming were found among the groups.

CT scan lesion site analysis was performed. The cortical and subcortical areas on CT scan that were examined for extent of lesion for each patient are shown in Fig. 1. No significant differences (Mann–Whitney U tests, $p < .01$ and beyond) were observed in the extent-of-lesion data for specific lesion site areas between the aphasia patients with no speech (Group 1) and those with stereotypies (Group 2). Therefore, the lesion site data from these two groups were combined, forming a no speech/stereotypies group ($n = 17$) for comparison with the nonfluent Broca's group ($n = 5$). (The lesion site data for patients who used only a few words and/or some overlearned phrases, Group 3, are discussed later.)

When the extent-of-lesion data for specific lesion site areas for each individual case were examined, no single neuroanatomical area alone could discriminate the 17 no speech/stereotypies cases from the 5 nonfluent Broca's cases. However, two lesion site areas combined produced no overlap between the no speech/stereotypies cases and the nonfluent Broca's cases. These two lesion site areas were two subcortical white matter areas including (1) the medial subcallosal fasciculus area (M Sc F) (mean lesion extent over slices B and B/W) and (2) the middle third periventricular white matter area (M 1/3 PVWM) (slice SM). The location of these two areas on CT scan is shown in the shaded areas in Fig. 10 (*top*).

A graph showing the extent of lesion in these two white matter areas, combined for the no speech/stereotypies cases versus the nonfluent Broca's cases shows no overlap between these two groups (Fig. 10, *bottom*). All the no speech/stereotypies cases had summed lesion extent scores above 7 and all the nonfluent Broca's cases had summed lesion extent scores below 6. No other lesion site combination could be used to discriminate these 22 cases into the two groups.

The mean lesion extent values in the M Sc F alone, were not adequate to discriminate these two very different groups of patients. The lesion extent values in the M 1/3 PVWM, *alone*, also were not adequate to discriminate these two very different groups of patients. Only when the lesion extent values were combined for these two lesion site areas (M Sc F at slices B and B/W plus M 1/3 PVWM at slice SM) were the two groups discriminated successfully on the basis of CT scan lesion extent values. The neuroanatomical connections contained within these two white matter pathway areas are discussed briefly here.

a. Medial subcallosal fasciculus area. The M Sc F area is a narrow white matter area surrounding the lateral angle of the frontal horn and contains a pathway through which fibers pass from the supplementary motor area (SMA) and the cingulate gyrus area 24 to the caudate. The subcallosal fasciculus first was described by Muratoff (1893) in the dog brain as the "fasciculus subcallosus." This structure is located under the corpus callosum. Dejerine (1895) diagrammed the structure in the human brain and labeled the medial portion as "substance grise sous-ependymaire" (Sge). The medial portion is very narrow and, in fact, is only one-tenth the distance from the lateral border of the frontal horn to the cortical mantle. (This distance represents only approximately 1 mm on a CT scan.) Yakovlev and Locke (1961) diagrammed these SMA and cingulate projections to the caudate in detail in the monkey brain. In their work, the most medial portion of the subcallosal fasciculus is labeled "stratum subcallosum" (St Sbc) (see Fig. 11).

Research by Benjamin and Van Hoesen (1982) using horseradish peroxidase injections in monkey brains showed strong reciprocal connections between cingulate gyrus area 24 and the SMA. The importance of the SMA in "the development of the intention-to-act" has been reviewed by Goldberg (1985). Research by Barnes, Van Hoesen, and Yeterian (1980) using the autoradiography technique in monkey brains showed that a major entry point for direct projections from the cingulate gyrus to the caudate

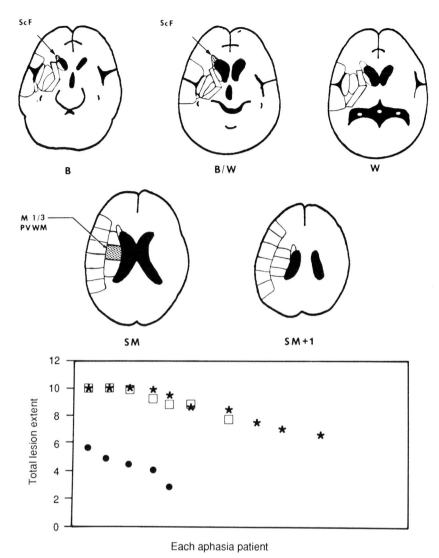

Figure 10 (Top) Location on CT scan slices of the two deep subcortical white matter areas that, when examined for total extent of lesion *combined*, discriminated the cases with no speech or only stereotypies versus those with nonfluent Broca's aphasia. These two deep subcortical white matter areas included (1) the medial subcallosal fasciculus (Sc F), mean lesion extent at slices B and B/W, and (2) the white matter deep to the lower motor/sensory cortex area for the mouth, the middle 1/3 PVWM at slice SM. (Bottom) Values for total lesion extent on the CT scan in the two deep subcortical white matter areas *combined*: (1) the medial Sc F area (mean lesion extent at slices B and B/W) plus (2) the white matter area deep to the lower motor/sensory cortex area for mouth, middle 1/3 PVWM at slice SM, for individual cases in three groups. Note that all cases with the most severe limitation in speech (□, Group 1 and ∗, Group 2) had total lesion extent values above 7; all cases with the least severe limitation in speech (●, Group 4, Broca's aphasia) had total lesion extent values below 6. The summed maximum lesion extent value on the graph represents maximum lesion extent ratings of 5 (entire area has solid lesion) in each of the two deep subcortical white matter areas combined. Reproduced with permission from Naeser et al. (1989).

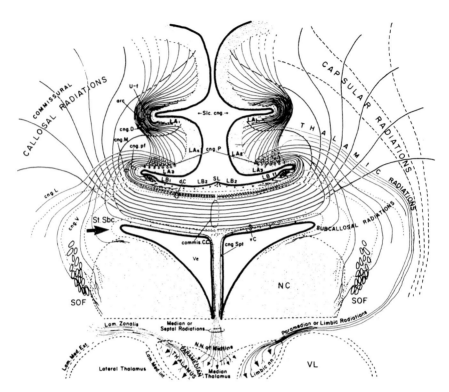

Figure 11 Drawing in coronal plane showing location of the medial subcallosal fasciculus (stratum subcallosum, St Sbc) in the lateral angle of the frontal horn (arrow). Note that the connections from the cingulate gyrus and supplementary motor area to the head of the caudate are located within the St Sbc area immediately lateral to the frontal horn. Reproduced with permission from Yakolev & Locke (1961).

(and indirect projections from the SMA to the caudate due to strong cingulate–SMA reciprocal connections) is in the most medial white matter surrounding the lateral angle of the frontal horn in its most rostral portion. Jürgens (1984) observed direct connections from the SMA to the caudate. These mesial frontal cortex projections then spread to the ventral and lateral portion of the caudate and to the lateral portion of the putamen.

Thus, lesion in the most medial white matter surrounding the lateral angle of the most rostral portion of the frontal horn (M Sc F) would interrupt pathways from the cingulate gyrus area 24 and SMA to the caudate and putamen, which would have an effect on initiation, preparation for speech movements, and limbic aspects of spontaneous speech.

b. Middle third periventricular white matter area. The M 1/3 PVWM area adjacent to the body of the lateral ventricle on CT scan slice SM is believed to contain, in part, the subcortical white matter fibers deep to the lower motor/sensory cortex area for mouth. These PVWM pathways are diagrammed coronally in Fig. 12. The motor cortex projections for the mouth have been shown in an anterograde staining study with rhesus monkeys, to project directly into the second quarter of the PVWM, adjacent to the body of the lateral ventricle (Schulz, Pandya, & Rosene, 1993). Thus, the M 1/3 PVWM area probably contains the motor/sensory projections for the mouth, immediately superior to their descent into the genu of the internal capsule.

In addition to containing the motor/sensory projections for the mouth, the M 1/3 PVWM area contains the body of the caudate nucleus and numerous other intra- and interhemispheric pathways. These pathways include, in part, (1) the descending pyramidal tract pathways for the leg and arm (Ross, 1980; Schulz et al., 1993), (2) the mid-callosal pathways, (3) additional medial subcallosal fasciculus pathways with connections from the SMA and cingulate gyrus to the body of the caudate (Dejerine, 1895; Muratoff, 1893; Yakovlev and Locke, 1961), (4) the occipito-frontal fasciculus (Dejerine, 1895), and (5) the superior lateral thalamic peduncle, which includes projections from the dorsomedial nucleus and the anterior nucleus to the cingulate (Mufson & Pandya, 1984) and projections from the ventrolateral nucleus to the motor cortex.

The lesion in the M 1/3 PVWM deep to the lower motor/sensory cortex area for mouth may interrupt the pathways necessary for motor execution as well as those pathways necessary for sensory feedback. Hence, we hypothesize that lesion in the two deep subcortical white matter pathway areas, the M Sc F and the M 1/3 PVWM, combined, effectively prevents any relevant spontaneous speech because no pathways are available for speech initiation, motor execution, or sensory feedback.

Note that the presence or absence of hemiplegia is not always a useful marker in predicting potential for long-term recovery of spontaneous speech (Naeser et al., 1989, Case 16). For example, the descending pyramidal tract pathways for the leg are most medial, within the second and third quarters of the PVWM area on CT scan, and immediately adjacent to the body of the lateral ventricle (slices SM and SM+1) (Naeser, Alexander, Stiassny-Eder, Galler, Hobbs, & Bachman, 1992; Schulz et al., 1993). The descending pyramidal tract pathways for the arm are slightly more anterior and lateral within the PVWM. Thus, if paralysis is caused by lesion in the PVWM, it will be related directly to the depth of the PVWM lesion adjacent to the body of the lateral ventricle, assuming absence of lesion in higher cortical motor pathways for the leg and arm and

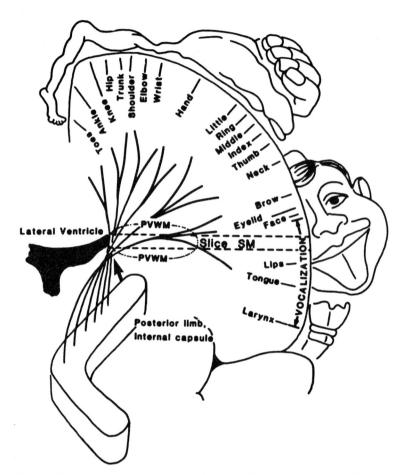

Figure 12 Coronal diagram showing location of descending pyramidal tract pathways in the deepest subcortical periventricular white matter (PVWM) area immediately adjacent to the body of the lateral ventricle (arrow). On CT scan, these descending pyramidal tract pathways are located in the second and third quarters of the PVWM on slices SM and SM + 1. On the CT scan slices inferior to these, the pyramidal tract pathways are located in the posterior limb of the internal capsule (CT scan slices W, B/W and B). Reproduced with permission from Naeser et al. (1989).

absence of lesion in lower subcortical motor pathways for the leg and arm (internal capsule and brainstem) (Naeser, Alexander, Stiassny-Eder, Galler, Hobbs, & Bachman, 1992).

A patient with no spontaneous speech may have lesion in the M Sc F and in more than half of the M 1/3 PVWM area, yet still spare the deepest portion of the M 1/3 PVWM area immediately adjacent to the body of the lateral ventricle, and have no paralysis. The CT scan of a patient without paralysis, but with no spontaneous speech, is shown in Fig. 8 (Naeser et al., 1989; Case 16). Thus, the severity of paralysis can be shown to have specific separate lesion sites, and the severity of spontaneous speech deficits also can be shown to have specific separate lesion sites. Therefore, recovery from paralysis is often a separate issue from recovery of spontaneous speech.

In summary, the cases with the least recovery of spontaneous speech, that is, those with no speech or only stereotypies (Groups 1 and 2), had combined lesion extent values for M Sc F plus M 1/3 PVWM above 7. Those cases with better recovery of spontaneous speech, that is, those with nonfluent Broca's aphasia (Group 4), had combined lesion extent values for M Sc F plus M 1/3 PVWM below 6. Those cases who fell in between these two groups in terms of severity of impairment of spontaneous speech, that is, those with a few words and/or some overlearned phrases, basically fell between these two groups in terms of combined lesion extent values (values around 6). Exceptional cases at either extreme were found in Group 3. A few case examples and CT scans are presented next.

Case examples

Case example for group 1: No speech or only a few irrelevant words. Case 3, H.J., is a 35-year-old man who, at 9 mo after stroke onset, still had no speech, although he could phonate and produce grunt-like sounds. He had a dense right hemiplegia with poor recovery (the second and third quarter PVWM lesion was immediately adjacent to the body of the lateral ventricle at slice SM). The CT scan in Fig. 13 shows a primarily subcortical infarct that included extensive lesion in the M Sc F at slices B and B/W and extensive lesion in the M 1/3 PVWM at slice SM. The total lesion extent in these two areas combined was 9.95 (see Fig. 13).

H.J. is an aphasia case with primarily subcortical lesion sites. This patient had no spontaneous speech output, and a moderate comprehension deficit was present (−0.21 on the BDAE Auditory Comprehension Z-Score 7 MPO) (see Tables 1 and 2). This moderate comprehension deficit is compatible with lesion in the anterior subcortical temporal isthmus area on slice B/W (see Section II,B).

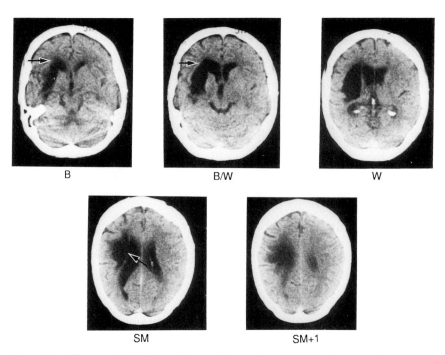

B   B/W   W

SM   SM+1

Figure 13 CT scan at 9 MPO for a 35-year-old man (Case 3) who had no speech at 7 MPO or even 2 years later (Group 1). A dense right hemiplegia was present. The left hemisphere lesion is on the left side of the CT scan. Lesion extent in the medial Sc F at slice B was rated 5, as was the lesion at slice B/W (see arrows), with a mean of 5. Lesion extent in the middle 1/3 PVWM at slice SM was rated 4.95 (see arrow); total lesion extent was 9.95. Note that the entire lesion is primarily subcortical. Reproduced with permission from Naeser et al. (1989).

Case example for Group 4: Nonfluent Broca's aphasia. Case 23, W.A., is a 50-year-old man who, at 7 months after stroke onset, produced nonfluent agrammatical speech that was compatible with Broca's aphasia (see Tables 1 and 2). Mild hemiparesis was present, and there was good recovery (no lesion was seen in the second and third quarter of the PVWM area immediately adjacent to the body of the lateral ventricle at slice SM or SM+1). The CT scan in Fig. 14 shows extensive lesion in the M Sc F at slices B and B/W but only minimal lesion in the M 1/3 PVWM at slice SM (small, patchy lesion). The total lesion extent in the two areas combined was 5.88 (see Fig. 14).

W.A. had a typical lesion distribution associated with longer-lasting Broca's aphasia, which we have observed repeatedly in our laboratory. (The Broca's aphasics who were included in this study were still nonfluent

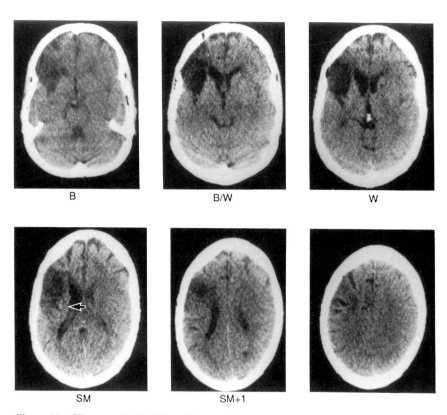

B   B/W   W

SM   SM+1

Figure 14 CT scan at 44 MPO for a 54-year-old man (Case 23) who had nonfluent agrammatic speech and Broca's aphasia at 7 MPO (Group 4). A mild hemiparesis was present and recovery was good. Lesion extent in the medial Sc F at slice B was rated 4; slice B/W was rated 3.75, with a mean of 3.88. Lesion extent in the middle 1/3 PVWM at slice SM was rated only 2; total lesion extent was 5.88. The arrow at slice SM shows a minimal lesion in the middle 1/3 PVWM which greatly reduced the combined total lesion extent to below 6, a value compatible with his mild limitation in speech. The mild hemiparesis with good recovery in this case was compatible with sparing of the deepest PVWM area immediately adjacent to the body of the lateral ventricle at slices SM and SM + 1. This deepest PVWM area contains, in part, the descending pyramidal tract pathways. Reproduced with permission from Naeser et al. (1989).

and agrammatical 7 months to 6 years following stroke onset.) This lesion distribution usually includes infarction in parts of Broca's area that extends across to the border of the frontal horn (including M Sc F, slices B and/or B/W) as well as superior lesion extension into lower motor cortex area for mouth (slices W and SM), which extends into the deep anterior

(A) 1/3 PVWM area and sometimes part of the M 1/3 PVWM area (slice SM). In some cases, the lower motor cortex area lesion is absent (slices W and SM). The deep subcortical A 1/3 PVWM lesion, however, is usually always present. The cortical portions of this lesion are compatible with lesion sites in longer-lasting Broca's aphasia cases, as previously published by Mohr et al. (1978).

Comparison of the CT scans for Case 3 (Fig. 13), who had no speech 9 MPO, with Case 23 (Fig. 14), who had functional nonfluent spontaneous speech 7 MPO, reveals that the less severe case 23 actually had more cortical damage (including Broca's area on slices B and B/W and the lower motor cortex area for mouth on slice SM) than the more severe case 3, who had no cortical lesion in either Broca's area or the lower motor cortex area for mouth. Comparison of the CT scans of these two cases suggests that the lesion extent within the two subcortical white matter areas (M Sc F and M 1/3 PVWM), not lesion extent within the cortex, is related to the severity of spontaneous speech output. Case 3, with no spontaneous speech, had complete lesion in the M Sc F at slices B and B/W and complete lesion in the M 1/3 PVWM at slice SM. Case 23, with functional nonfluent spontaneous speech, had lesion in more than half of the M Sc F at slices B and B/W but lesion in only a small part (less than half) of the M 1/3 PVWM at slice SM.

Broca's original Case, Leborgne.   The results from the 27 cases examined in the Naeser et al. (1989) study indicated that, when extensive lesion in the M Sc F deep to Broca's area is combined with extensive lesion in the M 1/3 PVWM deep to the lower motor/sensory cortex area for mouth, there is poor recovery of spontaneous speech with no speech or only stereotypies. These results are supported further by examination of the CT scan of Broca's original case, Leborgne.

Leborgne was 30 years old at the time of stroke onset and died 21 years later. His spontaneous speech was limited to the stereotypy "tan, tan." His auditory comprehension was reported to be good. He had a dense right hemiplegia. Broca attributed the poor speech to a lesion in the cortical region of the foot of the third left frontal convolution (Broca, 1861a,b). Broca himself, however, never observed the depth of the lesion in Leborgne's brain because it was never cut, as was common practice at that time. Recently, a CT scan was performed on the preserved brain 140 years after stroke onset (Castaigne, L'hermitte, Signoret, & Abelanet, 1980; Signoret, Castaigne, L'hermitte, Abelanet, & Lavoral, 1984) (see Fig. 15).

Figure 15 shows slices B/W, W, SM and SM+1 of Leborgne's brain (no slice B was available). Examination of the deep subcortical white matter surrounding the lateral angle of the left frontal horn reveals extensive lesion in the M Sc F at slice B/W. Because the lesion in the M Sc F is so

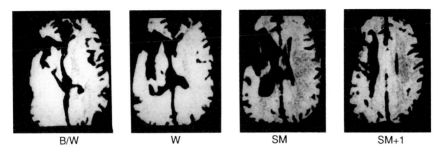

| B/W | W | SM | SM+1 |

Figure 15 CT scan of Broca's original case, Leborgne, who at age 51, 21 years after stroke onset, could produce only the stereotype "tan, tan." This case was similar to the Group 2 cases (only stereotypies) in the current study. A dense right hemiplegia was present. Lesion extent in the medial Sc F at slice B/W was rated 5; although slice B was not available we assumed that, because the lesion was so extensive on slice B/W, it was equally extensive on slice B (5); thus, the mean lesion extent of the medial Sc F was 5. Lesion extent in the middle 1/3 PVWM at slice SM was rated 4; total lesion extent was 9. This total lesion extent value of 9 in these two deep subcortical white matter lesion site areas was well within the range for cases with severe limitation in speech (total lesion extent values greater than 7). Reproduced with permission from Naeser et al. (1989).

extensive at slice B/W, a similarly extensive lesion is assumed to have been present in the M Sc F at slice B. Also, extensive lesion is seen in the M 1/3 PVWM at slice SM. The total lesion extent for the two areas combined was estimated to be 9, a value >7 and compatible with no long-term recovery of spontaneous speech. This patient was able to produce only the stereotypy "tan, tan" for 21 years after the stroke onset. The right hemiplegia may have been compatible with lesion in the deepest PVWM at slice SM+1 or possibly with some lesion in the posterior limb internal capsule at slice W.

The 1989 study by Naeser and associates focused on spontaneous speech and, although all cases in Groups 1 and 2 had severe limitation in spontaneous speech, not all these cases had complete cessation of speech, that is, 10 of 17 cases in Groups 1 and 2 could still repeat a few words and 4 of 17 could correctly name some pictures in response to visual confrontation. Research by Jürgens (Kirzinger & Jürgens, 1982) as well as others (Smith, Bourbonnais, & Blanchette, 1981) showed that lesion in the SMA has a direct effect on initiation of "spontaneous" motoric behavior patterns that are triggered internally, and not on those triggered directly by external stimuli. For example, Kirzinger and Jürgens (1982) observed that, after the SMA was ablated in squirrel monkeys and these monkeys were placed in isolation, the number of vocal "isolation calls" emitted from the monkeys was reduced, although the acoustic structure remained intact. Thus, the absence of internally-generated speech (spontaneous speech) in the presence of some externally-generated speech (word repetition and naming) may be compatible, in part, with lesion directly affecting projections from the SMA. Further, variation in word repetition and naming ability observed in those subjects who otherwise had no meaningful spontaneous speech may have been due, in part, to variation in the extent of the lesion in the projections from the SMA as well as in other areas. This possibility requires further study.

Results from this study suggest that careful examination of lesion in the M Sc F area and the M 1/3 PVWM area is a basic starting point for assessing potential for long-term recovery of spontaneous speech in severely nonfluent stroke patients with infarction in the various branches of the left middle cerebral artery (LMCA). When working with patients who have lesion outside the distribution of the LMCA, especially in the left anterior cerebral artery (LACA), one must examine different structures. For example, in cases with LACA infarcts, cortical lesion in the SMA and/or the cingulate gyrus area may combine with subcortical lesion in the M 1/3 PVWM to produce long-lasting impairment in speech, even when no lesion may be present in the M Sc F at slices B and B/W. Obviously, other cortical and/or subcortical lesion sites also may combine to produce severe limitation in spontaneous speech.

IV. SOME IMPLICATIONS FOR TREATMENT IN SEVERE APHASIA

We have completed two studies in which results from the CT scan recovery studies just described were applied to treatment studies with aphasia patients who had severe limitation in spontaneous speech. The first study focused on CT scan lesion sites in patients with limited spontaneous speech who were treated with Melodic Intonation Therapy (MIT). The second study focused on CT scan lesion sites in patients with no spontaneous speech who were treated with the nonverbal Computer-Assisted Visual Communication Program (C-VIC).

A. Good Response versus Poor Response to the Melodic Intonation Therapy Treatment Program

MIT is a treatment program for aphasia patients with limited spontaneous speech. In this retrospective study, CT scan lesion sites and good response versus poor response to MIT were examined for 8 chronic stroke patients (Frumkin, Naeser, Helm-Estabrooks, & Fitzpatrick, 1990).

The MIT program was designed to improve verbal expression in patients with severely limited or nonfluent speech (Albert, Sparks, & Helm, 1973; Sparks & Holland, 1976). The program uses phrases and sentences that are intoned slowly with continuous voicing, using simple high-note, low-note patterns based on normal speech prosody (Helm-Estabrooks, Nicholas, & Morgan, 1989a). Two studies have demonstrated that not all patients with limited verbal output respond positively to MIT (Helm, 1978; Sparks, Helm, & Albert, 1974). Positive response to MIT is defined as improvement in the number of words per phrase length as tested with the Cookie Theft Picture description from the BDAE (Goodglass & Kaplan, 1983).

Language data were collected from files on patients treated with MIT in the Audiology and Speech Pathology Service (Boston V.A. Medical Center). All patients were separated into two groups—good response to MIT and poor response to MIT. Good response to MIT was defined as an increase of at least two words per phrase length on the spontaneous speech characteristics rating scale as applied to the Cookie Theft Picture description, after a series of MIT treatments. Poor response to MIT was defined as no increase in the number of words per phrase length after a series of MIT treatments.

Data were reviewed for eight male stroke cases who were treated only with MIT, and who had chronic CT scans available for analysis. Each patient was right-handed and had suffered single-episode left hemisphere occlusive vascular stroke between the ages of 24 and 65 years ($x = 49$; SD 14.2). One patient (W.F.) had an additional small lesion in the right parietal lobe that was not extensive enough to be considered the primary cause of the aphasia. The CT scans used for CT scan lesion site analysis were performed between 3 and 36 MPO. All eight patients were treated with MIT during the chronic phase poststroke, beginning 3–51 MPO.

Four patients had good response with MIT; four patients had poor response with MIT. No significant differences were found between the GR cases and the PR cases in age at onset or MPO when the MIT treatments were begun. The mean age at onset for the GR cases was 49.5 years (SD 12.3) and MIT treatments began at a mean of 8.75 MPO (SD 6.9). The mean age at onset for the PR cases was 48.75 years (SD 17.75) and MIT treatments began at a mean of 16.25 MPO (SD 23.17).

The language characteristics of aphasia patients who are good candidates for successful treatment with MIT have been under development since the first published papers (Albert et al., 1973; Sparks et al., 1974).

These pre-MIT language characteristics were summarized by Helm-Estabrooks and Albert (1991):

1. poorly articulated, nonfluent, or severely restricted verbal output that may be confined to a nonsense stereotypy (e.g., "bika bika")
2. at least moderately preserved auditory comprehension, exceeding the 45th percentile on the BDAE rating scale
3. poor repetition, even for single words
4. poorly articulated speech, earning a rating of 3 or less for Articulatory Agility on the BDAE Profile of Speech Characteristics

These four language characteristics associated with good candidacy for successful treatment with MIT were refined over several years of experience with MIT at the Boston V.A. Medical Center. Several of the patients in the study being discussed were treated with MIT prior to final development of these four characteristics. In the current study, all patients treated with MIT had met at least three of these pre-MIT language characteristics. The decision to treat the patient with MIT was made by the speech pathologist treating the patient at that time; no CT scan lesion site information was used in the treatment decision.

The MIT treatment program is structured hierarchically and is divided into three levels. The decision to continue a patient in the MIT treatment program or to terminate the program was made by the speech pathologist based on the patient's scores at each level of the program (Helm-Estabrooks et al., 1989a). A wide range for the total number of MIT treatments provided (6–115 treatments) was observed across the GR and PR groups. This wide range was due, in part, to the fact that treatment was terminated if a patient could not complete Level I.

Pre-MIT. Mann–Whitney U tests were performed on the pre-MIT spontaneous speech data for the GR group and the PR group. The GR group had significantly better pre-MIT spontaneous speech scores for number of words per phrase length and for grammatical form than the PR group (see Table 3). Although the GR group had significantly better pre-MIT spontaneous speech scores than the PR group, note that each GR case and each PR case met at least three of the four pre-MIT language characteristics associated with good candidacy for successful treatment with MIT (Helm-Estabrooks & Albert, 1991).

Post-MIT. Mann–Whitney U tests also were performed on the post-MIT spontaneous speech data for the two groups. As would be expected, the GR group had significantly better post-MIT spontaneous speech scores than the PR group (see Table 3). Paired t-tests were performed on the spontaneous speech data at T_1 versus T_2 for the GR cases. The GR group showed significant improvement post-MIT in number of words per phrase length

Table 3 Spontaneous Speech Statistics for Good Response Group and Poor Response Group Treated with Melodic Intonation Therapy

| | Spontaneous speech scores | | | | | | Mann–Whitney U test comparisons | | | |
| | Pre-MIT | | Post-MIT | | Pre-Post Change | | Pre-MIT | | Post-MIT | |
	Mean	SD	Mean	SD	Mean	SD	Z corrected for ties	p-level 1-tail	Z corrected for ties	p-level 1-tail
Good response group										
No. words phrase length	3.0	2.3	5.8	1.9	+2.8	1.5	2.14	.016	2.38	.008
Articulatory agility	1.5	1.0	3.3	1.0	+1.8	1.9	1.52	.064	1.95	.025
Grammatical form	3.8	3.2	5.0	2.8	+1.3	1.9	2.12	.017	2.25	.012
Poor response group										
No. words phrase length	0.3	0.5	0.3	0.5	0	0				
Articulatory agility	0.5	1.0	0.8	1.5	+0.3	0.5				
Grammatical form	0.3	0.5	0.3	0.5	0	0				

($p < .006$) and articulatory agility ($p < .03$). The PR group showed no significant improvement post-MIT on any of the spontaneous speech scores.

1. CT data

CT scan lesion site analysis was performed. The cortical and subcortical areas on CT scan that were examined for extent of lesion are shown in Fig. 1.

Each of the four GR patients had a total extent-of-lesion value for the M Sc F area plus the M 1/3 PVWM area that was ≤7 (range 3.75–7) (see Table 4). Each of the four PR patients had a total extent-of-lesion value that was >7 (range 7.48–9.9). Note, in fact, that in this small study with eight patients, the GR patients could be separated from the PR patients on the basis of the M 1/3 PVWM extent of lesion rating alone, rather than on the basis of the lesion combination of M Sc F plus M 1/3 PVWM. All four GR cases had M 1/3 PVWM extent-of-lesion ratings <3; all four PR cases had M 1/3 PVWM extent-of-lesion ratings >3.

The extent of lesion in Wernicke's area could not be used to discriminate between the GR cases and the PR cases because, although four of four GR cases had lesion in less than half of Wernicke's area, two of four PR cases also had lesion in less than half of Wernicke's area (see Table 4).

All four GR cases had extent-of-lesion ratings <3 in the subcortical temporal isthmus area, whereas all four PR cases had extent-of-lesion ratings >3 in the subcortical temporal isthmus area. The GR group had pre-MIT BDAE Auditory Comprehension Z-Scores of −0.46, +0.46, +0.75, and +1.0. The PR group had pre-MIT BDAE Auditory Comprehension Z-Scores of −2.33, −1.48, and −0.4. The number of subjects with complete data was too small to permit statistical comparisons on auditory comprehension. However, prior to MIT treatment, four of the four GR cases were better than −0.5 on the BDAE Auditory Comprehension Z-Score, whereas only one of the three PR cases was better than −0.5 on the BDAE Auditory Comprehension Z-Score. The relatively greater deficit in auditory comprehension for most of the PR cases probably was related to lesion in the subcortical temporal isthmus area in all PR cases and/or Wernicke's area in two of the four PR cases.

In no cortical language areas could the extent of lesion be used to discriminate between all the GR cases and all the PR cases, including extent of lesion in Broca's area, Wernicke's area, the supramarginal and angular gyrus areas, or the SMA.

Case examples

The CT scan for one patient with good response to MIT is shown in Fig. 16. The CT scan for one patient with poor response to MIT is shown in Fig. 17.

Table 4 CT Scan Lesion Sites and Extent-of-Lesion Data for Patients with Good Response and Poor Response to Melodic Intonation Therapy

Case	CT scan (MPO)	Medial subcallosal fasciculus (mean B, B/W)	Mid. 1/3 PVWM (SM)	Total extent of lesion, (M Sc F + M 1/3 PVWM) (≤7 = GR)	Wernicke's area (mean B/W, W)	Temporal isthmus (mean B, B/W) (<3 = GR)	Occipital length asymmetry
Good response cases							
M.E.	3.5	3	2	5	1	0.5	R
W.A.	44	3.25	2.5	5.75	0	0	L
T.H.	77	1.75	2	3.75	2	1.25	L
M.J.	72	5	2	7	0	0	=
Poor response cases							
G.N.J.	4	4	4.5	8.5	1.87	4.5	=
S.F.	3.5	5	4.9	9.9	2.87	4	L
R.P.	18	3.63	3.85	7.48	4.5	3.75	=
W.F.	100	4.9	4.8	9.7	5	5	R

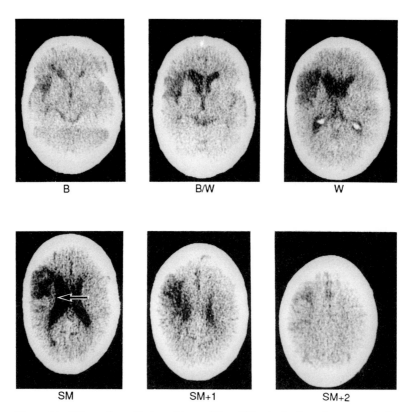

B   B/W   W

SM   SM+1   SM+2

Figure 16 CT scan for a 58-year-old man (Case ME) who had good response to the Melodic Intonation Therapy (MIT) treatment program beginning at 3 MPO. He improved from 1 to 3 words in phrase length after 3 months of MIT. The total extent of lesion value for the M Sc F *plus* the M 1/3 PVWM was 5. This total extent of lesion value was computed from (1) M Sc F at slice B = 2.5 (patchy lesion in less than half of the area, white arrow), M Sc F at slice B/W = 3.5 (patchy lesion in more than half of the area, white arrow), mean M Sc F at slices B and B/W = 3; and (2) M 1/3 PVWM at slice SM = 2 (small, patchy, partial lesion, black and white arrow). CT scan is 3.5 MPO. Note that this patient had almost no lesion in Wernicke's area on slices B/W and W (mean lesion extent value of 1), and almost no lesion in the subcortical temporal isthmus area on slices B and B/W (mean lesion extent value of 0.5).

The results from this study have expanded and revised the results from the Naeser & Helm-Estabrooks (1985) CT scan study with MIT. Results from this current study reveal that total extent of lesion in the M Sc F area plus the M 1/3 PVWM area discriminated completely between patients with good response to MIT and patients with poor response to MIT, in-

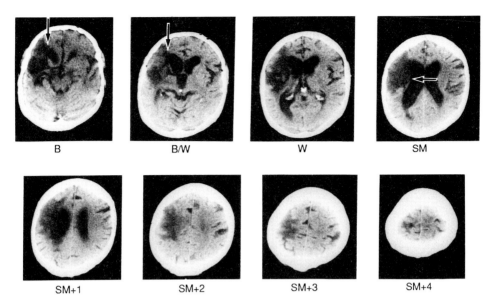

Figure 17 CT scan for a 65-year-old man (Case SF) who had poor response to the MIT treatment program beginning at 5 MPO. The total extent of lesion value for the M Sc F *plus* the M 1/3 PVWM was 9.9. This total extent of lesion value was computed from (1) M Sc F at slice B = 5 (entire area has solid lesion, white arrow), M Sc F at slice B/W = 5 (entire area has solid lesion, white arrow), mean M Sc F at slice B and B/W = 5; and (2) M 1/3 PVWM at slice SM = 4.9 (almost the entire area has solid lesion, black and white arrow). CT scan is 3.5 MPO. Note that this patient had lesion in less than half of Wernicke's area on slices B/W and W; the lesion was only small and partial (mean lesion extent value of 2.87). Lesion was present in more than half the subcortical temporal isthmus area, however, on slices B and B/W (mean, 4).

cluding even those patients treated as late as 4 years after stroke onset. The importance of these two subcortical white matter areas to the recovery (or nonrecovery) of spontaneous speech had not yet been recognized when our first CT scan study with MIT was published in 1985.

In summary, nonfluent aphasia patients who are the best candidates for successful treatment with MIT will meet at least three of the four pre-MIT language characteristics listed earlier in this section (Helm-Estabrooks & Albert, 1991). These patients will have total extent-of-lesion values <7 for M Sc F plus M 1/3 PVWM, as well as lesion in less than half of Wernicke's area and the subcortical temporal isthmus area. Future research may indicate that the results from this CT scan study with the MIT verbal treatment program may be applicable to candidacy for treatment with other verbal treatment programs.

B. Good Response versus Poor Response to the Nonverbal Computer-Assisted Visual Communication Treatment Program

The C-VIC treatment program enables patients with no spontaneous speech (or ability to read or write) to use pictures and icons on a computer screen to communicate needs and ideas. In a retrospective study, CT scan lesion sites and good response versus poor response to the C-VIC program were examined for 7 severe aphasia patients with no ability to speak, read, or write (Palumbo, Nicholas, Baker, Alexander, Frumkin, Naeser, 1992). These patients were treated with C-VIC beginning in the chronic phase after stroke.

Almost 20 years ago, the first systematic attempts to utilize a substituted language based on representational and arbitrary icons were reported (Baker, Berry, Gardner, Zurif, Davis, & Veroff, 1975; Gardner, Zurif, Berry, & Baker, 1976). More recently, the iconic C-VIC "language" that can be carried in and manipulated with a minicomputer was developed (Steele, Weinrich, Wertz, Kleczewka, & Carlson, 1989; Weinrich, Steele, Carlson, Kleczewska, Wertz, & Baker, 1989a; Weinrich, Steele, Kleczewska, Carlson, Baker, & Wertz, 1989b). These investigators have demonstrated that severely aphasic patients can manipulate the computer "mouse" and button-click necessary for operation, and can learn the rules of lexical organization. The patients learn to construct and comprehend complex sentences in the C-VIC pictorial "language." Not all severely aphasic patients, however, have been able to grasp the lexical and syntactic rules of the substituted language and use them to initiate communication independently.

In our study, CT scan lesion sites and GR versus PR to C-VIC were examined for 7 severe stroke patients (Palumbo et al., 1992). All 7 patients had suffered a left hemisphere cerebrovascular accident. The age at onset of stroke ranged from 43 to 65 years (mean, 56; SD 7.5). One case was left-handed. All 7 patients had severe right hemiplegia. One patient had a small right hemisphere infarction in addition to the major left hemisphere lesion.

The Boston Assessment of Severe Aphasia test (BASA) (Helm-Estabrooks, Ramsberger, Morgan, & Nicholas, 1989b) was performed immediately prior to C-VIC training and again at its termination. The BASA test was designed for severely aphasic patients and probes for even very small improvement in auditory comprehension or language production. Most patients also were tested with parts of the BDAE (Goodglass & Kaplan, 1983). Aphasia diagnosis prior to C-VIC treatment was severe aphasia with no spontaneous output—spoken or written—in conversation or picture description. Auditory comprehension also was impaired substantially. Table 5 summarizes language capacity.

Table 5 Patient Data and Language Test Scores for Patients Treated with the Computer-Assisted Visual Communication Treatment Program[a]

Case	Sex	Age at onset (years)	MPO when C-VIC started	Months in treatment		BDAE		BASA		Oral–gestural expression raw score (21)	C-VIC response phase II step 5 (PICA scale - 16)[b]
						No. words phrase length (7)	Auditory comprehension Z score	Auditory comprehension raw score (16)	Total raw score (61)		
Good response cases											
B.J.[c]	M	43	7	3	pre	0	−0.8	13	49	14	15.0
					post	1–2	+0.5	14	51	14	
D.J.	M	54	6	6	pre	0	−1.75	8	36	9	14.5
					post	0	NA[d]	7	39	14	
S.H.	F	65	4	18	pre	0	NA	7	26	3	14.0
					post	0	NA	10	37	5	
C.A.	M	49	72	9	pre	0	−0.63	13	40	7	12.5
					post	0	NA	13	40	8	
Poor response cases											
R.R.	M	59	21	28	pre	0	−0.61	6	25	3	9.0
					post	0	NA	6	38	13	
F.W.	M	59	7	7	pre	0	−1.6	9	31	6	8.5
					post	0	−0.33	13	41	7	
S.M.	M	60	60	7	pre	0	−1.74	4	24	7	8.0
					post	0	−1.61	8	29	7	

[a]Patients are rank ordered by response to Phase II, Step 5 of C-VIC program; see last column. Scores ≥ 13 reflect ability to initiate communication with the C-VIC program independently.

[b]The last step in Phase II, Step 5 of C-VIC, requires the patient to initiate a question or command independently. Scores of 8 or 9 reflect inability to initiate a question or command independently.

[c]Left-handed but aphasic from left hemisphere lesion.

[d]Information is not available.

All 7 cases had been treated previously with one or more traditional treatment programs without success, including verbal treatment programs such as MIT (Albert et al., 1973; Sparks & Holland, 1976), nonverbal treatment programs such as buccofacial visual action therapy, which trains patients with severe oral apraxia to produce representational gestures using the oral musculature (Ramsberger & Helm-Estabrooks, 1988), and limb–visual action therapy, which trains patients with severe aphasia and limb apraxia to produce representational purposeful gestures with the hand and arm (Helm-Estabrooks, Fitzpatrick, & Barresi, 1982).

1. C-VIC treatment program

All patients were treated with C-VIC in the chronic phase poststroke >3 MPO (range 4 MPO to 6 yr). The patients were seen as outpatients for half-hour treatment sessions, usually twice per week. All patients were able to match objects to pictured icons on a computer screen (and vice versa), and all were able to use the computer "mouse" easily with the left hand.

The C-VIC training consists of two phases (Baker & Nicholas, in press). In Phase I, patients are trained to use the computer "mouse" to carry out commands presented in C-VIC (comprehension); to answer questions; and, finally, to compose descriptions of simple acts (production). Phase II focuses on real-life communicative acts, including expressing needs, making requests (giving commands), and asking questions (see Fig. 18). Variability in duration of C-VIC treatment in this study reflected on-going program development as well as patient availability.

The quality of the communications generated by patients using C-VIC in Phase I and Phase II were rated by the speech–language pathologist using the Porch Index of Communicative Ability (PICA) rating scale, which ranges from 1 to 16 (Porch, 1967). A PICA score ≥13 represents independently initiated successful communication. Scores >13 were considered good C-VIC productions; scores <13 were considered poor C-VIC productions. To reach criterion at the end of Phase I, and to be considered a GR case at the end of Phase II, the patient's communications generated with C-VIC had to reach scores of at least 13 on the PICA scale. A patient with a Phase II C-VIC score <13 was considered a PR case. Three patients had GR, with Phase II scores ranging from 14 to 15; one patient had borderline GR, with a score of 12.5; three patients had PR, with scores ranging from 8 to 9 (see Table 5).

No significant correlation existed between the age at stroke onset and the Phase II C-VIC score ($r = -.482$), nor between the MPO when entering the C-VIC program and the Phase II C-VIC score ($r = -.352$). Also no significant correlation was found between the number of months a patient received the C-VIC program and the Phase II C-VIC score ($r = -.275$).

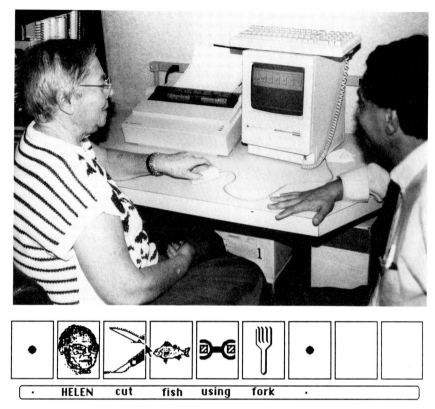

Figure 18 (Top) A severe nonverbal aphasia patient using the Computer-Assisted Visual Communication (C-VIC) program on the Macintosh computer. The patient has a right hemiplegia and controls the "mouse" with the nonparalyzed left hand to select pictures and icons on the computer screen. (Bottom) Example of communication generated by a nonverbal aphasia patient in Phase II of the C-VIC program. The patient's spouse has said that they ate at a fish restaurant over the weekend. The clinician asked, "When you were at the restaurant, who cut your food for you?" The patient generated the above response using C-VIC. The C-VIC program is customized to individual patient needs, including photos of family members, hospital staff, and so on. Note that the written English is not usually provided below each picture or icon, because this tends to confuse the patient who cannot read or write. The written English is provided here for purposes of illustration only.

One of the GR patients who had good response with C-VIC (Case S.H.) was able to remain at home with her spouse rather than transfer to a nursing home as a result of the new communication ability provided through C-VIC. The necessity of a nursing home had been considered prior to C-VIC training because of severe difficulties in communication

and management. As a result of this patient's success with the C-VIC program, a Macintosh computer was placed in the home and the patient was
able to use the system to communicate her needs to her husband, including when she felt a need for her prescriptions to be refilled and so on.

Even patients with poor response to C-VIC by PICA scoring were able
to use C-VIC for some interactions not possible with speech or writing.
Case R.R. was considered to have poor response to C-VIC because he was
not able to initiate communications independently with C-VIC following
Phase II training. He was, however, able to use C-VIC to answer specific
questions posed by another person. For example, R.R. has a Macintosh
computer in the home and can use it to respond to his wife's verbally
presented question, "What do you want for breakfast?"

The BASA scores had a general correspondence to C-VIC. The four
GR cases had pre-C-VIC overall BASA scores of 26–49/61 and auditory
comprehension BASA scores of 7–13/16. The three PR cases had pre-
C-VIC overall BASA scores of 24, 25, and 32/61 and auditory comprehension scores of 4, 6, and 9/16. Post-C-VIC testing showed significant improvements on the BASA in the following areas: overall BASA score
($p < .01$); Auditory Comprehension ($p < .05$); and Oral–Gestural Expression ($p < .05$).

2. CT data

CT scan lesion site analysis was performed. The cortical and subcortical
areas that were examined for extent of lesion are shown in Fig. 1. The CT
scans used for lesion localization were performed between 3 and 36 MPO.

No relationship was found between good or poor response to C-VIC
and lesion extent in any single neuroanatomical area analyzed on CT scan.
The Naeser et al. (1989) study had demonstrated that extensive lesion in
the M Sc F area plus the M 1/3 PVWM area combined was compatible
with no recovery of spontaneous speech. In fact, all 7 cases in the C-VIC
study had total extent-of-lesion values >7 for the M Sc F and M 1/3
PVWM combined (see Table 6). The GR and PR cases had complete overlap of total extent-of-lesion values in these two white matter areas.

Only one combination of additional lesion extension in two extra areas
completely discriminated between all GR cases and all PR cases treated
with C-VIC. These two extra areas included (1) extra area #1, the supraventricular area, including the SMA/cingulate gyrus area 24, and (2) extra
area #2, the temporal lobe area, including Wernicke's area or the subcortical temporal isthmus. The PR patients had extensive lesion (extent-of-
lesion value >3) in each of these two extra areas. The GR patients had
extensive lesion (extent-of-lesion value >3) in none or only one of these
two extra areas (see Table 6).

Table 6 CT Scan Lesion Sites and Extent-of-Lesion Values for Patients Treated with the Computer-Assisted Visual Communication Treatment Program

Case	CT scan (months postonset)	Medial subcallosal fasciculus (mean B, B/W)	Middle 1/3 PVWM (SM)	Total extent of lesion (M Sc F + M 1/3 PVWM) (>7 = Basic lesion, no recovery of spontaneous speech)	Extra area #1 Supraventricular		Wernicke's area (mean B/W, W)	Extra area #2 Temporal lobe	Right hemisphere lesion	Occipital length asymmetry
					Supplementary motor area	Cingulate gyrus area 24		Temporal isthmus (mean B, B/W)		
Good response cases										
B.J.[a]	12	5	4.9	9.9	0	0	0	0	No	=
D.J.	13	5	4.25	9.25	Deep	Deep	1	0	Yes, high right frontal	=
S.H.	6	4.9	4.75	9.65	0	0	3.25	4.37	No	R
C.A.	72	3.5	5	8.5	0	0	4.55	4.5	[b]	R
Poor response cases										
R.R.	49	4.75	4.9	9.65	Cortical & deep	Cortical & deep	2.5	4.5	No	L
F.W.	13	4.25	5	9.25	Cortical & deep	Cortical & deep	2.37	4.37	No	R
S.M.	60	2.37	4.75	7.12	Deep	Deep	4.9	5	No	L

[a] Left-handed by aphasic from left hemisphere lesion.
[b] Shunt in right ventricle.

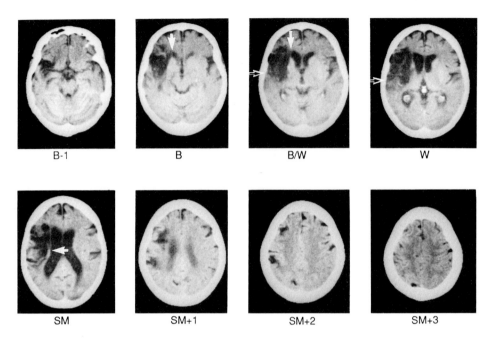

Figure 19 CT scan for patient SH, a 65-year-old woman who entered the C-VIC program at 4 MPO and had good response to the C-VIC treatment program. The basic lesion site pattern associated with no recovery of spontaneous speech was present in the M Sc F on slices B and B/W (white arrows) plus the M 1/3 PVWM at slice SM (white arrow). The total extent of lesion value for M Sc F plus M 1/3 PVWM was 9.65. In addition, extensive lesion was present in only *one* of the two extra areas. Lesion was present in extra area #2, the *temporal lobe* area including Wernicke's area on slices B/W and W (black and white arrows) and the temporal isthmus on slices B-1 and B. No lesion was present in extra area #1, the *supraventricular* area including SMA/cingulate gyrus on slices SM + 2 or SM + 3. CT scan is 6 MPO.

Case examples

The CT scan for one patient who had good response to C-VIC training is shown in Fig. 19. The CT scan for one patient who had poor response to C-VIC training is shown in Fig. 20.

The results from this study suggest that CT scan lesion site analysis may be useful in identifying severe nonverbal aphasia patients who probably will not recover spontaneous speech, but can be trained to communicate with the nonverbal C-VIC treatment program. Patients with total extent-of-lesion values >7 for the M Sc F area plus M 1/3 PVWM area are the appropriate patients for treatment with C-VIC.

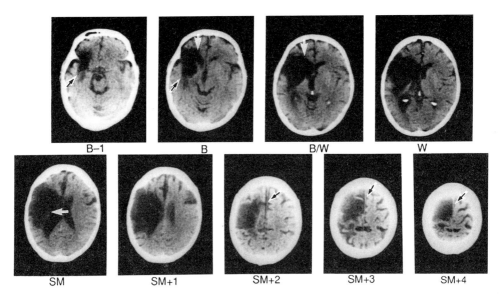

Figure 20 CT scan for patient RR, a 60-year-old man who entered the C-VIC program at 21 MPO and had poor response to the C-VIC treatment program. The basic lesion site pattern associated with no recovery of spontaneous speech was present in the M Sc F on slices B and B/W (white arrows) plus the M 1/3 PVWM at slice SM (white arrow). The total extent of lesion value for M Sc F plus M 1/3 PVWM was 9.65. In addition, extensive lesion was present in *both* of the two extra areas. Lesion was present in extra area #1, the *supraventricular* area including SMA/cingulate gyrus on slices SM + 2, SM + 3, and SM + 4 (black and white arrows); *and* lesion was present in extra area #2, the *temporal lobe* area including temporal isthmus on slices B—ru1 and B (black and white arrows). CT scan is 4 years after onset.

Further, patients with extensive lesion in both of the two extra areas, extra area #1, the supraventricular area, including the SMA/cingulate gyrus area 24, and extra area #2, the temporal lobe area, including Wernicke's area or the subcortical temporal isthmus, appear to be unable to initiate communication independently with C-VIC. They require assistance, such as a repeated cue or repeated instructions.

The poor response of some patients to C-VIC does not mean that these patients should not be trained to use C-VIC. The term "poor response" refers to communications that are rated below 13 on the PICA scale and inability to initiate communications independently with C-VIC at the Phase II level. The expectations of outcome with the C-VIC program may be lowered to accommodate patients who cannot initiate C-VIC messages independently, but can use the program with assistance to answer specific questions. Thus, practical use of C-VIC in the home, nursing home, or rehabilitation setting should be determined on a case-by-case basis.

Sarno and Levita (1981) have observed that the greatest recovery in severe aphasia patients occurs after 6–12 MPO. Therefore, with a severe nonverbal aphasia patient, a chronic CT scan could be obtained after 3 MPO and the results could be used to help make treatment decisions for the 6–12 MPO treatment period and beyond. Of course, other treatment approaches should be used earlier, including helping the patient use a basic communication board, drawing (Morgan & Helm-Estabrooks, 1987), or gesture (Skelly, Schinsky, Smith, Donaldson, & Griffin, 1974, 1975; Rao, 1986).

Careful analysis of a chronic CT scan may help reduce the overall cost of long-term rehabilitation of severe nonverbal aphasia patients by help-ing identify potential for recovery (or nonrecovery) of spontaneous speech. Several factors should be considered:

1. The complete borders of an infarct are best visualized on CT scans performed *after* 2–3 MPO. Acute CT scans that are performed earlier than 2–3 MPO do not reveal the complete borders of an infarct and are not useful in helping make predictions for long-term recovery (Palumbo & Naeser, in preparation).

2. To use the information in this chapter, the CT scan should be ob-tained at 20° to the canthomeatal line, without contrast, with 10-mm thick slices at 7-mm intervals, above the suprasellar cistern and through the ventricles. CT scans performed in this manner will con-form to the CT scan slice images shown in Fig. 1 in this chapter. The M Sc F area and the M 1/3 PVWM area, as well as Wernicke's area and the temporal isthmus, can be located easily for detailed extent-of-lesion analysis.

3. If the total extent of lesion for the M Sc F and the M 1/3 PVWM is <7, then the patient is likely to recover some spontaneous speech. If the extent of lesion is also <3 for Wernicke's area and the temporal isthmus, then the patient is likely to have good response with MIT or perhaps with another verbal treatment program.

4. If the total extent of lesion for the M Sc F and the M 1/3 PVWM is >7, then the patient is unlikely to recover spontaneous speech. For these patients, the nonverbal C-VIC treatment program or another nonverbal treatment program should be considered. If the extent of lesion is >3 for none or only one of the two extra areas observed in the C-VIC study, then the nonverbal patient is likely to have good response with C-VIC, and to be able to initiate communication inde-pendently with C-VIC. The two extra areas on CT scan that must be

examined regarding potential for good response with C-VIC are extra area #1, the supraventricular area, including the SMA/cingulate gyrus area 24, and extra area #2, the temporal lobe area, including Wernicke's area or the subcortical temporal isthmus. If the extent of lesion is >3 for each of these two extra areas, then the nonverbal patient is likely to have poor response with C-VIC and, although unable to initiate communication independently with C-VIC, may be able to answer simple questions with C-VIC.

ACKNOWLEDGMENTS

The author acknowledges the invaluable assistance of Carole Palumbo, Malee N. Prete, Neva Frumkin, Jessica Lydon, Patricia Emery, and Sulochana Naidoo for assistance in CT scan analysis and data collection, as well as Claudia Cassano and Roger Ray for assistance with manuscript preparation. We also thank the Radiology Service of the Boston V.A. Medical Center, including A. Robbins and R. N. Samaraweera, and the Medical Media Service of the Boston V.A. Medical Center, John Dyke and Mary Burke, for photography and illustrations.

This research was supported in part by the Medical Research Service of the Department of Veterans Affairs and by USPHS Grant DC00081 (NIDCD).

REFERENCES

Albert, M. L., Sparks, R., & Helm, N. (1973). Melodic intonation therapy for aphasia, *Archives of Neurology, 29*, 130–131.

Alexander, M. P., & LoVerme, S. R. (1980). Aphasia following left hemispheric intracerebral hemorrhage. *Neurology, 30*, 1193–1202.

Alexander, M. P., & Naeser, M. A. (1988). Cortical–subcortical differences in aphasia. In F. Plum (Ed.), *Language, communication and the brain* (pp. 215–228). New York: Raven Press.

Alexander, M. P., Naeser, M. A., & Palumbo, C. L. (1987). Correlations of subcortical CT lesion sites and aphasia profiles. *Brain, 110*, 961–-991.

Alexander, M. P., Naeser, M. A., & Palumbo, C. L. (1990). Broca's area aphasias: Aphasia after lesions including the frontal operculum. *Neurology, 40*, 353–362.

Baker, E., Berry, T., Gardner, H., Zurif, E., Davis, L., & Veroff, A. (1975). Can linguistic competence be dissociated from natural language functions? *Nature (London), 254*, 609–619.

Baker, E. H., & Nicholas, M. *Computer-assisted visual communication (C-VIC): An alternative communication system for severe aphasia.* American Journal of Speech–Language Pathology, A Journal of Clinical Practice, in press.

Barat, M., Constant, P. H., Mazaux, J. M., Caille, J. M., & Arné, L. (1978). Correlations anatomo-cliniques dans l'aphasie. Approt de la tomo densitometrie. *Revue Neurologique, 134*, 611–617.

Barnes, C. L., Van Hoesen, G. W., & Yeterian, E. H. (1980). Widespread projections to the striatum from the limbic mesocortices in the monkey. *Society for Neuroscience Abstracts, 6*, 271.

Benjamin, D., & Van Hoesen, G. W. (1982). Some afferents of the supplementary motor area (SMA) in the monkey. *Anatomical Record, 202*, 15A.

Bogen, J. E., & Bogen, G. M. (1976). Wernicke's region: Where is it? *Annals of the New York Academy of Sciences, 280*, 834–843.

Borod, J. C., Carper, M., Goodglass, H., & Naeser, M. (1984). Aphasic performance on a battery of constructional, visuo-spatial, and quantitative tasks: Factorial structural and CT scan localization. *Journal of Clinical Neuropsychology, 6*, 189–204.

Broca, P. (1861a). Perte de la parole; Ramollissement chronique et destruction partielle du lobe antérieur gauche du Cerveau. *Bulletin de la Société d'Anthropologie de Paris, 2*, 235–238.

Broca, P. (1861b). Remarques sur le siège de la faculté du langage articulé, suivies d'une observation d'aphémie (Perte de la parole). *Bulletins de la Société Anatomique de Paris, 36*, 330–357.

Cappa, S. F., Cavalloti, G., Guidotti, M., Papagno, C., & Vignolo, L. A. (1983). Subcortical aphasia: Two clinical-CT scan correlation studies. *Cortex, 19*, 227–241.

Castaigne, P., Lhermitte, F., Signoret, J. L., & Abelanet, R. (1980). Description et étude scan-nographique du cerveau de Leborgne: La découverte de Broca. *Revue Neurologique, 136*, 563–583.

Damasio, A. R., Damasio, H., Rizzo, M., Varney, N., & Gersh, F. (1982). Aphasia with non-hemorrhagic lesions in the basal ganglia and internal capsule. *Archives of Neurology, 39*, 15–20.

Damasio, H. (1981). Cerebral localization of the aphasias. In M. T. Sarno (Ed.), *Acquired aphasia* (pp. 27–50). New York: Academic Press.

Damasio, H., & Damasio, A. R. (1980). The anatomical basis of conduction aphasia. *Brain, 103*, 337–350.

DeArmond, S. J., Fusco, M. M., & Dewey, M. M. (1976). *Structure of the human brain: A photographic atlas* (2d Ed.). New York: Oxford University Press.

Dejerine, J. (1895). *Anatomie des centres nerveux* (Vol. 1). Paris: Rueff.

Freedman, M., Alexander, M. P., & Naeser, M. A. (1984). Anatomic basis of transcortical motor aphasia. *Neurology, 34*, 409–417.

Frumkin, N. L., Naeser, M. A., Helm-Estabrooks, N., & Fitzpatrick, P. M. (1990). Predicting Successful Treatment with Melodic Intonation Therapy (MIT) Using CT Scans. Presented at the 28th Annual Academy of Aphasia Meetings, Baltimore, Maryland, October 23.

Gardner, H., Zurif, E. B., Berry, T., & Baker, E. H. (1976). Visual communication in aphasia. *Neuropsychologia, 14*, 275–292.

Goldberg, G. (1985). Supplementary motor area structure and function. Review and hypoth-esis. *Behavioral and Brain Sciences*, 567–615.

Goodglass, H., & Kaplan, E. (1972). *The assessment of aphasia and related disorders*. Philadelphia: Lea and Febiger.

Goodglass, H., & Kaplan, E. (1983). *Assessment of aphasia and related disorders* (2d Ed.). Phila-delphia: Lea and Febiger.

Hanaway, J., Scott, W. R., & Strother, C. M. (1977). *Atlas of the human brain and the orbit for computed tomography*. St. Louis: Green.

Helm, N. A. (1978). Criteria for selecting aphasia patients for melodic intonation therapy. Paper presented at *Language Rehabilitation in Aphasia*. Annual meeting of the American Association for the Advancement of Science, Washington, DC.

Helm-Estabrooks, N. A., & Albert, M. L. (1991). *A manual of aphasia therapy*. Austin, Texas: Pro Ed.

Helm-Estabrooks, N., Fitzpatrick, P., & Barresi, B. (1982). Visual action therapy for global aphasia. *Journal of Speech and Hearing Disorders, 47*, 385–389.

Helm-Estabrooks, N., Nicholas, M., & Morgan, A. (1989a). *Melodic intonation therapy program.* San Antonio, Texas: Special Press.

Helm-Estabrooks, N., Ramsberger, G., Morgan, A., & Nicholas, M. (1989b). *Boston assessment of severe aphasia.* San Antonio, Texas: Special Press.

Hier, D. B., Davis, K. R., Richardson, E. P., & Mohr, J. P. (1977). Hypertensive putaminal hemorrhage. *Annals of Neurology, 11*, 152–159.

Hounsfield, G. N. (1973). Computerized transverse axial scanning (tomography): Description of system. *British Journal of Radiology, 46*, 1016–1025.

Jernigan, T. L., Zatz, L. M., & Naeser, M. A. (1979). Semiautomated methods for quantitating CSF volume on cranial computed tomography. *Radiology, 132*, 463–466.

Jürgens, U. (1984). The efferent and afferent connections of the supplementary motor area. *Brain Research, Amsterdam, 300*, 63–81.

Kertesz, A. (1983). Localization of lesions in Wernicke's aphasia. In A. Kertesz (Ed.), *Localization in neuropsychology.* Orlando, Florida: Academic Press.

Kertesz, A. (1979). *Aphasia and associated disorders: Taxonomy, localization and recovery.* New York: Grune and Stratton.

Kertesz, A., Harlock, W., & Coates, R. (1979). Computer tomographic localization, lesion size, and prognosis in aphasia and nonverbal impairment. *Brain and Language, 8*, 34–50.

Kirzinger, A., & Jürgens, U. (1982). Cortical lesion effects and vocalization in the squirrel monkey. *Brain Research, Amsterdam, 233*, 299–315.

Knopman, D. S., Selnes, O. A., Niccum, N., Rubens, A. B., Yock, D., & Larson, D. (1983). A longitudinal study of speech fluency in aphasia: CT correlates of recovery and persistent nonfluency. *Neurology, 33*, 1170–1178.

Matsui, T., & Hirano, A. (1978). *An atlas of the human brain for computerized tomography.* Tokyo: Igaku-Shoin.

Mazzocchi, F., & Vignolo, A. L. (1979). Localization of lesions in aphasia: Clinical CT scan correlations in stroke patients. *Cortex, 15*, 627–654.

Mohr, J. P., Pessin, M. S., Finkelstein, S., Funkenstein, H. H., Duncan, G. W., & Davis, K. R. (1978). Broca aphasia: Pathologic and clinical. *Neurology, 28*, 311–324.

Morgan, A., & Helm-Estabrooks, N. (1987). Back to the drawing board: A treatment program for nonverbal aphasic patients. In R. H. Brookshire (Ed), *Clinical aphasiology conference proceedings* (pp. 64–72). Minneapolis: BRK Publishers.

Mufson, E. J., & Pandya, D. N. (1984). Some observations on the course and composition of the cingulum bundle in the rhesus monkey. *Journal of Comprehensive Neurology, 225*, 31–43.

Muratoff, W. (1893). Secundäre Degeneration nach Durchschneidung des Balkens. *Neurologisches Centralblatt, 12*, 714–729.

Naeser, M. A. (1983). CT scan lesion size and lesion locus in cortical and subcortical aphasia. In A. Kertesz (Ed.), *Localization in neuropsychology,* New York: Academic Press.

Naeser, M. A., Alexander, M. P., Helm-Estabrooks, N., Levine, H. L., Laughlin, S. A., & Geschwind, N. (1982). Aphasia with predominantly subcortical lesion sites—description of three capsular/putaminal aphasia syndromes. *Archives of Neurology, 39*, 2–14.

Naeser, M. A., Alexander, M. P., Stiassny-Eder, D., Galler, V., Hobbs, J., & Bachman, D. (1992). Real versus sham acupuncture in the treatment of paralysis in acute stroke patients—A CT scan lesion site study, *Journal of Neurologic Rehabilitation, 6*, 163–173.

Naeser, M. A., Gaddie, A., Palumbo, C. L., & Stiassny-Eder, D. (1990). Late recovery of auditory comprehension in global aphasia: Improved recovery observed with subcortical temporal isthmus lesion versus Wernicke's cortical area lesion. *Archives of Neurology, 47*, 425–432.

Naeser, M. A., Hayward, R. W., Laughlin, S. A., & Zatz, L. M. (1981b). Quantitative CT scan studies in aphasia. I: Infarct size and CT numbers. *Brain and Language, 12*, 140–164.

Naeser, M. A., Hayward, R. W., Laughlin, S. A., Becker, J. M. T., Jernigan, T. L., & Zatz, L. M. (1981a). Quantitative CT scan studies in aphasia. II: Comparison of the right and left hemispheres. *Brain and Language, 12,* 165–189.

Naeser, M. A., & Hayward, R. W. (1978). Lesion localization in aphasia with cranial computed tomography and the Boston diagnostic aphasia exam. *Neurology, 28,* 545–551.

Naeser, M. A., & Helm-Estabrooks, N. (1985). CT scan lesion localization and response to melodic intonation therapy with nonfluent aphasia cases. *Cortex, 21,* 203–223.

Naeser, M. A., Helm-Estabrooks, N., Haas, G., Auerbach, S. & Srinivasan, M. (1987). Relationship between lesion extent in "Wernicke's area" on CT scan and predicting recovery of comprehension in Wernicke's aphasia. *Archives of Neurology, 44,* 73–82.

Naeser, M. A., Palumbo, C. L., Helm-Estabrooks, N., Stiassny-Eder, D., & Albert, M. L. (1989). Severe non-fluency in aphasia: Role of the medial subcallosal fasciculus plus other white matter pathways in recovery of spontaneous speech. *Brain, 112,* 1–38.

Nielsen, J. M. (1946). *Agnosia, apraxia, aphasia: Their value in cerebral localization* (2d Ed.). New York: Hoeber.

Palumbo, C. L., Nicholas, M. A., Baker, E. H., Alexander, M. P., Frumkin, N. L., & Naeser, M. A. (1992). CT Scan Lesion Sites Associated with Good Response to a Nonverbal Computer-Assisted Visual Communication Program (C-VIC) for Patients with Severe Aphasia. Presented at the Academy of Aphasia Meeting, Toronto, Canada, October 26.

Palumbo, C. L., & Naeser, M. N. Comparison of acute and subacute CT scans for usefulness in predicting recovery of spontaneous speech and/or auditory comprehension in aphasia patients. Ph.D. dissertation, Department of Behavioral Neuroscience, Boston University School of Medicine and Graduate School, in preparation.

Porch, B. E. (1967). *Porch index of communicative ability.* Palo Alto, California: Consulting Psychologists Press.

Ramsberger, G., & Helm-Estabrooks, N. (1988). Visual action therapy for bucco-facial apraxia. *Clinical Aphasiology Conference Proceedings.* San Diego: College Hill Press.

Rao, P. R. (1986). The use of Amer-Ind Code with aphasic adults. In R. Chapey (Ed.), *Language intervention strategies in adult aphasia* (pp. 360–367). Baltimore: Williams and Wilkins.

Ross, E. D. (1980). Localization of the pyramidal tract in the internal capsule by whole brain dissection. *Neurology, 30,* 59–64.

Sarno, M. T., & Levita, E. (1979). Recovery in aphasia during the first year post stroke. *Stroke, 10,* 663–670.

Sarno, M. T., & Levita, E. (1981). Some observations on the nature of recovery in global aphasia after stroke. *Brain and Language, 31,* 1–12.

Schulz, M. L., Pandya, D., & Rosene, D. (1993). The somatotopic arrangement of motor fibers in the periventricular white matter and internal capsule in the rhesus monkey. Ph.D. Thesis, Department of Behavioral Neuroscience, Boston University School of Medicine and Graduate School.

Selnes, O. A., Knopman, D. S., Niccum, N., Rubens, A. B., & Larsen, D. L. (1983). Computed tomographic scan correlates of auditory comprehension deficits in aphasia: A prospective study. *Annals of Neurology, 13,* 558–566.

Selnes, O. A., Niccum, N., Knopman, D. S., & Rubens, A. B. (1984). Recovery of single word comprehension: CT scan correlates. *Brain and Language, 21,* 72–74.

Signoret, J-L., Castaigne, P., Lhermitte, F., Abelanet, R., & Lavoral, P. (1984). Rediscovery of Leborgne's brain: Anatomical description with CT scan. *Brain and Language, 22,* 303–319.

Skelly, M., Schinsky, L., Smith, R., Donaldson, R., & Griffin, J. (1974). American Indian Sign (Amerind) as a facilitator of verbalization for the oral-verbal apraxic. *Journal of Speech and Hearing Disorder, 39,* 445–456.

Skelly, M., Schinsky, L., Smith, R., Donaldson, R., & Griffin, J. (1975). American Indian Sign: Gestural communication for the speechless. *Archives of Physical Medicine and Rehabilitation, 56,* 156–160.

Smith, A. M., Bourbonnais, D., & Blanchette, G. (1981). Interaction between forced grasping and a learned precision grip after ablation of the supplementary motor area. *Brain Research, 222,* 395–400.

Sparks, R., Helm, N., & Albert, M. (1974). Aphasia rehabilitation resulting from melodic intonation therapy. *Cortex, 10,* 303–316.

Sparks, R., & Holland, A. L. (1976). Method: Melodic intonation therapy for aphasia. *Journal of Speech and Hearing Disorders, 41,* 287–297.

Steele, R. D., Weinrich, M., Wertz, R. T., Kleczewska, M. K., & Carlson, G. S. (1989). Computer-based visual communication in aphasia. *Neuropsychologia, 27,* 409–426.

Vignolo, L. A. (1979). Lesions underlying defective performances on the Token Test: A CT scan study. In F. Boller (Ed.), *Auditory comprehension: Clinical and experimental studies with the Token Test* (pp. 161–167). Orlando, Florida: Academic Press.

Weinrich, M., Steele, R., Carlson, G. S., Kleczewska, M., Wertz, R. T., & Baker, E. H. (1989a). Processing of visual syntax in a globally aphasic patient. *Brain and Language, 36,* 391–405.

Weinrich, M., Steele, R., Kleczewska, M., Carlson, G. S., Baker, E. H., & Wertz, R. T. (1989b). Representation of "verbs" in a computerized visual communication system. *Aphasiology, 3,* 501–512.

Yakovlev, P. I., & Locke, S. (1961). Limbic nuclei of thalamus and connections of limbic cortex. III. Corticocortical connections of the anterior cingulate gyrus, the cingulum, and the subcallosal bundle in monkey. *Archives of Neurology, 5,* 364–400.

Chapter 10

Localization of Lesions in Transcortical Aphasia

Steven Z. Rapcsak and Alan B. Rubens

I. INTRODUCTION

In his influential paper on aphasia, Lichtheim (1885) described two patients with language profiles that did not fit easily into Wernicke's (1874) original model of aphasia. One had greatly reduced spontaneous speech but good language comprehension and was able to repeat. The other had fluent paraphasic speech and severely impaired comprehension, but could correctly repeat words and sentences he did not understand. Since Lichtheim was unable to account for the preservation of repetition on the basis of Wernicke's (1874) model of motor, sensory, and commissural aphasia ("Leitungsaphasie" or conduction aphasia), he proposed that the first syndrome was caused by an interruption of the pathway between the "center for the elaboration of concepts" and the "center of motor images" (i.e., Broca's area) whereas the second syndrome resulted from a disconnection between the "center of auditory images" (i.e., Wernicke's area) and the concept center (Fig. 1). The lesion in the first case led to the loss of volitional and conceptual control over speech production; in the second case, it separated word sounds from their meaning. The preservation of repetition in both cases was attributed to the sparing of Broca's and Wernicke's areas and their connecting association pathways. Although Broca's and Wernicke's areas were known to correspond to distinct perisylvian cortical regions, Lichtheim did not consider the concept center to be localizable to any single area and believed that it was diffusely represented in the cerebral cortex. His complex diagram (Fig. 2) emphasized the distributed cortical representation of the concept center and also illustrated the convergent radiation of white matter association pathways from multiple cortical regions on the language centers of the left perisylvian zone. Lichtheim thought that the lesions responsible for the two types of "inner

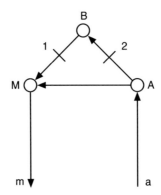

Figure 1 Lichtheim's (1885) model of aphasia. A, Center of auditory images (i.e., Wernicke's area); M, center of motor images (i.e., Broca's area); B, concept center; a, afferent auditory pathways; m, efferent motor pathways to the speech musculature. Lesion of Pathway 1 disconnects the concept center from Broca's area and produces transcortical motor aphasia. Lesion of Pathway 2 separates Wernicke's area from the concept center and produces transcortical sensory aphasia.

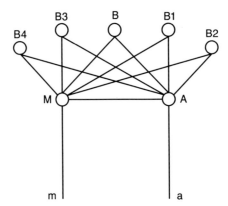

Figure 2 Another diagram by Lichtheim (1885) demonstrates the distributed cortical representation of the concept center and also illustrates the convergent radiation of white matter association pathways from multiple cortical regions on the perisylvian language centers. Lichtheim believed that lesions producing transcortical aphasia involved these white matter pathways at their points of confluence near A and M.

commissural aphasia" with preserved repetition involved these white matter radiations near their points of confluence, either at the base of the third frontal convolution or near the first temporal gyrus. Although Lichtheim did not have anatomical confirmation from his own case material, he quoted two reports from the literature with autopsy findings that seemed to bear out these predictions. Wernicke (1886, 1906) accepted Lichtheim's theoretical formulation, but suggested a change of nomenclature to emphasize the pathogenic role of damage to "transcortical" association pathways linking the concept center to the speech areas. Wernicke used the term "transcortical aphasia" to refer to aphasic syndromes with preserved repetition despite considerable loss of spontaneous speech (transcortical motor aphasia) or severe comprehension disturbance (transcortical sensory aphasia). He also briefly alluded to the possibility that, on rare occasions, one may encounter a combination of the two types of transcortical aphasia. This third transcortical syndrome—characterized by a severe reduction of spontaneous speech, poor comprehension, but intact repetition—was discussed in detail by Goldstein (1948) under the term "mixed transcortical aphasia." Goldstein thought that mixed transcortical aphasia was caused by widespread or multifocal pathology that produced a complete "isolation" of the perisylvian speech area. He stressed the pathological separation between the otherwise preserved "instrumentalities of speech" and "non-language mental performances," but he also considered the possibility that mixed transcortical aphasia may be the result of general cognitive impairment.

Although the clinical reality of transcortical aphasia was generally accepted, the pathophysiological mechanisms and anatomical correlates outlined in Lichtheim's original model have been questioned and challenged on various grounds by several investigators. Wernicke himself (1906) expressed some reservations about the anatomical evidence supporting the transcortical mechanism. Dejerine (1914) regarded transcortical aphasia with interest from a clinical point of view, but considered the interpretation of these syndromes highly debatable and thought no conclusive evidence existed that they were anatomically localizable. The transcortical mechanism was directly challenged by Bastian (1897), who suggested that repetition could be spared even in the face of damage to the cortical speech centers. He thought that damage to the speech centers raised the cortical excitability threshold to such a level that the centers could only be activated by strong external stimulation, as in the automatic act of repetition, but could not respond to weaker internal stimulation produced by "associations" or "voluntary recall." Henschen (1920–1922), Goldstein (1948), and Brown (1975) also considered partial damage to the speech areas to play a role in the pathogenesis of transcortical aphasia.

Whereas some of these investigators hypothesized that repetition was still within the functional capacity of the damaged language centers, others proposed that preserved repetition after left perisylvian damage either was mediated by the right hemisphere exclusively (Niessl von Mayendorf, 1911; Stengel, 1947), or was the result of the combined action of both hemispheres (Brown, 1975; Rubens, 1976).

Evidence for possible right hemisphere mediation of repetition in transcortical aphasia derives from several sources. In some cases, structural damage to the left perisylvian zone was quite extensive, making any contribution to repetition by the left hemisphere unlikely (Berthier et al., 1991; Stengel, 1947). In other cases with subcortical lesions, the perisylvian language zone appeared anatomically intact but positron emission tomography (PET) or single photon emission computerized tomography (SPECT) imaging demonstrated widespread ipsilateral cortical hypometabolism (Berthier et al., 1991; Perani, Vallar, Cappa, Messa, & Fazio, 1987), consistent with the hypothesis of Cambier, Elghozi, Khoury, and Strube (1980) that subcortical lesions may lead to "functional deactivation" of left hemisphere language areas. More direct evidence for right hemisphere participation in repetition derives from reports of transcortical aphasia in which repetition was abolished by a second lesion in the right hemisphere or following the injection of sodium amytal into the right internal carotid artery during the Wada test (Bando, Ugawa, & Sugishita, 1986; Berthier et al., 1991; Rapcsak, Krupp, Rubens, & Reim, 1990). Collectively, these findings suggest that after damage to left hemisphere language areas, the right hemisphere in some individuals may have the capacity to mediate repetition, although the minor hemisphere is much impaired in generating propositional speech. This formulation is consistent with observations that repetition may be better preserved than spontaneous speech after dominant hemispherectomy (Gott, 1973; Burklund & Smith, 1977). In our experience, most patients with left perisylvian lesions have impaired repetition in the acute phase. When the lesion involves Wernicke's area, a severe and persistent impairment of sentence repetition is frequently found (Selnes, Knopman, Niccum, & Rubens, 1985). These observations suggest that right hemisphere mediation of repetition, especially at the sentence level, is probably uncommon.

From this brief overview one can see that the "transcortical" mechanism outlined in the Lichtheim–Wernicke connectionist model of aphasia cannot account for all cases classified as "transcortical aphasia" based on taxonomic criteria. Lesions located outside or, less frequently, within the perisylvian language zone may give rise to aphasic syndromes with preserved repetition, but through different mechanisms and with somewhat different linguistic and behavioral profiles. With these caveats in mind, we now examine the clinicopathological correlates of the various transcortical aphasia syndromes.

II. TRANSCORTICAL MOTOR APHASIA

A. Clinical Description

The essential clinical feature of transcortical motor aphasia (TMA) is the striking dissociation between a marked reduction in the amount and complexity of spontaneous speech and a retained ability to produce speech in response to external stimulation in repetition, reading aloud, and confrontation naming (Rubens, 1976). In the acute phase, speech production is severely limited, often to the point of mutism. Patients with TMA generally do not speak unless spoken to. Answers to questions posed by the examiner are usually produced only after a long delay. Speech at times can only be initiated after considerable struggle. In TMA, verbal output is nonfluent, hesitant, poorly organized, and frequently perseverative. In milder or partially recovered cases, the main difficulty consists of an inability to propositionize. Patients rarely speak in full sentences and, although their laconic utterances may be relevant and even informative, they remain unable to generate a detailed narrative discourse. Verbal productions generally contain appropriate grammatical structure, although grammatical complexity is reduced. Articulation is normal and phonemic paraphasias are not prominent. Once initiated, serial speech is much better than spontaneous speech. Even when patients appear practically mute or produce only single words or short overlearned phrases, they can often repeat long and complicated sentences. Uninhibited echolalia is occasionally observed (Rubens, 1975). However, repetition is not necessarily automatic and parrot-like. Patients may spontaneously correct minor syntactic violations (e.g., errors of tense, number agreement, pronoun usage), although factually incorrect or semantically anomalous sentences are frequently repeated unaltered (Davis, Foldi, Gardner, & Zurif, 1978). Patients may show a compulsive tendency to complete open-ended sentences (Stengel, 1947). Auditory comprehension is well preserved for conversational speech in TMA but may break down when formally tested with grammatically complex sentences. Accurate assessment of auditory comprehension in some cases is made difficult by the prominent akinesia and the tendency to perseverate. Writing is impaired to the same degree as spontaneous speech. Patients are often completely unable to produce a written narrative on a given topic or in response to a picture. Writing to dictation may remain possible, but spelling errors and difficulty in forming legible graphemes are frequently noted. Reading aloud is often remarkably better than the ability to produce spontaneous speech. Reading comprehension is generally adequate for single words and simple sentences, but may be impaired for complex material or at the paragraph level. Patients can often name objects presented through a particular sensory channel more easily than they can evoke appropriate verbal labels during spontaneous speech. Word list generation (Thurstone & Thurstone, 1943) is a particularly difficult task for patients with TMA and frequently remains impaired even in otherwise recovered cases.

B. Lesion Localization

Lesions responsible for TMA involve the language-dominant frontal lobe, either along its superior and mesial aspect or over the dorsolateral convexity. Dorsolateral lesions are usually located in prefrontal and premotor regions anterior or superior to Broca's area. On occasion, TMA is seen with perisylvian lesions that involve Broca's area and/or the underlying white matter.

The left frontal lobe plays a highly complex role in human language and communication. Frontal language functions include activation–initiation, linguistic selection and formulation, monitoring, and motor execution (Alexander, Benson, & Stuss, 1989). These operations are mediated by distinct neuronal systems that may be damaged selectively or in various combinations by frontal pathology. Consequently, language profiles and associated neurobehavioral findings in patients with TMA may vary somewhat as a function of lesion size and location. In general, lesions closer to the superior and mesial frontal area are characterized by a severe impairment of speech initiation but no significant linguistic or articulatory disturbance, whereas lesions closer to Broca's area are more likely to produce additional linguistic impairment and motor speech deficits (Alexander, et al., 1989; Stuss & Benson, 1986).

1. Left superior and mesial frontal lesions involving the supplementary motor area

In summarizing their extensive experience from cortical electrical stimulation experiments in awake human subjects, Penfield and Roberts (1959) concluded that, in addition to Broca's and Wernicke's areas, a third cortical region was involved in speech mechanisms. This "superior speech area" corresponded to the supplementary motor area (SMA), a premotor region located on the mesial aspect of the superior frontal convexity directly anterior to the foot area of the primary motor cortex. Stimulation at this site in either hemisphere could produce speech arrest or involuntary vocalization (Brickner, 1940; Erickson & Woolsey, 1951; Penfield & Roberts, 1959; Penfield & Welch, 1951). Similar paroxysmal speech disturbances were also observed during focal epileptic discharges originating from the SMA, and a form of motor aphasia with relatively preserved repetition has been described in association with frontal parasagittal tumors, penetrating mesial premotor missile injuries, and cortical excision of the SMA (for reviews, see Jonas, 1981; Rubens, 1975, 1976; Rubens & Kertesz, 1983). Although abnormal vocal behavior has been reported with both left- and right-sided lesions, there appears to be an overall preponderance of left SMA pathology (Jonas, 1981). TMA is well documented after left anterior cerebral artery infarctions (Fig. 3), suggesting a major role for the dominant SMA in human vocalization (Alexander & Schmitt, 1980; Bogousslavsky, Assal, & Regli, 1987; Bogousslavsky & Regli, 1990; Damasio & Kassel, 1978; Damasio & Van Hoesen, 1980; Freedman, Alexander, & Naeser, 1984; Goldberg, Mayer, & Toglia, 1981; Jonas, 1981; Kertesz, Lesk, & McCabe, 1977; Környei, 1975; Masdeu, Schoene, & Funkenstein, 1978; Racy, Jannotta, & Lehner, 1979; Rubens, 1975; Von Stockert, 1974; Watson, Fleet, Gonzalez-Rothi, & Heilman, 1986). The participation of the SMA in speech production is supported further by regional cerebral blood flow (Ingvar & Schwartz, 1974; Larsen, Skinhøj, & Lassen, 1978) and PET (Petersen, Fox, Posner, Mintun, & Raichle, 1988) studies in normal subjects.

Acute vascular lesions involving the left SMA produce a characteristic clinical picture. Frequently, an initial stage of mutism may persist for several days. During this early period, some patients appear alert but do not spontaneously attempt to communicate (Damasio & Van Hoesen, 1980, 1983). Others struggle to initiate and sustain speech and are frustrated by their failures on language tasks (Masdeu et al., 1978; Rubens, 1975). As speech returns, some patients go through a stage in which they only mouth words silently or speak in a whisper or in a muffled monotone (Rubens, 1975). Even when patients generate only minimal propositional speech, they can sometimes repeat long sentences, produce serial speech, and read aloud effortlessly. Echolalia is not uncommon and, in some cases, compulsive repetition and forced completion could not be inhibited (Rubens, 1975). Although spontaneous verbal expression is severely compromised and consists of short and simplified utterances, agrammatism is not a feature, paraphasic errors are usually absent, and articulation is normal. Perseverative verbal and motor responses are common. Comprehension can approach normal, except for syntactically complex material. Confrontation naming may be relatively well preserved, but word list generation is severely impaired. Neurological findings that accompany acute left anterior cerebral artery occlusion include urinary incontinence and right hemiparesis, predominantly affecting the leg and the proximal right upper extremity. Sensory loss in the right lower extremity is frequently present. Movements of the right hand may be preserved and facial weakness may become prominent only during spontaneous emotional expression. Even in the absence of significant hemiparesis, motor neglect of the right extremities may be observed (Castaigne, Laplane, & Degos, 1972; Damasio & Van Hoesen, 1980, 1983; Laplane, Talairach, Meininger, Bancaud, & Orgogozo, 1977). Grasping and groping with the right hand is common and the alien hand sign may be present (Goldberg et al., 1981). Bilateral ideomotor apraxia has been described after left SMA lesions (Watson et al., 1986) and bimanual coordination is uniformly impaired. When the lesion involves the anterior corpus callosum, unilateral apraxia, agraphia, and tactile anomia of the left hand may be demonstrated.

The language production disturbance that follows SMA lesions may not be a true aphasia. Several investigators have proposed that the SMA is more likely to be involved in speech initiation than in linguistic formulation and elaboration (Alexander et al., 1989; Botez & Barbeau, 1971; Brown, 1987; Damasio & Van Hoesen, 1980; Freedman et al., 1984; Goldberg, 1985; Jonas, 1981, 1987; Stuss & Benson, 1986). Botez and Barbeau (1971) consider the SMA to be the highest cortical component of a neuronal system that functions as the "starting mechanism of speech" and includes the periaqueductal gray of the mesencephalon, the ventrolateral nucleus of the thalamus, and the anterior cingulate cortex. Cytoarchitectonically, the SMA represents a paralimbic expansion from ventrally adjacent limbic cortex (Sanides, 1970). The SMA is reciprocally interconnected with the anterior cingulate cortex and with dorsolateral premotor and primary motor cortices (Damasio & Van Hoesen, 1980). Goldberg (1985) proposed that, by virtue of its anatomical location between medial limbic cortex and primary motor cortex, the SMA plays an intermediary role between internally generated limbic drive for action and its external realization through the selection and execution of specific movement sequences and strategies. Thus, the SMA is involved in projecting limbic outflow onto motor executive regions and participates in translating intention into action (Goldberg, 1985). The SMA and related cingulate cortex are known to play an important role in the initiation of vocalization in primates (Jürgens, 1985; Jürgens & von Cramon, 1982; Myers, 1976; Robinson, 1976; Sutton, Trachy,

---

Figure 3 Lesion locations typically associated with transcortical motor aphasia. (A) Extensive mesial frontal infarction in the distribution of the left anterior cerebral artery. (B) T_2-weighted MRI scan demonstrating discrete mesial frontal infarction in the region of the left SMA. (C,D) Subcortical lesion anterolateral to the left frontal horn. (E,F) Dorsolateral frontal infarct located superior to Broca's area.

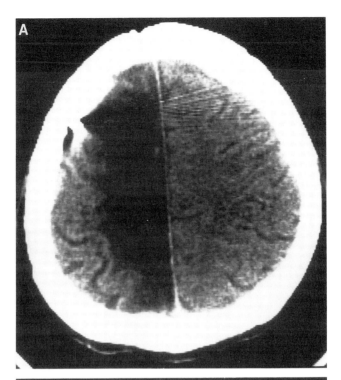

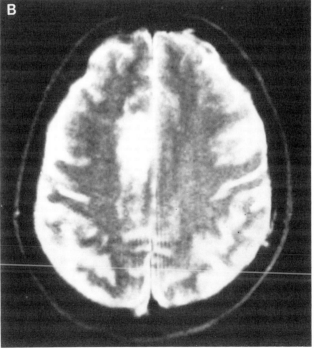

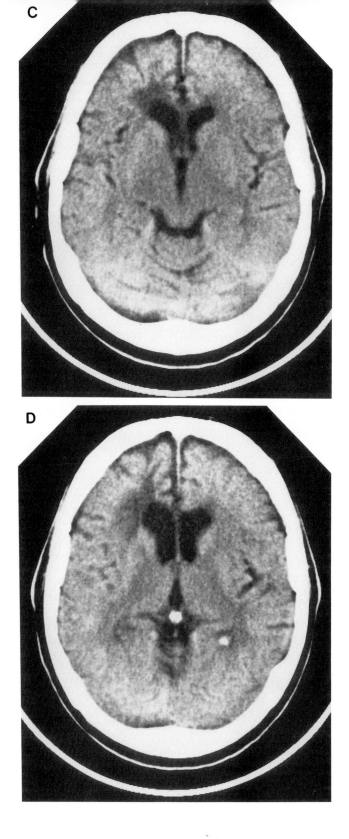

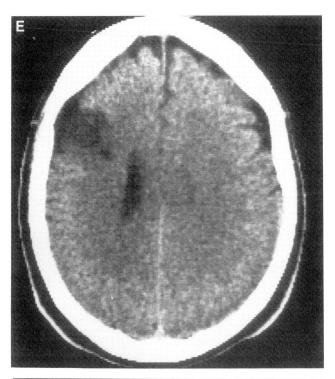

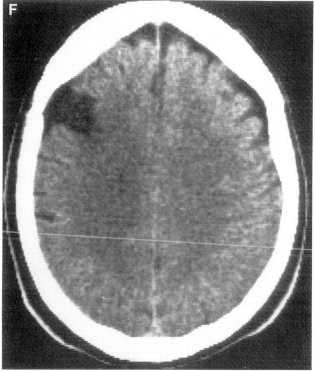

& Lindeman, 1985). Evidence from clinical case reports of TMA suggests that the SMA in humans also plays a crucial role in initiating volitional speech and in maintaining and modulating speech flow and volume (Botez & Barbeau, 1971; Damasio & Van Hoesen, 1980; Goldberg, 1985; Jonas, 1981, 1987; Rubens, 1975). The SMA is also likely to be involved in terminating the speech act and in inhibiting and suppressing involuntary or automatic vocalizations (Goldberg, 1985; Jonas, 1981, 1987; Rubens, 1975). In a sense, the role of the SMA in speech production might be that of a "go/no go" switch. According to Goldberg (1985), the SMA is involved primarily in the production of internally generated propositional speech, whereas the production of nonpropositional or automatic speech, such as repetition, is mediated by the lateral premotor system that operates in an input-dependent or "responsive" mode. The lateral premotor system in the human brain includes Broca's area, the anterior component of the traditional perisylvian reflex arc that mediates repetition in the Wernicke–Lichtheim connectionist model of language. After left SMA damage, the perisylvian language zone may become temporarily "disinhibited" and speech production may be driven exclusively by external input.

2. Dorsolateral frontal lesions

Early descriptions of TMA after dorsolateral frontal lesions located anterior or superior to Broca's area include reports by Marie and Foix (1917), Kleist (1934), and Luria (1970). A similar syndrome characterized by a remarkable reduction of spontaneous speech and severe impairment on tests of verbal fluency was described by Milner (1964) in patients with dominant frontal lobectomies that spared Broca's area. Regarding the mechanism of aphasia, Marie and Foix (1917) believed that the main difficulty in their patients consisted of an inability to condense ideas into phrases. Kleist (1934) emphasized the reduced drive to speak ("Antriebsmangel der Sprache") in patients with dorsolateral prefrontal lesions centered on Brodmann's area 9, just anterior and superior to Broca's area. Goldstein (1948) suggested that a lesion "located in the region between the frontal lobe and the motor speech area, which may not damage severely both areas themselves" resulted in a type of TMA that was characterized by an "impairment of the impulse to speak at all." Luria and Hutton (1977) distinguished two subtypes of TMA: perseverative and dynamic. In perseverative aphasia, the patient is only able to repeat single words or short sentences before perseverations interfere with repetition. Lesions in perseverative aphasia were frequently located in premotor regions superior to Broca's area (Luria, 1970). In dynamic aphasia, repetition of long and complicated sentences is possible but a severe reduction of spontaneous speech, amounting to an almost total loss of the ability to propositionize, is noted. Patients with dynamic aphasia generally do not try to communicate (Ardila & Lopez, 1984; Luria & Tsvetkova, 1967). According to Luria and Tsvetkova (1967), the basic defect in dynamic aphasia is an inability to convert the original idea into extended verbal expression because of a disturbance of the predicative function of inner speech. According to these authors, the role of inner speech in the transition from initial thought to verbal proposition is to provide the "linear scheme of the phrase." Other investigators have proposed that dynamic aphasia reflects a selective impairment of verbal planning (Costello & Warrington, 1989). Luria (1970) suggested that the lesion in dynamic aphasia was located in prefrontal regions anterior to Broca's area.

The association of TMA with dorsolateral frontal pathology sparing Broca's area is well documented in a series of studies that used radioisotope brain scan or computerized tomography (CT) localization in patients with circumscribed vascular lesions (Bogousslavsky & Regli, 1986; Damasio, 1981; Freedman et al., 1984; Naeser & Hayward, 1978; Rubens, 1976). Responsible ischemic lesions were located either in the watershed region between the anterior and middle cerebral arteries or completely within middle cerebral artery territory. Freedman and colleagues (1984) concluded that lesions in TMA are predominantly subcortical, with convergence of pathology in the white matter anterolateral to the left frontal horn (Fig. 3). Small lesions in this location may selectively interrupt connections between the SMA and Broca's area and may present with a relatively pure and often short-lived disturbance of speech initiation (Alexander et al., 1989; Damasio, 1981; Freedman et al., 1984; Stuss & Benson, 1986). Larger lesions that extend into the periventricular white matter adjacent to the body of the lateral ventricle deep to the precentral gyrus may be associated with dysarthria, whereas lesions that extend across the head of the caudate, the anterior limb of the internal capsule, the anterior putamen, and the insular region may be associated with comprehension impairment (Freedman et al., 1984). Extensive involvement of left prefrontal areas may lead to a total inability to formulate verbal expression, frequently accompanied by pervasive apathy and a general loss of cognitive control. This clinical picture corresponds to the "dynamic aphasia" of Luria (1970).

Patients with dorsolateral frontal infarctions anterior or superior to Broca's area may be mute in the beginning. When speech returns, verbal production is sparse, hesitant, and frequently perseverative. Semantic paraphasias may be present but phonemic paraphasias are not common. Hemiparesis is usually mild or transient. Watershed infarctions may give rise to predominantly proximal extremity weakness with sparing of the face.

TMA has been observed in the process of recovery from perisylvian lesions that resulted in partial or even complete destruction of Broca's area (Alexander, Naeser & Palumbo, 1990; Berthier et al., 1991; Freedman et al., 1984; Kertesz et al., 1977; Masdeu & O'Hara, 1983; Mohr et al., 1978; Rubens, 1976). Consistent with Lichtheim's prediction, TMA has also been described in association with a small white matter lesion located directly under Broca's area (Rothman, 1906). Goldstein (1948) proposed that TMA following damage to the motor speech area resulted from a "heightening of the threshold of motor speech performances." After acute frontal opercular lesions, most patients are mute and hemiparetic (Mohr et al., 1978). Mutism is short-lived and hemiparesis also rapidly improves, although lower facial weakness can persist. Buccofacial and limb apraxia may be present. Speech production after frontal opercular lesions is hesitant, effortful, and may be marked by numerous false starts resembling stuttering. Phonemic paraphasias are common and transient agrammatism may be seen. Dysarthria usually signifies extension of the lesion into the lower precentral gyrus or the white matter under it. Although repetition may be relatively preserved, the difference between spontaneous speech and repetition is often not as pronounced as in cases of TMA with lesions located outside the perisylvian zone (Goldstein, 1948). Patients with opercular lesions are anxious to communicate and are frequently frustrated by their failures.

The prognosis for TMA after vascular lesions is quite good; most cases evolve over weeks toward anomic aphasia and eventual full recovery (Kertesz & McCabe, 1977).

III. TRANSCORTICAL SENSORY APHASIA

A. Clinical Description

Transcortical sensory aphasia (TSA) is characterized by fluent speech, poor comprehension, and preserved repetition. Although spontaneous verbal production may be abundant, it is usually irrelevant, uninformative, semantically empty, and circumlocutory (Albert, Goodglass, Helm, Rubens, & Alexander, 1981; Rubens & Kertesz, 1983). Fluency in some cases is disrupted by frequent word-finding pauses. Semantic paraphasias are common and at times verbal output consists entirely of incoherent semantic jargon. Syntactic structure may be relatively intact. Patients generally appear unconcerned about their failure to produce coherent speech. On rare occasions, spontaneous speech can be relatively spared (Heilman, Rothi, McFarling, & Rottmann, 1981). In TSA, neologisms and phonemic paraphasias are much less frequent than in Wernicke's aphasia. Articulation is generally excellent and prosody may be preserved. Naming is usually significantly impaired but rare exceptions exist (Heilman et al., 1981). Comprehension in TSA is fragile and context dependent, but may not be as severely compromised as in Wernicke's aphasia (Albert et al., 1981). As originally noted by Lichtheim (1885), patients with TSA can repeat words and sentences they do not comprehend. In repeating, some patients correct minor syntactic violations but semantically anomalous sentences are repeated unaltered (Berndt, Basili, & Caramazza, 1987; Berthier et al., 1991; Coslett, Roeltgen, Gonzalez-Rothi, & Heilman, 1987; Davis et al., 1978; Heilman et al., 1981). Echolalia may be present. Although reading aloud may be possible, reading comprehension is usually impaired to the same degree as auditory comprehension. Patients with TSA cannot compose a coherent written narrative, but may be able to write to dictation words they fail to understand (Roeltgen, Gonzalez-Rothi, & Heilman, 1986).

Linguistically, the hallmark of TSA is the dissociation between impaired semantic processing and relatively preserved syntactic/phonological competence (Alexander, Hiltbrunner, & Fischer, 1989; Berndt et al., 1987; Kertesz, Sheppard, & MacKenzie, 1982; Rubens & Kertesz, 1983). In discussing the mechanism of TSA, Lichtheim (1885) stressed the disconnection between Wernicke's area and the concept center, resulting in a separation of word sounds from their meaning (Fig. 1). Similarly, Luria (1966) spoke of an "extinction of word meaning." If one accepts Teuber's (1968) definition of agnosia as a "normal percept that has somehow been stripped of its meaning," then the disconnection postulated by Lichtheim in TSA would result in an auditory comprehension defect that might be considered an agnosia for word meaning or "word meaning deafness" (Bramwell, 1897). However, the disconnection between Wernicke's area and the concept center in Lichtheim's model does not adequately explain the severe impairment of spontaneous speech and naming that is evident in most cases of TSA. To account for the prominent impairment of spontaneous speech and naming in sensory aphasia, Kussmaul (1877) postulated that for speech production the concept center projected to Wernicke's area first and subsequently activated Broca's area only indirectly (Fig. 4). According to Kussmaul's model, Wernicke's area and the concept center are reciprocally interconnected by two separate white matter association pathways. In TSA, both pathways are interrupted, creating a double disconnection of the semantic system from cortical regions involved in auditory analysis and speech production. On rare occasions, a selective impairment of the pathway from Wernicke's area to the concept center with sparing of the pathway in the reverse direction may give rise to TSA with relatively preserved spontaneous speech and naming (Heilman et al.,

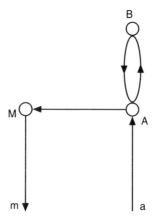

Figure 4 Kussmaul's (1877) model of aphasia. In speech production, the concept center projects to Wernicke's area first and activates Broca's area only indirectly. Wernicke's area and the concept center are reciprocally interconnected by two separate white matter association pathways.

1981). Alternatively, the more common syndrome of TSA with impaired speech production and comprehension may result from a central impairment of the semantic system itself (Alexander et al., 1989; Berndt et al., 1987; Rothi, 1990). Regardless of the mechanism of the semantic processing deficit (i.e., disconnection from semantics or central disruption of semantic competence), most investigators assume that the relative sparing of phonological and syntactic processing in TSA reflects the intrinsic capacity of the spared perisylvian language zone.

B. Lesion Localization

Lichtheim (1885) believed that the lesion responsible for disconnecting Wernicke's area from the concept center in TSA was located in the white matter near the first temporal convolution. Bastian (1897), however, considered TSA to result from partial damage to the auditory word center in the temporal lobe. Vix (1910) reported TSA with focal damage to the left temporo-occipital junction. Dejerine (1914) observed posterior cerebral atrophy in TSA and concluded that the syndrome did not have a specific anatomical localization. Henschen (1920–1922) also noted generalized cerebral atrophy in some cases of TSA, whereas in others the syndrome was produced by a focal lesion of the left temporal lobe. Henschen concluded that, although the term "transcortical aphasia" implied that the pathology was located outside the language zone, in most cases of TSA the lesion involved the posterior perisylvian speech area at least partially. Goldstein (1948) noted that similarly located lesions in the posterior temporal region may produce word deafness, Wernicke's aphasia, or TSA. He discussed several possible mechanisms in the pathogenesis of TSA, including partial damage to Wernicke's area, disconnection of Wernicke's area from the concept center, or a combination of the two. Luria (1966) considered damage to the left inferior temporo-occipital junction to play a critical role in producing the syndrome.

Modern studies using radioisotope and CT imaging have generally supported the temporo-parieto-occipital localization of lesions in TSA (Alexander et al., 1989; Bogousslavsky & Regli, 1986; Damasio, 1981; Heilman et al., 1981; Kertesz et al., 1977,1982). In the series by Kertesz and colleagues (1982), the most common lesion site for TSA was in the left inferior temporo-occipital region, either entirely within posterior cerebral artery territory or in the watershed area between posterior cerebral and middle cerebral artery territories. In a smaller number of cases, the lesions overlapped in the parieto-occipital convexity region (Fig. 5). The lesions in this latter group were generally somewhat lower and more posterior than those producing Gerstmann syndrome or alexia with agraphia. In the series of patients reported by Damasio (1981), the overlap of lesions in TSA centered on the posterior portions of the middle temporal gyrus (Brodmann's area 37) and the underlying white matter, with variable extension into the posterior portion of the superior temporal gyrus (Brodmann's area 22), the angular gyrus (Brodmann's area 39), and the visual association cortex (Brodmann's area 19). Alexander and associates (1989) proposed that the critical lesion in TSA following left posterior cerebral artery infarctions involved the deep temporo-occipital white matter extending up to and along the periventricular region of the proximal temporal horn, including the posterior half of the temporal isthmus and the adjacent posterolateral thalamus. Deep white matter lesions in this location disrupt thalamocortical and cortico-cortical pathways converging on inferolateral temporo-occipital association cortex (Brodmann's area 37).

Lesions in the left temporo-parieto-occipital region can give rise to a fascinating spectrum of complex neuropsychological syndromes (De-Renzi, Zambolin, & Crisi, 1987). Visual and/or tactile agnosia (Feinberg, Rothi, & Heilman, 1986; Ferro & Santos, 1984; Hahn, 1895; Larrabee, Levin, Huff, Kay, & Guinto, 1985; McCarthy & Warrington, 1986; Morin, Rivrain, Eustache, Lambert, & Cartheous, 1984; Pillon, Signoret, & L'hermitte, 1981), optic or tactile aphasia (Beauvois, Saillant, Meininger, & L'hermitte, 1978; Coslett & Saffran, 1989; Freund, 1889; Gil et al., 1985; Larrabee et al., 1985; L'hermitte & Beauvois, 1973), visual and tactile anomia (Poeck, 1984;

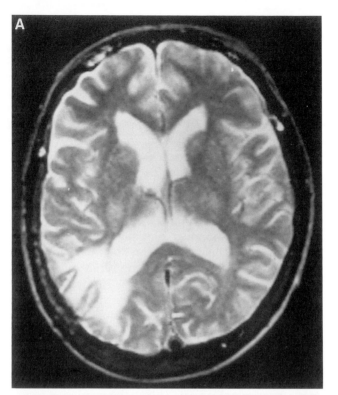

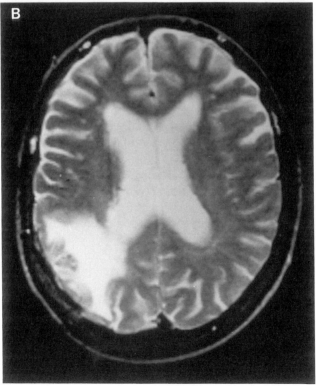

Rapcsak, Gonzalez-Rothi, & Heilman, 1987), anomic aphasia (Benson, 1979; Brown, 1972), color anomia (Geschwind & Fusillo, 1966), alexia without agraphia (Damasio & Damasio, 1983), and verbal amnesia (Benson, Marsden, & Meadows, 1974; De Renzi et al., 1987; Von Cramon, Hebel, & Schuri, 1988) have all been described following damage within this general area. Visual and/or tactile agnosia and verbal amnesia may coexist with TSA (Alexander et al., 1989; Kertesz et al., 1982). These observations suggest that left posterior temporo-parieto-occipital association areas play a crucial role in semantic functions and lexical retrieval. Sensory representations from different modalities converge on this region for semantic processing and verbal labeling, and lesions in this area may lead to modality-specific, multimodal, or supramodal deficits of recognition and naming. Alexander and associates (1989) proposed that left temporo-parieto-occipital association areas and their thalamic connections form a neuronal system that mediates semantic functions. Damage to this distributed system can disrupt lexical–semantic linguistic operations, recognition, and recall depending on the actual lesion configuration. Cortical lesions, especially those involving inferolateral temporo-occipital association cortex (Brodmann's area 37), may produce a central disturbance of semantic processing, whereas deep lesions involving cortico-cortical and thalamocortical white matter pathways converging on this region may isolate the semantic system (or systems) from specific sensory input or may interfere with thalamic activation of cortical semantic processing.

TSA is not infrequently observed in senile dementia of the Alzheimer type (SDAT) (Appell, Kertesz, & Fisman, 1982; Cummings, Benson, Hill, & Read, 1985). That TSA can occur in the setting of generalized cortical atrophy was often considered evidence against the anatomical specificity of this syndrome. However, Pick (1892) proposed that aphasic syndromes identical to those seen following focal lesions may occur in senile atrophy of the brain because of the "local accentuation of a diffuse process." The frequent occurrence of TSA in patients with SDAT may thus reflect the well-documented predilection of the degenerative process for the multimodal association cortex of the temporo-parieto-occipital region and the relative sparing, at least initially, of the immediate perisylvian language zone (Brun & Englund, 1981; Brun & Gustafson, 1976; Kemper, 1984).

In addition to the more common lesion location posterior and deep to Wernicke's area, TSA has also been described following perisylvian

Figure 5 (A and B) T_2-weighted MRI scan demonstrating metastatic tumor at the left temporo-parieto-occipital junction in a patient with transcortical sensory aphasia.

damage (Bando et al., 1986; Berndt et al., 1987; Berthier et al., 1991; Kertesz, Harlock, & Coates, 1979; Kertesz et al., 1977; Niessl von Mayendorf, 1911). Following posterior temporo-parietal perisylvian lesions, TSA may be observed in the process of recovery from Wernicke's aphasia (Kertesz & McCabe, 1977; Kertesz et al., 1982; Rubens & Kertesz, 1983). At least in some cases of TSA with perisylvian pathology, preserved repetition may have been mediated by the intact right hemisphere (Bando et al., 1986; Berthier et al., 1991; Niessl von Mayendorf, 1911).

Neurological deficits accompanying TSA include hemianopia and hemisensory loss. Significant hemiparesis is uncommon, except in cases in which the lesion in posterior cerebral artery territory involves the ipsilateral cerebral peduncle (Hommel et al., 1990). The prognosis in TSA following vascular lesions or trauma is generally excellent; many cases rapidly evolve through a stage of anomic aphasia to eventual complete recovery (Kertesz & McCabe, 1977; Kertesz et al., 1979, 1982; Rubens & Kertesz, 1983).

IV. MIXED TRANSCORTICAL APHASIA

A. Clinical Description

In mixed transcortical aphasia (MTA), meaningful spontaneous speech is dramatically reduced or is absent altogether. Verbal production in most cases is limited to meaningless, repetitive, and stereotypic utterances. Auditory comprehension is severely impaired. Comprehension of written language is also poor, although the ability to read aloud may be preserved (Heilman, Tucker, & Valenstein, 1976; Hübner, 1889; Ross, 1980; Rapcsak et al., 1990; Whitaker, 1976). In spite of the virtual absence of propositional speech, repetition is spared and is often compulsive in nature. Uninhibited echolalia and forced completion of open-ended sentences are frequently observed (Denny-Brown, 1963; Geschwind, Quadfasel, & Segarra, 1968; Goldstein, 1948; Hübner, 1889; Stengel, 1947; Whitaker, 1976). Automatic correction of minor syntactic violations has been described, but semantically anomalous sentences are not corrected (Whitaker, 1976). Repetition of affective prosody may be impaired (Speedie, Coslett, & Heilman, 1984). Serial speech can be relatively intact. Naming is severely defective, but rare exceptions may occur (Heilman et al., 1976). Articulation is usually excellent. Spontaneous writing is generally impossible, but some words can occasionally be written to dictation even though they are frequently not comprehended (Bogousslavsky, Regli, & Assal, 1988; Rapcsak et al., 1990). The ability to sing may be preserved (Geschwind et al., 1968).

Conceptually, the syndrome of MTA might be viewed as a combination of TMA and TSA (Wernicke, 1906). Functionally, MTA is characterized by a severe impairment of volitional cognitive–linguistic ability with preservation of nonvolitional–automatic language functions (Whitaker, 1976).

B. Lesion Localization

Goldstein (1948) suggested that MTA was caused by lesions that produced an anatomical "isolation" of the speech area posteriorly from parietal association cortex and anteriorly from frontal areas important in the production of volitional speech. Consistent with this hypothesis, MTA is often seen with diffuse or multifocal pathology affecting parietal and frontal association areas with relative sparing of the perisylvian language zone (Appell et al., 1982; Assal et al., 1983; Geschwind et al., 1968; Hübner, 1889; Mehler, 1988; Whitaker, 1976). Etiology in MTA is quite varied and includes degenerative dementia, carbon monoxide poisoning, and stroke. Acute MTA has been described in association with left internal carotid artery occlusion and simultaneous anterior precentral–central sulcus artery territory embolic infarction and posterior watershed infarction related to hemodynamic insufficiency (Bogousslavsky et al., 1988). The embolic frontal lesions were generally anterior and superior to Broca's area and the watershed lesions were located posterior to Wernicke's area. In these cases, lesions in anatomical locations characteristic for TMA and TSA combine to produce MTA. In other cases of MTA, following internal carotid occlusion or severe stenosis with hemodynamic insufficiency, a more-or-less confluent arc of infarction is seen (Fig. 6) involving the watershed territory (Speedie et al., 1984). MTA has also been described following extensive mesial frontoparietal infarction in the territory of the left anterior cerebral artery (Ross, 1980), and with infarction or hemorrhage located in the left parieto-occipital region (Speedie et al., 1984; Pirozzolo et al., 1981).

MTA has been observed following vascular lesions that directly involved the perisylvian language centers (Berthier et al., 1991; Brown, 1975; Kertesz et al., 1977; Rapcsak et al., 1990; Stengel, 1947; Trojano, Fragassi, Postiglione, & Grossi, 1988). In some cases the lesion was extensive, resulting in virtually complete destruction of the entire perisylvian language zone (Berthier et al., 1991; Stengel, 1947). MTA following perisylvian damage may evolve from an initial stage of global aphasia (Rapcsak et al., 1990). In some cases of MTA with perisylvian lesions, preserved repetition may have been mediated by the intact right hemisphere (Berthier et al., 1991; Rapcsak et al., 1990; Stengel, 1947).

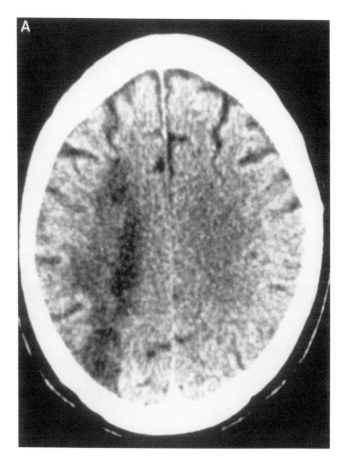

Figure 6   (A and B) Mixed transcortical aphasia associated with internal carotid occlusion and infarction of the watershed territory. The perisylvian region was spared.

The prognosis in MTA depends on the etiology. Variable recovery occurs in vascular cases (Kertesz & McCabe, 1977), but the prognosis may be quite good when the syndrome is seen after discrete anterior and posterior left hemisphere infarctions (Bogousslavsky et al., 1988).

V. TRANSCORTICAL APHASIA FOLLOWING LESIONS OF THE BASAL GANGLIA AND THALAMUS

Numerous reports exist of aphasia following ischemic or hemorrhagic lesions of the basal ganglia (Alexander & LoVerme, 1980; Alexander,

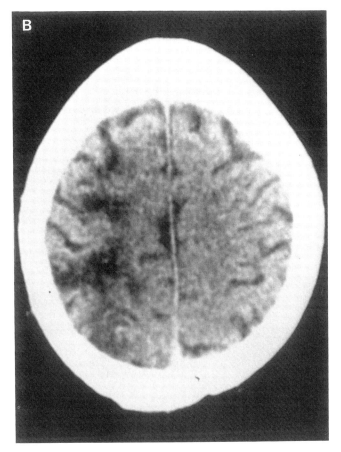

Figure 6 *(continued)*

Naeser, & Palumbo, 1987; Berthier et al., 1991; Cambier et al., 1980; Damasio, Rizzo, Varney, & Gersh, 1982; Brunner, Kornhuber, Seemuller, Suger, & Wallesch, 1982; Wallesch, 1985; Wallesch et al., 1983) and thalamus (for reviews, see Crosson, 1984; Jonas, 1982) in which the sparing of repetition was a conspicuous finding. Some of these cases met objective taxonomic criteria for the various transcortical aphasia syndromes, but atypical features were frequently noted. Although the exact pathophysiological mechanism of language disturbance in subcortical aphasia is still debated, these clinical observations clearly demonstrate that normal language operations require the dynamic interaction of anatomically distributed but functionally integrated neuronal networks that include both cortical and subcortical components. Functional integration of cortex, basal ganglia, and thalamus is mediated by complex neuronal circuitry that includes bidirectional thalamocortical projections as well as re-entrant basal ganglia–thalamocortical loops (Alexander, DeLong, & Strick, 1986). Damage to these distributed neuronal networks can occur at several different anatomical sites and may produce similar, although not necessarily identical, behavioral syndromes. In addition to lesions of the cortex, basal ganglia, and thalamus, damage to projection fibers linking these structures may occur in the internal capsule or in the deep white matter of the cerebral hemispheres and may play a critical role in the pathogenesis of the aphasic syndrome (Alexander et al., 1987; Damasio et al., 1982). Because of the anatomical proximity of white matter projection pathways, small differences in lesion size or site may have major consequences for the aphasia profile (Alexander et al., 1987; Naeser et al., 1982).

Disruption of certain functional circuits by subcortical lesions may have special relevance for the various transcortical aphasia syndromes. As we have seen, damage to the SMA or dorsolateral premotor and prefrontal cortex plays a critical role in the pathogenesis of TMA. These frontal regions are the cortical components of two parallel basal ganglia–thalamocortical re-entrant loops. The "motor" loop links SMA and dorsolateral premotor cortex with the putamen, which in turn projects back through the globus pallidus and ventrolateral nucleus of the thalamus to the SMA. The "complex" loop links dorsolateral prefrontal and premotor cortex to the caudate nucleus, and then projects back through the globus pallidus and ventral anterior and dorsomedial nuclei of the thalamus to dorsolateral prefrontal association cortex (Alexander et al., 1986). The "motor" and "complex" basal ganglia–thalamocortical loops are likely to participate in frontal lobe language functions, including the initiation and motor programming of speech. Damage to these neuronal circuits, at either a cortical or a subcortical level, may result in TMA (Fig. 7). In the pathogenesis of TSA, damage to a neuronal network that includes inferior temporo-parieto-occipital association cortex and thalamus might play a crucial role (Alexander et al., 1989).

In some patients with transcortical aphasia following subcortical lesions, major metabolic dysfunction was detected by PET or SPECT in ipsilateral cortical language areas (Berthier et al., 1991; Perani et al., 1987). Thus, although in terms of anatomy subcortical lesions are located outside the perisylvian zone, they may lead to "functional deactivation" of perisylvian language cortices (Cambier et al., 1980). Severe metabolic dysfunction of left hemisphere cortical language areas that was evident in some cases of transcortical aphasia after subcortical lesions suggests that spared

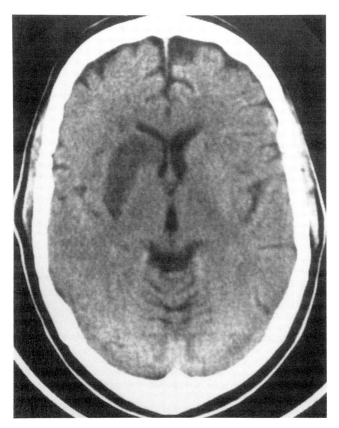

Figure 7 Large subcortical capsulo-putamino-caudate infarction associated with transcortical aphasia. Initially, spontaneous speech was virtually absent and auditory comprehension was also severely impaired. Repetition was intact and echolalia was observed. Comprehension rapidly improved, and by 3 wk post-onset the language profile was typical for transcortical motor aphasia.

repetition in these patients may have been mediated by the right hemisphere (Berthier et al., 1991; Cambier et al., 1980).

VI. CONCLUSION

Historically, the concept of transcortical aphasia arose from the need to account for the unexpected sparing of repetition in aphasic patients with impaired spontaneous speech production and language comprehension. The term "transcortical" carries with it strong anatomical and pathophysiological connotations that have not remained unchallenged.

In our experience, transcortical aphasia syndromes most typically are seen with lesions located outside the perisylvian language zone. Extraperisylvian lesions damage or disconnect distinct cortical or subcortical regions involved in semantic processing, linguistic formulation, and the production of volitional speech. After extra-perisylvian lesions, preserved repetition is probably mediated by the intact left perisylvian language

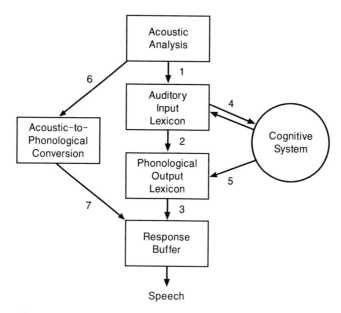

Figure 8   Linguistic model of repetition based on Patterson & Shewell (1987). Repeating via the semantic route (1,4,5,3) involves activation of an entry in the auditory input lexicon, followed by access to the corresponding semantic representation and subsequent retrieval of the appropriate phonological word form based on the semantic code. The lexical but nonsemantic route (1,2,3) bypasses the semantic system and relies on direct transcoding between corresponding representations in the auditory input and phonological output lexicons. Repetition by the nonlexical route (6,7) utilizes an algorithmic acoustic-to-phonological conversion procedure. Semantic and lexical but nonsemantic routes for repetition are also implicit in Lichtheim's connectionist model (Fig. 1). Repetition by an acoustic-to-phonological conversion procedure or "mimicry" was thought by Wernicke (1874) to play an important role in language acquisition.

zone. Transcortical aphasia is also seen following structural damage to the perisylvian language areas. In this setting, transcortical aphasia sometimes represents a stage of recovery from an aphasic syndrome that is more characteristic for the location of the perisylvian lesion. Following perisylvian damage, the functional capacity of the left hemisphere language system is reduced, but repetition may still be possible by residual left hemisphere mechanisms, by the compensatory action of the right hemisphere, or by a combination of both. Thus, although repetition may be preserved after extra-perisylvian or perisylvian lesions, it may not have the same neuronal substrate in all cases. Repetition may not be a unitary function with respect to linguistic mechanism either. According to information processing models of language (Morton, 1980; Patterson & Shewell, 1987), repetition may be accomplished by three different linguistic procedures: semantic, lexical but nonsemantic, and nonlexical (Fig. 8). A detailed discussion of these models is beyond the scope of this chapter, but evidence is accumulating that different aphasic patients may repeat by relying on different linguistic routines (Bernot et al., 1987; Caramazza, Miceli & Villa, 1986; Coslett et al., 1987; Katz & Goodglass, 1990; McCarthy & Warrington, 1986, 1987; Rothi, 1990). The neuronal systems involved in mediating repetition by the different routes remain to be elucidated.

Despite some within-group heterogeneity with respect to lesion location, pathogenesis, and linguistic mechanism, we feel that the term "transcortical aphasia" remains clinically useful. At the same time, we strongly believe that establishing the sparing of repetition and classifying the language disturbance as "transcortical aphasia" is only the first step. Advances in our understanding of the mechanisms underlying these intriguing syndromes will only come from investigations that combine structural and functional neuroimaging with detailed psycholinguistic assessment.

REFERENCES

Albert, M. L., Goodglass, H., Helm, N. A., Rubens, A. B., & Alexander, M. P. (1981). *Clinical aspects of dysphasia.* Wien: Springer-Verlag.

Alexander, G. E., DeLong, M. R., & Strick, P. L. (1986). Parallel organization of functionally segregated circuits linking basal ganglia and cortex. *Annual Review of Neuroscience, 9,* 357–381.

Alexander, M. P., Benson, D. F., & Stuss, D. T. (1989). Frontal lobes and language. *Brain and Language, 37,* 656–691.

Alexander, M. P., Hiltbrunner, B., & Fischer, R. S. (1989). Distributed anatomy of transcortical sensory aphasia. *Archives of Neurology, 46,* 885–892.

324 Steven Z. Rapcsak & Alan B. Rubens

Alexander, M. P., & LoVerme, S. R. (1980). Aphasia after left hemispheric intracerebral hemorrhage. *Neurology, 30,* 1193–1202.

Alexander, M. P., Naeser, M. A., & Palumbo, C. L. (1987). Correlations of subcortical CT lesion sites and aphasia profiles. *Brain, 110,* 961–991.

Alexander, M. P., Naeser, M. A., & Palumbo, C. L. (1990). Broca's area aphasias: Aphasia after lesions including the frontal operculum. *Neurology, 40,* 353–362.

Alexander, M. P., & Schmitt, M. A. (1980). The aphasia syndrome of stroke in the left anterior cerebral artery territory. *Archives of Neurology, 37,* 97–100.

Appell, J., Kertesz, A., & Fisman, M. (1982). A study of language functioning in Alzheimer's patients. *Brain and Language, 17,* 73–91.

Ardila, A., & Lopez, M. V. (1984). Transcortical motor aphasia: One or two aphasias. *Brain and Language, 22,* 350–353.

Assal, G., Regli, F., Thuillard, F., Steck, A., Deruaz, J.-P., & Perentes, E. (1983). Syndrome d'isolement de la zone du langage. Ètude neuropsychologique et pathologique. *Revue Neurologique, 139,* 417–424.

Bando, M., Ugawa, Y., & Sugishita, M. (1986). Mechanism of repetition in transcortical sensory aphasia. *Journal of Neurology, Neurosurgery and Psychiatry, 49,* 200–202.

Bastian, H. (1897). Some problems in connexion with aphasia and other speech defects. *Lancet, 1,* 933–942, 1005–1017, 1131–1137, 1187–1194.

Beauvois, M.-F., Saillant, B., Meininger, V., & Lhermitte, F. (1978). Bilateral tactile aphasia: A tacto-verbal dysfunction. *Brain, 101,* 381–401.

Benson, D. F. (1979). *Aphasia, alexia and agraphia.* London: Churchill Livingstone.

Benson, D. F., Marsden, C. D., & Meadows, J. C. (1974). The amnesic syndrome of the posterior cerebral artery occlusion. *Acta Neurologica Scandinavica, 50,* 133–145.

Berndt, R. S., Basili, A., & Caramazza, A. (1987). Dissociation of functions in a case of transcortical sensory aphasia. *Cognitive Neuropsychology, 4,* 79–107.

Berthier, M. L., Starkstein, S. E., Leiguarda, R., Ruiz, A., Mayberg, H. S., Wagner, H., Price, T. R., & Robinson, R. G. (1991). Transcortical aphasia. Importance of the nonspeech dominant hemisphere in language repetition. *Brain, 114,* 1409–1427.

Bogousslavsky, J., Assal, G., & Regli, F. (1987). Infarctus du territoire de l'artère cèrèbrale antèrieure gauche. II. Troubles du langage. *Revue Neurologique, 143,* 121–127.

Bogousslavsky, J., & Regli, F. (1986). Unilateral watershed cerebral infarcts. *Neurology, 36,* 373–377.

Bogousslavsky, J., & Regli, F. (1990). Anterior cerebral artery territory infarction in the Lausanne Stroke Registry: Clinical and etiologic patterns. *Archives of Neurology, 47,* 144–150.

Bogousslavsky, J., Regli, F., & Assal, G. (1988). Acute transcortical mixed aphasia: A carotid occlusion syndrome with pial and watershed infarcts. *Brain, 111,* 631–641.

Botez, M. I., & Barbeau, A. (1971). Role of subcortical structures and particularly the thalamus in the mechanism of speech and language. A review. *International Journal of Neurology, 8,* 300–320.

Bramwell, B. (1897). Illustrative cases of aphasia. *Lancet, 1,* 1256–1259.

Brickner, R. (1940). A human cortical area producing repetitive phenomena when stimulated. *Journal of Neurophysiology, 3,* 128–130.

Brown, J. W. (1972). *Aphasia, apraxia and agnosia.* Springfield, Illinois: Thomas.

Brown, J. W. (1975). The problem of repetition: A study of "conduction" aphasia and the "isolation" syndrome. *Cortex, 11,* 37–52.

Brown, J. W. (1987). The microstructure of action. In E. Perecman (Ed.), *The frontal lobes revisited* (pp. 251–272). New York: IRBN Press.

Brun, A., & Englund, E. (1981). Regional pattern of degeneration in Alzheimer's disease: Neuronal loss and histopathological grading. *Histopathology, 5,* 549–564.

Brun, A., & Gustafson, L. (1976). Distribution of cerebral degeneration in Alzheimer's disease: A clinico-pathological study. *Archiv für Psychiatrie und Nervenkrankheiten, 223*, 15–33.

Brunner, R. J., Kornhuber, H. H., Seemuller, E., Suger, G., & Wallesch, C. W. (1982). Basal ganglia participation in language pathology. *Brain and Language, 16*, 281–299.

Burklund, C. W., & Smith, A. (1977). Language and the cerebral hemispheres. Observations of verbal and nonverbal responses during 18 months following left ("dominant") hemispherectomy. *Neurology, 27*, 627–633.

Cambier, J., Elghozi, D., Khoury, M., & Strube, A. (1980). Rèpètition ècholalique et troubles sèvères de la comprèhension. Syndrome de dè-activation de l'hèmisphère gauche? *Revue Neurologique, 136*, 689–698.

Caramazza, A., Miceli, G., & Villa, G. (1986). The role of the (output) phonological buffer in reading, writing and repetition. *Cognitive Neuropsychology, 3*, 37–76.

Castaigne, P., Laplane, D., & Degos, J. D. (1972). Trois cas de nègligence motrice par lèsion frontale prè-rolandique. *Revue Neurologique, 126*, 5–115.

Coslett, H. B., Roeltgen, D. P., Gonzalez-Rothi, L., & Heilman, K. M. (1987). Transcortical sensory aphasia: Evidence for subtypes. *Brain and Language, 32*, 362–378.

Coslett, H. B., & Saffran, E. M. (1989). Preserved object recognition and reading comprehension in optic aphasia. *Brain, 112*, 1091–1110.

Costello, A. de L., & Warrington, E. K. (1989). Dynamic aphasia: The selective impairment of verbal planning. *Cortex, 25*, 103–114.

Crosson, B. (1984). Role of the dominant thalamus in language: A review. *Psychological Bulletin, 96*, 491–507.

Cummings, J. L., Benson, D. F., Hill, M. A., & Read, S. (1985). Aphasia in dementia of the Alzheimer type. *Neurology, 35*, 394–397.

Damasio, A. R., & Damasio, H. (1983). The anatomic basis of pure alexia. *Neurology, 33*, 1573–1583.

Damasio, A. R., Damasio, H. Rizzo, M., Varney, N., & Gersh, F. (1982). Aphasia with non-hemorrhagic lesions in the basal ganglia and internal capsule. *Archives of Neurology, 39*, 15–20.

Damasio, A. R., & Kassel, N. F. (1978). Transcortical motor aphasia in relation to lesions of the supplementary motor area. *Neurology, 28*, 396.

Damasio, A. R., & Van Hoesen, G. W. (1980). Structure and function of the supplementary motor area. *Neurology, 30*, 359.

Damasio, A. R., & Van Hoesen, G. W. (1983). Emotional disorders associated with focal lesions of the limbic frontal lobe. In K. M. Heilman & P. Satz (Eds.), *Neuropsychology of human emotion* (pp. 85–110). New York: Guilford Press.

Damasio, H. (1981). Cerebral localization of the aphasias. In M. T. Sarno (Ed.), *Acquired aphasia* (pp. 27–50). Orlando, Florida: Academic Press.

Davis, L., Foldi, N. S., Gardner, H., & Zurif, E. B. (1978). Repetition in the transcortical aphasias. *Brain and Language, 6*, 226–238.

Dejerine, J. (1914). *Sèmiologie des affections du système nerveux*. Paris: Masson.

Denny-Brown, D. (1963). The physiological basis of perception and speech. In L. Halpern (Ed.), *Problems of dynamic neurology* (pp. 30–62). Jerusalem: Hebrew University Press.

De Renzi, E., Zambolin, A., & Crisi, G. (1987). The pattern of neuropsychological impairment associated with left posterior cerebral artery infarcts. *Brain, 110*, 1099–1116.

Erickson, T. C., & Woolsey, C. N. (1951). Observations on the supplementary motor area of man. *Transactions of the American Neurological Association, 76*, 50–56.

Feinberg, T. E., Rothi, L. J. G., & Heilman, K. M. (1986). Multimodal agnosia after unilateral left hemisphere lesion. *Neurology, 36*, 864–867.

Ferro, J. M., & Santos, M. E. (1984). Associative visual agnosia: A case study. *Cortex, 20*, 121–134.

Freedman, M., Alexander, M. P., & Naeser, M. A. (1984). Anatomic basis of transcortical motor aphasia. *Neurology, 34*, 409–417.

Freund, C. S. (1889). Über optische Aphasie und Seelenblindheit. *Archiv für Psychiatrie und Nervenkrankheiten, 20*, 276–297.

Geschwind, N., & Fusillo, M. (1966). Color-naming defects in association with alexia. *Archives of Neurology, 15*, 137–146.

Geschwind, N., Quadfasel, F. A., & Segarra, J. M. (1968). Isolation of the speech area. *Neuropsychologia, 6*, 327–340.

Gil, R., Pluchon, C., Toullat, G., Micheneau, D., Rogez, R., & Lefevre, P. (1985). Disconnexion visuo-verbale (aphasie optique) pour les objets, les images, les couleurs et les visages avec alexie "abstractive". *Neuropsychologia, 23*, 333–349.

Goldberg, G. (1985). Supplementary motor area structure and function: review and hypothesis. *The Behavioral and Brain Sciences, 8*, 567–616.

Goldberg, G., Mayer, N. H., & Toglia, J. U. (1981). Medial frontal cortex infarction and the alien hand sign. *Archives of Neurology, 38*, 683–686.

Goldstein, K. (1948). *Language and language disturbances.* New York: Grune & Stratton.

Gott, P. S. (1973). Language after dominant hemispherectomy. *Journal of Neurology, Neurosurgery and Psychiatry, 36*, 1082–1088.

Hahn, E. (1895). Pathologisch-anatomische Untersuchung des Lissauer'schen Falles von Seelenblindheit. *Psychiatrische Klinik Breslau, 21*, 109–119.

Heilman, K. M., Rothi, L., McFarling, D., & Rottmann, A. L. (1981). Transcortical sensory aphasia with relatively spared spontaneous speech and naming. *Archives of Neurology, 38*, 236–239.

Heilman, K. M., Tucker, D. M., & Valenstein, E. (1976). A case of mixed transcortical aphasia with intact naming. *Brain, 99*, 415–426.

Henschen, S. E. (1920–1922). *Klinische und anatomische Beitrage zur Pathologie des Gehirns* (Vols. 5–7). Stockholm: Nordiska Bokhandeln.

Hommel, M., Besson, G., Pollak, P., Kahane, P., Le Bas, J. F., & Perret, J. (1990). Hemiplegia in posterior cerebral artery occlusion. *Neurology, 40*, 1496–1499.

Hübner, O. (1889). Über Aphasie. *Schmidt's Jahrbücher, Leipzig, 224*, 220–222.

Ingvar, D. H., & Schwartz, M. S. (1974). Blood flow patterns induced in the dominant hemisphere by speech and reading. *Brain, 97*, 273–288.

Jonas, S. (1981). The supplementary motor region and speech emission. *Journal of Communication Disorders, 14*, 349–373.

Jonas, S. (1982). The thalamus and aphasia, including transcortical aphasia: A review. *Journal of Communication Disorders, 15*, 31–41.

Jonas, S. (1987). The supplementary motor region and speech. In E. Perecman (Ed.), *The frontal lobes revisited* (pp. 241–250). New York: IRBN Press.

Jürgens, U. (1985). Implication of the SMA in phonation. *Experimental Brain Research, 58*, A12.

Jürgens, U., & von Cramon, D. (1982). On the role of the anterior cingulate cortex in phonation: A case report. *Brain and Language, 15*, 234–248.

Katz, R. B., & Goodglass, H. (1990). Deep dysphasia: Analysis of a rare form of repetition disorder. *Brain and Language, 39*, 153–185.

Kemper, T. (1984). Neuroanatomical and neuropathological changes in normal aging and dementia. In M. L. Albert (Ed.), *Clinical neurology of aging* (pp. 9–52). New York: Oxford University Press.

Kertesz, A., Harlock, W., & Coates, R. (1979). Computer tomographic localization, lesion size, and prognosis in aphasia and nonverbal impairment. *Brain and Language, 8*, 34–50.

Kertesz, A., Lesk, D., & McCabe, P. (1977). Isotope localization of infarcts in aphasia. *Archives of Neurology, 34*, 590–601.

Kertesz, A., & McCabe, P. (1977). Recovery patterns and prognosis in aphasia. *Brain, 100,* 1–18.

Kertesz, A., Sheppard, A., & MacKenzie, R. (1982). Localization in transcortical sensory aphasia. *Annals of Neurology, 39,* 475–478.

Kleist, K. (1934). *Gehirnpathologie.* Leipzig: Barth.

Környei, E. (1975). Aphasie transcorticale et ècholalie: Le problème de l'initiative de la parole. *Revue Neurologique, 131,* 347–363.

Kussmaul, A. (1877). *Die Störungen der Sprache.* Leipzig: Vogel.

Laplane, D., Talairach, J., Meininger, V., Bancaud, J., & Orgogozo, J. M. (1977). Clinical consequences of corticectomies involving the supplementary motor area in man. *Journal of Neurological Sciences, 34,* 310–314.

Larrabee, G. J., Levin, H. S., Huff, F. J., Kay, M. C., & Guinto, F. C. (1985). Visual agnosia contrasted with visual-verbal disconnection. *Neuropsychologia, 23,* 1–12.

Larsen, B., Skinhøj, E., & Lassen, N. A. (1978). Variations in regional cortical blood flow in the right and left hemispheres during automatic speech. *Brain, 101,* 193–209.

L'hermitte, F., & Beauvois, M.-F. (1973). A visual-speech disconnection syndrome. Report of a case with optic aphasia, agnosic alexia and color agnosia. *Brain, 96,* 695–714.

Lichtheim, L. (1885). On aphasia. *Brain, 7,* 433–484.

Luria, A. R. (1966). *Higher cortical functions in man.* New York: Basic Books.

Luria, A. R. (1970). *Traumatic aphasia.* The Hague: Mouton.

Luria, A. R., & Hutton, J. T. (1977). The modern assessment of the basic forms of aphasia. *Brain and Language, 4,* 129–151.

Luria, A. R., & Tsvetkova, L. S. (1967). The mechanism of "dynamic aphasia". *Foundations of Language, 4,* 296–307.

Marie, P., & Foix, C. (1917). Les aphasies de guerre. *Revue Neurologique, 24,* 53–87.

Masdeu, J. C., & O'Hara, R. J. (1983). Motor aphasia unaccompanied by faciobrachial weakness. *Neurology, 33,* 519–521.

Masdeu, J. C., Schoene, W. C., & Funkenstein, H. (1978). Aphasia following infarction of the left supplementary motor area. A clinicopathologic study. *Neurology, 28,* 1220–1223.

McCarthy, R., & Warrington, E. K. (1984). A two-route model of speech production. Evidence from aphasia. *Brain, 107,* 463–485.

McCarthy, R. A., & Warrington, E. K. (1986). Visual associative agnosia: A clinico-anatomical study of a single case. *Journal of Neurology, Neurosurgery and Psychiatry, 49,* 1233–1240.

McCarthy, R. A., & Warrington, E. K. (1987). The double dissociation of short-term memory for lists and sentences. Evidence from aphasia. *Brain, 110,* 1545–1563.

Mehler, M. F. (1988). Mixed transcortical aphasia in nonfamilial dysphasic dementia. *Cortex, 24,* 545–554.

Milner, B. (1964). Some effects of frontal lobectomy in man. In J. M. Warren & K. Akert (Eds.), *The frontal granular cortex and behavior* (pp. 313–334). New York: McGraw-Hill.

Mohr, J. P., Pessin, M. S., Finkelstein, S., Funkenstein, H. H., Duncan, G. W., & Davis, K. R. (1978). Broca aphasia: Pathologic and clinical. *Neurology, 28,* 311–324.

Morin, P., Rivrain, Y., Eustache, F., Lambert, J., & Courtheoux, P. (1984). Agnosie visuelle et agnosie tactile. *Revue Neurologique, 140,* 271–277.

Morton, J. (1980). Two auditory parallels to deep dyslexia. In M. Coltheart, K. Patterson, & J. C. Marshall (Eds.), *Deep dyslexia* (pp. 189–196). London: Routledge & Kegan Paul.

Myers, R. E. (1976). Comparative neurology of vocalization and speech: proof of a dichotomy. In S. R. Harnad, H. D. Steklis, & J. Lancaster (Eds.), *Origins and evolution of language and speech* (pp. 745–747). New York: New York Academy of Sciences.

Naeser, M. A., Alexander, M. P., Helm-Estabrooks, N., Levine, H. L., Laughlin, S. A., & Geschwind, N. (1982). Aphasia with predominantly subcortical lesion sites. Description of three capsular/putaminal aphasia syndromes. *Archives of Neurology, 39,* 2–14.

Naeser, M. A., & Hayward, R. W. (1978). Lesion localization in aphasia with cranial computed tomography and the Boston Diagnostic Aphasia Exam. *Neurology, 28*, 545–551.

Niessl von Mayendorf, E. (1911). *Die aphasischen Symptome und ihre kortikale Lokalization.* Leipzig: Barth.

Patterson, K., & Shewell, C. (1987). Speak and spell: Dissociations and word-class effects. In M. Coltheart, G. Sartori, & R. Job (Eds.), *The cognitive neuropsychology of language* (pp. 273–294). London: Erlbaum Associates.

Penfield, W., & Roberts, L. (1959). *Speech and brain mechanisms.* Princeton, New Jersey: Princeton University Press.

Penfield, W., & Welch, K. (1951). The supplementary motor area of the cerebral cortex: A clinical and experimental study. *AMA Archives of Neurology and Psychiatry, 66*, 289–317.

Perani, D., Vallar, G., Cappa, S., Messa, C., & Fazio, F. (1987). Aphasia and neglect after subcortical stroke. *Brain, 110*, 1211–1229.

Petersen, S. E., Fox, P. T., Posner, M. I., Mintun, M., & Raichle, M. E. (1988). Positron emission tomographic studies of the cortical anatomy of single-word processing. *Nature (London), 331*, 585–589.

Pick, A. (1892). On the relation between aphasia and senile atrophy of the brain. In D. A. Rottenberg & F. H. Hochberg (Eds.), *Neurological classics in modern translation* (pp. 35–40). New York: Hafner Press (1977).

Pillon, B., Signoret, J.-L., & Lhermitte, F. (1981). Agnosie visuelle associative. Role de l'hèmisphère gauche dans le perception visuelle. *Revue Neurologique, 137*, 831–842.

Pirozzolo, J. F., Kerr, K. L., Obrzut, J. E., Morley, G. K., Haxby, J. V., & Lundgren, S. (1981). Neurolinguistic analysis of the language abilities of a patient with a "double disconnection syndrome": A case of subangular alexia in the presence of mixed transcortical aphasia. *Journal of Neurology, Neurosurgery and Psychiatry, 44*, 152–155.

Poeck, K. (1984). Interhemispheric disconnection in optic anomia. *Neuropsychologia, 22*, 707–713.

Racy, A., Jannotta, F. S., & Lehner, L. H. (1979). Aphasia resulting from occlusion of the left anterior cerebral artery. Report of a case with an old infarct in the left rolandic region. *Archives of Neurology, 36*, 221–224.

Rapcsak, S. Z., Gonzalez-Rothi, L. J., & Heilman, K. M. (1987). Phonological alexia with optic and tactile anomia: A neuropsychological and anatomical study. *Brain and Language, 31*, 109–121.

Rapcsak, S. Z., Krupp, L. B., Rubens, A. B., & Reim, J. (1990). Mixed transcortical aphasia without anatomical isolation of the speech area. *Stroke, 21*, 953–956.

Robinson, B. W. (1976). Limbic influences on human speech. In S. R. Harnad, H. D. Steklis, & J. Lancaster (Eds.), *Origins and evolution of language and speech* (pp. 761–771). New York: New York Academy of Sciences.

Roeltgen, D. P., Gonzalez-Rothi, L., & Heilman, K. M. (1986). Linguistic semantic agraphia: A dissociation of the lexical spelling system from semantics. *Brain and Language, 27*, 257–280.

Rothi, L. J. G. (1990). Transcortical aphasias. In L. L. LaPointe (Ed.), *Aphasia and related neurogenic language disorders* (pp. 78–95). New York: Thieme Medical Publishers.

Rothman, M. (1906). Lichtheimsche motorische Aphasie. *Zeitschrift für Klinische Medizin, 60*, 87–121.

Ross, E. D. (1980). Left medial parietal lobe and receptive language functions: Mixed transcortical aphasia after left anterior cerebral artery infarction. *Neurology, 30*, 144–151.

Rubens, A. B. (1975). Aphasia with infarction in the territory of the anterior cerebral artery. *Cortex, 11*, 239–250.

Rubens, A. B. (1976). Transcortical motor aphasia. In H. Whitaker & H. A. Whitaker (Eds.), *Studies in neurolinguistics* (Vol. 1, pp. 293–306). New York: Academic Press.

Rubens, A. B., & Kertesz, A. (1983). The localization of lesions in transcortical aphasias. In A. Kertesz (Ed.), *Localization in neuropsychology* (pp. 245–268). New York: Academic Press.

Sanides, F. (1970). Functional architecture of motor and sensory cortices in primates in light of a new concept of neocortex evolution. In C. R. Noback & W. Montagna (Eds.), *The primate brain* (pp. 137–208). New York: Appleton.

Selnes, O. A., Knopman, D. S., Niccum, N., & Rubens, A. B. (1985). The critical role of Wernicke's area in sentence repetition. *Annals of Neurology, 17*, 549–557.

Speedie, L. J., Coslett, H. B., & Heilman, K. M. (1984). Repetition of affective prosody in mixed transcortical aphasia. *Archives of Neurology, 41*, 268–270.

Stengel, E. (1947). A clinical and psychological study of echo-reactions. *Journal of Mental Science, 93*, 598–612.

Stuss, D. T., & Benson, D. F. (1986). *The frontal lobes*. New York: Raven Press.

Sutton, D., Trachy, R. E., & Lindeman, R. C. (1985). Discriminative phonation in macaques: Effects of anterior mesial cortex damage. *Experimental Brain Research, 59*, 410–413.

Teuber, H.-L. (1968). Alteration of perception and memory in man. In L. Weiskrantz (Ed.), pp. 268–375. *Analysis of behavioral change*. New York: Harper & Row.

Thurstone, L. L., & Thurstone, T. (1943). *The Chicago test of primary mental abilities*. Chicago: Science Research Associates.

Trojano, L., Fragassi, N. A., Postiglione, A., & Grossi, D. (1988). Mixed transcortical aphasia. On relative sparing of phonological short-term store in a case. *Neuropsychologia, 26*, 633–638.

Vix, E. (1910). Anatomischer Befund bei transkortikaler sensorisher Aphasie. *Archiv fuer Psychiatrie und Nervenkrankheiten, 47*, 200–213.

Von Cramon, D. Y., Hebel, N., & Schuri, U. (1988). Verbal memory and learning in unilateral posterior cerebral infarction. *Brain, 111*, 1061–1077.

Von Stockert, T. R. (1974). Aphasia sine aphasia. *Brain and Language, 1*, 277–282.

Wallesch, C. W. (1985). Two syndromes of aphasia occurring with ischemic lesions involving the left basal ganglia. *Brain and Language, 25*, 357–361.

Wallesch, C. W., Kornhuber, H. H., Brunner, R. J., Kunz, T., Hollerbach, B., & Suger, G. (1983). Lesions of the basal ganglia, thalamus and deep white matter: Differential effects on language functions. *Brain and Language, 20*, 286–304.

Watson, R. T., Fleet, W. S., Gonzalez-Rothi, L., & Heilman, K. M. (1986). Apraxia and the supplementary motor area. *Archives of Neurology, 43*, 787–792.

Wernicke, C. (1874). *Der aphasische Symptomencomplex*. Breslau: Cohn & Weigert.

Wernicke, C. (1886). Einige neuere Arbeiten über Aphasie. *Fortschritte der Medizin, 4*, 377–463.

Wernicke, C. (1906). Der aphasische Symptomencomplex. *Deutsche Klinik am Eingange des 20 Jahrhunderts, 6*, 487–556.

Whitaker, H. (1976). A case of the isolation of the language function. In H. Whitaker & H. A. Whitaker (Eds.), *Studies in neurolinguistics* (Vol. 1, pp. 1–58). New York: Academic Press.

Chapter 11

Localization in Alexia

Sandra E. Black and Marlene Behrmann

I. INTRODUCTION

Two approaches have been used to study the localization of reading deficits caused by acquired brain damage. In one approach, the patient is selected on the basis of the behavior of interest and the deficits are analyzed in correlation with documentation of the anatomical damage in an attempt to draw inferences about cerebral localization. In this enterprise, well-studied single case examples usually are described as a first step in establishing the neuroanatomical locations linked to certain functional impairments. The generalizability of the observed brain–behavior relationships depends on subsequent confirmation in other individuals through either multiple single case studies or group series. A second approach involves classifying patients by lesion location and then determining their reading performance. Using this approach, negative cases can be documented and the frequency with which a certain behavior is associated with a particular site of anatomical damage can be assessed. With the advent of neuroimaging in the 1970s, researchers have been able to conduct such correlation studies in vivo. Consecutive series of patients with damage in the areas of interest or with selected behavioral deficits can be studied and followed over time. In the traditional neurological approach to acquired reading disorders, the behavioral analysis has emphasized associations and dissociations in performance in the language modalities such as writing and naming, in sensory-motor function such as hamianopsia, and in neuropsychological function such as memory. In this syndrome approach, alexia frequently is described in relation to these other potential deficits, for example, alexia and agraphia, pure alexia without hemianopsia, and so on. Occurrence of double dissociations, for example, impaired reading in the presence of normal writing and other language functions, has been interpreted to imply that separate functional processes are involved that may be localized in different brain regions. In such cases, the pattern of impairments has been attributed to damage in functionally important neuroanatomical regions or in the connections between them.

Within this neurological tradition, little attention was paid to psycholinguistic properties of words such as imageability, frequency, part of speech, and word length, although the ability to understand other visual symbols such as letters, numbers, and musical notation sometimes was documented (Hecaen, Ajuriaguerra, & David, 1952). Psycholinguistic parameters became important, however, when cognitive psychological methods began to be applied to the study of alexia (Marshall & Newcombe, 1966, 1973). Reading deficits began to be described in information processing terms, using models of reading derived from the study of normal individuals and incorporating neurolinguistic concepts. The analysis of dissociations and associations became more fine grained and sophisticated. With this approach, both speed and accuracy of single-word reading were quantitated and word recognition deficits were characterized in relation to spelling regularity, concreteness, and other characteristics. A wide variety of reading disorders has been identified in this manner (Coltheart, 1981, 1985; Ellis & Young, 1987; Shallice, 1988). The cognitive neuropsychological analysis of reading disorders, however, has been less conducive to neuroanatomical correlations than the modality-based approach. In fact, as behavioral analysis has become more detailed, brain–behavior correlations often seem more confusing; a one-to-one relationship between behavioral deficit and anatomical region is seldom definitive. This difficulty led to an eclipse of the localization method in the 1970s and 1980s in reading research, but interest has been reawakened in the last few years with advances in positron emission tomography (PET) scanning techniques, which have been used to study patterns of localized cerebral activation during the process of reading by normal individuals (Petersen, Fox, Posner, Mintun, & Raichle, 1988).

In this chapter, we first briefly review the structural and functional imaging methods currently in use for in vivo localization studies (Section II). We discuss the classical modality-based approach to localization in alexia and summarize case series in which reading behavior has been analyzed in relation to specific sites of damage (Section III). We then describe the cognitive neuropsychological approach to alexia, its theoretical framework, and current classification of reading deficits, including reference to possible neuroanatomical substrates (Section IV). Finally, we address some current controversial issues in this area, including the role of the right hemisphere in reading (Section IV, D) and the use of network modeling to simulate normal and impaired reading processes (Section V).

II. IMAGING TECHNIQUES

Since the last edition of this book, important advances have been made, in both structural and functional brain imaging, that permit individualized mapping of functional activation in normal individuals and precise localization of injury and dysfunction in brain-damaged individuals. These advances are summarized here since they have ushered in a new era in brain–behavior correlations and since reading is one of the first cognitive domains to which the latest brain activation techniques have been applied.

A. Structural Brain Imaging

With the arrival of computerized X-ray tomography (CT) in the mid-1970s, contemporaneous behavioral and neuroanatomical lesion analysis became possible for the first time. The mainstay of this localization approach was to trace or draw the brain lesion on standardized templates to identify the brain lesions involved and to allow comparisons in group studies (Kertesz, 1983). Differences in head size and in angle of orientation of scans often were not taken into account, detracting from the accuracy of the method. To permit stereotactic precision for neurosurgical procedures despite such variability, proportional methods were developed to account for individual differences (Vanier et al., 1985; Talairach & Tournoux, 1988). With CT scanning, tissue alterations must be sufficient to alter X-ray penetration and the lesions may not be visible if the scan is performed too soon after the injury. Also, the effects of brain edema can confound interpretation. Nevertheless, with proper procedures, in vivo neuroanatomy–behavior correlations now can be obtained routinely. (For discussion of methodology see Damasio & Damasio, 1989.)

Magnetic resonance imaging (MRI), which arrived in the early 1980s, provides a more sensitive technique for detecting early pathological alterations in the brain, as well as better contrast sensitivity between different brain tissue compartments (Pritchard & Brass, 1992). This technique is noninvasive and apparently without health hazard, making normative studies more feasible. Anatomical detail is also far superior, depending on the choice of pulse sequences. The limits of spatial resolution are near 1 mm. More importantly, technical improvements have shortened the time required for scanning, and computer software development now permits three-dimensional manipulation of brain images on a computer work station (Damasio & Frank, 1992). Quantitative morphometry of tissue compartments—for example, gray matter and white matter—as well as identification of major cerebral sulci on three-dimensional images,

provides unparalleled opportunity for precise individualized neuroanatomical correlations (Steinmetz & Seitz, 1991). Such studies already have
revealed considerable variability in the localization of the major cerebral
sulci, particularly the central sulcus, which demarcates the frontal and
parietal lobes (Steinmetz, Furst, & Freund, 1990). With respect to the
neuroanatomy of reading, right–left differences in the size of the occipital,
temporal, and inferior frontal lobes that are thought to reflect left
hemispheric specialization for language function now can be visualized
directly (Kertesz, Polk, Black, & Howell, 1992). Three-dimensional quantitative techniques are still in their early development and have not been
applied widely to individual case studies or group series, but the opportunity for highly accurate macroscopic lesion localization exists now as
never before.

B. Functional Brain Imaging

PET uses annihilation radiation generated when positrons are absorbed
in matter to provide an image of the spatial distribution of the radioactive
tracer in a selected plane of brain tissue. This technique takes advantage
of the close coupling of brain blood flow and metabolism (Raichle, 1989).
In the early 1980s, the use of [^{15}O]water to measure cerebral blood flow
with PET opened new possibilities for activation studies. With a short half-
life of 2 min, this tracer technique allows up to 10 measurements in a single
sitting within acceptable radiation exposure limits, enabling researchers
to perform several different experiments in the same individual. Improved anatomical localization procedures using stereotactic methods and
the use of the subtraction method, in which a change in local blood flow
is detected by comparison with a resting or control state, has allowed more
accurate localization of functional activation, to some extent overcoming
the inherent limits of spatial resolution which, in current devices, is in the
range of 10–20 mm (Raichle, 1989). Blood flow responses during primary
sensory activation, for example, when viewing flashing lights, are typically 20–40% above baseline (Fox, Mintun, Raichle, Miezin, Allman, & Van
Essen, 1986). Internal mental operations, however, may increase blood
flow only by 2–5% above control values (Raichle, 1989). By the subtraction
method, the control image is subtracted from the "activated" image so the
net effect of the functional activation appears as a focus of change. Intersubject averaging is used to increase signal-to-noise ratio in the active
regions and the significance of the increased signal is analyzed by statistical methods. In this manner, the low level responses that underlie higher
level cognitive processing can be discriminated from background noise,

and localization can exceed the spatial resolution of PET by an order of magnitude (Raichle, 1989).

The first groundbreaking attempt to apply this technique to the cognitive domain examined functional activation during the processing of words presented in auditory and visual modes to 17 normal right-handed volunteers (Petersen et al., 1988). Subjects first viewed a small cross on a computer screen as the baseline condition. In the passive word recognition, nouns appeared on the monitor at fixation at the rate of one per second or were heard through earphones on a digital tape recorder. No response was required. In the motor output condition, the subject read the word seen or repeated the word heard. In the final task, the person had to suggest a use for the word shown or heard, a process that involved semantic processing and selection (Petersen et al., 1988). With respect to the passive viewing of words, activation was noted in the left and right calcarine cortex and extrastriate areas in the lateral occipital region approaching the temporal junction. Passive listening activated the posterior superior temporal regions bilaterally. No overlap in activation was seen between the visual and auditory sensory tasks. When verbal responses were required, auditory and visual words both produced activation in the sensory-motor strip bilaterally and semantic processing activated the premotor region in the left lateral frontal region (Petersen et al., 1988). Somewhat surprisingly, no specific activation of parietal or Broca's area was detected in this experiment and no lateralization of activation was seen, although words were being used as stimuli. The authors argued that the nonoverlapping activations caused by auditory and visual word presentations suggested that, for word recognition, separate routes existed between auditory and visual inputs and the phonological, articulatory, and semantic coding areas (Petersen et al., 1988; Posner, Petersen, Fox, & Raichle, 1988).

In another series of PET experiments designed to explore orthographic effects more carefully, four different sets of stimuli were used, including single common nouns, pseudowords (letter strings that obey the rules of English orthography, e.g., flabe), unpronounceable consonant letter strings (e.g., JKLPM), and strings of letter-like false fonts (Petersen, Fox, Synder, & Raichle, 1990). Using the subtractive technique described earlier, lateral occipital (extrastriate) activation was noted with all four sets of stimuli. In addition, real words and pseudowords produced a left medial occipital extrastriate response. Subtraction of the real and pseudoword conditions showed no significant difference in the occipital region, but a significant activation in the left frontal cortex was seen in the real word relative to the pseudoword condition. The region activated was similar to that activated in the semantic processing tasks, suggesting that the lateral prefrontal region on the left is involved in semantic processing of single words. These findings provided additional support for the concept of direct visual access to word meaning and pronunciation, as is discussed in a subsequent section (Petersen et al., 1990).

These PET researches have highlighted functional brain activity during the normal process of reading, providing new insight into the neural circuitry of visible language. Some problems remain, however. A different locus of activation for word recognition emerged from more recent PET studies of reading conducted in another laboratory using a different imaging analysis technique (Howard et al., 1992). When subjects viewed letter-like fonts, increased ^{15}O uptake was detected bilaterally in the striate and extrastriate cortex. Viewing words caused increased blood flow in the posterior left middle temporal gyrus. Thus, different PET research centers have reported somewhat inconsistent results. Certain classical areas thought to participate in language processing have been conspicuously absent in these PET activations. For example, in the studies by Petersen and associates (1990), Broca's area was not activated in the verbal output task. The sensory-motor areas involved in tongue and mouth movement did show activation, but this response was likely the result of oral–facial movements not specific for language. Similarly, the inferior parietal region was not activated during any of the tasks, nor were subcortical thalamic and basal ganglia structures, which are known through lesion studies to participate in language circuitry. Some critics argue convincingly that the process of intersubject averaging may obscure individual activation patterns and render foci of language processing undetectable (Steinmetz et al., 1990; Steinmetz & Seitz, 1991). Intersubject variability can arise from individual variation in functional localization, as documented by intraoperative stimulation in subjects undergoing neurosurgical extirpation for epilepsy (Ojemann, 1983) or from morphometric variations in cerebral sulci, brain size, and shape, which contribute additional noise when anatomical standardization is performed. Intrasubject averaging therefore may be required to localize the low-level activation changes seen in higher cognitive functions properly.

Activation studies in brain-injured individuals are still in their infancy, but may become very important as a means of testing reserve function and mechanisms of recovery. Functional imaging studies in the resting state also provide rich information on the widespread functional deactivation that can result from focal brain damage. Information on such "remote" effects will, in the future, have to be factored in when considering brain–behavior correlations (Perani et al., 1988). Reduction of blood flow and presumably of function in the basal ganglia and thalamus, for example, frequently follow cortical injury and vice versa. Single photon emission computerized tomography (SPECT) is cheaper and more widely available than PET. The advent of newer brain perfusion agents such as HMPAO, which can be labeled using common radionuclear tracers such as technetium, allows tomographic brain imaging to use gamma camera technology that is readily available even in small hospital settings (Hellman and Tikofsky, 1990). Although the spatial resolution is poorer than that of PET, the technology is improving rapidly and, because of its accessibility, SPECT is likely to have more widespread use than PET. SPECT offers the possibility for large consecutive series as well as longitudinal studies. A final prospect is that of doing functional imaging with MRI, not only by topographical spectroscopy but also by imaging perfusion, diffusion, or even oxygenation changes in activated areas (Belliveau, Kennedy, McKenstry, & Buchbinder, 1991).

III. NEUROANATOMICAL APPROACHES TO STUDYING ACQUIRED ALEXIA

A. Background

For the purposes of this chapter, acquired alexia is defined, as in the previous edition of this book, as the "partial or complete loss of efficient reading for comprehension" (Greenblatt, 1983, p. 325) caused by brain damage in adults who were normal readers prior to the brain lesion. This type of deficit is distinguished from developmental dyslexia, a developmental reading problem in which a child may never learn to read normally.

The systematic analysis of reading disorders commenced with the work of the classical neurologists in the late 19th century. Neurologists such as Charcot and Wernicke noted that reading can be impaired selectively by cerebral lesions. In an analysis of 50 cases with sensory aphasia, Starr (1889) posited that reading comprehension was mediated by a tract from the occipital region to the angular gyrus and that alexia could result from occipital damage. The fundamental contribution in this clinical–pathological tradition, however, was the work of Dejerine. In 1891, Dejerine reported a patient with alexia and agraphia in the absence of other significant language impairment as the result of a stroke shown at autopsy to involve the left angular gyrus (Fig. 1). A year later, he described a patient who had alexia only, with no other significant language impairment, associated with a partial right field deficit (Dejerine, 1892). This patient's lesion involved the fusiform and lingual gyri of the left hemisphere. The patient also had a small lesion in the splenium which Dejerine

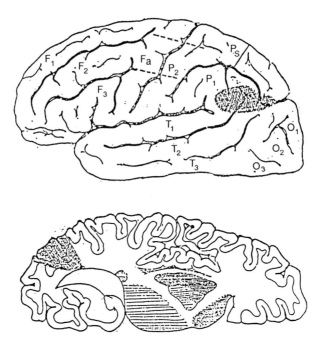

Figure 1 Dejerine's drawing of the left angular gyrus lesion (shaded) that caused alexia and agraphia in the case he reported (Dejerine, 1891).

did not regard as significant. Shortly before death, the patient developed agraphia; at autopsy, a more recent infarction of the angular gyrus was noted also (see Fig. 2). Dejerine explained the loss of reading ability with preserved writing, which he called pure alexia, as follows. He hypothesized that visual representations of words were located in the left angular gyrus, since this area was initially intact in the patient with pure alexia, explaining his ability to write. Because of the left occipital damage, however, visual information could not be transmitted to the angular gyrus to allow recognition of visually presented words. Accepting that visual information from the right occipital cortex was transmitted first to the left occipital cortex before traveling toward the angular gyrus, Dejerine argued that damage to the left primary visual cortex and visual association areas prevented input from the right occipital and left occipital lobes from reaching the left angular gyrus, where word recognition took place (Dejerine, 1892).

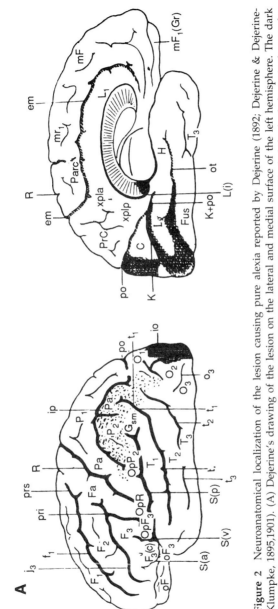

Figure 2 Neuroanatomical localization of the lesion causing pure alexia reported by Dejerine (1892; Dejerine & Dejerine-Klumpke, 1895,1901). (A) Dejerine's drawing of the lesion on the lateral and medial surface of the left hemisphere. The dark shading depicted the original lesion in the cuneus (C), lingual (TO₂) and fusiform (TO₁) gyri. The lightly shaded area in the angular gyrus (TC) and supramarginal gyrus (Pl) was the later lesion causing agraphia just prior to the patient's death. The medial section also shows the inferior involvement in the splenium of the corpus callosum (Dejerine, 1892). (B) The inferior surface of the lesion showing involvement of the lingual and fusiform gyri (Dejerine, 1892). (C) Horizontal section showing the occipital lesion (dark shading) extending into the white matter close to the occipital horn and the more recent parietal lesion (light shading) (Dejerine, 1892).

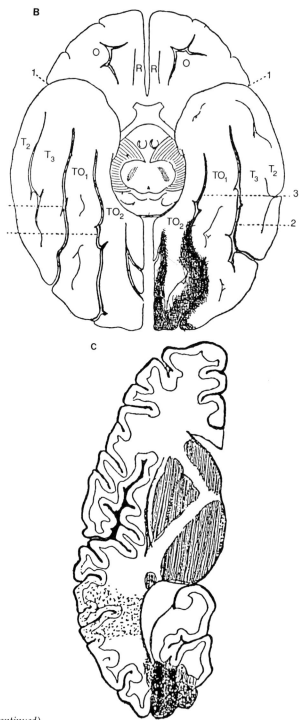

Figure 2 (*continued*)

Building on Dejerine's disconnection hypothesis, Geschwind articulated the modern neurological model of reading in 1965, in which he proposed that a sequence of anatomical regions is engaged to achieve comprehension and pronunciation of written words (Geschwind, 1965). First, visual information from the left visual association areas is transmitted to the angular gyrus where it is recoded phonologically. Semantic associations then are accessed in the nearby posterior temporal region (Wernicke's area). Finally, this information is transmitted through a white matter tract (the arcuate fasciculus) to Broca's area for motor speech output. This serial single-route model of word reading contrasts with the dual-route model of reading put forward by cognitive neuropsychological theorists, which has found some recent support in the PET activation studies mentioned earlier (Posner et al., 1988) and is discussed in Section IV.

According to this neurological model, the key neuroanatomical sites that constitute the neural network for reading are the left occipital region (visual processing), the left parietal–temporal region (for phonological and semantic coding), and the left inferior frontal region (for motor speech output). In a series of review articles, Benson and colleagues have described the features of the different types of alexia, which they classify in relation to damage in these key areas into posterior, central, and anterior alexia (Benson, 1977, 1985; Benson, Brown, & Tomlinson, 1971; Benson & Geschwind, 1969). Before discussing this classification, however, we briefly review some basic structure–function relationships in the human visual system.

In the primate nervous system, visual information from each visual field travels along the optic nerve to the lateral geniculate body of the opposite hemisphere. Fibers from the nasal portion of the retina of one eye cross at the optic chiasm to join with fibers from the temporal half of the retina of the other eye to synapse with neurons in the dorsal laminae of the ipsilateral lateral geniculate body. These cells then give rise to axons that run in the optic radiation lateral to the posterior horn of the lateral ventricle to the striate or calcarine cortex (Brodmann area 17) in the medial occipital lobe. Macular vision is represented posteriorly at the occipital pole and the more peripheral regions of the opposite hemifield are represented more anteriorly (Holmes, 1944). The upper visual quadrants project to the lower lip of the calcarine fissure and the lower quadrants to the upper lip (Holmes, 1944). The actual location and volume of the calcarine cortex is somewhat variable (Stensaas, Eddinton, & Dobelle, 1974) and, even in lower mammals, asymmetries occur in the number of cells in the right and left calcarine regions (Galaburda, Aboitiz, Rosen, & Sherman, 1986). In most right-handed humans, the left occipital pole protrudes more posteriorly, causing an indentation of the inner table of the skull that is referred to as a petalia (LeMay, 1977); these petalias are likely to represent language-related anatomical asymmetries. Neurons in the primary visual cortex (Brodmann area 17) project to the visual association areas (Brodmann areas 18 and 19). Studies of visual processing in the monkey have suggested the presence of over a dozen functionally distinct visual processing regions within these association areas, by which different features of visual stimuli such as color, form, and motion are processed in parallel (Van Essen & Maunsell, 1983). Brain–behavior studies in primates have established two separable paths for further visual processing: a ventral occipital–temporal system involved with object perception and a dorsal occipital–parietal system that mediates perception of spatial location (Mishkin, Ungerleider, & Macko, 1983) and/or motion (Van Essen & Maunsell, 1983) and prepares the subject for action on objects (Goodale & Milner, 1992) (see Fig. 3). A similar functional dichotomy is thought to occur in humans (Levine, 1982), but the white matter tracts involved in these projections have not been identified clearly. The inferior longitudinal fasciculus, in fact, consists of a series of projection fibers from areas 18 and 19 to the temporal and parietal cortices, now referred to as the occipital–temporal projection system (Tusa & Ungerleider, 1985). The white matter tract projecting to the parietal lobe is called the vertical longitudinal fasciculus (Greenblatt, 1983).

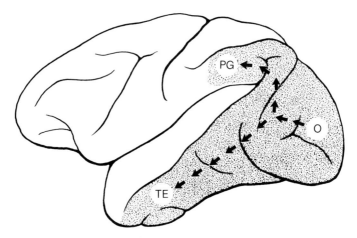

Figure 3 Lateral surface of a monkey's brain showing the two visual projection systems involved in visual perception. The superior occipital-parietal system (O–PG) mediates spatial localization and the inferior occipital–temporal region mediates visual object recognition (O–TE). Adapted with permission from Mishkin, Ungerleider, & Macko (1983).

With respect to localization of reading within the two visual projection systems, word recognition would be expected to be impaired primarily by damage to the ventral object-perception system. The dorsal system would likely be involved in the control of the eye movements, which are important for the saccadic and scanning movements involved in reading text; some reading deficits can arise from damage to the dorsal system (see Section IV,B). The studies of word recognition using PET activation suggest that two areas in the ventral visual pathway on the left—the inferior extrastriate cortex (Petersen et al., 1989) and the middle temporal region on the left (Howard et al., 1992)—are candidate sites for localization of visual lexicon.

The hemispheric interconnections implicated in the reading system also are not fully understood. Interhemispheric projections between the cortical visual areas have not been established in humans, but it is reasonable to suppose that patterns of interconnection are similar to those that have been mapped out in primates (Pandya & Seltzer, 1986). With respect to occipital interconnections, no direct projections exist between the primary visual cortices of the right and left hemispheres (Butler, 1977; Rockland & Pandya, 1981). The visual association areas project to homotypic cortex in the other hemisphere so, for example, area 18 on the right projects to area 18 on the left. In primates, the transcallosal visual association projections occupy the splenium with area 18 more inferior and area 19 more superior. Interparietal projections are more anterior in the body of the corpus callosum (Pandya & Seltzer, 1986) (see Fig. 4).

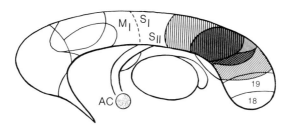

Figure 4 Topography of the commissural fibers in the corpus callosum of monkeys, modified from the diagram of Pandya & Seltzer (1986). The diagram shows the approximate location of interhemispheric connections from various brain regions. Lesion sites in some cases of pure alexia suggest that fibers interconnecting the right and left visual association cortex involved in reading pass through the inferior portion of the splenium (18 and 19). AC refers to the anterior commissure; M_I, S_I, and S_{II} refer to primary motor and sensory area interconnections.

According to contemporary neurological models of reading, the inter-hemispheric and intrahemispheric pathways that mediate word recognition are as follows. Visual information from the right visual field reaching the left calcarine cortex (area 17) is transmitted to the left visual association areas 18 and 19 and then is projected to the left angular gyrus (pathway abc in Fig. 5). Information from the left visual field reaches the primary visual cortex on the right (area 17), and subsequently the right visual association cortices (areas 18 and 19). To support the reading process, information is projected across the splenium to areas 18 and 19 on the left (path d in Fig. 5) and from there travels to the left angular gyrus (pathway c in Fig. 5). Another potential pathway would be from the right visual association area directly to the right angular gyrus (pathway e in Fig. 5) and across the posterior body of the corpus callosum to the left angular gyrus (pathway f in Fig. 5). According to Geschwind (1965), this latter pathway explains the recovery of reading that apparently occurs after complete occipital lobectomy. Henderson (1986) also has hypothesized a pathway (see g in Fig. 5), in which transcallosal fibers from the right travel directly to the left parietal–temporal language area without necessarily synapsing with visual association cortex on the left. Greenblatt (1990), on the other hand, would argue against such a pathway in favor of pathway d, which is more in keeping with current knowledge about interhemispheric connections in animals. Also note that, in this model, lesions confined to pathway b could cause a right field hemialexia and that a lesion confined to pathway c could disrupt reading even though vision in the right visual field would be intact.

B. Neuroanatomical Classification of Alexia

1. Posterior or occipital alexia

The reading disorder caused by damage in the left occipital region is called pure alexia or alexia without agraphia, since reading is affected often in the absence of any other language deficit (Benson, 1985; Benson & Geschwind, 1969). Typically in this syndrome, patients can write words but cannot read what they have just written. They can understand orally spelled words and can spell words out loud quite well. Indeed, many adopt a strategy of spelling letters of a word aloud to arrive at its pronunciation, a strategy called letter-by-letter reading (see Section IV,B). A number of neuropsychological deficits may co-occur. Patients may not be able to recognize or name letters, numbers, other symbols, or colors. A right hemianopsia or quadrantanopsia (usually superior) or loss of color vision in the right visual field also can occur (Damasio & Damasio, 1983; Geschwind

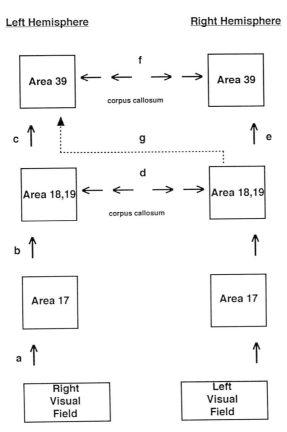

Figure 5 Key pathways for reading according to the disconnection hypothesis. According to this hypothesis, the pathways necessary for reading are a, b, c, and d. Damage to a+d, b+d, or c alone can cause pure alexia, at least in the acute phase. According to Geschwind (1965), reading may be regained through pathway e,f at a later stage, but additional studies suggest that an efficient reading system cannot be supported by this pathway alone. Henderson (1986) proposes an additional pathway g, which passes near, but does not require synapsis with, the visual association cortex. Greenblatt (1990), on the other hand, believes that the pathway through d must first synapse with areas 18 and 19 before proceeding to area 39. Since, anatomically, pathways c and g run close together by the lateral ventricle, both frequently are damaged at the same time. A lesion in pathway c would give pure alexia without hemianopsia. Pure alexia also can result from damage to a+g (Silver et al., 1988).

& Fusillo, 1966). Memory loss (Caplan & Hedley-White, 1974; DeRenzi, Zambolin, & Crisi, 1987) or visual agnosia (loss of object recognition) (Damasio & Damasio, 1983; Farah & Wallace, 1991) may occur as well.

a. Studies of the effects of occipital damage on reading.   With respect to localization, the neuroanatomical substrate for pure alexia has been studied intensively, in group series as well as in single cases. Reviewing some of this evidence is worthwhile, starting with the influential surgical series of Hecaen and colleagues (1952) that reported on the reading abilities of subjects who underwent occipital lobectomy for tumor excision. Six patients examined after left occipital lobectomy, including one left-hander, had severe alexia postoperatively. In three subjects followed over 1 yr, some recovery occurred but at least one patient was no longer able to read for pleasure. An additional subject who had a right occipital lobectomy had slow but accurate reading postoperatively. Based on the transient nature of the severe alexia in this series, Geschwind (1965), as mentioned earlier, argued that a pathway from the right visual association area to the right angular gyrus and across the corpus callosum to the left angular gyrus (pathway ef in Fig. 5) could mediate a return to normal reading. Given the limitations of the testing battery, however, recovery seems more likely to have been only partial in these patients. These patients probably had a form of spelling alexia in which speed of reading single words is increased greatly although word recognition may be regained (see Section IV,B). More recent evidence suggests, therefore, that information transfer along this alternative pathway may not suffice for fluent reading (Bub, Black, & Howell, 1989).

In a consecutive series of 708 patients with cerebral lesions verified at autopsy or operation, pure alexia occurred in 24 (10%) of the 241 patients with occipital lesions (Gloning, Gloning, & Hoff, 1968). Of these, 2 occurred with right hemisphere damage and 6 were bilateral. In the patients with pure alexia, 17 lesions involved the temporal–occipital region and 7 extended also into the parietal lobe. In addition, one temporal lesion, one parietal lesion, and one frontal–parietal lesion gave rise to alexia with agraphia. In the same series of posterior lesions, 3 of 241 subjects had visual agnosia, 1 had prosopagnosia, and 36 had color agnosia, usually associated with alexia. This case series is the largest ever reported on this subject, but had major shortcomings. Unfortunately, mixed etiologies such as tumor, stroke, trauma, and abscess were included in the series. Localization was made only within large lobar regions and the tests used to determine the presence or absence of neuropsychological symptoms were not specified.

In a more recent series, 16 consecutive patients with left posterior cerebral artery infarction were studied by CT scanning parallel to the orbitomeatal line (DeRenzi et al., 1987). Patients were asked to pronounce and comprehend a series of letters, words, and sentences and to perform tasks of object naming, color naming, and verbal memory. Of the individuals studied, 12 patients (75%) had a reading disorder that was mild in 2 and severe in 6. Some were unable to name or even recognize letters. None of the subjects overtly used a letter-by-letter strategy. Of the 4 subjects who did not have alexia, 1 was left-handed and another read accurately but laboriously, suggesting a subvocal letter-by-letter approach. Thus, 2 of 16 patients had no apparent reading difficulty. Writing was preserved in all subjects; 75% had a visual object-naming impairment, and verbal memory was deficient in all patients. On lesion analysis, 11 of the 12 patients with alexia had medial occipital damage at the level of the pineal gland, involving the lingual and parahippocampal gyri. Lesions frequently extended one or two slices below into the fusiform and hippocampal gyri. In 2 patients, no splenial damage was detected. The authors concluded that, in pure alexia, the connections from the right visual cortex to the left parietotemporal language areas could be interrupted anywhere in their course. The critical cortical structure mediating reading was the lingual gyrus. Extension to other medial association cortices, such as the fusiform inferiorly and the cuneate superiorly, made the alexia more severe but was not sufficient to disrupt reading if the lingual gyrus was spared. No one-to-one correlation of lesion to alexia was found in this series, however. One patient with an extensive left occipital lesion had only mild alexia and another with a large left posterior lesion had no obvious reading difficulty. One limitation of this study was that reading assessment was relatively insensitive, and that speed of reading was not documented, precluding recognition of a word-length effect in these subjects. Nevertheless this valuable consecutive group series suggested that the common neural substrate of pure alexia was the inferior medial occipital region.

In a consecutive series of 10 occipital infarct subjects (5 right and 5 left) in whom no overt reading deficit was present in the early stage poststroke, Black, Bub, and Behrmann (1987) used tachistoscopic presentation of single words to explore reading speed and accuracy. All patients had a hemianopsia that precluded testing of reading in the right visual field. In the 5 subjects with left occipital damage, all of whom had damage in the inferomedial occipital cortex and subjacent white matter, a subtle effect of word length of approximately 30 msec per letter was detected. This effect did not approach the typical length effect seen in pure alexia, which is usually on the order of 1 sec per letter. Instead, the deficit resembled the word-length effect seen in tachistoscopic left hemifield word presentations in normal subjects (Bub & Lewine, 1988). Thus, although no deficit existed in the accuracy of word reading, reading efficiency was impaired subtly.

b. Lesion analysis in pure alexia.   An important contribution to localization in pure alexia was the series reported by Damasio and Damasio
(1983). In this study, 16 patients, identified on a series of standardized
reading tests as having alexia without agraphia, underwent CT scanning
and careful lesion analysis as well as neuropsychological and neuro-
ophthalmological evaluation for color vision and naming, memory, visual
agnosia, and optic ataxia. The lesions associated with different alexic syndromes were classified into three types. Type I, found in patients with
right homonymous hemianopsia and color anomia, involved extensive
infarction of the posterior cerebral artery territory and included paraventricular damage in the white matter of the occipotemporal junction, in the
left half of the splenium and forceps major, and in inferior and superior
medial–occipital cortex. In Type II, associated with a right hemianopsia
without color anomia or verbal amnesia, paraventricular white matter
damage interrupting interhemispheric pathways was noted in conjunction with damage to the optic radiations, to the calcarine region, or to
both. Type III, alexia with an upper quadrantanopia and lower achromatopsia (color vision loss), demonstrated no color anomia or verbal amnesia. The inferior optic radiation and inferior visual association cortex were
involved, but paraventricular white matter damage was the critical lesion.
In some subjects the right visual field was spared. Extension into the adjacent association cortex and splenium could occur but was not essential.
The authors concluded that the critical substrate for pure alexia was in the
paraventricular white matter region. Such a lesion caused disconnection
of transcallosal fibers from the right occipital region and of fibers from the
left visual association cortex traveling to the left temporoparietal language
area. To date, this key anatomical localization for pure alexia, which originally was noted by Dejerine, has held up remarkably well to additional
scrutiny (see Fig. 6).

The localization of reading processes to the inferior visual association
cortex has been confirmed by several single-case studies, including Dejerine's original report (Dejerine, 1892; see also Ajax, Schenkenberg, &
Kosteljanetz, 1977; Caplan & Hedley-White, 1974; Vincent, Sadowsky,
Saunders, & Reeves, 1977). One report of a patient with pure alexia caused
by a hemorrhage in the left fusiform and inferior temporal gyrus and
subadjacent white matter concluded that the relevant transcallosal fibers
from the right visual association cortex pass inferior to the posterior horn
of the left lateral ventricle before reaching the left hemisphere language
areas (Henderson, Friedman, Teng, & Weiner, 1985). This case also indicated
that alexia could arise from inferolateral occipital–temporal damage rather
than medial occipital damage, an idea that has been confirmed in other
case reports (Caffarra, 1987; Greenblatt, 1990; Johansson & Fahlgren, 1979).

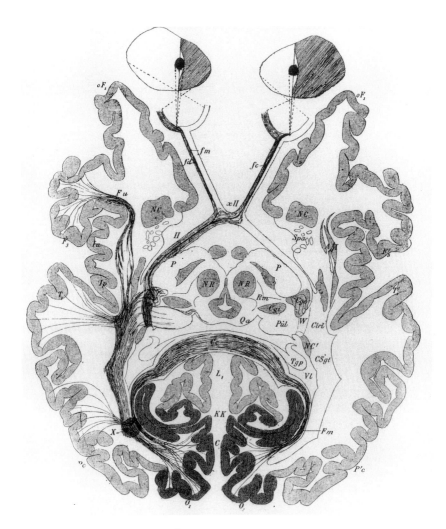

Figure 6 Dejerine's localization of the key lesion (marked *X*) causing pure alexia is shown in this drawing of a horizontal brain slice through the inferior occipital region (Dejerine & Dejerine-Klumpke, 1901). Damage to projection fibers in the periventricular white matter appears to be sufficient, a finding verified in the CT study by Damasio & Damasio (1983).

Right hemifield alexia without hemianopsia has been reported as caused by a left occipital lobe lesion involving the fusiform and lateral occipital gyrus (Castro-Caldas & Salgado, 1984). Words were presented tachitoscopically to the left and right visual field separately. The patient was unable to read in the right hemifield and identified 50% of the words on the left at an exposure of 100 msec. Word length was not explored in this assessment and the data provided suggested that reading was not entirely intact in the left visual field. This study illustrates one of the problems that bedevil research in this area. The definition of a reading disorder varies widely, as does the sophistication of testing of word comprehension and pronunciation, which can lead to quite different interpretations of the same data. In the case of posterior alexia, word length is an important variable since reading may be accurate but pathologically slow. Failure to manipulate this parameter in most neurological studies of this syndrome makes judging the validity of some localization inferences difficult since some patients with slow but accurate, letter-by-letter reading could be misclassified as nonalexic. Apparently, however, a similar lesion can produce different degrees of alexia in different subjects. More detailed brain–behavior correlation studies are required to determine the possible effects of different localizations of damage within the occipital region on reading performance.

In the previous edition of this book, Greenblatt (1983) proposed an anatomical classification for the lesions that can give rise to pure alexia, which he termed "pre-angular" alexia. In splenio-occipital alexia, damage to the splenium disconnects transcallosal input from the right visual association cortex, which has arrived from the left visual field. Damage in the left occipital cortex further disconnects input from the right visual field to the left language areas. If the lesion involves the calcarine cortex, the accompanying hemianopsia appears, but this symptom is not essential. In a callosal lesion, the ventral portion of the splenial fibers, which is critically involved in the reading process, is damaged (Ajax et al., 1977; Greenblatt, 1973). In occipital alexia, the lesion is confined to the occipital lobe and spares the splenial region. With lateral, medial, or complete occipital lobectomy, the left visual association cortex is removed partially or completely. Thus, even if splenial fibers are preserved, the left occipital association cortex with which transcallosal fibers can connect is insufficient. With partial resections, reading may improve but does not return to normal (Greenblatt, 1977,1983). In subangular alexia, a lesion in the white matter subjacent to the angular gyrus disrupts input from both the left and the right visual association cortex (Greenblatt, 1976).

According to Greenblatt's functional–anatomical model of the pre-angular pathways for reading, visual information relevant to reading crosses from the right visual association cortices in the ventral splenium to homologous association areas on the left (Greenblatt, 1977,1983,1990). Information then is projected from the left association cortices to the angular gyrus. Both medial and lateral inferior occipital–temporal visual association areas are involved in the reading process. If a lesion is confined to the lateral or medial cortex, then improvement can occur because some association cortex remains available for visual word processing. Since more association cortex is located laterally than medially, medial occipital lobectomy alexias have more potential for recovery.

Pure alexia also can arise from a combination of lesions that entirely spare the left occipital lobe (Silver et al., 1988). A patient with multiple cerebral emboli developed pure alexia due to infarction of the left lateral geniculate nucleus (LGN), which cut off input to the left visual cortex. He also had a splenial infarct that disconnected visual input from the right hemisphere. Lesions were confirmed by CT and MRI, and a PET scan showed a profound metabolic deficit in the left occipital region, indicating functional but not structural compromise of the occipital lobe. Interestingly, considerable improvement in occipital metabolism occurred 4.5 mo later, despite persisting spelling alexia. Thus, in the absence of visual input to the left occipital region (caused by the LGN lesion), disconnection of input from the right visual association area through the ventral splenium was sufficient to produce the pure alexic syndrome (Silver et al., 1988). Persistence of alexia after several months implies, contrary to Geschwind's (1965) hypothesis, that the transcallosal pathway from the right to left angular gyrus is not sufficient to restore functional reading.

Although the interpretation of the studies described here are all somewhat different, converging evidence exists for the following inferences about the localization of the visual processing necessary for word recognition. Lesion analysis suggests that visual association cortex implicated in the reading process is in the inferior occipitotemporal region, probably infracalcarine and infrasplenial, both medial and lateral. Certain white matter tracts are essential for transfer of visual word information to the language areas in the left parietotemporal region, presumably to contact lexical semantic and phonological representations. These tracts include interhemispheric fibers (pathway d, Fig. 5) through the ventral splenium from the visual association cortices on the right, which travel to the homologous inferior occipital–temporal association cortices (Brodmann areas 18, 37, and 19) on the left. Information then is conveyed to the temporal–parietal cortex, possibly including the angular gyrus, the posterior superior, and/or the middle temporal lobe areas (Brodmann areas 21, 22, and 40). Intrahemispheric connections occur via occipital–temporal and occipital–parietal projection systems, which pass by the posterior tip of the occipital horn of the lateral ventricle (pathway c, Fig. 5). A lesion in the periventricular white matter region at the temporal–

occipital–parietal junction would be sufficient to disconnect input from both the right and the left visual association cortex from the pertinent language areas and would cause pure alexia, since the projection system for word recognition appears to run through this region close to the lateral ventricle. If the calcarine cortex and optic radiations are spared, no concomitant hemianopsia need occur. Further, a splenial lesion is not essential and, if present in isolation, a splenial lesion or callosotomy will cause a left visual hemifield alexia (hemialexia) (Greenblatt, 1983). The minimal single lesion that causes pure alexia is in the white matter at the temporal–occipital–parietal junction (Damasio & Damasio, 1983; Dejerine & Dejerine-Klumpke, 1901; Greenblatt, 1983; Henderson, 1986).

Pure alexia appears, therefore, to be the quintessential disconnection syndrome. However, its manifestations and recovery may differ, if appropriately investigated, depending on the extent of involvement of the left visual association cortex as well as on individual factors. Unfortunately, the data currently available do not allow additional comment on this issue since the detailed experimental case studies on pure alexia have provided little anatomical information, and, in many of the neuroanatomically oriented studies, the investigation of the reading deficit has been insufficient. Hopefully this deficiency can be rectified in future case studies. Prospective series using modern neuroanatomical localization methods with MRI and functional imaging also will be necessary to determine how focally organized the visual association cortex is with respect to visual processing of other graphic symbols such as letters and numbers. The frequent dissociation between performance on reading numbers, letters, and words in pure alexia raises the possibility that different anatomical substrates may be involved. Careful behavioral and lesion analysis could contribute to a better understanding of these different types of form perception in humans.

2. Central alexia

a. Angular alexia. When damage occurs in the left perisylvian language area, a reading disorder frequently results. As first reported by Dejerine (1891), a lesion confined to the angular gyrus can produce alexia with agraphia in which oral language is relatively intact. Associated defects may include ideomotor apraxia, anomia, and elements of Gerstmann's syndrome which consists of agraphia, acalculia, right–left disorientation, and finger agnosia (Benson, 1985; Benson & Geschwind, 1969). Difficulty naming letters as well as naming words may occur. The psycholinguistic aspects of this reading and writing disorder have not been characterized, since this condition appears to be relatively rare. The localization may not be restricted to the angular gyrus, since the syndrome also has been reported with a lesion in the temporo-parietal junction (Kawahata & Nagata, 1988). Further, as described earlier, in PET activation studies of reading in normal adults, activation of the angular gyrus was not detected, as would have been predicted by lesion studies.

b. Aphasic alexia. More commonly with perisylvian damage, a reading disorder is present in parallel with the oral language deficit. For example, in patients with Wernicke's aphasia, comprehension of written words typically, although not always, is impaired to the same degree as auditory comprehension (Hecaen & Albert, 1978; Heilmann, Rothi, Campenella, & Wolfson, 1979; Hier & Mohr, 1977). The resulting reading deficit is described most meaningfully in relation to the lexical and phonological processes involved in reading, which are discussed in Section IV.

c. Anterior or frontal alexia. A subtype of aphasic alexia that has been singled out for special attention is associated with Broca's aphasia and usually arises from frontal lobe damage (Benson, 1977). The main features are impaired syntactic comprehension and letter anomia, also called literal alexia. Many aphasic subjects with Broca's aphasia can read and understand single words, in contrast to other alexic syndromes, but these patients are often unable to understand grammatical relationships within sentences. This deficit is referred to as "agrammatism," which usually manifests in speech as well as in oral and written comprehension (Caplan, 1987). In a series of 61 patients with Broca's aphasia, 51 had reading difficulty that was relatively severe in 34 (Benson, 1977). Benson (1977) suggested that gaze paresis, which prevented easy scanning of text from left to right and impaired sequential processing, exacerbated this reading disorder. In another study, 7 of 17 patients (41%) with Broca's aphasia, studied within 60 days of stroke onset, had a mild to moderate alexia on single-word tasks and sentences (Boccardi, Bruzzone, & Vignolo, 1984). Alexia was more frequent (72%) in patients with chronic Broca's aphasia, many of whom initially had global aphasia. Presumably, many of these patients had deep dyslexia (Section IV,C). No obvious differences were seen in the CT lesions, most of which involved frontal cortex, of the patients with and without alexia (Boccardi et al., 1984). Syntactic dyslexia caused by temporo-parietal lesions also has been reported, suggesting that frontal damage is not essential for this type of reading impairment (Rothi, McFarling, & Heilman, 1982). In fact, a psycholinguistic model-based approach to the reading disorders of Broca's and other aphasic alexias has proved to be more fruitful than the anatomical classification, as is described in the next section.

IV. COGNITIVE NEUROPSYCHOLOGICAL ANALYSIS OF ALEXIA

A. Background

The influential papers by Marshall and Newcombe (1966,1973), which launched the cognitive psychological approach to alexia, argued that normal reading is a complex componential skill that is susceptible to particular patterns of breakdown after damage. Because reading is a multifaceted behavior encompassing visual processing as well as linguistic analysis, deficits in any one of a number of processes could give rise to an acquired dyslexia. Soon many information processing models of single-word reading were put forward, most of which share the same general architecture (for example, Coltheart, 1987; Shallice, 1988). Figure 7 illustrates such a model, which sketches the components of reading that are central to the following discussion and is based loosely on the logogen model of Morton (1979; Morton & Patterson, 1980). When a written word is presented, detailed perceptual analysis of the orthography or visual form is undertaken. This stage of primary visual analysis is thought to be common to all forms of visual perceptual processing (Farah & Wallace, 1991; Friedman & Alexander, 1984). During this process, the primitive features of letters are analyzed and an abstract code (independent of case, style of writing, or font) is generated (Allport, 1979; Estes, 1975). This letter identification process generally occurs in a slow, serial, left-to-right manner in beginning readers but becomes automatic and parallel with increasing skill and expertise (LaBerge & Samuels, 1974; Seymour & MacGregor, 1984).

Once the primary visual analysis is complete, the abstract information is transmitted to the orthographic input lexicon, which stores visual word forms containing the constituent letters of a word and the order or position in which they occur. Once the orthographic word form has been activated, the word's meaning can be retrieved from the semantic lexicon; thereafter, the phonology or pronunciation of the word can be accessed. Because this "direct" route provides lexical mediation for word pronunciation, it also is referred to as the "addressed" route. An alternative "nonlexical" or "sublexical" route connecting orthography and phonology also exists and contains a set of rules to perform grapheme–phoneme correspondences on subcomponents of the input, thereby converting written spelling to sounds. Because its operation is on subcomponents of the input, this latter route is referred to as the route of "assembled" phonology (Coltheart, 1981,1985; Patterson & Morton, 1985). Researchers have argued that the existence of two routes is necessary for the following reason. If the only route for reading were from the orthographic word form to phonology via semantics, we would not be able to respond to novel words or nonwords

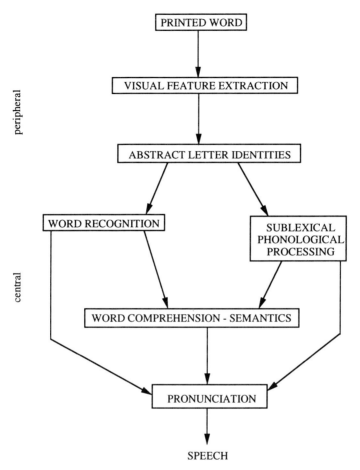

Figure 7 A processing diagram depicting the peripheral and central components involved in word reading.

(such as "wem" or "kalp"), which do not have an existing entry or representation in the lexical–semantic system. If, on the other hand, the only route were the one that led to pronunciation via sublexical componential analysis, we would not be able to read words that violate spelling–sound rules. Since we can read irregular or exception words such as "mortgage" and "choir," we must have a second route that does not operate via rules but uses a more holistic, gestalt form of processing. Controversy has arisen about the dual routes to reading. Some investigators claim that a third route proceeding directly from orthography to phonology is necessary to account for some data such as the ability to read aloud irregular words without comprehending them (Marin, Saffran, & Schwartz, 1980). Others argue that a single route alone is capable of dealing with both nonwords and irregular words (Seidenberg & McClelland, 1989; see Besner, Twilley, McCann, & Seergobin, 1990, for further discussion on the necessity of dual routes). For the sake of simplicity, we have chosen not to engage in the debate, but have elected to use a generic dual-route model for the purposes of our illustrations.

Brain damage can give rise to a deficit at any one of a number of possible loci shown in Fig. 7. The following section deals with the types of deficits, which we have grouped into two main categories. Damage to processes operating prior to the orthographic input lexicon, which prevent the attainment of the visual word form, can give rise to several different forms of alexia. These deficits are grouped together as "peripheral" alexias since they arise during the categorization of the stimulus as an orthographic entity (Shallice, 1988; Shallice & Warrington, 1980). Damage arising at stages beyond or "downstream" from the input lexicon causes "central" alexias, since it affects later stages of the reading process such as accessing the semantics or meaning of the words or obtaining their pronunciation. This broad distinction between peripheral and central variants will be adopted in our discussion. In each part, the behaviors are described first, followed by a summary of neuroanatomical data for each disorder. The term dyslexia is used in this section interchangeably with alexia.

B. Peripheral Alexias

1. Attentional dyslexia

According to Warrington and Shallice (1977; see also Shallice, 1988), patients with attentional dyslexia correctly read single words presented in isolation, but performance falters when multiple items are presented simultaneously. For instance, when several words appear simultaneously, letters from one word would migrate to the corresponding position in a second word. Thus, WIN FED would be read as FIN FED. These letter migration errors also have been observed with normal subjects under conditions of brief masked exposure of multiple words (Mozer, 1983; Shallice & McGill, 1978). The difficulty processing a single part of an array also is observed for single letters; although naming performance is almost perfect on individual letters, patients are unable to name the constituent letters of a visually presented word. The deficit in these cases is thought to arise from an inability to filter our irrelevant information, which then gains access to higher levels of processing and overloads the system. Thus, the presence of extraneous information in the visual field is not suppressed and interferes with processing of the relevant information. This phenomenon has been reported in two cases by Shallice and Warrington (1977), both of whom had tumors deep in the left parietal lobe, as well as in S.I., an adult with developmental dyslexia (Rayner, Murphy, Henderson, & Pollatsek, 1989). A group of developmentally dyslexic children also has been demonstrated to show characteristics similar to those of attentional dyslexia in adults (Geiger & Lettvin, 1987).

2. *Neglect dyslexia*

The primary symptom in neglect dyslexia is the failure to read information that appears on the side of space contralateral to the cerebral lesion. For example, after right hemisphere damage, patients may fail to read the left side of an open book or the beginning words of a line of text, or may make substitution (FARM → harm), addition (RIB → crib), or deletion (HAND → and) errors on the left side of a single word. Kinsbourne and Warrington (1962) described six cases of neglect dyslexia that followed right parietal lobe lesions, all of whom made errors on the leftmost letters of the words. This pattern of performance cannot be attributed to a field defect since the errors were seen even with presentations to the intact visual field. Although neglect dyslexia may co-occur with other manifestations of hemispatial neglect, "because visual neglect happens to compromise the reading process" (Ellis, Flude, & Young, 1987, p. 460), some reports show that neglect can be restricted to the reading system (Patterson & Wilson, 1990; Riddoch, Humphreys, Cleton, & Fery, 1990). Costello and Warrington (1987) documented the presence of left-sided neglect alexia with right-sided hemispatial neglect for other forms of visual processing, showing the independence of these two types of processes. In the last 5 years, considerable interest has arisen in this phenomenon; the pattern of performance has been analyzed in detail. Neglect dyslexia traditionally is interpreted as a disturbance in which visuospatial attention is distributed unevenly across the visual field, affecting the coding of visuospatial information at low levels of visual analysis. For example, when patients with neglect are given vertically presented words, they do not show neglect, suggesting that the attentional deficit affects the left side of space (Behrmann, Moscovitch, Black, & Mozer, 1990; Ellis et al., 1987,1991). This hypothesis is verified further by demonstrations that some patients neglect the left side of words when words are presented right-to-left as in mirror-reversed or backward writing (e.g., PEARL → earl and, backward, LRAEP → pear) (Ellis et al., 1987; Riddoch et al., 1990). Caramazza and Hillis (1990), on the other hand, have shown that a patient with right-sided neglect following a left-hemisphere lesion exhibited neglect for the right side of the word rather than for the information appearing on the right side of space. Irrespective of whether the item was printed horizontally, vertically, or even mirror reversed, the patient neglected the "right" or terminal letters of the word, suggesting that, for this patient, neglect was arising at an abstract level of representation where words are processed independent of input modalities.

With respect to anatomical localization, neglect dyslexia, like other forms of hemispatial neglect, usually is associated with middle cerebral artery territory damage in the right hemisphere that involves the frontal, parietal, and temporal lobes and subjacent deep structures and disrupts the diffuse distributed neural network underlying attention (Hier, Mondlock, & Caplan, 1983; Mesulam, 1987). Parietal damage appears to be the most common neural concomitant. In the few case studies of neglect dyslexia in which lesion data have been provided (Behrmann et al., 1990), the patients had moderate sized cortical–subcortical lesions involving the parietal cortex, basal ganglia, and centrum semiovale; in one case, the damage extended into the right frontal region.

3. Pure alexia

Pure alexia, so called because of the absence of any other neurobehavioral deficit, roughly corresponds to Dejerine's original "alexia without agraphia." Patients who retain letter naming ability characteristically read letter by letter, showing a monotonic increase in reading time as a function of the number of letters in a word (McCarthy & Warrington, 1991; Wolpert, 1924). This phenomenon is called spelling alexia. Reading accuracy may be good, provided that sufficient time is given for the patient to encode the letters. The severity of the deficit underlying the performance of different letter-by-letter readings can vary. The time required to read a three-letter word can range from 1–2 sec (Bub et al., 1989; Warrington & Shallice, 1980) to 17 seconds (Patterson & Kay, 1982). Letter-by-letter reading has been studied extensively in the last 10 years, resulting in several hypotheses about the underlying impairments. Whereas some researchers argue that the deficit is due to defective visual processing occurring at a stage of processing single letters (Arguin & Bub, 1992; Reuter-Lorenz & Brunn, 1990); not specific to words (Farah & Wallace, 1991; Friedman & Alexander, 1984), others argue that the deficit arises in accessing the input lexicon via parallel letter analysis (Howard, 1991; Patterson & Kay, 1982) or at an even higher level, in the word-form system itself (Warrington & Shallice, 1980).

Of particular interest is the finding that these patients may be able to attain some semantic knowledge of a word, even if it has been presented too briefly for explicit encoding of the constituent letters. For example, some patients can decide that the word refers to a "living thing or an object" or to an "animal or fruit" even though they cannot recognize the item overtly (Coslett, Saffran, Greenbaum, & Schwartz, 1993; Coslett & Saffran, 1989; Shallice & Saffran, 1986). The explanation offered to account for this ability is that these patients are utilizing processing mechanisms that are functionally and anatomically separate from those normally used for reading. According to Coslett and Saffran (1989,1992), the lesion in pure alexia prevents access from both visual cortices to lexical and semantic components in the left hemisphere, but the right visual cortex still has access to words in the more limited right hemisphere reading system. The preservation of semantic knowledge is seen only in covert reading in these patients, presumably because explicit identification requires the mediation of functional components of the dominant left hemisphere. The role of the right hemisphere in reading is discussed further in a subsequent section.

The lesion location in pure alexia has been discussed in some detail. In the cognitive neuropsychological case studies presented over the last several years, lesion analysis has not been emphasized. When this information has been provided, patients have had large inferior occipital lesions, as would be expected (Behrmann, Black, & Bub, 1990; Bub, Black, & Howell, 1989). The brain–behavior correlations, however, have been insufficient to comment on whether different degrees of severity of letter-by-letter reading or behavioral features, such as unconscious access to semantic information, are related in any way to lesion characteristics. Case studies should include templates or photographs of the brain lesions showing the sites of damage, so lesion analysis from multiple single-case studies could lead to better understanding of the neural substrate for the visual processing components involved in the early stages of reading.

C. Central Alexias

The two main types of central alexia correspond to deficits in one of the two reading routes described in the cognitive model presented in Section IV,A (see Fig. 7). Thus, surface alexics read via the nonlexical rule-based phonological route since the direct lexically mediated route is abolished, whereas the reverse is the case for patients with phonological alexia. The existence of this double dissociation in patients has been a major force in supporting the dual-route cognitive theory of reading.

1. Surface dyslexia

The hallmark feature of surface alexia is the regularization error, which arises from the overapplication of grapheme–phoneme rules used to convert sound to spelling. Although these patients can read words that obey spelling–sound rules (regular words such as "hand" and "fist") with normal speed and accuracy, they fail to read irregular words that violate the predicted form of pronunciation. For example, the patient may read the word "pint" to rhyme with "mint," giving "pint" the regular rather than its exceptional pronunciation. Thus, these patients show a striking dissociation between their intact ability to read regular words and their very poor performance on irregular words (Behrmann & Bub, 1992; Bub, Cancelliere, & Kertesz, 1985; Marshall & Newcombe, 1973; Shallice & Warrington, 1980; Shallice, Warrington, & McCarthy, 1983; Warrington, 1975). In another study, Patterson and Hodges (1992) argued that in progressive diseases (such as progressive aphasia or dementia), as word meaning deteriorates, the patients begin to demonstrate surface alexia. According to these investigators, when semantics deteriorates, the "glue" holding together the components of a word dissolves and the word must be read via the sublexical route.

With respect to neuroanatomical localization in surface dyslexia, a detailed lesion analysis is provided for each of the well-documented single cases published in a book devoted to the subject (Patterson, Marshall, & Coltheart, 1985). In this analysis, account was taken of angle of rotation of the scans and the lesion was drawn on templates from Dejerine & Dejerine-Klumpke (1895). Lesions then were projected onto the surface anatomy of the left hemisphere. A slice-by-slice anatomical description of the lesion was provided for each case. The chapter by Patterson and colleagues serves as a model for lesion analysis that hopefully will be followed in other compendia of single-case studies. The common areas that seem to be involved in all cases are the posterior superior and middle temporal gyri and underlying white matter. Occasionally, extension into the supramarginal and angular gyrus, and even the occipital region, is seen; one patient also had frontal and basal ganglia involvement. In three cases, the lesion was confined to the temporal lobe. Additional evidence for this localization was also forthcoming from a group study of 22 patients with acquired alexia due to focal lesions, who underwent linguistic analysis of reading and writing and were classified as having phonological or surface dyslexia. In the 6 subjects with surface dyslexia, the common site of overlap of lesions was in the posterior superior temporal gyrus (Roeltgen, 1983). The PET studies by Howard et al. (1992) found activation in the left middle temporal region in association with word rather than pseudoword reading, lending credence to possible localization of the visual lexicon in the posterior temporal–occipital region. Thus, converging evidence from single-case studies, a group series on surface dyslexia, and PET activation studies in normals suggests that the ability to read by the direct lexical route is dependent on structures located in the posterior superior and middle temporal region. The consistency of the brain–behavior relationships described in this disorder is certainly noteworthy.

2. Phonological dyslexia

The essence of this disorder is a severe impairment in the ability to read nonwords, coupled with the intact ability to read words. The phenomenon was described first in 1979 (Beauvois & Derouesne, 1979; Derouesne & Beauvois, 1985) and many subsequent reports have been made (Bub, Black, Howell, & Kertesz, 1987; Funnel, 1983; Warrington, 1983; Warrington and Shallice, 1980). The deficit in phonological dyslexia is thought to arise from damage to the nonlexical phonological route, which prevents the use of grapheme–phoneme rules. The intact direct route, however, enables the patients to read words (both regular and irregular) to a high level of accuracy.

The lesion location in phonological dyslexia is less clearly known than that in pure alexia or surface dyslexia. As frequently is the case in experimental single-case studies, little anatomical detail has been provided concerning the associated brain lesions. In the patient described by Bub, Black, Howell, and Kertesz (1983,1987) a left hemispheric infarction was found involving gray and white matter in the pre- and postcentral gyri, the insula, the supramarginal and superior temporal gyri, as well as the putamen and posterior internal capsule. In the group series described by Roeltgen (1983), 16 patients had phonological alexia. The areas of common overlap were the medial frontal region, the perisylvian region around the insula, the posterior middle temporal gyrus, and the inferolateral occipital region. In two subjects these regions were not involved, and in some subjects more than one region was affected. Lesions most commonly overlapped in the left mid-perisylvian region. In one subject with features of both surface and phonological dyslexia, the lesion localization was in the middle temporal, lateral occipital, and angular gyri. The essential neuroanatomical substrate of phonological dyslexia, therefore, is less clearly defined. More brain–behavior correlation studies are required to determine whether different profiles of patient performance reflect different sites of injury.

3. Deep dyslexia

Deep dyslexia also is thought to arise from damage to the indirect phonological route. Patients with this disorder show the same characteristics as patients with phonological dyslexia. The critical difference, however, that suggests an additional impairment is that these patients produce errors called semantic paralexias in which the response resembles the target word in meaning. Differences occur in the proportion of semantic errors produced by these patients; for example, patients H.W. (Caramazza & Hills, 1990) and K.E. (Hillis & Caramazza, 1990) produced in excess of 30% semantic errors. For other patients—G.R. (Glosser & Friedman, 1990), V.J. (Laine, Niemi, Niemi, & Koivuselka-Sallinen, and V.S. (Nolan & Caramazza, 1982)—5–10% of all their errors were semantic paralexias. A cluster of other features in deep dyslexia (Coltheart, 1980b; Coltheart et al., 1987) includes the production of visual and derivational paralexias, more difficulty with function than content words, poorer performance on abstract than on concrete words, and a part-of-speech effect in which nouns are read better than verbs, which are read better than functors.

The underlying impairment in deep dyslexia is much debated. One school of thought has argued that this disorder is the result of multiple impairments to a left hemisphere information processing system used for reading. For example, Morton and Patterson (1980) have argued that five separate impairments must co-occur to observe the pattern of deep dyslexia. This view is not particularly parsimonious; a more streamlined account has been given by those who argue that the co-occurrence of the various symptoms arises from the residual reading abilities of the right hemisphere when the left hemisphere is damaged extensively, since the anatomical substrate for deep dyslexia appears to be major cortical–subcortical damage in the left perisylvian language region. In the initial compilation of cases of deep dyslexia published by Coltheart et al. (1980b), CT scans were presented for 5 patients with the disorder (Marin, 1980). Only 1 slice was selected from each patient, but all showed massive infarction in the entire middle cerebral artery territory. On inspection of the scans provided, inferior frontal, inferior parietal, and superior temporal damage was present in all 5 cases, as was damage to subcortical white matter tracts and basal ganglia. Extensive left hemisphere damage in these and other reported patients has been interpreted as support for the hypothesis that deep dyslexia represents right hemisphere reading.

D. Role of the Right Hemisphere in Reading

The role of the right hemisphere in affective processing of language and the pragmatics of discourse is well accepted (Bryden & Ley, 1983). The right hemisphere appears to be specialized for the expression and perception of prosody and emotional content as well as the appreciation of humor, metaphor, and themes in discourse (Searleman, 1983). The extent to which the right hemisphere engages in linguistic processing, however, has been more controversial. Sources of data on this topic have included visual hemifield word presentations and dichotic listening in normals and callosotomy patients and studies of patients with unilateral brain lesions and

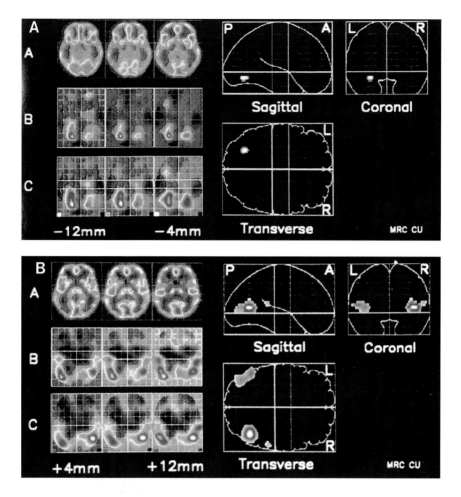

Color Plate 1 Visual cognition activation. (A) Coronal, sagittal, and transverse projections of statistical parametric maps obtained by planned comparison of means between conditions with viewing of colored versus viewing of gray-scale images. The stereotactic coordinates refer to the standard proportional grid of Talairach and Tournoux (1988), which defines the three-dimensional space into which all the subjects' brains have been rescaled (normalized). On the left, A, B, and C refer to transverse sections through the inferior occipital lobe at various levels. On the right, the contours of the brain in the transverse plane at the level of the AC–PC line (*bottom*), the midsagittal plane (*upper left*), and the coronal plane at the mid point of the AC–PC line (*upper right*), are shown on the respective grids. Activation of the left fusiform and lingual gyri and, to a lesser degree, the same area on the right is clearly demonstrated. (B) Similar data from the visual motion experiment, comparing brain activity during the viewing of objects in motion versus stationary objects. The map shows areas specifically subserving the perception of movement in the visual scene (human analog of V5). The areas are located on the convexity of the prestriate cortex at the junction of areas 19 and 37 of Brodmann. Reproduced with permission from Zeki, Lueck, and Watson (1991).

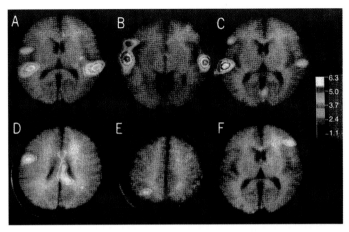

Color Plate 2 Activation of brain regions with ¹⁵O PET in auditory processing tasks. (A) Activation of Heschl's gyrus bilaterally with presentation of noise stimulation. (B, C) Activation of left and right superior temporal gyrus during presentation of syllable pairs compared with white noise. (D, E) Activation of left frontal and left parietal regions during performance of a rhyme judgment task (phonetic task) involving the sound pairs. (F) Activation of right frontal lobe during a pitch decision task involving the sound pairs. Reproduced with permission from Zatorre et al. (1992).

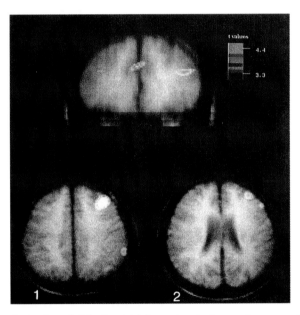

Color Plate 3 Activation of right frontal lobe cortex during performance of self-ordered pointing task. Top image is a coronal section from the self-ordered pointing minus control task subtraction, demonstrating activation within the right middorsolateral frontal cortex (area 46). Bottom horizontal sections show activation within the middorsolateral frontal cortex. Reproduced with permission from Petrides et al. (1993).

Subject 1

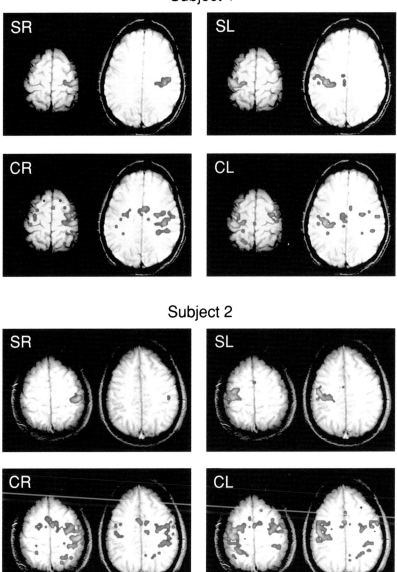

Subject 2

Color Plate 4 Functional images for the self-paced motor activation conditions—simple motor right hand (SR), simple left (SL), complex right (CR), and complex left (CL)—for two representative subjects. Two axial slices (centered 12 and 24 mm from the vertex of the brain) are presented for each condition. See text for details regarding the color scale used to indicate functional activity. The subject's right is on the reader's left. Reproduced with permission from Rao et al. (1993).

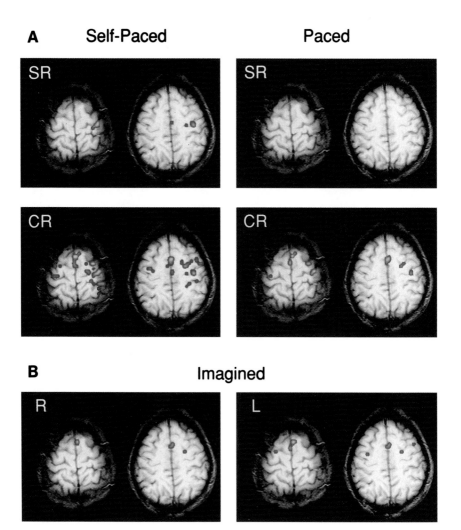

Color Plate 5 (A) Functional images for the self-paced and paced motor activation conditions, comparing simple right (SR) and complex right (CR) activation conditions. (B) Imagined complex motor activity of the right (R) and left (L) hands. The subject's right is on the reader's left. Reproduced with permission from Rao et al. (1993).

Subject A   Subject B

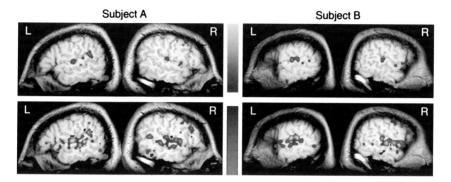

Color Plate 6 Activity during passive stimulation with white noise and speech sounds ("pseudo-words") in two normal subjects.

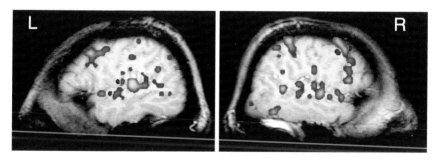

Color Plate 7 Activity during a task involving monitoring of pure tone sequences. Main areas of activation include bilateral primary and association auditory cortices of the superior temporal gyri, bilateral frontal lobes, and right supramarginal gyrus.

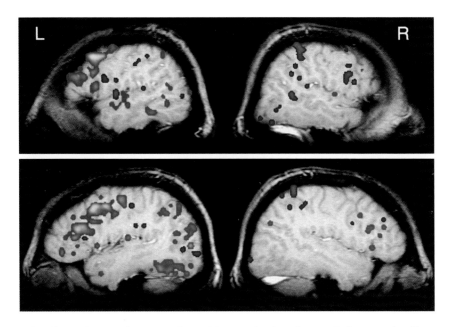

Color Plate 8 Activity during a task requiring monitoring of semantic features of auditory words. Control task was monitoring pure tone sequences. Activity coincident with the semantic task (red-yellow) is strongly lateralized to the left hemisphere and involves widespread polymodal association areas.

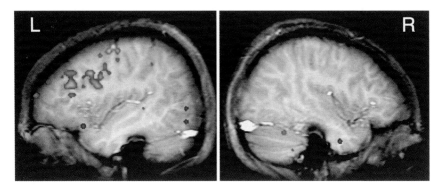

Color Plate 9 Left lateral frontal activity during silent retrieval of verbs associated with presented nouns.

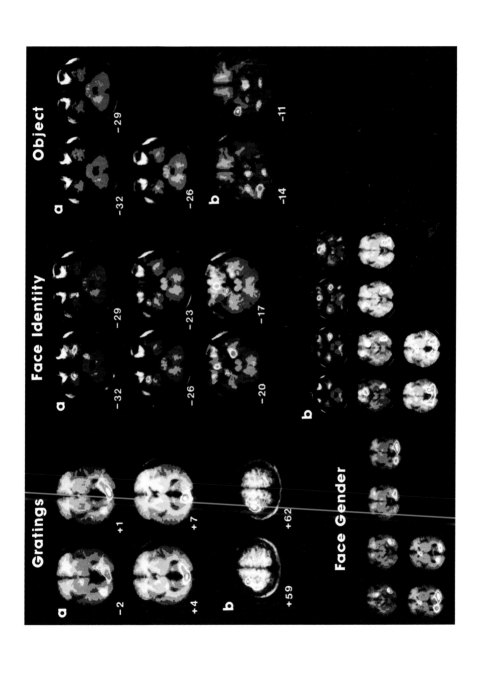

Gratings

Face Identity

Object

Face Gender

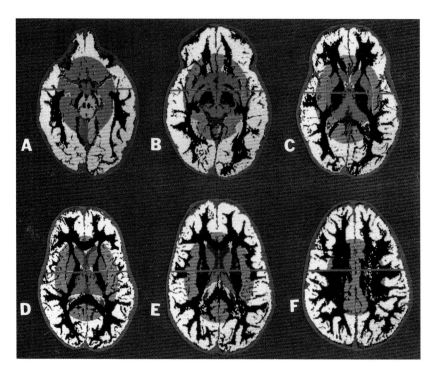

Color Plate 11 Representative, fully processed images. Pixels are classified and zones have been manually designated. The gray matter pixels have been color-coded to display the zone designations: peripheral cortex, white; mesial cortex, green; caudate, magenta; lenticular nucleus, orange; diencephalon, blue. CSF and white matter pixels in all zones are displayed here in red and black, respectively; however, these pixels are coded separately by zone so that regional measures may be computed. The yellow pixels within the subcortical zone are white matter with the signal characteristics of gray matter. (A–C) Sections completely classified as inferior to dividing plane. (D–F) Sections completely classified as superior to dividing plane. For both inferior and superior zones, the blue line shown within the sections indicates the position of the coronal dividing plane. The ellipsoid defining the mesial cortex has cardinal dimensions proportional to those of the supratentorial cranium and occupies 30% of its volume.

Color Plate 10 Foci of functional PET activation, superimposed on magnetic resonance images of normal subjects' brains performing tasks of grating discrimination, face gender categorization, face identification, and object recognition.

of patients who have had hemispherectomies (Coltheart, 1983; Hunter and Liederman, 1991; Schneiderman and Saddy, 1988; Searleman, 1983). On the whole, these studies have suggested a limited capacity for language processing in the right hemisphere, which can be summarized as follows. The right hemisphere in most right-handed individuals has limited phonetic and syntactic competence, both in comprehension and in speech production. Although individual variability may exist, speech production mediated by the right hemisphere in right-handers who have lost speech from left hemisphere damage is confined to automatic speech or singing and perhaps the pronunciation of some concrete nouns. Poor comprehension of propositional language is typical due to rudimentary syntactic processing and poor short-term verbal memory, but high frequency concrete nouns presented in both auditory and visual modalities usually can be understood. Researchers have speculated that these right hemisphere language functions play a role in recovery from aphasia. The abolition of aphasic speech by barbituate injection into the right carotid (Kinsbourne, 1971) and the relative increase in right hemisphere activation during language tasks in recovering aphasics measured by cortical evoked potentials (Papanicolaou, Bartlett, Deutsch, Lavin, & Eisenberg, 1988) lend support to this hypothesis.

Coltheart (1980b) and others (Saffran, Bogyo, Schwartz, & Marin, 1980) have argued strongly that the residual reading capacities demonstrated in patients with deep dyslexia reflect right-hemisphere processing of visible language. This claim is based primarily on the parallels between the reading performance of deep dyslexic patients and visual half-field studies in normal individuals and in patients who have undergone corpus callosum section for intractable epilepsy. In normal subjects, some tachistoscopic studies show a hemifield-by-concreteness interaction, indicating that the normal right hemisphere is better at reading concrete than abstract words. Visual half-field tachistoscopic presentations in callosotomy patient have shown that the right hemisphere in these patients is unable to derive phonology from print (Zaidel & Peters, 1981), but can process concrete words semantically (Zaidel, 1982). Semantic paralexias are seen also (Landis, Regard, Graves, & Goodglass, 1983; Zaidel, 1982). The finding that deep dyslexic patients show a left visual field advantage in tachistoscopic studies of word reading, in contrast to studies in normals, also has been interpreted to support the right hemisphere hypothesis (Saffran et al., 1980). In a Japanese deep dyslexic subject, Sasanuma (1980) showed that Kanji (ideographic) comprehension was relatively spared and attributed this result to right hemisphere processing, since a left visual field advantage for Kanji reading appears in normal subjects (Sasanuma, Itoh, Mori, & Kobayashi, 1977). Roeltgen (1987), on the other hand, reported a patient with deep dyslexia following a left hemisphere stroke who lost residual reading capacity after a second, more anterior left hemisphere stroke, suggesting that the deep dyslexic reading ability was dependent on previously intact left hemisphere structures. Thus, the right hemisphere hypothesis continues to be the subject of lively debate (Patterson & Besner, 1984a,b; Rabinowicz & Moskovitch, 1984; Zaidel & Schweiger, 1984).

One of the primary problems with the right hemisphere hypothesis has been generated by patients with pure alexia, who would be expected to show right hemisphere features in their reading but do not. To account for this discrepancy, a number of arguments have been put forward, including the possibility of interindividual differences in right hemispheric capacity for word recognition and differential effects of lesions on the transfer of information from the right to the left hemisphere in patients with pure alexia (Coltheart, 1983; Zaidel & Schweiger, 1984). Coltheart suggests that the right hemisphere is able to categorize letter strings orthographically into words and can derive meaning from highly imageable lexical items. Uncategorized letter information also can be derived and transferred to the left hemisphere. Since many pure alexic patients are unable to comprehend words until they spell them aloud, Coltheart suggests that uncategorized letter information can be transferred by the transcallosal route in these subjects, whereas categorical information cannot. Subsequent findings that at least some patients with pure alexia can perform lexical decision tasks successfully (Bub et al., 1989) and may derive semantic information from words they cannot read (Shallice & Saffran, 1986), however, suggests that some categorical information can be made available to the left hemisphere. Zaidel and Schweiger (1984) suggest that transcallosal inhibitory mechanisms could act to suppress the expression of right hemisphere linguistic abilities depending on different experimental conditions or on the location and degree of brain damage. Thus, pure alexics may show right hemisphere reading only when the left hemisphere releases control, which could occur through lesion factors or through manipulation of the task parameters so stimuli cannot be handled by the left hemisphere. These authors also suggest that the right and left hemispheres play complementary roles in the reading process and contribute different skills to the development of normal reading. In particular, the right hemisphere may facilitate "quick pattern recognition during speed reading" and "semantic–thematic orientation to a situation or narrative" (Zaidel & Schweiger, 1984, p. 362). The right hemisphere, therefore, may be important both for the acquisition of reading and for reading efficiency.

Two reports have provided additional converging evidence for a right hemisphere role in the reading performance of deep dyslexic patients. Patterson, Vargha-Khadem, and Polkey (1989) performed detailed reading analysis in two teenage patients who underwent hemispherectomy for seizure control after the acquisition of normal reading abilities. No reading abnormalities were seen in the patient with the right hemispherectomy. The reading performance of the left hemispherectomy patient showed striking parallels to deep dyslexia. The authors, who were formerly critical of the right hemisphere hypothesis, found this to be compelling evidence in its favor. In a xenon-inhalation regional cerebral blood flow study using subtractive techniques, Coltheart et al. (1992) found right hemisphere activation in word recognition tasks in a left hemisphere-damaged subject with deep dyslexia, in contrast to normal controls and to a left hemisphere-damaged subject with surface dyslexia. The same deep dyslexic subject showed increased activation in the left hemisphere during word production. Although the technique used was somewhat crude in comparison to PET, this study is the first to show on-line direct evidence for right hemisphere processing during word reading in deep dyslexia. The finding must be confirmed, but functional imaging provides an obvious technique with which to investigate further the role of the right hemisphere in acquired alexia.

E. Alexia in Ideographic Languages

The discussion in this chapter has centered on reading in phonetic language systems. In ideographic languages, brain organization for reading may be different. Alexia in Japanese subjects is of particular interest because they learn to read and write with both a phonetic alphabet (Kana) and ideographic symbols (Kanji). Studies in normal Japanese with tachistoscopic presentation of Kana and Kanji in the two visual fields suggests that Kanji is processed more efficiently by the right hemisphere and Kana by the left (Sasanuma et al., 1977). However, in two cases of occipital infarction, alexia for both Kana and Kanji occurred (Hirose, Kin, & Murakami, 1977; Kurachi, Yamaguchi, Inasaka, & Tori, 1979). If visual processing of Kanji and Kana were lateralized to the right and left hemisphere, respectively, one might expect dissociations in Kanji and Kana in pure alexic patients (Greenblatt, 1983). In Japanese patients who have undergone callosotomy, hemialexia occurs in the left visual field in both Kana and Kanji, although Kanji improves sooner than Kana (Sugishita, Iwata, Toyokura, Yoshiuka, & Yamada, 1978). Thus, left hemisphere processing appears to be important in both the phonetic and the ideographic symbol systems used in the Japanese language.

If brain damage occurs in Japanese subjects more anteriorly in the perisylvian language area, dissociations between reading and writing performance in Kana and Kanji become evident (Greenblatt, 1983). On the basis of the dual-route model of reading, one would predict that Kana would be processed by both lexical and nonlexical letter-to-sound routes and that Kanji would be processed by the direct lexical route. Dissociations along these lines have been reported with more anterior lesions (Soma, Sugishita, Kitamura, Maruyama, & Imanaga, 1989; Varney, 1984). The fact that dissociations are less likely to occur with left occipital damage, which mediates an earlier stage of processing, suggests that, with acquisition of reading skills, the early processing of ideographic symbols and of the letter arrays that form words may be similar. Therefore, seeing whether more complex ideograms require more time to process in Japanese patients with pure alexia, analogous to the word-length effect that would be expected in Kana, would be of interest.

V. ALTERNATIVE CONCEPTUALIZATION: CONNECTIONIST MODELS OF READING

The approach adopted in this chapter thus far has assumed that information processing is organized hierarchically so information is processed at one level (e.g., visual analysis) before being transmitted serially to the next stage (word forms), and so on. Although this approach has had much success in explaining the varieties of acquired dyslexias from a psychological standpoint, it does not easily coincide with our knowledge of neural processing nor does it specify precisely the type of computations (representation and processes) that are carried out at each stage of processing. In partial response to these shortcomings, an alternative formalism using artificial neural networks to model or simulate cognition and its breakdown after brain damage has gained much popularity. Artificial neural network models (also known as Parallel Distributed Processing or Connectionist models) assume that information processing takes place in parallel through the interaction of a large number of simple neuron-like processing units, each sending excitation or inhibition signals along connections between them. These models represent information as a pattern of activity distributed over several of these simple units. Learning to represent the information (e.g., produce the semantics in response to a visual input such as "cat") takes place through the gradual adjustment of the connection strengths (weights) joining the units. During training, the weights are modified until the system is able to represent any input pattern with a high degree of success. In simulating the performance of patients with brain damage, an analogous "lesion" (i.e., removing sets of units and/or connections) is introduced to the previous working model. (See Small, 1991, for a discussion of the correlation between the artificial lesion type and neurophysiology.) In many respects, these computational models are particularly well suited to modeling cognition after brain damage since they have inherent neurophysiological appeal. The types of computations performed by the neural-like units resemble the types of computations performed by the brain (Hinton & Shallice, 1991). In addition, these models degrade gracefully under damage (Hinton & Shallice, 1991; Patterson, 1990); as the model is damaged incrementally, so its performance slowly deteriorates, mirroring the decreased efficiency and speed seen in patients with brain damage. Several attempts have now been made to use networks to explore different forms of acquired dyslexia. Computationally explicit accounts of both peripheral and central forms of acquired dyslexia exist. For example, Mozer and Behrmann (1990,1992) have simulated neglect dyslexia in a model of word recognition and selective attention, and have shown that damage to the connections that direct attention to space gives rise to a rightward bias and neglect reading errors that are remarkably similar to the pattern seen in neglect dyslexia patients. In the same network, these researchers have been able to account for attentional dyslexia, which arises when the efficiency of the attentional mechanism is affected and irrelevant stimuli, which are also present in the visual field, gain access to higher levels of processing rather than being filtered out. Computational accounts of surface dyslexia (Patterson, 1990; Patterson, Seidenberg, & McClelland, 1989) and of deep dyslexia (Hinton & Shallice, 1991; Plaut & Shallice, 1991,1992) also exist. Although these efforts at modeling and providing explicit accounts of phenomena have shed light on the nature of the computations performed during reading, their exact relationship to brain functioning is still being determined.

VI. CONCLUSION

In the decade that has passed since the last edition of this book, considerable progress has been made in understanding the structural and functional architecture of the reading system in humans. In the decade ahead, the application of the three-dimensional MRI studies to acquired alexia will allow individualized neuroanatomical localization. The functional consequences of brain lesions can be quantitated by SPECT and PET, and potentially by MRI. Activation studies also will provide additional insight into normal mechanisms of reading. From the behavioral perspective, advances in microcomputer technology and software now make possible detailed individualized testing in subjects with acquired alexia. Many different types of experiments can be configured that include the speed and accuracy of reading as dependent measures. Model-based componential analysis of reading deficits is now a practical, if not an obligatory, approach. The challenge is to advance beyond the methodological limitations of the previous era, in which appropriate behavioral assessment frequently was lacking in the studies of careful lesion localization and in which sophisticated analysis of reading performance generally was not correlated with appropriate neuroanatomical measures. More attention also will have to be paid to the timing of analysis in relation to the onset of the injury and to patterns of recovery, if we are to understand acquired alexia properly and to design appropriate interventions to facilitate recovery. Interdisciplinary longitudinal studies of structural and functional lesion parameters in combination with theoretically relevant behavioral measures will be necessary to insure future progress in this field.

ACKNOWLEDGMENTS

The authors acknowledge personal support from Sunnybrook Health Science Centre's Research Program in Aging and Department of Medicine (S. E. Black) and the Medical Research Council of Canada (M. Behrmann). This work also was supported by grants from the Ontario Mental Health Foundation, the Heart and Stroke Foundation, and the Medical Research Council of Canada. We are grateful to Betty Ann Lemieux for assistance in preparation of the manuscript.

REFERENCES

Ajax, E. T., Schenkenberg, T., & Kosteljanetz, M. (1977). Alexia without agraphia and the inferior splenium. *Neurology, 27,* 685–688.

Allport, A. (1979). Word recognition in reading. In P. A. Kolers, M. E. Wrostad, & H. Bouma (Ed.), *Processing of visible language* (pp. 227–257). New York: Plenum Press.

Anderson, S. W., Damasio, A. R., & Damasio, H. (1990). Troubled letters but not numbers: Domain specific cognitive impairments following focal damage in frontal cortex. *Brain, 113,* 749–766.

Arguin, M., & Bub, D. N. (1992). Single-character processing in a case of pure alexia. *Neuropsychologia, 31,* 435–458.

Baxter, D., & Warrington, E. K. (1983). Neglect dysgraphia. *Journal of Neurology, Neurosurgery and Psychiatry, 46,* 1073–1078.

Beauvois, M. F., & Derouesne, J. (1979). Phonological alexia: Three dissociations. *Journal of Neurology, Neurosurgery and Psychiatry, 42,* 1115–1124.

Behrmann, M., Black, S. E., & Bub, D. (1990). The evolution of pure alexia: A longitudinal study of recovery. *Brain and Language, 39,* 405–427.

Behrmann, M., & Bub, D. N. (1992). Two routes, a single lexicon: Evidence from surface dyslexia and surface dysgraphia. *Cognitive Neuropsychology, 9,* 209–252.

Behrmann, M., Moscovitch, M., Black, S. E., & Mozer, M. (1990). Perceptual and conceptual factors in neglect dyslexia: Two contrasting case studies. *Brain, 113,* 1163–1883.

Belliveau, J. W., Kennedy, D. N., McKenstry, R. C., & Buchbinder, B. P. (1991). Functional mapping of the human visual cortex by magnetic resonance imaging. *Science, 254,* 716–718.

Benson, D. F. (1977). The third alexia. *Archives of Neurology, 34*, 327–331.

Benson, D. F. (1985). Alexia. In G. W. Bruyn, H. L. Klawans, & P. J. Vinken (Eds.), *Handbook of clinical neurology* (pp. 433–455). New York: Elsevier.

Benson, D. F., Brown, J., & Tomlinson, E. B. (1971). Variations of alexia: Word and letter blindness. *Neurology, 21*, 951–957.

Benson, D. F., & Geschwind, N. (1969). The alexias. In G. W. Bruyn & P. J. Vinken (Ed.), *Handbook of clinical neurology* (Vol. 4, pp. 112–140).

Besner, D., Twilley, L., McCann, R. S., & Seergobin, K. (1990). On the connection between connectionism and data: Are a few words necessary? *Psychological Review, 97(3)*, 432–446.

Black, S. E., Bub, D., & Behrmann, M. (1987). Asymmetries in lexical access: A study of patients with occipital damage. *Academy of Aphasia (Abstract)*.

Boccardi, E., Bruzzone, M. G., & Vignolo, L. A. (1984). Alexia in recent and late Broca's aphasia. *Neuropsychologia, 22*, 745–754.

Bryden, M. P., & R. G. Ley (1983). Right-hemispheric involvement in the perception and expression of emotion in normal humans. In K. M. Heilman & P. Satz (Eds.), *Neuropsychology of human emotion* (pp. 6–44). New York: Guilford Press.

Bub, A., Cancelliere, A., & Kertesz, A. (1985). Whole-word and analytic translation of spelling-to-sound in a non-semantic reader. In K. E. Patterson, M. Coltheart, & J. C. Marshall (Eds.), *Surface dyslexia* (pp. 15–34). Hillsdale, New Jersey: Erlbaum Associates.

Bub, D., Black, S. E., & Howell, J. (1989). Word recognition and orthographic context. *Brain and Language, 37*, 357–376.

Bub, D., Black, S. E., Howell, J., & Kertesz, A. (1983). A case of phonological dyslexia without grapheme–phoneme impairment. *Academy of Aphasia (Abstract)*.

Bub, D., Black, S. E., Howell, J., & Kertesz, A. (1987). Speech output processes and reading. In R. Job, G. Santori, & M. Coltheart (Eds.), *The cognitive neuropsychology of language* (pp. 79–110). Hillsdale, New Jersey: Erlbaum Associates.

Bub, D., & Lewine, J. (1988). Different modes of word recognition in the left and right visual fields. *Brain and Language, 33*, 161–188.

Butler, S. R. (1977). Interhemispheric transfer of visual information via the corpus callosum and anterior commissure in the monkey. In M. W. Hof, G. Berlucchi, & I. S. Russell (Eds.), *Structure and function of cerebral commissures* (pp. 343–357). Baltimore: University Park Press.

Caffarra, P. (1987). Alexia without agraphia or hemianopia. *European Neurology, 27*, 65–71.

Caplan, D. (1987). *Neurolinguistics and linguistic aphasiology.* Cambridge: Cambridge University Press.

Caplan, L. R., & Hedley-White, T. (1974). Cuing and memory dysfunction in alexia without agraphia: A case report. *Brain, 97*, 251–262.

Caramazza, A., & Hillis, A. E. (1990). Levels of representation, co-ordinate frames and unilateral neglect. *Cognitive Neuropsychology, 13*, 391–446.

Castro-Caldas, A., & Salgado, V. (1984). Right hemifield alexia without hemianopia. *Archives of Neurology, 41*, 84–87.

Cleton, P., Riddoch, M. J., Humphreys, G. W., & Fery, P. (1990). Levels of coding in neglect dyslexia. *Cognitive Neuropsychology, 7*, 479–518.

Coltheart, M. (1980a). Reading, phonological recoding and deep dyslexia. In M. Coltheart, K. E. Patterson, & J. C. Marshall (Eds.), *Deep dyslexia* (pp. 197–226). London: Routledge and Kegan Paul.

Coltheart, M. (1980b). Deep dyslexia: A review of the syndrome. In M. Coltheart, K. E. Patterson, & J. C. Marshall (Eds.), *Deep dyslexia* (pp. 22–47). London: Routledge and Kegan Paul.

Coltheart, M. (1981). Disorders of reading and their implications for models of normal reading. *Visible Language, 15,* 245–286.

Coltheart, M. (1983). The right hemisphere and disorders of reading. In A. W. Young (Ed.), *Functions of the right cerebral hemisphere* (pp. 171–201). London: Academic Press.

Coltheart, M. (1985). Cognitive neuropsychology and the study of reading. In M. I. Posner & O. S. M. Marin (Eds.), *Attention and performance XI.* Hillsdale, New Jersey: Erlbaum Associates.

Coltheart, M., Patterson, K. E., and Marshall, J. C. (Eds.) (1987). Deep dyslexia since 1980. In *Deep dyslexia* (pp. 407–451). London: Routledge and Kegan Paul.

Coltheart, M., Weekes, B., Savage, K., Simpson, L., Zurinsky, Y., & Gordon, E. (1992). Deep dyslexia and right hemisphere reading—A regional cerebral blood flow study (unpublished manuscript).

Coslett, H. B., & Saffran, E. M. (1989). Evidence for preserved reading in "pure" alexia. *Brain, 112,* 327–359.

Coslett, H. B., Saffran, E. M., Greenbaum, S., & Schwartz H. (1993). Reading in pure alexia. *Brain, 116,* 21–37.

Costello, A. D., & Warrington, E. K. (1987). Word comprehension and word retrieval in patients with localized cerebral lesions. *Brain, 101,* 163–185.

Costello, A. D., & Warrington, E. K. (1992). The dissociation of visual neglect and neglect dyslexia. *Journal of Neurology, Neurosurgery and Psychiatry, 50,* 1110–1116.

Damasio, A. R., & Damasio, H. (1983). The anatomic basis of pure alexia. *Neurology, 33,* 1573–1583.

Damasio, H., & Damasio, A. R. (1989). *Lesion analysis in neuropsychology.* New York: Oxford University Press.

Damasio, H., & Frank, R. (1992). Three-dimensional in vivo mapping of brain lesions in humans. *Archives of Neurology, 49,* 137–143.

Dejerine, J. (1891). Sur un cas de cecite verbale avec agraphie, suivi d'autopsie. *C. R. Societé du Biologíe, 43,* 197–201.

Dejerine, J. (1892). Contributions a l'etude anatomopathologique et clinique des differentes varietes de cecite verbale. *Memoires de la Societé Biologique, 44,* 61–90.

Dejerine, J., & Dejerine-Klumpke, A. (1895). *Anatomie des centres nerveux* (vol. 1). Paris: Rueff et Cie.

Dejerine, J., & Dejerine-Klumpke, A. (1901). *Anatomie des centres nerveux* (vol. 2). Paris: Rueff et Cie.

DeRenzi, E., Zambolin, A., & Crisi, G. (1987). The pattern of neuropsychological impairment associated with left posterior cerebral artery infarcts. *Brain, 110,* 1099–1116.

Derouesne, J., & Beauvois, M. F. (1985). The "phonemic" stage in the nonlexical reading process: Evidence from a case of phonological alexia. In K. E. Patterson, J. C. Marshall, & M. Coltheart (Eds.), *Surface dyslexia.* London: Erlbaum Associates.

Ellis, A. W., Flude, B., & Young, A. W. (1987). Neglect dyslexia and the early visual processing of letters in words and nonwords. *Cognitive Neuropsychology, 4,* 439–464.

Ellis, A. W., & Young, A. W. (1987). Human cognitive neuropsychology. Hillsdale, New Jersey: Elrbaum Associates.

Estes, W. K. (1975). The locus of inferential and perceptual processes in letter identification. *Journal of Experimental Psychology, 1,* 122–145.

Farah, M. J. (1991). Patterns of co-occurrence among associative agnosia: Implications for visual object representation. *Cognitive Neuropsychology, 8(1),* 1–20.

Farah, M. J., & Wallace, M. (1991). Pure alexia as a visual impairment: A reconsideration. *Cognitive Neuropsychology, 8,* 313–334.

Fox, P. T., Mintun, M. A., Raichle, M. E., Miezin, F. M., Allman, J. M., & VanEssen, D. C. (1986). Mapping human visual cortex with positron emission tomography. *Nature (London), 323*, 806–809.

Friedman, R. B., & Albert, M. L. (1985). Alexia. In K. M. Heilman & E. Valenstein (Eds.), *Clinical neuropsychology* (pp. 49–73). New York: Oxford University Press.

Friedman, R. B., & Alexander, M. P. (1984). Pictures, images, and pure alexia: A case study. *Cognitive Neuropsychology, 1*, 9–23.

Friedman, R., Ween, J. E., & Albert, M. L. (1992). Alexia. In K. M. Heilman & E. Valenstein (Eds.), *Clinical neuropsychology.* New York: Oxford University Press.

Funnell, E. (1983). Phonological processes in reading: New evidence from acquired dyslexia. *Biopsychology, 74*, 159–180.

Galaburda, A. M., Aboitiz, F., Rosen, G. D., & Sherman, G. F. (1986). Histological asymmetry in the primary visual cortex of the rat: Implications for mechanisms of cerebral asymmetry. *Cortex, 22*, 151–160.

Gazzaniga, M., & Hillyard, S. A. (1971). Language and speech capacity of the right hemisphere. *Neuropsychologia, 9*, 273–280.

Geiger, G., & Lettvin, J. Y. (1987). Peripheral vision in persons with dyslexia. *New England Journal of Medicine, 316*, 1238–1243.

Geschwind, N. (1965). Disconnexion syndromes in animals and man. *Brain, 88*, 17–294.

Geschwind, N., & Fusillo, M. (1966). Color-naming defects in association with alexia. *Archives of Neurology, 15*, 137–146.

Gloning, I., Gloning, K., & Hoff, H. (1968). *Neuropsychological symptoms and syndromes in lesions of the occipital lobe and the adjacent areas.* Paris: Gauthier-Villars.

Glosser, G., & Friedman, R. B. (1990). The continuum of deep/phonological dyslexia. *Cortex, 26*, 343–359.

Goodale, M. A., & Milner, A. D. (1992). Separate visual pathways for perception and action. *Trends in Neuroscience, 15(1)*, 20–25.

Greenblatt, S. H. (1973). Alexia without agraphia or hemianopsia: Anatomical analysis of an autopsied case. *Brain, 96*, 307–316.

Greenblatt, S. H. (1976). Subangular alexia without agraphia or hemianopsia. *Brain and Language, 3*, 229–245.

Greenblatt, S. H. (1977). Neurosurgery and the anatomy of reading: A practical view. *Neurosurgery, 1(1)*, 6–15.

Greenblatt, S. H. (1983). Localization of lesions in alexia. In A. Kertesz (Ed.), *Localization in neuropsychology* (pp. 323–356). New York: Academic Press.

Greenblatt, S. H. (1990). Left occipital lobectomy and the preangular anatomy of reading. *Brain and Language, 38*, 576–595.

Hecaen, H., & Albert, M. L. (1978). *Human neuropsychology.* New York: Wiley.

Hecaen, H., De Ajuriaguerra, J., & David, M. (1952). Les deficits fonctionnels apres lobectomie occipitale. *Monatsshrift fur Psychiatrie und Neurologie, 123*, 239–291.

Heilman, K. M., Rothi, L., Campanella, D., & Wolfson, S. (1979). Wernicke's and global aphasia without alexia. *Archives of Neurology, 36*, 129–133.

Heilman, K. M., Watson, R. T., & Valenstein, E. (1985). Neglect and related disorders. In K. M. Heilman & E. Valenstein (Eds.), *Clinical neuropsychology* (pp. 243–293). New York: Oxford University Press.

Hellman, R. S., & Tikofsky, R. S. (1990). An overview of the contribution of regional CBF studies in cerebrovascular disease. *Seminars in Nuclear Medicine, 20*, 303–324.

Henderson, V. W. (1984). Jules Dejerine and the third alexia. *Archives of Neurology, 41*, 430–432.

Henderson, V. W. (1986). Anatomy of posterior pathways in reading: A reassessment. *Brain and Language, 29,* 119–133.

Henderson, V. W., Friedman, R. B., Teng, E. L., & Weiner, J. M. (1985). Left hemisphere pathways in reading: Inferences from pure alexia without hemianopia. *Neurology, 35,* 962–968.

Heyman, S., Goldstein, H. A., Crowley, W., & Treves, S. (1980). The scintigraphic evaluation of hip pain in children. *Clinical Nuclear Medicine, 5,* 109–115.

Hier, D. B., & Mohr, J. P. (1977). Incongruous oral and written naming: Evidence for a subdivision of the syndrome of Wernick's aphasia. *Brain and Language, 4,* 115–126.

Hier, D. B., Mondlock, V., & Caplan, L. (1983). Behavioural abnormalities after right hemisphere stroke. *Neurology, 33,* 337–344.

Hillis, A. E., & Caramazza, A. (1990). The effects of attentional deficits in reading and spelling. In A. Caramazza (Ed.), *Cognitive neuropsychology and neurolinguistics* (pp. 211–276). Hillsdale, New Jersey: Erlbaum Associates.

Hinton, G. E., & Shallice, T. (1991). Lesioning an attractor network: Investigations of acquired dyslexia. *Psychological Review, 98(1),* 74–95.

Hirose, G., Kin, T., & Murakami, E. (1977). Alexia without agraphia associated with right occipital lesion. *Journal of Neurology, Neurosurgery, and Psychiatry, 40,* 225–227.

Holmes, G. (1944). The organization of the visual cortex in man. *Royal Society of London Proceedings, 132(Series B),* 348–361.

Howard, D. (1991). Letter-by-letter readers: Evidence for parallel processing. In G. W. Humphries & D. Besner (Eds.), *Basic processes in reading: Visual word recognition.* London: Erlbaum Associates.

Howard, D., Patterson, K., Wise, R., Brown, W. D., Friston, K., Weiller, C., & Frackowiak, R. (1992). The cortical localisation of the lexicons: PET evidence. *Brain, 115,* 1769–1782.

Hutner, N., & Liederman, J. (1991). Right hemisphere participation in reading. *Brain and Language, 41,* 475–495.

Johansson, T., & Fahlgren, H. (1979). Alexia without agraphia: Lateral and medial infarction of left occipital lobe. *Neurology, 29,* 390–393.

Kawahata, N., & Nagata, K. (1988). Alexia with agraphia due to the left posterior inferior temporal lobe lesion—Neuropsychological analysis and its pathogenetic mechanisms. *Brain and Language, 33,* 296–310.

Kertesz, A. (1983). *Localization in neuropsychology.* New York: Academic Press.

Kertesz, A., Polk, M., Black, S. E., & Howell, J. (1992). Anatomical asymmetries and functional laterality. *Brain, 115,* 589–605.

Kinsbourne, M. (1971). The minor hemisphere as a source of aphasic speech. *Transactions of the American Neurological Association, 96,* 141–145.

Kinsbourne, M., & Warrington, E. K. (1962). A variety of reading disability associated with right hemisphere lesions. *Journal of Neurology, Neurosurgery, and Psychiatry, 25,* 339–344.

Kirshner, H. S., & Webb, W. G. (1982). Word and letter reading and the mechanism of the third alexia. *Archives of Neurology, 39,* 84–87.

Kurachi, M., Yamaguchi, N., Inasaka, T., & Torii, H. (1979). Recovery from alexia without agraphia: Report of an autopsy. *Cortex, 15,* 297–312.

LaBerge, D. L., & J. Samuels (1974). Toward a theory of automatic word processing in reading. *Cognitive Psychology, 6,* 293–323.

Laine, M., Niemi, P., Niemi, J., & Koivuselka-Sallinen, P. (1990). Semantic errors in a deep dyslexic. *Brain and Language, 38,* 207–214.

Landis, T., Regard, M., Graves, R., & Goodglass, H. (1983). Semantic paralexia: A release of right hemispheric function from left hemispheric control?. *Neuropsychologia, 21(4),* 359–364.

LeMay, M. (1977). Asymmetries of the skull and handedness. *Journal of Neurological Sciences, 32*, 243–253.

Levine, D. (1982). Visual agnosia in monkey and in man. In M. A. Goodale & R. J. W. Mansfield (Ed.), *Analysis of Visual Behaviour*. Cambridge, Massachusetts: MIT Press.

Marin, O. S. M. (1980). CAT scans of five deep dyslexic patients. In K. Patterson, J. C. Marshall, & M. Coltheart (Eds.), *Deep dyslexia* (pp. 452–453). London: Routledge and Kegan Paul.

Marin, O. S. M., Saffran, E. M., & Schwartz, M. F. (1976). Dissociations of language in aphasia: Implications for normal function. *Annals of the New York Academy of Sciences, 280*, 868–884.

Marshall, J. C., & Newcombe, F. (1966). Syntactic and semantic errors in paralexia. *Neuropsychologia, 4*, 169–176.

Marshall, J. C., & Newcombe, F. (1973). Patterns of paralexia: A psycholinguistic approach. *Journal of Psycholinguistic Research, 2*, 175–199.

Marshall, J. C., & Patterson, K. E. (1983). Semantic paralexia and the wrong hemisphere: A note on Landis, Regard, Graves, and Goodglass. *Neuropsychologia, 21(4)*, 425–427.

McCarthy, R., & Warrington, E. K. (1990). *Cognitive neuropsychology*. New York: Academic Press.

Mesulam, M. M. (1990). Large scale neurocognitive networks and distributed processing for attention, language and memory. *Annals of Neurology, 28*, 597–613.

Mishkin, M., Ungerleider, L. G., & Macko, K. A. (1983). Object vision and spatial vision: Two cortical pathways. *Trends in Neuroscience, 6*, 414–417.

Morton, J. (1979). Word recognition. In J. Morton & J. C. Marshall (Eds.), *Psycholinguistics* (Series 2). London: Elek.

Morton, J., & Patterson, K. (1980). A new attempt at an interpretation, or, and attempt at a new interpretation. In M. Coltheart, K. E. Patterson, & J. C. Marshall (Eds.), *Deep dyslexia* (pp. 91–118). London: Routledge and Kegan Paul.

Mozer, M. C. (1983). Letter migration in word perception. *Journal of Experimental Psychology: Human Perception and Performance, 9*, 531–546.

Mozer, M. C., & Behrmann, M. (1990). On the interaction of selective attention and lexical knowledge: A connectionist account of neglect dyslexia. *Journal of Cognitive Neuroscience, 2(2)*, 96–123.

Mozer, M. C., & Behrmann, M. (1992). Reading with attentional impairments: A brain-damaged model of neglect and attentional dyslexias. In R. Reilly (Ed.), *Connectionism and Language*. London: Erlbaum Associates.

Nolan, K., & Caramazza, A. (1982). Modality-independent impairments in word processing in a deep dyslexic patient. *Brain and Language, 20*, 305–328.

Ojemann, G. A. (1983). Brain organization for language from the perspective of electrical stimulation mapping. *The Behavioural and Brain Sciences, 6*, 189–230.

Pandya, D. N., & Seltzer, B. (1986). The topography of commissural fibers. In M. Plito, F. Lepore, & H. H. Jasper (Eds.), *Two hemispheres—One brain: Functions of the corpus callosum* (pp. 47–73). Boston: Liss.

Papanicolaou, A. D., Bartlett, D. M., Deutsch, G., Lavin, H. S., & Eisenberg, H. M. (1988). Evidence for right-hemisphere involvement in recovery from aphasia. *Archives of Neurology, 45*, 1025–1029.

Patterson, K. (1990). Alexia and neural nets. *Japanese Journal of Neuropsychology, 6*, 90–99.

Patterson, K., & Besner, D. (1984a). Is the right hemisphere literate? *Cognitive Neuropsychology, 1*, 315–341.

Patterson, K., & Besner, D. (1984b). Reading from the left: A reply to Rabinowicz and Moscovitch and to Zaidel and Schweiger. *Cognitive Neuropsychology, 1*, 365–380.

Patterson, K., & Kay, J. (1982). Letter-by-letter reading: Psychological descriptions of a neurological syndrome. *Quarterly Journal of Experimental Psychology, 34A*, 411–441.

Patterson, K., Vargha-Khadem, F., & Polkey, C. E. (1989). Reading with one hemisphere. *Brain, 112*, 39–63.

Patterson, K., & Wilson, B. (1990). A rose is a nose: A deficit in initial letter identification. *Cognitive Neuropsychology, 13*, 447–478.

Patterson, K. E., & Hodges, J. (1992). Deterioration of word meaning: Implications for reading. *Neuropsychologia, 30*, 1025–1040.

Patterson, K. E., & Morton, J. (1985). From orthograph to phonology: An attempt at an old interpretation. In K. E. Patterson, J. C. Marshall, & M. Coltheart (Eds.), *Surface dyslexia* (pp. 335–359). London: Erlbaum Associates.

Patterson, K. E., Marshall, J. C., & Coltheart, M. (1985). Surface dyslexia: Neuropsychological and cognitive studies of phonological reading. London: Erlbaum Associates.

Patterson, K. E., Seidenberg, M. S., & McClelland, J. L. (1990). Connections and disconnections: Acquired dyslexia in a computational model of reading processes. In R. G. M. Morris (Ed.), *Parallel distributed processing: Implications for psychology and neuroscience* (pp. 131–181). London: Oxford University Press.

Perani, D., Vittorio, D. P., Lucignani, G., Gilardi, M. C., Pantano, P., Rossetti, C., Pozzilli, C., Gerundini, P., Fazio, F., & Lenzi, G. L. (1988). Remote effects of subcortical cerebrovascular lesions. A SPECT Cerebral Perfusion Study. *Journal of Cerebral Blood Flow and Metabolism, 8*, 560–567.

Petersen, S. E., Fox, P. T., Posner, M. I., Mintun, M., & Raichle, M. E. (1988). Positron emission tomographic studies of the cortical anatomy of single-word processing. *Nature (London), 331*, 585–589.

Petersen, S. E., Fox, P. T., Snyder, A. Z., & Raichle, M. E. (1990). Activation of extrastriate and frontal cortical areas by visual words and word-like stimuli. *Science, 249*, 1041–1044.

Plaut, D. C. (1991). Connectionist neuropsychology: The breakdown and recovery of behaviour in lesioned attractor networks. *Technical Report CMU-CS-91-185*, School of Computer Science, Carnegie Mellon University: Pittsburgh.

Plaut, D. C., & Shallice, T. (1991). Effects of abstractness in a connectionist model of deep dyslexia. In *Proceedings of the 13th Annual Conference of the Cognitive Science Society* (pp. 73–78). Hillsdale, New Jersey: Erlbaum.

Plaut, D. C., & Shallice, T. (1992). Analysis and further development of a connectionist model of deep dyslexia. *Cognitive Neuropsychology, 10(5)*, 377–507.

Posner, M. I., Petersen, S. E., Fox, P. T., & Raichle, M. E. (1988). Localization of cognitive operations in the human brain. *Science, 240*, 1627–1631.

Pritchard, J. W., & Brass, L. M. (1992). New anatomical and functional imaging methods. *Annals of Neurology, 32(3)*, 395–400.

Rabinowicz, B., & Moscovitch, M. (1984). Right hemisphere literacy: A critique of some recent approaches. *Cognitive Neuropsychology, 1*, 343–350.

Raichle, M. E. (1989). Developing a functional anatomy of the human brain with positron emission tomography. *Current neurology, 9*, 161–178.

Rayner, K., Murphy, L. A., Henderson, J., & Pollatsek, A. (1989). Selective attentional dyslexia. *Cognitive Neuropsychology, 6(4)*, 357–378.

Reuter-Lorenz, P., & Brunn, J. (1990). A prelexical basis for letter-by-letter reading: A case study. *Cognitive Neuropsychology, 7*, 1–20.

Riddoch, M. J., Humphreys, G. W., Cleton, P., & Fery, P. (1990). Interaction of attentional and lexical processes in neglect dyslexia. *Cognitive Neuropsychology, 7(5/6)*, 479–517.

Rockland, K. S., & Pandya, D. N. (1981). Cortical connections of the occipital lobe in the Rhesus monkey: Interconnections between areas 17, 18, 19 and the superior temporal sulcus. *Brain Research, 212,* 249–270.

Roeltgen, D. (1983). Proposed anatomic substrates for phonological and surface dyslexia. *Academy of Aphasia (Abstract).*

Roeltgen, D. P. (1987). Loss of deep dyslexic reading ability from a second left-hemispheric lesion. *Archives of Neurology, 44,* 346–348.

Rothi, L. J., McFarling, D., & Heilman, K. M. (1982). Conduction aphasia, syntactic alexia, and the anatomy of syntactic comprehension. *Archives of Neurology, 39,* 272–275.

Saffran, E., Bogyo, L. C., Schwartz, M. F., & Marin, O. S. M. (1980). Does deep dyslexia reflect right hemisphere reading? In M. Coltheart, K. Patterson, & J. C. Marshall (eds.), *Deep dyslexia* (pp. 381–406). London: Routledge and Kegan Paul.

Sasanuma, S. (1980). Acquired dyslexia in Japanese: Clinical features and underlying mechanisms. In M. Coltheart, K. Patterson, & J. C. Marshall (Eds.), *Deep dyslexia* (pp. 48–90). London: Routledge and Kegan Paul.

Sasanuma, S., Itoh, M., Mori, K., & Kobayashi, Y. (1977). Tachistoscopic recognition of Kana and Kanji words. *Neuropsychologia, 15,* 547–553.

Schneiderman, E., & Saddy, J. D. (1988). A linguistic deficit resulting from right hemisphere damage. *Brain and Language, 34,* 38–53.

Searleman, A. (1983). Language capabilities of the right hemisphere. In A. W. Young (Ed.), *Functions of the right cerebral hemisphere* (pp. 87–111). London: Academic Press.

Seidenberg, M., & McClelland, J. L. (1989). A distributed, developmental model of word recognition and naming. *Psychological Review, 96,* 523–568.

Seymour, P. H. K., & MacGregor, C. J. (1984). Developmental dyslexia: A cognitive experimental analysis of phonological, morphemic, and visual impairments. *Cognitive Neuropsychology, 1,* 43–82.

Shallice, T., & McGill, J. (1978). The origins of mixed errors. In J. Requin (Ed.), *Attention and Performance.* Hillsdale, New Jersey: Erlbaum Associates.

Shallice, T. (1988). From neuropsychology to mental structure. Cambridge: Cambridge University Press.

Shallice, T., & Saffran, E. (1986). Lexical processing in the absence of explicit work identification: Evidence from a letter-by-letter reader. *Cognitive Neuropsychology, 3,* 429–458.

Shallice, T., & Warrington, E. K. (1977). The possible role of selective attention in acquired dyslexia. *Neuropsychologia, 15,* 31–41.

Shallice, T., & Warrington, E. K. (1980). Single and multiple component central dyslexic syndromes. In M. Coltheart, K. E. Patterson, & J. C. Marshall (Eds.), *Deep dyslexia* (pp. 119–145). London: Routledge and Kegan Paul.

Shallice, T., Warrington, E. K., & McCarthy, R. (1983). Reading without semantics. *Quarterly Journal of Experimental Psychology, 35A,* 111–138.

Silver, F. L., Chawluk, J. B., Bosley, T. M., Rosen, M., Dann, R., Sergott, R. C., Alavi, A., & Reivich, M. (1988). Resolving metabolic abnormalities in a case of pure alexia. *Neurology, 38,* 730–735.

Small, S. L. (1991). Focal and diffuse lesions in cognitive models. In *Proceedings of the 13th Annual Conference of the Cognitive Science Society* (pp. 85–90). Hillsdale, New Jersey: Erlbaum.

Soma, Y., Sugishita, M., Kitamura, K., Maruyama, S., & Imanaga, H. (1989). Lexical agraphia in the Japanese language. *Brain, 112,* 1549–1561.

Starr, M. A. (1889). The pathology of sensory aphasia, with analysis of fifty cases in which Broca's centre was not diseased. *Brain, 12,* 82–101.

Steinmetz, H., Furst, G., & Freund, H. J. (1990). Variation of perisylvian and calcarine anatomic landmarks within stereotaxic proportional coordinates. *American Journal of Neuroradiology, 11*, 1123–1130.

Steinmetz, H., & Seitz, R. J. (1991). Functional anatomy of language processing: Neuroimaging and the problem of individual variability. *Neuropsychologia, 29*, 1149–1161.

Stensaas, S. S., Eddington, D. K., & Dobelle, W. H. (1974). The topography and variability of the primary visual cortex in man. *Neurosurgery, 40*, 747–755.

Sugishita, M., Iwata, M., Toyokura, Y., Yoshioka, M., & Yamada, R. (1978). Reading of ideograms and phonograms in Japanese patients after partial commissurotomy. *Neuropsychologia, 16*, 417–426.

Talairach, J., & Tournoux, P. (1988). *Co-planar stereotaxic atlas of the human brain.* New York: Thieme Medical Publishers.

Tusa, R. J., & Ungerleider, L. G. (1985). The inferior longitudinal fasciculus: A reexamination in humans and monkeys. *Annals of Neurology, 18*, 583–591.

Van Essen, D. C., & Maunsell, J. H. R. (1983). Hierarchical organization and functional streams in the visual cortex. *Trends in Neuroscience, 370*–375.

Vanier, M., & Caplan, D. (1985). CT scan correlates of surface dyslexia. In J. C. Marshall, M. Coltheart, & K. E. Patterson (Eds.), *Surface dyslexia: Neuropsychological and cognitive studies of phonological reading* (pp. 511–525). Hillsdale, New Jersey: Erlbaum Associates.

Vanier, M., Lecours, Ethier, A. R., Habib, R., Poncet, M., Millette, P. C., & Salomon, G. (1985). Proportional localization system for anatomical interpretation of computed cerebral tomograms. *Journal of Computer-Assisted Tomography, 9*, 715–724.

Varney, N. R. (1984). Alexia for ideograms: Implications for Kanji alexia. *Cortex, 20*, 535–542.

Vincent, F. M., Sadowsky, C. H., Saunders, R. L., & Reeves, A. G. (1977). Alexia without agraphia, hemianopia, or color-naming defect: A disconnection syndrome. *Neurology, 27*, 689–691.

Warrington, E. K. (1975). The selective impairment of semantic memory. *Quarterly Journal of Experimental Psychology, 27*, 635–657.

Warrington, E. K., & T. Shallice (1980). Word-form dyslexia. *Brain, 103*, 99–112.

Willemsen, R., van Dongen, J. M., Ginns, E. I., Sips, H. J., Schram, A. W., Tager, J. M., Barranger, J. A., & Reuser, A. J. J. (1987). Ultrastructural localization of glucocerebrosidase in cultured Gaucher's disease fibroblasts by immunocytochemistry. *Journal of Neurology, 234*, 44–51.

Young, A. W., Newcombe, F., & Ellis, A. (1991). Different disorders contribute to neglect dyslexia. *Cognitive Neuropsychology, 8*, 177–192.

Zaidel, E. (1982). Reading by the disconnected right hemisphere: An aphasiological perspective. In Y. Zotterman (Ed.), *Dyslexia: Neuronal cognitive and linguistic aspects.* Oxford: Pergamon Press.

Zaidel, E., & Peters, A. M. (1981). Phonological encoding and ideographic reading by the disconnected right hemisphere: Two case studies. *Brain and Language, 14*, 205–234.

Zaidel, E., & Schweiger, A. (1984). On wrong hypotheses about the right hemisphere: Commentary on K. Patterson and D. Besner, "Is the right hemisphere literate?". *Cognitive Neuropsychology, 1*, 351–364.

Chapter 12

Localization of Lesions in Agraphia

David P. Roeltgen

I. INTRODUCTION

Although Benedick (1865) was the first to apply the term "agraphia" to disorders of writing, Ogle (1867) was one of the first to address the issues of localization in agraphia. He found that aphasia and agraphia usually occurred together but occasionally were separable. Therefore he concluded that distinct cerebral centers existed for writing and speaking. Historically, discussions of the cerebral localization of agraphia primarily centered on the question of whether a "center" for writing exists. In addressing the issue of a center for writing, Nielson (1946) proposed that, although agraphia without associated neuropsychological signs (pure or isolated agraphia) was rare, when it occurred it was caused by a lesion of the frontal writing center (Exner's area) or the angular gyrus. Nielson theorized that Exner's area, the foot of the second frontal convolution, worked in association with the angular gyrus and the motor area to produce writing. He also proposed that information from the angular gyrus was carried to Exner's area by fibers that passed close to Broca's speech area. According to Nielson, these functional and anatomical connections accounted for the frequent association of agraphia and aphasia.

In recent decades, two distinct approaches to the study of acquired agraphia have evolved. One is an extension and elaboration of previous approaches such as the one used by Nielson (1946). The classification schemes developed from this approach have been labeled "neurological classifications" (Roeltgen & Heilman, 1985). Other neurological classifications of agraphia have included those of Leischner (1969), Hecaen and Albert (1978), Benson (1979), Kaplan and Goodglass (1981), and Benson and Cummings (1985). These classifications usually include five types of agraphia (Table 1).

Table 1 Neurological Classification of Agraphia

Pure agraphia
Aphasic agraphia
Agraphia with alexia
Apraxic agraphia
Spatial agraphia

In contrast to the neurological classifications are the classifications that can be described as neuropsychological (Roeltgen & Heilman, 1985), psychological, or linguistic (Roeltgen & Rapcsak, 1993) (Table 2). Figure 1 is a diagrammatic representation of a neuropsychological model developed from a neuropsychologically based analysis of agraphia (Roeltgen, 1993; Roeltgen & Heilman, 1985). Although certain agraphias, such as apraxic and spatial, as described by Roeltgen and colleagues (Roeltgen, 1993; Roeltgen & Heilman, 1985; Roeltgen & Rapcsak, 1993) and others (Margolin, 1984), are similar to the agraphias defined within the neurological classification, others are not. The lack of similarity is most evident for those agraphias that relate to poor spelling and poor word choice (aphasic agraphias in the neurological classifications and linguistic agraphia in the psychological classifications).

The terms and definitions best used for classifying agraphia are uncertain, and controversy exists within the field. In addition, controversy exists over the extent to which clinical–pathological correlations and attempts at localization of agraphia can be made (Margolin, 1984). Despite this concern, multiple studies within the past decade (Alexander, Friedman, Loverso, & Fischer, 1990; Auerbach & Alexander, 1981; Iwata, 1984; Kawamura, Hirayama, Hasegawa, Takahashi, & Yamaura, 1987; Kertesz,

Table 2 Neuropsychological Classification of Agraphia

Linguistic agraphias
  Phonological
  Lexical
  Semantic
  Deep
  Graphemic buffer disturbance
Motor or peripheral agraphias
  Apraxic
  Spatial
  Allographic store disturbance

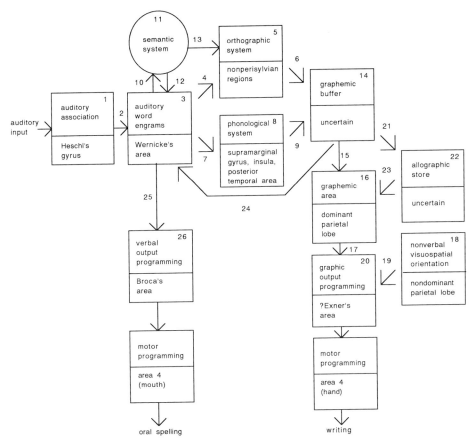

Figure 1 Diagrammatic representation of a neuropsychological model of writing and spelling. Reproduced with permission from Roeltgen (1993).

Latham, & McCabe, 1990; Laine & Marttila, 1981; Rapcsak, Arthur, & Rubens, 1988; Rapcsak & Rubens, 1990; Roeltgen, 1989a, 1991; Roeltgen & Heilman, 1983, 1984; Roeltgen, Rothi, & Heilman, 1986; Roeltgen, Sevush, & Heilman, 1983; Rosati & DeBastiani, 1981; Tanridag & Kirshner, 1985) have attempted clinical–pathological correlations. Given the controversies regarding classification of the agraphias, the concerns regarding the use of clinical–pathological correlation, and the large number of studies that have continued to analyze localization in agraphia, one important caveat must be made. Ideally, the lesion site (pathology) should act as an independent variable to support or refute the claim that a behavioral classification has significance. For example, if a consistent behavioral disturbance is associated with a consistent pathological site, such a correlation lends support to the usefulness of that behavioral classification.

II. NEUROLOGICAL CLASSIFICATION

A. Isolated or Pure Agraphia

Within the neurological classifications, the type of agraphia that has received the most interest over the past two decades is isolated or pure agraphia. Isolated or pure agraphia is a selective impairment of written communication (Rosati & DeBastiani, 1981). Patients with this disorder have agraphia in the absence of any other language disturbance. Note that the type of agraphia is not defined further and may include patients with disordered spelling, disordered handwriting, or both. Through the 1950s, the traditionally held position regarding the localization of pure agraphia was that the disorder resulted from lesions in the frontal lobe, specifically the lower end of the second frontal convolution known as Exner's area (Roeltgen, 1993; Rosati & DeBastiani, 1981). With the availability of neuro-imaging techniques, many more patients have been described with relatively pure agraphia and different lesion sites among the patients studied have been associated with this disturbance. With more frequent descriptions of this type of agraphia, multiple patterns of agraphia have been described. Based on studies such as these, researchers have concluded that a single center important for writing is unlikely. The case against a single site as the "center of writing" was emphasized by Chedru and Geschwind (1972), who described pure agraphia resulting from an acute confusional state. In those patients, the agraphia was characterized by poorly formed graphemes and inability to write on the line. When the letters were well formed, the patients produced spelling errors.

Although a single "center for writing" does not exist, reviewing the lesion sites associated with pure agraphia is still of interest.

1. Exner's area (second left frontal gyrus)

At least two studies have described relatively pure agraphia from lesions in, or close to, Exner's area. Vernea and Merory (1975) described a patient who made substitutions and rare omissions and showed striking improvement with slavish copying. Aimard and colleagues (Aimard, Devick, Lebel, Trouillas, & Boisson, 1975) described a patient who was interpreted to have a kinesthetic disorder, who produced well-formed letters but had impairment both in spontaneous writing and in copying. We have evaluated one patient with a lesion in this region (D. P. Roeltgen, unpublished study) who also had well-formed letters and spelling errors, and had a pattern of performance consistent with a disruption of the graphemic buffer (see Section III,A,6).

2. Posterior perisylvian region

Rosati and DeBastiani (1981) described a patient who had a lesion of the posterior perisylvian region that primarily involved the temporal lobe. Their patient produced misspellings and occasional extra strokes when forming graphemes. Spontaneous writing was poorer than copying. In a prospective study of agraphia, described in more detail later in the chapter, Roeltgen (1989a, 1991) studied a pure agraphic patient with a lesion in the posterior perisylvian area. The agraphia was equally prominent in oral spelling and in writing. Graphemes were well formed and, although an improvement in spelling was seen when the patient copied, no change in graphemic form when copying was detected. Linguistic analysis of this patient's agraphia (see Section III) indicated that he had an agraphia that combined elements of phonological agraphia and lexical agraphia (mixed phonological–lexical agraphia) (Fig. 2).

3. Superior parietal lobe

Basso and colleagues (Basso, Taborielli, & Vignolo, 1978) described two patients with pure agraphia. One had a lesion of the superior parietal lobule and the other had a lesion of the superior aspect of the postcentral gyrus. Both patients produced substitutions and omissions and improved with copying. In contrast, Kinsbourne and Rosenfeld (1974) and Auerbach and Alexander (1981) described two patients with lesions of the superior parietal lobe who produced poorly formed graphemes. The patient of Kinsbourne and Rosenfeld had relatively preserved oral spelling. The patient of Auerbach and Alexander, in addition to poorly formed graphemes, showed omissions, repetitions, and substitutions and showed these impairments in spontaneous writing and copying. This patient also showed better oral spelling than written spelling.

4. Left occipital lobe

Kapur and Lawton (1983) described a patient with a tumor in the occipital lobe. Their patient had preserved oral spelling but she was unable to write words. She produced well-formed but incorrect letters. The authors suggested that this impairment might be a form of motor memory deficit.

5. Posterior insula and posterior putamen

We (Roeltgen, 1989a; 1991) have evaluated a patient who had a lesion in this region (Fig. 3A). This patient had equal impairment in oral spelling

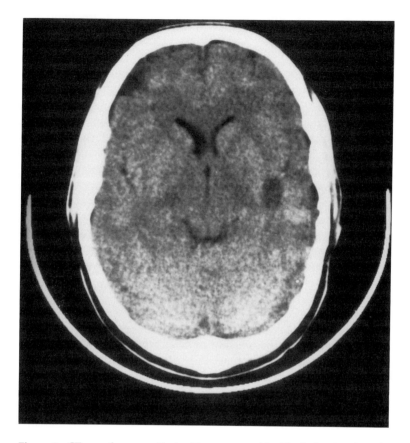

Figure 2 CT scan from a patient with pure agraphia. The lesion involves the posterior insula.

and writing and had relatively well-formed graphemes. His pattern of performance could be classified linguistically as phonological agraphia (see Section III,A,1).

6. Basal ganglia

Laine and Marttila (1981) described a patient with pure agraphia from a lesion of the caudate and the anterior limb of the internal capsule. Their patient made spelling errors on oral and written spelling. Written letters were well formed and performance improved with copying. Roeltgen evaluated two patients with lesions of the basal ganglia and internal capsule who had pure agraphia (Roeltgen, 1989a,1991). One of these patients,

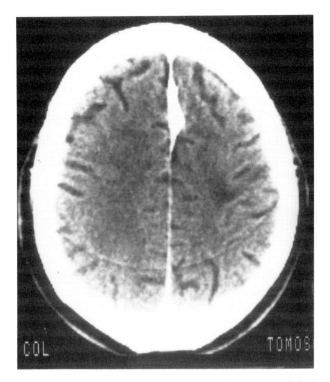

Figure 3   CT scan of a patient with phonological agraphia and alexia.

with a lesion of the mid and posterior basal ganglia, was unable to write but made frequent spelling errors when spelling aloud, and could be classified as having phonological agraphia (see Section III,A,1). The second had a lesion of the anterior basal ganglia and the internal capsule. That subject spelled aloud and wrote equally well. He produced well-formed graphemes and had a linguistic agraphia consistent with a disorder that has been called mixed phonological and lexical agraphia (see Sections III,A, 1 and 2).

7. Other subcortical structures

In the same prospective study as mentioned in the previous paragraphs, Roeltgen (1989a; 1991) studied one patient with a closed head injury who had a focal lesion of the left thalamus and showed pure agraphia. That patient had lexical agraphia (see Section III,A,2). Croisile and colleagues (Croisile, Laurent, Michel, & Trillet, 1990) described a patient with pure apraxic agraphia (see Section II,C,1), from a left centrum semiovale hemorrhage. Other studies have described significant agraphias from subcortical lesions (Kertesz et al., 1990; Tanridag & Kirshner, 1985), but most of the patients in those studies also had aphasia.

8. Comments

Pure agraphia has been associated with multiple lesion sites in the left hemisphere. Some lesion sites have been associated with multiple types of agraphia. In general, some pure agraphias can be classified as motor agraphias with poorly formed graphemes. Others can be classified as linguistic agraphias with well-formed graphemes and frequent misspellings. In addition, the type of linguistic agraphia in these patients also varies to a great extent. As will be evident in the following sections, strong evidence suggests that the presence or absence of aphasia or alexia is less important than the pattern of agraphia produced.

B. Agraphia with Aphasia and Agraphia with Alexia

1. Introduction

Although the review of lesion localization in pure agraphia demonstrates that behavioral analysis of the type of agraphia is important, a long and abundant literature on agraphia with aphasia and agraphia with alexia has addressed localization (Benson, 1979; Benson & Cummings, 1985; Kaplan & Goodglass, 1981; Marcie & Hecaen, 1979). Therefore, similar to the approach used for pure agraphia, this review summarizes both the agraphic behavior and the lesion locations.

2. Agraphia with aphasia

In general, descriptions of agraphia frequently indicate that the pattern of agraphia is similar to the oral output (Benson & Cummings, 1985), although discrepancies have been described (Roeltgen, 1993; Roeltgen & Heilman, 1985; Roeltgen et al., 1983).

a. Agraphia with Broca's aphasia. Some authors (Kaplan & Goodglass, 1981; Marcie & Hecaen, 1979) have described two distinct subtypes of agraphia in patients with Broca's aphasia. One subtype is characterized by difficulty in grapheme production and the second by agrammatism. Patients with the former make poorly formed graphemes and have difficulty spelling. Patients with the latter make well-formed graphemes but produce agrammatic sentence structure. In contrast, Benson and Cummings (1985) suggest that most individuals with nonfluent aphasia, including Broca's aphasia, have a characteristic writing disturbance that parallels the speech disturbance, with the exception that the agrammatism is more pronounced when writing than when speaking. According to Benson and Cummings, oral spelling and written spelling are both affected in these patients and their graphemic form is poor, even after the motor deficit that is usually present in these patients is taken into account. Interestingly, these patients have been described as having more difficulty spelling nonwords than real words (Marcie & Hecaen, 1979). This last result is similar to the results of linguistic analysis performed on patients with phonological agraphia, some of whom have had Broca's aphasia (Roeltgen et al., 1983) (see Section III,A,1). The lesions in patients who have agraphia and Broca's aphasia are similar to those found in patients with Broca's aphasia. Benson and Cummings conclude that patients who have agraphia with Broca's aphasia frequently have lesions that extend into subcortical structures, and consequently have an impairment of movement secondary to damage of these structures. This movement impairment produces impairment in the motor aspects of writing. Therefore, conclusions regarding writing behavior, especially the graphemic form which will be influenced significantly by motor impairment, must be made with caution in these patients.

b. Agraphia with conduction aphasia. This disorder was described by Marcie and Hecaen (1979) as an agraphia with correctly formed graphemes and poorly spelled words. In addition, real words are spelled better than nonwords. In the prospective study by Roeltgen (1989a; 1991), four subjects had agraphia with conduction aphasia. All had severely impaired spelling ability in both the oral and written modalities. Variability was seen among the subjects with respect to how well they produced graphemes. The lesions in these subjects were varied. One lesion was subcortical in the mid-basal ganglia and the anterior limb of the internal capsule. One lesion was in the posterior temporal region. One lesion was in the anterior parietal lobe and the mid-insula. Finally, one patient had two lesions, one in the mid-insula extending outward to the supramarginal gyrus and a second lesion in the posterior angular gyrus.

c. Agraphia with Wernicke's aphasia. These patients are described as having normal output with well-formed graphemes and a decreased production of substantive words and incorrect spellings (Benson & Cummings, 1985). Patients with this type of agraphia are typically thought to have lesions in the regions that are associated with Wernicke's aphasia.

d. Agraphia with transcortical sensory aphasia. Patients with this disorder have been described only rarely (Grossfeld & Clark, 1983). The agraphia in these patients is said to be similar to that seen in patients with

Wernicke's aphasia (Benson & Cummings, 1985). The lesion sites are typically in regions associated with transcortical sensory aphasia.

e. Agraphia with mixed transcortical aphasia. According to Benson and Cummings (1985), all patients with this type of aphasia will be agraphic, and typically have writing that is nonfluent and improves with copying. Roeltgen and associates (1986) described the syndrome of semantic agraphia (see Section III,A,3) in patients with mixed transcortical aphasia.

3. Agraphia with alexia

Agraphia with alexia often has been called parietal agraphia because these two symptoms, in the absence of aphasia, usually occur together in patients with parietal lobe lesions (Kaplan & Goodglass, 1981). Benson and Cummings (1985) suggest that the agraphia in these patients is similar to that seen in patients with fluent aphasia, and that frequently these patients also have a mild anomic aphasia. In contrast, other authors (Kaplan & Goodglass, 1981; Marcie & Hecean, 1979) suggest that patients with agraphia and alexia produce poorly formed graphemes. Those agraphias with disordered grapheme production (motor agraphias) include apraxic agraphia, which is described later in this chapter.

We (Roeltgen, 1989a,1991) have evaluated two patients who had agraphia with alexia without aphasia. One had a lesion deep to the superior parietal lobe that included the postcentral gyrus and extended into the precentral gyrus (Fig. 3). His agraphia was limited to nonwords, which he spelled poorly when writing and when spelling orally (phonological agraphia; see Section III,A,1). He produced well-formed graphemes. His reading impairment also was limited to nonwords (phonological alexia; see Chapter 11). A more typical lesion of alexia and agraphia is presented in Fig. 4. This patient also had severe difficulty spelling oral and written nonwords. However, he also had severe difficulty spelling real words and is best described as having a global agraphia. When writing, he produced well-formed graphemes. When reading, he had severe difficulty with nonwords and moderate difficulty with real words.

The heterogeneity of lesion sites and written language behaviors in the last two patients illustrates some of the difficulty of clinical–pathological correlation of agraphia, especially when using the type of classification described in this section of the chapter. Because both patients have agraphia with alexia (without aphasia), they can be grouped together superficially. However, their agraphias differ and the locations of their lesions differ, strongly suggesting that they should not be classified as having the same disorder.

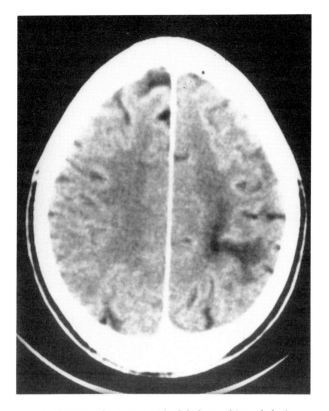

Figure 4   CT scan of a patient with global agraphia and alexia.

4. Summary

Although attempts to explain agraphia by its association with aphasia and alexia, or its occurrence as a pure syndrome, help explain some of the clinical and clinical–pathological results in patients with agraphia, inconsistencies are common. First, one subtype of agraphia may have the same behavioral disturbance as another type. For example, the agraphias produced by patients with Broca's aphasia with agraphia and characterized by disturbed grapheme production and by those with agraphia with alexia have similar descriptions (Kaplan & Goodglass; 1981; Marcie & Hecaen, 1979). Additionally, some patients with a specific agraphia, such as the agraphia with Broca's aphasia, may have writing and spelling disabilities that more closely resemble the agraphia of patients with

Wernicke's aphasia than the disabilities of patients with Broca's aphasia (Roeltgen et al., 1983). Consequently, lesion localization using the terminology described in this section has somewhat limited usefulness.

C. Peripheral Nonlanguage or Motor Agraphias

1. Apraxic agraphia

Apraxic agraphia is characterized by difficulty in forming graphemes when writing spontaneously and to dictation (Hecaen & Albert, 1978; Leischner, 1969; Marcie & Hecaen, 1979; Roeltgen, 1993). Although the written output usually improves when a copy is available, spelling may be disturbed as well. The localization of lesions in this syndrome has been reviewed by Alexander and colleagues (Alexander, Fisher, & Friedman, 1992). These investigators concluded that apraxic agraphia was a disorder of spatially learned movements and was associated with lesions of the superior parietal lobe. Apraxic agraphia also has been reported in patients with right hemispheric lesions (Roeltgen & Heilman, 1983), but these patients usually have had crossed aphasia, a language disorder induced by a right hemisphere lesion in a right-handed patient. Apraxic agraphia usually is associated with a more generalized limb apraxia, but occasionally can occur in patients without apraxia; this phenomenon has been termed "ideational agraphia" (Baxter & Warrington, 1986). The anatomical substrate for this syndrome appears to be similar to the anatomical substrate for apraxic agraphia with apraxia, the dominant parietal lobe.

Unilateral apraxic agraphia is a disorder in which the left hand produces unintelligible scrawls when the patient attempts to write. This syndrome usually is caused by a lesion of the corpus callosum (Bogen, 1969; Gersh & Damasio, 1981; Geschwind & Kaplan, 1962; Levy, Nebes, & Sperry, 1971; Liepmann & Maas, 1907; Rubens, Geschwind, Mahowald, & Mastri, 1977; Sugishita, Toyokura, Yoshioka, & Yamada, 1980; Watson & Heilman, 1983; Yamadori, Osumi, Ikeda, & Kanazawa, 1980). Some patients are able to type with their left hand (Watson & Heilman, 1983), whereas others cannot (Geschwind & Kaplan, 1962). Watson and Heilman (1983) suggested that the difference between their patient and that of Geschwind and Kaplan (1962) was that the genu of the corpus callosum was spared in Watson and Heilman's patient. In addition, Watson and Heilman have suggested that ideomotor praxis for writing and other motor tasks is transferred through the body of the corpus callosum, and lesions in this location lead to apraxic agraphia and apraxia. In contrast, they suggest that linguistic information is transferred through the posterior aspect of the corpus callosum, especially the splenium. Therefore, lesions involving the posterior callosum may result in a degree of aphasic agraphia or, using the term applied in this chapter, linguistic agraphia. This issue will be expanded later in this chapter in the discussion of localization of agraphia in Japanese patients.

2. Spatial agraphia

Patients with this type of agraphia typically produce duplicate strokes, have trouble writing in a horizontal line, write on only one side of the paper, and have intrusions of blank spaces between graphemes. This agraphia frequently is associated with the neglect syndrome (Benson, 1979; Ellis, Young, & Flude, 1987; Hecaen & Albert, 1978; Marcie & Hecaen, 1979). Ellis and colleagues (1987) have suggested that this syndrome can be fractionated into two symptom clusters. One symptom cluster is said to be associated with the left-sided neglect syndrome; the second symptom cluster, including the tendency to omit or repeat strokes, is said to be an impairment of visual and kinesthetic feedback. However, both sets of symptoms appear to be associated with lesions of the nondominant parietal lobe.

III. NEUROPSYCHOLOGICAL CLASSIFICATION

A. Linguistic Agraphias

Shallice (1981) and Beauvois and Derouesne (1981) provided the initial detailed linguistic studies of agraphia. Since then, data from their studies and others have been used to develop linguistic models similar to the one presented here. Ellis (1982) provided one of the first models and others have followed (Lesser, 1990; Margolin, 1984; Patterson, 1986; Rapcsak & Rubens, 1990; Roeltgen, 1993; Roeltgen & Heilman, 1985; Roeltgen & Rapcsak, 1993). These models, one of which is presented in Fig. 1, contain features similar to those first proposed by Ellis. All the models emphasize two dissociable systems or pathways that are used for the production of correct spellings: a phonological system and a lexical system. A third major component is the semantic system or "cognitive" system (Patterson, 1986). Finally, studies that have been conducted predominantly by Caramazza and colleagues (Caramazza, Miceli, Villa, & Romani, 1987; Hillis & Caramazza, 1989) have shown that, prior to the linguistic information being transferred into motor output, an apparent opportunity exists for the spellings to be disrupted after they have been chosen initially and correctly from the phonological or lexical systems. Caramazza and colleagues have argued that a disturbance of this ability is the result of a disruption of a graphemic buffer.

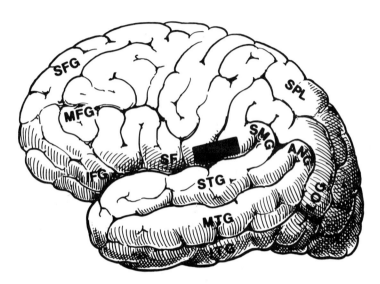

Figure 5 Lateral view of the left hemisphere. The blackened area is the region proposed as the anatomic substrate for phonological agraphia. SFG, Superior frontal gyrus; MFG, middle frontal gyrus; IFG, inferior frontal gyrus; SF, Sylvian fissure; SMG, supramarginal gyrus; SPL, superior parietal loblule; ANG, angular gyrus; LOG, lateral occipital gyrus; STG, superior temporal gyrus; MTG, middle temporal gyrus; ITG, inferior temporal gyrus. Reproduced with permission from Roeltgen (1993).

1. Phonological agraphia

The phonological system is described as a nonlexical system that requires phoneme to grapheme conversion (Roeltgen, 1993; Roeltgen & Rapcsak, 1993). Alternative descriptions of this process are phono-grapheme to letter (Lesser, 1990) or phonological to orthographic conversion (Margolin, 1984; Patterson, 1986). The phonological system is thought to be important for spelling pronounceable nonwords (e.g., nud) and novel real words, and is thought to be available as a back-up system for spelling orthographically regular real words if the orthographic system is dysfunctional. Phonological agraphia has been defined as impaired ability to spell nonwords with preserved ability to spell real words. Roeltgen et al. (1983) proposed that the anterior inferior supramarginal gyrus was the important anatomical substrate for phonological agraphia (Fig. 5). This conclusion was based on the analysis of four patients with phonological agraphia. The cerebral lesions from these patients, as determined by computerized tomography (CT), were depicted on a lateral view of the left hemisphere and the four lesions were overlapped. The anatomical region that was lesioned in all four patients was the anterior inferior supramarginal gyrus. However, Roeltgen and colleagues (Roeltgen, Rothi, & Heilman, 1982) described a patient with relatively isolated phonological agraphia who had a lesion confined to the insula, deep to the supramarginal gyrus. Based on these studies, as well as an additional study by Roeltgen and Heilman (1984), Roeltgen and colleagues concluded that the important anatomical substrate for phonological agraphia appeared to be the anterior inferior supramarginal gyrus or the insula deep to it (Roeltgen, 1993; Roeltgen & Heilman, 1985). Consistent with the hypothesis by Roeltgen and colleagues, other patients with phonological agraphia have had lesions in the same brain regions (Nolan & Caramazza, 1982; Shallice, 1981). In addition, a patient described by Baxter and Warrington (1985) had a lesion that appears, based on the CT scans available in their study, to be slightly posterior to the region implicated by Roeltgen and colleagues. In a study of a left-handed patient with phonological agraphia, Bola-Wilson, Speedie, and Robinson (1985) described a lesion of the frontal–parietal area that included that supramarginal gyrus, but the lesion was in the right hemisphere. Other studies (Roeltgen, 1989a,1991) indicate that a simple explanation consisting of a single anatomical region responsible for phonological agraphia is an oversimplification. The more complex interactions and clinical–pathological correlations are presented in Section III,A,5.

2. Lexical agraphia

The lexical system (Lesser, 1990; Roeltgen & Heilman, 1984), also called the lexical–orthographic system (Margolin, 1984) or orthographic system (Patterson, 1986; Rapcsak & Rubens, 1990), appears to utilize a whole-word retrieval process that may incorporate visual word images (Roeltgen et al., 1983). The lexical system is also likely to use whole word processing that is not based entirely on visual word images, but includes certain phonological components, word analogies, and mnemonic rules (Hatfield, 1985). The lexical system is used for spelling familiar words, especially those that are orthographically irregular (words that cannot be spelled using direct sound-to-letter correspondence rules, e.g., "comb") and ambiguous words (words with sounds that may be represented by multiple letters or letter clusters, e.g., "phone"). Lexical agraphia has been defined as impaired ability to spell irregular words relative to regular words and preserved ability to spell nonwords.

Roeltgen and Heilman (1984) attempted to delineate the anatomical substrate underlying lexical agraphia using the same methods that Roeltgen and colleagues used previously in attempting to delineate the

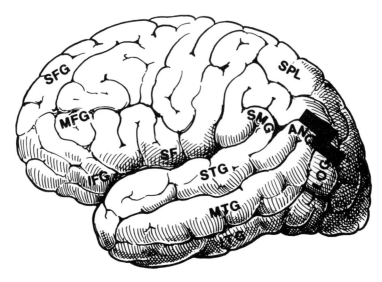

Figure 6 Lateral view of the left hemisphere. The blackened area indicates the region proposed as the anatomic substrate for lexical agraphia. The abbreviations are the same as those for Fig. 5. Reproduced with permission from Roeltgen (1993).

anatomical substrate of phonological agraphia. Overlap of the lesions from four patients with lexical agraphia indicated that the region commonly lesioned in all subjects was the posterior angular gyrus and the parieto-occipital lobule with extension subcortically into the white matter (Fig. 6).

Since Roeltgen and Heilman's 1984 study, some patients with lexical agraphia have been found to have lesions in or near the dominant angular gyrus (Alexander et al., 1990; Roeltgen, 1989a), but other patients have been found to have lesions away from that region. Other regions involved have been the right parietal lobe in a right-hander without aphasia (Rothi, Roeltgen, & Kooistra, 1987), the left posterior temporal region (middle temple gyrus) (Croisile, Trillet, Laurent, Latomby, & Schott, 1989), the left frontal region (Rapscak et al., 1988), the left caudate (Wolz & Roeltgen, 1989), and the left thalamus (Roeltgen, 1989b). Roeltgen and Rapscak (1993) reviewed reports of previous patients with lexical agraphia and added seven new cases. They concluded that patients with lexical agraphia had lesions that usually spared the immediate perisylvian region, especially the anterior supramarginal gyrus and insula (areas thought to be important for the production of phonological agraphia).

However, multiple lesion sites associated with lexical agraphia were identified. These researchers also noted a degree of linguistic heterogeneity that was consistent with the pathological heterogeneity. They suggested that the most striking cases appear to have lesions of the angular gyrus, but that the clinical and pathological heterogeneity were consistent with Hatfield's (1985) position that the lexical system might contain multiple cognitive components. Finally, additional studies (Roeltgen 1989a,1991) indicate that even the explanation offered by Roeltgen and Rapscak (1993) does not reflect the full complexity of the lexical system (see Section II,A,5).

3. Semantic agraphia

The incorporation of meaning into what is written has been termed semantic influence on writing (Roeltgen, 1993; Roeltgen & Heilman, 1985). The disruption of semantic ability or the disconnection of semantics from spelling has been termed semantic agraphia (Roeltgen et al., 1986). Patients with semantic agraphia lose their ability to spell and write with meaning. They may spell correctly (oral, written, or both) semantically incorrect homophones that have been dictated to them. Homophones are words that are pronounced the same but have different meanings, depending on the spelling ("bear" and "bare"). For example, when asked to write "bare" as in "the baby has a *bare* bottom," these patients may write "bear." They also may produce semantic jargon in sentence production (Rapscak & Rubens, 1990).

In their initial study of semantic agraphia, Roeltgen and colleagues (1986) concluded that their anatomical data did not appear to be as specific for semantic agraphia as the anatomical data for the syndromes of phonological and lexical agraphia. Their original patients with semantic agraphia had lesions involving (1) the cortical watershed area; (2) the caudate, the internal capsule, and the frontal subcortical region; (3) bilateral medial–frontal and medial–parietal areas; and (4) the thalamus. Therefore, Roeltgen and colleagues suggested that the lesions producing semantic agraphia did not overlap anatomically but occurred in areas frequently associated with transcortical aphasias with impaired comprehension. Consistent with the nonspecificity previously associated with semantic agraphia is the study by Rapscak and Rubens (1990), who described a patient with a left prefrontal lesion who had a disruption of semantic influence on writing. These results indicate that the clinical syndrome of semantic agraphia and the anatomical lesions associated with it are heterogeneous. Such a conclusion is not surprising because many studies that have examined the disruption of semantics in language have shown striking heterogeneity.

4. Deep agraphia

Patients have been described who have behavioral features similar to those of patients with phonological agraphia, that is, they have impaired ability to spell nonwords with preserved ability to spell real words. However, patients who have deep agraphia also make semantic paragraphias. Semantic paragraphias are spelled or written responses that have little phonological or visual resemblance to the stimulus word but have a semantic resemblance to the stimulus word (e.g., spelling "flight" when propeller is the stimulus). Bub and Kertesz (1982) and Hatfield (1985) have termed this syndrome "deep agraphia" because of the similarity this disorder has to the reading disorder deep dyslexia. Other patients with impaired ability to spell nonwords who produce semantic paragraphias also have been described (Assal, Buttet, & Jolivet, 1981; Marshall & Newcombe, 1966; Roeltgen et al., 1983, Patient 1; Van Lancker, 1990).

All the patients who had deep agraphia had lesions that included the supramarginal gyrus and insula. However, their lesions have been large, extending well beyond the circumscribed area thought to be important for the production of phonological agraphia. The syndrome of deep agraphia has been suggested to reflect right hemispheric writing mechanisms (Rapscak, Beeson, & Rubens, 1991) or, alternatively, residual left hemispheric mechanisms that include the left posterior angular gyrus, a region that has been spared in most patients with deep agraphia (Roeltgen & Heilman, 1982).

5. Prospective test of the linguistic components of the neuropsychological model

Roeltgen (1989a,1991) evaluated the linguistic components of the neuropsychological model described previously. He tested 43 consecutive patients with acquired left hemispheric lesions. He omitted subjects who were left-handed, who had low educational levels (usually less than 8th grade), and who did not speak English as the primary language. All patients were tested with the Battery of Linguistic Analysis for Writing and Reading (Roeltgen, Cordell, & Sevush, 1983) and, when the subjects had limited ability, portions of the Battery of Linguistics Analysis for Writing and Reading—Children's Version (Roeltgen, 1989b). These test batteries contain nonwords and matched sets of real words including orthographically regular and irregular words and homophones. The exact lists of words and nonwords are described in detail by Roeltgen and Blaskey (1992) and Roeltgen (1992). In addition to the patients with left hemispheric lesions, the test battery was given to a series of normal control subjects. All the patients with left hemispheric lesions had CT scans; the locus of each lesion was plotted on previously drawn templates using methods similar to those described by Damasio and colleagues (Damasio, Damasio, Rizzo, Varney, & Gersh, 1982). All normal control subjects were tested with the Wide Range Achievement Test-Revised (WRAT-R) (Jastak & Wilkinson, 1984); more than half the experimental group was given this test as well. This test is a single word reading and spelling test that can provide a grade level for a measurement of overall spelling and reading ability. Spelling impairment on words or nonwords for the patients with left hemispheric lesions was defined as impaired relative to the results from the normal control subjects.

The results from this study allow certain conclusions to be drawn. First, impairment in nonword spelling ability correlated highly with overall performance on the independent test of spelling achievement (WRAT-R). However, no subjects had selective impairment of nonword spelling to qualify as having pure phonological agraphia. Six patients had what Roeltgen has defined as "phonological-plus" agraphia. These patients had impaired nonword spelling ability and mild to moderate impairment of real word spelling. All five patients who had vascular lesions had involvement of the mid-perisylvian region (anterior supramarginal gyrus or insula). Seven additional patients had mild nonword spelling impairment (less than the mean, but within one standard deviation of the nonword spelling results from the normal control subjects) and had mild to moderate impairment of real word spelling. Six of these patients had vascular lesions and four of the six lesions were in the mid-perisylvian region. Importantly, none of these 13 patients showed a difference between regular and irregular word spelling that was more than one standard deviation greater than the mean difference found in the normal control subjects. Consequently, orthographic regularity appeared to have no effect on their spelling. Based on these results, although no subjects fulfilled the criteria for pure phonological agraphia, impairment in nonword spelling without an effect on orthographic regularity appears to be associated with mid-perisylvian lesions. This result is consistent with previous conclusions based on studies of phonological agraphia (Roeltgen & Heilman, 1984; Roeltgen et al., 1983).

In the prospective study by Roeltgen, one subject fulfilled the criteria for pure lexical agraphia. That subject had two lesions, one involving the anterior aspects of the internal capsule and anterior basal ganglia and the other involving subcortical white matter, deep to the angular gyrus, with extension into the posterior thalamus. Six other patients had excellent nonword spelling ability and showed moderate differences between abilities to spell regular and irregular words. These differences were greater than the average difference found in the normal control subjects. In this

Table 3 Acute Stroke vs. Chronic Stroke Patients[a]

Task	Chronic	Acute	p value
Spelling achievement (raw score)	19.6	17.2	<.5
Reading achievement (raw score)	52.5	54.3	<.9
Nonword spelling (number correct)	57.0	50.0	<.5
Nonword reading (number correct)	49.7	50.0	<.1
Lesion size (approximate)	273.7	201.2	<.5
Regular/irregular difference for spelling (% correct)	21.0	11.8	<.05
Regular/irregular difference for reading (% correct)	7.4	8.5	<.9
Aphasia quotient (Western Aphasia Battery)	87.8	83.2	<.5
Highest school grade completed	10.9	10.8	<.9
Age	60.4	56.7	<.5

[a] Prospective study performed by Roeltgen (1991).

way, these patients are similar to subjects previously described as having lexical agraphia. Four of these six patients had vascular lesions, one had a neoplasm, and one had a closed head injury with a focal lesion of the thalamus. Among the stroke patients, one had a lesion of the angular gyrus, one had a lesion of the posterior middle temporal gyrus, one had a lesion of the thalamus, and one had a lesion of the frontal pole.

The results from the prospective study by Roeltgen (1989a,1991) appear to support, in a general way, the previously described neuropsychological model of acquired agraphia. However, as has been discussed elsewhere by Roeltgen (1989a,1991), the discussion summarized here is based on methods used that were similar to those used in previous studies. Specifically, patients with specific behavioral syndromes were analyzed and the lesion sites recorded as the dependent variable. Such analysis does not account for a large number of patients. Many of these additional patients have lesions in the cerebral locations of interest, including the supramarginal gyrus and the angular gyrus. If the analysis is performed with the lesion sites as the independent variable and the linguistic behavior as the dependent variable, little consistent clinical–pathological correlation is obtained (Roeltgen, 1989a). Because of this inconsistency, Roeltgen (1991) analyzed the data using different variables, including chronicity of the lesion and the highest level of education obtained by the subjects. When all the subjects with vascular lesions are divided into groups of those with acute lesions (less than 6 wk) and those with chronic lesions (3 mo or longer), and comparisons are made for overall spelling achievement as well as performance on specific types of words, the only comparison that approaches statistical significance is the irregular word difference (difference between orthographically regular and irregular words) (Table 3). Interestingly, the patients with acute lesions show a regular–irregular difference that is less than that of the control subjects and the patients with chronic lesions show a regular–irregular word difference that is greater than that of the control subjects. In analysis of the effect of the level of education on performance, the correlation of spelling achievement and grade is .61, a correlation that is highly significant ($P < .005$).

A relationship also exists between nonword performance and level of education. Those patients with good nonword performance (within two standard deviations of the mean from the normal control subjects) had an average education of 12.4 years. In contrast to those patients, the patients with poor nonword performance had a mean education of 9.2 years. However, good education and good nonword performance are not sufficient to produce lexical agraphia. Two groups of subjects have good education and good nonword performance. One group fulfills the criteria for lexical agraphia because these subjects have a large regular–irregular word discrepancy and the subjects of the other group do not. Within the lexical agraphic group, all the patients had chronic lesions. Within the second group (good nonword performance, but no significant regular–irregular word difference) only 35% of the patients had chronic lesions. These patients are best described as having noncompensated lexical agraphia. This designation implies that the acuteness of the lesion limits the patient's capacity for utilizing preserved phonological skills to spell orthographically regular words.

Lesion site is also important in understanding why a patient develops lexical agraphia after a brain lesion. Nonperisylvian lesions are associated with lexical agraphia (compensated and noncompensated lexical agraphia; patients with good nonword performance); thus, nonperisylvian lesions appear to allow preserved nonword performance.

In summary, three major factors appear to impact on whether or not subjects have a particular type of agraphia. First, perisylvian lesions decrease nonword performance and are associated more frequently with phonological type agraphias (with or without large differences between orthographically regular and irregular word performance). Second, more years of education are associated with a strikingly better nonword performance and the more common occurrence of lexical agraphia. Third, chronicity of lesion appears to be associated with a large regular–irregular word difference and the more common occurrence of lexical agraphia.

The prospective study by Roeltgen (1989a,1991) also assessed subjects for performance on homophones. Frequent homophone confusions (consistent with semantic agraphia) were common in the patients with multi-

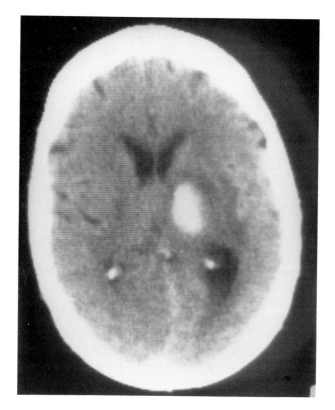

Figure 7 CT scan of a patient with relatively pure or isolated semantic agraphia. The hemorrhagic lesion involves the region of the
left thalamus.

ple types of agraphia. Consequently, impaired semantic access for spelling
production appears to be associated with a wide range of lesions. This
result is consistent with the opinion expressed by Roeltgen and colleagues
(1986). Only one subject had a relatively pure or isolated semantic
agraphia and that patient had a subcortical hemorrhagic lesion in the region of the thalamus (Fig. 7).

6. Agraphia from disturbances of the graphemic buffer

The output of the linguistic systems converges on a region that has been
termed the "graphemic" or "orthographic" buffer (Caramazza et al., 1987;
Hillis & Caramazza, 1989; Lesser, 1990; Margolin, 1984; Miceli, Silveri, &
Caramazza, 1985). According to Caramazza and colleagues, if the graphe-

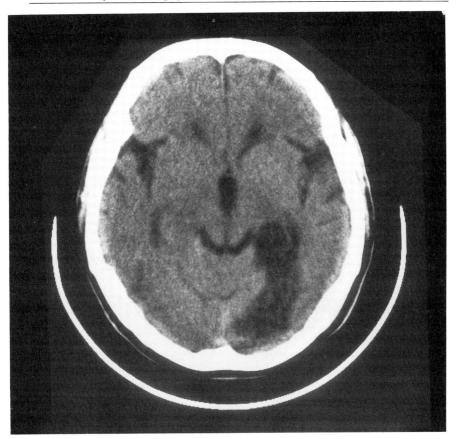

Figure 8 CT scan from a patient with agraphia due to disruption of the allographic store. The lesion involves the distribution of the posterior cerebral artery, and is more typically associated with pure alexia rather than agraphia with alexia.

mic buffer is disturbed, patients typically produce letter omissions, substitutions, insertions, and transpositions in nonwords and real words when spelling orally or when writing. Impairment appears in spontaneous writing, writing to dictation, written naming, and delayed copying. Although only a few reports have described patients with agraphia caused by impairment of the graphemic buffer, the lesion loci had varied widely and include the left frontal–parietal region (Hillis & Caramazza, 1989; Lesser, 1990), the left parietal region (Miceli et al., 1985), and the right frontal–parietal–temporal region (Hillis & Caramazza 1989). We (D. P. Roeltgen, unpublished observation; T. N. Laine, personal communication) have observed patients with apparent agraphia due to impairment of their graphemic buffer. These patients had relatively discrete lesions of the left posterior dorsolateral frontal lobe, the region typically labeled Exner's area. Based on the anatomical variability of the lesions in these six patients, any proposed clinical–pathological correlation would be speculation.

B. Motor or Peripheral Agraphias

1. Apraxic and spatial agraphias

These agraphias are similar to the agraphias described in the neurological classification in terms of the behavior as well as the lesion localization.

2. Agraphia due to a disruption of the allographic store

Ellis (1982), based on analysis of the slip of the pen (errors made by normal subjects when writing), hypothesized the existence of the allographic store, a system said to be important for directing the handwriting systems in the production of correct case (upper and lower) and style [script (cursive) or manuscript (print)]. A few reports have been made of patients with apparent impairment of the allographic store (Black, Bass, Behrmann, & Hacker, 1987; DeBastiani & Barry, 1989; Yopp & Roeltgen, 1987). These patients produced frequent case and style errors. The lesions in these three patients included the occipital lobe in one patient and the parietal–occipital lobe in two different sites. One lesion is illustrated in Fig. 8. Interestingly, this patient also had alexia and had a lesion more commonly associated with pure alexia than with agraphia and alexia. The large size of the lesions and the limited number of patients studied make clinical–pathological correlation difficult for this disorder. Although a posterior left hemispheric lesion is possible, any conclusions await further studies.

IV. LOCALIZATION OF AGRAPHIA IN JAPANESE PATIENTS

Two written language systems are used in Japanese: Kanji and Kana. Traditionally, Kanji is classified as "ideogram" writing. However, because the symbols have both semantic and phonetic value, a more precise description of the symbols would be the term "morphograms." Although the Japanese language contains approximately 10,000 of these symbols, which are used primarily for nouns and stems of verbs, adjectives, and adverbs, only approximately 1,900 of these usually are learned by the end of the 9th grade. Therefore, some errors in Kanji are common even in educated Japanese individuals (Iwata, 1984; Iwata, Sugishita, & Toyokura,

1981; Kawamura et al., 1987; Soma, Sugishita, Kitamura, Maruyama, & Imanaga, 1989; Tanaka, Yamadori, & Murata, 1987). In contrast, Kana has a limited number of syllabic characters that can be used for writing any of the words in the Japanese language. Kana symbols probably are described best as syllabograms, but frequently have been termed phonograms. Each Kana character has a one-to-one correspondence to a syllable.

Agraphias for Kanji and Kana have been reported, but the most frequently reported dissociation is impaired Kanji with normal or relatively normal Kana. Kanji agraphia usually is associated with left occipital pole lesions, a site that usually produces pure alexia in Indo-European languages. Kanji agraphia also has been ascribed to lesions of the posterior left temporal region, and of the junction of the posterior temporal region with the occipital lobe or the angular gyrus (Kawahata & Nagata, 1988) and the left inferior temporal gyrus (Kawamura et al., 1987). The selective occurrence of Kana agraphia is uncommon. The description that is closest to a case of pure Kana agraphia was given by Tanaka et al. (1987), who described a patient who produced frequent substitutions and occasional perseverations when writing Kana characters. Their patient had a lesion of the left inferior parietal lobe and a small lesion in the left corona radiata.

Patients described by Sugishita and colleagues (Sugishita et al., 1980) and Yamadori and colleagues (Yamadori et al., 1980) had lesions of the corpus callosum and showed a degree of linguistic agraphia with the left hand. These patients produced Kanji characters (morphograms) better than Kana characters (phonograms). Rothi and colleagues (Rothi et al., 1987) suggested that these results were consistent with the right hemisphere having the ability to support both linguistic and motor mechanisms for the more ideographic Kanji characters compared with the more phonological Kana characters.

V. SUMMARY

Acquired agraphia is a common disorder from cerebral lesions and can occur from lesions in multiple cerebral locations. In general, results from these studies, including the study of agraphia in confusional states by Chedru and Geschwind (1972), indicate that agraphia is a sensitive test for cerebral dysfunction. Clearly no single site is a "center for writing." However, certain associations of agraphic types, or patterns of agraphic performance, with certain cerebral regions do appear. Still, specific localization, given individual variability as well as the complex act being tested, must be done cautiously. Using the lesion site as an independent variable supports the contention that certain types of agraphia and certain patterns of performance—such as apraxic agraphia, spatial agraphia, and dissociations of phonological and lexical spelling—are significant behavioral constructs that appear to be important because of their fairly consistent association with particular cerebral lesions.

REFERENCES

Aimard, G., Devick, M., Lebel, M., Trouillas, M., & Boisson, D. (1975). Agraphie pure (dynamique?) origine frontale. *Revue Neurologique, 7,* 505–512.

Alexander, M. P., Fisher, R. S., & Friedman, R. (1992). Lesion localization in apraxic agraphia. *Archives of Neurology, 49,* 246–251.

Alexander, M. P., Friedman, R., LoVerso, F., & Fischer, R. (1990). Anatomic correlates of lexical agraphia. Presented at *The Academy of Aphasia,* Baltimore.

Assal, G., Buttet, J., & Jolivet, R. (1981). Dissociations in aphasia: A case report. *Brain and Language, 13,* 223–240.

Auerbach, S. H., & Alexander, M. P. (1981). Pure agraphia and unilateral optic ataxia associated with a left superior parietal lobule lesion. *Journal of Neurology, Neurosurgery and Psychiatry, 44,* 430–432.

Basso, A., Taborielli, A., & Vignolo, L. A. (1978). Dissociated disorders of speaking and writing in aphasia. *Journal of Neurology, Neurosurgery and Psychiatry, 41,* 556–563.

Baxter, D. M., & Warrington, E. K. (1985). Category-specific phonological dysgrahia. *Neuropsychologia, 23,* 653–666.

Baxter, D. M., & Warrington, E. K. (1986). Ideational agraphia: A single case study. *Journal of Neurology, Neurosurgery and Psychiatry, 49,* 369–374.

Beauvois, M. F., & Derouesne, J. (1981). Lexical or orthographic agraphia. *Brain, 104,* 21–49.

Benedikt, M. (1865). *Uber Aphasie, Agraphie und verwandte pathologische Zustände.* Wiener medizinishe Presse Pr. 6.

Benson, D. F. (1979). *Aphasia, alexia and agraphia.* New York: Churchill Livingstone.

Benson, D. F., & Cummings, J. L. (1985). Agraphia. In P. J. Vinken, G. W. Bruyn, H. L. Klawns, & J. A. M. Frederiks (Eds.), *Handbook of clinical neurology* (Vol. 45, pp. 457–472). Amsterdam: Elsevier.

Black, S. E., Bass, K., Behrmann, M., & Hacker, P. (1987). Selective writing impairment: A single case study of a deficit in allographic conversion. *Neurology, 37(3(1)),* 174.

Bogen, J. E. (1969). The other side of the brain. I. Dysgraphia and dyscopia following cerebral commissurotomy. *Bulletin of the Los Angeles Neurological Society, 34,* 3–105.

Bolla-Wilson, K., Speedie, L. J., & Robinson, R. G. (1985). Phonologic agraphia in a left-handed patient after a right-hemisphere lesion. *Neurology, 35,* 1778–1781.

Bub, D., & Kertesz, A. (1982). Deep agraphia. *Brain and Language, 17,* 146–165.

Caramazza, A., Miceli, G., Villa, G., & Romani, C. (1987). The role of the graphemic buffer in spelling: Evidence from a case of acquired dysgraphia. *Cognition, 26,* 59–85.

Chedru, F., & Geschwind, N. (1972). Writing disturbances in acute confusional states. *Neuropsychologia, 10,* 343–354.

Croisile, B., Laurent, B., Michel, D., & Trillet, M. (1990). Pure agraphia after deep left hemisphere haematoma. *Journal of Neurology, Neurosurgery and Psychiatry, 53,* 263–265.

Croisile, B., Trillet, M., Laurent, B., Latombe, D., & Schott, B. (1989). Agraphie lexicale par hematome temporo-parietal gauche. *Revue Neurologique (Paris), 145(4),* 287–292.

Damasio, A. R., Damasio H., Rizzo, M., Varney, N., & Gersch, F. (1982). Aphasia with non-hemorragic lesions in the basla ganglia and internal capsule. *Archives of Neurology, 39,* 15–20.

De Bastiani, K., & Barry, C. (1989). A cognitive analysis of an acquired dysgraphic patient with an "allographic" writing disorder. *Cognitive Neuropsychology, 6,* 25–41.

Ellis, A. W. (1982). Spelling and writing (and reading and speaking). In A. W. Ellis (Ed.), *Normality and pathology in cognitive functions* (pp. 113–146). London: Academic Press.

Ellis, A. W., Young, A. W., & Flude, B. M. (1987). "Afferent dysgraphia" in a patient and in normal subjects. *Cognitive Neuropsychology, 4(4),* 465–486.

Gersh, F., & Damasio, A. R. (1981). Praxis and writing of the left hand may be served by different callosal pathways. *Archives of Neurology, 38,* 634–636.

Geschwind, N., & Kaplan, E. F. (1962). A human cerebral disconnection syndrome. *Neurology, 12,* 675–685.

Grossfeld, M. L., & Clark, L. W. (1983). Nature of spelling errors in transcortical sensory aphasia: A case study. *Brain and Language, 18,* 47–56.

Hatfield, F. M. (1985). Visual and phonological factors in acquired dysgraphia. *Neuropsychologia, 23(1),* 13–29.

Hecaen, H., & Albert, M. L. (1978). *Human neuropsychology.* New York: John Wiley and Sons.

Hillis, A. E., & Caramazza, A. (1989). The graphemic buffer and attentional mechanisms. *Brain and Language, 36,* 208–235.

Iwata, M. (1984). Neuropsychological correlates of the Japanese writing system. *Trends in Neurosciences, 7(8),* 290–293.

Iwata, M., Sugishita, M., & Toyokura, Y. (1981). The Japanese writing system and functional hemispheric specialization. In S. Katsuki, T. Tsubaki, and Y. Toyokura (Eds.) *Neurology, Proceedings of the 12th World Congress of Neurology.* (pp. 55–62) Amsterdam-Oxford-Princeton: Excerpta Medica.

Jastak, D. E., & Wilkinson, G. S. (1984). Wide Range Achievement Test (revised). Wilmington, DE. Jastak Associates.

Kaplan, E., & Goodglass, H. (1981). Aphasia-related disorders. In M. T. Sarno (Ed.), *Acquired aphasia* (pp. 303–325). New York: Academic Press.

Kawahata, N., & Nagata, K. (1988). Alexia with agraphia due to the left posterior inferior temporal lobe lesion—Neuropsychological analysis and its pathogenetic mechanisms. *Brain and Language, 33,* 296–310.

Kawamura, M., Hirayama, K., Hasegawa, K., Takahashi, N., & Yamaura, A. (1987). Alexia with agraphia of Kanji (Japanese morphograms). *Journal of Neurology, Neurosurgery, and Psychiatry, 50,* 1125–1129.

Kapar, N., & Lawton, N. F. (1983) Dysgraphia for letters: A form of motor memory deficit? *Journal of Neurology, Neurosurgery and Psychiatry, 46,* 573–575.

Kertesz, A., Latham, N., & McCabe, P. (1990). Subcortical agraphia. *Neurology, 40(1),* 172.

Kinsbourne, M., & Rosenfeld D. B. (1974) Agraphia selective for written spelling, an experimental case study, *Brain and Language, 1,* 215–225.

Laine, T. N., & Marttila, R. J. (1981). Pure agraphia: A case study. *Neuropsychologia, 19,* 311–316.

Leischner, A. (1969). The agraphias. In P. J. Vinken & G. W. Bruyn (Eds.), *Disorders of speech, perception and symbolic behavior* (pp. 141–180). Amsterdam: North-Holland.

Lesser, R. (1990). Superior oral to written spelling: Evidence for separate buffers? *Cognitive Neuropsychology, 7(4),* 347–366.

Levy, J., Nebes, R. D., & Sperry, R. W. (1971). Expressive language in the surgically separated minor hemisphere. *Cortex, 71,* 49–58.

Liepmann, H., & Maas, O. (1907). Fall von linksseitiger Agraphie und Apraxie bei rechtsseitiger Lähmung. *Journal für Psychologie und Neurologie, 10*, 214–227.

Marcie, P., & Hecaen, H. (1979). Agraphia. In K. M. Heilman & E. Valenstein (Eds.), *Clinical
neuropsychology* (1st ed., pp. 92–127). New York: Oxford.

Margolin, D. I. (1984). The neuropsychology of writing and spelling: Semantic, phonological,
motor, and perceptual processes. *The Quarterly Journal of Experimental Psychology, 36A*,
459–489.

Marshall, J., & Newcombe, F. (1966). Syntactic and semantic errors in paralexia. *Neuropsychologia, 4*, 169–176.

Miceli, G., Silveri, M. C., & Caramazza, A. (1985). Cognitive analysis of a case of pure dysgraphia. *Brain and Language, 25*, 187–212.

Nielsen, J. M. (1946). *Agnosia, apraxia, aphasia: Their value in cerebral localization.* New York:
Hoeber.

Nolan, K. A., & Caramazza, A. (1982). Modality-independent impairments in word processing in a deep dyslexic patient. *Brain and Language, 16*, 236–264.

Ogle, J. W. (1867). Aphasia and agraphia. *Report of the Medical Research Counsel of St. George's
Hospital (London), 2*, 83–122.

Patterson, K. (1986). Lexical but nonsemantic spelling? *Cognitive Neuropsychology, 3(3)*,
341–367.

Rapcsak, S. Z., Arthur, S. A., & Rubens, A. B. (1988). Lexical agraphia from focal lesion of the
left precentral gyrus. *Neurology, 38*, 1119–1123.

Rapcsak, S. Z., Beeson, P. M., & Rubens, A. B. (1991). Writing with the right hemisphere.
Journal of Clinical and Experimental Neuropsychology, 13, 39.

Rapcsak, S. Z., & Rubens, A. B. (1990). Disruption of semantic influence on writing following
a left prefrontal lesion. *Brain and Language, 38*, 334–344.

Roeltgen, D. P. (1989a). Prospective analysis of a model of writing, anatomic aspects. Presented at *The Academy of Aphasia*, Sante Fe, New Mexico.

Roeltgen, D. P. (1989b). The battery of linguistic analysis for writing and reading—Children's
version. *Journal of Clinical and Experimental Neuropsychology, 11*, 57.

Roeltgen, D. P. (1991). Prospective analysis of writing and spelling. Part II: Results not related
to localization. *Journal of Clinical and Experimental Neuropsychology, 13(1)*, 48.

Roeltgen, D. P. (1992). Phonological error analysis, development and empirical evaluation.
Brain and Language, 43, 190–229.

Roeltgen, D. P. (1993). Agraphia. In K. M. Heilman & E. Valenstein (Eds.), *Clinical neuropsychology* (3d ed.) (pp. 39–55). New York: Oxford.

Roeltgen, D. P., & Blaskey, P. (1992). Processes, breakdowns and remediation in developmental disorders of reading and spelling. In D. Marjolin (Ed.), *Cognitive neuropsychology in
clinical practice* (pp. 298–326). New York: Oxford.

Roeltgen, D. P., Cordell, C., & Sevush, S. (1983). A battery of linguistic analysis for writing
and reading. *The INS Bulletin*, 31.

Roeltgen, D. P., & Heilman, K. M. (1982). Global aphasia with spared lexical writing. Presented at the *International Neuropsychological Society*, Pittsburgh.

Roeltgen, D. P., & Heilman, K. M. (1983). Apractic agraphia in a patient with normal praxis.
Brain and Language, 18, 811–827.

Roeltgen, D. P., & Heilman, K. M. (1984). Lexical agraphia, further support for the two system
hypothesis of linguistic agraphia. *Brain, 107*, 811–827.

Roeltgen, D. P., & Heilman, K. M. (1985). Review of agraphia and proposal for an anatomi-
cally-based neuropsychological model of writing. *Applied Psycholinguistics, 6*, 205–230.

Roeltgen, D. P., & Rapcsak, S. (1993). Acquired disorders of writing and spelling. In G.
Blanken (Ed.), *Linguistic disorders and pathologies.* Berlin: de Gruyter.

Roeltgen, D. P., Rothi, L. G., & Heilman, K. M. (1982). Isolated phonological agraphia from a focal lesion. Presented at *The Academy of Aphasia*, New Paltz, New York.

Roeltgen, D. P., Rothi, L. G., & Heilman, K. M. (1986). Linguistic semantic agraphia. *Brain and Language, 27*, 257–280.

Roeltgen, D. P., Sevush, S., & Heilman, K. M. (1983). Phonological agraphia: Writing by the lexical-semantic route. *Neurology, 33*, 733–757.

Rosati, G., & de Bastiani, P. (1981). Pure agraphia: A discreet form of aphasia. *Journal of Neurology, Neurosurgery and Psychiatry, 44*, 266–269.

Rothi, L. J. G., Roeltgen, D. P., & Kooistra, C. A. (1987). Isolated lexical agraphia in a right-handed patient with a posterior lesion of the right cerebral hemisphere. *Brain and Language, 30(1)*, 181–190.

Rubens, A. B., Geschwind, N., Mahowald, M. W., & Mastri, A. (1977). Posttraumatic cerebral hemispheric disconnection syndrome. *Archives of Neurology, 34*, 750–755.

Shallice, T. (1981). Phonological agraphia and the lexical route in writing. *Brain, 104*, 412–429.

Soma, Y. Sugishita, M., Kitamura, K., Maruyama, S., & Imanaga, H. (1989). Lexical agraphia in the Japanese language. *Brain, 112*, 1549–1561.

Sugishita, M., Toyokura, Y., Yoshioka, M., & Yamada, R. (1980). Unilateral agraphia after section of the posterior half of the truncus of the corpus callosum. *Brain and Language, 9*, 212–225.

Tanaka, Y., Yamadori, A., & Murata, S. (1987). Selective Kana agraphia: A case report. *Cortex, 23*, 679–684.

Tanridag, O., & Kirshner, H. S. (1985). Aphasia and agraphia in lesions of the posterior internal capsule and putamen. *Neurology, 35*, 1797–1801.

Watson, R. T., & Heilman, K. M. (1983). Callosal apraxia. *Brain, 106*, 391–404.

Wolz, W. E., & Roeltgen, D. P. (1989). Isolated lexical agraphia due to a basal ganglia lesion. *Journal of Clinical and Experimental Neuropsychology, 11*, 43.

Van Lancker, D. (1990). A case of deep dysgraphia attributed to right hemispheric function. Presented at *The Academy of Aphasia*, Baltimore.

Vernea, J. J., & Merory, J. (1975). Frontal agraphia (including a case report). *Proceedings of the Australian Association of Neurologists, 12*, 93–99.

Yamadori, A., Osumi, Y., Ikeda, H., & Kanazawa, Y., (1980). Left unilateral agraphia and tactile anomia. Disturbances seen after occulsion of the anterior cerebral artery. *Archives of Neurology, 37*, 88–91.

Yopp, K. S., & Roeltgen, D. P. (1987). Case of alexia and agraphia due to a disconnection of the visual input to and the motor output from an intact graphemic area. *Journal of Clinical and Experimental Neuropsychology, 9*, 92.

Chapter 13

Localization of Lesions in Limb and Buccofacial Apraxia

Leslie J. Gonzalez Rothi, Adele S. Raade, and Kenneth M. Heilman

. . . a large number of those paralyzed on the right side by cortical lesions, espe-cially those who are aphasic, have a weak or strong degree of apraxia in their left hand as well—to my mind, a very noteworthy fact which suggests that the left hemisphere is dominant, not only in speech, but also in action, and therefore earns, even if in considerably lesser degree, the name of Action Hemisphere as well as Speech Hemisphere. (Liepmann, 1905)

I. INTRODUCTION

Apraxia, a term originally used by Steinthal in 1871, is the inability to execute purposive movements properly as the result of neurological dys-function. Liepmann (1905) described the term "purposive movements" as ". . . those learned connections of elementary muscle actions, which either represent effects on the object world (such as brushing, knocking, tying) or are manifestations of mental events to others (movements of expres-sions)." In addition, we partly define apraxia by exclusion of other neuro-logical defects that may impair motor activity in such a way that skilled motor acts cannot be performed adequately. For example, movement ef-fects caused by weakness, deafferentation, akinesia, bradykinesia, and hy-pometria or impaired motor performance induced by tremors, dystonia, chorea, ballismus, athetosis, myoclonus, ataxia, or seizures are not termed apraxia. In addition to limb apraxia, many different behavioral disorders have been labeled as apraxic, for example, lid apraxia, dressing apraxia, gait apraxia, constructional apraxia, apraxic agraphia (see Coslett, Rothi, Valenstein, & Heilman, 1986), verbal apraxia (see Johns & Darley, 1970), and so on. These many forms of apraxia result from distinct and differing anatomical and behavioral mechanisms, however, and for the purposes of this chapter we exclude them from our discussion. In contrast, although the anatomical and behavioral mechanisms of limb and buccofacial apraxias are also distinguishable, our hypotheses about the mechanisms of these two disorders are complementary. For this reason, this chapter focuses on both limb apraxia and buccofacial apraxia.

Limb apraxia refers to the inability to perform skilled movements with the arm and/or hand specifically. In contrast, oral or *buccofacial apraxia* consists of the inability to perform skilled movements with the oral and facial structures (tongue, larynx, pharynx, velum, lips, cheeks, etc.). This chapter specifically concentrates on the pathological anatomy of these two apraxic disorders, each of which is discussed separately.

II. LIMB APRAXIA

A. Significance and Incidence

Apraxia, specifically limb apraxia, is considered a neuropsychological syndrome of theoretical but not clinical significance (Basso, Capitani, Della Sala, Laiacona, & Spinnler, 1985; Poeck, 1985). For theoreticians, the study of this disorder may lead to an understanding of how the brain mediates skilled purposive movement and how skill acquisition occurs. This information could be applied to improving the strategies associated with normal motor skill acquisition. However, the study of apraxia also may have pragmatic implications for patients with neurological disorders. In addition to the potential for providing information of localizing value, information regarding praxis processing ultimately may have behavioral implications. Disruption of a system that mediates skilled movement not only may affect the pantomime of tool use (the method typically used to test for ideomotor apraxia), but also may affect other forms of skilled limb movement such as intransitive gesture. Gesture of any kind is an important component of an individual's communicative strategy and may be especially important in patients with aphasia or dysarthria. For example, LeMay, David, and Thomas (1988) showed that aphasics, specifically Broca's aphasics, use more spontaneously generated hand gestures during speech than nonaphasics. These investigators suggest that this gesturing may represent a strategy on the part of the aphasic patient to compensate for what he or she recognizes as inadequate verbal communication. However, Borod, Fitzpatrick, Helm-Estabrooks, and Goodglass (1989) demonstrated that aphasic patients with limb apraxia do not use communicative gesture as much as those aphasics without apraxia, a result that may indicate that the availability or success of gesture as a compensatory strategy, as suggested by LeMay et al. (1988), may be limited by the presence of apraxia (Kertesz & Hooper, 1982). In addition to gestures that the aphasic patient might generate spontaneously, the clinician commonly looks to gesture in aphasia treatment for both restitutive purposes and substitutive purposes. In addition, patients with certain forms of apraxia may be impaired not only in pantomiming but also when using tools (Poizner, Soechting, Bracewell, Rothi, & Heilman, 1989). Might this inability explain the finding of Sundet, Finset, and Reinvang (1988) that, of all the neuropsychological deficits they documented in their group of left hemisphere-damaged patients, presence of apraxia most strongly predicted "poor outcome" for return to their homes? To summarize, our perspective is that the study of apraxia has both theoretical relevance and functional importance.

Reports of the incidence of ideomotor apraxia after left hemisphere lesions are varied. For example, whereas Poeck (1986) states that apraxia occurs in ". . . 80% of the patients who sustain a cerebrovascular accident (CVA) in the middle cerebral artery of the hemisphere dominant for language," Liepmann (1905) found in testing 41 patients with right hemiplegia (thus, the supposition that they were left hemisphere damaged) that 20 of them (49%) demonstrated a significant apraxia of the left hand. DeRenzi, Motti, and Nichelli (1980) also reported that 50% of 100 left hemisphere-damaged patients were apraxic with their left hands, whereas Pieczuro and Vignolo (1967) reported an incidence of 46%. Kertesz, Ferro, and Shewan (1984) found that of 152 acute left hemisphere stroke patients with aphasia, 54.6% had apraxia. Of 118 patients examined 3 or more months after onset of left hemisphere stroke and aphasia, 40% still had apraxia. In summary, researchers do not agree on the specific incidence of limb apraxia after left hemisphere damage. This disagreement may result from differences in subject selection: some authors reported the incidence of apraxia coexistent with aphasia, whereas others reported the incidence of apraxia subsequent to left hemisphere damage that may or may not have caused a concomitant aphasia. In addition, method of testing varies greatly among investigators; therefore, differences in incidence may result from these variations. Despite this inconsistency, apraxia certainly is quite common after left hemisphere dysfunction, and appears to be an enduring neuropsychological syndrome.

B. Limb Apraxia Syndromes

Liepmann pointed out almost 100 years ago that limb apraxia is not a homogeneous grouping but, in fact, is a grouping of forms that are qualitatively different from one another. Historically, we have respected this suggestion by following the multicomponent classification system, in part developed by Liepmann, that identifies a variety of syndromes such as ideomotor, ideational, limb kinetic, innervatory, and verbomotor disconnection apraxias. The reader can find a detailed discussion of the neuropsychology of these syndromes of limb apraxia in other reviews (e.g., Heilman & Rothi, 1985). We have attempted (on the basis of the performance of apraxic patients) to explicate more fully the psychological processes underpinning normal skilled limb praxis and, in turn, to formalize a cognitive neuropsychological model that more fully describes the psychological mechanisms of the different forms of limb apraxia (Rothi, Ochipa, & Heilman, 1991). The neuroanatomical implications of this approach represent new directions for apraxia and praxis research alike. Until new research is completed, we are left to review past neuroanatomical formulations in the context of this current cognitive neuropsychological perspective.

Roy and Square (1985) suggest that praxis processing is mediated by a two-part system involving both conceptual and production components. According to these authors, the praxis conceptual system involves three kinds of knowledge: knowledge of the functions of tools and objects, knowledge of actions independent of tools, and knowledge of the organization of single actions into sequences. The praxis production system involves the sensorimotor component of action knowledge, including the information contained in action programs and the translation of these programs into action. Thus, action is dependent on the interaction of conceptual knowledge related to tools and actions and the structural information contained in motor programs. Within this framework, ideomotor apraxia would result from a disruption of the praxis production system, whereas ideational apraxia may be viewed as a disruption of the praxis conceptual system.

1. Ideational apraxia

Ideational apraxia has been defined by a variety of defective performances involving actual tool use. In 1905, Liepmann described a patient who displayed a variety of praxis error types including tool substitution (using a razor for a comb) and inappropriate action selection (placing eyeglasses on his outstretched tongue). Although commonly described as the inability to perform a sequence of acts using tools and objects to achieve an intended goal (Liepmann, 1920; Poeck, 1983), ideational apraxics also fail on tasks of single-act object use (DeRenzi & Lucchelli, 1988; DeRenzi, Pieczuro, & Vignolo, 1968). That these errors were qualitatively different from errors produced by cases he termed ideomotor apraxic led Liepmann to conclude that they reflected distinctly different problems.

Importantly, Liepmann thought the errors just described reflected a difficulty in the ideation of tool use and, as a result, labeled this syndrome ideational apraxia.

Additional evidence for the dissociation of ideational apraxia and ideomotor apraxia has been provided by DeRenzi and colleagues (1968), who found impaired use of single objects in the absence of ideomotor apraxia in a subgroup of left hemisphere-damaged patients. Similar conclusions were drawn by C. Ochipa, L. J. G. Rothi, and K. M. Heilman (1992) with respect to the praxis abilities of patients with Alzheimer's disease. Ochipa, Rothi, and Heilman (1989) also described a left-handed patient who, subsequent to a unilateral right hemisphere lesion, was noted to use objects inappropriately in natural settings. His inability to use tools could not be explained solely by a production deficit since he could not match tools with the objects on which they were used, suggesting a more pervasive impairment in the appreciation of the functional relationship between tools and the objects which they act.

Patients with ideational apraxia usually have bilateral cerebral involvement such as infarcts, tumors, or Alzheimer's disease. Frequently, the computerized tomographic (CT) scan in these patients will show profound cortical atrophy, bilateral posteriorly placed infarcts, unilateral tumors, or, in the case of strokes, very large lesions as in the Ochipa group's (1989) case (Fig. 1).

2. Ideomotor apraxia

a. Callosal apraxia (unilateral limb apraxia). Liepmann and Maas (1907) postulated that, in right-handers, the guidance of both left- and right-sided skilled movement was the responsibility of a single mechanism of the left hemisphere. Liepmann (1905) believed that the acquisition of limb movement skill required the acquisition of "movement formulae" and of an "innervatory pattern" that would communicate the formula information to the appropriate primary motor areas. These "movement formulae" contained the "time–space–form picture of the movement" whereas the "innervatory pattern" assisted in adapting these memories to environmental conditions through the development and implementation of a motor program.

Callosal apraxia is a rare syndrome characterized by the inability to carry out verbal commands with the left hand with a preserved ability to carry out these same commands with the right hand. In the particular case described by Liepmann and Maas (1907), they proposed that a callosal lesion disconnected language as well as the "movement formulae" from right hemisphere innervations of the spinal motor neuron pool projecting to the muscles of the left hand.

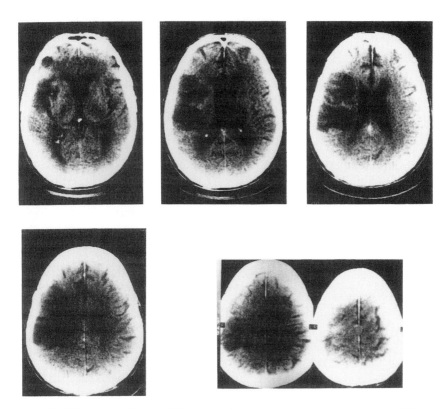

Figure 1   CT scan of a 67-year-old left-handed man performed 1 week after onset of ideational apraxia showing a large infarction involving the right middle cerebral artery.

Two clinical forms of callosal apraxia appear: those with impaired gesture imitation and those with normal gesture imitation. Liepmann and
Maas (1907) noted that their patient was unable to imitate pantomimed
acts with the left hand and was clumsy when using actual objects. In
contrast, Geschwind and Kaplan (1962) and Gazzaniga, Bogen, and Sperry
(1967) noted that, although their patients with callosal lesions could not
carry out commands correctly with their left hand, they could imitate and
use actual objects correctly. In cases of callosal apraxia in which patients
are able to imitate and use actual objects, the "movement formulae" have
been proposed to be less lateralized (Heilman, Rothi, & Kertesz, 1983).
Left-handed patients exist who have "movement formulae" in the right
hemisphere and language in the left hemisphere (Heilman, Coyle, Gonyea, & Geschwind, 1973; Valenstein & Heilman, 1979). If those patients

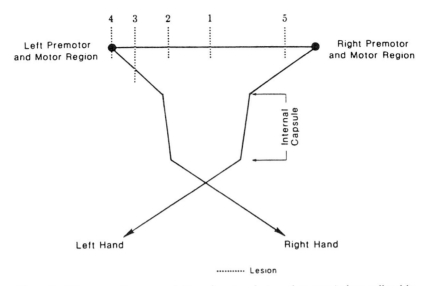

Figure 2 Diagrammatic representation of various lesions that may induce callosal lesions and anterior ideomotor apraxia: (1) callosal lesion involving ideomotor apraxia of the left hand without hemiparesis; (2) left hemisphere white matter lesions; (3) subcortical left hemisphere lesion inducing ideomotor apraxia of the left hand and right hemiparesis; (4) left cortical lesion inducing ideomotor apraxia of the left hand and right hemiparesis; and (5) right hemisphere subcortical lesion inducing ideomotor apraxia of the left hand without hemiparesis.

had a callosal section, the right hand would be deprived of information from the "movement formulae." Consequently, the right hand would perform poorly on command, in imitation, and in the use of actual objects. In contrast, the left hand would perform well in imitation and in the use of the actual objects; however, because that hand is disconnected from language, it would perform poorly on command. This specific pattern of deficits remains to be studied but, as can be seen from this discussion, the type of apraxic disturbance observed after a callosal lesion or section depends on the patterns of language and motor dominance of the individual patient. The variety of both real and potential disturbances of praxis that could be induced by callosal section is depicted in Fig. 2. Note that patients with callosal agenesis do not demonstrate limb apraxia (Sheremata, Deonna, & Romanul, 1978). They may be using their anterior commissure to transfer their skilled motor and language information, but these patients also are more likely to have stronger ipsilateral projections.

The most common natural lesions that induce callosal apraxia are infarctions in the distribution of the anterior cerebral artery. In these cases, most of the genu and body of the corpus callosum are damaged; however, the splenium usually is spared. Watson and Heilman (1983) described a right-handed patient who was not able to carry out purposeful movements with the left hand on command, in imitation, or in actual object use. Her right-hand performance was normal. The patient had an infarction of the anterior corpus callosum (Fig. 3). Callosal apraxia can be caused by trauma or tumor, by diseases such as cystic degeneration of the corpus callosum (Marchiafava Bignami disease), or by demyelinating diseases. Callosal apraxia also can be induced by cerebral commissurotomy for the treatment of intractable seizures.

b. Apraxia induced by lesions of the supplementary motor area. To perform a skilled movement, the arm must traverse a critical set of spatial loci in a specific temporal pattern; these space and time characteristics must be

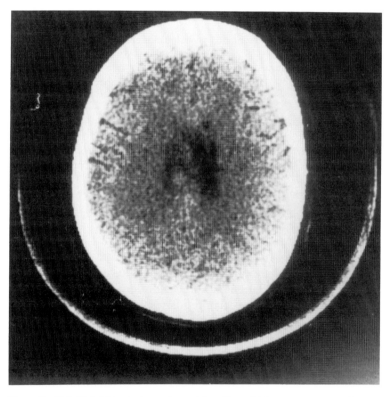

Figure 3   This high CT cut of a patient with unilateral ideomotor apraxia shows a lesion between the two ventricles in the body of the corpus callosum.

coordinated in very precise ways. Because the "movement formulae" have been posited to be coded in a three-dimensional supramodal code, before a skilled movement can take place this space–time representation must be transcoded into an innervatory pattern that adapts itself to the present environmental conditions. Although the final effectors are muscles activated by motor nerves, these motor nerves are controlled by corticospinal neurons found in the precentral gyrus of the frontal cortex (Brodmann's area 4). Although Geschwind (1965) suggested that convexity of premotor cortex may be important for praxis, apraxia has not been reported from a lesion limited to this area and its function remains to be elucidated.

In contrast, bilateral limb apraxia has been reported in two right-handed cases with lesions in the left supplementary motor area (SMA) (Watson, Fleet, Rothi, & Heilman, 1986). In both cases, the deficit was restricted to the performance of transitive gesture whereas intransitive gesture was normal. In the one case the authors were able to test extensively, the patient was able to comprehend and discriminate gestures. How do the contributions of these two motor areas (area 4 and SMA) differ? Watson et al. (1986) reviewed the point that adequate movement skill with a tool (as well as pantomime imitation and intransitive gesture) requires two forms of information: one is a centrally generated movement representation (the movement formula and its innervatory pattern) such as the one described by Liepmann and the other is on-line sensory feedback. In contrast, transitive pantomime on command may be influenced much less by on-line sensory feedback and, in turn, may rely much more crucially on the centrally generated movement representation. Because motor neurons in area 4 discharge after neurons in SMA relative to movement (Tanji & Kurata, 1982), because area 4 neurons are responsive to peripheral sensory stimulation whereas SMA is not (Brinkman & Porter, 1979; Wise & Tanji, 1981), and because SMA is said to influence area 4 during preparation for action (Tanji, Taniguchi, & Saga, 1980), Watson et al. (1986) proposed that the left SMA "translates" the information in the "motor formulae" into motor programs for both hands in much the same way as Liepmann described the activities of the "innervatory pattern." Watson et al. (1986) noted that SMA has both ipsilateral and contralateral innervations and therefore may be less dominant in the left hemisphere than other aspects of the praxis system. For this reason, the incidence of this form of limb apraxia may be lower than that of other forms.

c. Apraxia induced by anterior hemispheric lesions.   As we have discussed, Liepmann (1908) proposed that a lesion that affects the corpus callosum (Fig. 2, Lesion 1) induces left-sided apraxia (in right-handers) or hemiapraxia because the left hemisphere "movement formulae" and "innervatory patterns" do not have access to the right hemisphere motor areas that control movement of the left hand. The right hand, however, would not be paralyzed or apraxic. Liepmann (1908) noted that a variety of other lesions may produce similar disturbances. A lesion in the left hemisphere that affects the callosal fibers where they separate from the common white matter (the roof of the left lateral ventricle) also could induce an apraxia of the left hand (Fig. 2, Lesion 2). Many of the lesions also would induce a right hemiparesis since, in the centrum semiovale, the descending pyramidal fibers are in close proximity to the callosal fibers (Fig. 2, Lesion 3). Lesions of the premotor and motor cortex also would induce a similar clinical picture, that is, paresis of the right hand and apraxia of the left (Fig. 2, Lesion 4). Theoretically, a lesion of the right hemisphere that interrupts the callosal fibers on their way to the premotor cortex but does not involve the descending pyramidal fibers (Fig. 2, Lesion 5) also should induce an apraxia of the left arm. However, we know of no verified cases with a lesion in this distribution.

   d. Apraxia induced by posterior hemispheric lesions.   Using a neural model similar to that used by Wernicke (1874) to explain language processing, Geschwind (1965,1975) proposed that language elicits motor behavior in the following manner. Auditory stimuli reach Heschl's gyrus and undergo auditory analysis; then, they are processed in Wernicke's area (the posterior portion of the superior temporal gyrus), which appears to be important in language comprehension. Wernicke's area is connected to the premotor areas or motor association cortex by the arcuate fasciculus, and the motor association area on the left is connected to the primary motor area on the left. When one is asked to carry out a command with the right hand, this pathway is used. If one wishes to carry out a command with the left hand, a similar neural substrate is used, except that information is carried from the left motor association cortex to the right motor association cortex and then to the right motor cortex.

   Geschwind (1965,1975) suggested that disruptions in certain portions of Wernicke's arc and its connections to the right motor association cortex may explain most of the apraxic disturbances. When patients with lesions in Heschl's gyrus and Wernicke's area fail to carry out verbal commands, their difficulty may not be caused by a defect in skilled movement, but by a comprehension defect. Callosal lesions produce the type of apraxia that we have discussed already. Lesions that destroy the left motor association cortex (also discussed) usually are associated with a right hemiparesis, but they also destroy the callosal pathway from the left motor association cortex to the right motor association cortex.

   According to Geschwind's schema, lesions in the region of the arcuate fasciculus and the supramarginal gyrus should disconnect posterior language areas (important in language comprehension) from the motor association cortex (important in programming movements). Therefore, patients with lesions in this area should be able to comprehend commands but demonstrate difficulty in performing gestures on command. Unlike gesture-to-command that requires left hemisphere language processing, imitation of gesture should not require language processing. Because most right-handed apraxic patients do not have a right hemisphere lesion, they should be able to imitate but cannot. To explain this discrepancy, Geschwind (1965) proposed that the left arcuate fasciculus is dominant for these visuomotor connections.

An alternative hypothesis that may help explain the inability of these patients to carry out skilled movements both on command and in imitation is that these "movement formulae" are equally necessary for both gesture-to-command and gesture imitation tasks. We proposed (consistent with Liepmann's conceptualization) that these formulae are crucial to the programming of skilled movement with either hand and are stored in the dominant parietal lobe (Heilman & Rothi, 1985). In turn, these gesture memories interact with SMA to generate "innervatory patterns" that may gain access to motor areas for implementation. We wished to distinguish between dysfunction caused by destruction of the parietal areas and the apraxia induced by a disconnection of this parietal area from those motor association areas described previously. We tested the ability of patients with dominant (left hemisphere) parietal lesions to differentiate a correctly performed skilled act from a poorly performed one. We found that such patients had more difficulty making these discriminations than did patients with anterior lesions (Heilman, Rothi, & Valenstein, 1982). Patients with posterior left hemisphere lesions often failed to comprehend language. Despite an intact right hemisphere, these patients also failed to comprehend gestures (Rothi, Heilman, & Watson, 1985). Because of their posterior left hemisphere lesions, we proposed that the inability of these patients to comprehend gestures may be related to the destruction of these "movement formulae" in the left parietal lobe.

e. Pantomime agnosia. Agnosia is a recognition failure that cannot be attributed to an elemental sensory defect, a generalized cognitive defect such as dementia, a language disorder such as anomia, or a lack of prior knowledge of the stimulus (Bauer & Rubens, 1985). The gesture discrimination and recognition disorders we discussed previously cannot be considered agnosias because they are associated with disordered praxis production.

We (Rothi, Mack, & Heilman, 1986) described two patients who were not apraxic (i.e., did not have a production deficit) but who could not comprehend, recognize, or discriminate visually presented limb gestures. However, these patients could imitate these same gestures, suggesting that their recognition deficit was not being induced by a visual or perceptual deficit. Therefore, these patients appear to have pantomime agnosia. Both patients could recognize faces and real objects, suggesting that pantomime agnosia may be a specific disorder.

Both patients had ventral temporal occipital lesions (see Fig. 4). We propose that these lesions may have impaired the ability of visual percepts from gaining access to representations of learned skilled movements. Perhaps these patients could imitate because imitation, unlike recognition, does not require access to stored representations (i.e., subjects can imitate nonsense movements or movements they have never seen previously). Alternatively, these patients could gain access to stored representations; however, these representations could not access semantics.

f. Subcortical apraxia. Although the basal ganglia are known to participate in the control of limb movement, severe ideomotor apraxia for transitive movements has been reported only rarely in association with subcortical lesions. Agostoni, Coletti, Orlando, and Fredici (1983), Basso, Luzzatti, and Spinnler (1980), and Rothi, Kooistra, Heilman, and Mack (1988) have reported cases of ideomotor apraxia resulting from lesions of the left basal ganglia and/or thalamus. Most typically the nature of the limb apraxia from these lesions has been described as "mild" and characterized as "failure to execute." However, C. A. Kooistra, L. I. G. Rothi, L. Mack, and K. M. Heilman (unpublished data) have studied the error characteristics of the cases reported in the Rothi et al. (1988) abstract and found that the error pattern was quite similar in nature to the errors of cortically lesioned patients. These praxis errors included a preponderance of perseverations as well as spatial and movement errors. The lesions of the cases studied by Kooistra and associates involved the left lenticular nucleus, internal capsule, and subinsular white matter (see Fig. 5). The lesions were isolated and relatively discrete in size. No significant damage to intra- or interhemispheric white matter pathways was detected. In conclusion, although rarely reported, ideomotor apraxia does appear to result from subcortical lesions. This particular form of limb apraxia certainly warrants additional study.

3. Limb kinetic apraxia

Patients with limb kinetic apraxia have a unilateral (contralesional) disturbance of movement that can be seen when these patients are asked to pantomime acts, to imitate, or to use actual objects. Patients with this disorder improve less with imitation and actual object use than patients with forms of ideomotor apraxia. Unlike the patients with ideomotor apraxia who use body parts as objects, patients with limb kinetic apraxia may have a reduced number of movements or may move slowly or impre-

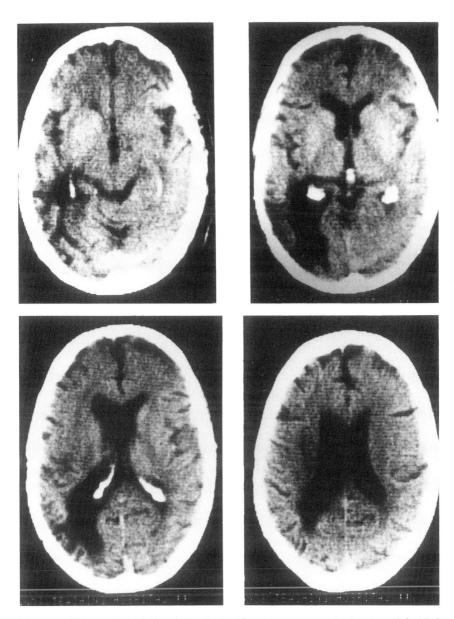

Figure 4 CT scan of a right-handed patient with pantomime agnosia showing a left-sided ventral temporal occipital lesion.

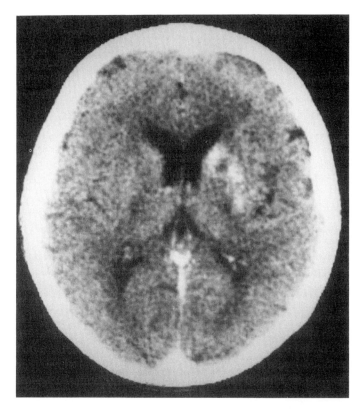

Figure 5 CT scan of a right-handed patient with limb apraxia from a left-sided subcortical lesion involving the internal capsule, lateral thalamus, lenticular nucleus, superior portions of the head of the caudate and the body of the caudate.

cisely (overshoot and undershoot). Usually, this disorder maximally affects movements of the hand.

Liepmann (1908) and Hecaen and Albert (1978) consider limb kinetic apraxia (melokinetic apraxia) to represent a form of movement disorder intermediate between paresis and apraxia. Kleist (1912) thought that this disorder resulted from an inability to link independent muscle groups that have a separate innervation correctly.

The precise anatomical localization of lesions that induce limb kinetic apraxia has not been determined. Liepmann (1908) thought that the seat of the disturbance was the cortex of the central convolutions. He also thought that the neighboring areas may be important. Pyramidal lesions of monkeys (Lawrence & Kuypers, 1968) may induce a clumsiness of the contralateral extremity that cannot be accounted for completely by weakness or changes in tone or posture. Although patients with lesions in motor and premotor cortex often demonstrate clumsy movements of the contralateral hand, they often exhibit spasticity and weakness and tend to assume a hemiparetic posture. Consequently, being certain that they are truly apraxic is often difficult.

4. Verbal–motor dissociation apraxia

In previous reports, verbal–motor dissociation apraxia was termed "ideational apraxia" (Heilman, 1973,1979). This choice of terms was unfortunate because this syndrome is not the same as the ideational apraxia described by Pick (1905) and Liepmann (1908).

As previously discussed, patients with ideomotor apraxia who fail to pantomime correctly often may improve their performance with imitation and the use of actual objects. However, compared with nonapraxics, ideomotor apraxics perform poorly on imitation and, at times, even in their use of actual objects. Patients with limb kinetic apraxia also often fail to imitate or use actual objects correctly. Heilman (1973) described three patients who were unlike others with ideomotor and limb kinetic apraxia; although they performed poorly with either hand on command, their performance in imitation and use of actual objects with either hand was excellent. Although these patients had fluent aphasia, they showed both verbally and nonverbally (by picking out the correct act performed by the examiner) that they comprehended the commands.

Liepmann (1908) interpreted the presence of preserved copying ability to suggest that the patients had lesions posterior to those that induced ideomotor and limb kinetic apraxia. He postulated that patients with preserved imitation were unable to arouse the "movement formulae." Similarly, Heilman (1973) proposed that patients with this form of apraxia had a disconnection between the areas decoding language and those containing the formulae for movements. Only one of the three patients described by Heilman (1973) had a positive radioisotope brain scan. This scan was abnormal in the region of, and deep to, the angular gyrus.

III. BUCCOFACIAL APRAXIA

Buccofacial apraxia (oral apraxia) was described in detail by Jackson (1932) and is defined as an inability to perform learned skilled movements of the lips, facial muscles, tongue, pharynx, larynx, and respiratory muscles. Similar to limb apraxia, the definition of buccofacial apraxia also is based on exclusion of other neurological deficits that may impair motor activity, for example, weakness or akinesia.

A. Incidence and Relevance

Consistent with limb apraxia, the relevance of buccofacial apraxia (other than for its lateralizing or localizing value) may be questioned by some authors. However, in the case of the commonly coexistent verbal communication problem, the presence of buccofacial apraxia may complicate the therapeutic process by rendering useless procedures that involve manipulation of the articulators on command (Square-Storer, 1989). Therefore, our perspective is that better understanding of the mechanism(s) underlying buccofacial apraxia may, at some point, have pragmatic relevance in the clinical setting.

As for limb apraxia, the reported incidence of buccofacial apraxia is variable. For example, one study by Mintz, Raade, and Kertesz (1989) reported that 65% of 112 right-handed subjects with a single left hemisphere lesion had buccofacial apraxia. In contrast, DeRenzi, Pieczuro, and Vignolo (1966) noted that 90% of Broca's aphasics demonstrated buccofacial apraxia. Liepmann (1908) noted that buccofacial apraxia often accompanied Broca's aphasia. Because buccofacial apraxia may occur in the absence of aphasia and Broca's aphasia may occur in the absence of buccofacial apraxia, the relationship is not likely to be causal. Therefore, although we have only a limited number of studies to which to refer, and although these studies are problematic, we are at the very least able to conclude that buccofacial apraxia is a relatively common syndrome subsequent to left hemisphere lesions.

The evolution of buccofacial apraxia has received scant attention in the literature to date. Ochipa and Rothi (1989b) reported a case of crossed buccofacial and limb apraxia in which the limb apraxia cleared with time whereas the buccofacial apraxia did not. Interestingly, Basso, Capitani, Della Sala, Laiacona, and Spinnler (1987) reported that, in a group of subjects in whom limb and buccofacial apraxias coexisted, the limb apraxia improved more often than the buccofacial apraxia. Therefore, in addition to being relatively common, buccofacial apraxia seems to be a relatively enduring syndrome in left hemisphere-damaged patients.

B. Localization of Lesions Producing Buccofacial Apraxia

Although buccofacial apraxia is associated most frequently with left hemisphere lesions, the disorder has on rare occasion been described in patients with right hemisphere lesions (Kramer, Delis, & Nakada, 1985;

Ochipa & Rothi, 1989a; Rapcsak, Rothi, & Heilman, 1987). Less consistent than the lateralization of buccofacial apraxia is the intrahemispheric localization of this disorder. One approach to studying this issue has been to refer to the aphasia types commonly associated with buccofacial apraxia and, therefore, to infer the localization of this form of apraxia. For example, DeRenzi et al. (1966) reported that buccofacial apraxia occurred far more often with Broca's aphasia than with any other aphasia type, suggesting a link to a frontal perisylvian lesion. However, DeRenzi et al. (1966) and others (Benson, Sheremata, Bouchard, Segarra, Price, & Geschwind, 1973; Kertesz, 1979,1985) have reported the co-occurrence of buccofacial apraxia with other forms of aphasia (although admittedly less frequently) that would imply a more posterior perisylvian lesion localization.

Using clinico-neuroradiological methods, Tognola and Vignolo (1980) investigated the brain lesions associated with buccofacial apraxia. CT scans for unilateral, left hemisphere-damaged patients with and without buccofacial apraxia were compared. The critical cortical areas in which lesions were found to correspond with the presence of buccofacial apraxia included the frontal and central (rolandic) opercula, the adjacent portion of the first temporal convolution, and the anterior portion of the insula. In addition, posterior (parietal) structures were not found to be crucial to the presence of buccofacial apraxia in their subjects. Raade, Rothi, and Heilman (1991) evaluated 15 left brain-damaged subjects for the presence of buccofacial apraxia. Their results revealed that the neuroanatomical structures significantly associated with the presence of buccofacial apraxia were inferior frontal gyrus, perisylvian central area, insula, and striatum. In addition, Mintz and colleagues (1989) examined the relationship between buccofacial apraxia (as defined by imitation errors) and lesion localization. Two regions and four specific structures were significant predictors of the occurrence of buccofacial apraxia: the central and the anterior subcortical areas and, specifically, the insula, the inferior frontal gyrus, the centrum semiovale, and the putamen. These studies support the importance of frontal cortical and subcortical gray matter structures in the presence of buccofacial apraxia (see Fig. 6).

In contrast, a clinicopathological study by Benson et al. (1973) provides evidence of the importance of posterior structures in buccofacial praxis processing in two cases of conduction aphasia with buccofacial apraxia. Analysis of the results of the Benson et al. (1973) study reveals that the posteriorly situated inferior parietal lobule, specifically the supramarginal and angular gyri, may play a critical role in buccofacial praxis. In addition, Heilman and co-workers (1983) reported cases of buccofacial apraxia from lesions located in the supramarginal and angular gyri that did not involve frontal cortical structures. Therefore, the anatomical structures critically

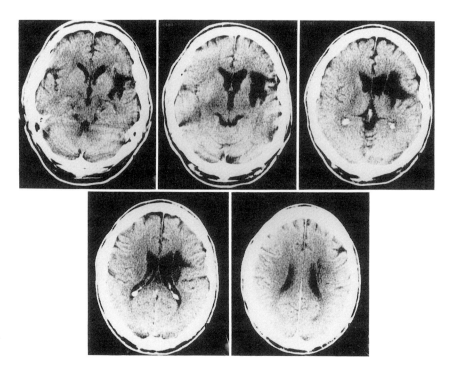

Figure 6 This patient with buccofacial apraxia demonstrated involvement of the following structures: posterior inferior frontal gyrus, perisylvian central cortex, insula, anterior superior temporal gyrus, striatum, anterior internal capsule, corona radiata, and anterior periventricular white matter.

associated with buccofacial apraxia are likely to include the anterior cortical and subcortical structures mentioned earlier, whereas involvement of the more posterior structures remains only a possibility.

REFERENCES

Agostoni, E., Coletti, A., Orlando, G., & Fredici, C. (1983). Apraxia in deep cerebral lesions. *Journal of Neurology, Neurosurgery and Psychiatry, 46*, 804–808.

Basso, A., Capitani, E., Della Sala, S., Laiacona, M., & Spinnler, H. (1985). Natural course of ideomotor apraxia: A study of initial severity and recovery. Paper presented at the annual meeting of the Academy of Aphasia.

Basso, A., Capitani, E., Della Sala, S., Laiacona, M., & Spinnler, H. (1987). Recovery from ideomotor apraxia: A study on acute stroke patients. *Brain, 110*, 747–760.

Basso, A., Luzzatti, C., & Spinnler, H. (1980). Is ideomotor apraxia the outcome of damage to well-defined regions of the left hemisphere? Neuropsychological study of CT correlation. *Journal of Neurology, Neurosurgery and Psychiatry, 43,* 357–369.

Bauer, R. M., & Rubens, A. B. (1985). Agnosia. In K. M. Heilman & E. Valenstein (Eds.), *Clinical neuropsychology* (pp. 187–242). New York: Oxford University Press.

Benson, F., Sheremata, W., Bouchard, R., Segarra, J., Price, D., & Geschwind, N. (1973). Conduction aphasia: A clinicopathological study. *Archives of Neurology, 28,* 339–346.

Borod, J. C., Fitzpatrick, P. M., Helm-Estabrooks, N., & Goodglass, H. (1989). The relationship between limb apraxia and the spontaneous use of communicative gesture in aphasia. *Brain and Cognition, 10,* 121–131.

Brinkman, C., & Porter, R. (1979). Supplementary motor area in the monkey: Activity of neurons during performance of a learned motor task. *Journal of Neurophysiology, 42,* 681–709.

Coslett, H. B., Rothi, L. J. G., Valenstein, E., & Heilman, K. M. (1986). Dissociations of writing and praxis: Two cases in point. *Brain and Language, 28,* 357–369.

De Renzi, E., & Lucchelli, F. (1988). Ideational apraxia. *Brain, 3,* 1173–1185.

De Renzi, E., Motti, F., & Nichelli, P. (1980). Imitating gestures: A quantitative approach to ideomotor apraxia. *Archives of Neurology, 37,* 6–10.

De Renzi, E., Pieczuro, A., & Vignolo, L. A. (1966). Oral apraxia and aphasia. *Cortex, 2,* 50–73.

De Renzi, E., Pieczuro, A., & Vignolo, L. A. (1968). Ideational apraxia: A quantitative study. *Neuropsychologia, 6,* 41–52.

Gazzaniga, M., Bogen, J., & Sperry, R. (1967). Dyspraxia following division of the cerebral commisures. *Archives of Neurology* (*Chicago*), *16,* 606–612.

Geschwind, N. (1965). Disconnexion syndromes in animals and man. *Brain, 88,* 237–294, 585–644.

Geschwind, N. (1975). The apraxias: Neural mechanisms of disorders of learned movements. *American Scientist, 63,* 188–195.

Geschwind, N., & Kaplan, E. (1962). A human disconnection syndrome. *Neurology, 12,* 675–685.

Hecaen, H., & Albert, M. L. (1978). *Human Neuropsychology.* New York: Wiley.

Heilman, K. M. (1973). Ideational apraxia: A re-definition. *Brain, 96,* 861–864.

Heilman, K. M. (1979). Apraxia. In K. M. Heilman & E. Valenstein (Eds.), *Clinical neuropsychology* (pp. 159–155). London: Oxford University Press.

Heilman, K. M., Coyle, J. M., Gonyea, E. F., & Geschwind, N. (1973). Apraxia and agraphia in a left-hander. *Brain, 96,* 1–28.

Heilman, K. M., & Rothi, L. J. G. (1985). Apraxia. In K. M. Heilman & E. Valenstein (Eds.), *Clinical neuropsychology* (pp. 131–150). New York: Oxford University Press.

Heilman, K. M., Rothi, L. J. G., & Kertesz, A. (1983). Localization of apraxia-producing lesions. In A. Kertesz (Ed.), *Localization in neuropsychology* (pp. 371–392). New York: Academic Press.

Heilman, K. M., Rothi, L. J., & Valenstein, E. (1982). Two forms of ideomotor apraxia. *Neurology, 32,* 342–346.

Jackson, H. (1932). Remarks on non-protrusion of the tongue in some cases of aphasia. In J. Taylor (Ed.), *Selected writings* (Vol. II). London: Hodder and Stoughton. (Original work published in 1878.)

Johns, D. F., & Darley, F. L. (1970). Phonemic variability in apraxia of speech. *Journal of Speech and Hearing Research, 13,* 556–583.

Kertesz, A. (1979). *Apraxia, aphasia and associated disorders.* New York: Grune and Stratton.

Kertesz, A. (1985). Apraxia and aphasia: Anatomical and clinical relationship. In E. A. Roy (Ed.), *Neuropsychological studies of apraxia and related disorders* (pp. 163–178). Amsterdam: Elsevier Science Publishers.

Kertesz, A., Ferro, J. M., & Shewan, C. M. (1984). Apraxia and aphasia: The functional–anatomical basis for their dissociation. *Neurology, 34*, 40–47.

Kertesz, A., & Hooper, P. (1982). Praxis and language. The extent and variety of apraxia in aphasia. *Neuropsychologia, 20*, 275–286.

Kleist, K. (1912). Der Gang und der gegenwürtige Stand der Apraxie-Forschung. *Ergebnisse der Neurologie und Psychiatrie, 1*, 342–452.

Kramer, J. H., Delis, D. C., & Nakada, T. (1985). Buccofacial apraxia without aphasia due to a right parietal lesion. *Annals of Neurology, 18*, 512–514.

Lawrence, D. G., & Kuypers, H. G. J. M. (1968). The functional organization of the motor systems in the monkey. *Brain, 91*, 1–36.

LeMay, A., David, R., & Thomas, A. P. (1988). The use of spontaneous gesture by aphasic patients. *Aphasiology, 2*, 137–145.

Liepmann, H. (1905). The left hemisphere and action. *Münchener medizinische Wochenschrift*, 48–49. (In D. Kimura, *Translations from Liepmann's essays on apraxia* (1980), Research Bulletin #506, Department of Psychology, The University of Western Ontario, Canada.)

Liepmann, H. (1908). *Drei Aufsätze aus dem Apraxie-Gebiet*. Berlin: Karger.

Liepmann, H. (1920). Apraxie. *Ergebnisse bei Gesamter Medizin, 1*, 515–543.

Liepmann, H. & Maas, O. (1907). Fall von linksseitiger Agraphie und Apraxie bei rechtsseitiger Lähmung. *Zeitschrift fuer Psychologie und Neurologie, 10*, 214–227.

Mintz, T., Raade, A., & Kertesz, A. (1989). Lesion size and localization in buccofacial apraxia: A retrospective study. Paper presented at the meeting of the Canadian Association of Speech-Language Pathologists and Audiologists (CASLPA), Toronto, Ontario.

Ochipa, C., & Rothi, L. J. G. (1989a). Buccofacial apraxia recovery in a patient with atypical cerebral dominance. *ASHA, 31*, 73.

Ochipa, C., & Rothi, L. J. G. (1989b). Recovery and evolution of a subtype of crossed aphasia. *Aphasiology, 3*, 465–472.

Ochipa, C., Rothi, L. J. G., & Heilman, K. M. (1992). Conceptual apraxia in Alzheimer's disease. *Brain, 115*, 1061–1071.

Ochipa, C., Rothi, L. J. G., & Heilman, K. M. (1989). Ideational apraxia: A deficit in tool selection and use. *Annals of Neurology, 25*, 190–193.

Pick, A. (1905). *Studien über motorische Apraxie und ihre nahestehende Erscheinungen*. Leipzig: Deuticke.

Pieczuro, A., & Vignolo, L. A. (1967). Studio sperimentale sull'aprassia ideomotoria. *Sistema Nervoso, 19*, 131–143.

Poeck, K. (1983). Ideational apraxia. *Journal of Neurology, 230*, 1–5.

Poeck, K. (1985). Clues to the natures of disruptions to limb praxis. In E. A. Roy (Ed.), *Neuropsychological studies of apraxia and related disorders* (pp. 99–109). Amsterdam: Elsevier–North Holland.

Poeck, K. (1986). The clinical examination for motor apraxia. *Neuropsychologia, 24*, 129–134.

Poizner, H., Soechting, J. F., Bracewell, M., Rothi, L. J., & Heilman, K. M. (1989). Disruption of hand and joint kinematics in limb apraxia. *Society for Neuroscience Abstracts, 15*, 481.

Raade, A. S., Rothi, L. J. G., & Heilman, K. M. (1991). The relationship between buccofacial and limb apraxia. *Brain and Cognition, 16*, 130–146.

Rapcsak, S. Z., Rothi, L. J. G., & Heilman, K. M. (1987). Apraxia in a patient with atypical cerebral dominance. *Brain and Cognition, 6*, 450–463.

Rothi, L. J. G., Heilman, K. M., & Watson, R. T. (1985). Pantomime comprehension and ideomotor apraxia. *Journal of Neurology, Neurosurgery and Psychiatry, 48*, 207–210.

Rothi, L. J. G., Kooistra, C., Heilman, K. M., & Mack, L. (1988). Subcortical ideomotor apraxia. *Journal of Clinical and Experimental Neuropsychology, 10*, 48.

Rothi, L. J. G., Mack, L., & Heilman, K. M. (1986). Pantomime agnosia. *Journal of Neurology, Neurosurgery and Psychiatry, 49*, 451–454.

Rothi, L. J. G., Ochipa, C., & Heilman, K. M. (1991). A cognitive neuropsychological model of limb praxis. *Cognitive Neuropsychology, 8*, 443–458.

Roy, E. A., & Square, P. A. (1985). Common considerations in the study of limb, verbal and oral apraxia. In E. A. Roy (Ed.), *Advances in psychology: Neuropsychological studies of apraxia and related disorders* (pp. 111–162). Amsterdam: North-Holland.

Sheremata, W. A., Deonna, R. W., & Romanul, F. C. (1978). Agenesis of the corpus callosum and interhemispheric transfer of information. *Neurology, 23*, 390.

Square-Storer, P. (1989). Traditional therapies for apraxia of speech reviewed and rationalized. In P. Square-Storer (Ed.), *Acquired apraxia of speech in aphasic adults* (pp. 145–161). London: Taylor and Francis.

Steinthal, P. (1871). *Abriss der Sprachwissenschaft*. Berlin: Dümmlers Verlagsbuchh.

Sundet, K., Finset, A., & Reinvang, I. (1988). Neuropsychological predictors in stroke rehabilitation. *Journal of Clinical and Experimental Neuropsychology, 10*, 363–379.

Tanji, J., & Kurata, K. (1982). Comparison of movement-related neurons in two cortical motor areas of primates. *Journal of Neurophysiology, 48*, 633–653.

Tanji, J., Taniguchi, K., & Saga, T. (1980). Supplementary motor area: Neuronal response to motor instructions. *Journal of Neurophysiology, 43*, 60–68.

Tognola, G., & Vignolo, L. A. (1980). Brain lesions associated with oral apraxia in stroke patients: A clinico-neuroradiological investigation with the CT scan. *Neuropsychologia, 18*, 257–272.

Valenstein, E., & Heilman, K. M. (1979). Apraxic agraphia with neglect-induced paragraphia. *Archives of Neurology, 26*, 506–508.

Watson, R. T., Fleet, W. S., Rothi, L. G. J., & Heilman, K. M. (1986). Apraxia and the supplementary motor area. *Archives of Neurology, 43*, 787–792.

Watson, R. T., & Heilman, K. M. (1983). Callosal apraxia. *Brain, 106*, 391–403.

Wernicke, C. (1874). *Der aphasische Symptomenkomplex*. Breslau: Cohn & Weigart.

Wise, S. P., & Tanji, J. (1981). Supplementary and precentral motor cortex: Contrast in responsiveness to peripheral input in the hindlimb area of the unanesthetized monkey. *Journal of Comparative Neurology, 195*, 433–451.

Chapter 14

Lesion Localization in Visual Agnosia

Janet Jankowiak and Martin L. Albert

I. INTRODUCTION

Visual agnosia is one of a group of rare selective disorders of visual recognition that have sparked intense investigation over the past decade. The interest in determining lesion localization in cases of visual agnosia is better to understand which areas of the brain may be responsible for processing and integration of visual input. Definitive knowledge of lesion localization can be determined best by autopsy, but few have been reported in the literature. Most data on lesion localization in visual agnosia are available from other static studies (computerized tomography, CT; magnetic resonance imaging, MRI). However, the understanding of the anatomical basis of complex visual processing requires compilation and coordination from static studies as well as dynamic neuroimaging studies (positron emission tomography, PET; single photon emission CT, SPECT; functional MRI), the latter of which are still in early stages of exploration. Hence our conclusions necessarily must be tentative and undoubtedly will be modified with the introduction of new information concerning functional neuroanatomy and chemical neuroanatomy over the next decade.

II. CLINICAL DEFINITION OF VISUAL AGNOSIA

Teuber's definition of agnosia as "a percept stripped of its meaning" has the appealing merit of brevity and clarity (Teuber, 1968). Visual agnosia, thus, refers to the ability to perceive visual stimuli without appreciating their meaning. Freud (1891) defined visual agnosia as the inability to recognize objects visually in the absence of a primary visual deficit, general intellectual impairment, or aphasia. Patients with the disorder are,

Localization and Neuroimaging
in Neuropsychology

nonetheless, able to recognize objects when they are presented in a tactile or auditory modality.

Lissauer (1890) usually is given credit for the first thorough clinical description of visual agnosia, which he called "mindblindness." He described an elderly man who suddenly developed a right homonymous visual field defect without associated aphasia or disturbance of motor or sensory function. He had difficulty dressing (presumably a dressing apraxia) and could not recognize family members on sight (prosopagnosia), find his way about a familiar neighborhood (topographagnosia), identify familiar objects presented visually (visual agnosia), identify colors (color agnosia or color anomia), or read, although he could write intelligent letters to friends (alexia without agraphia). On the basis of this case and others seen in his clinic, Lissauer proposed two possible forms of visual agnosia.

1. "Apperceptive" mindblindness referred to a defect of identification caused by distortion of visual perception. Patients with this form could not describe the features of a visually presented stimulus, reproduce it by drawing, or match it to similar alternatives, although they could see it. Normally in this stage of visual recognition, physical characteristics of a stimulus are processed and integrated into a structured percept (DeRenzi, Faglioni, Grossi, & Nichelli, 1991; DeRenzi, Scotti, & Spinnler, 1969). Marr (1982) divides this stage into two levels of visual representation, both "viewer centered," to arrive at a "two-and-a-half-dimensional sketch" of the object. The first level—"the primal sketch"—incorporates light intensity and geometry whereas the second level takes account of the orientation, depth, and discontinuities of the visible surfaces.

2. "Associative" mindblindness was characterized by an inability to identify visual stimuli despite intact visual perception. Patients with this form could describe the stimulus, draw it, and match it but, despite achieving a fully structured percept, failed to comprehend what they saw (Benson & Greenberg, 1969). Normally, in this stage, a fully specified structured description of a stimulus evokes associations, stored in semantic memory, that give it its meaning (DeRenzi et al., 1991). For Marr (1982), recognition of an object requires a third step which yields a "three-dimensional model representation" and provides an "object-centered" description that is compared with memory stores of like descriptions.

The term "visual agnosia" has replaced mindblindness but the dichotomy proposed by Lissauer has remained, although at times viewed controversially. In fact, the very concept of visual agnosia as a distinct entity has been questioned. In 1953, Bay proposed that the apparently selective impairment in visual object recognition was due to a combination of two more general deficits: (1) All the reported cases he reviewed suffered from at least subtle impairments in elementary visual functions. (2) In addition, all the cases demonstrated general intellectual decline. Bender and Feldman (1972) further supported this notion after reviewing 20 years of case records from their clinical practice. Despite a large population sample, they stated that they did not find a single case of "true visual agnosia without concomitant defects in the fields of vision or perception, mental dysfunction, or dysphasia." Rather, they found a spectrum of patients with severe mental and mild visual defects at one end and mild mental and severe visual defects at the other end. They selected four patients to describe in detail as evidence of their conclusions. All four had a slowly progressive course of deterioration most consistent with a dementing process. Almost two decades later, Mendez, Mendez, Martin, Smyth, and Whitehouse (1990a) and Mendez, Tomsak, and Remler (1990b) studied 30 patients with dementia of the Alzheimer type and found that, although patients had preserved visual acuity and color recognition, they all had disturbances in figure–ground analysis and 57% had difficulties with visual recognition of real objects (visual agnosia). Patients with more severe dementia had the most complex visual disturbances. Thus, these researchers concluded that a range of complex visual disturbances is common in Alzheimer's disease. Benson, Davis, and Snyder (1988) also noted visual agnosia as an early feature in their study of five patients with a progressive cognitive decline associated with posterior cortical atrophy. We surmise, therefore, that the Bay (1953) and Bender and Feldman (1972) cases may have included a mixture of medical problems, such as stroke, Alzheimer's disease, or head injury.

Despite the skepticism regarding a "pure" visual agnosia, Ettlinger (1956) argued, after performing a series of visual discrimination tests on a group of brain damaged patients, that some patients with severe elementary visual impairments were not agnosic. Further, as Efron (1968) pointed out, the simple occurrence of a visual defect in a patient with visual agnosia is not proof that the failure of recognition is due to that defect. More recently, Farah (1990) defined visual agnosia as "an impairment in the higher visual processes necessary for object recognition, with relative preservation of elementary visual functions." She cited a range of preserved and impaired visual capabilities that vary from one case of visual agnosia to another. The specific role of elementary visual impairment in visual agnosia has not been resolved (Benson & Greenberg, 1969; Levine & Calvanio, 1989). However, from numerous case reports that document established etiologies and onsets that can be correlated with specific focal lesions, visual agnosia has become accepted as a distinct disorder (Alexander & Albert, 1983; Hecaen & Albert, 1978).

A. Apperceptive Visual Agnosia

1. Clinical features

The failure to recognize visually presented stimuli in cases of apperceptive visual agnosia is believed to be the result of distortion of visual perception despite relatively preserved primary visual functions. These patients generally have intact visual acuity, brightness discrimination, and elementary color vision, yet are impaired severely in describing the features of a visual stimulus, copying it, or matching it to similar objects.

Different authors have focused on different attributes of their respective patients' presentation, leading to the creation of possible subgroups of apperceptive visual agnosia. Some patients seem to present simultaneously features of apperceptive and associative visual agnosia and consequently, are difficult to classify. Table 1 lists cases defined as apperceptive visual agnosia; etiologies and lesion localizations are listed also.

The deficit in apperceptive visual agnosia commonly appears to be at the level of shape or form discrimination, although the patient often is able to trace the shape with hand or head movements. However, the tracing process frequently is termed "slavish" and the patient may be distracted completely by irrelevant lines, being unable to discriminate figure from ground. Nonetheless, the patient may be able to recognize readily real objects in which color and size clues help compensate for defective perception of form (Sparr, Jay, Drislane, & Venna, 1991). Performance generally improves when objects are shown in context rather than in isolation. However, responses are usually inferences made by piecing together color, size, textures, and contextual cues, and patients may express uncertainty about the correctness of their answers. Responses occasionally may appear confabulatory or perseverative, probably because perceptual clues have been misinterpreted (Kertesz, 1979). However, some of these patients have been able to guess correctly the functional category of the object despite an inability to identify the specific target. For example, Campion and Latto's (1985) patient asked "Is it a tool?" when shown a pair of scissors.

Carbon monoxide poisoning represents a "classic" etiology of apperceptive visual agnosia. Some patients with this etiology present with specific clinical features that would suggest their inclusion in a common subgroup (Mendez, 1988; Sparr et al., 1991). For example, these patients are said to view objects in a "piecemeal" fashion, using feature-by-feature analysis. When viewing complex visual patterns, they attempt to reduce

Table 1 Reported Cases of Apperceptive Visual Agnosia: Lesion Localization and Etiology

Reference	Lesion localization	Etiology
Bauer & Verfaellie (1988)	CT: bilateral temporal–occipital; Angio: 50% stenosis L vertebral	Infarction
Benson & Greenberg (1969)	Pneumoencephalogram: bilateral posterior ventricular dilatation	Carbon monoxide
Campion & Latto (1985)	CT: diffuse low-density L occipital (striate and prestriate); R prestriate (also frontal—not mentioned)	Carbon monoxide
DeRenzi et al. (1991)		
1	MRI: lateral R temporo-occipital, medial R temporal, bilateral parasagittal frontal	Trauma
Kertesz (1979)	CT: L deep occipital, R frontal	Trauma
Larrabee et al. (1985)	CT: L occipital, R anterior limb internal capsule	Infarction
Mendez		
1	CT: bilateral basal ganglia; MRI: absent splenium corpus callosum	Carbon monoxide
2 (evolved over time); L handed	CT: L occipital; R occipital	Mitochondrial encephalopathy, lactic acidosis, and stroke-like syndrome (MELAS)
Sparr et al. (1991)	CT: bilateral occipital atrophy; MRI: bilateral occipital atrophy	Carbon monoxide
Warrington & James (1988)		
1	CT: L occipital hematoma; midline shift to L	Hemorrhage
2	CT: R parietal cyst	Metastatic squamous cell carcinoma
3	CT: R posterior temporal–parietal; ring enhancing after contrast	Grade IV astrocytoma

the complexity by focusing on smaller and more manageable components. However, since some patients still fail to recognize even these elemental components, they consequently cannot reintegrate them into a coherent whole (Sparr et al., 1991). Other patients are capable of performing adequately on tasks of visual matching and copying in a slow, laborious manner if given enough time (Mendez, 1988). This feature-by-feature analysis greatly compromises both speed and accuracy in visuoperceptual tasks. For example, performance is usually poor on tests that distort or obscure salient features of the visual image, for example, overlapping figures (e.g., Poppelreuter test) or identification of fragmented representations of letters or objects (e.g., Gollins test).

Investigators have speculated that at least some patients with apperceptive visual agnosia are unable to achieve an immediate global impression of a complex pattern and revert to processing in a slower serial fashion, thus analyzing systematically feature by feature. This pattern suggests a disturbance in the initial phase of visual recognition, which is believed to occur through parallel processing. Normally, parallel processing by multiple units results in the simultaneous treatment of a visual stimulus, which facilitates an early extraction of the overall spatial relationships within the useful field of view. These units have multiple, large, overlapping receptive fields, creating a trade off between resolution and accuracy. With neuronal damage and loss of receptive fields, resolution decreases and the patient reverts to a slower process of serial analysis of smaller fields of view. In addition, researchers postulate that the duplicate processing provided by the overlapping receptive fields normally aids in "cleaning up" the visual image by selecting the "correct" connections and inhibiting incorrect ones. Thus, the essence of a pattern is extracted and the background "noise" is suppressed. When parallel processing is disturbed, irrelevant components of a visual input may not be ignored; thus, background noise and figure–ground problems interfere (Mendez, 1988).

Whether processing by a systematic feature-by-feature analysis and reversion to serial processing are unique to patients with apperceptive visual agnosia is debatable. Clearly, this behavior is not limited to patients who suffered carbon monoxide intoxication. Bauer and Verfaellie (1988) presented a patient with prosopagnosia and possible apperceptive visual agnosia whose "descriptions of her percepts indicated fragmented vision with serial scanning of details." Although the focus of their study was on the psychophysiological distinction between processing of familiar and unfamiliar faces, they argued that her visual recognition defects were "apperceptive with some of the characteristics of 'simultanagnosia.'" They did not attempt to specify the nature of her visual agnosia, but simply noted that "she named real objects with hesitation, but was nearly totally incapable of naming line drawings or photographs."

The patient described by Riddoch and Humphreys (1987a) reportedly showed features of both apperceptive and associative visual agnosia, which they termed "integrative visual agnosia." His visual object identification was based on serial independent local feature descriptions that he could not integrate into a meaningful whole. Performance was consequently very slow because of feature-by-feature analysis. These authors postulated that their patient's inability to identify more than one object at a time was the result of a deficiency in switching attention between simultaneously available stimuli.

Other authors have suggested that morphological errors are a sign of "perceptual underspecification" (Charnallet, Carbonnel, & Pellat, 1988). Their patient was particularly interesting because his visual recognition deficit was restricted to the right visual field only—a hemiagnosia. He also showed features of apperceptive and associative visual agnosia in that visual field, but his core deficit was termed "a form recognition disorder" (see Table 4).

Other patients have been described with features of both apperceptive and associative visual agnosia; etiologies have been variable. Although Mendez (1988) classified both of his patients (one with carbon monoxide poisoning and the other with mitochondrial encephalopathy, lactic acidosis, and stroke-like syndrome, MELAS) as having "an apperceptive profile," he noted that they could copy and match (features of associative visual agnosia). Nonetheless, he considered them to be a subgroup of apperceptive visual agnosia because, if given enough time, these patients performed adequately on these tasks by a slow serial comparison of smaller segments. Kertesz (1979) described a patient with what he described as a "dual deficit of perception and recognition" after a severe head trauma. He argued that her ability to draw better from memory than to copy was suggestive of apperceptive visual agnosia, whereas most of her other findings were more characteristic of associative visual agnosia.

Yet another group of patients has been described, none of whom seems to fit well into the apperceptive–associative visual agnosia dichotomy. Warrington and James (1988) described three patients: one with a unilateral right occipital hematoma, another with metastatic squamous carcinoma to the right parietal lobe, and a third with a grade IV malignant astrocytoma in the right posterior temporo-parietal region. All three were classified as apperceptive visual agnosics although they could copy and perform match-to-sample tasks. This classification was based on their severely impaired performance on tasks that manipulated perception by degrading or distorting salient features of the visual target. Despite their impaired visual perception, these patients were not impaired seriously in everyday life, a result that was attributed to preservation of visual semantic knowledge as tested by tasks that (1) compared relative sizes of animals to objects, (2) required pairing of items that were physically dissimilar but had a common function and name (e.g., two types of dogs), and (3) forced decisions regarding semantic attributes of groups of animals and objects. These investigators concluded that apperceptive visual agnosia is a deficit in "perceptual categorization" (i.e., inability to match objects based on physical features) that occurs at a postsensory and presemantic stage of visual analysis. Unlike other apperceptive visual agnosics described, these patients had intact visual shape discrimination. According to these authors, patients whose object recognition deficits are characterized by failure in tasks of shape discrimination should be classified as "pseudoagnosics."

In addition to patients who seem to present with simultaneous features of both apperceptive and associative visual agnosia are others who seem to evolve from one form to another. Larrabee, Levin, Huff, Kay, and Guinto (1985) described the case of a patient who initially (10 days after infarction) had an apperceptive visual agnosia, based on her inability to name common objects presented visually, to copy drawings, or visually to match common objects to samples. By 48 days after infarction, she could match visually pairs of common objects, suggesting that she had evolved to an associative visual agnosic. At 2 yr postonset, she was able to name real objects although she continued to have difficulty naming pictures of objects. This case supports a notion that the deficits seen in visual agnosia may fall along a continuum rather than into discrete clinical entities and may evolve over time.

2. Lesion localization

Lesion localization in apperceptive visual agnosia tends to be posterior in the cerebral hemispheres, involving occipital, parietal, or posterior temporal regions bilaterally. Small focal or unilateral lesions rarely, if ever, produce this syndrome. Pathological changes are scattered widely, as might occur with carbon monoxide intoxication. Before the era of CT and MRI, anatomical correlations were even more speculative. Benson and Greenberg's (1969) patient had a radioactive isotope brain scan that was normal. A pneumoencephalogram revealed bilateral posterior ventricular dilatation that would be consistent with posterior hemispheric atrophy. In addition, a persistent bilateral slow wave pattern in the parieto-occipital regions was found on electroencephalograms (EEGs). These authors postulated that diffuse cerebral edema had occurred early in the course, causing a herniation syndrome with bilateral compression of the posterior cerebral arteries. This vascular insult presumably accentuated the anoxic effects of their patient's carbon monoxide poisoning. These investigators further speculated that, because carbon monoxide poisoning produces a laminar necrosis, it could separate interacting striatal neurons and thereby interfere with form perception. They argued that form perception may be more dependent than other attributes of visual discrimination (e.g., perception of color, light intensity, or direction) on the interaction of several neurons, connected by fibers presumably running through a single specific cortical level of the calcarine cortex.

Benson and Greenberg (1969) also suggested a second mechanism for the deficit in form perception found in their patient with carbon monoxide poisoning, based on diffuse cerebral insufficiency. They predicted multifocal disseminated cortical lesions, also called "salt and pepper infarctions." Almost two decades later, Campion and Latto (1985) postulated a similar mechanism to explain the apperceptive visual agnosia manifested by their patient after carbon monoxide intoxication. A CT scan revealed diffuse lesions, with a prominent low density area above the calcarine sulcus on the left that extended into the striate and prestriate cortex. A smaller low density patch was noted in the prestriate cortex of the right hemisphere. Again, because carbon monoxide in known to cause disseminated multifocal infarctions of cortical tissue as a result of hypoxemia, these researchers postulated that the multiple infarcts produced minute scotoma that "peppered" the entire visual field. This condition would result in a "masking" of object contours, preventing the patient from appreciating a global overview or details.

A T_1-weighted MRI on another patient who developed apperceptive visual agnosia after carbon monoxide poisoning showed findings similar to those seen on her noncontrasted CT (Sparr et al., 1991). Bilateral occipital atrophy was noted, but no focal pathology. A T_2-weighted study showed patchy areas of increased signal over the occipital sulci. Gliosis in the immediately adjacent subcortex could not be ruled out definitively.

An initial CT scan of Mendez's (1988) patient, who also suffered carbon monoxide intoxication, demonstrated bilateral nonenhancing lucencies in the basal ganglia. A subsequent CT scan revealed resolution of these abnormalities. An MRI scan showed absence of the splenium of the corpus callosum, an incidental finding. No other findings on the MRI were mentioned, nor was the time at which it was done relative to the anoxic event. Generally, dysgenesis of the corpus callosum is clinically asymptomatic; the author did not feel that this condition could explain the patient's visuoperceptual deficits or reversion to serial processing.

Depending on the etiology of apperceptive visual agnosia, neuroimaging studies may show variability of lesion localization, although the presence of bilateral posterior lesions remains fairly constant. In the patient with MELAS (Mendez, 1988), a CT scan showed a left occipital non-enhancing hypolucency following the acute onset of a generalized seizure with a residual right homonymous hemianopsia. However, an EEG suggested bilateral involvement by the presence of bioccipital spike and wave discharges. Over the following month, the right visual field defect resolved and reappeared on two occasions. The patient then developed a transient left homonymous hemianopsia with a corresponding new right occipital hypolucency on CT scan. Subsequent CT scans reportedly "returned to normal." However, 8 mo after his original presentation, the patient developed transient blindness. A CT then revealed bilateral occipital hypolucencies, which apparently persisted. EEGs showed diffuse slowing and epileptiform discharges emanating from the occipital regions. No additional anatomical correlates were given, although the author attributed the patient's apperceptive visual recognition deficits to damage to primary visual cortical areas.

Patients whose etiology is trauma also tend to have diffuse or multifocal injury, with frequent frontal involvement. On a CT scan, Kertesz's (1979) patient, who demonstrated features of apperceptive and associative visual agnosia, showed an area of decreased absorption in the left occipital region from the occipital horn to the periphery, as well as a left frontal hypolucency. The pattern was consistent with a coup–contrecoup injury. Ventricular dilatation was noted also, consistent with cerebral atrophy. The MRI of another patient who suffered head trauma, resulting in shape discrimination difficulties, showed "lesions (possibly of ischemic nature) in the lateral part of the right temporo-occipital region, in the medial aspect of the right temporal lobe and in both parasagittal frontal regions" (DeRenzi et al., 1991).

When cerebral infarction is the etiology for the visual recognition deficits, patients tend to have features of both apperceptive and associative visual agnosia. The patient with "integrative visual agnosia" described by Riddoch and Humphreys (1987a) demonstrated "extensive infarctions extending forward in both occipital lobes in the distribution of the posterior cerebral arteries." The patient of Bauer and Verfaellie (1988) with prosopagnosia and possible apperceptive visual agnosia also demonstrated bilateral occipito-temporal infarctions on a CT scan. An angiogram showed 50% stenosis of the left vertebral artery; no mention was made of the right vertebral artery. Charnallet et al.'s (1988) patient with right visual hemiagnosia showed an area of hypodensity (ischemic infarction) in the left lingual and fusiform gyri on a nonenhanced CT scan. MRI confirmed the left inferior occipital lesion, which was situated under the calcarine fissure. In addition, the scan revealed two other lesions: a small punctiform hypersignal on the medial side of the left cuneus above the inferior occipital lesion and a larger punctiform lesion in the homologous white matter of the right hemisphere. The discovery of a second contralateral lesion by MRI demonstrates clearly that CT scans are not always sufficient to document the nature or extent of lesion.

The three patients described by Warrington and James (1988), who were classified as apperceptive visual agnosics but did not have any deficits in form recognition, all reportedly had unilateral right sided lesions on CT scans. The first patient had an intracerebral hematoma in the right occipital region. Mass effect and slight shift of the midline to the left was reported. A repeat CT scan approximately 1 yr later showed complete resorption of the hematoma with a residual cystic cavity. Psychological testing of the patient apparently began around the time of the first CT scan, raising the question of what effect the midline shift and compression of the left hemisphere may have had. The second patient had a large cystic lesion in the right parietal lobe, as demonstrated on CT. A craniotomy was performed with resection of a metastatic squamous carcinoma and subsequent cranial irradiation. Psychological studies on this patient were conducted preoperatively. Again, the effect the probable surrounding edema may have had on testing is unknown. In addition, the CT may not have revealed a left-sided lesion that may have been discovered with an MRI scan. The third patient had a right posterior temporo-parietal low density lesion with ring enhancement noted on a contrast-enhanced CT scan. Surrounding edema with midline shift to the left was noted. The patient's psychological testing was conducted prior to a craniotomy that confirmed a grade IV malignant astrocytoma. As with the first patient, one could question what effect the edema and midline shift, compressing the left hemisphere, may have had on behavior.

The patient of Larrabee et al. (1985) who evolved from an apperceptive to an associative visual agnosic had an initial CT scan performed 7 days postonset. This scan demonstrated decreased attenuation in the left occipital lobe and the anterior limb of the right internal capsule. A repeat scan at 20.1 mo was reportedly consistent with cerebral infarctions in these locations. However, because the patient was a known diabetic, she also may have had other lacunar infarctions, probably distributed bilaterally.

In summary, lesion localization in patients with apperceptive visual agnosia tends to be diffuse or multifocal and to involve the posterior cerebral hemispheres. When elements of associative visual agnosia are present, the lesion may be more focal and usually involves posterior cerebral artery (PCA) territory bilaterally. Some tendency exists for right PCA involvement to predominate. To date no autopsies or studies using PET of patients with apperceptive visual agnosia have been reported.

B. Associative Visual Agnosia

1. Clinical features

Classically, associative visual agnosia is defined as a selective impairment in the recognition of visually presented objects despite adequate visual perception. To be included in this category, the patient should meet the following criteria: (1) be unable to recognize objects when seen, including an inability to name them, demonstrate their function by gesture, or group them into their semantic category; (2) be able to show recognition of the object through modalities other than vision, such as tactile or auditory; (3) demonstrate visual perception adequate to discriminate one object from another; and (4) exhibit no aphasia, dementia, clouding of consciousness, or other sensory or motor abnormality. Traditionally, these patients can copy objects or pictures they cannot recognize and match pairs of objects as being the same or different.

Reports of patients with associative visual agnosia are perhaps even more heterogeneous than those with apperceptive visual agnosia. Review of the literature reveals that virtually no two patients with visual agnosia are identical. As Benson and Greenberg (1969) state, "the number of suggested mechanisms [for visual agnosia] very nearly equals the number of recorded cases." Because visual agnosia for objects rarely exists in isolation, different authors tend to focus on different aspects of the presenting features; consequently, the same patient may be cited by various authors as an example of a different syndrome. For example, visual agnosia frequently is seen with prosopagnosia. This latter syndrome has been the target of considerable investigation, particularly over the past decade, partly because of the striking functional significance of a patient's inability to recognize familiar faces visually. Consequently, many cases of prosopagnosia are reported that clearly have not explored the patient's full ability to recognize other visual stimuli (e.g., real objects, photographs, line drawings, objects viewed from unusual perspectives). By minimizing these other aspects of visual recognition, some authors have concluded that prosopagnosia exists in isolation and that the processing of human faces may be attributed to a unique process with specific anatomical localization (Baylis, Rolls, & Leonard, 1985; Bodamer, 1947; Perrett, Smith, Potter, Mistlin, Head, Milner, & Jeeves, 1984; Tzavaras, Hecaen, & LeBras, 1970; Yin, 1970). Other authors have suggested that prosopagnosia is not specific to human faces (Bauer & Trobe, 1984; Davidoff & Landis, 1990) but may be broadened to encompass stimuli belonging to visually "ambiguous" categories (Damasio, Damasio, & Van Hoesen, 1982). Table 2 lists various visual recognition and cognitive deficits that have been found associated with visual agnosia.

Table 2 Visual Recognition and Cognitive Deficits Associated with Visual Agnosia

Reference	Apperceptive VA	Associative VA	VA otherwise classified	Prosopagnosia	Topographagnosia	Alexia	Color disturbance	Other cognitive deficits
Albert et al. (1975a)	+			+	?[a]	–	+	Amnesia
Aptman et al. (1977); L handed			+	+	NM[b]	+	–	VIQ: 97; PIQ: 58; neglect L VF
Assal et al. (1984)			+ (zooagnosia)	+	+	–	–	Retrograde amnesia
Bauer (1982)			NS[c]	+	?		–	Amnesia; poor interpretation facial expression
Bauer & Verfaellie (1988)	+			+	NM	+	+	Poor visual memory; simultanagnosia
Benke (1988)		+		+	–	+	+	Amnesia; poor visual categorization
Benson & Greenberg (1969)	+			+	NM	+	–	
Benton & Van Allen (1972)			?	+	NM	–	NM	Amnesia
Beyn & Knyazeva (1962)			NS	+	?	NM	+	Simultanagnosia; visual amnesia; poor visual imagery
Bornstein et al. (1969)			Zooagnosia	+	–	–	–	Mild amnesia
Bruyer et al. (1983)			NS	+	–	–	+	Simultanagnosia; mild STM loss
Campbell et al. (1990)								
1			NM	+	+	–	NM	–
2			?	+ (developmental)	+	–	–	
Campion & Latto (1985)	+			NM	NM	NM	+	Mild amnesia, resolved
Charnallet et al. (1988)			+NS	NM	NM	+	–	
Cohn et al. (1977)								
1			"Grossly intact"	+	?	–	+	Severe amnesia
2			NM	+	NM	NM	NM	Amnesia; disoriented to time

continues

Table 2 (*continued*)

Reference	Apperceptive VA	Associative VA	VA otherwise classified	Prosopagnosia	Topographagnosia	Alexia	Color disturbance	Other cognitive deficits
Damasio et al. (1982)								
1			NM	+	NM	NM	+	NM
2			NS	+	NM	+	+	NM
3			NM	+	NM	+	-	NM
Damasio & Damasio (1983)								
4		+		+	NM	+	+	NM
7			NS	NM	NM	+	+ (L. hemiachromatopsia)	NM
Davidoff & Wilson (1985)		+		?	NM	?	NM	Tactile agnosia; VIQ: 94
DeHaan et al. (1987)			NS	+	NM	-	-	Amnesia
DeRenzi (1986)								
1			NM	+	+	NM	NM	
2			NS	+	+	?-	NM	
DeRenzi et al. (1991)								
1	+			+	NM	Paralexic	NM	?-
2			+	+	+	-	NM	VIQ: 100; PIQ: 92; poor spatial memory
3	-	-	-	+	+	-	NM	VIQ: 89; PIQ: 87
Farah et al. (1989)		+		+	-	-	+	?-
Farah et al. (1991)		+		+	?	-	-	Amnesia
Feinberg et al. (1986)			NS	-	NM	+	?; Color anomia	Amnesia; tactile agnosia
Gomori & Hawryluk (1984)			+	+	+	-	+	Anterograde amnesia; ? simultanagnosia
Iwata (1990)	?			+	-	-	NM	-
Jankowiak et al. (1992)		+		+	+	-	+	Mild verbal memory encoding problem
Kawahata & Nagata (1989)		+		+	NM	+	+	Amnesia; poor visuoconstruction
Kertesz (1979)	+	+		+	NM	+	+	Amnesia; simultanagnosia; Balint's syndrome; poor visual memory

Study							Associated deficits
Landis et al. (1986)							
1		NM	+	+	-	NM	Dressing apraxia
2		NS; zooagnosia	+	+	-	NM	Mild amnesia; anosagnosia
3		NM	+	+	?	+	Mild anosagnosia
4		NM	+	+	-	-	Anosagnosia; poor nonverbal memory
5		? NS	+	+	-	-	Anosagnosia
6		NM	+	NM	NM	NM	Poor visual memory
Landis et al. (1988)		NS	+	+	-	-	L neglect
Larrabee et al. (1985)							
1	+		+	+	+	+	Apraxia
Levine (1978)		NS	? +	+	-	-	Amnesia
Levine et al. (1985)							
1		+	+	-	-	+	Mild amnesia; poor object/color imagery
Mack & Boller (1977)	+		+	NM	-	+	Amnesia
Macrae & Trolle (1956)		NS; ? zooagnosia	+	+	+	Color anomia	Mild amnesia; poor visual imagery; dyscalculia
Malone et al. (1982)							
1		?	+	?	+	+	
2		? NS	+	?	+	-	Amnesia
Marks & DeVito (1987)							
1		NS	+	?	+	+	? Simultanagnosia
2		NS	-	?	+	+	? Optic ataxia; mild anomia
McCarthy & Warrington (1986)	+		-	NM	+	+	Mild anomia
Mendez (1988)							
1	+		?	NM	+ . . . resolved	-	Poor STM, resolved
2	+		?	NM	+	-	Psychomotor slowing
Michel et al. (1986)		NS (mild)	+	-	-	-	
Milner et al. (1991)		+ (visual form agnosia)	+	NM	? +	-	Amnesia; acalculia
Newcombe et al. (1989)		NS	+	?	-	+	Poor semantic memory
Pallis (1955)		Zooagnosia	+	+	+ . . . resolved	+	

continues

Table 2 *(continued)*

Reference	Apperceptive VA	Associative VA	VA otherwise classified	Prosopagnosia	Topographagnosia	Alexia	Color disturbance	Other cognitive deficits
Riddoch & Humphreys (1987a)			+ integrative VA	+	+	+	+	
Rubens & Benson (1971)		+		+	NM	+	+	Amnesia
Sparr et al. (1991)	+			+	+	Slow laborious	–	Poor visual imagery
Striano et al. (1981)			NS	+	NM	? –	NM	Tactile agnosia; ? simultanagnosia
Tagawa et al. (1990)			?	+	+	NM	+	
Taylor & Warrington (1971)		+		+	NM	?	+	Progressive amnesia and aphasia
Tranel & Damasio (1988)								
1			NM	+	NM	–	+	
2			NM	+	NM	–	+	
3			NM	+	NM	–	–	
4			NM	+	NM	–	+	
Tranel et al. (1988)								
3			NM	+	NM	NM	–	Amnesia
Warrington (1975)								
1			+	NM	NM	+/–	NM	
2			+	NM	NM	+/–	NM	Progressive amnesia
3			+	NM	NM	+/–	NM	
Warrington & James (1988)								
1	+			NM	NM	–	–	
2	+			NM	NM	–	–	Poor verbal memory
3	+			NM	NM	–	–	
Whiteley & Warrington (1977)								
1			NS	+	–	–	+	
2			NS	+	–	–	–	
3			NS	+	NM	NM	–	

[a] ?, probable occurrence, but not sufficient information given.
[b] NM, not mentioned.
[c] NS, visual agnosia present, but type not specified.

A detailed clinical description of a patient with associative visual agnosia is provided by Rubens and Benson (1971). Their patient suffered an acute hypotensive episode, followed by right homonymous hemianopsia and significant deficits in visual recognition. He had prosopagnosia, alexia without agraphia, and color agnosia. He was labeled as having associative visual agnosia because he had great difficulty naming visually presented materials, from real objects to drawings of objects, but could name them using other sensory modalities. He could not describe or demonstrate the use of any of the objects he could not name, nor could he group objects by semantic categories unless he could name them first. Visual perception was demonstrated to be adequate by his ability to copy and match line drawings that he could not recognize. He also was able to pick out predesignated geometrical shapes in a hidden figure test, suggesting that he could "perceive" all parts of a design at a glance and not simply piece by piece. He was noted to have a "two-way defect" of visual recognition, that is, an inability to name when a target was presented as well as an inability to choose from an array of targets when the name was offered. According to Geschwind and Fusillo (1966), this two-way defect indicates a visual–verbal disconnection.

The two-way color naming defect, in conjunction with alexia without agraphia and a right homonymous hemianopsia, led Rubens and Benson to speculate that their patient had a lesion involving the medial left occipital lobe and the posterior corpus callosum. The fact that most patients with such a lesion do not demonstrate prosopagnosia or visual agnosia suggested that the patient may have had a silent infarct affecting the right posterior hemisphere at an earlier time. These researchers also postulated that, unlike split-brain patients who demonstrate normal recognition by the nondominant hemisphere of stimuli presented to their right visual field, this patient only had a partially interrupted callosum. The remaining intact callosum may have permitted continued dominance by the left hemisphere over the right hemisphere. The intact right hemisphere "may be receiving and acting upon accurate percepts (as demonstrated by the superb ability to copy drawings, etc.), but when visual–verbal associations are required, the result is failure."

Other cases of associative visual agnosia in which the presumed etiology appeared to be cerebral infarction are those reported by Albert, Reches, and Silverberg (1975a), McCarthy and Warrington (1986), and Mack and Boller (1977). Severe head trauma also has been reported as the etiology (Farah, Hammond, Mehta, & Ratcliff, 1989; Farah, McMullen, & Meyer, 1991; Kawahata & Nagata, 1989). Table 3 lists case reports of associative visual agnosia; etiologies and lesion localizations are listed. Table 4 lists a variety of cases of visual agnosia in which the type was not specified

Table 3 Reported Cases of Associative Visual Agnosia: Lesion Localization and Etiology

Reference	Lesion localization	Etiology
Albert et al. (1975a)	Brain scan: increased uptake bilateral inferior and posterior occipital	? Infarction
Albert et al. (1979)	Autopsy: bilateral temporo–occipital, R inferior frontal gyrus	Multiple infarctions
Benke (1988)	CT: L parieto–occipital	Intracerebral hemorrhage
Benson et al. (1974)	Autopsy: L PCA occlusion; splenium corpus callosum; bilateral globus pallidus	Infarction
Damasio & Damasio (1983)	CT: bilateral temporo–occipital	Infarction
Davidoff & Wilson (1985)	CT: ventricular dilatation, especially L occipital horn	Trauma
Farah et al. (1989)	CT: bilateral temporo–occipital, R inferior frontal	Trauma
Farah et al. (1991)	CT: L temporal edema	Trauma
Iwata (1990)	CT: bilateral inferior temporal (area 37)	Infarction
Jankowiak et al. (1992)	MRI: bilateral occipital, inferior and middle temporal gyri, angular gyri; L superior cerebellum	Trauma
Kawahata & Nagata (1989)	CT: bilateral temporo–occipital hematomas; MRI: bilateral temporo–occipital hematomas; PET: bilateral temporo–occipital hematomas	Trauma
Kertesz (1979)	CT: L deep occipital, R occipital	Trauma
Larrabee et al. (1985) #1	CT: L occipital, R anterior limb internal capsule	Infarction
Levine et al. (1985)	CT: R anterior temporal to R inferior frontal; small L temporo–occipital	Trauma
Mack & Boller (1977)	CT: L occipital, temporo–occipital, parieto–occipital, R inferior occipital, temporo–occipital	Infarction
McCarthy & Warrington (1986)	CT: L medial occipital, posterior temporal; MRI: L inferior medial occipital	Infarction
Ratcliff & Newcombe (1982)	Brain scan: increased uptake R parieto–occipital CT (10 yrs post): bilateral occipital, mild temporo–parietal, R>L	Presumptive herpes encephalitis Probable infarction
Rubens & Benson (1971)	Pneumoencephalogram: mild ventricular enlargement, L>R	Presumed hypotensive infarction
Taylor & Warrington (1971)	L carotid angiogram: ventricular enlargement; Air encephalogram: cerebral and cerebellar atrophy	"Cerebral atrophy, aggravated by alcoholism'' (probable progressive dementia)

Table 4 Unspecified Visual Agnosia: Lesion Localization and Etiology

Reference	Lesion localization	Etiology
Aptman et al. (1977); L handed	CT: R parieto–temporal; bilateral occipital; Angio: occlusion R PCA	Infarction
Assal et al. (1984)	CT: bilateral temporo–occipital junction; bilateral medial occipital	SDH, infarction
Bauer (1984)	CT: bilateral temporo–occipital, R>L	Trauma
Beyn & Knyazeva (1962)	No study: "probable bilateral occipital"	Emboli in PCAs
Bruyer et al. (1983)	CT: bilateral occipital; Cerebral blood flow: R parieto–occipital	Infarction
Campbell et al. (1990)		
1	CT: R medial temporo–occipital	Infarction
Charnallet et al. (1988)	CT: L lingual and fusiform; MRI: bilateral inferior occipital	Infarction
Cohn et al. (1977)		
1	Autopsy: bilateral PCA territory, R>L	Infarction
2	Autopsy: bilateral PCA territory	Infarction
Damasio et al. (1982)		
2	Autopsy: L occipital, temporo–occipital junction; R inferior occipital, temporal-occipital junction	Infarction
Damasio & Damasio (1983)	No mention R hemisphere in either study	
4	CT: L mesial occipital	Infarction
7	CT: L mesial occipital, mesial temporo–occipital	Infarction
DeHaan et al. (1987)	CT (at time of injury): "generalized cerebral edema"	Trauma
DeRenzi (1986)		
2	CT: R parahippocampus, occipital, splenium on R side	Infarction

continues

Table 4 (*continued*)

Reference	Lesion localization	Etiology
DeRenzi et al. (1991)		
2	CT: Abscess R temporal	Trauma
Feinberg et al. (1986)	CT: L temporo–occipital, inferior parietal	Infarction
Gomori & Hawryluk (1984)	CT: bilateral temporo–occipital; L mesial occipital; L forceps splenium corpus callosum	Colloid cyst 3rd ventricle; infarction
Landis et al. (1986)		
1	CT: R PCA territory	Infarction
2	CT: R PCA territory	Infarction
3	CT: R PCA territory	Infarction
4	CT: R parieto–occipital epidural hematoma	Trauma
5	CT: R temporo–parieto–occipital tumor	Glioblastoma
6	CT: R parieto–occipital (after resection)	Oliodendroglioma
Landis et al. (1988)	Autopsy: R PCA territory, areas 17 & 18, splenium corpus callosum; old L deep lateral parieto–occipital, old R inferior frontal	Infarction
Levine (1978)	CT: R occipital, temporo–occipital (after resection)	Meningioma
Macrae & Trolle (1956)	Pneumonencephalogram: normal; EEG: bilateral parietal slowing	Trauma
Malone et al. (1982)		
1	CT: bilateral occipital, R parietal; Angio: narrowed R PCA	Infarction
2	CT: bilateral parieto–occipital	Hemorrhagic infarction s/p ligation L vertebral art s/p pseudoaneurysm secondary to trauma
Marks & DeVito (1987)		
1	CT: inferior L occipital, deep temporal, R inferior temporo–occipital	Infarction
2	CT: R occipital, posterior parietal and temporal	Infarction

Study	Imaging/findings	Diagnosis
Michel et al. (1986)	CT (several days): R occipital hematoma; CT (1 year): R fusiform gyrus	Hemorrhage
Milner et al. (1991)	CT (2 mo): no abnormality; MRI (17 d): bilateral lentiform nuclei, ? L temporo–occipital; MRI (13 mo): bilateral globus pallidus, bilateral occipital and parieto–occipital; SPECT (1 mo): hypoperfusion L posterior parietal and temporo–occipital, bilateral frontal; SPECT (8 mo): bilateral parieto–occipital hypoperfusion	Carbon monoxide intoxication
Pallis (1955)	Vertebral angio: defective filling R PCA	Infarction
Ratcliff & Newcombe (1982)	[Angio: normal; EEG: severe bilateral & R occipital abnormality; Brain scan: abnormal uptake R parieto–occipital]; CT: bilateral occipital, mild parietal and temporal, R>L	? Encephalitis vs infarction (by CT 10 yr later)
Riddoch & Humphreys (1987a)	CT: bilateral occipital (PCA territory)	Infarction
Striano et al. (1981)	CT: R temporo–parieto–occipital, L parieto–occipital	Infarction
Tagawa et al. (1990)	Angio: Occlusion R PCA; L PCA territory poorly visualized; CT: bilateral occipital; PET: decreased blood flow and O_2 metabolism, bilateral occipital	Infarction
Tranel & Damasio (1988)		
1	CT/MRI: bilateral white matter temporo–occipital, L>R	Infarction
2	CT/MRI: bilateral inferior mesial occipital	Trauma/infarction
3	CT/MRI: L mesial temporal, R temporo–occipital	Herpes simplex encephalitis
4	CT/MRI: bilateral posterior hippocampus, R inferior mesial occipital	Infarction
Tranel et al. (1988)		
3	CT/MRI: bilateral anterior, mesial temporal	Herpes simplex encephalitis
Whiteley & Warrington (1977)		
1	CT: bilateral occipital, L>R	Infarction
2	CT: R lateral occipital hematoma	Hemorrhage
3	CT: R parieto–occipital tumor, midline shift to L	Glioma, grade III astrocytoma

or did not conform to the apperceptive–associative dichotomy. Etiologies and lesion localizations are listed also.

The case report of L. H., a patient with associative visual agnosia and prosopagnosia following a severe head injury (Farah et al., 1989), is particularly interesting because he appeared to have greater impairment in visual recognition of living things than of nonliving things. The authors conducted an experiment with L. H. and 12 normal controls to determine whether this difference was simply the result of a difference in task difficulty (i.e., are living things harder to recognize because they are more complex and visually confusable, one to another, than nonliving things?) or the result of a category-specific difference. Because prosopagnosic patients make more errors in the "living thing" category, prosopagnosia has been argued actually to be a mild form of visual object agnosia. The performance by L. H. was comparable to that of the controls on questions of semantic knowledge about nonliving things that did not require visual information (e.g., "Were wheelbarrows invented before 1920?"). His scores were within the range of the normal controls on questions of semantic knowledge about living things that did not require visual information (e.g., "Are roses given on Valentine's day?") and on questions of semantic knowledge about nonliving things that required visual information (e.g., "Is a canoe wider in the center?"). His performance was impaired significantly on questions of semantic knowledge about living things that required visual information (e.g., "Are the hind legs of a kangaroo larger than the front legs?"). From the performances of the normal controls, these researchers concluded that sheer difficulty "does not account for the difference between a prosopagnosic patient's knowledge of the visual appearance of living and nonliving stimuli." L. H. had a selective deficit in semantic memory for visual information about living things that appeared to be both modality specific and category specific. The authors suggested, therefore, that "the representation of visual stimuli that fall under the general category of 'living things' may require particular capacities not needed for other categories of stimuli . . . but not necessarily an entirely distinct system." They commented further that the category "living things" may not be strictly literal, based on the fact that some prosopagnosics have difficulty recognizing makes of automobiles. Rather, the classification of living versus nonliving may be based more on differences in the aspects of shape that are necessary for the differential classification and not on "differences in aliveness *per se*."

M. B., another patient with associative visual agnosia and prosopagnosia following severe head injury, was presented by Farah et al. (1991). Like patient L. H., M. B. had more difficulty recognizing living things than nonliving things. To determine the reason for this selective impairment, both L. H. and M. B. were given a battery of tests to explore three hypotheses: (1) living things are more visually complex than nonliving things, (2) living things are visually more similar to one another, and (3) a generic category response is often sufficient for nonliving things but a more specific, within-category response is necessary for living things (e.g., when shown a desk chair, the response "chair" is considered acceptable, whereas when shown a robin, the acceptable response is "robin" and not "bird"). Even when factors such as visual complexity, familiarity, and similarity to the most similar other object were factored out, M. B. and L. H. made consistently more errors in the living than in the nonliving category. Both made errors that were predominantly visual (i.e., errors in which the named object was visually similar to the correct object), although they also made some purely semantic errors (errors in which the named object was related to the correct object in the sense of its taxonomy or function).

From these results, the authors concluded that "there is some neural system that is relatively more important for recognizing living things than nonliving things ... (which) can be selectively impaired by brain damage." In addition, the predominance of visual errors suggested that the locus for the recognition deficit may be in the visual system. The investigators postulated further that higher visual centers are specialized for face perception as well as perception of more general biological forms. The distinction between living things and nonliving things therefore would likely be related to the characteristic *shapes* of each. However, the presence of some purely semantic errors suggested that some semantic knowledge about living things also may be impaired in the recognition deficit. The brain region subserving semantic knowledge about living things was postulated to be separated at least partly from the region subserving face recognition, based on the fact that some patients show impairment in knowledge about living things but are not prosopagnosic. An example was the patient presented by Warrington and Shallice (1984) who had a lesion probably "too anterior to affect the visual system." Farah et al. (1991) argued that the closed head injuries suffered by M. B. and L. H. would be compatible with posterior lesions causing prosopagnosia, as well as with more anterior involvement. Thus, they concluded that, although they could not specify the locus of the impairment definitively, a semantic impairment could involve damage to modality-specific visual semantics and that visual perceptual as well as visual semantic representations may be involved. Warrington and McCarthy (1983) already had suggested that visual information may be more essential to our knowledge of living things than to our knowledge of nonliving things.

In many cases of associative visual agnosia, a hierarchy of impairment is apparent (Farah et al., 1991; Gomori & Hawryluk, 1984; Kawahata &

Nagata, 1989; Striano, Grossi, Chiacchio, & Fels, 1981). Real objects usually are best recognized, perhaps because of the additional information provided by color, texture, and shape (Kertesz, 1979; Newcombe & Ratcliff, 1975; Sparr et al., 1991). Next in the hierarchy is recognition of photographs. Recognition of line drawings of objects is even more difficult. In a few cases, even more demanding tasks using noncanonical representations have been requested.

In this last situation, photographs of real objects are taken from unusual angles or using lighting from angles that distort perspectives. Warrington and Taylor (1973) had patients who demonstrated difficulty recognizing objects photographed from an unconventional angle whereas recognition of pictures of objects presented in a canonical view was intact. Subsequent studies by Whiteley and Warrington (1977) revealed this same deficit in three other patients with a "perceptual categorization" defect that they classified with apperceptive visual agnosia. Landis, Regard, Bliestle, and Kleihues (1988) presented a patient with prosopagnosia, topographagnosia, and visual agnosia for objects seen from noncanonical views, all following an infarction. She named correctly most objects or pictures presented in a typical representation, but was unable to identify even real objects when shown noncanonically. These same misidentified objects were identified correctly in canonical presentations. In another case, the hierarchy of visual recognition deficits seemed to evolve over time (Striano et al., 1981). This patient presented initially with prosopagnosia, tactile agnosia, and an unspecified type of visual agnosia for complex real life objects and pictures, following a stroke. By 25 days postonset, he could identify complex objects in real life but not pictures that were composed of various parts. By 45 days postonset, he was able to recognize even complex pictures, but was still unable to recognize scenes if they were shown from an unusual perspective (noncanonical view).

Levine (1978) has postulated that a hierarchy of deficits also exists in visual imagery that parallels the visual perceptual problems. According to Sergent and Corballis (1990), "visual image generation is the process by which information stored in long-term memory is reactivated to give rise to a visual representation of an object's physical attributes, so that the object can be revisualized and inspected in the absence of direct sensory stimulation." The patient described by Levine (1978) had prosopagnosia, visual disorientation (visual misreaching, impaired visual counting of stimuli, and erratic performance on tests of distance discrimination), and a visual object agnosia that had features of apperceptive and associative forms after resection of the right occipital lobe for a recurrent vascular tumor. She could match and copy in a crude fashion but had perceptual deficits, including a disturbance of depth perception. She showed a hierarchy of difficulties, having the most difficulty identifying more complex objects such a human faces, followed by pictures of animals, then common objects, and finally letters, for which she was "almost 100% accurate." Color naming and matching, on the other hand, were excellent. Her ability to name the characteristic colors corresponding to objects named (e.g., sky, blood, inside of a watermelon) was excellent. Like her visual recognition, her descriptions and drawings of objects from memory were impaired moderately. Her description of faces from memory was extremely poor. This parallel hierarchy of deficits between visual imagery and perception led Levine to postulate that visualization, or "inner seeing," is a process that must utilize at least some of the same neural pathways as perception itself.

Levine and Calvanio (1989) also emphasized a parallel between visual perceptual deficits and visual imagery. These researchers asserted that no documented cases of prosopagnosia with preserved visual imagery could be found, if certain pitfalls in the assessment of visual imagery were observed. They refer to the patient L. H., also discussed by Farah et al. (1989; Farah, Hammond, Levine, & Calvanio, 1988). L. H., a prosopagnosic with impaired color recognition and probable associative visual agnosia, had no difficulties in depth perception or other aspects of visual spatial orientation. He was found to have a selective impairment in visual imagery for objects and colors, that is, he had lost the inner knowledge of how previously familiar objects and colors looked. On the other hand, he had excellent spatial imagery. The authors noted three ways that visual imagery may be misinterpreted: (1) if spatial imagery, for example, description of a route, is used as evidence of entirely preserved visual imagery, (2) if the patient's introspective report that his visual imagery is intact is taken as proof (the patient may be unaware of the defect), and (3) if other skills can compensate or mask deficits on a test of visual imagery, for example, "verbal descriptions may employ verbally encoded knowledge that can be recalled without generating a visual image" (Pylyshin, 1973) and "drawing from memory may utilize spatiomotor skills without requiring generation of a fully configured visual image" (Goldstein & Gelb, 1918).

Although previous studies may not have explored visual imagery in the manner specified by Levine and Calvanio, the statement by a patient of Pallis (1955) seems reliable: "I can shut my eyes and can well remember what my wife looked like or the kids." The patient stated further that his dreams were less frequent, "but just as vivid as before." This latter statement was taken as evidence that the patient did not have the "syndrome described by Charcot (1883) and Wilbrand (1887), characterized by a failure to conjure up visual images and by a loss of the visual components of dreaming." Farah et al. (1988) emphasized that mental imagery comprises two functionally independent subsystems, one visual and the other spatial, that are neurologically dissociable.

An additional exploration of the relationship between visual perception and visual imagery was possible with our patient (M. D.), who demonstrated a unique discrepancy between his impaired ability to recognize objects visually and his maintained ability to generate visual images of these objects. Details of M. D.'s associative visual agnosia are described elsewhere (Jankowiak, Kinsbourne, Shalev, & Bachman, 1992) and will be summarized here.

2. Case Report

M. D. is a 29-year-old right-handed man who sustained a gunshot wound at age 19, resulting in a bitemporo-occipital lesion (Fig. 1). He was left with an associative visual agnosia, prosopagnosia, topographagnosia, and mild difficulty naming colors, particularly in the right visual field. Reading was functional but slow. Detailed neuropsychological testing demonstrated normal intelligence with adequate language skills although verbal memory testing showed some problem with encoding.

A series of experiments was conducted to assess an apparent discrepancy between his impaired ability to recognize objects and his maintained ability to generate visual images of these objects. In one experiment, line drawings of objects from the Snodgras and Vanderwart (1980) picture series were presented tachistoscopically for 100 msec and then for 1000 msec. M. D. was asked to draw what he had seen and then to name or describe it. At the 100-msec exposure, he was able to produce recognizable sketches of 5 of 8 items, despite inability to name or describe any of them (Fig. 2). Even at 1000 msec, when his sketches were more detailed, he could draw some items (even with reverse orientation) that he could not identify (Fig. 3). This experiment demonstrated that M. D. has preserved visual pattern perception even with brief (100 msec) presentations. Of particular interest was his preserved ability to draw stimuli correctly that he was, at times, quite unable to recognize.

Given unlimited time to copy pictures of animals from Snodgras and Vanderwart (1980) presented in free field, M. D. produced accurate drawings even though he could only identify 1 of 9 samples more specifically than as "animals" (Fig. 4). His approach was never piecemeal and he correctly named the parts of the animal as he drew. At a later session, he was asked to describe animals that he had identified incorrectly but had drawn well. For example, he had drawn a bear yet had called it a pig. He described a bear as a "four-legged animal with a large head, black or white fur, sharp teeth, and a short tail, if any, which when standing on hind legs could be 7–8 feet tall." A pig he described as "short, stocky, smelly, with a

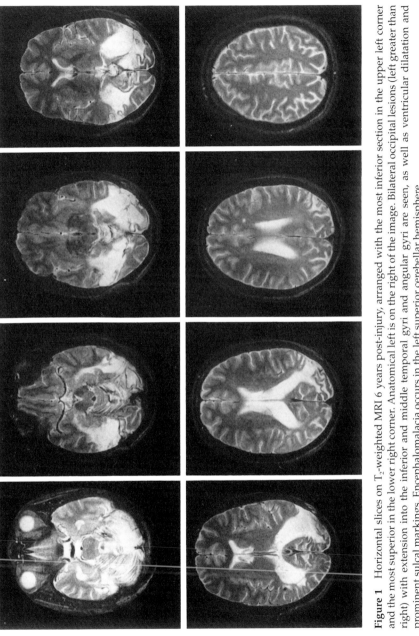

Figure 1 Horizontal slices on T_2-weighted MRI 6 years post-injury, arranged with the most inferior section in the upper left corner and the most superior in the lower right corner. Anatomical left is on the right of the image. Bilateral occipital lesions (left greater than right) with extension into the inferior and middle temporal gyri and angular gyri are seen, as well as ventricular dilatation and prominent sulcal markings. Encephalomalacia occurs in the left superior cerebellar hemisphere.

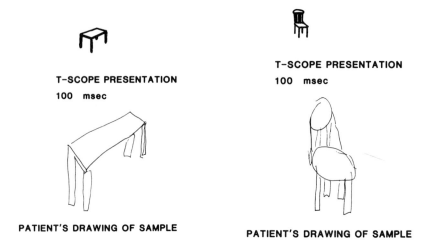

T–SCOPE PRESENTATION

100 msec

T–SCOPE PRESENTATION
100 msec

PATIENT'S DRAWING OF SAMPLE

PATIENT'S DRAWING OF SAMPLE

Figure 2   M. D.'s sketches of (1) table and (2) chair drawn after 100-msec exposure, despite his inability to identify the stimulus.

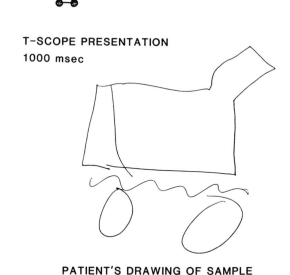

T–SCOPE PRESENTATION
1000 msec

PATIENT'S DRAWING OF SAMPLE

Figure 3   M. D.'s sketch of a baby carriage drawn after 1000-msec exposure, despite his inability to identify target.

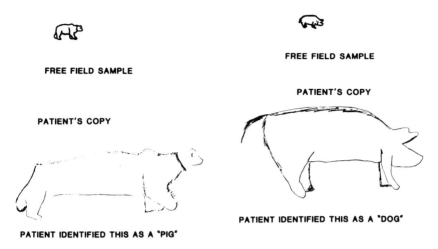

Figure 4   M. D.'s copy of a (1) bear and (2) a pig, both of which he misidentified.

snout nose and a squiggly tail, grey smooth skin, about knee high in height and weighing 100–150 pounds."

On a visual object decision task, M. D. correctly classified 40 of 53 (75%) real object drawings and correctly rejected 47 of 51 (92%) pseudoreal object drawings. The pseudoreal category included 39 drawings of "pseudoreal" objects with distorted features and 12 drawings with either added or substituted real features. This experiment demonstrated that M. D. generally knew when an object was real even if he did not know what it was. He rejected pseudo-objects even when the component features were realistic, suggesting that he used the overall gestalt in the recognition process and did not base his conclusions on just one or two features.

In another experiment, M. D. was able to sort pictures from Snodgras and Vanderwart (1980) into appropriate categories such as fruits, vegetables, insects, or animals despite an inability to identify 15 of 33 correctly. Of note, the stimuli within each category were visually markedly different (e.g., pineapple and grapes, lettuce and carrot, butterfly and caterpillar, kangaroo and stag). Thus, this experiment demonstrated that unrecognized stimuli nonetheless held enough meaning for M. D. to enable him to categorize them correctly.

On a standardized imagery questionnaire (Eddy & Glass, 1981) that compares visual imagery (termed high imagery) and nonvisual imagery (low imagery), M. D. did as well as two age-matched controls. In addition, he responded correctly to 45 of 46 questions that tapped size imagery and self-corrected the one error on retesting at a later time. His verbal descriptions of animate and inanimate objects were quite adequate. These findings all indicated that M. D. was able to generate mental visual images.

M. D. seemed to have difficulty in the process of matching the visual target with his intact internal image of the same object. According to the model of visual object identification proposed by Kosslyn, Flynn, Amsterdam, and Wang (1989), a "pattern activation" subsystem exists within which the central representation of patterned visual input is compared with patterns stored in memory of objects previously seen. Within this theoretical framework, M. D. was capable of adequate representation both of input and of items from memory, but was impaired selectively with respect to the matching process within the pattern activation subsystem. Using a neural network model, this matching abnormality might be thought of as an unstable neural response of patterns stored in memory. Normally, when rival interpretations are competing for control of the central representation of patterned visual input, "difference amplifying" laterally inhibitory feedback interactions ensure a correct identification. However, if these "winner-take-all" interactions are inactivated, similar but incorrect (although not unrelated) responses might gain control intermittently. On the other hand, when the process occurs endogenously, for example, when the patient visualizes a remembered object, no rivalry of competing interpretations occurs and deficient lateral inhibition would not be of consequence.

As in certain cases of apperceptive visual agnosia, associative visual agnosia also may evolve over time. Benke (1988) described a patient with associative visual agnosia, prosopagnosia, achromatopsia, alexia, and a persistent amnestic syndrome following an intracerebral hematoma that was removed by craniotomy. When shown common objects or line drawings, she initially could provide only a vague description of the item, its shape, and of what it reminded her. One year later, she could pantomime the use of most functional objects, which sometimes even cued her naming efforts. This improvement was interpreted as an evolution of associative visual agnosia toward optic aphasia (a visual–verbal disconnection syndrome in which patients cannot name a visually presented object but can demonstrate its use by gesture, thus suggesting intact visual recognition) (Beauvois, 1982; L'hermitte & Beauvois, 1973; Riddoch & Humphreys, 1987b).

3. Lesion localization

Lesion localization in associative visual agnosia is somewhat less heterogeneous than in cases of apperceptive visual agnosia, although inconsistencies, incomplete information, and variations from case to case render all conclusions hazardous (see Tables 3, 4). Many case reports do not provide confirmed clinico-anatomical data. For the patient described by Rubens and Benson (1971), who had suffered an acute hypotensive episode, relatively little anatomical information was available initially. A radionuclide brain scan was negative. An EEG revealed diffuse symmetrical 5–7 cps slowing. A pneumoencephalogram demonstrated mild ventricular enlargement, more prominent on the left. Bilateral carotid arteriograms were interpreted as normal. Based on the patient's clinical findings of a right homonymous hemianopsia without macular sparing, color agnosia, and alexia without agraphia, these researchers speculated that he would have a lesion in the left medial occipital cortex and splenium of the corpus callosum.

Rubens and Benson's patient came to autopsy 3 yr later following a rapidly spreading pharyngeal carcinoma (Benson, Segarra, & Albert, 1974). Gross examination revealed an occluded left PCA from a point distal to the junction with the posterior communicating artery and including the posterior temporal branch. A vascular lesion ran the length of the fusiform gyrus along the undersurface of the left temporo-occipital lobe. The right occipital lobe showed a loss of bulk. On coronal sections, moderate ventricular enlargement anteriorly, with the left slightly greater than the right, was detected. Old infarctions were noted in the right globus pallidus and superior aspects of the left globus pallidus. Infarction was seen in the white matter of the left parahippocampal cortex and anterior third of the fusiform lobe, which expanded as it extended posteriorly to involve the periventricular white matter and ended at the pial surface of the lingual gyrus. An elongated cystic lesion involved the white matter of the right fusiform gyrus. Necrosis of the lower half of the splenium of the corpus callosum was noted. The occipital lesions interrupted several fiber pathways. On the left, the inferior longitudinal fasciculus (ILF), which connects the fusiform gyrus to temporal lobe structures, was involved. Optic radiation and hippocampal connections were severed by the left hippocampal infarct. The lesions were less extensive but almost symmetrical on the right, with involvement of the right ILF and other occipito-temporal connections. However, right optic radiation fibers were spared. Damage to the inferior forceps of the corpus callosum interrupted interhemispheric connections between right and left areas 18. Of note, the calcarine cortex was spared bilaterally, although on the left it was almost totally isolated. On the right, only major outflow tracts to the temporal lobe were interrupted. Thus, almost all visual interhemispheric connections were destroyed.

Benson et al. (1974) attributed their patient's associative visual agnosia to the combination of infarctions of the left medial occipital region (which prevented the left visual cortex from participating in visual activities) and splenium (which produced a visual–verbal disconnection.) However, these lesions are found typically in patients with alexia without agraphia who do not have visual agnosia. These investigators postulated that their patient could perform visual–motor tasks (e.g., copying) because of intact pathways connecting right visual association cortex and right motor association cortex, but failed to recognize the nature of the visual stimuli because of bilateral disruption of visual–limbic connections (visual association cortex–ILF–hippocampus). Tactile and auditory recognition were retained because of intact association pathways to language areas. This specific combination of lesions involving primary visual pathways of the dominant hemisphere with lesions of the splenium of the corpus callosum and the ILF of the nondominant hemisphere was credited for both the patient's associative visual agnosia and his prosopagnosia.

A second autopsy report of a patient with associative visual agnosia and prosopagnosia is that by Albert, Soffer, Silverberg, and Reches (1979). Their patient additionally had a left hemianopic color blindness, topographagnosia, an amnestic syndrome, and a right superior quadrantanopsia (Albert, Reches, & Silverberg, 1975a,b). However, despite his striking visual recognition deficits for nonverbal stimuli, this patient was not alexic. Autopsy revealed moderate to severe arteriosclerosis of the circle of Willis and a hypoplastic left posterior communicating artery. Multiple infarctions were found bilaterally. On the right were lesions of the parahippocampus, including the right ILF and white matter underneath the inferior lip of the calcarine fissure and lingual gyrus; the insula, claustrum, white matter of the temporal lobe, and inferior frontal gyrus also had lesions. On the left, mirror infarcts were seen in the parahippocampal gyrus; lesions also were noted in the collateral sulcus, lingual gyrus, fusiform gyrus, and pulvinar. The corpus callosum was spared. The authors proposed that a visual–limbic disconnection was responsible for their patient's associative visual agnosia, that is, bilateral lesions of the ILF, the major pathway that connects visual cortical areas 18 and 19 to temporal lobe and limbic system. Visual information could be processed for perception in the visual association cortices, as evidenced by his ability to describe, copy, and match visual stimuli, but could not be processed for meaning because the stimuli would not excite visual memory stores in the limbic system. Unlike the patient of Benson et al. (1974), this patient had sparing of the corpus callosum (and was not alexic), which argues against its necessity in visual agnosia or prosopagnosia.

Relatively few other cases of associative visual agnosia have come to autopsy, but the report by Kawahata and Nagata (1989) demonstrates the capabilities of modern technology. Their patient is described as having associative visual agnosia, prosopagnosia, alexia, agraphia, color anomia, visuospatial constructional problems, and impaired recent verbal memory following a head injury. A CT scan initially showed bilateral temporo-occipital subcortical hematomas with surrounding edema. A CT scan 6 mo later showed extensive low density areas in the posterior–inferior temporal lobes and lateral occipital lobes bilaterally, with no involvement of parietal lobes or the corpus callosum. An MRI scan 18 mo postinjury revealed extensive bilateral temporo-occipital low intensity lesions, involving subcortical and deep basal white matter. Lingual, fusiform, and posterior inferior temporal gyri were involved bilaterally, including both ILFs. The medial occipital lobes, left angular gyrus, and splenium of the corpus callosum all were spared. A PET study showed severe depression of cerebral blood flow and oxygen metabolism in the same impaired regions demonstrated by CT and MRI. Bilateral carotid angiograms ruled out stenosis or occlusion, but did demonstrate mildly decreased blood flow in bilateral temporo-parietal regions. These studies in a live patient are consistent with the autopsy reports that have demonstrated intrahemispheric disconnection between visual cortices and the temporal lobes, that is, bilateral ILF disruption, postulated to be responsible for associative visual agnosia.

Damasio et al. (1982) analyzed 10 postmortem reports of prosopagnosia and compared them with CT reports from 3 of their own patients with prosopagnosia and varying degrees of visual agnosia, as well as with CT scans of 19 control patients. The controls did not have prosopagnosia but had other neuropsychological syndromes often associated with prosopagnosia such as achromatopsia, alexia, visual field defects, or visuospatial defects. For these authors, prosopagnosia is not specific for human faces; they contend that the emphasis on a dissociation of facial recognition and object recognition is misleading. These researchers argue that, if objects were tested in the same manner used for testing familiar faces, the objects would not be recognized by most true prosopagnosics. Instead, the dissociation should be between the recognition of the generic conceptual class to which the object belongs (which is preserved for the prosopagnosic but not for the visual agnosic) and the recognition of the historical context of a given object for the patient (which is impaired in both types of patient). Consequently, both disorders may be considered memory disorders because "an ongoing correct visual percept" fails "to evoke its appropriate, previously acquired context." From their anatomical correlation studies, these investigators concluded that lesions associated with prosopagnosia are located in the mesial occipito-temporal region and are functionally symmetrical.

Although many other studies demonstrate bilateral temporo-occipital lesions associated with associative visual agnosia (see Table 3), cases were reported of presumably unilateral lesions producing the disorder. However, these early case reports did not have the benefit of the sophisticated neuroimaging techniques currently available and may have missed a contralateral lesion. For example, Macrae and Trolle (1956) reported a patient with prosopagnosia, unspecified visual agnosia, mild topographagnosia, and mild dyslexia following head trauma. Visual fields were intact, but opticokinetic nystagmus was reduced moderately to the left. He had tachistoscopic tunnel vision for colors and depth, especially in the left homonymous field; flicker fusion was depressed slightly on the left. A pneumoencephalogram was normal; an EEG showed bilateral parietal slowing, especially on the left. Based on the co-existing clinical findings (especially visual), these researchers postulated that the patient had a left temporo-parietal lesion. However, as Damasio et al. (1982) point out, although the absence of a right visual field defect has been considered sufficient to exclude the presence of a left hemisphere lesion this may not be true if the optic radiations and striate cortex are spared.

Even with CT scans, cerebral lesions may be missed. A common problem is that an early CT, done at the time of lesion onset, may show no abnormality. One example was the patient of Riddoch and Humphreys (1987a), who suffered a perioperative stroke that resulted in "integrative visual agnosia," prosopagnosia, topographagnosia, and achromatopsia. The initial CT scan done shortly after the stroke did not define the lesion. However, the CT scan was repeated 5–8 wk later and then revealed bilateral occipital infarctions in the distribution of the PCAs. However, in the case of DeRenzi's (1986) two patients with prosopagnosia and some degree of topographagnosia (no mention was made of assessment for visual agnosia or alexia), CT scans revealed unilateral lesions in the distribution of the right PCA only. The CT scans were done "approximately 20 days" postonset, and therefore may not have revealed the full extent of the lesion. According to M. Naeser (personal communication, 1991), a CT scan must be done at least 2 mo postonset to appreciate the full extent of the lesion. In addition, CT scanning may miss lesions that are detected subsequently by MRI. Charnallet et al. (1988) offered this precaution when dealing with anatomical interpretations based solely on CT scans. Their patient, with a right visual hemiagnosia and no visual field deficits, had a hypodensity in the left lingual and fusiform gyri on a CT scan. An MRI scan demonstrated the same left inferior occipital lesion under the calcarine fissure, as well as two additional lesions. One was in the white matter on the inner side of the cuneus on the left; a second more pronounced lesion was seen in the homologous white matter in the right cuneus. Although the corpus callosum was spared, the MRI demonstrated bilateral involvement of fibers of the forceps major, especially on the right. The additional lesions revealed by MRI support disconnection hypotheses of associative visual agnosia.

Other cases in which apparently unilateral lesions resulted in associative visual agnosia have been published. McCarthy and Warrington (1986) described a patient with associative visual agnosia, alexia, difficulties in color recognition and a right homonymous hemianopsia following a stroke. CT scan demonstrated a medial left occipital infarction that possibly involved the posterior temporal lobe. MRI scan showed a lesion consistent with a left PCA infarction, involving gray and white matter in the lower part of the left occipital lobe, particularly the medial occipital cortex and fusiform and lateral occipital gyri. The corpus callosum was spared. The fact that MRI did not reveal any right hemispheric or corpus callosal involvement led the authors to conclude that no disconnection between the hemispheres had occurred. This case supports the notion that "the integrity of regions in the posterior left hemisphere is essential for crucial and distinct components of visual object recognition."

Fewer reports exist of right unilateral lesions causing associative visual agnosia, but the autopsy report by Landis et al. (1988) is particularly interesting. This patient had prosopagnosia, topographagnosia, a left homonymous hemianopsia, and a visual agnosia for noncanonical views, even for real objects. She died 10 days postonset of her symptoms from a pulmonary embolism. Autopsy revealed a large recent infarction in the territory of the right PCA, "extending from the level of the splenium of the corpus callosum to the occipital pole." The right parahippocampal gyrus was involved, as well as the entire visual cortex (areas 17 and 18). In addition, two old vascular lesions were noted: one in the right inferior frontal gyrus at the level of the anterior commissure and the other, a cortical microinfarct, in a deep lateral parieto-occipital sulcus of the left cerebral hemisphere. From this case, the authors argued that the relevance of the old left posterior lesion to the patient's visual recognition disturbance was still unclear, although the lesions definitely were not functionally symmetrical, as Damasio et al. (1982) had stressed. However, the presence of a right posterior lesion with impaired recognition of objects or faces from atypical perspectives is well documented (DeRenzi, Bonacini, & Faglioni, 1989; Sergent & Corballis, 1989; Whiteley & Warrington, 1977).

Although investigators generally have established that visual agnosia is caused by unilateral or, more often, bilateral posterior hemispheric lesions (usually involving occipito-temporal connections), a number of cases report frontal involvement (see Tables 1, 3). Larrabee et al. (1985) speculated that deep frontal lesions may contribute to agnosia based on evidence that projections from basal forebrain nuclei provide most of the cholinergic input to the cerebral cortex. They postulated that a lesion in the deep frontal white matter may interrupt cholinergic fibers projecting to the posterior cerebral cortex. In the case of their patient, with lesions in the left occipital lobe and in the anterior limb of the right internal capsule, the right anterior lesion might have disrupted fibers projecting to the right occipital lobe, thus producing the effect of bilateral occipital lesions. The authors refer to the cholinergic deficiency noted in Alzheimer's disease and the occurrence of visual agnosia in the course of that illness (Mendez et al., 1990b).

To date, localization studies in visual agnosia have been dependent on static anatomical studies. The introduction of positron emission tomography (PET) provides a means of studying cerebral metabolism while a patient is performing a specific task (e.g., visually identifying an object). In normal subjects, categorization of photographs of common objects activated the following areas: left lateral occipito-temporal regions (Brodmann areas 19, 37, and 20), left superior parietal lobule (area 7), supramarginal gyrus (area 40) bilaterally, left middle temporal region (area 21), and the gyrus rectus (area 11) (Sergent, Ohta, & MacDonald, 1992a). Activation of left lateral occipital cortex also was found during categorization of line drawings of objects (Sergent, Zuck, Levesque, & MacDonald, 1992b), confirming the robustness of the earlier findings (Sergent et al., 1992a). Of interest, no involvement of right hemispheric regions specifically activated during face-identification tasks (i.e., right fusiform, parahippocampal, and anterior temporal areas) was detected. However, posterior areas of the left hemisphere (areas 37 and 21) and the gyrus rectus (area 11) were activated in both tasks. The absence of anterior temporal activation in object categorization illustrates a fundamental difference between object and face recognition, that is, that object recognition does not require the activation of biographical information. However, objects that are personal, historical monuments, or familiar places may fail to be recognized by prosopagnosic patients, especially if the anterior temporal cortex is involved (Ellis, Young, & Critchley, 1989). Sergent and colleagues (1992a) postulate that object recognition may occur predominantly in the left posterior hemisphere, with no need for a right hemisphere contribution, at least for canonical views of objects. However, a right posterior lesion responsible for prosopagnosia also may disturb the recognition of objects presented in noncanonical perspective. Studies with PET also have been found to reveal deficits not noted on CT or MRI. For example, a prosopagnosic patient was described as having a lesion in the right fusiform gyrus, sparing the parahippocampal gyrus. PET revealed absent metabolic activity in the right parahippocampus, implying that even when an area seems to be structurally intact, damage to posterior adjacent areas may deprive that region of afferent information and render it *functionally* disturbed. SPECT also may be useful in early detection of functional impairment before CT or MRI demonstrate abnormality. Milner and colleagues (1991) present a case of "visual form agnosia" following carbon monoxide intoxication in which neither CT nor MRI initially demonstrated the posterior cortical lesions that were evident by SPECT at an early stage and only later by MRI, but not by CT.

In summary, lesion localization in associative visual agnosia most often involves bilateral posterior hemispheric lesions in PCA territory. Lesion size may play a role since a certain critical size must not be exceeded because a massive lesion may produce more global devastation of visual perception (Benton, 1990). It seems likely that lesions must be focally precise, interrupting fibers of the ILF that connect visual systems with limbic memory stores. PET studies suggest that object recognition occurs predominantly in the left posterior hemisphere, at least for canonical views of objects; the right posterior hemisphere is involved with interpretation of noncanonical representations.

III. CONCLUSIONS

Cerebral processing of complex visual information operates by means of two related visual association networks: a "what" system for identifying an object and a "where" system for locating it. A set of dorsal (occipito-parietal) pathways transmits visual information from peripheral and central visual field representations in the striate and parastriate cortices to association areas of the parietal lobe, especially areas 7 and 39 (Iwata, 1990). This set of pathways is involved in visuospatial analysis (the "where" system), so lesions in this system result in optic ataxia and impaired recognition of spatial relationships. Clark, Geffen, and Geffen (1989) postulate that the high concentration of catecholamines found in the inferior parietal lobule may be important in facilitating the disengagement of attention needed in rapidly switching visual focus of attention. Because the inferior parietal lobule has extensive reciprocal limbic connections with the cingulate gyrus, and subsequently the hypothalamus, this dorsal system is implicated in emotional arousal and in rapid selective orientation to stimuli that have motivational significance for the organism (Bauer, 1984). This system operates in a redundant *parallel* fashion to permit a cursory preliminary "preattentive" analysis of where to focus attention (Mega, Ween, & Albert, 1991).

A set of ventral (occipito-temporal) pathways transmits visual information from foveal representation areas to the inferior temporal lobe, area 37. Object identification (the "what" of visual processing) involves inferior temporal structures. At this stage, attention is deployed *serially* across an array of stimuli in a selective (foveated) manner, by shifting a "spotlight" from one focus to the next, to achieve an integrated representation (Coslett & Saffran, 1991). This set of pathways connects visual information with stored visual memories, so a lesion in this system causes disturbances of object recognition (visual agnosia) (Iwata, 1990).

As we have seen from the review of lesion localization data in this chapter, lesions producing apperceptive visual agnosia are generally more diffuse than those producing associative visual agnosia. The latter syndrome most likely requires a precise lesion that interrupts the ILF, especially on the left, thereby interfering with ventral (occipito-temporal) visual networks.

REFERENCES

Albert, M. L., Reches, A., & Silverberg, R. (1975a). Associative visual agnosia without alexia. *Neurology, 25,* 322–326.

Albert, M. L., Reches, A., & Silverberg, R. (1975b). Hemianopic colour blindness. *Journal of Neurology, Neurosurgery, and Psychiatry, 38,* 546–549.

Albert, M. L., Soffer, D., Silverberg, R., & Reches, A. (1979). The anatomic basis of visual agnosia. *Neurology, 29,* 876–879.

Alexander, M. P., & Albert, M. L. (1983). The anatomical basis of visual agnosia. In A. Kertesz (Ed.), *Localization in neuropsychology* (pp. 393–415). New York: Academic Press.

Aptman, M., Levin, H., & Senelick, R. C. (1977). Alexia without agraphia in a left-handed patient with prosopagnosia. *Neurology, 27,* 533–536.

Assal, G., Favre, C., & Anderes, J. P. (1984). Non-reconnaissance d'animaux familiers chez un paysan. *Revue Neurologique, 140,* 580–584.

Bauer, R. M. (1982). Visual hypoemotionality as a symptom of visual-limbic disconnection in man. *Archives of Neurology, 39,* 702–708.

Bauer, R. M. (1984). Autonomic recognition of names and faces in prosopagnosia: A neuropsychological application of the guilty knowledge test. *Neuropsychologia, 22,* 457–469.

Bauer, R. M., & Trobe, J. D. (1984). Visual memory and perceptual impairments in prosopagnosia. *Journal of Clinical Neuro-Ophthalmology, 4,* 39–46.

Bauer, R. M., & Verfaellie, M. (1988). Electrodermal discrimination of familiar but not unfamiliar faces in prosopagnosia. *Brain and Cognition, 8,* 240–252.

Bay, E. (1953). Disturbances of visual perception and their examination. *Brain, 76,* 515–550.

Baylis, G. C., Rolls, E. T., & Leonard, C. M. (1985). Selectivity between faces in the response of a population of neurons in the cortex in the superior temporal sulcus of the monkey. *Brain Research, 342,* 91–102.

Beauvois, M. F. (1982). Optic aphasia: A process of interaction between vision and language. *Philosophical Transactions of the Royal Society of London, Series B: Biological Sciences, 298,* 35–47.

Bender, M. B., & Feldman, M. (1972). The so-called "visual agnosias". *Brain, 95*, 173–186.

Benke, T. (1988). Visual agnosia and amnesia from a left unilateral lesion. *European Neurological, 28*, 236–239.

Benson, D. F., Davis, R. J., & Snyder, B. D. (1988). Posterior cortical atrophy. *Archives of Neurology, 45*, 789–793.

Benson, D. F., & Greenberg, J. P. (1969). Visual form agnosia. A specific defect in visual discrimination. *Archives of Neurology, 20*, 82–89.

Benson, D. F., Segarra, J., & Albert, M. L. (1974). Visual agnosia–prosopagnosia. *Archives of Neurology, 30*, 307–310.

Benton, A. (1990). Facial recognition 1990. *Cortex, 26*, 491–499.

Benton, A. L., & Van Allen, M. W. (1972). Prosopagnosia and facial discrimination. *Journal of Neurological Sciences, 15*, 167–172.

Beyn, E. S., & Knyazeva, G. R. (1962). The problem of prosopagnosia. *Journal of Neurology, Neurosurgery, and Psychiatry, 25*, 154–158.

Bodamer, J. (1947). Die Prosop-Agnosie. *Archiv für Psychiatrie und Nervenkrankheiten, 179*, 6–54.

Bornstein, B., & Kidron, D. P. (1959). Prosopagnosia. *Journal of Neurology, Neurosurgery, and Psychiatry, 22*, 124–131.

Bornstein, B., Sroka, H., & Munitz, H. (1969). Prosopagnosia with animal face agnosia. *Cortex, 5*, 164–169.

Bottini, G., Cappa, S. F., & Vignolo, L. A. (1991). Somesthetic-visual matching disorders in right and left hemisphere-damaged patients. *Cortex, 27*, 223–228.

Bruyer, R., Laterre, C., Seron, X., Feyereisen, P., Strypstein, E., Pierrard, E., & Rectem, D. (1983). A case of prosopagnosia with some preserved covert remembrance of familiar faces. *Brain and Cognition, 2*, 257–284.

Campbell, R., Heywood, C. A., Cowey, A., Regard, M., & Landis, T. (1990). Sensitivity to eye gaze in prosopagnosic patients and monkeys with superior temporal sulcus ablation. *Neuropsychologia, 28*, 1123–1142.

Campion, J., & Latto, R. (1985). Apperceptive agnosia due to carbon monoxide poisoning. An interpretation based on critical band masking from disseminated lesions. *Behavioural Brain Research, 15*, 227–240.

Charcot, J. M. (1883). Un cas de suppression brusque et isolee de la vision mentale des signes et des objets (formes et couleurs). *Le Progres Medical, 11*, 568–571.

Charnallet, A., Carbonnel, S., & Pellat, J. (1988). Right visual hemiagnosia: A single case report. *Cortex, 24*, 347–355.

Clark, C. R., Geffen, G. M., & Geffen, L. B. (1989). Catecholamines and the covert orientation of attention in humans. *Neuropsychologia, 27*, 131–139.

Cohn, R., Neumann, M. A., & Wood, D. H. (1977). Prosopagnosia: A clinicopathological study. *Annals of Neurology, 1*, 177–182.

Coslett, H. G., & Saffran, E. (1991). Simultanagnosia. To see but not two see. *Brain, 114*, 1523–1545.

Damasio, A. R., & Damasio, H. (1983). The anatomic basis of pure alexia. *Neurology, 33*, 1573–1583.

Damasio, A. R., Damasio, H., & Van Hoesen, G. W. (1982). Prosopagnosia: Anatomic basis and behavioral mechanisms. *Neurology, 32*, 331–341.

Davidoff, J., & Landis, T. (1990). Recognition of unfamiliar faces in prosopagnosia. *Neuropsychologia, 28*, 1143–1161.

Davidoff, J., & Wilson, B. (1985). A case of visual agnosia showing a disorder of presemantic visual classification. *Cortex, 21*, 121–134.

DeHaan, E. H. F., Young, A., & Newcombe, F. (1987). Faces interfere with name classification in a prosopagnosic patient. *Cortex, 23,* 309–316.

DeRenzi, R. (1986). Prosopagnosia in two patients with CT scan evidence of damage confined to the right hemisphere. *Neuropsychologia, 24,* 385–389.

DeRenzi, E., Bonacini, M. G., & Faglioni, P. (1989). Right posterior brain-damaged patients are poor at assessing the age of a face. *Neuropsychologia, 27,* 839–848.

DeRenzi, E., Faglioni, P., Grossi, D., & Nichelli, P. (1991). Apperceptive and associative forms of prosopagnosia. *Cortex, 27,* 213–221.

DeRenzi, E., Scotti, G., & Spinnler, H. (1969). Perceptual and associative disorders of visual recognition. Relationship to the side of the cerebral lesion. *Neurology, 19,* 634–642.

Eddy, J. K., & Glass, A. L. (1981). Reading and listening to high and low imagery sentences. *Journal of Verbal Learning and Verbal Behavior, 20,* 333–345.

Efron, R. (1968). What is perception? *Boston Studies in Philosophy of Science, 4,* 137–173.

Ellis, A. W., Young, A. W., & Critchley, E. M. R. (1989). Loss of memory for people following temporal damage. *Brain, 112,* 1469–1484.

Ettlinger, G. (1956). Sensory deficits in visual agnosia. *Journal of Neurology, Neurosurgery, and Psychiatry, 19,* 297–307.

Farah, M. J. (1990). *Visual agnosia: Disorders of object recognition and what they tell us about normal vision.* Cambridge, Massachusetts: MIT Press.

Farah, M. J., Hammond, K. M., Levine, D. N., & Calvanio, R. (1988). Visual and spatial mental imagery: Dissociable systems of representation. *Cognitive Psychology, 20,* 439–462.

Farah, M. J., Hammond, K. M., Mehta, Z., & Ratcliff, G. (1989). Category-specificity and modality-specificity in semantic memory. *Neuropsychologia, 27,* 193–200.

Farah, M. J., McMullen, P. A., & Meyer, M. M. (1991). Can recognition of living things be selectively impaired? *Neuropsychologia, 29,* 185–193.

Feinberg, T. E., Gonzalez Rothi, L. J., & Heilman, K. M. (1986). Multimodal agnosia after unilateral left hemisphere lesion. *Neurology, 36,* 864–867.

Freud, S. (1891). *On aphasia* (E. Stengel, trans., 1953). London: Image Publishing.

Geschwind, N., & Fusillo, M. (1966). Color-naming defects in association with alexia. *Archives of Neurology, 15,* 137–146.

Goldstein, K., & Gelb, A. (1918). Psychologische Analysen hirnpathologischer Falle auf Grund von Untersuchungen Hirnverletzter. *Zeitschrift fur die Gesamte Neurologie und Psychiatrie, 41,* 1–142.

Gomori, A. J., & Hawryluk, G. A. (1984). Visual agnosia without alexia. *Neurology, 34,* 947–950.

Hecaen, H., & Albert, M. L. (1978). *Human neuropsychology.* New York: John Wiley & Sons.

Iwata, M. (1990). Visual association pathways in human brain. *Tohoku Journal of Experimental Medicine, 161,* 61–78.

Jankowiak, J., Kinsbourne, M., Shalev, R., & Bachman, D. (1992). Preserved visual imagery and categorization in a case of associative visual agnosia. *Journal of Cognitive Neuroscience, 4,* 119–131.

Kawahata, N., & Nagata, K. (1989). A case of associative visual agnosia: neuropsychological findings and theoretical considerations. *Journal of Clinical and Experimental Neuropsychology, 11,* 645–664.

Kertesz, A. (1979). Visual agnosia: The dual deficit of perception and recognition. *Cortex, 15,* 403–419.

Kosslyn, S. M., Flynn, R. A., Amsterdam, J. B., & Wang, G. (1989). Components of high-level vision: A cognitive neuroscience analysis and accounts of neurological syndromes. *Cognition, 34,* 203–277.

Landis, T., Cummings, J. L., Christen, L., Bogen, J. E., & Imhof, H-G. (1986). Are unilateral right posterior cerebral lesions sufficient to cause prosopagnosia? Clinical and radiological findings in six additional patients. *Cortex, 22*, 243–252.

Landis, T., Regard, M., Bliestle, A., & Kleihues, P. (1988). Prosopagnosia and agnosia for noncanonical views. *Brain, 111*, 1287–1297.

Larrabee, G. J., Levin, H. S., Huff, F. J., Kay, M. C., & Guinto, F. C., Jr. (1985). Visual agnosia contrasted with visual-verbal disconnection. *Neuropsychologia, 23*, 1–12.

Levine, D. N. (1978). Prosopagnosia and visual object agnosia: A behavioral study. *Brain and Language, 5*, 341–365.

Levine, D. N., & Calvanio, R. (1989). Prosopagnosia: A defect in visual configural processing. *Brain and Cognition, 10*, 149–170.

Levine, D. N., Warach, J., & Farah, M. (1985). Two visual systems in mental imagery: Dissociation of "what" and "where" in imagery disorders due to bilateral posterior cerebral lesions. *Neurology, 35*, 1010–1018.

L'Hermitte, F., & Beauvois, M. F. (1973). A visual-speech disconnection syndrome. Report of a case with optic aphasia, agnosic alexia and colour agnosia. *Brain, 96*, 695–714.

Lissauer, H. (1890). Ein Fall von Seelenblindheit nebst einem Beitrage zur Theorie derselben. *Archiv fur Psychiatrie und Nervenkrankheiten, 21*, 222–270.

Mack, J. L., & Boller, F. (1977). Associative visual agnosia and its related deficits: The role of the minor hemisphere in assigning meaning to visual perceptions. *Neuropsychologia, 15*, 345–349.

Macrae, D., & Trolle, E. (1956). The defect of function in visual agnosia. *Brain, 79*, 94–110.

Malone, D. R., Morris, H. H., Kay, M. C., & Levin, H. S. (1982). Prosopagnosia: A double dissociation between the recognition of familiar and unfamiliar faces. *Journal of Neurology, Neurosurgery, and Psychiatry, 45*, 820–822.

Marks, R. L., & DeVito, T. (1987). Alexia without agraphia and associated disorders: Importance of recognition in the rehabilitation setting. *Archives of Physical Medicine and Rehabilitation, 68*, 239–243.

Marr, D. (1982). *Vision: A computational investigation into the human representation and processing of visual information.* San Francisco: Freeman.

McCarthy, R., & Warrington, E. K. (1986). Visual associative agnosia: A clinico-anatomical study of a single case. *Journal of Neurology, Neurosurgery, and Psychiatry, 49*, 1233–1240.

Mega, M. S., Ween, J. E., & Albert, M. L. (1991). Higher visual functions. In S. Lessel & J. Van Dalen (Eds.), *Current neuro-ophthalmology* (pp. 85–100). Chicago: Year Book Medical.

Mendez, M. F. (1988). Visuoperceptual function in visual agnosia. *Neurology, 38*, 1754–1759.

Mendez, M. F., Mendez, M. A., Martin, R., Smyth, K. A., & Whitehouse, P. J. (1990a). Complex visual disturbances in Alzheimer's disease. *Neurology, 40*, 439–443.

Mendez, M. F., Tomsak, R. L., & Remler, B. (1990b). Disorders of the visual system in Alzheimer's disease. *Journal of Clinical Neuro-Ophthalmology, 10*, 62–69.

Meyer, J. S. (1991). Editorial. Does diaschisis have clinical correlates? *Mayo Clinic Proceedings, 66*, 430–432.

Michel, F., Perenin, M. T., & Sieroff, E. (1986). Prosopagnosie sans hémianopsie après lésion unilatérale occipito-temporale droite. *Revue Neurologique, 142*, 545–549.

Milner, A. D., Perrett, D. I., Johnston, R. S., Benson, P. J., Jordon, T. R., Heeley, D. W., Bettucci, D., Mortara, F., Mutari, R., Terazzi, E., & Davidson, D. L. W. (1991). Perception and action in "visual form agnosia." *Brain, 1114*, 405–428.

Newcombe, F., & Ratcliff, G. (1975). Agnosia: A disorder of object recognition. In F. Michel & B. Schott (Eds.), *Les syndromes de disconnexion calleuse chez l'homme* (Colloque International de Lyon, 1974). Lyon: Hopital Neurologique de Lyon.

Newcombe, F., Young, A. W., & DeHaan, E. H. F. (1989). Prosopagnosia and object agnosia without covert recognition. *Neuropsychologia, 27*, 179–191.

Pallis, C. A. (1955). Impaired identification of faces and places with agnosia for colours. Report of a case due to cerebral embolism. *Journal of Neurology, Neurosurgery, and Psychiatry, 18*, 218–224.

Perrett, D. I., Smith, P. A. J., Potter, D. D., Mistlin, A. J., Head, A. S., Milner, A. D., & Jeeves, M. J. (1984). Neurones responsive to faces in the temporal cortex: Studies of functional organization, sensitivity to identity and relation to perception. *Human Neurobiology, 3*, 197–208.

Pylyshin, Z. W. (1973). What the mind's eye tells the mind's brain: A critique of mental imagery. *Psychological Bulletin, 80*, 1–24.

Ratcliff, G., & Newcombe, F. (1982). Object recognition: Some deductions from the clinical evidence. In A. W. Ellis (Ed.), *Normality and pathology in cognitive functions* (pp. 147–171). London: Academic Press.

Riddoch, M. J., & Humphreys, G. W. (1987a). Case of integrative visual agnosia. *Brain, 110*, 1431–1462.

Riddoch, M. J., & Humphreys, G. W. (1987b). Visual object processing in optic aphasia: A case of semantic access agnosia. *Cognitive Neuropsychology, 4*, 131–185.

Rubens, A. B., & Benson, D. F. (1971). Associative visual agnosia. *Archives of Neurology, 24*, 305–316.

Sergent, J., & Corballis, M. C. (1989). Categorization of disoriented faces in the cerebral hemispheres of normal and commissurotomized subjects. *Journal of Experimental Psychology: Human Perception and Performance, 15*, 701–710.

Sergent, J., & Corballis, M. C. (1990). Generation of multipart images in the disconnected cerebral hemispheres. *Bulletin of the Psychonomic Society, 28*, 309–311.

Sergent, J., Ohta, S., & MacDonald, B. (1992a). Functional neuroanatomy of face and object processing. A positron emission tomography study. *Brain, 115*, 15–36.

Sergent, J., Zuck, E., Levesque, M., & MacDonald, B. (1992b). Positron emission tomography study of letter and object processing: Empirical findings and methodological considerations. *Cerebral Cortex, 2*, 68–80.

Snodgrass, J. G., & Vanderwart, M. (1980). A standardized set of 260 pictures: Norms for name agreement, image agreement, familiarity, and visual complexity. *Journal of Experimental Psychology: Human Learning and Memory, 6*, 174–215.

Sparr, S. A., Jay, M., Drislane, F. W., & Venna, N. (1991). A historic case of visual agnosia revisited after 40 years. *Brain, 114*, 789–800.

Striano, S., Grossi, D., Chiacchio, L., & Fels, A. (1981). Bilateral lesion of the occipital lobes. *Acta Neurologica, 3*, 690–694.

Tagawa, K., Nagata, K., & Shishido, F. (1990). Occipital lobe infarction and positron emission tomography. *Tohoku Journal of Experimental Medicine (Suppl.), 161*, 139–153.

Taylor, A., & Warrington, E. K. (1971). Visual agnosia: A single case report. *Cortex, 7*, 152–161.

Teuber, H. L. (1968). Alteration of perception and memory in man. In L. Weiskrantz (Ed.), *Analysis of behavioral change*. New York: Harper and Row.

Tranel, D., & Damasio, A. R. (1988). Non-conscious face recognition in patients with face agnosia. *Behavioural Brain Research, 30*, 235–249.

Tranel, D., Damasio, A. R., & Damasio, H. (1988). Intact recognition of facial expression, gender, and age in patients with impaired recognition of face identity. *Neurology, 38*, 690–696.

Tzavaras, A., Hecaen, H., & LeBras, H. (1970). Le problème de la spécificite du déficit de la reconnaissance du visage humain lors des lésions hémisphériques unilaterales. *Neuropsychologia, 8*, 403–416.

Warrington, E. (1975). The selective impairment of semantic memory. *Quarterly Journal of Experimental Psychology, 27,* 635–657.

Warrington, E. K., & James, M. (1988). Visual apperceptive agnosia: A clinico-anatomical study of three cases. *Cortex, 24,* 13–32.

Warrington, E. K., & McCarthy, R. (1983). Category specific access dysphasia. *Brain, 106,* 859–878.

Warrington, E. K., & Shallice, T. (1984). Category specific impairments. *Brain, 107,* 829–854.

Warrington, E. K., & Taylor, A. M. (1973). The contribution of the right parietal lobe to object recognition. *Cortex, 9,* 152–164.

Whiteley, A. M., & Warrington, E. K. (1977). Prosopagnosia: A clinical, psychological, and anatomical study of three patients. *Journal of Neurology, Neurosurgery, and Psychiatry, 40,* 395–403.

Wilbrand, H. (1887). *Die Seelenblindheit als Herderscheinung.* Wiesbaden: Bergmann.

Yin, R. K. (1970). Face recognition by brain-injured patients: A dissociable ability? *Neuropsychologia, 8,* 395–402.

Chapter 15

Cognitive and Neural Structures in Face Processing

Justine Sergent

I. INTRODUCTION

The idea that the brain is organized into distinct areas of relative functional autonomy and specialization is a basic principle of cognitive neuroscience. This principle is supported and illustrated by the selectivity of behavioral and cognitive deficits that result from focal cerebral injury, and governs the research aimed at understanding brain–behavior relationships. This principle implies that the brain realizes its functions by the joint activation of several of its component structures, each performing specific operations. One of the goals of cognitive neuroscience is uncovering the operations that are performed by each of these structures and how these operations are coordinated and related to one another to produce adapted behavior. However, the logic of the underlying processes cannot be operationalized at a neurophysiological level. Indeed, strictly speaking, no difference may exist at such a neurophysiological level between a rat brain and a human brain. The understanding of brain–behavior relationships must rely on a clear and comprehensive description of the processes at a psychological level.

Psychological and cognitive functions are no longer viewed as made of unitary processes (e.g., reading, writing, face recognition) but as composed of several subprocesses, organized in specific ways (e.g., in parallel or in succession, independently or interactively; see McClelland, 1979). The cognitive architecture of mental functions is characterized best by a compartmented organization of interactive components. Therefore, through a decomposition of a given function into its component operations—thus providing a theoretical framework specifying the nature, the goals, the logical order, and the interactive relationships of the processing steps to be performed for the realization of that function—a better specification of

Localization and Neuroimaging in Neuropsychology

473

the functional organization of the cerebral cortex can be achieved. The understanding of brain–behavior relationships then can be conceived as an enterprise that aims at mapping a fractionated set of mental operations underlying cognitive functions onto their corresponding cerebral structures, and at understanding the relationships among these structures.

Such an enterprise is confronted with a variety of theoretical and methodological problems. First and foremost, we have no guarantee that the brain divides its functions according to the psychological principles we apply to describe the component subprocesses that make up mental functions. A conceptual decomposition of a cognitive function does not necessarily correspond to the cerebral fractionation of its realization. For example, what is considered a single mental operation may, in fact, be implemented in the brain by the interactive recruitment of several structures. In addition, even if evidence supports a structural and functional *modular* organization of the brain, considerable evidence also points to the close interdependence of cerebral structures (e.g., Felleman & Van Essen, 1991). The modularity of the brain is embedded within a highly *interactive* neural network (Sergent, 1984). However, despite the current state of knowledge of structures and connections of the primate brain that underlie visual cognition—32 distinct visual areas have been identified in the monkey brain, with more than 300 connecting pathways among these areas (Felleman & Van Essen, 1991)—the methods currently available for uncovering human brain–behavior relationships impose some limitations on the level of sophistication that can be achieved, especially when dealing with higher order cognitive processes.

The main source of information about the neurobiological substrates of cognition in humans is the study of brain-damaged patients, and relies on two complementary approaches. One is the anatomical–clinical method that seeks to establish correlations between the site of a lesion and the patterns of successes and failures in behavioral tasks in an attempt to identify the contribution of the damaged cerebral area to cognition (e.g., McCarthy & Warrington, 1990). This method generally relies on group studies and is more concerned with the locus of the lesion responsible for specific impairments than with the nature of the processes normally underlying the defective function. This source of information is important for the localization of lesions responsible for specific symptoms, but may not be sufficient to explain how the observed deficits come to the fore. The other approach is computationally oriented, relies primarily on single case studies, and uses the study of brain-damaged patients to infer the functional architecture of cognition from the pattern of disruptions that characterizes the cognitive breakdown, generally irrespective of the actual site and nature of the cerebral lesion (e.g., Caramazza & Badecker, 1989; Shallice, 1988). Thus, this approach provides important information about the decomposition of cognitive functions, but is less concerned with the underlying organization of their neurobiological substrates. In spite of the current debate over the respective merits of these two approaches, and notwithstanding their inherent limitations, both yield relevant and complementary information and, together, allow a more exhaustive and comprehensive specification of the mapping of mental functions onto cerebral structures.

A more recent approach to uncovering the functional organization of the human brain consists of positron emission tomography (PET) measurements of regional changes in cerebral blood flow (CBF) during the performance of behavioral or mental tasks. With the advance in activation techniques (Raichle, Martin, Herscovitch, Mintun, & Markham, 1983), data analyses (Mintun, Fox, & Raichle, 1989), and stereotactic mapping of physiological changes (Evans, Marrett, Collins, & Peters, 1988; Fox, Perlmutter, & Raichle, 1985), local CBF increases that underlie sensory and motor processing can be detected reliably and translated into PET images of the activated cerebral structures, highlighting the areas participating in the realization of a given function (e.g., Fox et al., 1986; Lueck et al., 1989). The introduction of paired-image subtraction and intersubject averaging (Fox, Mintun, Reiman, & Raichle, 1988) has enhanced the sensitivity of PET measurements of CBF and has opened the way to the study of anatomical–functional correlations of higher order cognitive processes, the neural substrates of which are distributed more broadly and are activated less intensely than are primary sensory and motor cerebral areas. The use of short-lived isotopes (such as ^{15}O), allowing measurement of successive tasks in the same subjects, and of the subtraction method, by which measurements made in a control task are subtracted from those made in a functionally activated state (Raichle, 1990) to "isolate" the cerebral areas concerned with the operations that differentiate the two tasks, has given investigators the opportunity to map the functional organization of the brain that underlies cognition (e.g., Petersen, Fox, Posner, Mintun, & Raichle, 1989; Posner, Petersen, Fox, & Raichle, 1988).

When considered within this general context, the study of the neurobiological substrates of cognition may proceed with fewer degrees of freedom since findings from different sources may provide converging evidence and constrain one another's interpretation. In addition, this multidisciplinary procedure offers the possibility of better identifying the strengths and weaknesses of each approach by comparing areas of convergence and divergence between these different sources. This research strategy, when applied to the study of face recognition, then may provide an opportunity to specify the cerebral areas underlying this function.

II. NEUROLOGICAL STUDIES OF FACE RECOGNITION

Inquiry into the processes that underlie the perception and recognition of faces has led to the development of computational models that describe the various operations that must be implemented for a face to become meaningful and be identified. Faces convey a variety of information about individuals; some of this information is *derived visually* in the sense that it can be accessed on the sole basis of the physical attributes of the physiognomy irrespective of the identity of the individual (e.g., gender, age, emotion), whereas other information is *derived semantically* in the sense that it can be accessed only after the perceived representation of the face "makes contact" with a corresponding stored representation from which biographical information about the individual can be reactivated (Sergent, 1989; Young & Bruce, 1991). This distinction implies a hierarchical organization of the processes underlying face recognition, since some preliminary perceptual and mnesic operations must be performed before pertinent biographical information can be accessed. Such a theoretical model has found confirmation in the study of brain-damaged patients, particularly those rare prosopagnosic patients who find themselves unable to recognize persons they know through the visual inspection of their faces (Damasio, 1985; Damasio, Damasio, Tranel, & Brandt, 1991; Sergent & Poncet, 1990; Sergent & Signoret, 1992a; Young and Bruce, 1991).

The study of these patients has shown that prosopagnosia, although a functionally well-delineated impairment—an inability to experience a feeling of familiarity at the view of faces of known persons and, consequently, to identify these faces—may result from lesions at different brain sites and may manifest itself differently across patients. Some patients display a complete inability to perform any operation on faces, even to tell apart men from women, whereas others have no difficulty in processing the gender, age, or emotion of faces and are deficient only when the identity of the face is a critical factor. In addition, the deficit in face recognition is highly selective in most patients and does not extend to the recognition of objects, although some but not all patients have difficulty in identifying objects of the same category when they are physically similar (e.g., feline animals) and in recognizing objects presented from an unusual viewpoint. Although the location of the cerebral lesion varies considerably across these patients, prosopagnosia has been suggested to be the result of a bilateral posterior damage, involving principally the mesial occipitotemporal junction (Damasio, 1985). However, the pattern of damage does not conform to this general rule in all patients; in some cases, the lesion is unilateral and does not invade the posterior cortex. In addition, in most patients, the lesion is extensive so more cortical tissue than area specifically involved in the processing of faces may be destroyed, making it difficult to identify the cortical regions specifically involved in face recognition. Moreover, because structural integrity as identified by radiological data does not guarantee functional normality, due to distant detrimental effects of brain damage, whether the damaged areas are those that normally underlie the processing of faces or whether the lesion also disturbs the processing of adjacent intact regions cannot be determined unequivocally. For all these reasons, the examination of face and object recognition in normal subjects through PET measurements of CBF would provide important data that would bear on still controversial issues.

III. THE PROBLEM OF FACE RECOGNITION

The face enjoys an important and unique status in human lives. As an essential medium of interpersonal relationships and communication, the face conveys and reveals a wide variety of information about an individual. The extraction and interpretation of this information require elaborate and refined perceptual skills that few other categories of objects call for. These skills have attained a very high degree of proficiency, although they are acquired without formal training; we are not aware of the intricate operations that must be implemented for their processing. These operations obviously are taken for granted, as illustrated by Searle's (1984) suggestion that recognizing a face is "as simple and automatic as making footprints in the sand" (p. 52) and by the disbelief of most people when they learn that face recognition can be abolished selectively in some brain-damaged patients.

Processing faces is so natural and automatic that realizing the nature of being deprived of the ability to process faces, as prosopagnosics are, may be difficult. Although the comparison is not entirely exact, Fig. 1 may provide some idea of the nature of this inability. Although we are quite capable of seeing that these faces are all different, we would be unable to achieve a reliable and durable representation of each of them, primarily because we have not acquired and incorporated the basic knowledge that would guide our processing of such simian faces and that would specify the relevant information about their pattern of variations. As a brief test, if we were to match the test faces to the target faces, we would have to look at each detail of each face, not knowing which features or combination of features carry the pertinent information for recognition. We would spend much time on tiny aspects instead of forming a configuration of each face that we would keep in memory for further comparison, as we normally do with human faces. However, this inability does not apply to

Target Faces

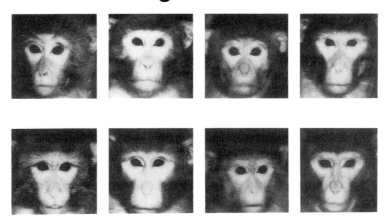

Test Faces

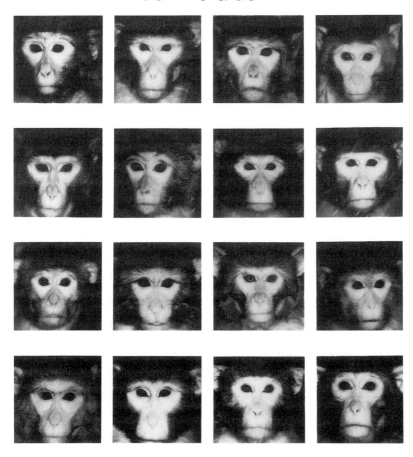

everyone. Some researchers who earn their living with regular contact with monkeys have developed the necessary skills to discriminate and encode the facial features of these monkeys reliably. Indeed, these monkeys are part of a laboratory colony, each with its own name, and their caretaker can identify each one readily regardless of its expression or viewpoint. Nonetheless, for individuals who have not been exposed to monkeys regularly, such faces are very difficult to process. When prosopagnosic patients are presented with these simian faces, they are as much at a loss at processing them as they are at processing human faces; these patients indicate that the simian faces make the same impression on them as human faces have since the onset of their prosopagnosia.

This analogy suggests that underlying our astonishing efficiency in processing human faces is the integration, within our processing structures, of a series of functional operations capable of quickly and reliably extracting the relevant physiognomical invariants that uniquely describe a face. This task is not easy, however, because faces are not static and can take so many different expressions or mimics that their appearance is ever changing, requiring highly proficient mechanisms to detect the constants across different views of the same face. In addition, all faces are shaped on the same basic format, or first-order isomorphism, as suggested by Diamond and Carey (1986), and differ from one another by subtle variations in their general configuration. However, confusions among faces, although they do occur (e.g., Young, Hay, & Ellis, 1985), are rather infrequent, indicating that we have at our disposal very efficient discriminatory skills that are specific to the processing of faces, rather than not only other objects but also faces that do not belong to our race. Our proficiency with faces of our own race does not make us expert at processing faces of other races. We need long and repeated exposure to a given class of persons or objects to achieve a high level of efficiency. Also remarkable is that we have almost no conscious knowledge of the particular features on which we rely to derive specific information from facial representations. Although this is true of the *identification* of faces, it also applies to other operations on faces when identity is not a relevant factor. For instance, although very few people can describe specifically what differentiates a female from a male face, nearly everyone can tell them apart readily just by looking at them; this is also the case in estimating the age of a face or recognizing the emotion expressed by a face. These abilities illustrate that our processing structures have incorporated a considerable amount of functional knowledge about faces, the regularities and diversities of their

Figure 1 Sample of simian faces.

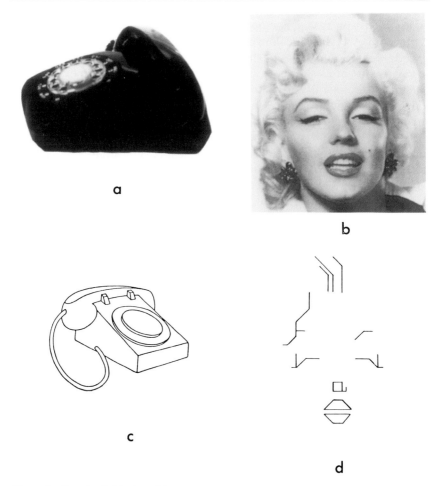

Figure 2 Sample of objects and faces.

variations, and the meaning of these variations to such a degree that one fails to realize the intricate operations that must be at work when processing faces.

IV. SPECIFICITY OF FACE PROCESSING

Perception and recognition of faces differ from the processing of other visual objects in several ways. If we look at the object shown in Fig. 2a, we readily recognize a telephone; information related to the functional properties of this object category then can be accessed, irrespective of the unique characteristics of this particular telephone. If we look at the object in Fig. 2b, we do not see merely a face but recognize a unique individual; the perception of this face can lead to the evocation of a host of information about this individual. Thus, one way faces stand out among visual objects is that most objects are processed at the basic-object category level (e.g., Rosch, Mervis, Gray, Johnson, & Poyes-Braem, 1976) whereas faces are treated by considering each instance as different within the category of faces. In addition, even if we were not familiar with the face of Marilyn Monroe, a large variety of information about the individual still could be accessed on the sole basis of the physical attributes of the face. We could, for instance, make inferences about the age, gender, race, and emotion; we could judge the beauty of the face, its charm, its cuteness, and so on. Thus, visually we derive valuable information relevant to the bearer of the face. Another way in which faces differ from other objects, therefore, seems to be the large variety of categories into which faces can be entered; the same facial attributes are treated and combined differently, depending on the information one wishes to access about an individual (Sergent, 1989).

Although faces often are thought to be morphologically more complex than other objects, this is not necessarily so. When a face is stripped to its bare essential, in a very schematic drawing, it still conveys the properties that make it a unique individual. A face can be described with as few elements as most other objects, as illustrated in Fig. 2c,d. Two main factors seem to make face recognition a complex and almost unique process. One is the large number of faces with which we are confronted and the necessity to detect, from a configuration common to all faces, the subtle differences that make each face unique; the other is the need to retrieve pertinent information about the individual for a face to become meaningful. These two main operations, in conjunction with the establishment of a relationship between a facial representation and its pertinent memories, seem to be essential to the recognition of faces. As will be suggested next, for one of these operations to be defective is sufficient for the whole face-recognition system to become disabled.

V. COGNITIVE STRUCTURES IN FACE PROCESSING

Prosopagnosia is a rare neurologically based deficit characterized by the inability to experience a feeling of familiarity at the view of faces of known individuals and, therefore, to identify these faces. This disorder occurs in the absence of severe intellectual, perceptual, and memory impairments and, although associated deficits may be detected in these patients, none is more dramatic than the failure to identify faces of known persons, including those of relatives and even the patient's own face. One of the main difficulties in understanding the underlying nature of this deficit derives from the diversity of the impairments displayed by these patients. Although the functional deficit is the same across patients, it may result from a breakdown at different stages in the operations that must be performed from the initial perception of a face to its recognition. This variability can be illustrated by examining the nature of the impairment of four prosopagnosic patients.

A specification of the decomposition of the operations inherent in the perception and recognition of faces provides a useful framework for examining the various steps that must be implemented for a face to be identified. Faces lend themselves to a variety of operations corresponding to the extraction of different types of information about an individual. Figure 3 presents some of the steps leading to the recognition of a face. This figure is adapted from the model suggested by Bruce and Young (1986) and should not be regarded as a theoretical model of face recognition, but as a schematic functional description of some of the operations by which a face acquires meaning and eventually can be identified. A first step indicated here is that of the structural encoding, which is the end product of visual sensory processing of the incoming information and which broadly corresponds to Marr's (1982) two-and-a-half description. From such a description, a series of operations can be performed to access information on the basis of the visual properties of the face, irrespective of its identity. The ability of the patients to perform such operations may provide some indications about the functional level of their deficits. Considering the performance of prosopagnosic patients shown in Table 1, one patient (R. M.) was unable to extract from the physical attributes of the face the information related to its gender (as was P. M.), its age, and its emotion. In fact, R. M. was defective in all aspects of face processing, including matching two identical views of the same faces presented simultaneously in front of him. Another patient (P. C.) performed above chance but nonetheless was impaired compared with controls, whereas the performance of the fourth patient (P. V.) was not significantly different from that of controls. These findings suggest that not only is prosopagnosia a selective deficit, but a good deal of selectivity exists even within the prosopagnosic disturbance. In other words, prosopagnosia is not a homogeneous impairment. In addition, the perceptual deficit that results from a breakdown of the structural encoding stage and that characterizes R. M.'s prosopagnosia is specific to the processing of faces and does not necessarily extend to the discrimination and recognition of other categories of objects. For instance R. M., who collected miniature cars as a hobby and owned old cars, could identify correctly 172 of 210 cars by the name of their makes and models and by their year of fabrication (Sergent & Signoret, 1992b).

A next step in the processing of faces indicated in Fig. 3 consists of extracting the invariant physiognomic attributes that uniquely describe a face. This step does not involve the recognition of the face as such, but the ability to determine, for example, whether or not two different views of the same face are the same or different. The patients were tested on a series of matching tasks in which faces photographed from different viewpoints or at different ages were presented; the patients had to put together the

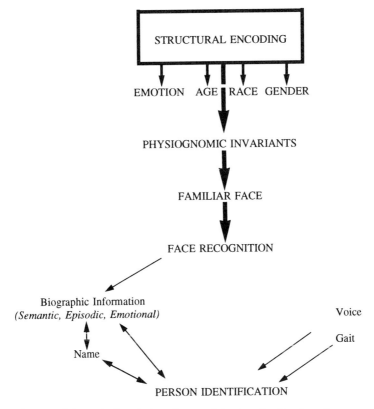

Figure 3 Schematic description of some of the steps leading from face perception to person identification.

faces of the same individuals. Such tasks cannot be performed simply by comparing the visual information contained in the faces and requires perceptual operations by which the facial features that are unique to a face are extracted. The results are shown in Table 1 (physiognomic invariants) and indicate that two patients (R. M. and P. M.) performed at chance, whereas patients P. C. and P. V. could achieve above-chance matching. Whereas the deficit of R. M. lies at the structural encoding level, that of P. M. reflects an inability to derive the configuration that is unique to a face. A next step shown in Fig. 3 consists of determining whether a face is familiar or not, which requires that some contact be made between the perceived face and the representation of that face in memory. At this level all prosopagnosics are defective; the performance of the other two patients was at chance at this level of processing. As the diversity in the patterns of deficits displayed by these four patients suggests, the prosopagnosic disturbance may result from a breakdown at different levels of processing, and the deficits associated with this disturbance vary among patients.

VI. FUNCTIONAL NEUROANATOMY OF FACE PROCESSING

These different patterns of disruption across patients reflect different underlying breakdowns and suggest that the different stages that compose the face-recognition system may be disabled selectively and therefore must be represented in different locations in the cerebral structures. However, several difficulties exist in inferring the actual anatomical locus of

Table 1 Performance of Prosopagnosic Patients and Control Subjects on Tasks of Object and Face Processing

| | Patient | | | | Controls |
Task	P. C.	P. M.	P. V.	R. M.	(range)
WAIS verbal IQ	114	108	102	103	
Memory quotient	109	111	98	112	
Object recognition	52/54	54/54	52/54	52/54	50–54
Gender categorization	38/50[a]	31/50[b]	43/50	18/50[b]	43–50
Age estimation	17/20	16/20	17/20	4/20[b]	16–20
Face emotion	12/24	8/24[b]	18/24	6/24[b]	18–24
Physiognomic invariants	14/24	4/24[b]	17/24	2/24[b]	16–24
Face recognition	5/100	6/100	2/100	0/100	94–100

[a]Underlined: Performance lower than lowest performance of controls.
[b]Performance not different from chance.

the various stages that compose face recognition from the performance of prosopagnosic patients since their lesions generally invade large cerebral territories and even may affect, functionally, adjacent cortical regions that are structurally intact.

Advances in brain-imaging techniques for measuring blood flow within cerebral structures have made possible the visualization of the neuronal substrates of cognitive abilities in normal subjects. Such techniques provide the opportunity to infer the neuroanatomy of a given function without interference from the dynamic effects of a cortical lesion on the functioning of the whole brain. The basic procedure consists of injecting a subject with radioactive material that binds to a physiologically active compound and serves as a blood flow tracer by detecting, through a tomograph, the gamma rays that are emitted following the decay of positrons. Much progress has been made in the study of cognition using ^{15}O, the half-life of which is very short and allows repeated tasks to be performed on the same subject (Raichle et al., 1983). Posner and his colleagues (1988) have developed a technique consisting of several complementary tasks that differ from one another in one or a few cognitive operations to "isolate" the cerebral areas specifically activated by these operations.

This basic procedure was used to examine the neural substrates of face and object processing in normal subjects, with the [^{15}O]water-bolus technique. In addition, a magnetic resonance imaging (MRI) scan of each subject's brain was obtained to perform a mapping of the physiological activation measured in the PET study onto the actual cerebral structures of the subjects, following the method developed by Evans and his colleagues (1989).

The study consisted of 6 tasks run consecutively, in a different order for each subject, with 3 control and 3 experimental tasks (Sergent, Ohta, & MacDonald, 1992). The control tasks consisted of a passive fixation of the lit screen, a passive viewing of unfamiliar faces, and a two-choice reaction-time task requiring the subjects to discriminate sine wave gratings of varying spatial frequencies as a function of their vertical or horizontal orientation. The experimental tasks were an object-categorization task, consisting of the presentation of common objects (either living or non-living); a gender-categorization task, in which the subjects had to decide whether the face was that of a man or a woman; and a face-identity task, consisting of an occupational categorization of famous faces, a task that requires the recognition of the face. The subjects responded with the right index or middle finger by pressing one of two buttons of the computer mouse that lay on their abdomen, to insure that they actually were performing the requested tasks and to control for response accuracy. The 3 experimental tasks, as well as the grating-discrimination task, were exactly the same in terms of procedure, monitor apparatus, and stimuli as those carried out on the prosopagnosic patients so the neural substrates of the operations involved in these tasks could be compared in normal and neurological subjects. Thus, the main foci of activation in the normal subjects' brains are shown in Color Plate 10 and the radiological data of the patients are presented in Fig. 4.

In each of the experimental and grating tasks, strong activation of the sensorimotor area of the left hemisphere was seen, corresponding to the

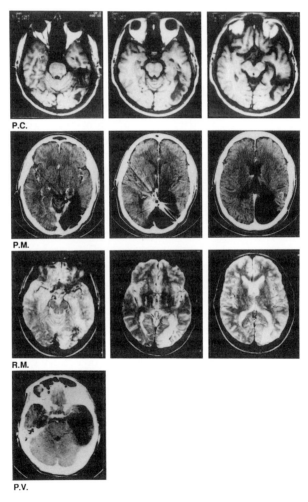

Figure 4   Radiological data from four prosopagnosic patients.

representation of the right hand which was used to make the two-choice response. This response is illustrated for Grating Condition B (Color Plate 10). As shown in the plate, the foci of activation obtained from the PET study are seen superimposed on the MRI of the subject's brain in the latero-dorsal region of the left frontal and parietal lobes. Although the passive viewing of faces was intended as a control task, it did not result in significantly different activation from the gratings condition and, compared with the passive fixation condition, both the passive face-viewing and the grating-discrimination tasks produced significant activation of areas 17 and 18 in the occipital cortex and resulted in a larger involvement of the lingual gyrus in the right than in the left hemisphere (Color Plate 10, Grating Condition A). In the comparison of the passive fixation and the gender-categorization task, cerebral activation was observed significantly in the ventro-medial region of the right hemisphere and involved the lingual gyrus and the posterior part of the fusiform gyrus. In this and the following PET images, the sequence of slices from top left to bottom right corresponds to ventral regions toward dorsal cerebral regions. As can be seen for the gender-categorization task (Color Plate 10), bilateral activation at the level of the striate cortex occurred, but much of the processing underlying this task took place within the lingual gyrus of the right hemisphere, with no corresponding activation of the left hemisphere beyond area 18.

Comparing this pattern of activation in normal subjects performing the gender-categorization task with the radiological data from the two prosopagnosic patients (P. M. and R. M.) whose disturbance included the categorization of faces as male and female (see Table 1) may be informative. In the case of P. M. (Fig. 4), the entire right occipital cortex and posterior ventral temporal cortex were ablated in 1972; she has been prosopagnosic since then. Her lesion affects the cerebral structures that were activated during the PET study in the normal subjects. Similarly, in the case of R. M., whose impairment encompasses all operations on faces, the lesion also affects the ventro-medial area of the right occipital and posterior temporal cortex up to the fusiform gyrus, but extends somewhat more anteriorly than the damage of P. M. (Fig. 4). In both patients, the lesion invades those areas activated in normal subjects during the gender-categorization task. In contrast, the MRI of P. C.'s brain—one of the two patients who were able to tell men from women, to perform the extraction of physiognomic invariants, and to match different views of the same face—indicates a more anterior lesion in the white matter surrounding the fusiform gyrus without affecting the cortex of this gyrus. Patient P. C., although impaired, was performing above chance at tasks that required him to extract the physiognomic invariants, suggesting that his functional deficit reflected an inability to activate the semantic information from the perception of faces. Therefore, the processes underlying the reactivation of pertinent memories must take place in cerebral areas anterior to the fusiform gyrus.

This hypothesis was confirmed in the PET study by considering the pattern of activation associated with the face-identification task. As shown in Color Plate 10 (Face Identity B), the same pattern of activation as in the gender-discrimination task is apparent in the posterior region of the right hemisphere during the face-identity task. However, activation is now located more anteriorly, specifically in the right parahippocampal gyrus, whereas no such activation can be detected on the left side. Two additional findings are worth noting with respect to the activation associated with the face-identity task. One concerns the activation of the orbital region of the frontal cortex. This activation also was present during the object-recognition task and therefore may not be specific to the processing of faces as such, but the actual operations served by this region in the processing of faces cannot be identified at this time. The other finding is the activation of the left fusiform gyrus, which was not involved in the face gender-categorization task and therefore must participate in the processing of face identity. However, this area also was activated during the object recognition task, again suggesting that its participation may not be specific to the processing of faces and may reflect the visual analysis of complex objects. Face Identity Condition A (Color Plate 10) presents the results of the subtraction of the gender-categorization from the face-identity task, and suggests several foci of activation specifically associated with the processing of face identity. Clearly apparent here is the activation of the right parahippocampal gyrus, as already mentioned, but activation is seen also in the most anterior parts of the temporal lobes of both hemispheres. As a comparison, the computerized tomography (CT) scan of patient P. V. (Fig. 4), whose deficit involved only the recognition of familiar faces but did not markedly affect any other aspect of face processing, indicates a lesion that had destroyed the anterior half of her right temporal lobe as well as the pole of the left temporal lobe as a result of encephalitis. This pattern of brain damage concurs with the pattern of activation observed in the normal subjects during the face-identity task. The type of prosopagnosia that characterizes patient P. V.'s disturbance is therefore different from the essentially perceptual impairment discussed earlier with respect to patients R. M. and P. M., and suggests an inability to evoke the pertinent stored information about an individual without which a face cannot become meaningful and be identified.

In contrast, the object-categorization task resulted in no activation of the anterior temporal cortex relative to the grating-discrimination condition (see Color Plate 10). In fact, this task did not result in any significant activation of the cerebral structures in the right hemisphere that were involved specifically in the processing of faces. At a functional level, therefore, one does not need to access specific information related to the particular instance of an object to recognize it, consistent with the different processing requirements made by the recognition of faces and of objects. Instead, the categorization of objects resulted in significant activation of the lateral occipito-temporal cortex of the left hemisphere, specifically in the fusiform gyrus, as well as in the activation of the left temporal area 21 in the middle temporal gyrus. The activation of the posterior left hemisphere concurs with the evidence suggesting that this area plays a crucial role in the processing of the semantics of visual objects, the disruption of which results in visual agnosia of the associative type (Kertesz, 1979; McCarthy and Warrington, 1990). Thus, not only do the recognition of faces and objects make different processing demands, but their respective representations are not inscribed in the same areas of the brain.

VII. IMPLICATIONS

The current findings may help us understand the anatomical and functional architecture underlying the processing of faces and its dissociation from the processing of objects. The recognition of faces requires the involvement of the ventral areas of the right hemisphere. As shown by the PET and radiological data from the patients, the right hemisphere seems to be both necessary and sufficient to sustain face recognition. Three cortical areas that subserve specific operations appear to be essential to this function. The right lingual and fusiform gyri perform the perceptual operations by which the configurational properties of a face are encoded and the physiognomic invariants are extracted. The right parahippocampal gyrus, but not the hippocampus itself, seems to play a crucial role in the reactivation of pertinent memories associated with a given facial representation. The anterior temporal lobes of both hemispheres seem to contain the biographical information that must be reactivated for a given face to become meaningful and, thus, be identified. This information is not used uniquely for the processing of faces, however, and seems to be involved in any operation that calls for the evocation of personal memories. In fact, patient P. V., whose lesion involved the anterior temporal cortex, was unable to recognize such famous monuments as the Arc de Triomphe and the Eiffel Tower in addition to her prosopagnosia. At present, whether or not the left temporal cortex plays an indispensable role in this process is not clear since evidence reported by Ellis, Young, and Critchley (1989) suggests that damage to the right anterior temporal cortex may be sufficient to produce an impaired recollection of personal memories in response to the view of a face or the evocation of a name.

None of the cortical areas specifically active during the face-identity task were activated in the object-recognition task, which instead relied on structures in the posterior left hemisphere. This result is consistent with the frequently observed dissociation of visual agnosia and prosopagnosia. In fact, these two deficits are associated only in patients with bilateral damage; when the lesion responsible for prosopagnosia is restricted to the right hemisphere or does not involve the posterior left hemisphere, as in the four patients presented here, no deficit in object recognition is seen. Note also that no activation of the superior temporal sulcus nor of the ventro-lateral infero-temporal cortex could be detected in the PET study. In these areas cells that are selectively responsive to faces have been found in the monkey brain (Rolls, 1989). Either these areas are not indispensable to the process of face recognition or no direct anatomical correspondence exists between the simian and the human neural substrates of face processing, despite the frequent references to the work on monkeys in the literature on humans. In fact, damage to these areas in humans does not result in prosopagnosia.

This overview of the functional neuroanatomy of face and object processing suggests a striking convergence of the results derived from brain-damaged patients with specific functional deficits at different stages of face processing on the one hand and, on the other hand, the PET findings from normal subjects performing the same tasks as the patients. The results provide strong evidence of structural and functional dissociation of face and object processing and, within face processing itself, of a decomposition into specific operations that has a clear correspondence at the anatomical level. The results also confirm the crucial role of the ventral cortex in determining "what" is perceived, as opposed to "where," and suggest an essential contribution of the right hemisphere to the processing of faces.

VIII. CONCLUSIONS

No neuroimaging technique is free of methodological and theoretical difficulties, and each has its own limitations that could prevent exclusive reliance on its findings to infer brain–behavior relationships unequivocally. Nonetheless, the concordance of the conclusions derived from correlations between cognitive deficits and location of lesions in the four prosopagnosic patients and from the activated cerebral areas in normal subjects performing the same tasks as the patients suggests that each technique yields pertinent evidence for the understanding of brain–behavior relationships. Together, such measurements thus may provide convergence as well as complementarity of viewpoints. For instance, the results of the PET study, by virtue of the subtraction technique, yield more precise information about the location of the critical areas recruited for the realization of a specific operation than do radiological data from patients whose lesions may invade larger areas than would be necessary to disrupt this operation. However, if the realization of this operation engages more than one cerebral area, radiological data from the patients may become indispensable to specify the area of destruction that is critical in the disruption of that operation. On the other hand, the visualization of activated cerebral areas in *normal* subjects provides information that is not influenced by secondary or distant effects of a lesion in a damaged brain.

The benefits associated with the PET approach should not mask its inherent limitations nor the risk of promoting a modern type of phrenology that would disregard evidence of interactive and distributed processing carried out in cerebral structures (Sergent, 1990). PET findings often are suggested to offer evidence of a modular organization of the brain, since activation clearly is localized in specific areas as a function of the particular demands of a cognitive task. However, such a finding of local activation is, in itself, the result of the subtraction method that aims at eliminating all activation but that uniquely related to one operation among the many operations that constitute a given task. This procedure hides the interactions among various cerebral areas underlying the realization of a given task. In addition, because the PET technique is a "noisy" procedure, stringent statistical tests are used, with the result that only a few of the areas in which an increased activation can be detected are considered significant. Whether the local changes that do not reach statistical reliability are actually insignificant and do not reflect a true contribution to the realization of the operation under investigation is unknown (see Sergent, Zuck, Levesque, & MacDonald, 1993b).

Another limitation of the PET technique derives from the static and cumulative nature of the data it produces, which conceals the temporal and relational components of cerebral information processing. For instance, the present data cannot determine whether the involvement of a cortical structure (e.g., the left fusiform gyrus in the face-identification task) is conditional to earlier operations in another structure (e.g., the right lingual and fusiform gyri). The fact that the four prosopagnosics described earlier had an intact left fusiform gyrus suggests that this area cannot, on its own, perform the operations required for face identification; its activation in normal subjects may reflect a participation of this area that is not sufficient to sustain the recognition of faces and may not even be necessary. Clearly, the results of PET measurements of CBF need to be confronted with findings from other sources. However, even if this technique is not self-sufficient, it provides a significant contribution to the understanding of brain–behavior relationships.

ACKNOWLEDGMENTS

The work reported in this chapter was supported by the L.B. Foundation, the Medical Research Council of Canada, the National Institute of Mental Health, and the Natural Sciences and Engineering Research of Canada.

REFERENCES

Bruce, V., & Young, A. W. (1986). Understanding face recognition. *British Journal of Psychology*, *77*, 305–327.

Caramazza, A., & Badecker, W. (1989). Patient classification in neuropsychological research. *Brain and Cognition*, *10*, 256–295.

Damasio, A. (1985). Prosopagnosia. *Trends in Neuroscience*, *8*, 132–135.

Damasio, A., Damasio, H., Tranel, D., & Brandt, J. P. (1991). The neural regionalization of neural access: Preliminary evidence. *Cold Spring Harbor Symposium of Quantitative Biology*, *55*, 1039–1047.

Diamond, R., & Carey, S. (1986). Why faces are and are not special: An effect of expertise. *Journal of Experimental Psychology: General*, *115*, 107–117.

Ellis, A. W., Young, A. W., & Critchley, E. M. R. (1989). Loss of memory for people following temporal damage. *Brain*, *112*, 1469–1484.

Evans, A. C., Beil, C., Marrett, S., Thompson, C. J., & Hakim, A. (1988). Anatomical–functional correlation using an adjustable MRI-based region of interest atlas with positron emission tomography. *Journal of Cerebral Blood Flow and Metabolism*, *8*, 513–530.

Evans, A. C., Marrett, S., Collins, L., & Peters, T. M. (1989). Anatomical-functional correlative analysis of the human brain using three-dimensional imaging systems. *Proceedings of the International Society of Optical Engineering (SPIE)*, *1092*, 264–274.

Felleman, D. J., & Van Essen, D. C. (1991). Distributed hierarchical processing in the primate cerebral cortex. *Cerebral Cortex*, *1*, 1–47.

Fox, P. T., Mintun, M. A., Raichel, M. E., Miezin, F. M., Allman, J. M., & Van Essen, D. C. (1986). Mapping human visual cortex with positron emission tomography. *Nature (London)*, *323*, 806–809.

Fox, P. T., Mintun, M. A., Reiman, E. M., & Raichle, M. E. (1988). Enhanced detection of focal brain responses using intersubject averaging and change-distribution analysis of subtracted PET images. *Journal of Cerebral Blood Flow and Metabolism*, *8*, 642–653.

Fox, P. T., Perlmutter, J. S., & Raichle, M. E. (1985). A stereotactic method of anatomical localization for positron emission tomography. *Journal of Computer Assisted Tomography*, *9*, 141–153.

Kertesz, A. (1979). Visual agnosia: The dual deficit of perception and recognition. *Cortex, 15,* 403–419.

Lueck, C. J., Zeki, S., Friston, K. J., Deiber, M. P., Cope, P., Cunningham, V. J., Lammerstma, A. A., Kennard, C., & Frackowiak, S. J. (1989). The colour center in the cerebral cortex of man. *Nature (London), 340,* 386–389.

Marr, D. (1982). *Vision.* San Francisco: Freeman.

McCarthy, R. A., & Warrington, E. K. (1990). *Cognitive neuropsychology. A clinical introduction.* London: Academic Press.

McClelland, J. L. (1979). On the time relations of mental processes: An examination of systems of processes in cascade. *Psychological Review, 86,* 287–330.

Mintun, M., Fox, P. T., & Raichle, M. E. (1989). A highly accurate method of localizing neuronal activity in the human brain with positron emission tomography. *Journal of Cerebral Blood Flow and Metabolism, 9,* 96–103.

Petersen, S. E., Fox, P. T., Posner, M. I., Mintun, M., & Raichle, M. E. (1989). Positron emission tomographic studies of the processing of single words. *Journal of Cognitive Neuroscience, 1,* 153–170.

Posner, M. I., Petersen, S. E., Fox, P. T., & Raichle, M. E. (1988). Localization of cognitive operations in the human brain. *Science, 240,* 1627–1631.

Raichle, M. E. (1990). Images of the functioning human brain. In H. Barlow, C. Blakemore, & M. Weston-Smith (Eds.), *Images and understanding* (pp. 284–296). Cambridge: Cambridge University Press.

Raichle, M. E., Martin, W. R. W., Herscovitch, P., Mintun, M. A., & Markham, J. (1983). Brain blood flow measured with intraveinous H_2O. II. Implementation and validation. *Journal of Nuclear Medicine, 24,* 790–798.

Rolls, E. T. (1989). The representation and storage of information in neuronal networks in the primate cortex and hippocampus. In R. Durbin, C. Miall, & G. Mitchison (Eds.), *The computing neuron* (pp. 73–90). Amsterdam: Addison-Wesley.

Rosch, E., Mervis, C. B., Gray, W. D., Johnson, D. M., & Poyes-Braem, P. (1976). Basic objects in natural categories. *Cognitive Psychology, 8,* 382–439.

Searle, J. (1984). *Minds, brains and science: The 1984 Reith lectures.* London: British Broadcasting Corporation.

Sergent, J. (1984). Inferences from unilateral brain damage about normal hemispheric functions in visual pattern recognition. *Psychological Bulletin, 96,* 99–115.

Sergent, J. (1989). Structural processing of faces. In A. W. Young & H. D. Ellis (Eds.), *Handbook of research on face processing* (pp. 57–91). Amsterdam: North-Holland.

Sergent, J. (1990). The neuropsychology of visual image generation: Data, method and theory. *Brain and Cognition, 13,* 98–129.

Sergent, J., Ohta, S., & MacDonald, B. (1992). Functional neuroanatomy of face and object processing: A positron emission tomography study. *Brain, 115,* 15–36.

Sergent, J., & Poncet, M. (1990). From covert to overt recognition of faces in a prosopagnosic patient. *Brain, 113,* 989–1004.

Sergent, J., & Signoret, J. L. (1992a). Functional and anatomical decomposition of face processing: Evidence from prosopagnosia and PET study of normal subjects. *Philosophical Transactions of the Royal Society, London, Series B. 335,* 55–62.

Sergent, J., & Signoret, J.-L. (1992b). Varieties of functional deficits in prosopagnosia. *Cerebral Cortex, 2,* 375–388.

Sergent, J., Zuck, E., Lévesque, M., & MacDonald, B. (1992). A positron emission tomography study of letter and object processing: Empirical findings and methodological considerations. *Cerebral Cortex, 2,* 68–80.

Shallice, T. (1988). *From neuropsychology to mental structure.* Cambridge: Cambridge University Press.

Young, A. W., & Bruce, V. (1991). Perceptual categories and the computation of "grandmother." *European Journal of Cognitive Psychology, 3,* 5–49.

Young, A. W., Hay, D. C., & Ellis, A. W. (1985). The faces that launched a thousand slips: Everyday difficulties in recognizing people. *British Journal of Psychology, 76,* 495–523.

Chapter 16

Localization of Lesions in Neglect and Related Disorders

Kenneth M. Heilman, Robert T. Watson, and Edward Valenstein

I. INTRODUCTION

Because the major intent of this book is to correlate behavioral abnormalities and their pathological substrates, this chapter concentrates on the pathological anatomy of unilateral neglect and related disorders. However, before discussing the anatomical aspects of the neglect syndrome, we define and discuss the symptoms and signs. The reader can find a detailed discussion of the pathophysiology of neglect in several reviews (e.g., Heilman, Valenstein, & Watson, 1993).

II. DEFINITIONS

Patients with unilateral neglect and related disorders fail to report, respond, or orient to stimuli presented to the side contralateral to a central nervous system lesion. A patient is not considered to have neglect if he or she fails to respond to stimuli because of a primary motor or sensory defect. The symptoms of neglect may occur under a variety of stimulus and performance conditions. Different behavioral signs and symptoms may be seen during the evolution of, and recovery from, neglect. In addition, not all patients with neglect will have the complete spectrum of signs and symptoms associated with the neglect syndrome.

Neglect may occur in two spatial domains, environmental or personal, and neglect may be related to one or more of three basic mechanisms, disorders of sensory attention, disorders of action or motor intention, and representational or memory disorders. Based on the domain and the basic mechanisms, many specific disorders have been described, including inattention or sensory neglect; extinction to simultaneous stimuli; hemispatial

Localization and Neuroimaging in Neuropsychology

495

or unilateral spatial neglect; personal neglect or asomatognosia; intentional or motor neglect including akinesia, hypokinesia, motor extinction, and impersistence; and associated signs such as allesthesia, anosognosia (denial of hemiplegia or illness), and anosodiaphoria.

III. SIGNS AND SYMPTOMS

A. Inattention (Sensory Neglect) and Extinction

Patients with sensory inattention are unaware of contralesional stimuli. Depending on the sensory modality tested, inattention may be spatial (visual, auditory) or personal (e.g., tactile). Distinguishing between inattention and primary sensory disturbances in the visual or somesthetic modalities may be difficult. However, a patient with visual or somesthetic inattention may be able to report a stimulus when their attention is drawn to it. Since inattention may occur in a hemispatial field, having subjects gaze toward ipsilesional hemispace may improve their awareness of stimuli presented to the contralesional visual field (Kooistra & Heilman, 1989; Nadeau & Heilman, 1991). Caloric stimulation of the ear may improve unawareness of contralesional tactile stimuli (Vallar, Sterzi, Bottini, Cappa, & Ruscani, 1990).

Patients with inattention often improve so they can detect, respond to, and lateralize unilateral stimuli successfully. With frequent repetition of stimuli, some patients will show inattention again (fatigue phenomenon). Even when not fatigued, however, when some patients are given bilateral simultaneous stimuli, they fail to report the stimulus presented contralateral to the damaged hemisphere. This phenomenon, extinction to double simultaneous stimulation, can be visual, tactile, or auditory. Extinction also may be seen with stimuli given on the same side of the body (or space). Although extinction may be more common and more severe when stimuli are presented contralesionally, extinction may be seen even when both stimuli are presented on the ipsilesional side (Rapcsak, Watson, & Heilman, 1987).

B. Spatial Neglect

When performing spatial tasks, patients may neglect the contralesional space. Spatial neglect may be tested using a variety of tasks. Spatial neglect can be demonstrated when drawing a daisy or other objects. Often, the patient with spatial neglect will draw only half a daisy (Fig. 1). The line-bisection and crossing-out tasks may be more sensitive for demon-

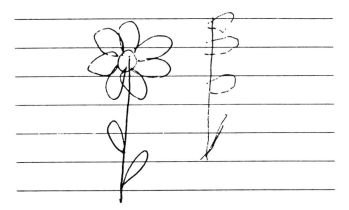

Figure 1 Examiner's drawing of daisy on left; patient's drawing on right.

strating spatial neglect. In the former task, a long horizontal line is drawn and the patient is asked to bisect the line. The patient with spatial neglect tends to bisect the line toward the normal (ipsilesional) side (Fig. 2). In the cancellation task, a patient is presented a sheet of paper with short lines randomly distributed over the entire sheet. When a patient with spatial neglect is asked to mark out (cancel) all lines, he or she will fail to mark out some of the lines on the neglected side of space.

Spatial neglect also can be identified in the activities of daily living. A patient may fail to eat food from one side of the plate. He or she may fail to read one side of a word or sentence (neglect-induced paralexia) or type

Figure 2 Line bisection tasks demonstrating the hemispatial effect on hemispatial neglect. On first trial, line is in right hemispace (right side); before bisecting the line the patient reads the letter *O* (left start). On second trial, the line is placed in left hemispace (left side), and the patient reads the letter *O* (left start) before bisecting line.

on one side of the keyboard (neglect-induced paragraphia) (Valenstein & Heilman, 1978).

Although spatial neglect is most severe in the space contralateral to the lesioned hemisphere (Heilman & Valenstein, 1979), patients with spatial neglect also may demonstrate neglect in ipsilateral hemispace (Albert, 1973; Heilman & Valenstein, 1979; Weintraub & Mesulam, 1987).

Although horizontal neglect is the most common type of spatial neglect seen in the clinic, some patients may neglect lower (Rapcsak, Cimino, & Heilman, 1988) or upper (Shelton, Bowers, & Heilman, 1990) vertical space. In addition, Shelton et al. (1990) reported a patient who neglected near radial (peripersonal) space and Mennemeier, Wertman, and Heilman (1993) reported a patient who neglected far peripersonal space. Radial and vertical neglect can be tested using tasks such as line bisection and cancellation. However, to test for radial or vertical neglect, the stimulus (e.g., the line) must be oriented in a vertical or radial direction.

C. Personal Neglect (Asomatognosia)

Patients with personal neglect may fail to recognize that their own limbs belong to them. They also may fail to groom and dress the neglected parts of their body.

D. Intentional Neglect (Motor Neglect)

Although unawareness of stimuli may cause a failure to respond, patients with intentional neglect may fail to respond to stimuli of which they are aware. Four types of intentional disorders will be discussed: akinesia, hypokinesia, motor extinction, and impersistence.

Akinesia is a motor initiation disorder that is not caused by a lesion in the corticospinal motor system, but is induced by dysfunction in the systems that are responsible for activating the motor neurons (e.g., premotor). This disorder may involve almost any part of the body or the entire body. Akinesia also may be directional (Butter, Rapcsak, Watson, & Heilman, 1988) or spatial (Meador, Watson, Bowers, & Heilman, 1986). Whereas patients with spatial akinesia fail to act in contralesional hemispace, patients with directional akinesia fail to move in a direction contralateral to a hemispheric lesion. This directional akinesia may occur with the eyes (gaze palsy), head, and arms. Directional akinesia may be associated with a motor bias so the eyes, head, or arm may deviate toward the lesioned hemisphere (DeRenzi, Colombo, Faglioni, & Gilbertoni, 1982; Heilman, Bowers, & Watson, 1984).

Patients with mild forms of akinesia may be able to move a contra-lesional limb or move in a contralesional direction, but only after a delay. This slowness in initiating a response is called *hypokinesia* (Heilman et al., 1993; Meador et al., 1986).

Patients who are able to move both their right and left limbs indepen-dently may be impaired when attempting simultaneous movements and either move only the ipsilesional extremity or move the contralesional extremity only after they have moved the ipsilesional extremity. These disorders have been termed *motor extinction* (Valenstein & Heilman, 1981; Viader, Cambier, & Pariser, 1982).

Motor impersistence is defined by the inability to sustain an act or main-tain a posture. Impersistence may be observed with a variety of body parts including the limbs, the eyes, the jaw, and even the tongue. Like akinesia, impersistence may be directional (Kertesz, Nicholson, Cancelliere, Kassa, & Black, 1985) or spatial (Roeltgen, Roeltgen, & Heilman, 1989).

E. Associated Signs

When stimulated on the contralesional side, patients with *allesthesia* may indicate they were stimulated on the ipsilesional side (Obersteiner, 1882).

Patients with a hemiparesis or hemianopia may be unaware of or may deny their disability. This unawareness or denial has been termed *anosog-nosia* (Babinski, 1914). Awareness of a disability with diminished concern has been termed *anosodiaphoria* (Critchley, 1966).

Although allesthesia, anosognosia, and anosodiaphoria often are asso-ciated with neglect, these associated signs may be seen with other condi-tions such as spinal cord disease (allesthesia) and blindness (anosognosia).

IV. ANATOMY OF NEGLECT

Ideally, one would like to correlate each of the signs and symptoms of neglect (discussed previously) with a specific lesion locus. However, many signs and symptoms coexist and factors other than the anatomical locus of the lesion influence behavior, including the speed of the ablative pro-cess, the nature of the ablation, the time between ablation and testing, premorbid processing strategies, and the ability of the unaffected brain to compensate. Therefore, we have divided the localization discussion into three sections. In the first section, based primarily on knowledge that has been gained from the ablative paradigm, we discuss the systems that sub-serve spatially directed attention, arousal, and motor activation or inten-tion. We also discuss how lesions in different parts of this system cause neglect. In the second section we discuss how elements of the neglect syndrome may be fractionated as a function of their anatomical localization. In the third section we discuss right–left asymmetries.

A. Localization of Lesions that Induce Neglect

Neglect has been associated with lesions in many parts of the central nervous system, including cortical polymodal sensory convergence areas such as the inferior parietal lobe and the dorsolateral frontal lobe.

The posterior temporo-parietal convergence areas may be necessary to analyze sensory input, and to determine its spatial location and its significance. Subcortical systems including the mesencephalic reticular formation (MRF) and portions of the thalamus probably are necessary for mediating arousal. Most of these cortical and subcortical areas receive heavy projections from the limbic system and the frontal lobes. Whereas limbic input may be important in determining biological significance, frontal lobe input may be important in determining significance as it relates to goal-directed behavior. These corticolimbic and reticular areas, in conjunction with portions of the frontal lobes and basal ganglia, are also important for initiating responses. The regions of the central nervous system associated with neglect are all highly interconnected and all can be construed, in part, as subserving attention–arousal–activation (intention) functions.

1. Cortical lesions

a. Parietal lobe. In 1876, John Hughlings Jackson (Taylor, 1932) described a patient who neglected the left side of the page when reading. Autopsy disclosed a glioma of the right posterior temporal lobe. Subsequently, many investigators noted that neglect usually is associated with lesions in the region in which the inferior parietal lobe meets the temporal and occipital lobes (temporo-parieto-occipital junction; posterior portion of Brodmann's areas 39 and 40).

In the first edition of this book, we (Heilman, Valenstein, & Watson, 1983) discussed the lesions of 10 patients with hemispatial neglect and multimodal extinction who had cortical and subcortical lesions as determined by computerized tomography (CT). All the lesions were in the right hemisphere. The CT scans of one patient are shown in Fig. 3. Figure 4 shows a lateral view of the right hemisphere. We have projected and superimposed the CT scans of these 10 patients on this lateral projection. Figure 4 illustrates that, when damaged, the inferior parietal lobule—in conjunction with the temporo-parieto-occipital junction—is a critical area for inducing neglect. Vallar and Perani (1987) published a similar compos-

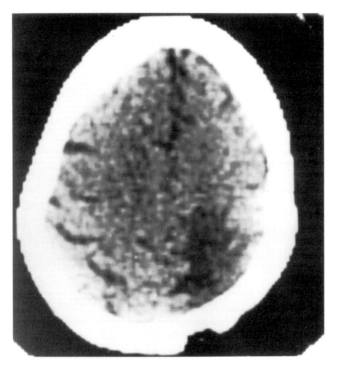

Figure 3   A section of a computerized tomographic scan (CT) demonstrating hypodense area in parieto-occipital region suggesting an infarction.

ite contour map of 8 patients with severe neglect; the lesions were in the same location. Vallar and Perani (1987) also published a composite contour map of 10 right-hemisphere-damaged patients who did not have neglect. This map demonstrates that the temporo-parieto-occipital junction is free of lesions.

In the monkey, researchers have demonstrated that unimodal sensory association areas converge on polymodal association areas, including the prefrontal cortex (periarcuate, prearcuate, orbitofrontal) and both banks of the superior temporal sulcus (STS) (Pandya & Kuypers, 1969). Unimodal association areas may reach the inferior parietal lobe (IPL) after a synapse in polymodal convergence areas (e.g., prefrontal cortex and both banks of the STS) (Mesulam, Van Hesen, Pandya, & Geschwind, 1977). Polymodal convergence areas may subserve polymodal sensory synthesis, and polymodal sensory synthesis may be important in "modeling"

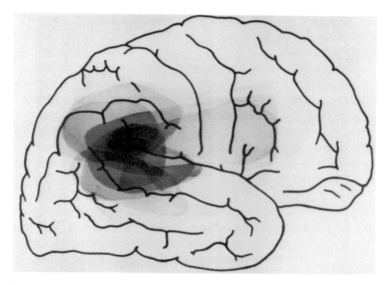

Figure 4   Lateral view of right hemisphere where we have projected and superimposed the CT scans of 10 patients with the neglect syndrome.

(detecting stimulus novelty), determining stimulus significance, and developing spatial representations.

Polymodal (e.g., STS) and supramodal (IPL) areas have important limbic and frontal connections. The polymodal cortices (STS) project to the
cingulate gyrus (a portion of the limbic system), and the cingulate gyrus
projects to the IPL. Prefrontal cortex, STS, and IPL have strong reciprocal
connections. Stimulus significance is determined in part by the needs of
the organism (motivational state). Limbic system (cingulate) input into
brain regions important for determining stimulus significance (STS and
IPL) may provide these posterior neocortical regions with information
about biological needs. Since the frontal lobes do play a critical role in
goal-oriented behavior and in developing sets, the frontal lobes might
provide polymodal STS and supramodal IPL with input about needs related to goals that are neither directly stimulus dependent nor motivated
by an immediate biological need.

Investigators have defined the properties of individual neurons in the
inferior parietal lobule (area 7) of the monkey (Goldberg & Robinson,
1977; Lynch, 1980; Motter & Mountcastle, 1981; Robinson, Goldberg, &
Stanton, 1978). Unlike cells in the primary sensory cortex, the activity of
many neurons in the inferior parietal lobule correlates best with stimuli of

Figure 5 Radioisotope scan of patient with neglect syndrome demonstrating a large dorsolateral frontal lobe lesion.

importance to the animal, whereas similar stimuli that are unimportant are associated with either no change or a lesser change in neuronal activity.

b. Dorsolateral frontal lobe. Silberpfenning (1941) reported "pseudo-hemianopsia" in two patients with frontal lesions. We described three patients with extinction to simultaneous visual stimuli, two of whom also had somesthetic extinction and hemispatial neglect (Heilman & Valenstein, 1972). Radioisotope brain scan determined that these patients had lesions in the dorsolateral frontal lobe (Fig. 5). The lesions appear to involve Brodmann's areas 8 and 9, as well as area 46, which includes the frontal eye fields.

Damasio, Damasio, and Chang (1980) reported a case of hemispatial neglect with inattention to auditory and visual stimuli induced by dorsolateral frontal lesion, as determined by CT scan. Van Der Linden, Seron, Gillet, and Bredart (1980) also reported neglect from frontal lesions.

Although the frontal lobe of monkeys is not as well developed as that of humans, a lesion in the region of the arcuate gyrus, which is cytoarchitectonically similar to area 8, also induces neglect (Welch & Stuteville,

1958). In humans, however, whether lesions restricted to the inferior portion of area 8 would induce neglect is not known. Rats with frontal lesions also demonstrate neglect (Corwin, Kanter, Watson, Heilman, Valenstein, & Hashimoto, 1986). The connections of the dorsolateral frontal lobes are important in understanding its role in attention and intention to multimodal sensory and limbic inputs. The dorsolateral frontal lobe has reciprocal connections with visual, auditory, and somesthetic association cortex (Chavis & Pandya, 1976). This area also is connected reciprocally with STS, another site of multimodal sensory convergence, and with the intraparietal sulcus, an area of somatosensory and visual convergence. The dorsolateral frontal lobe has reciprocal connections with subcortical areas including the dorsomedial nucleus (DM), the adjacent centromedian–parafascicularis (CM–PF) complex (Akert & Von Monakow, 1980; Kievet & Kuypers, 1977), and the MRF (Kuypers & Lawrence, 1967). Nonreciprocal projections to the caudate also exist. The dorsolateral frontal lobe also receives input from the limbic system, primarily from the anterior cingulate gyrus (Baleydier & Mauguiere, 1980).

The neocortical sensory association and sensory convergence area connections to the dorsolateral frontal lobe may provide the frontal lobe with information about the external stimuli that may call the individual to action. The limbic connections with the cingulate gyrus may provide the frontal lobe with motivation information. Connections with the MRF may be important in arousal.

The connections with the sensory association cortex, limbic, and reticular formation make the dorsolateral frontal lobes ideal candidates for implementing a response to a stimulus to which the subject is attending. Although this area may not be critical for mediating *how* to respond (e.g., providing instruction for the spatial trajectory and temporal patterns), it may control *when* one responds and when one does not respond. Physiological studies appear to support this hypothesis. Recordings from single cells in the posterior frontal arcuate gyrus (dorsolateral frontal lobe of monkeys) reveal visually responsive neurons that show enhanced activity time-locked to the onset of stimulus and preceding eye movements. Attending to an object without an associated eye or limb movement is not associated with enhancement (Goldberg & Bushnell, 1981). This response pattern differs from that of IPL neurons, which respond to visual input independent of behavior. Therefore, the IPL neurons appear to be responsible for mediating selective spatial attention independent of behavior (Bushnell, Goldberg, & Robinson, 1981) and frontal eye-field neurons are linked to behavior, but only to movements that have motivation significance.

c. Medial frontal lobe. Heilman and Valenstein (1972) described three patients with the neglect syndrome from right medial frontal lobe lesions in the region of the cingulate gyrus and supplementary motor area. The location was determined by postmortem examination (Fig. 6) in one patient and by radioisotope scans in two. Damasio et al. (1980) described two patients with neglect syndrome induced by left anterior cingulate–supplementary motor area lesion, as demonstrated by CT.

The anterior cingulate of monkeys is similar to that of humans. To confirm that focal lesions of the anterior cingulate gyrus can induce neglect, Watson, Heilman, Cauthen, and King (1974a) made such lesions in monkeys and induced contralateral neglect.

We already have discussed the close anatomical relationships between the anterior cingulate gyrus, the dorsolateral frontal lobe, and the inferior parietal lobe, and suggest that the connection provides these areas with motivational information.

2. Subcortical lesions: thalamus and mesencephalic reticular formation

In cats (Reeves & Hagamen, 1971) and in monkeys (Watson, Heilman, Miller, & King, 1974b), MRF lesions induce profound and enduring contralateral neglect. In cats, stimulation of the MRF is associated with behavioral arousal and desynchronization of the electroencephalogram (EEG), a physiological measure of arousal (Moruzzi & Magoun, 1949). Unilateral stimulation of the MRF induces greater EEG desynchronization in the

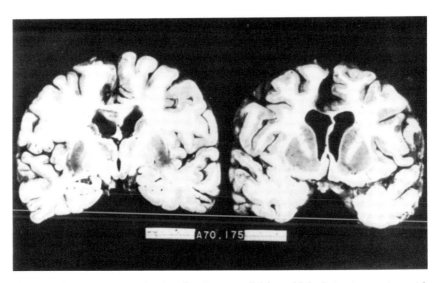

Figure 6 Postmortem examination showing a medial frontal lobe lesion in a patient with neglect syndrome.

ipsilateral than in the contralateral hemisphere (Moruzzi & Magoun, 1949). Arousal is a physiological state of heightened neuronal excitability that prepares the organism for processing. Whereas bilateral MRF lesions result in coma, unilateral lesions result in contralateral neglect.

Many of the neurons that ascend from the mesencephalic reticular activating system and its environs are monoaminergic. The locus coeruleus norepinephrine system projects diffusely to the cortex. Although this norepinephrine system should appear to be ideal for mediating cortical arousal (Jouvet, 1977), destruction of most of the locus coeruleus does not affect behavioral or electrophysiological (EEG) arousal profoundly (Jacobs & Jones, 1978).

Whereas the dopaminergic system may be critical for mediating action–intention, dopamine does not appear to be important in arousal because blockade of dopamine syntheses or dopamine receptors does not appear to affect desynchronization (Robinson, Vanderwolf, & Pappas, 1977).

Shute and Lewis (1967) described an ascending cholinergic reticular system. Stimulation of the mesencephalic reticular system not only induces an arousal response but also increases the rate of acetylcholine release from the neocortex (Kanai & Szerb, 1965). Acetylcholine makes some neurons more responsive to sensory input (McCormick, 1989). Cholinergic agonists induce neocortical desynchronization and electrophysiological signs of arousal. Cholinergic antagonists abolish desynchronization (Bradley, 1968). Therefore, cholinergic input may be responsible for behavioral arousal. The cholinergic projections from the nucleus basalis may be responsible for increasing neuronal responsitivity (Sato, Hata, Hagihara, & Tsumoto, 1987). The cuneiform area of the mesencephalon could influence the nucleus basalis via the peripeduncular area of the mesencephalon, which receives projections from the cuneiform nucleus and projects to the nucleus basalis (Arnault & Roger, 1987).

However, a neglect-like syndrome has not been reported in patients with destruction of the nucleus basalis. Schiebel and Schiebel (1967) suggested that the mesencephalic reticular system projects to the cortex in a diffuse polysynaptic fashion. Steriade and Glenn (1982) demonstrated that the centralis lateralis and paracentralis thalamic nuclei project to widespread cortical regions. However, an alternative means exists by which the mesencephalic activating system may influence cortical processing of sensory stimuli. Sensory information that reaches the cortex must be relayed through specific thalamic nuclei. The nucleus reticularis thalami (NR) is a thin reticular nucleus that envelops the thalamus and projects to the thalamic relay nuclei. This thalamic reticular nucleus appears to inhibit thalamic relay to the cortex (Schiebel & Schiebel, 1966). The mesencephalic reticular system projects to the NR. Rapid mesencephalic reticular system stimulation inhibits the inhibitory NR and thereby enhances thalamic transmission to the cerebral cortex (Singer, 1977). Unilateral lesions of MRF may induce neglect not only because the cortex, in the absence of MRF-mediated arousal, is not prepared for processing sensory stimuli but also because the thalamic relay nuclei are being inhibited by the NR.

Lesions of the sensory thalamic relay nuclei, thalamocortical projections, or primary sensory cortex may induce a sensory defect rather than neglect. The primary cortical sensory areas project to the modality-specific (unimodal) association cortex. The association cortex synthesizes multiple features of complex stimuli. Lesions of the unimodal association cortex may induce perceptual deficits in a single modality. These unimodal or modality-specific association areas also may be detecting stimulus novelty (modeling) (Sokolov, 1963). When a stimulus is neither novel nor significant, corticofugal projections to the NR may allow habituation to occur by selectively influencing thalamic relay. However, if a stimulus is novel or significant, corticofugal projections may inhibit the NR and thereby allow the thalamus to relay additional sensory input. This capacity for selective control of sensory input is supported by the observation that stimulation of specific areas within the NR related to specific thalamic nuclei results in abolition of corresponding (visual, auditory, tactile) cortically evoked responses (Yingling & Skinner, 1977).

Polymodal cortical areas may have a more general inhibitory action on the NR and may provide further arousal after cortical analysis. These convergence areas also may project directly to the MRF, which may induce a general hemispheric state of arousal because of diffuse multisynaptic connections to the cortex, via the thalamus or the basal forebrain cholinergic system, or may increase thalamic transmission by its connections with the NR. The evidence that polymodal cortex is important in arousal derives from neurophysiological studies showing that stimulation of select polymodal sites induces a generalized arousal response. In the monkey, these sites include the prearcuate region (dorsolateral frontal lobe) and both banks of the STS (Segundo, Naguet, & Buser, 1955), which is a precursor to the human inferior parietal lobe. When similar sites are ablated, EEG evidence suggests ipsilateral hypoarousal (Watson, Andriola, & Heilman, 1977).

Watson and Heilman (1979) described three cases of the neglect syndrome induced by thalamic hemorrhage (Fig. 7). Some of the patients had hemispatial neglect, a poor orienting response to the contralesional side, limb akinesia, multimodal extinction to simultaneous stimulation, and anosognosia. Postmortem examination of one patient showed that, although edema and mass effect compressed the ipsilateral internal capsule, no abnormalities of the cortex existed (Fig. 8). However, these thalamic lesions

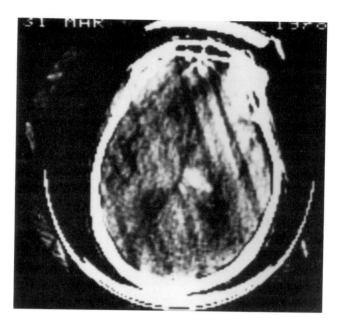

Figure 7 Hyperdense CT abnormality suggesting right thalamic hemorrhage.

may have also compressed the MRF. The thalamic hemorrhages were too large to help determine the part of the thalamus that may be critical for inducing neglect. However, Orem, Schlag-Rey, and Schlag (1973) induced unilateral visual neglect in cats by unilateral intralaminar thalamic lesions. Watson, Miller, and Heilman (1978) induced multimodal neglect in monkeys with CM–PF lesions. Neither anterior nor medial dorsal thalamic lesions induced neglect. We (Watson, Valenstein, & Heilman, 1981) also described a patient who had the neglect syndrome induced by a discrete thalamic infarction (Fig. 9) that involved portions of the posterior ventral nucleus, most of the medial nuclear group including the CM–PF, and possibly the anterior inferior aspect of the pulvinar. We proposed that medial thalamic lesions may induce predominantly motor neglect (akinesia) because the CM–PF is associated anatomically with motor systems such as the ventrolateral nucleus (VL) of the thalamus, frontal lobe, and neostriatum. Unilateral lesions of the CM–PF may, therefore, induce contralateral akinesia, whereas bilateral lesions of the CM–PF or its connections induce akinetic mutism (Segarra & Angelo, 1970).

Velasco and Velasco (1979) demonstrated "motor inattention" in humans with VL lesions. Hassler (1979) induced hemi-inattention in humans

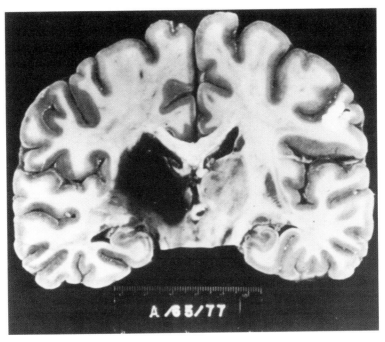

Figure 8 Right thalamic hemorrhages as determined by postmortem examination in patient with the neglect syndrome.

by pallidothalamic lesions, that is, by interrupting the pathway from the globus pallidus to the VL. Watson and associates (1981) proposed that lesions of the CM–PF or its primary connections with the VL, prefrontal cortex, or neostriatum may disrupt preparation for a meaningful response to stimuli of behavioral significance. The CM–PF may mediate this preparation to respond primarily through a pathway from the CM–PF to prefrontal cortex to the NR. The previously described cortico-cortical connections between inferior parietal lobule and dorsolateral frontal lobes, as well as connections of the MRF (tonic arousal) and the CM–PF to the NR, serve as anatomical interfaces for sensorimotor integration and aid in understanding why patients with neglect usually have both attentional and intentional deficits.

3. Basal ganglia and dopaminergic system

The caudate, putamen, globus pallidus, and substantia nigra constitute major portions of the basal ganglia. Hier, Davis, Richardson, and Mohr

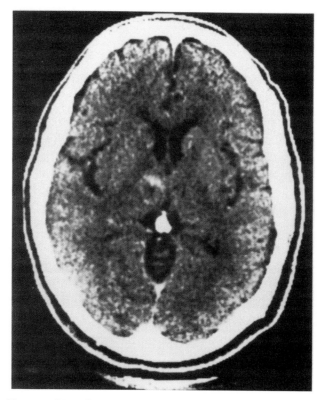

Figure 9 Hypodense CT scan suggesting thalamic infarction in a
patient with the neglect syndrome.

(1977) noted that patients with right putaminal hemorrhage often had
unilateral neglect. We also noted the association between unilateral neglect and putaminal hemorrhages (Fig. 10). Because these lesions often
compress the internal capsule—thereby inducing a contralateral hemiplegia and hemisensory loss—tests for inattention, akinesia, and extinction may be invalid. However, we have examined several of these patients
who were inattentive to auditory stimuli and who had a poor contralesional orienting response to the side opposite the lesion or, less often,
allesthesia.

Damasio et al. (1980) described two patients with inattention and hemispatial neglect in association with infarction of the putamen and caudate.
Valenstein and Heilman (1981) described a patient with a lesion mainly in
the caudate nucleus. This patient did not show hemi-inattention, sensory

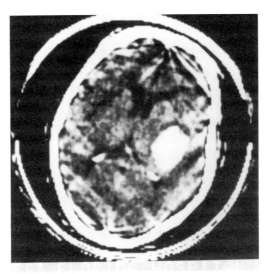

Figure 10 Hyperdense CT abnormality in lenticular region suggesting right ventricular (putaminal) hemorrhage.

extinction, or hemispatial neglect to simultaneous stimuli. He did, however, have a contralesional limb akinesia. This left-sided limb akinesia was increased dramatically by bilateral simultaneous movement (motor extinction).

Therefore, a neglect syndrome has followed lesions of all structures of the major basal ganglia. These structures are connected directly to areas previously described as important in the induction of neglect (e.g., CM–PF, VL, dorsolateral frontal lobe, and inferior parietal lobule).

In animals, unilateral lesions of the ascending dopaminergic pathways cause unilateral neglect. Three related dopaminergic pathways have been defined. The nigrostriatal pathway originates in the pars compacta of the substantia nigra (SN) and terminates in neostriatum (caudate and putamen). The mesolimbic and mesocortical dopaminergic pathways originate principally in the ventral tegmental area (VTA) of the midbrain and terminate in the limbic areas of the basal forebrain (nucleus accumbens septi and olfactory tubercle) and the cerebral cortex (frontal and cingulate cortex), respectively (Lindvall, Bjorklund, Morre, & Stenevi, 1974; Unger-stedt, 1971).

These dopaminergic fibers course through the lateral hypothalamus (LH). Whereas bilateral lesions in the LH of rats induce an akinetic state (Teitelbaum & Epstein, 1962), unilateral LH lesions cause unilateral neglect (Marshall, Turner, & Teitelbaum, 1971). Neglect occurs with 6-hydroxydopamine lesions of LH that selectively damage dopaminergic fibers (Marshall, Richardson, & Teitelbaum, 1974). Unilateral damage to the same dopaminergic fibers in the midbrain also cause unilateral neglect (Ljungberg & Ungerstedt, 1976).

The frontal and cingulate cortex receive dopaminergic input from the ventral tegmental area (Brown, Crane, & Goldman, 1979) and the neocortex projects strongly to the striatum. This corticostriatal projection is, at least in part, glutaminergic (Divac, Fonnum, & Storm-Mathison, 1977). Stimulation of cat cortex causes a release of dopamine in the striatum and substantia nigra (Nieoullon, Cheramy, & Glowinski, 1978). Although the pharmacological effects of cortical lesions have not been well described or understood, after unilateral frontal lesions (Corwin et al., 1986) rats demonstrated contralateral neglect that was reversed by the dopamine agonist apomorphine. This apomorphine treatment of neglect is blocked by the prior administration of the dopamine blocker spiropiridol. Fleet, Valenstein, Watson, and Heilman (1987), in an open trial in humans, demonstrated that dopamine agonist therapy also reduced neglect.

4. White matter

Since the networks that mediate attention and intention (motor activation) include both cortical structures (i.e., frontal and parietal cortex) and subcortical nuclei (e.g., thalamus, mesencephalic reticular system, basal ganglia) and since these areas are connected by white matter pathways, one would expect that white matter lesions also should be associated with neglect. Stein and Volpe (1983) reported neglect after right frontal lobe subcortical lesions. Ferro and Kertesz (1984) reported that neglect was associated with posterior internal capsule infarctions.

5. Networks for sensory attention and motor intention

Based on the behavioral, anatomical, and physiological research discussed, we have posited two highly interactive networks that mediate spatially direct attention and intention (Heilman, 1979; Heilman & Watson, 1977; Watson et al., 1981; Heilman et al., 1993). These networks are similar to the one proposed by Mesulam (1981).

The attentional model we have proposed is summarized in Fig. 11. Unilateral inattention is associated with unilateral mesencephalic reticular system lesions, because loss of inhibition of the ipsilateral NR by the mesencephalic reticular system decreases thalamic transmission of sensory input to the cortex, because the mesencephalic reticular formation cannot activate cortical neurons and these cortical neurons are not prepared for sensory processing, or both. Lesions of the primary or association cortices

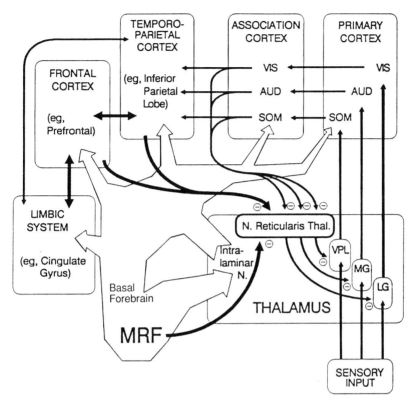

Figure 11 Schematic representation of pathways important in sensory attention and tonic arousal. See text for details. MRF, mesencephalic reticular formation; VIS, visual; AUD, auditory; SOM, somasthetic. Within the thalamus, sensory relay nuclei are indicated: VPL, ventralis posterolateralis; MG, medial geniculate; LG, lateral geniculate. Reproduced with permission from Heilman, Valenstein, & Watson (1993).

cause contralateral unimodal sensory loss or inability to synthesize contralateral unimodal sensory input. Corticothalamic collaterals from the sensory association cortex to the NR may serve unimodal attention and habituation. Lesions of multimodal sensory convergence areas that project to the mesencephalic reticular system and the NR may induce contralateral inattention because the subject cannot process or be aroused to multimodal contralateral stimuli. A lesion of the inferior parietal lobule, because of its reciprocal connections with polymodal areas (prefrontal lobes, STS) and the limbic system (cingulate gyrus), may impair a person's ability to code the spatial coordinates and determine the significance of a stimulus.

The frontal lobes have strong projections to the striatum. Whereas the dorsolateral frontal lobe projects primarily to the caudate, the supplementary motor area projects to the putamen and the cingulate gyrus projects to the ventral striatum (Alexander, DeLong, & Strick, 1986). These striatal nuclei project to the internal portion of the globus pallidus, which projects to specific thalamic nuclei [e.g., ventral anterior (VA), ventral lateralis (VL), and medial dorsal (MD)]. These thalamic nuclei then project back to the same frontal areas from which these frontal, basal ganglial, and thalamic loops initiate (e.g., dorsolateral frontal lobe, supplementary motor area, and cingulate gyrus).

Intentional disorders such as akinesia may be associated with diseases that affect the frontal lobes (dorsolateral and medial), basal ganglia, and thalamus. Intentional disorders also may be induced by an interruption of the dopaminergic system that projects to both the striatum and the frontal lobes. Thalamic lesions of VA/VL or the medial nuclei such as the CM–PF also can induce akinesia.

Intentional disorders such as akinesia also can be associated with diseases that affect the white matter that connects the frontal lobes with these subcortical structures. Intentional disorders including akinesia have been reported with temporo-parietal lobe lesions as well (Valenstein, Van Den Abell, Watson, & Heilman, 1982).

Based on the evidence cited, the frontal lobes appear to play a central role in an intentional network (Fig. 12). The dorsolateral frontal lobes receive projections from the inferior parietal lobes (polymodal association cortex) and from primary association cortices. The frontal lobe also has strong reciprocal connections to the cingulate gyrus (a portion of the limbic system) and the medial thalamic nuclei, and nonreciprocal connections with the striatum, all of which have been discussed.

The inferior parietal lobe may provide the frontal lobe with stored knowledge (cognition) and a spatial map. The cingulate (limbic) connections may provide the frontal lobe with motivational information. Sensory association areas may provide the frontal lobes with information about external stimuli that may call the organism to action. Afferents from the reticular system, including the thalamus, may be important for modulating arousal and activation.

B. Anatomical–Behavioral Fractionation of the Neglect Syndrome

As we have discussed, the neglect syndrome has multiple components (i.e., sensory inattention or unawareness, defective motor intention or in-

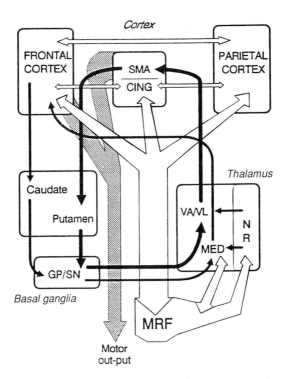

Figure 12 Schematic representation of pathways important for motor activation and preparation to respond. See text for details. Two basal ganglia "loops" are indicated: one (thick black arrows) from the supplementary motor area (SMA) to putamen to globus pallidus (GP)/substantia nigra (SN) to VA/VL (ventralis anterior and ventralis lateralis) of thalamus back to the SMA; the other (thin black arrows) from prefrontal cortex to caudate to GP/SN to the medial thalamic nuclei (MED) back to prefrontal cortex. NR, nucleus reticularis thalami; CING, cingulate gyrus. Reproduced with permission from Heilman, Valenstein, & Watson (1993).

action). These components may occur in different spatial domains (i.e., environmental, personal) and may have different patterns in these spatial domains (e.g., horizontal, vertical, radial). Although the neuronal networks mediating attention, intention, and spatial and personal representations share components, they do not overlap completely. Certain portions of a network may be more critical to one activity (e.g., motor intention) than to another activity (e.g., sensory attention). Therefore, elements of the neglect syndrome may fractionate.

One of the first experiments to study fractionation of neglect systematically was performed in monkeys with frontal lesions. Monkeys with dorsolateral frontal lobe lesions were thought to have sensory neglect because, following ablation of this area, the animals were not weak but failed to respond to contralesional stimuli. However, Watson et al. (1978) trained monkeys to respond with the hand opposite the side on which they were stimulated. When stimulated on the contralesional side, the animals responded normally with their ipsilesional hand. However, when stimulated on the normal ipsilesional side, they often failed to respond correctly with their contralesional hand, suggesting that the dorsolateral frontal lobe lesion induced motor neglect rather than sensory neglect.

In humans, Zingerle (1913) reported a patient with a frontal lesion who had akinesia or motor neglect of the centralateral limbs. Castaigne, Laplane, and Degos (1972) reported three cases with frontal damage that had limb akinesia or motor neglect. Meador et al. (1986) reported a patient with a medial frontal lobe lesion who had a limb akinesia that was worse in contralesional space that in ipsilesional space. Ocular directional akinesia has been reported with dorsolateral frontal lesions (Butter et al., 1988) and, as discussed, basal ganglia and thalamic lesions may be associated with motor extinction and motor neglect. Patients with frontal lesions also are more likely to have motor impersistence (Kertesz et al., 1985).

Although, based on the motor-anterior sensory-posterior dichotomy, researchers would like to see inattention and sensory extinction associated with posterior lesions, this dichotomy has not been substantiated completely. When Valenstein et al. (1982) tested monkeys with temporo-parietal ablation using the same crossed-response task used by Watson et al. (1978), they also found these monkeys to have motor neglect. In addition, a directional ocular motor bias (DeRenzi et al., 1982) and limb akinesia may be associated with temporo-parietal lesions (Castaigne, Laplane, & Degos, 1970; Critchley, 1966). Patients with frontal lesions may have sensory inattention (e.g., see Butter et al., 1988) and may have extinction (Heilman & Valenstein, 1972).

With respect to spatial neglect, Coslett, Bowers, Fitzpatrick, Haws, and Heilman (1990) and Bisiach, Geminiani, Berti, and Rusconi (1990) used different paradigms in an attempt to determine whether spatial neglect was induced primarily by a motor intentional (e.g., directional or hemispatial akinesia) or a sensory attentional deficit. Although both groups found that patients with frontal lesions were more likely to have intentional deficits and those with posterior lesions were more likely to have attentional deficits, their studies did not have sufficient numbers of subjects to draw definite conclusions.

Whereas lower vertical neglect is associated with dorsal parietal occipital lesions (Rapcsak et al., 1988), upper vertical neglect is associated with ventral temporo-occipital lesions (Shelton et al., 1990). Neglect of far peripersonal space also is associated with ventral temporo-occipital lesions (Shelton et al., 1990).

C. Hemispheric Asymmetries of Neglect

Although several of the earlier investigators noted that neglect was associated more often with right than with left hemisphere lesions (Brain, 1941; McFie, Piercy, & Zangwell, 1950), Battersby, Bender, and Pollack (1956) thought that this high association of right hemisphere lesions and neglect was related to a sampling artifact caused by the exclusion of aphasic subjects. However, more recent studies confirm that lesions in the right hemisphere are associated more often with elements of the neglect syndrome. We briefly discuss each of these elements.

We mentioned earlier that inattention is often difficult to dissociate from a primary sensory loss. However, Kooistra and Heilman (1989) and Nadeau and Heilman (1991) reported patients with visual inattention and Vallar et al. (1990) reported somesthetic inattention. All these patients had right hemisphere lesions. Using hemispheric barbituate anesthesia (Wada testing), Meador, Loring, Lee, Brooks, Thompson, and Heilman (1988) demonstrated that sensory extinction is more common with right than with left hemispheric anesthesia.

To account for a hemispheric asymmetry of attention in humans, we suggested that the right hemisphere attentional networks are more likely than those in the left hemisphere to have bilateral spatial receptive fields (Heilman & Van Den Abell, 1980). Thus, the networks in the left hemisphere would be activated predominantly by novel or significant stimuli in the right hemispace or hemifield, but networks in the right hemisphere would be activated by novel or significant stimuli on either side. In addition, although the left hemisphere directs attention rightward, the right hemisphere may direct attention in both directions. Therefore, when the left hemisphere is damaged, the right can attend to ipsilateral stimuli and direct attention rightward but the left hemisphere cannot attend to ipsilateral stimuli or direct attention leftward after right-sided damage. If the right hemisphere has bilateral spatial attentional fields, the right hemisphere of normals should desynchronize to stimuli presented in either field, whereas the left hemisphere should desynchronize only to right-sided stimuli. We presented lateralized visual stimuli to normal subjects while recording the EEG found that the right parietal lobe desynchronized equally to right- or left-sided stimuli whereas the left parietal lobe desynchronized mainly to right-sided stimuli (Heilman & Van Den Abell, 1980). A similar phenomenon has been demonstrated using positron emission tomography (Pardo, Fox, & Raichle, 1991). These studies provide evidence for a special role of the right hemisphere in attention and also may help explain why inattention is associated more often with right hemisphere lesions.

The right hemisphere also appears to play a dominant role in the mediation of arousal. Patients with right temporo-parietal lesions, when compared with patients with left temporo-parietal lesions, have a reduced arousal response as measured by galvanic skin response (Heilman, Schwartz, & Watson, 1978). Yokoyama, Jennings, Ackles, Hood, and Boller (1987) obtained similar results using heart rate as a measure of arousal. Using EEG, we also found that patients with right hemisphere lesions showed more theta and delta activity over the nonlesioned hemisphere than patients with left hemisphere lesions.

With respect to motor intentional disorders, patients with right hemisphere lesions more often demonstrated a contralateral limb akinesia than patients with left hemisphere lesions (Coslett & Heilman, 1989). However, akinesia is not always limited to the contralateral extremities. Warning stimuli may prepare an individual for action and thereby reduce reaction times (Lansing, Schwartz, & Lindsley, 1959). Patients who have a motor activation deficit may not be prepared to respond. Although they attributed their findings to a loss of topographical sense, Howes and Boller (1975) demonstrated that patients with right hemisphere lesions had slower reaction times than those with left hemisphere lesions, and found that the right hemisphere lesions associated with slowing were not larger than the left hemisphere lesions. Because patients with right hemisphere lesions have been shown to have reduced behavioral evidence of motor activation or intention, we have postulated that, in humans, the right hemisphere also may be dominant in mediating action intentional processes. Whereas the left hemisphere prepares the right extremities for action, the right hemisphere prepares extremities on both sides. Therefore, with left hemisphere lesions, left-sided limb akinesia is minimal, but with right hemisphere lesions, left-limb akinesia is severe. In addition, because the right hemisphere is more involved than the left hemisphere in activating ipsilateral extremities, more ipsilateral hypokinesia will be seen with right hemisphere lesions than with left hemisphere lesions.

To learn whether the right hemisphere is dominant for action–intention, we presented lateralized warning stimuli to normal subjects and measured reaction times to subsequent central (reaction time) stimuli. We found that stimuli projected to the right hemisphere reduced reaction times of the right hand more than warning stimuli projected to the left hemisphere reduced left-handed times; warning stimuli projected to the right hemisphere reduced reaction times of the right hand even more than did warning stimuli projected directly to the left hemisphere. These results support the hypothesis that the right hemisphere dominates motor activation or intention (Heilman & Van Den Abell, 1979). Because motor bias, directional akinesia, and hypokinesia are more common with right than with left hemisphere lesions, the right hemisphere also may be able to prepare movements in both the right and the left direction, but the left hemisphere may prepare only for rightward movements.

REFERENCES

Akert, K., & Von Monakow, K. H. (1980). Relationship of precentral, premotor, and prefrontal cortex to the mediodorsal and intralaminer nuclei of the monkey thalamus. *Acta Neurobiologiae Experimentalis (Waszawa), 40,* 7–25.

Albert, M. D. (1973). A simple test of visual neglect. *Neurology, 23,* 658–664.

Alexander, G. E., DeLong, M. R., & Strick, P. L. (1986). Parallel organization of functionally segregated circuits linking basal ganglia and cortex. *Annual Review of Neuroscience, 9,* 357–381.

Arnault, P., & Roger, M. (1987). The connections of the peripeduncular area studied by retrograde and anterograde transport in the rat. *Journal of Comparative Neurology, 258,* 463–478.

Babinski, J. (1914). Contribution a l'etude des troubles mentaux dans l'hemiplegie organique cerebrale (agnosognosie). *Revue Neurologique, 27,* 845–847.

Baleydier, C., & Mauguiere, F. (1980). The duality of the cingulate gyrus in monkey— Neuroanatomical study and functional hypothesis. *Brain, 103,* 525–554.

Battersby, W. S., Bender, M. B., & Pollack, M. (1956). Unilateral spatial agnosia (inattention) in patients with cerebral lesions. *Brain, 79,* 68–93.

Bisiach, E., Geminiani, G., Berti, A., & Rusconi, M. L. (1990). Perceptual and premotor factors of unilateral neglect. *Neurology, 49,* 686–694.

Bradley, P. B. (1968). The effect of atropine and related drugs on the EEG and behavior. *Progress in Brain Research, 28,* 3–13.

Brain, W. R. (1941). Visual disorientation with special reference to lesions of the right cerebral hemisphere. *Brain, 64,* 224–272.

Brown, R. M., Crane, A. M., & Goldman, P. S. (1979). Regional distribution of monoamines in the cerebral cortex and subcortical structures of the rhesus monkey: Concentrations and in vivo synthesis rates. *Brain Research, 168,* 133–150.

Bushnell, M. C., Goldberg, M. E., & Robinson, D. L. (1981). Behavioral enhancement of visual responses in monkey cerebral cortex: I. Modulation of posterior parietal cortex related to selected visual attention. *Journal of Neurophysiology, 46,* 755–772.

Butter, C. M., Rapcsak, S. Z., Watson, R. T., & Heilman, K. M. (1988). Changes in sensory inattention, direction hypokinesia, and release of the fixation reflex following a unilateral frontal lesion: A case report. *Neuropsychologia, 26,* 533–545.

Castaigne, P., Laplane, D., & Degos, J. D. (1970). Trois cas de négligence motrice par lésion retro-rolandique. *Revue Neurologique, 122,* 233–242.

Castaigne, P., Laplane, D., & Degos, J. D. (1972). Trois cas de négligence motrice par lésion frontale pre-rolandique. *Revue Neurologique, 126,* 5–15.

Chavis, D. A., & Pandya, D. N. (1976). Further observations on corticofrontal connections in the rhesus monkey. *Brain Research, 117,* 369–386.

Corwin, J. V., Kanter, S., Watson, R. T., Heilman, K. M., Valenstein, E., & Hashimoto, A. (1986). Apomorphine has a therapeutic effect on neglect produced by unilateral dorso-medial prefrontal cortex lesions in rats. *Experimental Neurology, 36,* 683–698.

Coslett, H. B., Bowers, D., Fitzpatrick, E., Haws, B., & Heilman, K. M. (1990). Directional hypokinesia and hemispatial inattention in neglect. *Brain, 113,* 475–486.

Coslett, H. B., & Heilman, K. M. (1989). Hemihypokinesia after right hemisphere strokes. *Brain and Cognition, 9,* 267–278.

Critchley, M. (1966). *The parietal lobes.* New York: Hafner.

Damasio, A. R., Damasio, H., & Chang, C. H. (1980). Neglect following damage to frontal lobe or basal ganglia. *Neuropsychologia, 18,* 123–132.

De Renzi, E., Colombo, A., Faglioni, P., & Gilbertoni, M. (1982). Conjugate gaze paralysis in stroke patients with unilateral damage. *Archives of Neurology, 39,* 482–486.

Divac, I., Fonnum, F., & Storm-Mathison, J. (1977). High affinity uptake of glutamate in terminals of corticostriatal axons. *Nature (London), 266,* 377–378.

Ferro, J. M., & Kertesz, A. (1984). Posterior internal capsule infarction associated with neglect. *Archives of Neurology, 41,* 422–424.

Fleet, W. S., Valenstein, E., Watson, R. T., & Heilman, K. M. (1987). Dopamine agonist therapy for neglect in humans. *Neurology, 37,* 1765–1771.

Goldberg, M. E., & Bushnell, M. C. (1981). Behavioral enhancement of visual responses in monkey cerebral cortex. Modulation in frontal eye fields specifically related saccades. *Journal of Neurophysiology, 46,* 773–787.

Goldberg, M. E., & Robinson, D. C. (1977). Visual responses of neurons in monkey inferior parietal lobule. The physiological substrate of attention and neglect. *Neurology (Abstr.), 27,* 350.

Hassler, R. (1979). Striatal reputation of adverting and attention directing induced by pallidal stimulation. *Applied Neurophysiology, 42,* 98–102.

Heilman, K. M. (1979). Neglect and related disorders. In K. M. Heilman & E. Valenstein (Eds.), *Clinical neuropsychology.* New York: Oxford University Press.

Heilman, K. M. (1991). Anosognosia: Possible neuropsychological mechanisms. In G. Priga-tano & D. Schacter (Eds.), *Awareness of defect after brain injury.* New York: Oxford University Press.

Heilman, K. M., & Valenstein, E. (1972). Frontal lobe neglect in man. *Neurology, 22,* 660–664.

Heilman, K. M., & Valenstein, E. (1979). Mechanisms underlying hemispatial neglect. *Annals of Neurology, 5,* 166–170.

Heilman, K. M., & Van Den Abell, T. (1979). Right hemisphere dominance for mediating cerebral activation. *Neuropsychologia, 17,* 315–321.

Heilman, K. M., & Van Den Abell, T. (1980). Right hemisphere dominance for attention: The mechanisms underlying hemispheric asymmetries of inattention (neglect). *Neurology, 30,* 327–330.

Heilman, K. M., & Watson, R. T. (1977). The neglect syndrome—A unilateral defect of the response. In S. Hardned, R. W. Doty, L. Goldstein, J. Jaynes, & G. Kean Thamer (Eds.), *Lateralization in the nervous system.* New York: Academic Press.

Heilman, K. M., Bowers, D., & Watson, R. T. (1984). Performance on a hemispatial pointing task by patients with neglect syndrome. *Neurology, 33,* 661–664.

Heilman, K. M., Schwartz, H. D., & Watson, R. T. (1978). Hypoarousal in patients with neglect syndrome and emotional indifference. *Neurology, 28,* 229–232.

Heilman, K. M., Valenstein, E., & Watson, R. T. (1983). Localization of neglect. In A. Kertesz (Ed.), *Localization in neuropsychology*. New York: Academic Press.

Heilman, K. M., Valenstein, E., & Watson, R. T. (1993). Neglect and related disorders. In K. M. Heilman & E. Valenstein (Eds.), *Clinical neuropsychology*. New York: Oxford University Press.

Hier, D. B., Davis, K. R., Richardson, E. T., & Mohr, J. P. (1977). Hypertensive putaminal hemorrhage. *Annals of Neurology, 1*, 152–159.

Howes, D., & Boller, F. (1975). Evidence for focal impairment from lesions of the right hemisphere. *Brain, 98*, 317–332.

Jacobs, B. L., & Jones, B. E. (1978). The role of central monoamine and acetylcholine systems in sleep wakefulness states. Mediation or modulation? In I. I. Butcher (Ed.), *Cholinergic–monoaminergic interactions of the brain*. New York: Academic Press.

Jouvet, M. (1977). Neuropharmacology of the sleep waking cycle. In L. L. Iverson, S. D. Iverson, & S. H. Snyder (Eds.), *Handbook of psychopharmacology*. New York: Plenum Press.

Kanai, T., & Szerb, J. C. (1965). Mesencephalic reticular activating system and cortical acetylcholine output. *Nature (London), 205*, 80–82.

Kertesz, A., Nicholson, I., Cancelliere, A., Kassa, K., & Black, S. E. (1985). Motor impersistance: A right-hemisphere syndrome. *Neurology, 35*, 662–666.

Kievet, J., & Kuypers, H. G. J. M. (1977). Organization of the thalamo-cortical connections to the frontal lobe in the rhesus monkey. *Experimental Brain Research, 29*, 299–322.

Kooistra, C. A., & Heilman, K. M. (1989). Hemispatial visual inattention masquerading as hemianopsia. *Neurology, 39*, 1125–1127.

Kuypers, H. G. J. M., & Lawrence, D. G. (1967). Cortical projections to the red nucleus and the brain stem in the rhesus monkey. *Brain Research, 4*, 151–188.

Lansing, R. W., Schwartz, E., & Lindsley, D. B. (1959). Reaction time and EEG activation under alerted and nonalerted conditions. *Journal of Experimental Psychology, 58*, 1–7.

Lindvall, O., Bjorklund, A., Morre, R. Y., & Stenevi, U. (1974). Mesencephalic dopamine-neurons projecting to the neocortex. *Brain Research, 81*, 325–331.

Ljungberg, T., & Ungerstedt, U. (1976). Sensory inattention produced by 6-hydroxydopamine-induced degeneration of ascending dopamine neurons in the brain. *Experimental Neurology, 53*, 585–600.

Lynch, J. C. (1980). The functional organization of posterior parietal association cortex. *The Behavioral and Brain Sciences, 3*, 485–534.

Marshall, J. F., Richardson, J. S., & Teitelbaum, P. (1974). Nigrostriatal bundle damage and the lateral hypothalamic damage. *Journal of Comparative Physiology and Psychology, 87*, 808–830.

Marshall, J. F., Turner, B. H., & Teitelbaum, P. (1971). Sensory neglect produced by lateral hypothalamic damage. *Science, 174*, 523–525.

McCormick, D. A. (1989). Cholinergic and noradrenergic modulation of thalamocortical processing. *Trends in Neuroscience, 12*, 215–221.

Mcfie, J., Piercy, M. F., & Zangwell, O. L. (1950). Visual spatial agnosia associated with lesions of the right hemisphere. *Brain, 73*, 167–190.

Meador, K., Loring, D. W., Lee, G. P., Brooks, E. E., Thompson, W. O., & Heilman, K. M. (1988). Right cerebral specialization for tactile attention as evidenced by intracarotid sodium amytal. *Neurology, 38*, 1763–1766.

Meador, K., Watson, R. T., Bowers, D., & Heilman, K. M. (1986). Hypometria with hemispatial and limb motor neglect. *Brain, 109*, 293–305.

Mennemeier, M., Wertman, E., & Heilman, K. M. (1992). Neglect of near peripersonal space: evidence for multidirectional attention systems in humans. *Brain, 115*, 37–50.

Mesulam, M. (1981). A cortical network for directed attention and unilateral neglect. *Annals of Neurology, 10*, 309–325.

Mesulam, M., Van Hesen, G. W., Pandya, D. N., & Geschwind, N. (1977). Limbic and sensory connections of the inferior parietal lobule (area PG) in the rhesus monkey: A study with a new method for horseradish perosidase histochemistry. *Brain Research, 136*, 393–414.

Moruzzi, G., & Magoun, H. W. (1949). Brainstem reticular formation and activation of the EEG. *Electroencephalography and Clinical Neurophysiology, 1*, 455–473.

Motter, B. C., & Mountcastle, V. B. (1981). The functional properties of the light sensitive neurons of the posterior parietal cortex studies in waking monkeys: Foveal sparing and opponent vector organization. *Journal of Neuroscience, 1*, 3–26.

Nadeau, S. E., & Heilman, K. M. (1991). Gaze-dependent hemianopia without hemispatial neglect. *Neurology, 41*, 1244–1250.

Nieoullon, A., Cheramy, A., & Glowinski, J. (1978). Release of dopamine evoked by electrical stimulation of the motor and visual areas of the cerebral cortex in both caudate nuclei and in the substantia nigra in the cat. *Brain Research, 15*, 69–83.

Obersteiner, H. (1882). On allochiria—A peculiar sensory disorder. *Brain, 4*, 153–163.

Orem, J., Schlag-Rey, M., & Schlag, J. (1973). Unilateral visual neglect and thalamic intralaminar lesions in the cat. *Experimental Neurology, 40*, 784–797.

Pandya, D. M., & Kuypers, H. G. J. M. (1969). Cortico-cortical connections in the rhesus monkey. *Brain Research, 13*, 13–36.

Pardo, J. V., Fox, P. T., & Raichle, M. E. (1991). Localization of a human system for sustained attention by positron emission tomography. *Nature (London), 349*, 61–64.

Rapcsak, S. Z., Cimino, C. R., & Heilman, K. M. (1988). Altitudinal neglect. *Neurology, 38*, 277–281.

Rapcsak, S. Z., Watson, R. T., & Heilman, K. M. (1987). Hemispace-visual field interactions in visual extinction. *Journal of Neurology, Neurosurgery, and Psychiatry, 50*, 1117–1124.

Reeves, A. G., & Hagamen, W. D. (1971). Behavioral and EEG asymmetry following unilateral lesions of the forebrain and midbrain of cats. *Electroencephalography and Clinical Neurophysiology, 30*, 83–86.

Robinson, D. L., Goldberg, M. E., & Stanton, G. B. (1978). Parietal association cortex in the primate sensory mechanisms and behavioral modulations. *Journal of Neurophysiology, 41*, 910–932.

Robinson, T. E., Vanderwolf, C. H., & Pappas, B. A. (1977). Are the dorsal noradrenergic bundle projections from the locus coeruleus important for neocortical or hippocampal activation? *Brain Research, 8*, 75–98.

Roeltgen, M. G., Roeltgen, D. P., & Heilman, K. M. (1989). Unilateral motor impersistence and hemispatial neglect from a striatal lesion. *Neuropsychiatry, Neuropsychology, and Behavioral Neurology, 2*, 125–135.

Sato, H., Hata, Y., Hagihara, K., & Tsumoto, T. (1987). Effects of cholinergic depletion on neuron activities in the cat visual cortex. *Journal of Neurophysiology, 58*, 781–794.

Schiebel, M. E., & Schiebel, A. B. (1966). The organization of the nucleus reticularis thalami: A Golgi study. *Brain Research, 1*, 43–62.

Schiebel, M. E., & Schiebel, A. B. (1967). Structural organization of nonspecific thalamic nuclei and their projection toward cortex. *Brain, 6*, 60–94.

Segarra, J. M., & Angelo, J. N. (1970). Presentation I. In A. C. Benton (Ed.), *Behavioral changes in cerebrovascular disease*. New York: Harper.

Segundo, J. P., Naguet, R., & Buser, P. (1955). Effects of cortical stimulation on electrocortical activity in monkeys. *Journal of Neurophysiology, 18*, 236–245.

Shelton, P. A., Bowers, D., & Heilman, K. M. (1990). Peripersonal and vertical neglect. *Brain, 113*, 191–205.

Shute, C. C. D., & Lewis, P. R. (1967). The ascending cholinergic reticular system, neocortical olfactory and subcortical projections. *Brain, 90,* 497–520.

Silberpfenning, J. (1941). Contributions to the problem of eye movements. Disturbances of ocular movements with pseudohemianopsia in patients with frontal lobe tumors. *Corfinia Neurologica, 4,* 1–13.

Singer, W. (1977). Control of thalamic transmission by corticofugal and ascending reticular pathways in the visual system. *Physiological Reviews, 57,* 386–420.

Sokolov, Y. N. (1963). *Perception and the conditioned reflex.* Oxford: Pergamon Press.

Stein, S., & Volpe, B. T. (1983). Classical "parietal" neglect syndrome after subcortical right frontal lobe infarction. *Neurology, 33,* 797–799.

Steriade, M., & Glenn, L. (1982). Neocortical and caudate projections of intralaminer thalamic neurons and their synaptic excitation from the midbrain reticular core. *Journal of Neurophysiology, 48,* 352–370.

Taylor, J. (Ed.) (1932). *Selected writings of John Hughlings Jackson.* London: Hodder and Stoughton.

Teitelbaum, P., & Epstein, A. N. (1962). The lateral hypothalamic syndrome: Recovery of feeding and drinking after lateral hypothalamic lesions. *Psychological Review, 69,* 74–90.

Ungerstedt, U. (1971). Striatal dopamine release after amphetamine or nerve degeneration revealed by rotational behavior. *Acta Physiologica Scandinavica, (367) 82,* 49–68.

Valenstein, E., & Heilman, K. M. (1978). Apraxic agraphia with neglect induced paragraphia. *Archives of Neurology, 36,* 506–508.

Valenstein, E., & Heilman, K. M. (1981). Unilateral hypokinesia and motor extinction. *Neurology, 31,* 445–448.

Valenstein, E., Van Den Abell, T., Watson, R. T., & Heilman, K. M. (1982). Nonsensory neglect from parietotemporal lesions in monkeys. *Neurology, 32,* 1198–1201.

Vallar, G., & Perani, D. (1987). The anatomy of spatial neglect in humans. In M. Jeannerod (Ed.) *Neurophysiological and neuropsychological aspects of spatial neglect.* Elsevier: North Holland.

Vallar, G., Sterzi, R., Bottini, G., Cappa, S., & Ruscani, M. C. (1990). Temporary remission of left hemianesthesia after vesticular stimulation. A sensory neglect phenomena. *Cortex, 26,* 123–131.

Van Der Linden, M., Seron, X., Gillet, J., & Bredart, S. (1980). Heminégligence par lésion frontale droite. *Acta Neurologica Belgica, 80,* 298–310.

Velasco, F., & Velasco, M. (1979). A reticulthalamic system mediating proprioceptive attention and tremor in man. *Neurosurgery, 4,* 30–36.

Viader, F., Cambier, J., & Pariser, P. (1982). Phénomène d'extinction motrice gauche. *Revue Neurologique, 138,* 213–217.

Watson, R. T., Andriola, M., & Heilman, K. M. (1977). The EEG in neglect. *Journal of the Neurological Sciences, 34,* 343–348.

Watson, R. T., & Heilman, K. M. (1979). Thalamic neglect. *Neurology, 29,* 690–694.

Watson, R. T., Heilman, K. M., Cauthen, J. C., & King, F. A. (1974a). Neglect after cingulectomy. *Neurology, 23,* 1003–1007.

Watson, R. T., Heilman, K. M., Miller, B. D., & King, F. A. (1974b). Neglect after mesencephalic reticular formation lesions. *Neurology, 24,* 294–298.

Watson, R. T., Miller, B. D., & Heilman, K. M. (1978). Nonsensory neglect. *Annals of Neurology, 3,* 505–508.

Watson, R. T., Valenstein, E., & Heilman, K. M. (1981). Thalamic neglect: The possible role of the medial thalamus and nucleus reticularis thalami in behavior. *Archives of Neurology, 38,* 501–507.

Weintraub, S., & Mesulam, M. M. (1987). Right cerebral dominance in spatial attention: Further evidence based on ipsilateral neglect. *Archives of Neurology, 44*, 621–625.

Welch, K., & Stuteville, P. (1958). Experimental production of neglect in monkeys. *Brain, 81*, 341–347.

Yingling, C. D., & Skinner, J. E. (1977). Gating of thalamic input to cerebral cortex by nucleus reticularis thalami. In J. E. Desmedt (Ed.), *Progress in clinical neurophysiology* (Vol. 1). New York: Karger.

Yokoyama, K., Jennings, R., Ackles, P., Hood, P., & Boller, F. (1987). Lack of heart rate changes during an attention demanding task after right hemisphere lesions. *Neurology, 37*, 624–630.

Zingerle, H. (1913). Ueber Störungen der Wahrnehmung des eigenen Körpers bei organischen Gehirnerkrankungen. *Monatsschrift für Psychiatrie und Neurologie, 34*, 13–36.

Chapter 17

Localization of Lesions in Constructional Impairment

Andrew Kirk and Andrew Kertesz

I. INTRODUCTION

Constructional impairment is a frequent outcome of brain damage. Despite decades of study, much controversy still surrounds the cerebral localization of constructional abilities and the cognitive mechanisms presumed to underlie them. A historical approach to the subject illustrates that the understanding of constructional impairment has not followed a steady evolution. Instead, some remarkable reversals in prevailing opinion have occurred, particularly regarding the interhemispheric localization of these abilities.

Although Liepmann (1912) and Poppelreuter (1917) made early contributions to the field, it was Kleist (1922,1934) who coined the term "constructional apraxia" to denote "a disturbance which appears in formative activities (such as assembling, building, drawing, etc.) in which the spatial form of the task is missed, although there is no apraxia of the single movements." Kleist considered this deficit to lie between visuoperceptual function and executive motor function, so patients with perceptual deficits or ideomotor apraxia would not be considered to have constructional apraxia. Since that time, however, general usage of the term has expanded, so patients displaying any abnormalities on constructional tests often are described as having constructional apraxia.

Drawing is the most frequently used clinical test of constructional ability, but a broad variety of two- and three-dimensional constructional tasks has been developed and even drawing tasks vary. Copying (Benton, 1962), drawing from memory or to verbal request (Critchley, 1953; Warrington, 1969), and drawing of real objects or of simple or complex geometric figures all have been used clinically and in experimental studies. Although Kleist appears to have considered constructional apraxia a unitary deficit affecting all types of construction "such as assembling, building, drawing, etc.," Benton and Fogel (1962) found that the correlations between brain-damaged patients' performance on four different constructional tasks were not high. Warrington, James, and Kinsbourne (1966) recommended that the labels applied to constructional deficits should be descriptive and should not prejudge the underlying mechanisms. These investigators suggested the use of terms such as "drawing disability" in preference to "constructional apraxia." In this chapter, we use the term "constructional impairment" to refer to the entire class of deficits discussed. When appropriate, we use a more specific term such as "drawing impairment."

Research of constructional impairment has centered around a few major questions. (1) What are the interhemispheric differences in incidence and severity of constructional impairment? (2) What are the qualitative differences between constructions of patients with right and left hemisphere lesions, and can we infer from these differences information about the normal contributions of each hemisphere to construction or about the mechanisms underlying constructional impairment? (3) Can constructional abilities be localized intrahemispherically? (4) What types of constructional impairment are seen in patients with diffuse cerebral disease, and how do these deficits relate to those seen after focal disease?

II. INTERHEMISPHERIC DIFFERENCES IN INCIDENCE AND SEVERITY OF CONSTRUCTIONAL IMPAIRMENT

Early reports placed constructional impairment in the context of diffuse cerebral disease (L'hermitte & Trelles, 1933; Mayer-Gross, 1935) or, noting its frequent association with the Gerstmann syndrome, of focal left hemisphere disease, particularly in the posterior parietal region (Head, 1926; Poppelreuter, 1917). At the time, investigators generally assumed that the left hemisphere was dominant for most cognitive abilities. Lange (1936) and Dide (1938), however, reported patients with right hemisphere disease who manifested constructional impairment.

Subsequent to these early reports, a series of investigations by Zangwill and colleagues began a trend that would result in a complete reversal of opinion concerning interhemispheric localization of constructional abilities, whereby impairment would come to be considered a sign of right rather than left hemisphere disease. Paterson and Zangwill (1944) published a detailed study of two patients with right posterior lesions, both of whom were impaired at constructional tasks. McFie, Piercy, and Zangwill (1950) reported eight additional patients with right hemisphere lesions and gross constructional impairment. In 1957, Ettlinger, Warrington, and Zangwill described another 10 such cases. Hécaen and co-workers reported similar findings (Hécaen, Ajuriaguerra, & Massonet, 1951; Hécaen, Penfield, Bertrand, & Malmo, 1956).

Until 1960, the literature consisted of a series of descriptions of patients with constructional impairment. Piercy, Hécaen, and Ajuriaguerra (1960) made the first attempt to study the incidence and severity of constructional impairment by directly comparing patients with right and left hemisphere lesions. These investigators retrospectively examined 8 years of clinical records from a neurosurgical unit. Every patient had been examined by one of the authors and the records of patients thought to display "constructional apraxia" were studied in greater detail. The incidence of constructional impairment in this study was 22.3% in patients with right-sided lesions but only 11.6% in those with left-sided lesions, a difference significant at $p < .01$. Piercy et al. (1960) also found that the severity of impairment of copying simple geometrical figures was greater in patients with right rather than left hemisphere lesions.

Piercy and Smyth (1962) studied 37 consecutive patients with unilateral lesions involving the parietal lobe. Patients were studied on copying drawn designs, copying matchstick patterns, copying block designs, object assembly, and drawing to verbal request. Whereas 68.4% of patients with right parietal lesions exhibited constructional impairment, the deficit was seen in only 38.9% of those with left parietal damage. Again, the impairment was more severe in the right parietal group.

Costa and Vaughan (1962) examined 36 patients with unilateral cerebral lesions and found that patients with right hemisphere lesions performed more poorly on the Wechsler Adult Intelligence Scale (WAIS) Block Design subtest than did left hemisphere patients. Several additional studies also demonstrated that constructional impairment was more frequent and more severe in patients with right rather than left hemisphere lesions (Arrigoni & De Renzi, 1964; Benton, 1962,1967,1968; Benton & Fogel, 1962; Black & Bernard, 1984; De Renzi & Faglioni, 1967; Stiles-Davis, Janowsky, Engel, & Nass, 1988). However, several other studies did not demonstrate major differences in interhemispheric incidence and severity of constructional impairment (Arena & Gainotti, 1978; Benson & Barton, 1970; Benton, 1973; Black & Strub, 1976; Colombo, De Renzi, & Faglioni, 1976; Dee, 1970; Gainotti, Messerli, & Tissot, 1972; Mack & Levine, 1981; Villa, Gainotti, & De Bonis, 1986; Warrington et al., 1966). and some studies found drawings by patients with left hemisphere lesions to be worse than those of patients with right-sided lesions (Kimura & Faust, 1987; Kirk & Kertesz, 1989,1993a). Also of interest is that Gazzaniga, Bogen, and Sperry (1965) reported that, following callosal section, complex drawings were done better by the left hand (right hemisphere). Several plausible explanations for these discrepancies exist.

Benton (1962) pointed out that many early studies used unspecified criteria to identify patients with constructional impairment. For example, Piercy et al. (1960) retrospectively studied patients who had been diagnosed clinically as having constructional impairment. No information was given on how this distinction had been made. Benton (1962) suggested that tests of graded difficulty with clearly specified scoring criteria be adopted and that patients be considered as having constructional impairment only if their scores fell below those of a given percentage of a group of normal controls.

The second difficulty in comparing the results of various studies is that different constructional tasks are used. Benton and Fogel (1962) studied 80 patients with unilateral lesions on two- and three-dimensional constructional tasks. The correlation between performance on the two types of tasks was significant but by no means overwhelming ($\phi = .20; p < .05$). Further, when normal performance was defined by reference to that of normal controls, 14 patients who performed adequately on drawing were impaired in the three-dimensional tasks and 8 who scored normally on three-dimensional construction were impaired in copying drawings. Only 12 patients performed inadequately on both tasks. Benton (1967) found that patients with right hemisphere lesions were more impaired than those with left hemisphere lesions in copying designs and in three-dimensional construction with blocks, whereas left and right patients were impaired similarly on the WAIS Block Design subtest. Dee (1970), in a study of 86 patients with unilateral hemispheric lesions, also found that many patients who could copy drawn designs could not perform three-dimensional block construction and that many patients could perform the three- but not the two-dimensional task. Researchers also have demonstrated that patients with left hemisphere lesions perform better on tests of copying than on tests of spontaneous drawing, whereas provision of a model does not improve the performance of right hemisphere patients (Gainotti et al., 1972; Paterson & Zangwill, 1944). Clearly, the choice of constructional task employed has an important effect on the outcome of a study.

A third source of variability among studies is their relative inclusion or exclusion of left hemisphere patients with aphasia. Benton (1973) studied performance on a three-dimensional constructional task by patients with right hemisphere lesions and by three groups with left hemisphere lesions: nonaphasics, expressive aphasics, and receptive aphasics. He found that defective constructional performance was most frequent in the patients with receptive aphasia (50%) whereas left hemisphere patients who were nonaphasic or who had expressive aphasia were impaired relatively infrequently (13%). Patients with right hemisphere lesions had an intermediate frequency of constructional impairment (36%). Similarly, Arena and Gainotti (1978) found that impaired performance in copying drawings was much more frequent in aphasic than in nonaphasic patients with left hemisphere lesions. The correlation between constructional performance and comprehension score (Token Test) in this study was also significant. These authors suggested that, in some studies, patients with impaired comprehension may have been underrepresented out of concern that they would not understand the task required of them. This bias may have led to an underestimate of the frequency and severity of constructional impairment resulting from left hemisphere disease. DeRenzi (1982) pointed out that, over a 20-year period, similar incidences had been reported for constructional impairment in patients with right hemisphere disease, whereas the incidence in those with left hemisphere lesions had risen steadily. He attributed this effect to the fact that early studies had, perhaps unwittingly, excluded many aphasics whereas more recent studies often explicitly retained patients with impaired comprehension.

A fourth factor that must be considered in comparing studies is the type of lesion studied. Some series (e.g., Arrigoni & De Renzi, 1964; Kimura & Faust, 1987; Villa et al., 1986; Warrington et al., 1966) have studied groups of patients with different types of lesions, mixing neoplastic and vascular pathology. A tumor may not necessarily cause the same deficits seen after a stroke in the same area (Anderson, Damasio, & Tranel, 1990). Restricting study to patients with strokes (Kirk & Kertesz, 1989) avoids confounding effects of tumor growth, peritumoral edema, and surgical intervention, which may cause dysfunction remote from the site of major involvement. If patients with cerebrovascular lesions are studied, carrying out testing at comparable times since onset is also important to minimize the influence of acute effects such as diaschisis and edema as well as the chronic effects of recovery (Kirk & Kertesz, 1989,1993a; Swindell, Holland, Fromm, & Greenhouse, 1988).

A fifth potential source of variability among studies is the effect of lesion size. Investigators have suggested that otherwise unselected patients with right hemisphere disease may, on average, have larger lesions than those with left hemisphere disease. Although this assertion is controversial, few studies have included data on lesion volume. Benson and Barton (1970) estimated lesion size using radioisotope brain scan, but did not find a consistent relationship between lesion size and severity of constructional impairment. Black and Bernard (1984) estimated lesion size from operative neurosurgical information and did not find it to be an important determinant of severity of constructional impairment. Kirk and Kertesz

(1989) measured lesion volume on computerized tomography (CT) scan in 69 patients with single strokes and found significant correlations between lesion size and degree of constructional impairment in patients with right but not left hemisphere lesions.

A final factor that may complicate studies of interhemispheric differences in construction is intrahemispheric lesion location. This issue is discussed at length in a subsequent section.

III. QUALITATIVE INTERHEMISPHERIC DIFFERENCES IN CONSTRUCTIONAL IMPAIRMENT

Since construction is undoubtedly a complex activity involving numerous potentially separable processes, studies of relative interhemispheric incidence and severity of impairment may, in effect, be comparing apples with oranges if different types of impairment caused by damage to different underlying mechanisms are seen in each hemispheric group. Under these circumstances, clinical lesion localization may be aided more by the type of disturbance than by the overall severity of impairment. Extensive support exists for the idea that construction is affected differently by lesions in each hemisphere. Note that many of the confounding factors discussed in the last section also mar most studies of qualitative aspects of drawing. However, certain qualitative interhemispheric differences emerge consistently. Their study has contributed to the development of hypotheses concerning the different contributions of each hemisphere to constructional ability.

Paterson and Zangwill (1944), in their description of two patients with right hemisphere lesions, emphasized the "piecemeal approach" in which "patients typically drew complex objects . . . detail by detail with little apparent grasp of the whole and poor articulation of the parts." McFie et al. (1950) described similar characteristics and attributed constructional impairment in patients with right hemisphere lesions to hemispatial neglect and to "a disorganization of discriminative spatial judgment (planotopokinesia)." Duensing (1953) stated that the drawings of patients with left hemisphere damage were hesitant and simplified, whereas patients with right-sided lesions failed to portray spatial relationships between parts of the drawing adequately. He also noted that left hemisphere patients' drawings improved when a model was provided for direct copying but that right hemisphere patients' drawings were not improved by provision of a model. Duensing suggested that right hemisphere damage resulted in a "spatioagnostic" form of constructional impairment and that left hemisphere lesions caused an "ideational–apraxic" deficit. This theory of a dichotomy of hemispheric function, in which right hemisphere disease produces a perceptual deficit and left hemisphere disease an executive impairment of construction, has been the subject of much investigation since its inception.

McFie and Zangwill (1960) reported eight patients with impaired construction caused by left hemisphere lesions and emphasized that their drawings were simplified, including fewer details while maintaining accurate spatial relationships between components. Whereas right hemisphere patients were said to produce fragmented "exploded diagrams," patients with left-sided lesions produced drawings essentially correct in outline but lacking in detail. These authors suggested that right hemisphere disease resulted in "spatial agnosia" whereas left hemisphere lesions caused "constructional apraxia." Piercy et al. (1960) noted that patients with right hemisphere disease produced drawings that were complex, even when unrecognizable. Right hemisphere patients' drawings also were noted to be marked by hemispatial neglect, increased number of lines, and abnormal orientation on the page. Researchers suggested that right hemisphere lesions caused a perceptual deficit and that constructional impairment in left hemisphere disease was the result of ideomotor apraxia. However, in the same year, Ajuriaguerra, Hécaen, and Angelergues (1960) demonstrated that patients with ideomotor apraxia are not necessarily impaired at construction. Bay (1962) used simplification of clay models made by aphasics to argue that failure to form adequately differentiated concepts is an important deficit in some patients with focal left hemisphere disease.

Since the initial comparisons of drawings by patients with left and right hemisphere disease, several investigators have compared drawings by these two groups of patients in an attempt to better define the qualitative differences between them. The results of these studies have been compatible, for the most part, with the earliest clinical reports. Arrigoni and DeRenzi (1964), in comparing left and right brain-damaged patients' performances at copying drawings, found that the left hemisphere group tended to simplify whereas patients with right-sided lesions displayed hemispatial neglect and gross alterations in spatial relationships between components of the drawings. DeRenzi and Faglioni (1967), using a similar test, found similar results. In another study of copying drawings, Warrington et al. (1966) found that drawings by patients with left-sided lesions included fewer details, and a tendency to widen angles and include more right angles, whereas right hemisphere lesioned patients drew more acute angles and produced less symmetrical drawings. Gainotti & Tiacci (1970) compared the performance of 100 right and 100 left brain-damaged patients on a test of design copying and quantified specific qualitative aspects of drawing to compare the two groups statistically. Left hemisphere patients reduced the size of the figures, simplified, increased the number of right angles, and had difficulty representing angles whereas right hemisphere patients displayed neglect, a piecemeal approach, abnormal orientation of the drawings on the page, alterations in spatial relationships, an increased number of lines, and inclusion of irrelevant script or other material. Collignon and Rondeaux (1974) noted that left hemisphere patients drew slowly and laboriously, drew more right angles, and simplified whereas right lesioned patients had difficulty representing spatial relationships between components, oriented drawings abnormally, and displayed hemispatial neglect. Ducarne and Pillon (1974) noted more right angles and simplification in left brain-damaged patients' drawings and hemispatial neglect and complexity in right hemisphere patients' drawings. Gasparrini, Shealy, and Walters (1980), in a study of drawing objects in response to verbal request, noted that left brain-damaged patients drew smaller figures that tended to be displaced toward the upper left quadrant of the page. Kimura and Faust (1987) found that left hemisphere damaged patients produced smaller drawings with fewer lines. Stiles-Davis et al. (1988), in a study of four 5-year-old children with congenital brain injury, found that the children with left hemisphere injury drew normally but that those with right-sided damage were impaired and displayed abnormal spatial relationships between components of the drawings. Swindell et al. (1988) found that left hemisphere patients' drawings were simplistic and often placed in the upper left quadrant. Right lesioned patients produced detailed but scattered and fragmented drawings, often with evidence of hemispatial neglect. Kirk and Kertesz (1989), in a comparison of normal controls and stroke patients tested at uniform times since onset, with left and right groups not significantly different in lesion volume, found that patients with left hemisphere lesions produced simplified tremulous drawings with fewer details and fewer angles, but a greater proportion of right angles. These patients had difficulty representing angles clearly and their drawings tended to be displaced toward the left side of the page. Right hemisphere patients' drawings displayed hemispatial neglect and impaired representation of spatial relationships between components. Compared with normal controls, left and right hemisphere patients displayed perseveration and impaired representation of three-dimensional perspective in their drawings.

Despite differences among the studies just presented, clearly several common themes emerge. The reported qualitative differences, although far from absolute, are clinically helpful in distinguishing between patients with right and left hemisphere lesions. Figures 1 and 2 show examples of drawings by patients with left and right hemisphere strokes. These studies

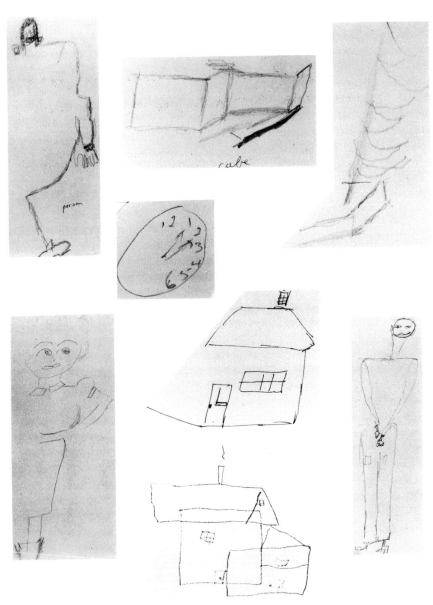

Figure 1 Examples of drawings by patients with left hemisphere strokes.

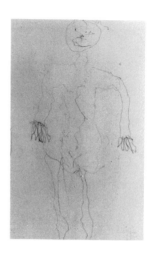

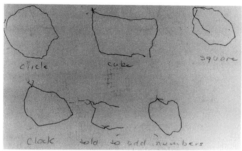

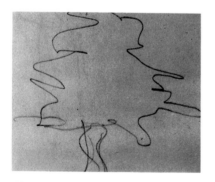
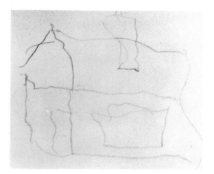

Figure 2   Examples of drawings by patients with right hemisphere strokes.

of qualitative aspects of drawing have given rise to debate over the fundamental mechanisms underlying constructional impairment of left or right hemisphere origin. The first hypothesis to be studied experimentally was Duensing's (1953) theory that right hemisphere constructional impairment is the result of perceptual impairment whereas left hemisphere deficit is related to an executive–motor problem.

Several studies have addressed the question of whether a visuoperceptual problem is responsible for constructional impairment with right but not left hemisphere disease. If this were the case, we would expect that right hemisphere patients with constructional impairment should perform more poorly on a visuoperceptual test than right hemisphere patients without constructional impairment. Similarly, right hemisphere patients with constructional impairment should perform worse on a visuoperceptual test than left hemisphere patients with constructional impairment. Piercy and Smyth (1962), Arrigoni and DeRenzi (1964), DeRenzi and Faglioni (1967), Dee (1970), Gainotti and Tiacci (1972), and Arena and Gainotti (1978), using a variety of perceptual tasks, all found that visuoperceptual dysfunction was associated with constructional impairment regardless of the hemisphere that was damaged. Costa and Vaughan (1962), Warrington and James (1967), Mack and Levine (1981), and Kirk and Kertesz (1989) all found that the relationship between construction and perception was stronger in patients with right hemisphere lesions, although in the study by Kirk and Kertesz (1989) the correlation between the two abilities was also significant in the left brain-damaged group. This body of literature shows that no clear-cut distinction can be made between the two hemispheric groups but that, instead, impaired visuoperceptual function probably plays some role in constructional impairment regardless of lesion laterality.

Several studies have investigated other possible explanations for constructional impairment caused by left hemisphere damage. As noted earlier, Ajuriaguerra et al. (1960) demonstrated that many patients with ideomotor apraxia perform normally on constructional tasks. Kimura and Faust (1987), however, suggested that patients with apraxia may be more impaired at construction, although Kirk and Kertesz (1989) could not confirm a significant relationship between construction and praxis. Warrington et al. (1966) suggested that left hemisphere damaged patients' constructional impairment might be explained by the presence of a planning disturbance. They noted that left but not right hemisphere patients' drawings improved with systematic practice at copying components of the complete drawing. Hécaen and Assal (1970) found that left but not right brain-damaged patients' drawings of cubes improved when they were provided with landmarks marked on the paper. These authors used this finding as evidence to support the existence of a planning disorder in left hemisphere patients. However, these results were not confirmed in later studies (Collignon & Rondeaux, 1974; Pillon, 1981) and Gainotti, Miceli, and Caltagirone (1977) found no significant improvement in either group with the use of landmarks. Although right hemisphere patients used fewer of the landmarks provided, this phenomenon occurred only on the left side of the drawing and presumably was the result of hemispatial neglect. Bay (1962), Gainotti, Silveri, Villa, and Caltagirone (1983), and Kirk and Kertesz (1989) all have suggested that at least some patients with left hemisphere damage may have a disorder of concept formation that parallels comprehension disturbance and may contribute to difficulty constructing real objects from memory.

Another explanation for left hemisphere patients' errors, which has received surprisingly little attention, is the possibility that use of a hemiparetic right hand or a nondominant left hand is responsible for many low level executive errors. In a study of normal subjects, Dee and Fontenot (1969) found no significant differences between error scores of right- and left-handed copying of stimuli in the Visual Retention Test (Benton, 1962). This study shows only that this particular test, scored on a limited number of items, is not sensitive to the differences between left- and right-handed copying. Archibald (1978) found that patients with left hemisphere lesions who were forced by hemiparesis to draw with the nonpreferred left hand made more errors of simplification in copying the Rey–Osterrieth figure than did those patients who were able to use the preferred right hand. Kirk and Kertesz (1989) found that severity of hemiparesis was a more important predictor of severity of drawing impairment than was size or location of lesion. Severity of hemiparesis was significantly correlated with difficulty representing angles, number of angles produced, tremulousness, number of details drawn, and overall severity of drawing impairment. In a later study, Kirk and Kertesz (1993a) found that left hemisphere patients able to draw with the right hand were superior, on many aspects of drawing, to those who used the left hand. Low level motor deficits are probably an important, but not the only, cause of constructional impairment in patients with left hemisphere lesions. The nature of the test used obviously influences the relative effect of hemiparesis on constructional ability.

Despite decades of study, characterization of the fundamental nature of the constructional deficits in patients with left and right hemisphere lesions remains elusive. Paterson and Zangwill's (1944) statement, "Whereas constructive defects are commonly associated with disorders of visual perception, the relative predominance of agnostic and apraxic elements varies from case to case and no single generalization will cover the diversity of clinical pictures," could have been written today.

IV. INTRAHEMISPHERIC LOCALIZATION OF LESIONS RESPONSIBLE FOR CONSTRUCTIONAL IMPAIRMENT

Early investigators regarded constructional impairment as a sign of left or right parietal disease. In fact, since constructional impairment was not expected with anterior lesions, some studies were limited to patients with parietal or parieto-occipital lesions (e.g., Piercy & Smyth, 1962). Piercy et al. (1960) thought that lesions responsible for constructional impairment were more restricted to posterior regions in right than in left hemisphere patients. DeRenzi and Faglioni (1967) found that constructional impairment tended to occur in patients with right posterior rather than anterior lesions (as determined by presence of visual field defect). However, in 1968, Benton reported that the incidence of constructional impairment in patients with right frontal lesions was comparable to its incidence in patients with right hemisphere lesions as a whole. Conflicting data concerning the importance of intrahemispheric lesion location have emerged since that time.

Newcombe (1969) found that patients with parietal lesions performed worse on block design. Benson and Barton (1970), using radioisotope brain scans to localize lesions, did not find significant differences in constructional ability among patients with anterior and posterior lesions. Taylor and Warrington (1973) found no significant difference among constructional abilities of right posterior, anterior, and temporal patients. Black and Strub (1976), in a study of patients with missile wounds, found that posterior lesions in either hemisphere caused more severe constructional impairment than did lesions in anterior regions. Kertesz, Harlock, and Coates (1979) failed to find any specific localization of right hemisphere strokes in patients with constructional impairment. Kertesz and Dobrowolski (1981) did not find a strong relationship between right hemisphere lesion location and severity of constructional impairment and suggested that intellectual functions may be distributed more diffusely in the right hemisphere than in the left. Kirk and Kertesz (1989) found no significant differences in severity of drawing impairment with different lesion locations within each hemisphere. Grossman (1988) reported that right central (mainly parietal) lesions resulted in impaired representation of shape, right posterior lesions caused impaired representation of color, and right anterior lesions impaired color, shape, and relative size in drawings. Few of these studies controlled for the size of the lesions in each location and, in most studies, the number of patients with lesions in each area was small.

Despite controversy over the relationship between intrahemispheric lesion location and severity of impairment, Luria and Tsvetkova (1964) suggested that different mechanisms might be responsible for constructional impairment in patients with anterior and posterior lesions. These authors suggested that faulty analysis of spatial relationships underlies the constructional impairment of patients with parieto-occipital lesions, whereas defective planning causes the impairment of patients with frontal lesions. Several studies have shown that visuospatial impairment often accompanies constructional impairment in patients with posterior lesions (e.g., Arena & Gainotti, 1978; DeRenzi & Faglioni, 1967) although whether this relationship is not also seen with frontal disease is not entirely clear. In support of the hypothesis that a disorder of planning is responsible for constructional impairment in frontal lesions, L'hermitte, Derousné, and Signoret (1972) showed that patients with frontal lobe lesions are helped more than those with posterior lesions in copying the Rey figure if the task is broken down into steps so patients are not forced to develop their own plan or strategy for carrying out the task. Pillon (1981) found similar results.

Subcortical cerebral damage is recognized increasingly as a cause of aphasia (Alexander & Loverme, 1980; Alexander, Naeser, & Palumbo, 1987; Basso, Della Sala, & Farabola, 1987; Hier, Davis, Richardson, & Mohr, 1977; Kirk and Kertesz, 1993b; Naeser, Alexander, Helm-Estabrooks, Levine, Laughlin, & Geschwind, 1982) and other disorders traditionally deemed cortical. Constructional impairment is said to be common in disorders such as Parkinson's disease, in which the pathology is mainly subcortical (e.g., Boller, Passafiume, Keefe, Rogers, Morrow, & Kim, 1984), although drawing has not been studied systematically and in detail in Parkinson's disease. Relatively little attention has been directed toward constructional impairment with subcortical lesions. Basso et al. (1987), in a study of subcortical aphasia, reported that impairment of copying is about as frequent in patients with left subcortical lesions as it is said to be in patients with unselected left hemisphere lesions. However, right subcortical patients were not tested, no direct comparison was made with cortical patients, and no information was given on drawing characteristics. Kirk and Kertesz (1993), in a study of 125 patients with left and right cortical and subcortical lesions matched for lesion volume on CT, found that, although previously reported interhemispheric differences were confirmed in patients with subcortical lesions, no significant differences were found in either overall drawing impairment or specific drawing characteristics between patients with cortical and subcortical lesions of similar size.

V. CONSTRUCTIONAL IMPAIRMENT IN DIFFUSE CEREBRAL DISEASE

Constructional impairment long has been recognized as a frequent and often early sign of dementia (Ajuriaguerra, Muller, & Tissot, 1960; Perez, Rivera, Meyer, Gay, Taylor, & Mathew, 1975; Sjögren, Sjögren, & Lindgren,

1952). Many of the very early reports of constructional impairment described patients with dementia. Mayer-Gross (1935) drew attention, in patients who probably suffered from multi-infarct dementia, to the "closing-in" phenomenon, a tendency for patients to draw directly on the model in copying tasks. Aguriaguerra and colleagues (1960) and Gainotti (1972) have demonstrated that this finding is uncommon in patients with focal disease but that it is common in dementia. In particular, Gainotti (1972) found that the phenomenon became more frequent in advanced stages of dementia and that a progression in qualitative aspects of the phenomenon occurred with advance of the disease. In that study, the progression of the "closing-in" phenomenon with worsening of dementia appeared to mirror the loss of the same phenomenon in normal children with advancing development.

Benton and Fogel (1962) documented a high incidence of constructional impairment in a group of patients with unspecified "diffuse or bilateral lesions." Villa et al. (1986) found that coexistent mental deterioration is an important determinant of the degree of constructional impairment in patients with unilateral lesions. McFie and Zangwill (1960), Piercy and Smyth (1962), Warrington et al. (1966), and Gainotti et al. (1972) all have suggested that a stronger relationship may exist between constructional impairment and generalized mental deterioration in patients with left rather than right hemisphere lesions. In a small number of patients with Alzheimer's disease, researchers have found that drawing may be more impaired in patients with more marked right than left parietal hypometabolism on positron emission tomography (Friedland et al., 1988).

In a study comparing the drawings of patients with Alzheimer's disease with those of normal controls matched for age and education, Kirk and Kertesz (1991) found that drawings by a group of patients with early Alzheimer's disease (mean estimated disease duration 29.0 mo) displayed fewer angles, impaired perspective and spatial relationships, simplification, and overall impairment. This result represents a combination of the deficits seen following both right and left hemisphere lesions. Neglect, tremulousness, and perseveration were not prominent in the Alzheimer patients' drawings. Brantjes and Bouma (1991) found that drawings by patients with Alzheimer's disease contained more errors of omission, simplification, and confabulation than did control drawings. Some types of perseveration also were significantly more frequent in the Alzheimer patients' drawings.

VI. CONCLUSIONS

Constructional tasks are complex and clearly require many different abilities. An attempt to localize *the* lesion that impairs construction may be somewhat akin to an attempt to localize *the* lesion that impairs ability to drive an automobile. Instead the existence of several (perhaps many) different types of constructional impairment is likely (Benton, 1983). Techniques of cognitive neuropsychology, through detailed case studies, have yielded some insights into mechanisms underlying some cognitive deficits such as agraphia and alexia (e.g., Beauvois & Dérouesné, 1981; Bub & Kertesz, 1982; Kirk, Blonder, Wertman, & Heilman, 1991; Shallice, 1981). These techniques are beginning to be applied to constructional impairment (Delis, Kiefner, & Fridlund, 1988; Delis, Kramer, & Kiefner, 1988; Griffiths, Cook, & Newcombe, 1988; Grossi, Orsini, Modafferi, & Liotti, 1986; Roncato, Sartori, Masterson, & Rumiati, 1987) and, although interesting dissociations have been reported, their implications for localization are not yet clear.

Many complicating factors discussed in this chapter make comparing studies and drawing overall conclusions from the literature difficult. Although a growing number of studies suggests otherwise, most investigations suggest that right rather than left hemisphere lesions result in greater impairment of constructional abilities and that parietal lesions cause more impairment than lesions elsewhere. Qualitative differences exist between the drawings of patients with right and left hemisphere lesions—this is the most consistent finding in the literature. Visuoperceptual deficits seem to contribute to constructional impairment following both right and left hemisphere insult, although their effect may be more pronounced in individuals with right hemisphere lesions. A comprehensive theory of construction and the brain is not likely to be proposed in the near future. However, detailed studies of a variety of constructional tasks and their component activities performed by patients with accurate anatomical and functional lesion localization, such as that afforded by magnetic resonance imaging and positron emission tomography, are likely to continue to improve our understanding of these complex abilities.

REFERENCES

Ajuriaguerra, J., Hécaen, H., & Angelergues, R. (1960). Les apraxies, variétés cliniques et latéralisation lésionelle. *Revue Neurologique, 102*, 494–566.

Ajuriaguerra, J., Muller, M., & Tissot, R. (1960). A propos de quelques problèmes posés par l'apraxie dans les demences. *Encephale, 49*, 275–401.

Alexander, M. P., & Loverme, S. R. (1980). Aphasia after left hemispheric intracerebral hemorrhage. *Neurology, 30*, 1193–1202.

Alexander, M. P., Naeser, M. A., & Palumbo, C. L. (1987). Correlations of subcortical CT lesion sites and aphasia profiles. *Brain, 110*, 961–991.

Anderson, S. W., Damasio, H., & Tranel, D. (1990). Neuropsychological impairments associated with lesions caused by tumor or stroke. *Archives of Neurology, 47*, 397–405.

Archibald, Y. M. (1978). *Simplification in the drawings of left hemisphere patients—A function of motor control?* Paper presented at the meeting of the Academy of Aphasia, Chicago.

Arena, R., & Gainotti, G. (1978). Constructional apraxia and visuoperceptive disabilities in relation to laterality of cerebral lesions. *Cortex, 14*, 463–473.

Arrigoni, G., & De Renzi, E. (1964). Constructional apraxia and hemispheric locus of lesion. *Cortex, 1*, 170–197.

Basso, A., Della Sala, S., & Farabola, M. (1987). Aphasia arising from purely deep lesions. *Cortex, 23*, 29–44.

Bay, E. (1962). Aphasia and non-verbal disorders of language. *Brain, 85*, 411–426.

Beauvois, M. -F., & Dérouesné, J. (1981). Lexical or orthographic agraphia. *Brain, 104*, 21–49.

Benson, D. F., & Barton, M. I. (1970). Disturbances in constructional ability. *Cortex, 6*, 19–46.

Benton, A. L. (1962). The visual retention test as a constructional praxis task. *Confinia Neurologica, 22*, 1–16.

Benton, A. L. (1967). Constructional apraxia and the minor hemisphere. *Confinia Neurologica, 27*, 1–17.

Benton, A. L. (1968). Differential behavioral effects in frontal lobe disease. *Neuropsychologia, 6*, 53–60.

Benton, A. L. (1973). Visuoconstructive disability in patients with cerebral disease: Its relationship to side of lesion and aphasic disorder. *Documenta Ophthalmologica, 34*, 67–76.

Benton, A. L. (1983). Visuoperceptual, visuospatial, and visuoconstructive disorders. In K. M. Heilman & E. Valenstein (Eds.), *Clinical neuropsychology* (pp. 151–179). New York: Oxford University Press.

Benton, A. L., & Fogel, M. L. (1962). Three dimensional constructional praxis. *Archives of Neurology, 7*, 347–354.

Black, F. W., & Bernard, B. A. (1984). Constructional apraxia as a function of lesion locus and size in patients with focal brain damage. *Cortex, 20*, 111–120.

Black, F. W., & Strub, R. L. (1976). Constructional apraxia in patients with discrete missile wounds of the brain. *Cortex, 12*, 212–220.

Boller, F., Passafiume, D., Keefe, N. C., Rogers, K., Morrow, L., & Kim, Y. (1984). Visuospatial impairment in Parkinson's disease: Role of perceptual and motor factors. *Archives of Neurology, 41*, 485–490.

Brantjes, M., & Bouma, A. (1991). Qualitative analysis of the drawings of Alzheimer patients. *The Clinical Neuropsychologist, 5*, 41–52.

Bub, D., & Kertesz, A. (1982). Deep agraphia. *Brain and Language, 17*, 146–157.

Collignon, R., & Rondeaux, J. (1974). Approche clinique des modalités de l'apraxie constructive secondaire aux lésions corticales hémisphériques gauches et droites. *Acta Neurologica Belgica, 74*, 137–146.

Colombo, A., De Renzi, E., & Faglioni, P. (1976). The occurrence of visual neglect in patients with unilateral cerebral disease. *Cortex, 12*, 221–231.

Costa, L. D., & Vaughan, H. G. (1962). Performance of patients with lateralized cerebral lesions. 1. Verbal and perceptual tests. *Journal of Nervous and Mental Disease, 134*, 162–168.

Critchley, M. (1953). *The parietal lobes.* London: Edward Arnold.

Dee, H. L. (1970). Visuoconstructive and visuoperceptive deficit in patient with unilateral cerebral lesions. *Neuropsychologia, 8*, 305–314.

Dee, H. L., & Fontenot, D. J. (1969). Use of the non-preferred hand in graphomotor performance: A methodological study. *Confinia Neurologica, 31*, 273–280.

Delis, D. C., Kiefner, M. G., & Fridlund, A. J. (1988). Visuospatial dysfunction following unilateral brain damage: Dissociations in hierarchical and hemispatial analysis. *Journal of Clinical and Experimental Neuropsychology, 10,* 421–431.

Delis, D. C., Kramer, J. H., & Kiefner, M. G. (1988). Visuospatial functioning before and after commissurotomy. Disconnection in hierarchical processing. *Archives of Neurology, 45,* 462–465.

De Renzi, E. (1982). *Disorders of space exploration and cognition.* London: Wiley and Sons.

De Renzi, E., & Faglioni, P. (1967). The relationship between visuo-spatial impairment and constructional apraxia. *Cortex, 3,* 327–342.

Dide, M. (1938). Les désorientation temporo-spatielles et la prépondérance de l'hémisphere droit dans les agnoso-akinésies proprioceptives. *Encephale, 33,* 276.

Ducarne, B., & Pillon, B. (1974). La copie de la figure complexe de Rey dans les troubles visuo-constructifs. *Journal de Psychologie Normale et Pathologique, 4,* 449–469.

Duensing, F. (1953). Raumagnostische und ideatorisch-apraktische Störung des gestaltenden Handelns. *Deutsche Zeitschrift fuer Nervenheilkunde, 170,* 72–94.

Ettlinger, G., Warrington, E., & Zangwill, O. L. (1957). A further study of visuo-spatial agnosia. *Brain, 80,* 335–361.

Friedland, R. P., Koss, E., Haxby, J. V., Grady, C. L., Luxenberg, J., Schapiro, M. B., & Kaye, J. (1988). Alzheimer disease: Clinical and biological heterogeneity. *Annals of Internal Medicine, 109,* 298–311.

Gainotti, G. (1972). A quantitative study of the "closing-in" symptom in normal children and in brain-damaged patients. *Neuropsychologia, 10,* 429–436.

Gainotti, G., Messerli, P., & Tissot, R. (1972). Troubles du dessin et lésions hémisphériques rétrorolandiques unilatérales gauches et droites. *Encéphale, 61,* 245–264.

Gainotti, G., Miceli, G., & Caltagirone, C. (1977). Constructional apraxia in left brain-damaged patients: A planning disorder? *Cortex, 12,* 109–113.

Gainotti, G., Silveri, M. C., Villa, G., & Caltagirone, C. (1983). Drawing objects from memory in aphasia. *Brain, 106,* 613–622.

Gainotti, G., & Tiacci, C. (1970). Patterns of drawing disability in right and left hemispheric patients. *Neuropsychologia, 8,* 379–384.

Gasparrini, B., Shealy, C., & Walters, D. (1980). Differences in size and spatial placement of drawings of left versus right hemisphere brain-damaged patients. *Journal of Consulting and Clinical Psychology, 48,* 670–672.

Gazzaniga, M. S., Bogen, J. E., & Sperry, R. W. (1965). Observations on visual perception after disconnexion of the cerebral hemispheres in man. *Brain, 88,* 221–236.

Griffiths, K. M., Cook, M. L., & Newcombe, R. L. G. (1988). Cube copying after cerebral damage. *Journal of Clinical and Experimental Neuropsychology, 10,* 800–812.

Grossi, D., Orsini, A., Modafferi, A., & Liotti, M. (1986). Visuoimaginal constructional apraxia: On a case of selective deficit of imagery. *Brain and Cognition, 5,* 255–267.

Grossman, M. (1988). Drawing deficits in brain-damaged patients' freehand pictures. *Brain and Cognition, 8,* 189–205.

Head, H. (1926). *Aphasia and kindred disorders of speech.* London: Cambridge University Press.

Hécaen, H., Ajuriaguerra, J., & Massonet, J. (1951). Les troubles visuoconstructifs par lésions pariéto-occipitales droites. Rôle des perturbations vestibulaires. *Encéphale, 1,* 122–179.

Hécaen, H., & Assal, G. (1970). A comparison of constructive deficits following right and left hemispheric lesions. *Neuropsychologia, 8,* 289–303.

Hécaen, H., Penfield, W., Bertrand, C., & Malmo, R. (1956). The syndrome of apractognosia due to lesions of the minor cerebral hemisphere. *Archives of Neurology and Psychiatry, 75,* 400–434.

Hier, D. B., Davis, K. R., Richardson, E. P., & Mohr, J. P. (1977). Hypertensive putaminal hemorrhage. *Annals of Neurology, 1*, 152–159.

Kertesz, A., & Dobrowolski, S. (1981). Right-hemisphere deficits, lesion size and location. *Journal of Clinical Neuropsychology, 3*, 283–299.

Kertesz, A., Harlock, W., & Coates, R. K. (1979). Computer tomographic localization, lesion size and prognosis in aphasia and nonverbal impairment. *Brain and Language, 8*, 34–50.

Kimura, D., & Faust, R. (1987). Spontaneous drawing in an unselected sample of patients with unilateral cerebral damage. In D. Ottoson (Ed.), *Duality and unity of the brain* (pp. 114–146). New York: Macmillan.

Kirk, A., Blonder, L. X., Wertman, E., & Heilman, K. M. (1991). Phonolexical agraphia: Superimposition of acquired lexical agraphia on developmental phonological dysgraphia. *Brain, 114*, 1977–1996.

Kirk, A., & Kertesz, A. (1989). Hemispheric contributions to drawing. *Neuropsychologia, 27*, 881–886.

Kirk, A., & Kertesz, A. (1991). On drawing impairment in Alzheimer's disease. *Archives of Neurology, 48*, 73–77.

Kirk, A., & Kertesz, A. (1992). Recovery from drawing impairment after hemispheric stroke. Paper presented at the twentieth annual meeting of the International Neuropsychological Society, San Diego.

Kirk, A., & Kertesz, A. (1993a). Subcortical contributions to drawing. *Brain and Cognition, 21*, 57–70.

Kirk, A., & Kertesz, A. (1993b). Cortical and subcortical aphasias compared. *Aphasiology*, in press.

Kleist, K. (1922). Kriegsverletzungen des Gehirns. In O. Schjerning (Ed.), *Handbuch der ärztlichen Erfahrungen im Weltkriege 1914–1918.* (pp. 343–370). Leipzig: Barth.

Kleist, K. (1934). *Gehirnpathologie.* Leipzig: Barth.

Lange, J. (1936). Agnosien und Apraxien. In F. Bumke & O. Foerster (Eds.), *Handbuch der neurologie* (Vol. 6, pp. 807–860). Berlin: Springer-Verlag.

L'hermitte, F., Derousné, J., & Signoret, J. L. (1972). Analyse neuropsychologiques du syndrome frontal. *Revue Neurologique, 127*, 415–440.

L'hermitte, J., & Trelles, J. O. (1933). Sur l'apraxie pure constructive. *Encephale, 28*, 413–444.

Liepmann, H. (1912). Anatomische Befunde bei Aphasischen und Apraktischen. *Neurologisches Zentralblatt, 31*, 1524–1530.

Luria, A. R., & Tsvetkova, L. S. (1964). The programming of constructive activity in local brain injuries. *Neuropsychologia, 2*, 95–108.

Mack, J. L., & Levine, R. N. (1981). The basis of visual constructional disability in patients with unilateral cerebral lesions. *Cortex, 17*, 515–532.

Mayer-Gross, W. (1935). Some observations on apraxia. *Proceedings of the Royal Society of Medicine, 28*, 1203–1212.

McFie, J., Piercy, M. F., & Zangwill, O. L. (1950). Visual spatial agnosia associated with lesions of the right cerebral hemisphere. *Brain, 73*, 167–190.

McFie, J., & Zangwill, O. L. (1960). Visual-constructive disabilities associated with lesions of the left cerebral hemisphere. *Brain, 82*, 243–259.

Naeser, M. A., Alexander, M. P., Helm-Estabrooks, N., Levine, H. L., Laughlin, S. A., & Geschwind, N. (1982). Aphasia with predominantly subcortical lesion sites: Description of three capsular/putaminal syndromes. *Archives of Neurology, 39*, 2–14.

Newcombe, F. (1969). *Missile wounds of the brain.* Oxford: Oxford University Press.

Paterson, A., & Zangwill, O. L. (1944). Disorders of visual space perception associated with lesions of the right cerebral hemisphere. *Brain, 67*, 331–358.

Perez, F. I., Rivera, V. M., Meyer, J. S., Gay, J. R. A., Taylor, R. L., & Mathew, N. T. (1975). Analysis of intellectual and cognitive performance in patients with multi-infarct dementia, vertebrobasilar insufficiency with dementia, and Alzheimer's disease. *Journal of Neurology, Neurosurgery, and Psychiatry, 38*, 533–540.

Piercy, M., Hécaen, H., & Ajuriaguerra, J. (1960). Constructional apraxia associated with unilateral cerebral lesions—Left and right sided cases compared. *Brain, 85*, 775–789.

Piercy, M. F., & Smyth, V. O. G. (1962). Right hemisphere dominance for certain non-verbal intellectual skills. *Brain, 85*, 775–789.

Pillon, B. (1981). Troubles visuo-constructifs et méthodes de compensation: Résultats de 85 patients atteints de lésions cérébrales. *Neuropsychologia, 19*, 375–383.

Poppelreuter, W. (1917). *Die psychische Schädingungen durch Kopfschuss im Kriege* (Vol. 1). Leipzig: Voss.

Roncato, S., Sartori, G., Masterson, J., & Rumiati, R. (1987). Constructional apraxia: An information processing analysis. *Cognitive Neuropsychology, 4*, 113–129.

Shallice, T. (1981). Phonological agraphia and the lexical route in writing. *Brain, 104*, 413–429.

Sjögren, T., Sjögren, H., & Lindgren, A. G. H. (1952). Morbus Alzheimer and morbus Pick. *Acta Psychiatrica et Neurologica Scandinavica, 82 (Suppl.)*, 1–152.

Stiles-Davis, J., Janowsky, J., Engel, M., & Nass, R. (1988). Drawing ability in four young children with congenital unilateral brain lesions. *Neuropsychologia, 26*, 359–371.

Swindell, C. S., Holland, A. L., Fromm, D., & Greenhouse, J. B. (1988). Characteristics of recovery of drawing ability in left and right brain-damaged patients. *Brain and Cognition, 7*, 16–30.

Taylor, A. M., & Warrington, E. K. (1973). Visual discrimination in patients with localized cerebral lesions. *Cortex, 9*, 82–93.

Villa, G., Gainotti, G., & De Bonis, C. (1986). Constructive disabilities in focal brain-damaged patients. Influence of hemispheric side, locus of lesion and coexistent mental deterioration. *Neuropsychologia, 24*, 497–510.

Warrington, E. K. (1969). Constructional apraxia. In P. J. Vinken & G. W. Bruyn (Eds.), *Handbook of clinical neurology* (Vol. 4., pp. 69–83). Amsterdam: North Holland.

Warrington, E. K., & James, M. (1967). Disorders of visual perception in patients with localized cerebral lesions. *Neuropsychologia, 5*, 253–266.

Warrington, E. K., James, M., & Kinsbourne, M. (1966). Drawing disability in relation to laterality of cerebral lesion. *Brain, 89*, 53–82.

Chapter 18

Subcortical Lesions and Cognitive Deficits

Stefano Cappa and Claus-W. Wallesch

I. INTRODUCTION

Reports of neuropsychological disorders (usually aphasia) in patients with lesions involving the subcortical structures of the brain can be found in the anatomical–clinical literature of the 19th century (for a summary, see Henschen, 1922). The number of reports, however, increased considerably only after the introduction of the computerized tomography (CT) scan to neurological practice in the early 1970s. The aim of these case descriptions generally was to demonstrate that disorders of the functions traditionally ascribed to the cerebral cortex ("higher cortical functions" such as language or perception) actually could follow cerebral lesions centered on the thalamus and the basal ganglia. In accordance with the paradigm then prevailing in neuropsychological research, the main focus of these reports was on syndrome description and on the comparison of these subcortical syndromes with their cortical counterparts. Attempts to integrate these findings into neural models of the correlates of cognitive functions were rare, with the possible exception of the pioneering work of Luria (1966,1977).

The advances in the fields of cognitive neuropsychology and of neuroscience in general have led, in the last few years, to a thorough revision of the research questions and the methodology in this area. A number of important issues have emerged, which can be summarized briefly as follows.

At the psychological level, the analysis of patterns of cognitive impairment that occur after subcortical lesions must become more fine grained, that is, researchers must go beyond the syndrome approach in the quest for differential aspects between "cortical" and "subcortical" impairments, making reference to the available cognitive models for language and other functions (Shallice, 1988).

Localization and Neuroimaging
in Neuropsychology

545

At the neural level, the physiopathological interpretation of subcortical cognitive disorders carries important implications for neural models of cognitive functions. Investigators have suggested that, at least in cases caused by vascular lesions, neuropsychological impairment after subcortical damage is actually the result of coexisting cortical dysfunction (Skyhoj-Olsen, Bruhn, & Oberg, 1986). This point of view denies the subcortical structures any relevant role in cognition, and is incompatible with recent models that include subcortical components within the neural networks that underlie cognitive functions (Cappa & Vallar, 1992; Crosson, 1985; Mesulam, 1990; Wallesch & Papagno, 1988). A closely connected question is whether damage to the gray nuclei plays any relevant role, in comparison to white matter damage, which practically always coexists in natural lesions and which may result in neuropsychological impairments by means of the time-honored "disconnection" mechanism (Alexander, 1989).

A careful consideration of these issues is crucial to any effort to build neural models of cognitive functions that include the participation of subcortical structures. In this chapter, after an anatomical–physiological summary, we provide a selective review of the available empirical data on subcortical cognitive disorders, summarized according to the area of neuropsychological impairment, and of their implications for neural models of cognitive functions. The review, in general, is limited to evidence from focal vascular lesions and does not include memory disorders. Comprehensive discussions of the cognitive disorders found in extrapyramidal diseases and of the neuropathology of amnesia are available (for example, see Vallar, Cappa, & Wallesch, 1992; Chapter 20).

II. ANATOMY AND PHYSIOLOGY

In this chapter, only the three major subcortical nuclei of the cerebral hemispheres are considered: (1) the lenticular (lentil-shaped) nucleus underneath the insula, (2) the caudate nucleus that consists of a head situated between the frontal horn of the lateral ventricle and the anterior limb of the internal capsule and a long and slim body and tail that stretch along the lateral wall of the lateral ventricle, and (3) the thalamus, an aggregation of more than 30 separate nuclei that forms a large part of the lateral wall of the third ventricle (Figs. 1 and 2).

The lentil shape of the lenticular nucleus matches the funnel shape of the internal capsule that contains the major spinal and brainstem afferents and efferents of the cerebral cortex. The nucleus can be subdivided into two parts—a lateral one, the putamen, and a medial one, the pallidum.

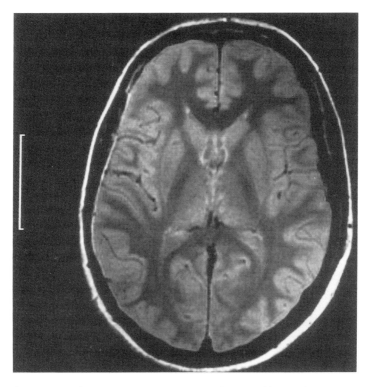

Figure 1 Axial magnetic resonance images of a normal brain, showing the deep gray nuclei and their spatial relationship. (Courtesy of R. Gasparotti, Institute of Radiology, University of Brescia.)

The latter is a common efferent relay both for the putamen and the caudate. Putamen and caudate nuclei have a similar internal structure and together form the neostriatum.

The thalamic nuclei are grouped together by Y-shaped white matter sheets, the internal medullary laminae, into five groups: the anterior, medial, ventral, posterior, and "nonspecific" intralaminar and midline nuclei (Walker, 1938; Williams & Warwick, 1975). With respect to their cortical projections, the thalamic nuclei have been subdivided into specific (projecting on circumscribed cortical regions) and nonspecific nuclei. Specific cortical projections connect the nucleus ventralis anterior with the premotor cortex, the nucleus medialis dorsalis with the prefrontal, and the anterior nuclei with the cingulate and pericingulate cortex (Williams & Warwick, 1975).

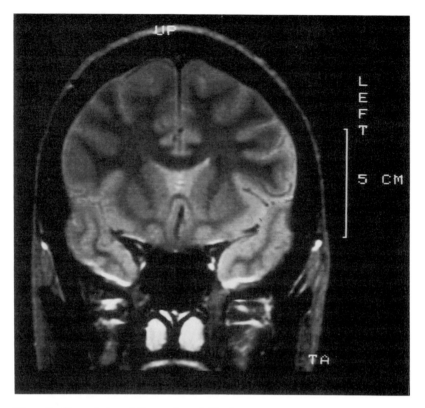

Figure 2 Coronal view of the specimen in Fig. 1.

The cerebral cortex of each hemisphere projects in topographically organized fashion on the neostriatum, the associative areas of the prefrontal, temporal, and parietal lobe (mainly to the caudate nucleus), and the motor cortex (mainly to the putamen) (Goldman & Nauta, 1977; Künzle, 1975). Apart from cortical afferents, the neostriatum receives projections from the midbrain substantia nigra and from nonspecific thalamic nuclei (Parent, 1986). The cortical and thalamic afferents are excitatory. The nigrostriatal projection is dopaminergic, and its mode of action is probably more intricate than is reflected in the excitatory–inhibitory dichotomy (Chiodo & Berger, 1986). All in all, this area seems to have a modulatory inhibitory effect (see subsequent discussion).

Neostriatal efferents project to the pallidum and to the substantia nigra and are inhibitory. The pallidum receives mainly neostriatal afferents and projects on the nuclei of the ventral (nuclei VA and VL), mediodorsal, and nonspecific thalamus. The pallidal projections are assumed to be inhibitory (Penney & Young, 1983; Uno, Ozawa, & Yoshida, 1978). Ventral–mediodorsal thalamus and frontal cortex are connected bidirectionally by excitatory projections, thus forming reverberating loop systems (Alexander, DeLong, & Strick, 1986) that are dampened by the pallidal projection on the thalamus. Indirectly, the midbrain dopaminergic projections may be powerful modulators of forebrain performance.

Alexander and colleagues (1986) proposed a parallel organization of functionally segregated circuits linking basal ganglia and cortex to form the functional basis of cortical–subcortical interaction. These researchers distinguished five loop systems: the motor, the oculomotor, the dorsolateral prefrontal, the lateral orbitofrontal, and the anterior cingulate. The latter three are assigned nonmotor "cognitive" and emotional functions.

In the neostriatum, the corticostriatal, as well as the nigrostriatal dopaminergic, projections synapse with an intricate system of neostriatal cells—the spiny I and II neurons. The spiny I cells—which constitute the large majority of neostriatal neurons (96%)—are supposed to form a collateral inhibiting network that suppresses neuronal activity outside the immediate region of afferent cortical excitation (Groves, 1983; Katayama, Miyazaki, & Tsubokawa, 1981; Penney & Young, 1983). Thus, this structure provides "means for sharpening the diffuse pattern of input from cerebral cortex and brainstem" on target cells in the next station of the neuronal chain, the pallidum (Groves, 1983). This focusing of output seems to depend on nigrostriatal dopaminergic projection. Focusing in the neostriatum results in an equally focused inhibition of the pallidum, which again acts via GABA-ergic inhibition on the ventral thalamus. At this level, then, the inhibition of the pallidal inhibitor results in a disinhibition of thalamocortical positive feedback loops. Regulation of regional frontal cortex activity thus is achieved by release or suppression, at the level of the thalamus, of interactive thalamocortical activation mechanisms. The dopaminergic projection in this model plays a role in filtering and focally activating (by inhibition of inhibitors) the cortical projections on the neostriatum, thus allowing only those impulses that pass through the filter to release the thalamocortical activator (Penney & Young, 1983). However, the activity of nigrostriatal dopaminergic neurons seems to be quite unaltered in different functional states, including movement (Groves, 1983), indicating that this projection indeed imposes a tonic effect on its target, as would a preset filter or amplifier.

In geometrical terms, the projections from cortex to neostriatum, from the neostriatum to the pallidum, and further to the thalamus and substantia nigra have been demonstrated to be highly convergent and functionally probably highly integrative (Percheron, Yelnik, & Francois, 1984a,b), although the different functional circuits appear to be kept separate. The within-circuit integration, however, includes projections from those cortical regions that project on the same neuron population in the striatum, which appear to be those cortical regions that are cortico-cortically connected (Yeterian & van Hoesen, 1978) and functionally related. According to Penney and Young (1983), this architecture serves the function of behavior selection and maintenance.

Neurophysiological data suggest that the basic functional unit of the cortex is a barrel shaped "module" containing some 10^4 neurons (Mountcastle, 1957; Phillips, Zeki, & Barlow, 1984). Phillips and Porter (1977) propose that the thalamic afferents transmit "features of a central command" onto the frontal motor cortex. According to Goldman-Rakic (1984), the modular machinery in the prefrontal cortex could permit combinations and recombinations among inputs that would constitute a highly adaptive and plastic mechanism for information processing. Eccles (1977) assumed a competitive inhibition of neighboring and functionally parallel modules at the level of the cortex to govern behavior. If thalamic afferents transmit features of a central command and enhance the activity of some modules but not others, then the thalamic projection plays a key role in the competition of cortical modules.

Anatomically and physiologically, it seems plausible that the thalamus and, indirectly via the thalamus, the basal ganglia participate in those actions of the cerebral cortex that underlie higher mental functions. In the following discussion, we review the clinical literature for supporting evidence.

III. SUBCORTICAL APHASIA

In classical aphasiology, the term "subcortical aphasia" denoted the assumed disconnection effect of a fiber tract lesion on the function of a cortical language center. Today, the term is used for language disturbances resulting from lesions of the deep nuclei of the forebrain, that is, the basal ganglia (caudate and lentiform nucleus) and the thalamus. Because of the different functional roles of basal ganglia and thalamus, we discuss the respective aphasia syndromes resulting from lesions of these structures separately.

In this chapter, only effects of naturally occurring lesions are considered. For language disturbances resulting from stereotaxic surgery, the reader is referred to the review by Wallesch and Papagno (1988). Further, only such cases are analyzed in whom a cortical lesion detectable by CT has been excluded. Most naturally occurring lesions in the deep nuclei are vascular. Therefore, some knowledge of vascular anatomy is a prerequisite for the understanding of subcortical aphasia.

The putamen (with the exception of its anterior pole), the lateral pallidum, and the caudate nucleus (with the exception of the rostral and basal part of the head) are supplied by lenticulostriate branches of the middle cerebral artery (Lazorthes, Gouaze, & Salomon, 1976). Most infarctions in these structures and their vicinity result from (in many instances temporary) occlusions of the main stem of this artery, so effects of ischemia on the cerebral cortex must be taken into account.

The rostral and basal aspects of the head of the caudate and the frontal poles of the putamen and pallidum are supplied from the anterior cerebral artery, mainly through the recurrent artery of Heubner. Most of the pallidum receives its supply from the anterior choroid artery, which usually stems directly from the internal carotid (Krayenbühl & Yasargil, 1965). The area of supply of this artery includes the cortex only at the basal surface of the temporal lobe. Therefore, the effects of anterior choroidal infarction are especially important for the distinction between subcortical and cortical dysfunction in the pathogenesis of subcortical aphasia syndromes.

Great interindividual differences exist with respect to thalamic vascularization. According to Lazorthes et al. (1976), in 60–70% of individuals the thalamus receives its principal blood supply from the posterior communicating artery and a lesser amount from the posterior choroid artery. In the other individuals, its main supply stems from branches of the posterior cerebral artery. In addition, the watershed between the internal carotid and the vertebrobasilar supply systems varies interindividually.

A. Basal Ganglia Aphasia

The early literature contains a small number of cases of ("motor") aphasia resulting from infarctions in the lenticular nucleus (Moutier, 1908). Kleist (1934) and his pupil Merzbach (1928) related the symptoms of stereotypy and iteration to lesions in the basal ganglia (for a review, see Wallesch, 1990).

After the introduction of the CT scan in the mid-1970s, a large number of reports of cases of aphasia resulting from caudate and putaminal infarctions and hemorrhages was published (e.g., Cambier, Elghozi, & Strube, 1979; Hier, Davies, & Richardson, 1977). In one of the first systematic studies, Damasio, Damasio, Rizzo, Varney, and Gersh (1982) described an atypical aphasia syndrome occurring with subcortical infarctions involving the head of the caudate, the anterior limb of the internal capsule, and the rostral putamen. The most constant symptom was semantic paraphasia.

Fluency, presence of phonemic paraphasia, and deficits of repetition and comprehension varied among patients. This subcortical aphasia syndrome, with the core symptom of semantic paraphasia, in patients with infarctions in the area of supply of the anterior lenticulostriate branches of the middle cerebral artery had been described earlier by Barat, Mazaux, Bioulac, Giroire, Vital, and Arné (1981) and was confirmed by Wallesch (1985), who pointed out that, with this type of lesion, cortical dysfunction cannot be excluded since such lesions usually result from occlusion of the main stem of the middle cerebral artery (Weiller, Ringelstein, Reiche, Thron, & Buell, 1990). On the other hand, the syndrome contains features that are unusual for anterior cortical infarctions, namely the presence of frequent paraphasia and dissociations of impaired auditory and less impaired written language comprehension (Wallesch, 1985). The latter finding supports the explanation of Damasio et al. (1982) that a compromised connection between the temporal auditory association cortex and the head of the caudate might account for these patients' comprehension deficits. Aphasia following left anterior lenticulostriate infarction is usually transient. Caplan and colleagues (1990) reported language abnormalities, stuttering, or stammering in only 2 of 10 patients with chronic left caudate infarctions (compare Ludlow, Rosenberg, Salazar, Grafman, & Smutok, 1987, for acquired stuttering with caudate lesions).

The stroke syndrome of striatocapsular infarction was investigated prospectively by Donnan, Baldin, Berkovic, Longley, and Saling (1991). These authors found aphasia in the acute stages in 18 of 20 cases, and related its presence to cortical dysfunction.

Another syndrome of subcortical aphasia, nonfluent aphasia with largely or partially preserved repetition and usually mild comprehension deficit, first was described by Sterzi and Vallar (1978) and later was confirmed by Brunner, Kornhuber, Seemueller, Suger, and Wallesch (1982) and Wallesch (1985). This type of "transcortical" subcortical aphasia usually occurs with lesions of the lenticular nucleus or the knee of the internal capsule.

Cappa, Cavallotti, Guidotti, Papagno, and Vignolo (1983) reported two investigations on subcortical aphasia. The first included six consecutive cases with left-sided strictly subcortical lesions, of whom two were aphasic. In one, the lesion extended to the temporal isthmus. The other aphasic patient exhibited fluent empty speech with a posterior putaminal lesion. The lesions of three nonaphasic subjects also were situated in the posterior part of the lenticular nucleus (one additional patient suffered from a small hemorrhage). In the second study, the authors selected, from 250 CT scans of patients assessed for aphasia, 6 patients with lesions confined to subcortical structures. Two of them had been found not aphasic in standard assessment. Of the other four, the lesion extended to the temporal isthmus in one case and into the thalamus in another. One patient with a small hematoma involving the head of the caudate and the anterior limb of the internal capsule exhibited mild nonfluent aphasia with excellent repetition and mild agraphia. Comprehension was well preserved. The authors noted the presence of semantic paraphasia. Finally, the authors described mild fluent aphasia with semantic paraphasia in a patient with a small hemorrhage in the lateral putamen.

Probably the largest and most thoroughly studied series of 25 patients with left subcortical lesions was reported by Puel and associates (1984). A number of cases match even strict criteria. The authors found atypical fluent aphasia with semantic paraphasia with an anterior putamino-capsular hemorrhage and nonfluent aphasia with phonemic and semantic paraphasia in patients suffering from a hemorrhage involving the head of the caudate nucleus and parts of the anterior limb of the internal capsule. Semantic paraphasia was also noticed, with a small infarction involving the pallidum, the posterior limb of the internal capsule, and the body of the caudate. In patients in whom the CT lesion was restricted to the white matter, Puel et al. (1984) found semantic jargon aphasia with an apparent anterior choroidal infarction involving the knee and posterior limb of the internal capsule, and nonfluent paraphasic aphasia with an anterior limb infarct. Hemorrhages confined to the head of the left caudate have been reported to result in nonfluent aphasia, and small putaminal hemorrhages to result in fluent aphasia (Cappa et al., 1983; Puel et al., 1984).

As discussed earlier, anterior choroidal infarctions are crucial to the consideration of possible participation by the basal ganglia in language processes. This artery supplies the knee and posterior limb of the internal capsule, parts of the pallidum, only temporobasal cortex, and normally not the thalamus (although some individual variation is seen here). Reports of presumably selected cases (Cambier, Graveleau, Decroix, El-ghozi, & Masson, 1983; Cappa & Sterzi, 1990; Wallesch, 1985) demonstrated nonfluent aphasia with semantic paraphasia or transcortical motor aphasia to occur with left-sided anterior choroidal artery infarction. Rapid, although not always total, improvement was seen. In Wallesch's and Cappa and Sterzi's cases, the lesions involved both the pallidum and the knee of the internal capsule, as well as parts of the posterior limb of varying extent. In three previously unreported cases of Wallesch, with left anterior choroidal artery infarction without aphasia, the lesion spared both pallidum and the knee of the internal capsule.

Decroix, Graveleau, Masson, and Cambier (1986) reported 10 patients with left-sided infarction in the territory of supply of this vessel. In two

cases, these researchers noted "thalamic aphasia" with loss of verbal fluency, difficulty in organizing speech, and rare semantic paraphasias, but
sparing of repetition and comprehension. The authors speculated that an
interruption of thalamocortical connections, particularly between pulvinar and post-rolandic language cortex, accounted for these deficits. In
another three patients, language deficits of short duration were noted (reduction in fluency and initiating speech, unspecified language deficit, initial loss of speech, difficulty in organizing conversation, and impairment
in a word fluency task). The remaining five cases exhibited nonlanguage
disturbances. Of six patients with left anterior choroidal infarction in the
series of Graff-Radford, Damasio, Yamada, Eslinger, and Damasio (1985),
three exhibited mild language deficits.

In summary, left anterior choroidal artery infarction does not result in
aphasia in all instances. Aphasia seems to be infrequent and especially
short-lived in cases in whom only the posterior limb of the internal capsule is involved, and more pronounced when pallidum and knee of the
capsule are lesioned. With few exceptions (Decroix et al., 1986), the prognosis of aphasia seems good and language symptoms are only transient
(Cappa & Sterzi, 1990).

Table 1 outlines the aphasic syndromes that occur with small ischemic
lesions in the left basal ganglia and their vicinity.

Much debate has occurred over whether the presence of aphasia in subcortical stroke indicates a participation of the basal ganglia in language

Table 1 Acute Syndromes of Aphasia with Left Ischemic Lesions in the Region
of the Basal Ganglia

Territory	Likelihood of cortical involvement	Language pathology
Anterior lenticulostriate arteries	Yes	Fluent or nonfluent, semantic paraphasia, phonemic paraphasia, auditory comprehension deficit, repetition preserved
Anterior choroideal arteries, with pallidum and knee of internal capsule	No	Nonfluent, preserved repetition and comprehension
Anterior choroideal arteries, posterior limb only	No	Minimal or no aphasia

processing. Sryhöj Olsen et al. (1986) found, in five aphasic patients with subcortical lesions, a focal cortical low-flow area in regional cerebral blood flow, whereas in five nonaphasic patients no focal low flow was seen. The authors interpret their results to indicate that, in "subcortical aphasia," cortical hypoperfusion is the underlying cause of the language dysfunction. However, both groups were not parallel in a number of aspects. Four of five nonaphasic patients had exclusively white matter lesions that did not involve the nuclei, and four of five aphasic subjects showed evidence of middle cerebral artery occlusion. Perani, Vallar, Cappa, Messa, and Fazio (1987) also found cortical hypoperfusion in aphasic patients with left subcortical (two thalamic, two basal ganglia) lesions. Since the thalamus is not supplied by the middle cerebral artery, ischemia-caused cortical hypoperfusion seems rather unlikely with thalamic pathology. Perani et al. (1987) suggest instead a functional cortical depression. This mechanism resembles von Monakow's (1914) concept of diaschisis and would indicate only that the lesioned structure is linked functionally to the hypoactive one.

B. Thalamic Aphasia

Based on observations during and after stereotaxic surgery, a participation in language processing has been claimed for the anterior and lateral parts of the ventral thalamus (nuclei VA and VL) and for the pulvinar, the largest of the posterior group of nuclei (for review and analysis, see Wallesch & Papagno, 1988). VA and VL are the main thalamic projection sites of pallidal and cerebellar fibers. As has been discussed, these sites are interconnected bidirectionally with the motor and premotor areas of the frontal lobes, including Broca's and the supplementary motor areas, and are the target of the cortico-striato-pallido-cortical projection. The pulvinar is linked bidirectionally with the retro-rolandic cortex, with the exception of the occipital pole and those parts of the temporal cortex that belong to the limbic system. The pulvinar is the specific thalamic nucleus for most of the temporo-parietal association cortex including Wernicke's area.

Language pathology after thalamic hemorrhages and infarctions has been reviewed by Jonas (1982) and by Crosson (1984). Both authors stress the absence of deficits of language comprehension and repetition in the vast majority of cases. Language production seems to be affected mainly at the word level. Paraphasia or even jargon, perseveration, and word finding difficulties are observed frequently. A predominance of semantic over phonemic paraphasia has been reported (Alexander & LoVerme, 1980; Cappa & Vignolo, 1979). Aspontaneity of speech production has been observed occasionally. Syntactic errors seem to be rare. Jonas (1982) drew attention to the disparity between the severe disruption of propositional and the relative preservation of nonpropositional speech.

An analysis of the role of specific thalamic nuclei in language functions is hampered by the fact that, in the most common type of pathology— hemorrhages, the lesion frequently exceeds the boundaries of the thalamus. Patients with larger hemorrhages tend to suffer from impairments of consciousness, either because of damage to the "unspecific" thalamic nuclei that are parts of the ascending reticular activating system or because of ventricular hemorrhage or occlusion. In a series of left thalamic infarction cases, Graff-Radford et al. (1985) found aphasia only in those three patients in whom the area of supply of the tuberothalamic artery that includes the ventral nuclei was affected. Details of the language pathology in three patients with left ventral thalamic lesions were described by Gorelick, Hier, Benevento, Levitt, and Tan (1984) and Bogousslavsky, Regli, and Assal (1986). The patients exhibited reduced speech production, phonemic and semantic paraphasia, anomia, and relatively preserved syntax and repetition. A patient of Archer, Ilinsky, Goldfader, and Smith (1981) with a left ventral thalamic infarction was fluently aphasic but otherwise similar. Ventral thalamic lesions in the dominant hemisphere without language symptoms have been described by Mohr (1983).

The following symptom constellation is characteristic of ventral thalamic aphasia: fluent or nonfluent speech production; prominent, mainly semantic, paraphasia; and well-preserved comprehension. In addition, better than expected repetition and rapid fluctuations (within minutes) in degree or even type of symptomatology may indicate thalamic pathology.

Aphasia following circumscribed ventral thalamic lesions is transient. Wallesch, Kornhuber, Brunner, Kunz, Hollerbach, and Suger (1983) were unable to find lateralized linguistic deficits more than 6 months afer thalamic stroke. Thalamic aphasia with circumscribed lesions outside the ventral nuclei is rare. Crosson, Parker, Kim, Warren, Kepes, and Tulley (1986) described an anatomically verified case of left pulvinar infarction who was fluently aphasic with anomia and semantic paraphasia and relatively preserved repetition and comprehension. Other fluently aphasic patients with posterior lesions have been reported by Reynolds, Turner, Harris, Ojemann, and Davis (1979) and Alexander and LoVerme (1980); negative cases have been presented by Graff-Radford et al. (1985) and Cappa, Papagno, Vallar, and Vignolo (1986).

The patient of Crosson et al. (1986) was impaired only mildly when discussing familiar topics, but speech deteriorated into jargon with unfamiliar themes. The authors explain this finding as a disorder of semantic monitoring. However, such a phenomenon may only indicate a less specific disorder of attention. Frequently, aphasia in patients with posterior thalamic lesions is confounded with attentional deficits or fluctuating states of consciousness. "When rendered fully alert, the patient appeared virtually intact in language function, . . . but quickly lapsed into a state of unwonted logorrhoic paraphasia resembling delirium" (Mohr, Watters, & Duncan, 1975, p. 3). The findings of Graff-Radford et al. (1985) indicate that disorders of attention are characteristic for infarctions in the territory of the interpeduncular profundus artery, which supplies mainly unspecific thalamic nuclei.

IV. SUBCORTICAL APRAXIA

Several cases of ideomotor apraxia in association with aphasia in patients with subcortical lesions have been reported (Basso & Della Sala, 1986; DeRenzi, Faglioni, Scarpa, & Crisi, 1986). Persistent apraxia after rapid recovery from aphasia was present in a patient with a white matter lesion (Sanguineti, Agostoni, Ajello, Apale, Bogliun, & Tagliabue, 1989).

A large-scale retrospective study (Della Sala & Basso, 1992) involved 35 cases of deep left hemispheric stroke; of these, 7 had ideomotor apraxia that was associated with oral apraxia in 6. Only 1 patient, whose performance on the apraxia tests was just below the normal cut-off value, was not aphasic. Of the 7, 5 patients had lesions limited to the white matter, without any relevant involvement of the thalamus and the basal ganglia. This study suggests that apraxia in subcortical lesions might be the consequence of interruption of long fiber pathways connecting anterior and posterior cortical regions. This result is in agreement with the traditional neurological model of cerebral organization for praxis (Heilman & Gonzalez Rothi, 1985; Liepmann, 1908). A major problem, as in the case of aphasia, is the interpretation of negative cases, that is, patients whose lesions on the CT scan are comparable in site and size to those found in apraxic patients, but who are nevertheless not impaired on apraxia tests.

In the case of aphasia, "negative" cases have been found to be associated with minor degrees of hypoperfusion in ipsilateral cortex (Perani et al., 1987; Sryhöj Olsen et al., 1986). Functional studies of regional cerebral blood flow or metabolism assessed in the resting state, however, do not make a major contribution to the understanding of the respective roles of subcortical damage and of coexistent cortical dysfunction in the pathogenesis of apraxia since, as reported earlier, aphasia co-occurs practically always, preventing any specific correlation. On the other hand, studies of regional modifications of blood flow or metabolism assessed with positron emission tomography while the subject is engaged in motor tasks will provide important information, as suggested by the investigations that have been reported in normal subjects (Colebatch, Deiber, Passingham, Friston, & Frackowiak, 1991) and in patients (Chollet, Di Piero, Wise, Brooks, Dolan, & Frackowiak, 1991) engaged in relatively simple motor tasks.

No studies have been done on the qualitative aspects of subcortical apraxia or on the presence of any distinctive feature with cortical disorders. However, note that with a few notable exceptions, the detailed qualitative analysis of disorders of intentional movement is only beginning (Freund & Hummelsheim, 1985; Poizner, Mack, Verfaellie, Gonzalez Rothi, & Heilman, 1990).

V. SUBCORTICAL NEGLECT

An extensive number of reports has established clearly that unilateral neglect (ULN), operationally defined as a failure to explore the hemispace contralateral to the cerebral lesion, can be found in association with right subcortical lesions involving the thalamus (Watson & Heilman, 1979), the basal ganglia (Damasio, Damasio, & Chang Chui, 1980), or white matter structures such as the anterior (Viader, Cambier, & Pariser, 1982) or the posterior (Cambier et al., 1983) limb of the internal capsule. If we consider large-scale studies on samples of right hemisphere damaged patients screened for the presence of ULN (Vallar, Rusconi, Geminiani, Berti, & Cappa, 1991; Vallar & Perani, 1986), the percentage of patients presenting with the disorder in the recent stage after a stroke (1–30 days) is smaller with subcortical than with cortical lesions (about 25% compared with more than 50%). This difference is accounted for, in part, by the lower frequency of occurrence of ULN with lesions limited to the basal ganglia or to the white matter. Thalamic neglect, on the other hand, is apparently both frequent and severe (see Cappa & Vallar, 1992 for a discussion).

ULN has been reported with several different stroke syndromes. In the case of ischemic pathology, ULN frequently is observed in the so-called "large striato-capsular infarction" because of occlusion of lenticulo-striate arteries. As discussed in Section IIIA, this type of stroke is secondary to middle cerebral artery pathology and may be expected to be associated with some degree of cortical hypoperfusion (Donnan et al., 1991; Weiller et al., 1990). ULN, however, also is found with right anterior choroidal artery stroke (Cambier et al., 1983), which generally is considered a "lacunar" syndrome (Bruno, Graff-Radford, Biller, & Adams, 1989), and in caudate infarction, which also is associated with small vessel disease (Caplan et al., 1990). In both these cases, any direct reduction of cortical perfusion would be unexpected. ULN in thalamic infarction correlated with lesions in the posterolateral region or in the paramedian artery territory (Bogousslavsky, Regli, & Uske, 1988; Graff-Radford et al., 1985). Hemorrhagic lesions in the putaminal (Hier, Davis, Richardson, & Mohr, 1977) and posterior thalamic (Hirose et al., 1985) region have been associated with the occurrence of ULN. Caudate hemorrhages, on the other hand, usually are associated with disorders of consciousness and confusional states, which may preclude the assessment of directed attention (Pedrazzi, Bogousslavsky, & Regli, 1990; Stein et al., 1984).

The contribution of studies of subcortical neglect to the current debate about the pathophysiological mechanisms and fractionation of the syndrome (Bisiach & Vallar, 1988) is at its beginning. A particular type of neglect that has a strong link to subcortical pathology is motor neglect (Castaigne, Laplane, & Degos, 1970,1972). This unusual condition can be defined as a reduction of the spontaneous motility of the affected side (usually the left), reversible with verbal incitations and accompanied by motor extinction (absence of movement of the left hand after bilateral stimulation in the absence of tactile extinction). Motor neglect has been reported in association with a thalamic hematoma (Schott, Laurent, Mauguiere, & Chazot, 1981) and with a hemorrhage centered on the head of the caudate nucleus (Valenstein & Heilman, 1981). These lesions disrupt the thalamo-frontal connections, which probably are important for the voluntary generation of motor acts (Marsden, 1982; Viader et al., 1982). An interesting activation study with the ^{133}Xe method has been reported in a case of motor neglect associated with a lesion in the anterior limb of the right internal capsule (de la Sayette, Bouvard, Chapon, Rivaton, Viader, & Lechevalier, 1989): no contralateral rolandic activation was observed when the patient palpated an object with the left hand.

Another important issue that is raising considerable interest in the neglect literature is the distinction between "intentional" and attentional/ representational aspects of ULN (Heilman, Bowers, Branch Coslett, Whelan, & Watson, 1985). Directional hypokinesia, that is, difficulty in executing movements toward the contralesional side, has been shown to play an important role in the physiopathology of ULN (Bisiach, Geminiani, Berti, & Rusconi, 1990; Branch Coslett, Bowers, Fitzpatrick, Haws, & Heilman, 1990; Tegner & Levander, 1991). In the neural model of ULN proposed by Heilman, Watson, and Valenstein (1985) and by Mesulam (1981,1990) the basal ganglia participate in an anterior loop that is involved in the generation of movements directed to the contralateral space, whereas the thalamus is involved, mainly in association with the parietal cortex, in the attentional and sensory aspects of spatial behavior. The evidence available from human studies to support this anatomical–physiological distinction is, at the moment, extremely limited (see, for example, Bisiach, Berti, &

Vallar, 1985). However, a more detailed investigation of single cases with tests that go beyond a simplified clinical assessment of the presence or absence of ULN is likely to provide the relevant information in the near future.

Another interesting issue is the fractionation, within the syndrome of ULN, of disorders of extrapersonal and personal space exploration and of anosognosia (Bisiach, Perani, Vallar, & Berti, 1986a; Bisiach, Vallar, Perani, Papagno, & Berti, 1986b). The full-fledged "dyschiria" syndrome (Bisiach & Vallar, 1988) usually is found only with thalamic lesions, whereas anosognosia frequently is lacking in cases associated with basal ganglia and white matter damage (see Cappa & Vallar, 1992 for discussion).

Some investigators have suggested (Colombo, De Renzi, & Gentilini, 1982) that involvement of the thalamus in cases of ULN associated with cortical lesions is a negative prognostic factor for recovery. A study in patients with chronic neglect, however, has not shown consistent thalamic involvement in the CT scan of this patient group (Cappa, Guariglia, Messa, Pizzamiglio, & Zoccolotti, 1991).

VI. SUBCORTICAL BEHAVIORAL DISORDERS

The first observations of severe behavioral disorders in patients with subcortical lesions were made in cases of "psychic akinesia" in which pseudo-obsessional activities were associated with bilateral pallidal lesions caused by carbon monoxide poisoning or toxic encephalopathy (Laplane, Baulac, Pillon, & Panayotopoulou-Achimestos, 1982; Laplane, Baulac, Wildlocher, & Dubous, 1984; Laplane, Wildlocher, Pillon, Baulac, & Binoux, 1981). The clinical picture in these patients was characterized by a severe reduction of all aspects of motor and verbal behavior.

In contrast, a case of severe behavioral disturbances associated with bilateral lesions of the head of the caudate reported by Richfield, Twyman, and Berent (1987) was characterized by restlessness, disinhibition, and impulsivity. Other patients with similar lesions, however, have been reported to be severely abulic (Pozzilli, Passafiume, Bastianello, D'Antona, & Lenzi, 1987; Trillet, Croisile, Tourniaire, & Schott, 1990; Williams, Owen, & Heath, 1988).

In a large series of patients with unilateral caudate lesions (Mendez, Adams, & Skoog Lewandowski, 1989), a generalized impairment on "frontal lobe" tests was found. Apathy and affective disorders frequently were associated with dorsolateral lesions, whereas disinhibition was more prominent in patients with ventromedial caudate damage. In the series by Caplan et al. (1990), abulia was more frequent in cases with a left caudate lesion; extension to the internal capsule on CT was not necessary for the presence of this behavioral disorder.

VII. CONCLUSIONS

Research on cognitive disorders associated with subcortical lesions is coming of age. After a period necessarily dominated by a "look and see" approach, the majority of studies is now based on explicit models of cognitive functions. These refinements on the psychological side often are combined with comparable attention to the developments in the field of neuroscience, a focus that is not always found in other areas of neuropsychological research. By definition, research in this area remains strongly anchored to the quest for the neurological correlates of cognitive functions. The study of subcortical lesion may prove to be one of the privileged avenues for this enterprise.

REFERENCES

Alexander, G., DeLong, M., & Strick, P. (1986). Parallel organization of functionally segregated circuits linking basal ganglia and cortex. *Annual Review of Neuroscience, 9*, 357–381.

Alexander, M. P. (1989). Clinical-anatomical correlations of aphasia following predominantly subcortical lesions. In F. Boller & J. Grafman (Eds.), *Handbook of neuropsychology* (Vol. 29, pp. 47–66). Amsterdam: Elsevier Science Publishers.

Alexander, M. P., & LoVerme, S. R. (1980). Aphasia after left intracerebral hemorrhage. *Neurology, 30*, 1193–1202.

Archer, C. R., Ilinsky, I. A., Goldfader, P. R., & Smith, K. R. (1981). Aphasia in thalamic stroke: CT stereotaxic localization. *Journal of Computer Assisted Tomography, 5*, 427–432.

Barat, M., Mazaux, J. M., Biaulac, B., Giroire, J. M., Vital, C. L., & Arne, L. (1981). Troubles du langage de type aphasique et lesions putamino-caudees. *Revue Neurologique, 137*, 343–356.

Basso, A., & Della Sala, S. (1986). Ideomotor apraxia arising from a purely deep lesion. *Journal of Neurology, Neurosurgery and Psychiatry, 49*, 458.

Bisiach, E., Berti, A., & Vallar, G. (1985). Analogical and logical disorders of space. In M. I. Posner & O. S. M. Marin (Eds.), *Attention and performance XI* (pp. 239–249). Hillsdale, New Jersey: Erlbaum Associates.

Bisiach, E., Geminiani, G., Berti, A., & Rusconi, M. L. (1990). Perceptual and premotor factors of unilateral neglect. *Neurology, 40*, 1278–1281.

Bisiach, E., Perani, D., Vallar, G., & Berti, A. (1986a). Unilateral neglect: Personal and extrapersonal. *Neuropsychologia, 24*, 759–767.

Bisiach, E., & Vallar, G. (1988). Hemineglect in humans. In F. Boller & J. Grafman (Eds.), *Handbook of neuropsychology* (Vol. 1, pp. 195–222). Amsterdam: Elsevier Science.

Bisiach, E., Vallar, G., Perani, D., Papagno, C., & Berti, A. (1986b). Unawareness of disease following lesions of the right hemisphere: Anosognosia for hemiplegia and anosognosia for hemianopia. *Neuropsychologia, 24*, 471–482.

Bogousslavsky, J., Regli, F., & Assal, G. (1986). The syndrome of unilateral tuberothalamic artery territory infarction. *Stroke, 17*, 434–441.

Bogousslavsky, J., Regli, F., & Uske, A. (1988). Thalamic infarcts: Clinical syndromes, etiology and prognosis. *Neurology, 38*, 837–848.

Branch Coslett, H., Bowers, D., Fitzpatrick, E., Haws, B., & Heilman, K. M. (1990). Directional hypokinesia and hemispatial inattention in neglect. *Brain, 113*, 475–486.

Brunner, R. J., Kornhuber, H. H., Seemüller, E., Suger, G., & Wallesch, C. W. (1982). Basal ganglia participation in language pathology. *Brain and Language, 16*, 281–299.

Bruno, A., Graff-Radford, N. R., Biller, J., & Adams, H. P. (1989). Anterior choroidal artery territory infarction: A small vessel disease. *Stroke, 20*, 616–619.

Cambier, J., Elghozi, D., & Strube, E. (1979). Hemorrhagie de la tete du noyau caude gauche. Desorganisation du discours et de l'expression graphique, perturbations des series gesturelles. *Revue Neurologique, 135*, 763–774.

Cambier, J., Graveleau, P., Decroix, J. P., Elghozi, D., & Masson, M. (1983). La syndrome de l'artere choroidienne anterieure. Etude neuropsychologique de 4 cas. *Revue Neurologique, 139*, 553–559.

Caplan, L. R., Schmahmann, J. D., Kase, C. S., Feldmann, E., Baquis, G., Greenberg, J. P., Gorelick, P. B., Helgason, C., & Hier, D. B. (1990). Caudate infarcts. *Archives of Neurology, 47*, 133–143.

Cappa, S. F., Cavallotti, G., Guidotti, M., Papagno, C., & Vignolo, L. (1983). Subcortical aphasia: Two clinical-CT scan correlation studies. *Cortex, 19*, 227–241.

Cappa, S. F., Guariglia, C., Messa, C., Pizzamiglio, L., & Zoccolotti, P. L. (1991). Computed tomography correlates of chronic unilateral neglect. *Neuropsychology, 5*, 195–204.

Cappa, S. F., Papagno, C., Vallar, G., & Vignolo, L. A. (1986). Aphasia does not always follow left thalamic hemorrhage. *Cortex, 22*, 639–647.

Cappa, S. F., & Sterzi, R. (1990). Infarction in the territory of the anterior choroideal artery: A cause of transcortical motor aphasia. *Aphasiology, 4*, 213–217.

Cappa, S. F., & Vallar, G. (1992). Neuropsychological disorders after subcortical lesions: Implications for neural models of language and spatial attention. In G. Vallar, S. F. Cappa, & C.-W. Wallesch (Eds.), *Neuropsychological disorders associated with subcortical lesions* (pp. 7–41). Oxford: Oxford University Press.

Cappa, S. F., & Vignolo, L. A. (1979). "Transcortical" features of aphasia following left thalamic hemorrhage. *Cortex, 15*, 121–130.

Castaigne, P., Laplane, D., & Degos, J. D. (1970). Trois cas de negligence motrice par lesion retrorolandique. *Revue Neurologique, 122*, 233–242.

Castaigne, P., Laplane, D., & Degos, J. D. (1972). Trois cas de negligence motrice par lesion frontal prerolandique. *Revue Neurologique, 126*, 5–14.

Chiodo, L. A., & Berger, T. W. (1986). Interactions between dopamine and amino-acid induced excitation and inhibition in the striatum. *Brain Research, 375*, 198–203.

Chollet, F., Di Piero, V., Wise, R. J. S., Brooks, D. J., Dolan, R. J., & Frackowiak, R. S. J. (1991). The functional anatomy of motor recovery after stroke in humans: A study with positron emission tomography. *Annals of Neurology, 29*, 63–71.

Colebatch, J. G., Deiber, M.-P., Passingham, R. E., Friston, K. J., & Frackowiak, R. S. J. (1991). Regional cerebral blood flow during voluntary arm and hand movements in human subjects. *Journal of Neurophysiology, 65*, 1392–1401.

Colombo, A., De Renzi, E., & Gentilini, M. (1982). The time course of visual hemi-inattention. *Archiv fuer Psychiatrie und Nervenkrankheiten, 29*, 644–653.

Crosson, B. (1984). Role of the dominant thalamus in language: A review. *Psychological Bulletin, 96*, 491–517.

Crosson, B. (1985). Subcortical functions in language: a working model. *Brain and Language, 25,* 257–292.

Crosson, B., Parker, J. C., Kim, A. K., Warren, R. L., Kepes, J. J., & Tulley, R. (1986). A case of thalamic aphasia with postmortem verification. *Brain and Language, 29,* 301–314.

Damasio, A. R., Damasio, H., & Chang Chui, H. (1980). Neglect following damage to frontal lobe and basal ganglia. *Neuropsychologia, 18,* 123–132.

Damasio, A. R., Damasio, H., Rizzo, M., Varney, N., & Gersh, F. (1982). Aphasia with non-hemorrhagic lesions in the basal ganglia and internal capsule. *Archives of Neurology, 39,* 15–20.

Decroix, J. P., Graveleau, P., Masson, M., & Cambier, J. (1986). Infarction in the territory of the anterior choroideal artery. *Brain, 109,* 1071–1085.

de la Sayette, V., Bouvard, G., Chapon, F., Rivaton, F., Viader, F., & Lechevalier, B. (1989). Infarct of the anterior limb of the right internal capsule causing left motor neglect: Case report and cerebral blood flow study. *Cortex, 25,* 147–154.

Della Sala, S., & Basso, A. (1992). Subcortical localisation of ideomotor apraxia: A review and an experimental study. In G. Vallar, S. F. Cappa, & C.-W. Wallesch (Eds.), *Neuropsychological disorders associated with subcortical lesions* (pp. 357–380). Oxford: Oxford University Press.

De Renzi, E., Falgioni, P., Scarpa, M., & Crisi, G. (1986). Limb apraxia in patients with damage confined to the left basal ganglia and thalamus. *Journal of Neurology, Neurosurgery and Psychiatry, 49,* 1030–1038.

Donnan, G. A., Bladin, P. F., Berkovic, S. F., Longley, W. A., & Saling, M. M. (1991). The stroke syndrome of striatocapsular infarction. *Brain, 114,* 51–70.

Eccles, J. C. (1977). In K. R. Popper & J. C. Eccles (Eds.), *The self and its brain.* Berlin: Springer.

Freund, H.-J., & Hummelsheim, H. (1985). Lesions of premotor cortex in man. *Brain, 108,* 697–733.

Goldman, P. S., & Nauta, W. J. H. (1977). An intricately patterned prefronto-caudate projection in the Rhesus monkey. *Journal of Comparative Neurology, 171,* 369–385.

Goldman-Rakic, P. S. (1984). Modular organization of the prefrontal cortex. *Trends in Neuroscience, 7,* 419–424.

Gorelick, P. B., Hier, D. B., Benevento, L., Levitt, S., & Tan, W. (1984). Aphasia after left thalamic infarction. *Archives of Neurology, 41,* 1296–1298.

Graff-Radford, N. R., Damasio, H., Yamada, T., Eslinger, P. J., & Damasio, H. (1985). Non-hemorrhagic thalamic infarction. Clinical, neuropsychological and electrophysiological findings in four anatomical groups defined by computerized tomography. *Brain, 108,* 485–516.

Groves, P. M. (1983). A theory of the functional organization of the neostriatum and the neostriatal control of voluntary movement. *Brain Research Reviews, 5,* 109–132.

Heilman, K. M., Bowers, D., Branch Coslett, H. B., Whelan, H., & Watson, R. T. (1985). Directional hypokinesia. *Neurology, 35,* 855–859.

Heilman, K. M., & Gonzalez Rothi, L. (1985). Apraxia. In K. M. Heilman & E. Valenstein (Eds.), *Clinical neuropsychology* (pp. 131–152). Oxford: Oxford University Press.

Heilman, K. M., Watson, R. T., & Valenstein, E. (1985). Neglect and related disorders. In K. M. Heilman & E. Valenstein (Eds.), *Clinical neuropsychology* (2d ed., pp. 243–293). New York: Oxford University Press.

Henschen, S. E. (1922). *Klinische und anatomische Beitraege zur Pathologie des Gehirns.* Stockholm: Nordiska Bockhandeln.

Hier, D., Davies, K., & Richardson, E. P., Jr. (1977). Hypertensive putaminal hemorrhage. *Annals of Neurology, 1,* 152–159.

Hirose, G., Kosoegawa, H., Saeki, M., Kitagawa, Y., Oda, R., Kanda, S., & Matsuihira, T. (1985). The syndrome of posterior thalamic hemorrhage. *Neurology, 35,* 998–1002.

Jonas, S. (1982). The thalamus and aphasia, including transcortical aphasia: A review. *Journal of Communication Disorders, 15,* 31–41.

Katayama, Y., Miyazaki, S., & Tsubokawa, N. (1981). Electrophysiological evidence favoring intracaudate axon collaterals of GABAergic caudate output neurons in the cat. *Brain Research, 216,* 180–186.

Kleist, K. (1934). *Gehirnpathologie.* Leipzig: Barth.

Krayenbühl, H., & Yasargil, M. G. (1965). *Die zerebrale Angiographie.* Stuttgart: Thieme.

Künzle, H. (1975). Bilateral projections from precentral motor cortex to the putamen and other parts of the basal ganglia. *Brain Research, 88,* 195–210.

Laplane, D., Baulac, M., Pillon, B., & Panayotopoulou-Achimestos, I. (1982). Perte de l'autoactivation psychique. Activité compulsive d'allure obsessionelle. Lésion lenticulaire bilatérale. *Revue Neurologique, 138,* 137–141.

Laplane, D., Baulac, M., Wildlocher, D., & Dubois, B. (1984). Pure psychic akinesia with bilateral lesions of basal ganglia. *Journal of Neurology, Neurosurgery and Psychiatry, 47,* 377–385.

Laplane, D., Wildlocher, D., Pillon, B., Baulac, M., & Binoux, F. (1981). Comportement compulsif d'allure obsessionelle par nécrose circonscrite bilatérale pallido-striatale. Encephalopatie par piqure de guepe. *Revue Neurologique, 137,* 269–276.

Lazorthes, G., Gouaze, A., & Salomon, C. (1976). *Vascularisation et circulation de l'encephale.* Paris: Masson.

Liepmann, H. (1908). *Drei Aufsaetze aus dem Apraxiegebiet.* Berlin: Karger.

Ludlow, C. L., Rosenberg, J., Salazar, A., Grafman, J., & Smutok, M. (1987). Site of penetrating brain lesions causing acquired stuttering. *Annals of Neurology, 22,* 60–66.

Luria, A. R. (1966). *Higher cortical functions in man.* New York: Basic Books.

Luria, A. R. (1977). On quasi-aphasic speech disturbances in lesions of deep structures of the brain. *Brain and Language, 4,* 432–459.

Marsden, C. D. (1982). The mysterious motor function of the basal ganglia. *Neurology, 32,* 514–539.

Mendez, M. F., Adams, N. L., & Skoog Lewandowski, K. (1989). Neurobehavioral changes associated with caudate lesions. *Neurology, 39,* 349–354.

Merzbach, A. (1928). Die Sprachiterationen und ihre Lokalisation bei Herderkrankungen des Gehirns. *Journal der Psychologie und Neurologie, 36,* 210–319.

Mesulam, M.-M. (1981). A cortical network for directed attention and unilateral neglect. *Annals of Neurology, 10,* 309–325.

Mesulam, M.-M. (1990). Large-scale neurocognitive networks and distributed processing for attention, language and memory. *Annals of Neurology, 28,* 597–613.

Mohr, J. P. (1983). Thalamic lesions and syndromes. In A. Kertesz (Ed.), *Localization of lesion in neuropsychology* (pp. 269–293). New York: Academic Press.

Mohr, J. P., Watters, W. C., & Duncan, G. W. (1975). Thalamic hemorrhage and aphasia. *Brain and Language, 2,* 3–17.

Mountcastle, V. B. (1957). Modality and topographic properties of single neurons of cat somatic sensory cortex. *Journal of Neurophysiology, 20,* 615–622.

Moutier, F. (1908). *L'aphasie de Broca.* Paris: Steinheil.

Parent, A. (1986). *Comparative neurobiology of the basal ganglia.* New York: Wiley.

Pedrazzi, P., Bogousslavsky, J., & Regli, F. (1990). Hématomes limités à la tete du noyau caudé. *Revue Neurologique, 146,* 726–738.

Penney, J. B., & Young, A. B. (1983). Speculations on the functional anatomy of basal ganglia disorders. *Annual Review of Neuroscience, 6,* 73–94.

Perani, D., Vallar, G., Cappa, S., Messa, C., & Fazio, F. (1987). Aphasia and neglect after subcortical stroke. *Brain, 110*, 1211–1229.

Percheron, G., Yelnik, J., & Francois, C. (1984a). A Golgi analysis of the primate globus pallidus. III. Spatial organization of the striatopallidal complex. *Journal of Comparative Neurology, 227*, 214–227.

Percheron, G., Yelnik, J., & Francois, C. (1984b). The primate striato-pallido-nigral system: An integrative system for cortical information. In J. S. McKenzie, R. E. Kemm, & L. M. Wilcock (Eds.), *The basal ganglia* (pp. 205–226). New York: Plenum.

Phillips, C. G., & Porter, R. (1977). *Corticospinal neurones.* London: Academic Press.

Phillips, C. G., Zeki, S., & Barlow, H. B. (1984). Localization of function in cerebral cortex architecture. *Brain, 107*, 327–362.

Poizner, H., Mack, L., Verfaellie, M., Gonzalez Rothi, L. J., & Heilman, K. M. (1990). Three-dimensional computer graphic analysis of apraxia. *Brain, 113*, 85–101.

Pozzilli, C., Passafiume, D., Bastianello, S., D'Antona, R., & Lenzi, G. L. (1987). Remote effects of caudate hemorrhage: A clinical and functional study. *Cortex, 23*, 341–349.

Puel, M., Demonet, J. F., Cardebat, D., Bonafe, A., Gazounaud, Y., Guiraud-Chaumeil, B., & Rascol, A. (1984). Aphasies sous-corticales. *Revue Neurologique, 140*, 695–710.

Reynolds, A. F., Turner, P. T., Harris, A. B., Ojemann, G. A., & Davis, L. E. (1979). Left thalamic hemorrhage with dysphasia: A report of five cases. *Brain and Language, 7*, 62–73.

Richfield, E. K., Twyman, R., & Berent, S. (1987). Neurological syndrome following bilateral damage to the head of the caudate nuclei. *Annals of Neurology, 22*, 768–771.

Sanguineti, I., Agostoni, E., Ajello, U., Apale, P., Bogliun, G., & Tagliabue, M. (1989). Aphasia and apraxia caused by ischemic damage to the white substance of the dominant hemisphere. *Italian Journal of Neurological Sciences, 10*, 97–100.

Schott, B., Laurent, B., Mauguière, F., & Chazot, G. (1981). Négligence motrice par hématome thalamique droit. *Revue Neurologique, 137*, 447–455.

Shallice, T. (1988). *From neuropsychology to mental structure.* Cambridge: Cambridge University Press.

Sryhöj Olsen, T., Bruhn, P., & Öberg, R. G. E. (1986). Cortical hypoperfusion as a possible cause of "subcortical aphasia." *Brain, 109*, 393–410.

Stein, R. W., Kase, C. S., Hier, D. B., Caplan, L. R., Mohr, J. P., Hemmati, M., & Henderson, K. (1984). Caudate hemorrhage. *Neurology, 34*, 1549–1554.

Sterzi, R., & Vallar, G. (1978). Frontal lobe syndrome as a disconnection syndrome: Report of a case. *Acta Neurologica, 33*, 419–425.

Tegnér, R., & Levander, M. (1991). Through a looking glass. A new technique to demonstrate directional hypokinesia in unilateral neglect. *Brain, 114*, 1943–1951.

Trillet, M., Croisile, B., Tourniaire, D., & Schott, B. (1990). Perturbations de l'activité motrice volontaire et lésions des noyaux caudés. *Revue Neurologique, 146*, 338–344.

Uno, M., Ozawa, N., & Yoshida, M. (1978). The mode of participation of pallido-thalamic transmission investigated with intracellular recording from cat thalamus. *Experimental Brain Research, 33*, 493–507.

Valenstein, E., & Heilman, K. M. (1981). Unilateral hypokinesia and motor extinction. *Neurology, 31*, 445–448.

Vallar, G., Cappa, S. F., & Wallesch, C.-W. (1992). *Neuropsychological disorders associated with subcortical lesions.* Oxford: Oxford University Press.

Vallar, G., & Perani, D. (1986). The anatomy of unilateral neglect after right hemispheric stroke lesions. A clinical/CT scan correlation study in man. *Neuropsychologia, 24*, 609–622.

Vallar, G., Rusconi, M. L., Geminiani, G., Berti, A., & Cappa, S. F. (1991). Visual and nonvisual neglect after unilateral brain lesions: Modulation by visual input. *International Journal of Neuroscience, 61*, 229–239.

Viader, F., Cambier, J., & Pariser, P. (1982). Phénomene d'extinction motrice gauche. *Revue Neurologique, 138*, 213–217.

von Monakow, C. (1914). *Die Lokalisation im Grosshirn.* Wiesbaden: Bergmann.

Walker, E. (1938). *The primate thalamus.* Chicago: Chicago University Press.

Wallesch, C. W. (1985). Two syndromes of aphasia occurring with ischemic lesions involving the left basal ganglia. *Brain and Language, 25*, 357–361.

Wallesch, C. W. (1990). Repetitive verbal behaviour: Functional and neurological considerations. *Aphasiology, 4*, 133–154.

Wallesch, C. W., Kornhuber, H. H., Brunner, R. J., Kunz, T., Hollerbach, B., & Suger, G. (1983). Lesions of the basal ganglia, thalamus and deep white matter: Differential effects upon language functions. *Brain and Language, 20*, 286–304.

Wallesch, C. W., & Papagno, C. (1988). Subcortical aphasia. In F. C. Rose, R. Whurr, & M. A. Wyke (Eds.), *Aphasia* (pp. 256–287). London: Whurr.

Watson, R. T., & Heilman, K. M. (1979). Thalamic neglect. *Archives of Neurology, 38*, 501–506.

Weiller, C., Ringelstein, B., Reiche, W., Thron, A., & Buell, O. (1990). The large striatocapsular infarct. *Archives of Neurology, 47*, 1085–1091.

Williams, A. C., Owen, C., & Heath, D. A. (1988). A compulsive movement disorder with cavitation of caudate nucleus. *Journal of Neurology, Neurosurgery and Psychiatry, 51*, 447–448.

Williams, P. M., & Warwick, R. (1975). *Functional neuroanatomy of man.* Edinburgh: Churchill Livingstone.

Yeterian, E. M. & Van Hoesen, G. W. (1978). Cortico-striate projections in the rhesus monkey: The organization of certain cortico-caudate connections. *Brain Research, 139*, 43–63.

Chapter 19

Frontal Lesions and Function

Andrew Kertesz

I. INTRODUCTION

The evolutionary increase in size of frontal lobes in the human has suggested to anatomists and anthropologists, including Gall, Bouillaud, and Broca in the 19th century, that language made frontal lobe function and size different. The famous case of Phineas Gage, described by Harlow (1848, 1868), influenced thinking about the frontal lobes considerably. These reports chronicle the changes in the personality of a previously conscientious, industrious foreman, who walked away from an injury after a tamping iron penetrated his skull through the maxilla with an exit wound in the medial dorsal frontal region. Subsequently, Gage was noted to be irresponsible, inattentive, superficial, and unbound by social rules. Additional case descriptions, such as one by Welt (1888), confirmed the personality changes that result from frontal lobe pathology. Her patient became inappropriately jocular, aggressive, and malicious after a penetrating injury to the orbitofrontal regions (bilateral rectus gyrus and right mesial inferior frontal gyrus lesion on autopsy).

Electrical stimulation and ablation of the frontal lobes in animals did not produce dramatic changes like those of the motor and visual areas in the studies of 19th century physiologists. Bianchi (1895), however, emphasized the personality changes in animals after bilateral ablations. The subjects became impulsive and hyperactive and lost affection and sociable behavior. His concept of frontal lobe function, "serializing and synthesizing groups of representations," was the forerunner of many modern formulations. Fulton and Jacobsen (1935) described a delayed response task in which animals were allowed to see the placement of the bait, but a delay and a visual distraction occurred before they were allowed to respond. This response was impaired characteristically in frontal lobe damaged animals, and was interpreted variously as an impairment of short-term memory, a difficulty in carrying a representation, or distractibility. These deficits also were found in humans (Freedman & Oscar-Berman,

Localization and Neuroimaging
in Neuropsychology

1986; Hebb & Penfield, 1940). Jacobsen (1936) also described a chimpanzee who developed "experimental neurosis," but after a bilateral frontal lobotomy the animal became placid and did not fly into a rage after failing an experiment. Shortly afterward, Moniz (1936) began to use the lobotomy to alleviate severe anxiety and psychosis in patients, and received the Nobel Prize for this.

Frontal lobe injuries produced variable personality disorders: some patients were depressed, apathetic, and lacked concern whereas others were childish, and still others were psychopathic, antisocial "hysteroids" (Feuchtwanger, 1923). In a few patients perseveration, "stickiness" and inability to handle more than one environmental event at the same time were prominent features. Behaviorally, frontal lobe patients have been divided into two types. The apathetic type is slow and lethargic, and lacks initiative and spontaneity (abulia). This syndrome is characteristic of massive frontal lobe damage, at times related to specific mesofrontal lobe involvement and at other times to dorsolateral convexity damage (Kleist, 1934; Luria, 1973). The other contrasting type of impairment is characterized by restlessness, impulsivity, hyperkinesis, and even explosiveness. Patients with orbitofrontal lesions seem to fit this description. Rarely do these types occur in pure form. Description of a frontal injury usually includes a mixture of symptoms that could be the result of impairment at one or the other location.

Fairly large lesions of the frontal lobes, however, often result in little, if any, deficit that can be documented easily with objective and specific neuropsychological tests. Changes in behavior, such as inattention, apathy, perseveration, inflexibility, lack of judgment, insight, and irritability, may be difficult to measure. Stable vascular lesions are relatively infrequent outside the middle cerebral artery territory, and many of the clinical studies were done on cortical resections of epileptic brains, trauma or tumor patients, or leukotomies for psychiatric illness. Frontal lobe tumors are relatively silent until they become massive or bilaterally situated. Closed head injury often produces significant frontal and temporal damage, but with rather diffuse pathology. Frontal lobe resections for epilepsy are easier to demarcate and define, but the psychological functions often are altered by premorbid pathology. These patients usually have dorsolateral damage. The second group after neurosurgery are leukotomized patients who have pathology in the orbital medial white matter (Benson, Stuss, Naeser, Weir, Kaplan, & Levine, 1981). These lesions were bilateral. Often significant premorbid symptomatology had to be considered in evaluating the residual symptoms.

The analysis of behavior in frontal lobe damage is complicated by the variability of the size of the lesion, the type of pathology, the extent of subcortical versus cortical damage, the region of the cortex affected, the time course of the disease, and the extent of the disconnection from other cortical areas (see case descriptions that follow). Therefore, the description and definition of frontal lobe syndromes vary greatly according to these factors. No such entity as a single frontal lobe syndrome or frontal lobe disease exists, although this terminology is used frequently. Clinicians may even say "the patient looks frontal," which may mean any of a multitude of variable symptoms. Lumping together all frontal lesions produces almost as much error as considering only hemispheric differences in any particular function and ignoring lesion size, location, and etiology.

II. ANATOMY AND PHYSIOLOGY

The frontal lobes constitute about one-third of the surface of each hemisphere. In front of the rolandic fissure, the motor cortex (Brodmann area 4) has the most distinct cytoarchitectonics. In front of this primary motor area is the premotor area or motor association area that comprises Brodmann areas 6, 8, 44, and 45, and usually includes the cortical eye fields and the supplementary motor area on the medial surface of the hemisphere. The remaining frontal cortex anterior to the premotor cortex is called the prefrontal cortex. Many of the behavior changes considered characteristic of frontal lobe damage are related to lesions in this area (Brodmann areas 46, 9, and 10—dorsofrontal; 11, 12, 25, 32, and 47—ventral or orbitofrontal). The medial surface of the frontal lobe is bordered by the corpus callosum and consists of the cingulate gyrus, which is considered to be part of the limbic system (Brodmann areas 24 and 32, and the medial aspects of areas 6, 8, 9, and 10).

The extensive connections of this complex association cortex have been reviewed by Goldman-Rakic (1987) and the functional studies in humans, by Stuss and Benson (1984). Nauta (1971) suggested that the prefrontal cortex has a unique relationship with the cortical areas of interoceptive and exteroceptive sensory domains and plays the role of synthesizing inner and outer sensory worlds. The prefrontal cortex appears to be the ultimate target of sensory cascades, as shown by the cortico-cortical connections discovered using the Nauta method (Jones & Powell, 1970; Pandya & Kuypers, 1979). Abundant input from the sensory cortex arrives at this area, directly and through the interlaminar nucleus of the thalamus, as well as through the hypothalamus, hippocampus, medulla septum, and midbrain tegmentum. The striatum has the strongest connections with the premotor cortex and the limbic system with the medial and orbitofrontal cortex. The bidirectional association with limbic and reticular activity and structures implies a major role in arousal, motivation, and affect. Various synthetic functions in experiments, such as "spatial memory," "response inhibition," "short-term memory," "polymodal integration," "planning," and "the temporal structuring of behavior," were allocated to various regions of the prefrontal cortex (Fuster, 1989; Jacobsen, 1936; Mishkin, 1964). Goldman-Rakic (1987) argued that the prefrontal cortex is necessary for regulating behavior that is guided by representations of internalized models of reality, but not for behavior directly guided by external stimuli.

Major subcortical input to the prefrontal cortex is from the dorsal medial nucleus of the thalamus. The lateral parvocellular portion projects to the upper frontal pole, and the medial magnocellular portion projects to the orbitofrontal cortex (Nauta, 1971; Walker, 1935). The dorsolateral frontal cortex has reciprocal connections to other association areas and diencephalic regions, with efferents to brainstem monoamine sites (Goldman-Rakic, 1987). Orbitofrontal cortex has more interconnection with limbic sites (Mishkin, 1964); basal forebrain, the medial septal area, the diagonal band of Broca, and the nucleus basalis of Meynert send cholinergic efferents (Mesulam & Mufson, 1984). Impaired delayed response (DR) that has memory and spatial components has been associated with lesions on the lateral convexity (Bauer & Fuster, 1976; Butters & Pandya, 1969; Jacobsen, 1936). The periarcuate cortex has been associated with compound discrimination and association (Goldman & Rosvold, 1970; Petrides & Iversen, 1978). Orbitofrontal damage produces reduced aggression and increased aversive reactions (Bowden, Goldman, & Rosvold, 1971), as well as deficits of response inhibition as evidenced from the delayed alteration (DA) task (Fuster, 1989; Warren & Akert, 1964).

Double anterograde tracing studies, in which one tracer was injected into the caudal principal sulcus of a monkey and the second tracer into the posterior parietal cortex, revealed a distributed network of spatial cognition (Goldman-Rakic, 1988). Investigators postulated that this area in the prefrontal cortex (caudal principal sulcus) is crucial for the working memory of visuospatial cues and behavioral performance when it is guided by remembered spatial locations rather than by external cues. The dorsal prefrontal cortex is connected to the neostriatum, substantia nigra, and medial dorsal nucleus of the thalamus, which is considered a positive feedback loop to trigger motor actions by disinhibition of the thalamocortical neurons. The functional aspects of this network were confirmed by 2-deoxyglucose metabolic studies (Friedman, Janas, & Goldman-Rakic, 1990). The metabolic activity in prefrontal cortex that is activated for working memory shows alternating patches of high and low metabolic activity. Also, some cytoarchitectonic similarities exist between some areas of the prefrontal cortex and the parahippocampal cortex to support a functional relationship.

The distinction between the convexity, orbitofrontal cortex, and mesial prefrontal cortex as functional units continues to be maintained, mostly based on animal experiments. The human equivalents of this distinction are less clear, as discussed in the following section.

III. COGNITIVE DOMAINS AND LOCALIZATION

A. Intelligence

Formal intelligence tests in patients with frontal lesions did not yield evidence of intellectual deficit (Feuchtwanger, 1923), even in substantial frontal lobectomies as in the cases of Hebb (1939). Subsequent series by Teuber (1964) and Milner (1964), and another one in Vietnam veterans (Black, 1976), confirmed the relative preservation of the abilities measured by general tests of intelligence. Postleukotomy patients showed higher IQs with larger lesions (Fig. 1). Presumably, this effect was related to the recovery from their schizophrenic symptoms after surgery (Stuss & Benson, 1983). Apparently one can lose a large part of the frontal lobe and still have a normal IQ. Nevertheless, certain deficits of intelligence were described with dorsolateral lesions (Malmo, 1948; McFie & Thompson, 1972); subtests of the Wechsler Adult Intelligence Scale–Revised (WAIS-R)— Digit Span, Picture Arrangement, and Block Design—were found to be impaired in frontal lesions in more recent studies (Janowsky, Shimamura, Kritchevsky, & Squire, 1989).

The contradiction in the concept that the frontal lobes control the highest forms of behavior in humans and the finding that intelligence test performance is not affected after frontal damage is a result of what is being measured. Changes in affect, emotional response, personality, and the control of social behaviors are not quantified by tests of instrumental cognition. Similarly, high level problem solving and the ability to plan ahead differ from memory-based responses.

B. Cognitive Flexibility and Perseveration

In contrast to relatively preserved cognitive abilities of visuospatial and verbal function, certain areas of cognition typically have been shown to be disturbed with frontal lesions: these are supramodular functions of cognition, such as abstraction, concept formation, and category shifting. The Wisconsin Card Sorting Test (Grant & Berg, 1948) is a test of cognitive

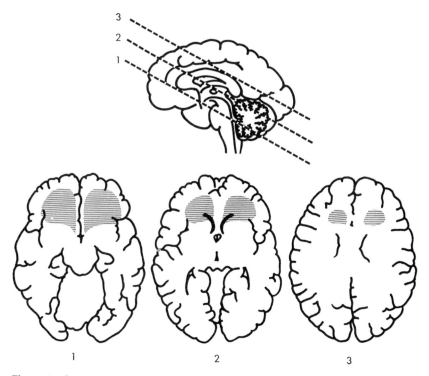

Figure 1 Composite diagram of frontal leukotomies in schizophrenics. Reproduced with permission from Stuss & Benson (1983).

flexibility in which the subjects must discover themselves the shifting criteria of sorting by number, color, or form on the basis of the limited feedback of right or wrong for each choice. Milner (1963) found that patients with left dorsolateral frontal lesions performed worse on this task than patients with resection elsewhere. Some patients were observed to verbalize sorting strategies, although they did not shift. Perseveration is a feature that occurs with both right and left frontal lobe lesions. Leukotomy patients also had impaired performance on the card sorting task (Stuss, Kaplan, Benson, Wier, Naeser, & Levine, 1981). Drewe (1984) suggested that the critical area may be the medial and not the dorsolateral convexity.

Several investigators (Cohen, 1959; Teuber, 1964; Yacorzynski & Davis, 1945) have found that visual perceptual shifting is also a problem for patients with frontal lobe lesions, because these patients had greater difficulty than those with more posterior lesions in shifting perspective on stimuli such as the Necker cube. Similarly, Ricci and Blundo (1990) showed that frontal lobe lesions cause a deficit in the ability to perceive more than one image in a reversible figure (e.g., Rubin's vase). More recently, Meenan and Miller (1992) reported that the right frontal lobe is more important than the left for this type of cognitive shifting. They tested patients who had undergone focal cortical resections for the relief of epilepsy and found that only removals from the right frontal lobe interfered with the ability to see a second image in a reversible figure; patients with left frontal, right temporal, or left temporal lobe removals performed as well as normal control subjects. Miller (1990) found a similar deficit in the ability to shift strategy on a visuospatial problem-solving task that is modeled on Guilford's (1967) Matchstick test. Again, only patients with right-sided frontal lesions showed a selective deficit in ability to switch strategies.

The Stroop test (1935) measures the ability to inhibit interference from a conflict of perceptual categories. Perret (1974) originally reported left hemisphere damage, especially left frontal involvement, to result in the greatest impairment. Leukotomy patients had little difficulty with the Stroop test (Stuss et al., 1981). Conditional associative learning tasks also were found to be specifically impaired after frontal lobe lesions in humans and in animals (Petrides, 1985). Retrieval from the set of alternative responses in the presence of various cues seems to be impaired, although discrimination between the cues is intact.

One of the most common symptoms of frontal lobe pathology is perseveration, which is defined as any situation in which elements of previous tasks or behaviors fuse or interfere with subsequent behavior, or when behavior cannot be terminated (Luria, 1966). Recurrent perseveration seems to typify the impairment caused by frontal lobe damage (Vilkki, 1989). Continuous perseveration, which is described as an inappropriate prolongation or repetition of a behavior without interruption, has been found to be most common in patients with basal ganglia damage (Sandson & Albert, 1984). Luria's "go–no go" test is designed to test perseveration and response inhibition. Drewe (1975) found that mesial frontal lesions impair the ability to monitor "go–no go" responding. Nevertheless, the specificity of frontal lobe for perseveration and for executive functions has been doubted on the basis of patients with similar symptoms who have lesions in other areas of the brain (Teuber, 1972).

C. Planning, Problem Solving, and Executive Function

The executive aspects of cognitive control such as planning, sequential organization, and selectivity are considered prime frontal lobe functions. The difficulty in planning and organizing activities is tested by carrying out maze-learning tests or shopping lists and itineraries to be carried out in an optimal sequence. The major factors resulting in poor maze-learning are related to impulsivity, perseveration, and a failure to respond to feedback (Canavan, 1983; Milner, 1964,1965). Patients with right frontal excisions showed impairment mainly on nonverbal tasks (Petrides & Milner, 1982).

Luria's (1966) theory of frontal lobes regarded their function as a system for the programming, regulation, and verification of activity. A similar model developed by Norman and Shallice (1986) is expressed in information processing terms. This formulation suggests that certain action schemata are triggered through the perceptual system and the selection of these schemata for action is regulated by a lateral inhibitory mechanism, which is called contention scheduling, and also by a supervisory system, which modulates the activation level of the schemata. Researchers proposed that damage to the supervisory system would explain some of the disorders seen after prefrontal lesions, such as difficulty with attention, inability to concentrate, and responding to irrelevant stimuli and environment. According to this model, the supervisory system would be required in situations that would involve planning or decision making, error correction, trouble shooting, novel sequences of actions, technically difficult or dangerous situations, or actions that required the overcoming of a strong habitual response. Measuring this function has turned out to be difficult. Shallice (1982) gave a problem-solving task (Tower of London), consisting of switching colored rings among pegs to achieve a goal pattern, and found that patients with anterior lesions were affected more significantly. (However, one of their unpublished studies with more selected patients with normal IQ did not show any difference between anterior and posterior patients.) Similar tests of problem solving include scheduling a set of simple shopping activities in real time. Longer duration, well-learned, multistage activities such as going to a restaurant are processed by memory organization packages (MOP) (Schank, 1982). According to Shallice and Burgess (1991), in some patients further fractionation of the supervisory system is possible: (1) plan formulation and modification, (2) marker creation or triggering, (3) evaluation and goal articulation, and (4) MOP organization. Marker creation is a process related to the maintenance of intentions.

Impairment of cognitive judgment with frontal lobe lesions was quantitated by Shallice and Evans (1978). Patients with frontal lobe lesions on either side had bizarre estimations of amounts, distances, and durations. Smith and Milner (1984) asked patients to estimate the prices of various items represented by toys. Right frontal lobectomy patients performed worse than other brain-damaged groups. Shallice attributed these difficulties with novel stimuli as support for the existence of a "supervisory system." Impulsive behavior seems to interfere with cognition in frontal damage. Studies by Miller and Milner (1985; Miller, 1985, 1992) illustrated the impulse control problems of patients who have undergone frontal lobectomy. These deficits were most clearly evident when the patient was responding manually rather than orally.

D. Attention

Attention is considered to be impaired characteristically in frontal lobe patients. Distractibility and inattention are features of brain damage in general, but are particularly prominent in frontal lobe disease. External events often trigger a lapse of attention. Much of the evidence concerning lesions and attention derives from animal literature, but patients with brain tumor and trauma (Goldstein, 1944; Hecaen, 1964) and frontal lobotomy (Rylander, 1939) also demonstrate these deficiencies.

Attention often is categorized as divided, focused, or sustained. Divided attention refers to the ability to process multiple sources of information simultaneously. Focused attention is defined as the ability to inhibit irrelevant information. Sustained attention refers to the ability to maintain attention over time. The paced auditory serial addition task (PASAT) is an example of a rather demanding sustained attention task (Gronwall, 1977). Other clinical tests of attention include the mental control tests of the WAIS; serial seven subtraction; the digit span, forward and backward; the Knox cube imitation test; and the attention–concentration index of the Wechsler Memory Scale–Revised (WMS-R). These tests have no apparent external interference factors. The Trail-Making test and the Stroop test, on the other hand, have specifically designed interference components.

Attention deficit often is examined by a signal detection task in which patients have particular difficulty in detecting a unique stimulus interspersed among repeated presentations of other stimuli (Salmaso & Denes, 1982). The presentation of an attention-generating stimulus results in a variety of physiological changes such as increased galvanic response or changes in the EEG with suppression of the alpha rhythm. After repeated stimulus presentation, habituation occurs and these physiological changes disappear. Patients with cerebral lesions habituate more easily, regardless of the location of the lesion, and their orienting reflex to novel stimuli is diminished. Luria and Homskaya (1964) found that frontal lobe damaged patients could not maintain the orienting reflex on verbal commands, in contrast to patients with more posterior lesions. Attention deficit, at times, also is manifested by inconsistent performance on short-term or immediate memory tests, as assessed by the digit span (Black & Strub, 1978). Studies of neglect also suggested that the right hemisphere frontal regions impair attention (Kertesz & Dobrowolski, 1981).

The electrophysiological equivalent of attention can be measured by event-related potentials (Woods & Knight, 1986). Distractibility is related to the inability to inhibit responding to irrelevant events in the internal and external environment. The P30 component of the auditory or somato-sensory evoked response is disinhibited by prefrontal damage (Knight, Scabini, & Woods, 1989). Focused sustained auditory attention generates negative evoked responses at 50 msec, often lasting 500 msec; these responses have been called "processing negativity." Right frontal damage produces absent attention negativity to contralateral ear stimuli with a concomitant decrease in the selective attention detection capacity in that ear (Knight, Hillyard, Woods, & Neville, 1981). P300 potentials are generated by phasic attention capacity. Involuntary phasic attention appears to be impaired particularly from prefrontal lesions (Knight, 1990).

E. Memory

Generally, memory impairments are not salient features in patients with frontal lobe lesions; recent and remote memory and learning have been demonstrated to be intact after frontal lobe lesions (Hecaen & Albert, 1978; Janowsky et al., 1989). On the other hand, short-term memory or working memory that is necessary to integrate perceptual input and organize motor and cognitive sequences is considered to be affected characteristically (Baddeley & Wilson, 1988; Goldman-Rakic, 1987).

One of the theories of frontal lobe function is the integration of behavior with information from the recent past and in the near future (Fuster, 1985). The feedback circuit of perception and action has a temporal element, in all sequential action that is characterized by deliberation and choice. Short-term memory and preparatory motor set play essential roles in this integrative function and are presumed to be represented in the dorsolateral prefrontal cortex on the basis of DR experiments mentioned earlier. The application of these tasks to several human populations revealed impairment in both DR and DA paradigms in Alzheimer's and frontal lesioned patients. DA tasks only were affected in Huntington's and Korsakoff's patients (Freedman & Oscar-Berman, 1986), representing mainly ventromedial lesions or presumable cholinergic deficiency. Parkinson's patients with dementia only had impairment in DR tasks, presumably representing dopaminergic deficiency. Visuospatial and attentional factors appear to be critical for DR performance, and short-term memory and response inhibition are needed for DA performance (Freedman & Oscar-Berman, 1986).

Patients with frontal lobe lesions do not have amnesia, but have poor performance on the attention–concentration index of the WMS-R (short-term memory). These patients also are impaired significantly in recalling words in the Rey Auditory Verbal Learning Task (Janowsky et al., 1989). However, their recognition memory appears to be intact, suggesting that some encoding has taken place but a problem with retrieval may exist. Conditional associative learning tasks also were found to be impaired specifically after frontal lobectomy in humans and in animals (Petrides, 1985). Retrieval from the set of alternative responses in the presence of various cues seems to be impaired, although discrimination between the cues is intact. Petrides and Milner (1982) and Wiegersma, Vander Scheer, and Human (1990) also showed that patients with frontal lesions perform poorly on self-ordered tasks that require them to organize and retain their own responses. Leonard and Milner (1991) showed that retention of kinesthetic–location information depends on frontal lobe mechanisms, particularly in the right hemisphere.

Frontal lobe impaired patients have particular difficulty in making judgment about the recency of information. The significance of this test has been the subject of debate, but it is considered to indicate subtle short-term memory impairment. Verbal recency judgments were more impaired by left frontal lesions and nonverbal judgments by right frontal lesions (Ladavas, Umilta, & Provinciali, 1979; Milner, 1971). Frontal lobe damage also results in problems in sequencing recalled words or objects (impaired memory for temporal order) (Shimamura, Janowsky, & Squire, 1990). Impaired memory for temporal order occurred even when the memory for items was good. Memory for the source of information also was impaired (Janowsky et al., 1989). Frontal lobe damage also interfered with meta-memory, the feeling of knowing something. Patients were asked to say whether they could or could not answer multiple choice questions about an item they could not recall (Janowsky et al., 1989). Prospective memory is a term used to describe the deficits of planning, monitoring, organization, and initiation. A deficit in prospective memory is seen in frontal lobe patients with no impairment in declarative memory or learning. This concept overlaps with terms such as dysexecutive syndrome (Baddeley & Wilson, 1988) and disinhibition.

Frontal lobe disease also impairs memory functioning through enhanced proactive interference. Although no deficit is seen on commonly used memory tests such as the WMS and the Recurring Figures test, interference from previous trials was noted in patients with dorsolateral resections (Milner, 1964) and in leukotomy patients (Stuss, Kaplan, & Benson, 1982). This phenomenon seems to be the effect of inflexible behavior and perseveration of "central sets." Frontal lobe patients also fail to show normal release from proactive interference (Freedman & Cermak, 1986; Moscovitch, 1982). However, Goldstein, Levin, Boake, and Lohrey (1989) could not confirm this phenomenon. Frontal lobe memory disorders differ from true amnesia that typically is seen with bilateral temporal lesions. The deficit has been described as "forgetting to remember" (Hecaen & Albert, 1978). Inability to follow external cues and rules, inattention, difficulty in sequencing, and susceptibility to interference and distraction are the mechanisms that are presumed to be involved.

F. Language

Two areas of the frontal cortex affect language behavior. The left inferior posterior frontal lobe has an important role in articulation; the medial supplementary motor area on the left functions in the initiation of speech. The language syndrome that is associated most commonly with left frontal lobe impairment is, of course, Broca's aphasia, but the lesion in persistent Broca's aphasia often involves the central inferior rolandic cortex and even the inferior parietal regions cortically, as well as a substantial part of the insula and subcortical region (Kertesz, Harlock, & Coates, 1979; Mohr, Pessin, Finkelstein, Funkenstein, Duncan, & Davis, 1978). Nevertheless, some patients have transient disturbance of articulation and speech output with preserved comprehension and only Broca's area is involved at the foot of the premotor cortex of areas 44 and 45. However, inferior rolandic and subcortical regions will compensate, so recovery tends to be rapid if the cortical deficit is restricted in Broca's area. These patients are often left with significantly decreased word fluency and some verbal apraxia (Kertesz, 1988).

The second type of language deficit, which is seen with lesions of the supplementary motor area or the premotor cortex in the medial aspect of the left hemisphere, is called transcortical motor aphasia (Arseni & Botez, 1961; Goldstein, 1948; Kornyey, 1975; Rubens, 1975). This deficit is characterized by significantly reduced fluency with relatively normal repetition. The extent of this dissociation, particularly the extent of reduced fluency, is determined with variable degrees of rigor. Therefore, the lesion localization is divergent to some extent since variable patients tend to be included in some series. A variable naming deficit is common, and the writing output is also poor in these patients. The importance of the supplementary motor area on the left was recognized by Penfield and Roberts (1959), who renamed it the "supplementary speech area" because of the frequent speech arrest that was found during stimulation of this region. Lesions anterior and superior to Broca's area also were described in transcortical motor aphasia (Freedman, Alexander, & Naeser, 1984). Investigators generally agree that the supplementary motor area and its connections to Broca's area are important in initiating speech. The lack of speech initiation often is considered part of a general hypokinetic syndrome associated with frontal lobe lesions. The term "adynamic aphasia" also has been used to describe this behavior.

Word fluency is impaired in many lesions of the brain, especially in the left hemisphere. Deficits in word fluency occur in aphasic syndromes and in dementia. When frontal lesions are examined, a more consistent impairment is seen from left side lesions (Milner, 1964). When dorsolateral resection and an orbitofrontal leukotomy groups were compared, word fluency was affected mainly by the dorsolateral resections (Stuss et al., 1981). Frontal lobe lesions impair "letter word fluency" predominantly, rather than word association in semantic categories (Milner, 1964). Left medial supplementary motor lesions particularly result in decreased word list generation (Alexander & Schmitt, 1980). Ramier and Hecaen (1970) suggested a deficit in initiation that may result after lesions of either frontal lobe. The linguistic component of the deficit, as tested by word fluency, would be left hemisphere sensitive.

The right hemisphere corollary of the word fluency paradigm is design fluency for the production of drawings (Jones-Gotman & Milner, 1977). However, the specificity of design fluency for frontal lobe damage has been questioned as has that of word fluency. Design fluency deficit can be seen after lesions in several cortical areas other than the frontal lobe. Right frontal convexity lesion also may disturb emotional prosody and contribute to relative flatness and inefficiency of communication of patients with right frontal lesions.

Difficulty with verbal abstraction and abstract reasoning in a verbal mode has been studied with the Proverb Interpretation Test, especially with bilateral frontal lobe disturbance (Benton, 1968). Luria (1966) emphasized the uncoupling of verbal and motor behaviors, although some of the experimental studies subsequently investigating Luria's theory showed this hypothesis to be only partially valid (Drewe, 1975). Patients with frontal lobe damage may give an appropriate verbal response to a question of judgment, yet in actual behavior they are unable to carry out actions according to the same principle. Some of these patients may verbalize the consequences of actions, yet act as though they were not aware of them. Patients with leukotomies on the other hand do not show any action–verbalization dissociation in tests in which the cue stimulus conflicted with verbal instruction (Benson & Stuss, 1982).

G. Motor Behavior

Prefrontal deficits may consist of difficulty in initiating movements, completing motor sequences, or inhibiting inappropriate responses (Luria, 1966). Premotor lesions resulted in deficits of dexterity and reproducing gestures, and deeper subcortical lesions resulted in perseveration. Liepmann (1905) suggested three types of perseveration: intentional (similar to utilization behavior), clonic, and tonic (similar to grasp reflex). Dexterity and finger tapping speed are altered by precentral involvement. Buccolingual and facial apraxia often are seen after left premotor cortex involvement (Geschwind, 1965). Sympathetic or ideomotor apraxia also was considered to be related to pathology at the origin of callosal motor connections, but subcortical and posterior cortical lesions also can produce this disorder (Kertesz & Ferro, 1984). Kinetic apraxia of the magnetic type and grasp reflex are considered to be related to medial frontal cortex pathology (Denny-Brown, 1958). Denny-Brown conceptualized frontal lobe function as avoidance behavior, in contrast to the approach behavior subserved by the parietal lobes. A grasp reflex, for instance, would represent a loss of avoidance behavior and excessive approach behavior without the inhibitory effect of the frontal lobes. Gegenhalten or "holding against" represents a semivoluntary increase in resistance to passive movements or a failure of inhibiting resistance. This phenomenon is described often, although it is difficult to quantitate or even differentiate from rigidity, as a clinical characteristic of diffuse from frontal dysfunction. The so-called "frontal motor tasks" such as motor sequences, response alternation, sequential production of the letters "m" and "n," and conflicting motor tasks are impaired characteristically by dorsolateral lesions with negative results from leukotomy (Benson & Stuss, 1982; Luria, 1966).

The difficulty initiating action with the arm, although strength and coordination may be intact, is called akinesia. Akinesia or hypokinesia may be present in the limbs and the eyes, and may be unilateral (Bianchi, 1895; Heilman, Bowers, Coslett, Whelan, & Watson, 1985). Hypometria and motor neglect also have been described. These dysfunctions may occur even in one hemispace selectively; when the same limb is tested in the other hemispace (side of the patient), the movements are normal (Meador, Watson, Bowers, & Heilman, 1986). Motor impersistence often is related to frontal lobe lesions, but has been described in association with lesions more posteriorly, especially in the right hemisphere (Kertesz, Nicholson, Cancelliere, Kassa, & Black, 1985). With damage to the supplementary motor area may come a release of automatisms and primitive synergies of the arm and hand. The alien hand syndrome can be considered a defective response inhibition (Goldberg & Bloom, 1990). Another form of impaired response inhibition is called allochiria, in which the ipsilesional limb moves instead of the contralateral paralyzed one when movement is requested or when the limb is stimulated. Motor perseveration also is seen as a result of frontal lesions and can be detected clinically on drawing and writing tasks. Verfaellie, Bowers, and Heilman (1988) demonstrated that attention could be dissociated from intention using a reaction time paradigm and showed a right hemisphere dominance for intention. Akinesia involving frontal lobe mechanisms may occur in Parkinson's disease and in frontal white matter disease, such as arteriosclerotic encephalopathy or hydrocephalus. Premotor cortex has a role in preparing saccades of eye movements; contralateral gaze palsy is common with frontal lesions. However, recovery from this deficit occurs in stages; akinesia of the eye movements, especially for saccades, persists whereas other movements, such as pursuit, remain intact. A more subtle deficit, an inability to inhibit reflexive saccades after frontal lobectomy, was found by Guitton, Buchtel, and Douglas (1985).

Utilization behavior also is related to lesions of the frontal lobes (Lhermitte, 1983). This effect consists of automatic and inappropriate utilization of objects in front of the patient. Lhermitte explained uninhibited utilization or imitation behavior on the basis of imbalance between frontal inhibitory and parietal perceptual mechanisms. This explanation has been reformulated to use the supervisory system model (Shallice, Burgess, Schon, & Baxter, 1989). This phenomenon has some similarity to the grasp reflex, forced groping (Schuster & Pineas, 1926), and the "alien hand syndrome." Impairment of initiating movement with the lower extremity results in gait apraxia or the so-called magnetic gait in which the feet appear to be sticking to the ground (Denny-Brown, 1958). This behavior often is associated with hydrocephalus, which affects the medial and basal parts of the frontal lobe.

IV. PERSONALITY CHANGES AND SOCIOPATHY

Personality disorders in frontal lobe deficits are considered characteristic. Attempts have been made to establish syndromes that correspond to localization (Blumer and Benson, 1975; Kleist, 1934). Blumer and Benson (1975) divided behavioral changes in patients with frontal lesions into two types. The pseudodepressed group is characterized by apathy, unconcern, reduced sexual interests, reduced emotion, and inability to plan ahead. These patients typically have dorsolateral symptomatology. The pseudopsychopathic group shows excessive jocularity (Witzelsucht), sexual disinhibition, increased motor activity, self-indulgence and inappropriate social behavior; these patients have orbitofrontal pathology. More commonly, however, patients with large frontal lesions will have a mixture of both personality types, resulting in paradoxical descriptions of personality disorder, that is, being apathetic, irritable, and euphoric at the same time (Geschwind, 1965). Primates (Kimble, Bagshaw, & Primbram, 1965) and patients with frontal lobe lesions have defects in autonomic orienting responses (Luria, 1973). Acquired sociopathy in a patient with orbitofrontal tumor resection was associated with reduced skin conductance in response to emotionally and socially relevant stimuli (Damasio, Tranel, & Damasio, 1990). Investigators suggested that somatic (automatic) markers of the social significance of stimuli were not available to assist in complex decision making in this patient, leading to socially inadequate behavior.

A. Frontal Lobes in Head Injury

Many frontal lobe lesions are related to head injury. In addition to the penetrating head injuries that are prominent in the literature, moderate to severe closed head injuries also cause significant damage to the frontal lobes. These lesions range from focal edema to hemorrhagic contusions that later absorb to leave hypodensities (see Fig. 3). Magnetic resonance imaging (MRI) has shown that the frontal region is the most common location of focal lesions even after mild to moderate closed head injury (Levin et al., 1987). These patients fail on a number of tasks with perseverative errors, verbal dysfluency, and interference with short-term memory resulting in the dysexecutive syndrome described by Baddeley and Wilson (1988). Positron emission tomography (PET) scanning has shown that hypometabolic areas extend beyond the boundaries of the morphological lesions detected by CT or MRI (Langfitt et al., 1986). Lateral ventricular enlargement occurs in a surprisingly large number of survivors of moderate to severe head injury (Levin, Grossman, Sarwar, & Meyers, 1981). Anosmia is a relatively frequent complication of orbitofrontal head injury. These patients tend to have psychosocial problems rather than deficits on neuropsychological testing (Martzke, Swan, & Varney, 1991). In most of these studies, neuropsychological deficits other than frontal lobe functions seem to be affected as well. This result, of course, reflects the frequency of temporal lobe involvement in closed head injuries, as well as the relative lack of specificity of the tests that are used to characterize frontal lobe dysfunction. Many of these tasks measure more than one function, to which many anatomical structures may contribute.

B. Frontal Lobe Tumors

Many classical syndromes of frontal lobe dysfunction have been described with frontal lobe tumors or after a neurosurgical removal of the tumor and some of the surrounding tissue (lobectomy). However, the variable relationships of the tumor to the brain anatomy, the amount of distortion, the accommodation of function, the degree of infiltration, and the degree of distance (pressure) effect from the actual tumor location make this material difficult to evaluate. More recent imaging techniques such as the CT scan and MRI result in earlier diagnosis of these tumors; some of the classical descriptions of frontal lobe tumor syndromes now are seen less often. Such is the Foster Kennedy (1911) syndrome that is characteristic of an olfactory groove meningioma. This syndrome consists of anosmia and blindness as a result of optic atrophy on the side ipsilateral to the subfrontal tumor, and contralateral papilledema. Presumably the compression of the sheet of the optic nerve on the side of the tumor prevents the raised intracranial pressure to produce the papilledema. Tumors in the lateral and dorsal aspects of the frontal lobes usually are accompanied by focal epilepsy. Head turning (aversive seizures), grasping movements, and jack-knifing of the body have been described with frontal lobes seizures, in addition to focal motor and generalized seizures (Penfield & Jasper, 1954). The closer the tumors are to the precentral gyrus, the more likely they are to produce contralateral pyramidal signs. Parasagittal tumors are most likely to produce abulia, akinesia, and transcortical motor aphasia. Basal tumors, in addition to olfactory disorders, produce the autonomic disturbance and sociopathy, as well as personality changes, inappropriate jocularity, and so on (Damasio et al., 1990; Rylander, 1939; Welt, 1888). Frontal release signs such as the palmomental reflex, the pouting reflex, the grasp reflex, utilization behavior, the alien hand sign, akinesia, apathy, early disturbance of consciousness, and incontinence are often the presenting symptoms in large diffuse tumors affecting the white matters of the frontal lobes and spreading across the corpus callosum, at times assuming the appearance of a "butterfly" on autopsy or on neuroimaging.

C. Schizophrenia: A Frontal Lobe Disease?

Lately the notion that schizophrenia involves the impaired function of the frontal lobes has been revived. Cerebral blood flow studies indicate frontal hypofunction that is particularly apparent when certain cognitive tasks such as card sorting are performed (Ingvar & Franzen, 1974; Weinberger, Berman, & Zec, 1986). The frontal hypometabolism in schizophrenia is considered to be related to diminished subcortical dopaminergic neuronal activity.

D. Mood Disorders

From the beginning of the century, depressive states have been reported after strokes affecting the frontal lobes. Gainotti (1972) found that left hemisphere lesions produced depressive catastrophic reaction whereas right hemisphere lesions were associated with jocularity and euphoria. The severeity of depression in the left hemisphere correlates significantly with the proximity of lesions to the frontal lobes and is considered to be related to the interruption of the dopaminergic network between the basal ganglia and the dorsofrontal regions (Starkstein, Robinson, & Price, 1987). Secondary mania is associated with lesions of orbitofrontal and limbic lesions, almost always in the right hemisphere (Rylander, 1939; Starkstein et al., 1990). Associations between site of frontal lobe injury and mood regulation disorders also are made by Grafman, Vance, Weingartner, Salazar, and Amin (1986), based on the study of a series of veterans with penetrating brain wounds.

E. Frontal Lobe Dementia

Certain degenerative disorders such as Pick's disease at times affect the frontal lobes disproportionately. Focal frontal lobe dementias have been described without the typical Pick pathology. A group of patients with a striking change in personality with relatively minor memory changes also was characterized by neuroimaging changes such as selective reduction of HMPAO–^{99}Tc uptake in the anterior cerebral hemispheres (Neary et al., 1987). Other groups describe similar features, and autopsy showed frontotemporal atrophy without Alzheimer pathology (Brun, 1987; Gustafson, 1987). The presentation of these patients is somewhat variable; sometimes lack of initiative and concern, neglect of personal responsibilities, and impaired occupational performance are the earliest symptoms. Subsequently, rigidity and inflexibility of thinking and impaired judgment occur. Some patients may present as overactive, restless, highly distractible, and overly disinhibited. Others may appear fatuous and superficially jocular and may exhibit stereotypic behavior, such as singing or humming a favorite rhythm, producing puns, or reciting verbatim a repertoire of phrases. Some patients may develop apathy, inertia, aspontaneity, and emotional blunting; sometimes these behaviors are not clearly dissociable. Eventually a language disturbance and mutism may occur, but comprehension may be relatively preserved. A few of these cases are associated with motor neuron disease (Neary, Snowden, Mann, Northen, Goulding, & MacDermott, 1990). The pathology in a few cases that came to autopsy showed mild status spongiosis and gliosis, but the typical changes of Pick's disease, such as Pick bodies are infrequent.

V. ILLUSTRATIVE CASES

Case 1

This 55-year-old woman was seen 1 and 3 wk after her stroke. She had a language deficit that initially was characterized by decreased spontaneous speech, but good repetition. She did not initiate conversation and often did not respond when asked questions. She remained unconcerned about her problems, even denying them when asked. She was abulic and her affect was flat. Her voice was low volume and monotonous. She required a lot of probing and encouragement to provide responses. It was difficult to be certain that one had engaged her full attention in any one task.

On the Wisconsin Card Sorting Test, she was unable to establish even a single response set, but merely ordered the cards sequentially. Performance on the Stroop test was also poor. She also showed apraxia and motor impersistence, and provided bizarre impulsive answers on the environmental sounds and prosopagnosia tests. Her language deficit initially was classified as transcortical motor aphasia. Her CT scan at that time showed subcortical frontal and striatal lesions in the anterior cerebral artery distribution (Fig. 2). She refused to return for a follow-up visit.

Comment

This woman presents with a clinical picture of transcortical motor aphasia, abulia, apraxia, and motor impersistence characteristic of left mesial frontal damage. Her CT lesion is a deep white matter disconnection of the mesial frontal cortex, as well as a partial striatal stroke with damage to the anterior forceps of the corpus callosum (severe apraxia). Her cognitive rigidity, impulsiveness, and bizarre answers could be related to the striatal damage in addition to the mesial white matter destruction. Although these deep embolic infarcts are not uncommon, the variation of the anatomical damage and the complexity of the combination of clinical symptoms contributes to the difficulty of the lesion study of frontal lobe functions in humans.

Case 2

This 23-year-old factory worker sustained a closed head injury on November 18, 1989. Subsequently, he was in a coma for 2.5 weeks (Glasgow

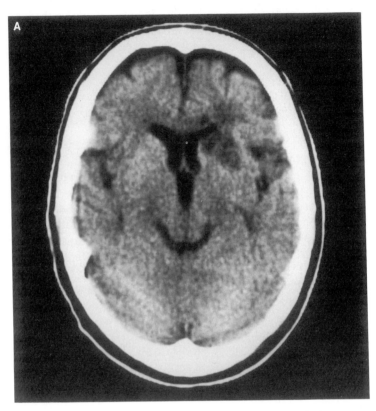

Figure 2 CT scan and MRI of Case 1 (a stroke). (A) Left frontal white matter hypodensities involving the outflow tract of the corpus callosum, anterior capsule, caudate, and head of the putamen. (B) Frontal white matter hypodensity at a higher cut on the CT. (C) Left frontal hyperintensity in the white matter on a "spin echo" MRI.

Coma Scale: 7). He regained consciousness gradually. At that time, he was reported to have amnesia, expressive and receptive aphasia, and bilateral frontal contusions (Fig. 3). While in the hospital, he exhibited temper outbursts, verbal threats, wandering behavior, and urinary incontinence. He was evaluated 14 weeks postonset. In a testing situation, he was cooperative although he required prompting to complete tasks. Intellectually his scores were in the low average range (full scale IQ 72; verbal IQ 80; performance IQ 62). His scores on the Information, Digit Span, Arithmetic, and Similarity tests were borderline. His vocabulary was within low average range and his comprehension score was impaired significantly. His

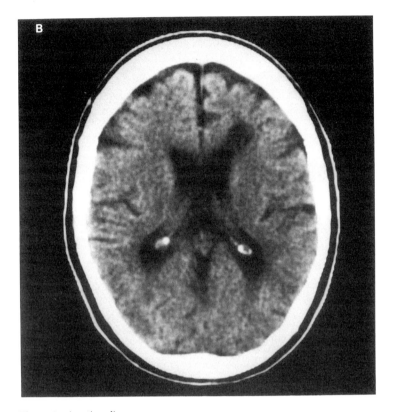

Figure 2 (*continued*)

drawings were incomplete and lacked relevant detail. Raven's matrices were low (16). Simple calculations, Block Design, and Paired Associate learning were within normal limits.

On the WMS-R, his scores all were impaired significantly, especially on visual tasks; on the other hand, verbal learning was intact. Delayed recall scores reflected a floor effect. Attention–concentration index was low. On formal language testing, he had a mild anomia (Western Aphasia Battery IQ 90.3) with excessively rambling circumlocutory speech. Semantic word fluency was low (8). On the Wisconsin Card Sorting Test, his performance was poor and perseverative. He was not able to shift from sorting by color. He could verbalize other sorting strategies but did not act on them. In addition to cognitive rigidity, he also showed inattention, social disinhibition, and inappropriate jocularity.

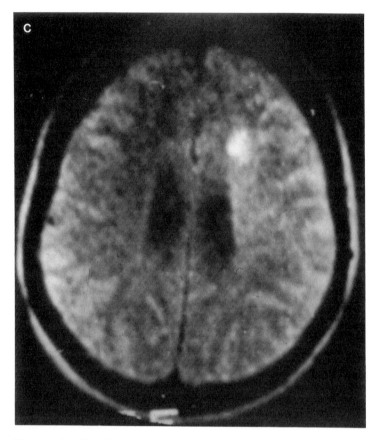

Figure 2   (*continued*)

Comment

This case illustrates the clinical picture after extensive bilateral frontal lobe damage secondary to head injury. The neuropsychological tests document severe cognitive rigidity, poor attention and concentration ("dorsolateral function"). This interferes with some aspects of his memory performance, but he had preserved verbal learning (diencephalic memory). His personality changes fit the picture of "pseudopsychopathic type," indicating orbitofrontal involvement. The CT localization (Fig. 3) suggests extensive white matter damage in both frontal lobes, probably beyond the fluid-filled destruction of tissue remaining after the hemorrhagic contusion. This lesion is evidenced further by the enlargement of

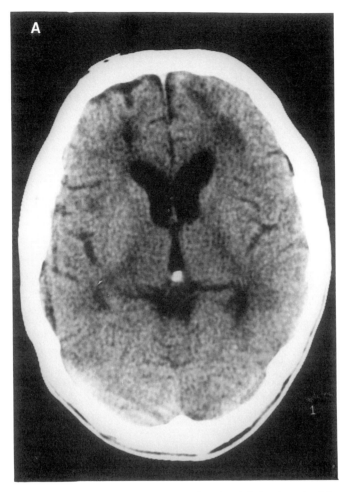

Figure 3 (A) Posttraumatic CT scan showing inferior frontal white matter hypodensities (Case 2). Lower cut through the third ventricle. (B) CT scan of Case 2. Bilateral posttraumatic frontal white matter hypodensities at a higher level. (C) CT scan of Case 2. Bilateral cortical and subcortical posttraumatic lesions in the convexity of the frontal lobes.

the lateral ventricles, especially the frontal horns. One can assume extensive disconnection of both dorsolateral and orbitofrontal premotor systems from the rest of the brain. The consequences are even more devastating than those of cortical resection for epilepsy.

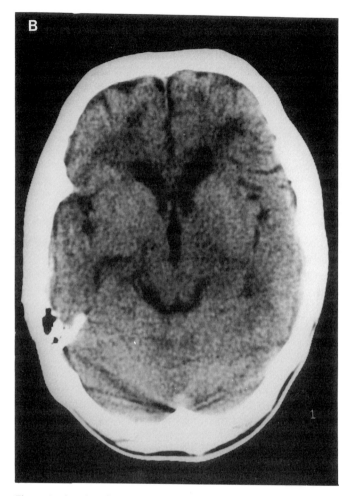

Figure 3 *(continued)*

VI. SUMMARY

The localization of lesions in the frontal lobe syndrome is complex and the clinical picture is variable. At least three location-related syndromes can be distinguished although the overlap between them is considerable and controversy exists over their independent existence or even their definition. Table 1 summarizes the major symptoms in each of the syndromes.

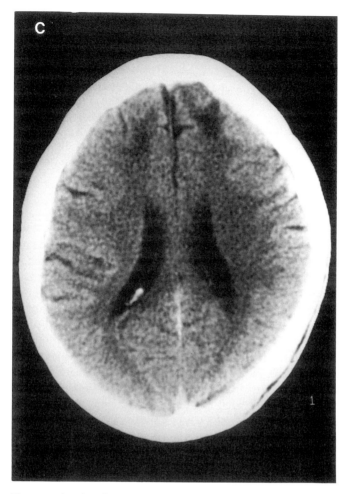

Figure 3 (*continued*)

1. Dorsolateral lateral frontal lesions produce difficulty in problem solving, planning, cognitive flexibility, short-term memory, judgment, and attention, all of which popularly are grouped under "executive function." Right-sided lesions are more likely to produce contralateral hemispatial neglect, motor impersistence, and the alien hand syndrome. Left-sided lesions may be responsible for diminished word fluency, verbal apraxia, or anomia. Advanced dorsolateral disease, especially when bilateral, produces frontal release signs (snout, grasp, palmomental reflexes), difficulty

Table 1 Frontal Symptomatology

Dorsolateral	Orbitofrontal	Dorsomedial
Perseveration	Inattention	Mutism
Rigidity	Distractibility	Apathy, slowness
Concreteness	Euphoria	Abulia, laziness
Verbal–action	Impulsiveness	Aspontaneity
Dissociation	Disinhibition	Verbal dysfluency (L)
Oral apraxia	Jocularity	Design dysfluency (R)
Broca's aphasia (L)	Irresponsibility	Transcortical motor aphasia
Impersistence (R)	Inappropriateness	Reduced emotions

with sequential motor activity, and, at times, utilization behavior. Tests of cognitive flexibility such as the Wisconsin Card Sorting Test and the Stroop test, tests of problem solving such as the Tower of London, and tests of delayed response or some aspects of memory also may be affected.

2. Dorsomedial lesions involving the supplementary motor and speech area and cingulate gyrus often result in an apathetic abulic state and, on the dominant side, produce transcortical motor aphasia. Large bilateral lesions may produce an akinetic mute state. In milder cases, motor programming deficits are seen as apractic disturbances.

3. Orbitofrontal damage results in personality changes, emotional lability, inattention, decreased impulse inhibition, poor social integration, inappropriate jocularity (Witzelsucht), bursts of anger, lack of judgment, and other asocial behaviors. Response inhibition, delayed alternating tasks, and autonomic responsiveness to emotional stimuli may be affected.

More often these syndromes appear in combination or in an incomplete form. Nevertheless, throughout the years of accumulated clinical and experimental evidence, they have become more or less established and continue to serve as guideposts.

REFERENCES

Alexander, M. P., & Schmitt, M. A. (1980). The aphasia syndrome of stroke in the left anterior cerebral artery territory. *Archives of Neurology, 37,* 97–100.

Arseni, C., & Botez, M. I. (1961). Speech disturbances caused by tumors of the supplementary motor area. *Acta Psychiatrica Scandinavica, 36,* 379–299.

Baddeley, A., & Wilson, B. (1988). Frontal amnesia and the dysexecutive syndrome. *Brain and Cognition, 7,* 212–230.

Bauer, R. H., & Fuster, J. M. (1976). Delayed-matching and delayed-response deficit from cooling dorsolateral prefrontal cortex in monkeys. *Journal of Comparative and Physiological Psychology, 90*, 293–302.

Benson, D. F., & Stuss, D. T. (1982). Motor abilities after frontal leukotomy. *Neurology, 32*, 1353–1357.

Benson, D. F., Stuss, D. T., Naeser, M. A., Weir, W. S., Kaplan, E. F., & Levine, H. L. (1981). The long term effects of prefrontal leukotomy. *Archives of Neurology, 38*, 165–169.

Benton, A. L. (1968). Differential behavioral effects on frontal lobe disease. *Neuropsychologia, 6*, 53–60.

Bianchi, L. (1895). The functions of the frontal lobes. *Brain, 18*, 397–522.

Black, F. W. (1976). Cognitive deficits in patients with unilateral war-related frontal lobe lesions. *Journal of Clinical Psychology, 32*, 366–372.

Black, F. W., & Strub, R. L. (1978). Digit repetition performance in patients with focal brain damage. *Cortex, 14*, 12–21.

Blumer, D., & Benson, D. F. (1975). Personality changes with frontal and temporal lobe lesions. In D. F. Benson & D. Blumer (Eds.), *Psychiatric aspects of neurologic disease* (pp. 151–170). New York: Grune & Stratton.

Bowden, D. M., Goldman, P. S., & Rosvold, H. E. (1971). Free behavior of rhesus monkeys following lesions of the dorsolateral and orbital prefrontal cortex in infancy. *Experimental Brain Research, 12*, 265–274.

Brun, A. (1987). Frontal lobe degeneration of non-Alzheimer type. I. Neuropathology. *Archives of Gerontology and Geriatrics, 6*, 193–208.

Butters, N., & Pandya, D. N. (1969). Retention of delayed-alternation: Effect of selective lesion of sulcus principalis. *Science, 165*, 1271–1273.

Canavan, A. G. M. (1983). Stylus-maze performance in patients with frontal lobe lesions: Effects of signal valency and relationship to verbal and spatial abilities. *Neuropsychologia, 21*, 375–382.

Cohen, L. (1959). Perception of reversible figures after brain injury. *Archives of Neurology and Psychology, 81*, 765–775.

Damasio, A. R., Tranel, D., & Damasio, H. C. (1990). Somatic markers and the guidance of behavior: Theory and preliminary testing. In H. S. Levin, H. M. Eisenberg, & A. L. Benton (Eds.), *Frontal lobe function and dysfunction* (pp. 217–229). New York: Oxford University Press.

Denny-Brown, D. (1958). The nature of apraxia. *Journal of Nervous and Mental Disorders, 126*, 9–33.

Drewe, E. A. (1975). An experimental investigation of Luria's theory on the effects of frontal lesions in man. *Neuropsychologia, 13*, 421–429.

Drewe, E. A. (1984). The effect of type and area of brain lesion in Wisconsin Card Sorting test performance. *Cortex, 10*, 159–170.

Feuchtwanger, E. (1923). *Die Funktionen des Stirnhirns: Ihre Pathologie und Psychologie.* Berlin: Springer-Verlag.

Freedman, M., Alexander, M. P., & Naeser, M. A. (1984). Anatomic basis of transcortical motor aphasia. *Neurology, 34*, 409–417.

Freedman, M., & Cermak, L. S. (1986). Semantic encoding deficits in frontal lobe disease and amnesia. *Brain and Cognition, 5*, 108–114.

Freedman, M., & Oscar-Berman, M. (1986). Comparative neuropsychology of cortical and subcortical dementia. *Canadian Journal of Neurological Sciences, 13*, 410–414.

Friedman, H. R., Janas, J., & Goldman-Rakic, P. S. (1990). Enhancement of metabolic activity in the diencephalon of monkeys performing working memory tasks: A 2-deoxyglucose study in behaving rhesus monkeys. *Journal of Cognitive Neuroscience, 2*, 18–31.

Fulton, J. F., & Jacobsen, C. F. (1935). The functions of the frontal lobes: A comparative study in monkeys, chimpanzees and man. *Advances in Modern Biology, 4*, 113–123.

Fuster, J. M. (1985). The prefrontal cortex, mediator of cross-temporal contingencies. *Human Neurobiology, 4*, 169–179.

Fuster, J. M. (1989). *The prefrontal cortex* (2d ed.). New York: Raven Press.

Gainotti, G. (1972). Emotional behavior and hemispheric side of the brain. *Cortex, 8*, 41–55.

Geschwind, N. (1965). Disconnections syndromes in animals and man. *Brain, 88*, 237–294, 585–644.

Goldberg, G., & Bloom, K. K. (1990). The alien hand sign: localization, lateralization and recovery. *American Journal of Physical Medicine and Rehabilitation, 69*, 228–238.

Goldman, P. S., & Rosvold, E. (1970). Localization of function within the dorsolateral prefrontal cortex of the rhesus monkey. *Experimental Neurology, 27*, 291–304.

Goldman-Rakic, P. S. (1987). Circuitry of the prefrontal cortex and the regulation of behavior by representational memory. In F. Plum (Ed.), *Handbook of physiology: The nervous system* (Vol. 5, pp. 373–417). Bethesda, Maryland: American Physiological Society.

Goldman-Rakic, P. S. (1988). Topography of cognition: Parallel distributed networks in primate association cortex. *Annual Review of Neuroscience, 11*, 137–156.

Goldstein, F. C., Levin, H. S., Boake, C., & Lohrey, J. H. (1989). Conceptual encoding following severe closed head injury. *Cortex, 25*, 541–554.

Goldstein, K. (1944). The mental changes due to frontal lobe damage. *Journal of Psychology, 17*, 187–208.

Goldstein, K. (1948). *Language and language disturbances.* New York: Grune & Stratton.

Grafman, J., Vance, S. C., Weingartner, H., Salazar, A. M., & Amin, D. (1986). The effects of lateralized frontal lesions on mood regulation. *Brain, 109*, 1127–1148.

Grant, A. D., & Berg, E. A. (1948). A behavioral analysis of reinforcement and ease of shifting to new responses in a Weigl-type card sorting problem. *Journal of Experimental Psychology, 38*, 404–411.

Gronwall, D. (1977). Paced auditory serial addition task: A measure of recovery from concussion. *Perceptual and Motor Skills, 44*, 367–373.

Guildford, J. P. (1967). *The nature of human intelligence.* New York: McGraw-Hill.

Guitton, D., Buchtel, H. A., & Douglas, R. M. (1985). Frontal lobe lesions in man cause difficulties in suppressing reflexive glances and in generating goal-directed saccades. *Experimental Brain Research, 58*, 455–472.

Gustafson, L. (1987). Frontal lobe degeneration of non-Alzheimer type. II. Clinical picture and differential diagnosis. *Archives of Gerontology and Geriatrics, 6*, 209–223.

Harlow, J. M. (1848). Passage of an iron bar through the head. *Boston Medical and Surgical Journal, 39*, 389–393.

Harlow, J. M. (1868). Recovery from the passage of an iron bar through the head. *Publications of the Massachusetts Medical Society, 2*, 327–347.

Hebb, D. O. (1939). Intelligence in man after large removals of cerebral tissue: Report of four left frontal lobe cases. *Journal of General Psychology, 21*, 73–87.

Hebb, D. O., & Penfield, W. (1940). Human behavior after extensive bilateral removal from the frontal lobes. *Archives of Neurology and Psychiatry, 44*, 421–438.

Hecaen, H. J. (1964). Mental symptoms associated with tumors of the frontal lobe. In J. M. Warren & K. Akert (Eds.), *The frontal granular cortex and behaviour* (pp. 335–352). New York: McGraw-Hill.

Hecaen, H. J., & Albert, M. L. (1978). *Human neuropsychology.* New York: Wiley.

Heilman, K. M., Bowers, D., Coslett, B., Whelan, H., & Watson, R. T. (1985). Directional hypokinesia: Prolonged reaction times for leftward movements in patients with right hemisphere lesions and neglect. *Neurology, 35*, 855–868.

Ingvar, D. H., & Franzen, G. (1974). Distribution of cerebral activity in chronic schizophrenia. *Lancet, 2,* 1484–1486.

Jacobsen, C. C. (1936). Studies of cerebral function in primates. *Comparative Psychology and Monographs, 13,* 1–68.

Janowsky, J. S., Shimamura, A. P., Kritchevsky, M., & Squire, L. R. (1989). Cognitive impairment following frontal lobe damage and its relevance to human amnesia. *Behavioral Neuroscience, 103,* 548–560.

Jones, E. G., & Powell, T. P. (1970). An anatomical study of converging sensory pathways within the cerebral cortex of the monkey. *Brain, 93,* 793–820.

Jones-Gotman, M., & Milner, B. (1977). Design fluency: The invention of nonsense drawings after focal cortical lesions. *Neuropsychologia, 15,* 643–652.

Kennedy, F. (1911). Retrobulbar neuritis as an exact diagnostic sign of certain tumors and abscesses in the frontal lobes. *American Journal of the Medical Sciences, 142,* 335.

Kertesz, A. (1988). What do we learn from recovery from aphasia? In S. G. Waxman (Ed.), *Advances in neurology—Functional recovery in neurological disease* (Vol. 47, pp. 175–196). New York: Raven Press.

Kertesz, A., & Dobrowolski, S. (1981). Right hemisphere deficits, lesion size and location. *Journal of Clinical Neuropsychology, 3,* 282–299.

Kertesz, A., & Ferro, J. M. (1984). Lesion size and location in ideomotor apraxia. *Brain, 107,* 921–933.

Kertesz, A., Harlock, W., & Coates, R. (1979). Computer tomographic localization, lesion size and prognosis in aphasia and nonverbal impairment. *Brain and Language, 8,* 34–50.

Kertesz, A., Nicholson, I., Cancelliere, A., Kassa, K., & Black, S. E. (1985). Motor impersistence. *Neurology, 35,* 662–666.

Kimble, D. P., Bagshaw, M. H., & Primbram, K. H. (1965). The GSR of monkeys during orienting and habituation after selective partial ablations of the cingulate and frontal cortex. *Neuropsychologia, 3,* 121–128.

Kleist, K. (1934). *Kriegsverleteungen des Gehirns.* Leipzig: Barth.

Knight, R. T. (1990). ERPs in patients with focal brain lesions. *Electroencephalography and Clinical Neurophysiology (Abs.), 75,* 72.

Knight, R. T., Hillyard, S. A., Woods, D. L., & Neville, S. J. (1981). The effects of frontal cortex lesions on event-related potentials during auditory selective attention. *Electroencephalography and Clinical Neurophysiology, 52,* 571–582.

Knight, R. T., Scabini, D., & Woods, D. L. (1989). Prefrontal cortex gating of auditory transmission in humans. *Brain Research, 504,* 338–342.

Kornyey, E. (1975). Aphasie transcorticale et echolalie: Le probleme de l'initiative de la parole. *Revue Neurologique, 131A,* 347–363.

Ladavas, E., Umilta, C., & Provinciali, L. (1979). Hemisphere-dependent performances in epileptic patients. *Epilepsia, 20,* 493–502.

Langfitt, T. W., Obrist, W. D., Alavi, Grossman, R. I., Zimmerman, R., Jaggi, J., Uzzell, B., Reivich, M., & Patton, D. R. (1986). Computerized tomography, magnetic resonance imaging, and positron emission tomography in the study of brain trauma. Preliminary observations. *Journal of Neurosurgery, 64,* 760–767.

Leonard, G., & Milner, B. (1991). Contribution of the right frontal lobe to the encoding and recall of kinesthetic distance information. *Neuropsychologia, 29,* 47–58.

Levin, H. S., Amparo, E., Eisenberg, H. M., Williams, D. S., High, W. M., Jr., McArdle, C. B., & Weiner, R. L. (1987). Magnetic resonance imaging and computerized tomography in relation to the neurobehavioral sequelae of mild and moderate head injuries. *Journal of Neurosurgery, 66,* 706–713.

Levin, H. S., Grossman, R. G., Sarwar, M., & Meyers, C. A. (1981). Linguistic recovery after closed head injury. *Brain and Language, 12*, 360–374.

Lhermitte, F. (1983). "Utilization behaviour" and its relation to lesions of the frontal lobes. *Brain, 106*, 237–255.

Liepmann, H. (1905). *Ueber Störungen des Handelns bei Gehirnkranken.* Berlin: Karger.

Luria, A. R. (1966). *Higher cortical functions in man.* London: Tavistock.

Luria, A. R. (1973). The frontal lobes and the regulation of behavior. In K. H. Pribram & A. R. Luria (Eds.), *Psychophysiology of the frontal lobes* (pp. 3–26). New York: Academic Press.

Luria, A. R., & Homskaya, E. D. (1964). Disturbance in the regulative role of speech with frontal lobe lesions. In J. M. Warren & K. Akert (Eds.), *The frontal granular cortex and behaviour* (pp. 353–371). New York: McGraw-Hill.

Malmo, R. B. (1948). Psychological aspects of frontal gyrectomy and frontal lobotomy in mental patients. *Research Publications—Association for Research in Nervous and Mental Diseases, 27*, 537–564.

Martzke, J., Swan, C. M., & Varney, N. R. (1991). Neuropsychological and neuropsychiatric characteristics of patients with post-traumatic damage of orbital frontal cortex. *Neuropsychology, 5*, 213–226.

McFie, J., & Thompson, J. A. (1972). Picture arrangement: A measure of frontal lobe function? *British Journal of Psychiatry, 121*, 547–552.

Meador, K., Watson, R. T., Bowers, D., & Heilman, K. M. (1986). Hypometria with hemispatial and limb motor neglect. *Brain, 109*, 293–305.

Meenan, J. P., & Miller, L. A. (1992). The perception of reversible figures after unilateral temporal or frontal lobectomy. Paper presented at the annual meeting of the Canadian Society for Brain Behaviour and Cognitive Science, Quebec City, Quebec.

Mesulam, M-M., & Mufson, E. J. (1984). Neural inputs into the nucleus basalis of the substantia innominata (CH4) in the Rhesus monkey. *Brain, 107*, 253–274.

Miller, L. (1985). Cognitive risk-taking after frontal or temporal lobectomy—I. The synthesis of fragmented visual information. *Neuropsychologia, 23*, 359–369.

Miller, L. (1990). Effects of cortical lesions on visuo-spatial problem-solving abilities. *International Journal of Neuroscience, 51*, 124–125.

Miller, L. (1992). Impulsivity, risk-taking, and the ability to synthesize fragmented information after frontal lobectomy. *Neuropsychologia, 30*, 69–79.

Miller, L., & Milner, B. (1985). Cognitive risk-taking after frontal or temporal lobectomy—II. The synthesis of phonemic and semantic information. *Neuropsychologia, 23*, 371–379.

Milner, B. (1963). Effects of different brain lesions on card sorting. *Archives of Neurology, 9*, 90–100.

Milner, B. (1964). Some effects of frontal lobectomy in man. In J. M. Warren & K. Akert (Eds.), *The frontal granular cortex and behaviour* (pp. 313–334). New York: McGraw-Hill.

Milner, B. (1965). Visually guided image learning in man: Effects of bilateral hippocampal, bilateral frontal, and unilateral cerebral lesions. *Neuropsychologia, 3*, 317–338.

Milner, B. (1971). Interhemispheric differences in the localization of psychological processes in man. *British Medical Bulletin, 27*, 272–277.

Mishkin, M. (1964). Perseveration of central sets after frontal lesions in monkeys. In J. M. Warren & K. Akert (Eds.), *The frontal granular cortex and behaviour* (pp. 219–241). New York: McGraw Hill.

Mohr, J. P., Pessin, M. S., Finkelstein, S., Funkenstein, H. H., Duncan, G. W., & Davis, K. R. (1978). Broca: Aphasia, pathologic and clinical. *Neurology, 28*, 311–324.

Monriz, E. (1936). Premiers essais de psycho-chirurgie. Technique et résultats. *Lisboa Medica, 12*, 152.

Moskovitch, M. (1982). Multiple dissociations of function in amnesia. In L. Cermak (Ed.), *Human memory and amnesia* (pp. 337–370). Hillsdale, New Jersey: Erlbaum.

Nauta, W. J. H. (1971). The problems of the frontal lobe: A reinterpretation. *Journal of Psychiatric Research, 8,* 167–187.

Neary, D., Snowden, J. S., Mann, D. M. A., Northen, B., Goulding, P. J., & Macdermott, N. (1990). Frontal lobe dementia and motor neurone disease. *Journal of Neurology, Neurosurgery, and Psychiatry, 53,* 23–32.

Neary, D., Snowden, J. S., Shields, R. A., Burjan, A. W. I., Northern, B., MacDermott, N., Prescott, M. C., & Testa, H. J. (1987). Single photon emission tomography using 99m Tc-AM-PM in the investigation of dementia. *Journal of Neurology, Neurosurgery and Psychiatry, 50,* 1101–1109.

Norman, D., & Shallice, T. (1986). Attention to action: Willed and automatic control of behaviour. Center for human information processing (Technical Report No. 99). (Reprinted in revised form in R. J. Davidson, G. E. Schwartz, & D. Shapiro (Eds.), *Consciousness and Self-Regulation* (Vol. 4). New York: Plenum Press.)

Pandya, D. N., & Kuypers, H. G. J. M. (1979). Cortico-cortical connections in the rhesus monkey. *Brain Research, 13,* 13–36.

Penfield, W., & Jasper, H. H. (1954). *Epilepsy and the functional anatomy of the human brain.* Boston: Little, Brown.

Penfield, W., & Roberts, L. (1959). *Speech and brain mechanisms.* Princeton, New Jersey: Princeton University Press.

Perret, E. (1974). The left frontal lobe of man and the suppression of habitual responses in verbal categorical behaviour. *Neuropsychologia, 12,* 323–330.

Petrides, M. (1985). Deficits on conditional associative-learning tasks after frontal- and temporal-lobe lesions in man. *Neuropsychologia, 23,* 601–614.

Petrides, M., & Iversen, S. (1978). The effect of selective anterior and posterior association cortex lesions in the monkey on performance of a visual-auditory compound discrimination test. *Neuropsychologia, 16,* 527–537.

Petrides, M., & Milner, B. (1982). Deficits of subject-ordered tasks after frontal and temporal-lobe lesions in man. *Neuropsychologia, 20,* 249–262.

Ramier, A. M., & Hecaen, H. (1970). Role respectif des atteintes frontales et de la lateralisation lesionnelle dans les deficits de lat "fluence verbale". *Revue Neurologique, 123,* 17–22.

Ricci, C., & Blundo, C. (1990). Perception of ambiguous figures after focal brain lesions. *Neuropsychologia, 28,* 1163–1173.

Rubens, A. B. (1975). Aphasia with infarction in the territory of the anterior cerebral artery. *Cortex, 11,* 239–250.

Rylander, G. (1939). *Personality changes after operations on the frontal lobes. A clinical study of 32 cases.* Copenhagen: Munskgaard.

Salmaso, D., & Denes, G. (1982). The frontal lobes on an attention task: A signal detection analysis. *Perceptual and Motor Skills, 45,* 1147–1152.

Sandson, J., & Albert, M. L. (1984). Varieties of perseveration. *Neuropsychologia, 22,* 715–732.

Schank, R. C. (1982). *Dynamic memory.* Cambridge: Cambridge University Press.

Schuster, P., & Pineas, H. (1926). Weitere Beobachtungen über Zwangsgreifen und Nachgreifen und deren Beziehungen zu ähnlichen Bewegungsstörungen. *Deutsche Zeitschrift der Nervenheilkunde, 94,* 29–42.

Shallice, T. (1982). Specific impairments of planning. *Philosophical Transactions of the Royal Society of London. Series B: Biological Sciences (London), 298,* 199–209.

Shallice, T., & Burgess, P. W. (1991). High-order cognitive impairments and frontal lobe lesions in man. In H. L. Levin, H. M. Eisenberg, & A. L. Benton (Eds.), *Frontal lobe function and dysfunction* (pp. 125–138). New York: Oxford University Press.

Shallice, T., Burgess, P. W., Schon, F., & Baxter, D. M. (1989). The origins of utilization behaviour. *Brain, 112,* 1587–1598.

Shallice, T., & Evans, M. E. (1978). The involvement of the frontal lobes in cognitive estimation. *Cortex, 14,* 294–303.

Shimamura, A. P., Janowsky, J. S., & Squire, L. R. (1990). Memory for the temporal order of events in patients with frontal lobe lesions and amnesic patients. *Neuropsychologia, 28,* 803–14.

Smith, M. L., & Milner, B. (1984). Differential effects of frontal-lobe lesions on cognitive estimation and spatial memory. *Neuropsychologia, 22,* 697–705.

Starkstein, S. E., Mayberg, H. S., Berthier, M. L., Fedoroff, P., Price, T. R., Dannals, R. F., Wagner, H. N., Leiguarda, R., & Robinson, R. G. (1990). Mania after brain injury: Neuroradiological and metabolic findings. *Annals of Neurology, 27,* 652–659.

Starkstein, S. E., Robinson, R. G., & Price, T. R. (1987). Comparison of cortical and subcortical lesions in the production of post-stroke mood disorders. *Brain, 110,* 1045–1059.

Stroop, J. R. (1935). Studies of interference in serial verbal reactions. *Journal of Experimental Psychology, 18,* 643–662.

Stuss, D. T., & Benson, D. F. (1983). Frontal lobe lesions and behavior. In A. Kertesz (Ed.), *Localization in neuropsychology* (pp. 429–454). New York: Academic Press.

Stuss, D. T., & Benson, D. F. (1984). Neuropsychological studies of the frontal lobes. *Psychological Bulletin, 95,* 3–28.

Stuss, D. T., Kaplan, E. F., & Benson, D. F. (1982). Long-term effects of prefrontal leucotomy: Cognitive functions. In R. N. Malatesha & L. C. Hartlage LC (Eds.), *Neuropsychology and cognition* (Vol. II, pp. 252,271). The Hague: Martinus Nijhoff.

Stuss, D. T., Kaplan, E. F., Benson, D. F., Wier, W. S., Naeser, M. A., & Levine, H. L. (1981). Long-term effects of prefrontal leucotomy—An overview of neurospychologic residuals. *Journal of Clinical Neuropsychology, 3,* 13–32.

Teuber, H. L. (1964). The riddle of frontal lobe function in man. In J. M. Warren & K. Akert (Eds.), *The frontal granular cortex and behavior* (pp. 410–444). New York: McGraw-Hill.

Teuber, H. L. (1972). Unity and diversity of frontal lobe functions. *Acta Neurobiologiae Experimentalis, 32,* 615–656.

Verfaellie, E., Bowers, D., & Heilman, K. M. (1988). Hemispheric asymmetries in mediating intention but not selection attention. *Neuropsychologia, 26,* 521–531.

Vilkki, J. (1989). Perseveration in memory for figures after frontal lobe lesions. *Neuropsychologia, 27,* 1101–1104.

Walker, A. E. (1935). The retrograde cell degeneration in the thalamus of macacus rhesus following hemidecortication. *Journal of Comparative Neurology, 62,* 407–419.

Warren, J. M., & Akert, K. (1964). *The frontal granular cortex and behavior.* New York: McGraw-Hill.

Weinberger, D. R., Berman, K. F., & Zec, R. F. (1986). Physiologic dysfunction of dorsolateral prefrontal cortex in schizophrenia. I: Regional cerebral blood flow evidence (rCBF). *Archives of General Psychiatry, 43,* 114–125.

Welt, L. (1888). Über Charakterveränderungen des Menschen infolge von Läsionen des Stirnhirns. *Deutsche Archiv für Klinische Medizin, 42,* 339–390.

Wiegersma, S., van der Scheer, E., & Human, R. (1990). Subjective ordering short-term memory and the frontal lobes. *Neuropsychologia, 28,* 95–98.

Woods, D. L., & Knight, R. T. (1986). Electrophysiological evidence of increased distractibility after dorsolateral prefrontal lesions. *Neurology, 36,* 212–216.

Yacorozynski, G. K., & Davis, L. (1945). An experimental study of the functions of the frontal lobes in man. *Psychosomatic Medicine, 7,* 97–107.

Chapter **20**

Structural Basis of Memory

Terry L. Jernigan and Laird S. Cermak

I. INTRODUCTION

In this chapter we summarize the evidence from magnetic resonance imaging (MRI) studies that indicates that specific brain structures are implicated in human memory. Past strategies of fractionating memory into independent memory systems, such as explicit and implicit memory, and concomitant localization attempts are highlighted. However, we also emphasize that contemporary memory models seem to be shifting from a multiple-systems approach to a processing approach that is orthogonal to memory systems. The implications of this theoretical shift for modern localization studies are explored and ways of mapping MRI findings onto processing accounts are suggested. We hope that this change in strategy can provide a framework for future MRI studies that will conform more closely with contemporary information processing and neural systems models.

II. CLASSICAL LOCALIZATION STUDIES

Before the advent of modern brain imaging techniques, theories about the sites of memory processing in the human brain were based on neuropathology performed at autopsy in patients who, in life, had had clinical memory impairments. This approach relied on the definition of the amnesic syndrome current at the time and, of course, on the sensitivity of the available pathological methods. Inferences regarding localization rested on the assumption that little significant damage to the brain had occurred in the interval between clinical characterization and subsequent death, fixation, and measurement. These lesion studies, coupled with several excellent reports of cases involving surgical resections, revealed most of the information that we now regard as the classical anatomy of human memory. Sir Charles Symonds (1966) and J. B. Brierley (1977) summarized many of these individual cases in which damage had occurred bilaterally to mesial temporal lobe or diencephalic structures and in which the patients had selective loss of recent and, in some cases, remote memory. Among temporal lobe structures, the hippocampus and parahippocampal gyrus were implicated most often, whereas the roles of the amygdala and the uncus were less clear because surgical lesions confined solely to these areas usually seemed to spare memory functions. Diencephalic lesions often were reported in alcoholic patients who became amnesic (Korsakoff patients) and in patients who were amnesic following infarction, tumor, or trauma. These lesions seldom were limited to a single diencephalic structure; however, midline structures such as the mammillary bodies, adjacent hypothalamic structures, the anterior nuclear group of the thalamus, and the dorsomedial thalamic nucleus frequently were affected in these postmortem cases. Victor, Adams, and Collins (1971) suggested a critical role in amnesia for dorsomedial nucleus damage; however, Spiegel, Wycis, Orchinik, and Freed (1955) reported that bilateral stereotactic lesions of this nucleus often produced only transient amnesic symptoms. Other pathological studies have confirmed a critical role for the hippocampus in amnesia (Zola-Morgan, Squire, & Amaral, 1986) and have provided additional evidence that neither mammillary body nor dorsomedial nucleus damage is required for diencephalic amnesia to occur (Mair, Warrington, & Weiskrantz, 1979; Mayes, Meudell, Mann, & Pickering, 1988).

III. MODERN LOCALIZATION STUDIES

A. Computerized Tomography

Advances in medical imaging technologies have widened substantially the scope of research correlating behavioral impairments with the site of brain damage. Although the resolution of in vivo structural imaging methods still lags far behind that of modern histological techniques, the potential for simultaneous behavioral and anatomical assessments and for repeated observations over time are important advantages of imaging techniques. We can now examine the brains of memory-impaired patients when their cognitive impairments are well defined. Most studies using these methods have, like the classical histological studies, attempted to detect lesions in patients with circumscribed amnesic syndromes. Deep vascular lesions within the diencephalon are the lesions reported most frequently in these patients, because small cystic infarcts, particularly those occurring within deep structures, are well visualized with computerized tomography (CT). Thus, this technique was the first employed to map lesions in living patients with memory disorders.

An early report by Alexander and Freedman (1984) described a group of amnesic patients who suffered anterior communicating artery (ACoA) aneurysm rupture. These patients suffered various cognitive and personality changes but their defining characteristic was a delayed recall deficit. The structures most consistently showing damage on the patients' CTs were the medial septal nuclei, the paraventricular nucleus of the anterior hypothalamus, and the medial forebrain bundle. The authors speculated that the memory deficits they had observed might be the result of damage to the cholinergic projection from septal nuclei to the hippocampal region or the result of this damage in combination with hypothalamic damage. Damasio, Graff-Radford, Eslinger, Damasio, and Kassell (1985) also used CT to study patients with amnesia as a result of ACoA aneurysms or vascular resection. Like Alexander and Freedman, these authors found damage to basal forebrain structures. They speculated that such damage produces amnesia by inducing malfunction in the mesial temporal lobe structures with which the basal forebrain has strong connections. Damasio et al. (1985) noted that their patients showed impaired association of the stimuli within an event and poor temporal encoding of that event, but seemed to encode the individual stimuli within an event relatively well, unlike patients with mesial temporal lobe damage. However, direct behavioral comparisons between their patients and patients with mesial temporal lobe damage were not performed. A more recent CT study of patients with ACoA ruptures (Irle, Wowra, Kunert, Hampl, & Kunze, 1992) emphasized that the most severe memory impairments were observed in patients who, in addition to basal forebrain lesions, had damage to the ventral striatum, notably the caudate nucleus. Behaviorally, the patients had conspicuous recall impairment but milder recognition deficits and additional attentional disturbance, so inferences about memory functions per se were difficult to make.

CT findings from a group of patients with vascular lesions in the thalamus were summarized by von Cramon, Hebel, and Schuri (1985), who found that lesions associated with memory impairment frequently involved the mammillo-thalamic tract (which connects the anterior thalamic nucleus and the hippocampus) and the lamina medullaris interna (an amygdalo-thalamic tract). These researchers speculated that this type of amnesia might represent a disconnection of the amygdala and hippocampus from their thalamic and cortical projections.

B. Magnetic Resonance Imaging

More recent studies of thalamic amnesia have exploited the increased anatomical specificity obtained with MRI. Nichelli, Bahmanian-Behbahani,

Gentilini, and Vecchi (1988) focused on the spared memory abilities of a patient with amnesia resulting from bilateral thalamic infarctions. The lesion involved the dorsomedial nucleus, the midline nuclei, and the internal medullary lamina bilaterally, and the mammillo-thalamic tract on the right side. Severe deficits of recall and recognition memory were present. However, the patient exhibited apparently normal learning on mirror tracking and mirror reading tasks, and on the learning of a mathematical rule.

Graff-Radford, Tranel, van Hoesen, and Brandt (1990) summarized the results of a group of patients with MRI-determined lesions in the medial thalamus. These researchers concluded that the memory deficits in these patients probably were attributable to the involvement of fiber tracts connecting the hippocampus and the amygdala to the thalamus. They reviewed evidence from their study and others that involvement of one of these tracts alone was insufficient to produce a dense amnesia. Graff-Radford and colleagues also performed anatomical studies in the monkey that revealed that such tracts course adjacent to each other where the inferior thalamic peduncle passes through the ventral anterior nucleus of the thalamus. Thus, a small bilateral lesion in this location would affect both tracts. These authors attempted to address dissociability of memory systems in three ways. First, they assessed remote memory and showed that some but not all of their patients had measurable impairments. From this result, they concluded that no localization inferences could be drawn regarding remote memory impairment. The researchers tested motor learning and concluded that it was intact in their patients because the patients improved on pursuit rotor and mirror tracing tasks. Finally, they suggested that diencephalic amnesia disproportionately affected temporal processing (that is, the ability to use temporal cues in recall) since providing their patients with temporal cues substantially improved their recall performance.

A similar patient, reported by Malamut, Graff-Radford, Chawluk, Grossman, and Gur (1992), showed marked impairments on a number of recall and recognition tasks (although one figural recognition task showed no impairment) but normal performance on stem completion priming. Remote memory appeared to be intact. The lesion for this patient involved both the mammillo-thalamic tract and the inferior thalamic peduncle. The authors concluded that remote memory and stem-completion priming are mediated by structures other than those damaged in their patient.

MRI also has been used to examine two cases of amnesia with lesions in the mesial temporal lobe. Conlon, Kertesz, and Mount (1988) reported a case of dense amnesia with multiple cognitive and behavioral symptoms in a patient who had suffered herpes encephalitis. Unfortunately, specific memory processes could not be studied because the patient's cerebral damage and his behavioral effects were quite extensive. Shimauchi, Wakisaka, and Kinoshita (1989) described a case with bilateral glioblastoma in the mesial temporal lobes. Their case was described as amnesic, without aphasia, but verbal intelligence was only 63. Also, although the lesion was described as limited to the hippocampus, the figure they presented suggested that the amygdala and entorhinal and parahippocampal gyri were involved also.

Although these cases seem to confirm that severe impairment on memory tests can result either from damage confined to mesial and anterior parts of the diencephalon or from damage to structures in the mesial temporal lobe, growing evidence from animal studies suggests that, for clinically significant amnesia to occur, diencephalic lesions must involve the projections to both the more posteriorly lying hippocampal and parahippocampal structures and the more anteriorly lying amydala and periamygdalar structures (Mishkin, 1982). Studies by Squire and associates (see Squire, 1992, for review) suggest a limited role in amnesia for amygdala damage, but an important role for the adjacent entorhinal cortex. Graff-Radford et al. (1990) traced the connections between the amygdala and thalamus that lie within the inferior thalamic peduncle, and suggested that involvement of this amygdalo-thalamic tract was important in producing diencephalic amnesia. Their tracer stained not only the amygdala but parts of the adjacent entorhinal cortex. Thus, to date, precisely which temporal lobe structures must be involved to produce amnesia is not entirely clear.

Questions obviously also remain regarding the equivalence of diencephalic and mesial temporal lobe amnesias. Clearly many similarities exist, but whether or not equivalent effects on the encoding, elaborative, and retrieval processes required by explicit memory tests are seen for both amnesias has yet to be determined. This uncertainty is due in part to the paucity of studies directly contrasting these two groups of patients. The little evidence that does exist suggests that skill learning and priming are unimpaired as often in diencephalic amnesics as in patients with mesial temporal lobe damage, whereas both groups have recall and recognition deficits.

C. Imaging Studies of Memory-Impaired Patients without Focal Lesions

Some patients with prominent memory impairments—such as alcoholic amnesics with Korsakoff's syndrome, patients who have suffered brain damage from anoxia or ischemia, and early Alzheimer patients—have few if any detectable lesions on CT or MRI. However, careful anatomical analysis of their MR images reveals evidence for volume loss in particular structures in these patients, loss that presumably is caused by neuronal death or reduced cell size (e.g., from loss of synapses). MRI allows investigators to measure such volume losses in different structures thought to contribute to memory processing. Several studies have taken this approach.

In one early study using MRI (Press, Amaral, & Squire, 1989), three amnesic patients were compared with four normal controls. One of the patients became amnesic after a cardiac arrest associated with a seizure; cause of the amnesia in the other two patients was unknown. The areas of the hippocampal formation, parahippocampal gyrus, temporal lobe, and third and lateral ventricles were measured on three adjacent 5-mm thick sections. Investigators found that the hippocampal formation was reduced in size significantly in the patients relative to the controls. Although some of the patients had enlargement of the ventricles and reduction in the size of the parahippocampal gyrus, group differences on these measures did not reach significance. In a later study (Squire, Amaral, & Press, 1990), which included four alcoholic Korsakoff patients and four non-Korsakoff patients, the same methods were used to examine the temporal lobes; a separate protocol was included for visualizing the mammillary bodies. In this study, the Korsakoff patients did not differ from controls in the size of their hippocampal formations; however, they did show reductions in the size of the mammillary bodies. Other diencephalic, striatal, and cortical structures were not examined specifically, but T_2-weighted images were inspected for high signal abnormalities. Only a few small abnormalities were reported, some of which were in the striatum.

Two studies also have examined the hippocampal formation in small groups of Alzheimer patients and controls. Seab et al. (1988) measured the cross-sectional areas of the hippocampus, the lenticular nucleus, the whole brain, and the cerebrospinal fluid (CSF) spaces in Alzheimer patients and controls. A measure of the hippocampal area normalized to lenticular nucleus area showed no overlap between the groups. Kesslak, Nalcioglu, and Cotman (1991) also observed a large group difference between eight Alzheimer patients and seven controls on an estimate of the volume of the hippocampus and parahippocampal gyrus. These authors speculated that the observed volume losses in these mesial temporal lobe structures may produce the prominent early memory deficits in Alzheimer's disease. Unfortunately, no information was provided about performances on different kinds of memory tasks, and only a small group of structures was examined.

Jernigan, Schafer, Butters, and Cermak (1991d) examined 8 Korsakoff patients, 12 age- and sex-matched nonamnesic alcoholic controls, and 13 similarly matched nonalcoholic controls using MRI. Volumes of multiple cortical and subcortical structures then were estimated with interactive morphometric analyses of the images. MRI was performed with a 1.5-Tesla superconducting magnet. Two spatially registered images were obtained simultaneously for each section, using an asymmetrical multiple-echo sequence (TR = 2000 msec, TE = 25, 70 msec) to obtain images of the entire brain in the axial plane. Detailed descriptions of the image-analytic methods are contained in several articles (Jernigan et al., 1991a; Jernigan, Press, & Hesselink, 1990; Jernigan et al., 1991d). Image data sets were assigned (random) numeric codes, and all analyses were conducted blind to any subject characteristics. Briefly, each pixel location within a section of the imaged brain was classified (on the basis of its signal values in the two images of that section) as most resembling CSF, gray matter, white matter, or signal hyperintensity (tissue abnormality). Then, consistently identifiable landmarks in the corpus callosum and the interhemispheric fissure were designated by trained image analysts, using a stylus-controlled cursor on the displayed "pixel-classified" images. Using these points as anchors, the processed image data then were transformed spatially so that the locations of all points within the brain images could be expressed in anatomically standardized three-dimensional coordinates. Regional boundaries separating different brain structures were determined either entirely by manual designation or, when boundaries could not be determined reliably on a visual basis, by combining visual and stereotactic criteria. Pixels designated as lying within a given structure were summed over all sections in which the structure appeared. In Color Plate 11, a representative set of processed images is displayed; the pixels within the measured structures are color coded. Ventricular and cortical sulcal CSF spaces, diencephalic and basal ganglia structures, and eight separate regions within the cerebral cortex were measured.

Both alcoholic groups showed significant increases in the fluid-filled spaces relative to controls; however, only the ventricular increase was significantly greater in Korsakoff patients than in the nonamnesic alcoholics. These CSF increases in the alcoholic groups were associated with decreased volumes of gray matter structures. The gray matter losses in this small group of nonamnesic alcoholics were slight, and only a few reached statistical significance. A later study, which examined a larger group of nonamnesic alcoholics, confirmed that these mild volume losses were reliable, that they were distributed over both cortical and subcortical structures, and that they were correlated with CSF increases (Jernigan et al.,

1991b). The gray matter losses in the eight Korsakoff patients generally were more pronounced than in the nonamnesic alcoholic controls; in three regions, these losses were significantly greater in the Korsakoff patients. These three regions were an anterior diencephalic region that included hypothalamic, septal and anterior thalamic gray matter; a mesial temporal lobe region including the amygdala, uncus, hippocampus, and parahippocampal gyrus; and a neocortical region including the orbitofrontal cortices. In all three regions, the losses in the alcoholic Korsakoff patients significantly exceeded those in both other groups. Thus, unlike Squire et al. (1990), we found evidence for damage to mesial temporal lobe structures as well as diencephalic structures in Korsakoff patients. The discrepancy may be related to the methods used. MR morphometry was based on a double-echo T_2-weighted protocol in the Jernigan et al. study and on a T_1-weighted protocol in the Squire et al. study. Although no a priori reason exists to expect a difference using T_1-weighted rather than T_2-weighted protocols, an effect of protocol differences cannot be ruled out. Also, the sample size was considerably larger in the Jernigan et al. study. Finally, the mesial temporal lobe measure used in the Jernigan et al. study included all the structures lying on the mesial surface throughout the entire extent of the temporal lobe. Thus, a small reduction distributed over several of the structures in this region may have been easier to detect. Alternatively, the Korsakoff patients' losses may have been greatest in structures within this region that were not measured by Squire et al. In any event, the results of the Jernigan et al. study suggest that diencephalic, mesial temporal lobe, and frontal cortical damage may contribute to the memory impairments of Korsakoff patients.

Another study using morphometric techniques identical to those used in our study of Korsakoff patients compared groups of patients with Alzheimer's disease or Huntington's disease with controls (Jernigan, Salmon, Butters, & Hesselink, 1991c). Differences were found in each group: Alzheimer patients had widespread cortical volume losses including highly significant losses in the volume of the mesial temporal lobe structures; Huntington patients had particularly dramatic striatal losses. However, some unexpected differences were observed also. The Huntington patients also had significant volume losses in mesial temporal lobe structures. The absolute volume losses in these structures were smaller in the younger Huntington patients than in the Alzheimer patients, but the age-adjusted losses (relative to age-matched controls) were not significantly different between the two groups. Another unexpected result was that significant volume losses in subcortical structures were found in the Alzheimer patients, including basal ganglia and diencephalic changes.

IV. SOME ISSUES RELEVANT TO THE LOCALIZATION OF MEMORY PROCESSES

Throughout the history of the study of memory and memory disorders, a push–pull effect has always existed between two major theoretical viewpoints. One of these perspectives tends to cast memory as a sort of storage system within the brain. According to this view, the brain "handles" information by placing it into one of the various proposed memories. The information resides there until some form of retrieval is desired. However, the alternative view has proposed that memory is not a storage bin as much as it is a form of "processing." According to this view, the way in which information is processed rather than the site of storage eventually allows retrieval. This concept has led to theories of encoding in which the proposed "depth" of encoding produces the probability of retrieval (Craik & Lockhart, 1972). At times in recent history the systems theories have predominated; at other times the processing theories have been at the forefront. When systems theories dominated, neurophysiological approaches prospered. When processing approaches dominated, neurophysiological correlates generally took a back seat. To understand this imbalance in perspective, a brief review of the recent history of memory theories is necessary. This discussion is followed by an explanation of how anatomical approaches may track processing theory comfortably. We propose that this will be increasingly necessary as the pendulum returns to processing approaches in the 1990s.

During the 1960s, systems theories clearly predominated. Researchers largely concluded that alcoholic Korsakoff amnesics' anterograde amnesia reflected an impaired "long-term" memory. Most investigators (Baddeley & Warrington, 1970) felt that amnesic patients had an intact "short-term" memory since they could repeat short strings of numbers and retain material for very brief time intervals. The problem for the patient was in long-term memory, for which clearly information was not available. During the 1970s, this unavailability became the major topic of research. Determining whether information loss was the result of an impaired retrieval mechanism or of an inability to process the material into a retrievable form at the encoding stage became a matter of some concern. Some investigators, such as Warrington and Weiskrantz (1970,1974) thought the problem represented a pure retrieval deficit. These researchers proposed that material had to be represented in the brain somewhere, but the patient could not seem to find it. Other workers, including Cermak (1979), believed that the disorder was produced by an encoding deficit, an initial inability to process the material in such a way that later retrieval would be possible.

Processing theory received a real impetus in 1972 with the publication of a theory by Craik and Lockhart called "Levels of Processing." This theory proposed that the depth to which information is processed initially determines the probability that an item will be recalled later. Depth was defined loosely in terms of the cognitive analysis performed on the material. However, investigators demonstrated in several different ways that, if a subject was instructed to analyze the semantic characteristics of a word, he or she was more likely to remember that word later than if asked to analyze only the phonemic characteristics. These observations spawned a decade of research that was led by processing theorists, who were eager to show that the nature of initial processing determined the life-span of memory for a piece of information. This hypothesis was in contrast with explanations based on whether the information had been encoded in short-term memory or long-term memory. Investigators showed that semantic analysis produced a longer life-span than did phonemic analysis. The relevance to amnesia was obvious. In a series of experiments based on the Craik and Lockhart thesis, we (Cermak & Reale, 1978) were able to demonstrate that alcoholic Korsakoff patients were unable to profit from deeper analysis of incoming information in the same way as normal individuals. We showed that the patients did not analyze the semantic features of verbal information automatically, and that, even when they could be induced to perform such analysis, the product of such analysis was not a lasting memory.

As we entered the decade of the 1980s, the pendulum certainly had swung from dominance of systems theories to dominance of processing theories. Even Warrington and Weiskrantz (1978) in one of their later articles together concluded that the extent to which amnesics could "cognitively manipulate" verbal information during storage and retrieval was deficient relative to normal controls. These authors still emphasized the retrieval aspects of amnesic disability, but contended that a major contributing factor to the lack of retrieval routes available to the amnesic was his or her inability to produce them at the time of input. Interestingly, a clinical curiosity that was long known to clinicians, when formalized and studied under the general rubric of "priming," led to the beginning of the swing back to systems theory. Eventually, this change in focus culminated in the introduction of two new memory systems into the literature of amnesia and of normal memory, namely "explicit memory" and "implicit memory."

One of the first instances of priming an amnesic patient's memory was reported by Gardner, Boller, Moreines, and Butters (1973) who called it the "out of the blue" phenomenon. These investigators presented a list of words to the patient and asked for recall, but then proceeded to an ostensibly quite different task. This second task required that the patient free associate to the name of a category of words (e.g., automobiles). The researchers found that the patient could be induced to respond to this category cue with an exemplar from the just-presented list at a rate much higher than that anticipated from free association norms. In other words, the patient would respond to the cue of automobile with the word "Buick," which happened to be in the previous list, as one of his first associates. Clearly, other exemplars normally would be first associates but the patient had in some manner been "primed" by the inclusion of that particular exemplar in the list, although he had not been able to recall it during testing.

Priming became more formalized in a much later study by Graf, Squire, and Mandler (1984), who utilized a procedure that became known as "word-stem completion." Basically, the examiner presented a list of words to the patient under instructions to try to remember them for later recall. Then, at the time of recall, the patient was shown a three-letter cue (the first three letters of the word) and asked to complete the word. Under one set of instructions, the patient was told to complete the word with one that had been in the list. Under a second set of instructions, he was told simply to respond with the first word that came to mind and no reference to the list was made. Graf and colleagues found that the amnesic patients performed significantly worse than the normal controls under the usual cued recall conditions, but performed normally (i.e., filled in the word stems with words that came from the just-presented list) when simply asked to put down the first word that came to mind. Thus, when the instructions for the test were phrased in terms of a recall test, the patient was impaired; however, when no reference to the to-be-retained list was made, the patients' performance was normal. This result seemed to demonstrate that some memory for the items was present.

A similar phenomenon was reported by Cermak, Talbot, Chandler, and Wolbarst (1985) using a completely different procedure. In their task, patients also were shown a list of words and then asked to try to remember them, this time using a recognition procedure. As expected, recognition was at chance level, far below the level of the normal controls. However, the patients then were asked to try to identify words presented on a computer screen at durations that were below the patients' thresholds. Duration was increased for each word until the patient could identify the word on the screen. Some of the words were words that had been presented during the just-completed memory test but, of course, they had not been recognized as such by the patients. However, these same words had an effect on performance during the perceptual identification task. The words presented previously during the memory test were identified by the amnesic patients at shorter durations than were the new words. Further, the extent of the "priming" effect for the just-presented words was as great as that seen for normal controls. Thus, the phenomenon of "priming" became established across paradigms. Other paradigms rapidly emerged in the literature, such as those influencing "spelling bias" of homonyms and reading speed of inverted text, but word-stem completion and perceptual identification were the two most frequently used paradigms.

Trying to provide a theoretical explanation for this phenomenon of "priming" in amnesia became important, so several theories emerged to fill the void. One of the most popular was a theory that presented memory as divided into the two components "procedural learning" and "declarative learning." Championed by Squire and Cohen (1984), this theory proposed that amnesics have normal procedural memory but impaired declarative memory. Procedural memory was defined as "knowing how" to do a task, whereas declarative memory was defined as "knowing that" a certain fact had been learned. Cohen and Squire (1980) proposed that amnesics can learn and retain "procedures" needed to perform a task but cannot acquire "declarative" knowledge of this act.

An alternative memory system theory that could incorporate priming phenomena was the "episodic" and "semantic" memory distinction. Initially described by Tulving (1972,1983), episodic memory was defined as the ability to retain and retrieve an item within the context in which it was presented originally, whereas semantic memory was thought to be a context-free system in which facts and knowledge were retained without reference to the context in which they were learned. It was suggested (Cermak, 1984; Kinsbourne & Wood, 1975) that amnesics have relatively normal semantic but below-normal episodic retrieval. Impaired explicit task performance occurred because the patient was asked to provide an item in context ("Was the item on the just-presented word list?"); such contextual information was not available to the amnesic because it would be in episodic memory. The normal priming performance of amnesics, on the other hand, was supported by their normal semantic memory because the word's presence on the original to-be-retained list was considered to produce "activation" of that item's representation in semantic memory. This activation sensitized the item's representation, making it more easily accessible the next time the stimulus was presented or partially presented. Thus, the initial presentation "primed" the accessibility of that same item on an implicit task such as perceptual identification or word-stem completion. The two different memory systems theories, therefore, used the same outcome as evidence for their validity.

Finally, a third systems dichotomy was introduced to explain the distinction between explicit memory and priming. This distinction was based more on the notion of differential processing than were the former two dichotomies and received far less attention in the amnesia literature. Basically, this hypothesis suggested that two forms of processing exist when a subject attempts to remember information. Mandler (1980) and Jacoby (1984) were among the first to suggest this idea and to propose a mechanism that could contribute to amnesic patients' normal performance on implicit tasks. The emphasis in this theory is on processing during the episode itself; at the time of initial stimulus presentation, a subject performs some processing of which he or she is aware, which is retained to aid during reconstruction of the episode on an explicit memory task, and some processing of which he or she is unaware. Naturally, the "aware" processing is not available to the amnesic. The second type of process occurs automatically and includes such aspects of processing as perceptual analysis of the visually presented word. This type of processing does not aid reconstruction at the time of recall, but probably facilitates faster processing on an implicit measure such as perceptual identification, although the patient is never aware that this processing is taking place. This "unaware" level of processing may be available to amnesics and would support their normal implicit performance. Thus, intact processing in amnesic patients could exist on a purely automatic perceptual level and may be sufficient to support implicit, but not explicit, memory.

At that time, at least three dichotomies of memory existed, each using the performance of amnesics on implicit priming tasks to substantiate a belief in two distinct memory systems. However, by the end of the 1980s, many more such theories had been developed, including direct–indirect, associative–habit, vertical–horizontal, and, of course, simply explicit–implicit. The problem was that no real way to differentiate among all these dichotomies existed since all predicted performance of amnesics in the same direction. However, Roediger (1990) suggested a solution to this dilemma by proposing that processing abilities actually can cut across memory systems and make different predictions within memory systems depending on which type of processing is utilized. The processing dimensions that he considers to be orthogonal to the explicit–implicit dimension are called data driven and conceptually driven. Data-driven processes are those that occur automatically as soon as the stimuli are presented and include perceptual, phonemic, and acoustic processes. Conceptual processes are initiated by the subject and include semantic, reflective, and organizational processes. These processing abilities are proposed to be orthogonal to explicit–implicit memory systems because they exist within each component. In other words, subjects have data-driven implicit or explicit performance and, potentially, conceptually driven implicit or explicit performance. The mistake investigators have been making (according to Roediger) is that they have been exploring only two cells of this

2 × 2 contingency table, namely, the data-driven implicit cell and the conceptually driven explicit cell. Roediger suggests that researchers also should devise tasks that study data-driven explicit performance and conceptually driven implicit performance. These tasks might demonstrate that amnesic patients are impaired not along the explicit–implicit dimension but along the data-driven–conceptually driven dimension. Of course, inventing creative paradigms that test this hypothesis is difficult, but Roediger suggests two such tests: a sound-alike cue given during recall as a data-driven explicit task and a trivia-like question delivered during the priming portion of an experiment as a conceptually driven implicit task.

The results of this new analysis must be determined with amnesic patients, but early indications seem to be showing a difference between task performance of amnesics when they must rely on perceptual or conceptual processing (Blaxton, 1992; Cermak, Verfaellie, Butler, & Jacoby, 1993; Cermak, Verfaellie, Sweeney, & Jacoby, 1992; Verfaellie & Treadwell, 1993). We have suggested elsewhere (Cermak, 1993; Cermak and Verfaellie, 1992) that, when these perceptual processes are automatic, they are preserved in amnesia, whereas the conceptual processes, being more strategic, generally are impaired in amnesics. The terminology is not identical, but the meaning is the same as that proposed by Roediger. A processing-ability dimension might cut across memory systems; amnesics may be capable of one type of processing (automatic) but not the other (strategic). However, the expectations of systems theories must be pitted against those of processing theories in the performance of amnesics on tests that directly assess all cells of the 2 × 2 matrix proposed by Roediger. This comparison will enable investigators, in the the future, to determine whether the effects on different memory systems account for the behavior of amnesics on memory tasks or whether the ability (or inability) to perform specific types of processing, regardless of the memory task, accounts for their performance. If processing ability is deficient, which is our prediction, the pendulum of emphasis on systems vs processing will return in the 1990s to the side of processing as an explanation of memory disorders. Therefore, MRI researchers should acknowledge this shift in emphasis and begin to look for processing deficits rather than, or in addition to, systems deficits.

V. LOCALIZATION OF PROCESSES: A MULTIVARIATE APPROACH

Jernigan and Ostergaard (1993) have conducted a localization study that examined memory processing as measured by priming and recognition memory tasks, and correlated processing ability with the amount of damage present in specific brain structures. These investigators examined the caudate nuclei, the mesial temporal lobes, and the posterior temporo-occipito-parietal (posterolateral) cortex, using the MRI measures previously described. These structures were chosen because the mesial temporal lobes often have been implicated in explicit memory processes (such as recognition), and both caudate and posterolateral cortex have been implicated indirectly in priming deficits. A group of patients was recruited within which a large degree of variability on priming and recall memory measures existed. The 30 subjects included Alzheimer patients, Huntington patients, Korsakoff patients, patients with amnesia due to anoxic episodes, and normal subjects spanning the age range of the patients. This study was not conceptualized as one of the behavioral and anatomical abnormalities in various diagnostic groups; instead, the subjects were considered to constitute a single group spanning the full range from normal to highly abnormal on the memory measures and the anatomical indices. It was hoped that behavioral and anatomical heterogeneity within the group could be exploited to uncover underlying brain-processing relationships. We reasoned that, if the extent of damage to a specific structure is associated with the degree of impairment of a specific memory-related process, this relationship should be observable regardless of the diagnosis of the patient. Repetition priming and recognition memory were measured with a tachistoscopic identification task, as described previously. Identification threshold was measured for some of the words from a previously presented list and for some new words. Finally, each patient was asked if the word had been on the studied list. The measure of repetition priming was the difference between the mean identification threshold for all "old" word trials and the mean for all "new" word trials. The measure of recognition memory was the number of correct classifications of words as on or off the list (i.e., number of hits plus number of correct rejections). Relationships then were examined between the anatomical measures and three measures of performance on the memory tasks—mean identification threshold for all "new" word trials, repetition priming, and recognition memory.

For the recognition memory measure, only temporal limbic loss showed a significant specific association with poor recognition memory in a multiple regression analysis. A similar regression analysis for the priming measure revealed significant contributions by both the caudate and the temporal limbic measures. However, the effects were opposite to those predicted by "systems" theories: caudate volume loss was associated with increased priming, whereas temporal limbic loss was associated with decreased priming. For the "new" word threshold, the regression analysis suggested that caudate losses were related most strongly to inefficient lexical processing (as reflected in increased threshold).

The results of this study suggested that, relative to caudate and posterior neocortical damage, volume loss in the structures of the mesial temporal lobe was related specifically to poor recognition memory. Damage to the caudate, on the other hand, seemed to affect perceptual or lexical processing of the words. An analysis of the relationship between priming, identification threshold for "new" words, and recognition memory in this study revealed that the correlation between priming and recognition memory was virtually zero; however, recognition memory and "new" word threshold both contributed significantly to the prediction of priming in a multiple regression analysis; good recognition memory and slowed "new" word processing were associated with increased priming. Of particular interest was the fact that slower word identification predicted larger priming effects. Jernigan and Ostergaard found that impairment in the perceptual or lexical processing of the stimuli, associated with caudate damage, actually seemed to increase the priming effects of some of the subjects within the sample. A significant correlation between poorer processing of the "new" items and increased priming was observed. On the other hand, when temporal limbic losses occurred in isolation, priming effects were decreased. This result could have important implications for other findings of "intact" priming in patients who may have slowed perceptual or lexical function. Clearly, "baseline" levels on tasks such as these must be considered in the assessment of priming. Priming variability would seem to be composed of a significant memory-related component and an independent perceptual or lexical processing component. Limbic damage appears to affect the memory component and thereby reduces measured priming. However, when damage in other areas (in this case, of caudate nuclei) affects lexical or perceptual processing, measured priming effects may increase and, therefore, mask the reduced priming resulting from the limbic damage. Jernigan and Ostergaard (1993) obtained evidence that striatal damage affected the efficiency of stimulus processing in their subjects. A review of anatomical studies of memory-impaired patients with different etiologies revealed that, when the striatum is examined, damage frequently is found in these patients (Irle et al., 1992; Jernigan et al., 1991c,d; Press et al., 1989; Shimamura, Jernigan, & Squire, 1988). Further, damage to other structures also may contribute to inefficient stimulus processing and thereby augment priming effects. Therefore, such augmentation may have masked the decrementing effects of mesial temporal lobe damage in other studies of priming in amnesics. If so, normal priming in amnesics may not indicate sparing of an independent memory system.

Of course the results of this study, and the validity of the interpretations, should be confirmed in other priming studies. If they are, theories about the localization of priming effects may have to be revised. Multiple cognitive processes may contribute, although to different degrees, to both priming and recognition performance; much as data-driven and conceptually driven processes contribute to both explicit and implicit memory. Future structural imaging studies of memory-impaired patients may help delineate the nature and neural bases of these processes. The time-honored tradition of searching for patients with minimal lesions and highly circumscribed deficits will continue to play an important role in determining localization. However, the Jernigan and Ostergaard (1993) study suggests that studies of patients with milder or less discrete impairments and more diffuse damage can be informative.

VI. CONCLUSIONS AND PROSPECTS FOR THE FUTURE

Anatomical studies of memory-impaired subjects have focused attention on a few structures deep within the cerebral hemispheres that almost certainly play crucial roles in the processes of remembering. However, work in cognitive psychology is defining processing dimensions that clearly play important roles in memory performance. Controversy exists over which of these processing dimensions represent separate forms of memory per se and which modulate memory processes, either at encoding or at retrieval. However, future localization studies may help reveal the contributions of different brain systems to these processes. Several exciting functional imaging methods are being used now to continue localization studies of normal memory. These studies may be especially helpful in modeling the possible parallel and overlapping roles of different systems. Indeed, functional imaging techniques with MRI may provide a completely noninvasive method for conducting activation studies in normal and in memory-impaired subjects. However, concluding that only patients with dramatic and highly circumscribed memory disorders can provide insight into the neural bases of memory would be erroneous. In fact, from a processing perspective, every exemplar is informative since each places or reinforces certain constraints on models that map lesion sites onto behavioral outcomes. Further, multivariate analyses of imaging and behavioral data from patients with multiple abnormalities may be particularly useful for demonstrating the interactions between different brain systems that support memory processing.

REFERENCES

Alexander, M. P., & Freedman, M. (1984). Amnesia after anterior communicating artery aneurysm rupture. *Neurology, 34,* 752–757.

Baddeley, A. D., & Warrington, E. K. (1970). Amnesia and the distinction between long and short-term memory. *Journal of Verbal Learning and Verbal Behavior, 9,* 176–189.

Blaxton, T. A. (1989). Investigating dissociations among memory measures: Support for a transfer of appropriate processing framework. *Journal of Experimental Psychology: Learning, Memory, and Cognition, 15,* 657–668.

Blaxton, T. A. (1992). Dissociations among memory measures in memory-impaired subjects. Evidence for a processing account of memory. *Memory and Cognition, 20,* 549–562.

Brierley, J. B. (1977). Neuropathology of amnesic states. In C. W. M. Whitty & O. L. Zangwill (Eds.), *Amnesia* (pp. 199–223). London: Butterworths.

Butters, N., Heindel, W. C., & Salmon, D. P. (1990). Dissociation of implicit memory in dementia: Neurological implications. *Bulletin of the Psychonomic Society, 28(4),* 359–366.

Cave, C. B., & Squire, L. R. (1992). Intact and long-lasting repetition priming in amnesia. *Journal of Experimental Psychology: Learning, Memory, and Cognition, 18,* 509–520.

Cermak, L. S. (1979). Amnesic patients' level of processing. In L. S. Cermak & F. I. M. Craik (Eds.), *Levels of processing in human memory* (pp. 119–140). Hillsdale, New Jersey: Erlbaum Associates.

Cermak, L. S. (1984). The episodic/semantic distinction in amnesia. In L. R. Squire & N. Butters (Eds.), *Neuropsychology of memory* (pp. 55–62). New York: Guilford Press.

Cermak, L. S. (1993). Automatic vs. controlled processing and the implicit task performance of amnesic patients. In P. Graf & M. E. J. Masson (Eds.), *Implicit memory: New directions in cognition, development and neuropsychology* (pp. 287–302). Hillsdale, New Jersey: Erlbaum Associates.

Cermak, L. S., O'Connor, M., & Talbot, N. (1986). Biasing of alcoholic Korsakoff patients' semantic memory. *Journal of Clinical and Experimental Neuropsychology, 8,* 543–555.

Cermak, L. S., & Reale, L. (1978). Depth of processing and retention of words by alcoholic Korsakoff patients. *Journal of Experimental Psychology: Human Learning and Memory, 4,* 165–174.

Cermak, L. S., Talbot, N., Chandler, K., & Wolbarst, L. R. (1985). The perceptual priming phenomenon in amnesia. *Neuropsychologia, 23,* 615–622.

Cermak, L. S., & Verfaellie, M. (1992). The role of fluency in the implicit and explicit task performance of amnesic patients. In L. R. Squire & N. Butters (Eds.), *The neuropsychology of memory* (revised) (pp. 36–45). New York: Guilford Press.

Cermak, L. S., Verfaellie, M., Butler, T., & Jacoby, L. L. (1993). Attributions of familiarity in amnesia: Evidence from a fame judgment task. *Neuropsychology,* in press.

Cermak, L. S., Verfaellie, M., Sweeney, M., & Jacoby, L. L. (1992). Fluency vs. conscious recollection in the word completion performance of amnesic patients. *Brain and Cognition, 20,* 367–377.

Cohen, N. J., & Squire, L. R. (1980). Preserved learning and retention of pattern-analyzing skill in amnesia: Dissociation of knowing how and knowing that. *Science, 210,* 207–210.

Conlon, P., Kertesz, A., & Mount, J. (1988). Kluver Bucy syndrome with severe amnesia secondary to herpes encephalitis. *Canadian Journal of Psychiatry, 33(8),* 754–756.

Craik, F. I. M., & Lockhart, R. S. (1972). Levels of processing: A framework for memory research. *Journal of Verbal Learning and Verbal Behavior, 11,* 671–684.

Damasio, A. R., Graff-Radford, N. R., Eslinger, P. J., Damasio, H., & Kassell, N. (1985). Amnesia following basal forebrain lesions. *Archives of Neurology, 42,* 263–271.

Diamond, R., & Rozin, P. (1984). Activation of existing memories in anterograde amnesia. *Journal of Abnormal Psychology, 93*, 98–105.

Dunn, J. C., & Kirsner, K. (1988). Discovering functionally independent mental processes: The principle of reversed association. *Psychological Review, 95*, 91–101.

Gabrieli, J. D. E., Cohen, N. J., Huff, F. J., Hodgeson, J., & Corkin, S. (1984). Consequences of recent experience with forgotten words in amnesia. *Society for Neuroscience Abstracts, 10*, 383.

Gabrieli, J. D. E., Milberg, W., Keane, M. W., & Corkin, S. (1990). Intact priming of patterns despite impaired memory. *Neuropsychologia, 28*, 417–428.

Gardner, H., Boller, F., Moreines, J., & Butters, N. (1973). Retrieving information from Korsakoff patients: Effects of categorical cues and reference to the task. *Cortex, 9*, 165–175.

Graf, P., Squire, L. R., & Mandler, G. (1984). The information that amnesic patients do not forget. *Journal of Experimental Psychology: Learning, Memory, and Cognition, 10*, 164–178.

Graff-Radford, N. R., Tranel, D., van Hoesen, G. W., & Brandt, J. P. (1990). Diencephalic amnesia. *Brain, 113*, 1–25.

Haist, F., Musen, G., & Squire, L. R. (1991). Intact priming of words and nonwords in amnesia. *Psychobiology, 19*, 275–285.

Heindel, W. C., Salmon, D. P., & Butters, N. (1991). The biasing of weight judgments in Alzheimer's and Huntington's disease: A priming or programming phenomenon? *Journal of Clinical and Experimental Neuropsychology, 13(2)*, 189–203.

Heindel, W., Salmon, D., Shults, C., Walicke, P., & Butters, N. (1989). Neuropsychological evidence for multiple implicit memory systems: A comparison of Alzheimer's, Huntington's and Parkinson's disease patients. *Journal of Neuroscience, 2*, 582–587.

Hintzman, D. L. (1980). Simpson's paradox and the analysis of memory retrieval. *Psychological Review, 87*, 398–410.

Hintzman, D. L. (1990). Human learning and memory: Connections and dissociations. *Annual Review of Psychology, 41*, 109–139.

Hintzman, D. L., & Hartry, A. L. (1990). Item effects in recognition and fragment completion: Contingency relations vary for different subsets of words. *Journal of Experimental Psychology: Learning, Memory, and Cognition, 16*, 955–968.

Irle, E., Wowra, B., Kunert, H. J., Hampl, J., & Kunze, S. (1992). Memory disturbances following anterior communicating artery rupture. *Annals of Neurology, 31*, 473–480.

Jacoby, L. L. (1984). Remembering and awareness as separate issues. In L. R. Squire & N. Butters (Eds.), *Neuropsychology of memory* (pp. 145–156). New York: Guilford Press.

Jacoby, L. L., & Dallas, M. (1981). On the relationship between autobiographical memory and perceptual learning. *Journal of Experimental Psychology: General, 110*, 306–340.

Jacoby, L. L., & Witherspoon, D. (1982). Remembering without awareness. *Canadian Journal of Psychology, 36*, 300–324.

Jernigan, T. L. (1990). Techniques for imaging brain structure: Neuropsychological applications. In A. A. Boulton, G. B. Baker, & M. Hiscock (Eds.), *Neuromethods* (Vol. 17, pp. 81–105). Clifton, New Jersey: Humana Press.

Jernigan, T. L., Archibald, S. L., Berhow, M. T., Sowell, E. R., Foster, D. S., & Hesselink, J. R. (1991a). Cerebral structure on MRI, Part I: Localization of age-related changes. *Biological Psychiatry, 29*, 55–67.

Jernigan, T. L., Butters, N., DiTraglia, G., Schafer, K., Smith, T., Irwin, M., Grant, I., Schuckit, M., & Cermak, L. S. (1991b). Reduced cerebral grey matter observed in alcoholics using magnetic resonance imaging. *Alcoholism: Clinical and Experimental Research, 15*, 418–427.

Jernigan, T. L., & Ostergaard, A. L. (1993). Word priming and recognition memory both affected by mesial temporal lobe damage. *Neuropsychology, 7*, 1–13.

Jernigan, T. L., Salmon, D. P., Butters, N., & Hesselink, J. R. (1991c). Cerebral structure on MRI, Part II: Specific changes in Alzheimer's and Huntington's diseases. *Biological Psychiatry, 29*, 68–81.

Jernigan, T. L., Schafer, K., Butters, N., & Cermak, L. S. (1991d). Magnetic resonance imaging of alcoholic Korsakoff patients. *Neuropsychopharmacology, 4(3)*, 175–186.

Jernigan, T. L., Press, G. A., & Hesselink, J. R. (1990). Methods for measuring brain morphologic features on magnetic resonance images: Validation and normal aging. *Archives of Neurology, 47*, 27–32.

Keane, M. M., Gabrieli, J. D. E., Fennema, A. C., Growdon, J. H., & Corkin, S. (1991). Evidence for a dissociation between perceptual and conceptual priming in Alzheimer's disease. *Behavioral Neuroscience, 105(2)*, 326–342.

Kesslak, J. P., Nalcioglu, O., & Cotman, C. W. (1991). Quantification of magnetic resonance scans for hippocampal and parahippocampal atrophy in Alzheimer's disease. *Neurology, 41*, 51–54.

Kinsbourne, M., & Wood, F. (1975). Short-term memory processes and the amnesic syndrome. In D. D. Deutsch & J. A. Deutsch (Eds.), *Short-term memory* (pp. 257–291). New York: Academic Press.

Mair, W. P. G., Warrington, E. K., & Weiskrantz, L. (1979). Memory disorders in Korsakoff's psychosis. *Brain, 102*, 749–783.

Malamut, B. L., Graff-Radford, N., Chawluk, J., Grossman, R. I., & Gur, R. C. (1992). Memory in a case of bilateral thalamic infarction. *Neurology, 42*, 163–169.

Mandler, G. (1980). Recognizing: The judgment of previous occurrence. *Psychological Review, 87*, 252–271.

Mandler, G. (1991). Your face looks familiar but I can't remember your name: A review of dual process theory. In W. E. Hockley & S. Lewandowsky (Eds.), *Relating theory and data: Essays on human memory in honor of Bennet B. Murdock* (pp. 207–225). Hillsdale, New Jersey: Erlbaum Associates.

Martone, M., Butters, N., Payne, M., Becker, J. T., & Sax, D. S. (1984). Dissociations between skill learning and verbal recognition in amnesia and dementia. *Archives of Neurology, 41*, 965–970.

Mayes, A. R. (1988). *Human organic memory disorders.* Cambridge: Cambridge University Press.

Mayes, A. R., Meudell, P. R., Mann, D., & Pickering, A. (1988). Location of lesions in Korsakoff's syndrome: Neuropsychological and neuropathological data on two patients. *Cortex, 24*, 367–388.

Milner, B., Corkin, S., & Teuber, H.-L. (1968). Further analysis of the hippocampal amnesic syndrome: 14-year follow-up study of H. M. *Neuropsychologia, 6*, 215–234.

Mishkin, M. (1982). A memory system in the monkey. *Philosophical Transactions of the Royal Society of London, B, 298*, 83–95.

Nichelli, P., Bahmanian-Behbahani, G., Gentilini, M., & Vecchi, A. (1988). Preserved memory abilities in thalamic amnesia. *Brain, 111*, 1337–1353.

Norris, D. (1984). The effects of frequency, repetition and stimulus quality in visual word recognition. *Quarterly Journal of Experimental Psychology. A, Human Experimental Psychology, 36*, 507–518.

Olton, D. S. (1989). Inferring psychological dissociations from experimental dissociations: The temporal context of episodic memory. In H. L. Roediger & F. I. M. Craik (Eds.), *Varieties of memory and consciousness: Essays in honour of Endel Tulving* (pp. 161–177). Hillsdale, New Jersey: Erlbaum Associates.

Ostergaard, A. L. (1992). A method for judging measures of stochastic dependence: Further comments on the current controversy. *Journal of Experimental Psychology: Learning, Memory and Cognition, 18(2)*, 413–420.

Ostergaard, A. L., & Jernigan, T. L. (1993). Are word priming and explicit memory mediated by different brain structures? In P. Graf and M. Masson (Eds.), *New directions in cognition, development and neuropsychology.* Hillsdale, New Jersey: Erlbaum Associates.

Parkin, A. J. (1982). Residual learning capability in organic amnesia. *Cortex, 18,* 417–440.

Polster, M. R., Nadel, L., & Schacter, D. L. (1991). Cognitive neuroscience analyses of memory: A historical perspective. *Journal of Cognitive Neuroscience, 3,* 95–117.

Press, G. A., Amaral, D. G., & Squire, L. R. (1989). Hippocampal abnormalities in amnesic patients revealed by high-resolution magnetic resonance imaging. *Nature (London), 341,* 54–57.

Roediger, H. L., III (1984). Does current evidence from dissociation experiments favor the episodic/semantic distinction? *The Behavioral and Brain Sciences, 7,* 252–254.

Roediger, H. L., III (1990). Implicit memory: Retention without remembering. *American Psychologist, 45,* 1043–1056.

Scarborough, D. L., Cortese, C., & Scarborough, H. S. (1977). Frequency and repetition effects in lexical memory. *Journal of Experimental Psychology: Human Perception and Performance, 3,* 1–17.

Schacter, D. L. (1990). Perceptual representation systems and implicit memory: Toward a resolution of the multiple memory systems debate. In A. Diamond (Ed.), *The development and neural bases of higher cognitive functions* (pp. 543–571). New York: The New York Academy of Sciences.

Schacter, D. L., Cooper, L. A., Tharan, M., & Rubens, A. B. (1991). Preserved priming of novel objects in patients with memory disorders. *Journal of Cognitive Neuroscience, 3,* 117–130.

Seab, J. P., Jagust, W. J., Wong, S. T. S., Roos, M. S., Reed, B. R., & Budinger, T. F. (1988). Quantitative NMR measurements of hippocampal atrophy in Alzheimer's disease. *Magnetic Resonance in Medicine, 8,* 200–208.

Shimamura, A. P. (1985). Problems with the finding of stochastic independence as evidence for multiple memory systems. *Bulletin of the Psychonomic Society, 23,* 506–508.

Shimamura, A. P. (1986). Priming effects in amnesia: Evidence for dissociable memory function. *Quarterly Journal of Experimental Psychology, 38A,* 619–644.

Shimamura, A. P., Jernigan, T. L., & Squire, L. R. (1988). Korsakoff's syndrome: Radiological (CT) findings and neuropsychological correlates. *Journal of Neuroscience, 8(11),* 4400–4410.

Shimauchi, M., Wakisaka, S., & Kinoshita, K. (1989). Amnesia due to bilateral hippocampal glioblastoma. MRI finding. *Neuroradiology, 31,* 430–432.

Spiegel, E. A., Wycis, H. T., Orchinik, L. W., & Freed, H. (1955). The thalamus and temporal orientation. *Science, 121,* 771–772.

Squire, L. R. (1986). Mechanisms of memory. *Science, 232,* 1612–1619.

Squire, L. R. (1987). *Memory and brain.* Oxford: Oxford University Press.

Squire, L. R. (1992). Memory and the hippocampus: A synthesis from findings with rats, monkeys, and humans. *Psychological Review, 99(2),* 195–231.

Squire, L. R., Amaral, D. G., & Press, G. A. (1990). Magnetic resonance imaging of the hippocampal formation and mammillary nuclei distinguish medial temporal lobe and diencephalic amnesia. *Journal of Neuroscience, 10(9),* 3106–3117.

Squire, L. R., & Cohen, N. J. (1984). Human memory and amnesia. In G. Lynch, J. L. McGaugh, & N. M. Weinberger (Eds.), *Neurobiology of learning and memory* (pp. 3–64). New York: Guilford Press.

Squire, L. R., & Zola-Morgan, S. (1988). Memory: Brain systems and behavior. *Trends in Neurosciences, 11,* 170–175.

Symonds, C., Sir (1966). Disorders of memory. *Brain, 89(4),* 625–644.

Tulving, E. (1972). Episodic and semantic memory: Different retrieval mechanisms. In E. Tulving & W. Donaldson (Eds.), *Organization of memory* (pp. 381–403). New York: Academic Press.

Tulving, E. (1983). *Elements of episodic memory.* Oxford: Oxford University Press.

Tulving, E. (1985). How many memory systems are there? *American Psychologist, 40,* 385–398.

Tulving, E., Hayman, C. A. G., & MacDonald, C. A. (1991). Long-lasting perceptual priming and semantic learning in amnesia: A case experiment. *Journal of Experimental Psychology: Learning, Memory, and Cognition, 17,* 595–617.

Tulving, E., & Schacter, D. L. (1990). Priming and human memory systems. *Science, 247,* 301–306.

Verfaellie, M., & Treadwell, J. (1993). Status of recognition memory in amnesia. *Neuropsychology 7,* 5–13.

Victor, M., Adams, R. D., & Collins, G. H. (1971). *The Wernicke–Korsakoff syndrome.* Oxford: Blackwell.

von Cramon, D. Y., Hebel, N., & Schuri, U. (1985). A contribution to the anatomical basis of thalamic amnesia. *Brain, 108,* 993–1008.

Warrington, E. K., & Weiskrantz, L. (1968). New method of testing long-term retention with special reference to amnesic patients. *Nature (London), 217,* 972–974.

Warrington, E. K., & Weiskrantz, L. (1970). Amnesic syndrome: Consolidation or retrieval? *Nature (London), 228,* 628–630.

Warrington, E. K., & Weiskrantz, L. (1974). The effect of prior learning on subsequent retention in amnesic patients. *Neuropsychologia, 12,* 419–428.

Warrington, E. K., & Weiskrantz, L. (1978). Further analysis of the prior learning effect in amnesic patients. *Neuropsychologia, 16,* 169–177.

Weiskrantz, L., & Warrington, E. K. (1970). Comparison of so-called amnesic states in monkey and man. *Brain Research, 24,* 555–556.

Winnick, W. A., & Daniel, S. A. (1970). Two kinds of response priming in tachistoscopic recognition. *Journal of Experimental Psychology, 84,* 74–81.

Zola-Morgan, S., & Squire, L. R. (1990). The neuropsychology of memory: Parallel findings in humans and nonhuman primates. In A. Diamond (Ed.), *The development and neural bases of higher cognitive functions* (pp. 434–456). New York: The New York Academy of Sciences.

Zola-Morgan, S., Squire, L. R., & Amaral, D. G. (1986). Human amnesia and the medial temporal region: Enduring memory impairment following a bilateral lesion limited to field CA1 of the hippocampus. *Journal of Neuroscience, 6(10),* 2950–2967.

Chapter 21

Neuroimaging in Dementia

Kevin F. Gray[1] and Jeffrey L. Cummings

I. INTRODUCTION

Dementia is a major public health challenge, not only for clinicians, but also for society as a whole. Studies suggest that up to 4 million Americans suffer from severe dementia; an additional 1–5 million patients have mild to moderate intellectual impairment (U.S. Congress Office of Technology Assessment, 1987). Community surveys indicate that as many as 50% of people over age 85 have a dementia syndrome (Evans et al., 1989). Given the growing size of the elderly population, the number with severe dementia will increase by 60% by the year 2000 and by 100% by 2020 if current trends continue.

A wide variety of disorders have been identified as causes of dementia (Cummings & Benson, 1992), including primary brain diseases and diseases of other organ systems that lead to secondary brain dysfunction. Many of these conditions have available therapies, so distinguishing patients with potentially reversible conditions from those with degenerative dementias is crucial. Some irreversible dementias may have important interventions (e.g., vascular dementia), and accurate diagnosis is necessary in the degenerative disorders to provide prognostic and genetic counseling. Neuroimaging plays a critical role in the diagnostic evaluation of the dementia patient; vascular dementia, hydrocephalus, neoplasms, subdural hematomas, and some infectious and degenerative processes can be identified with contemporary imaging technology.

This chapter describes the diagnostic application of neuroimaging in the evaluation of dementia. The imaging technologies are described; a discussion of current neuroimaging findings reported in specific dementia syndromes follows. Finally, principles guiding the use of neuroimaging in dementia are summarized. Only those syndromes with well-established

---

[1]Present address: University of Texas, Southwestern Medical School, 5323 Harry Hines Blvd., Dallas, TX 75235-9070.

Localization and Neuroimaging in Neuropsychology

neuroimaging alterations are discussed in detail. Imaging results are correlated with the known pathology of the individual dementing disorders.

II. NEUROIMAGING TECHNIQUES

A. Computerized Tomography

Computerized tomography (CT) became available in the mid-1970s as a dramatically improved X-ray technology that allows noninvasive visualization of brain structures in vivo. An X-ray source is directed at the brain, and scintillation detectors record the amount of radiation passing through particular areas of the brain. This information then is processed by a computer, which assigns a relative numerical attenuation value to each specific scanned volume element (voxel). These numerical values are translated onto a "gray scale"—a graduated spectrum of gray shades ranging from black to white, in which black represents low-density low-attenuation elements such as cerebrospinal fluid (CSF) and white represents high-density high-attenuation elements such as bone. Each voxel with its gray scale value then is transferred to a corresponding two-dimensional pixel (picture element) on plain X-ray film. Resolution is limited by "partial-volume" ("volume-averaging") artifact that represents the major technical limitation of CT. This effect occurs in voxels containing areas of widely differing attenuation that are averaged to an intermediate value, making precise delineation or measurement difficult.

Clinically, CT has been used primarily to detect intracranial mass lesions such as tumors and hematomas, but cerebral infarctions, regional cerebral atrophy, ventricular enlargement, and white-matter alterations also can be studied. Iodinated radio-opaque contrast dye frequently is used to enhance structures with compromised blood–brain barriers and to improve the diagnosis of tumors, infarctions, certain infections, and demyelinating disease. Contrast enhancement also improves detection of vascular malformations and aneurysms. CT technology is relatively inexpensive and widely available (Burns, 1990).

B. Magnetic Resonance Imaging

Magnetic resonance imaging (MRI) has been available for clinical application since 1980. This structural imaging technique is derived from measuring the behavior of protons when influenced by a magnetic field. Protons in a static magnetic field align in the direction of that field. A brief radiofrequency pulse is administered that transiently shifts this alignment by 90°. As the protons return to their previous alignment ("relaxation"), they emit energy; the differing rate of energy signal decay between tissues confers contrast that can be exploited for imaging purposes. Three components of the signal are utilized: T_1-weighted, T_2-weighted, and proton-density measures. T_1-weighted images clearly distinguish between gray and white matter, producing high-quality anatomical displays. T_2-weighted images are of poorer anatomical quality but are best for detecting pathological lesions, particularly of the white matter. Proton-density images, although showing limited contrast among soft tissues, have excellent contrast between soft tissues and CSF, allowing precise outlining of the ventricular system and the gray matter–CSF interface.

MRI does not employ ionizing radiation, may be used safely on the same patient repeatedly, and thus is well suited for longitudinal studies of disease. This technique is superior to CT for identifying ischemic lesions, cerebral tumors, demyelination and other pathologies, but is less sensitive for detecting calcification (Besson, 1990). MRI is more expensive than CT and is less well tolerated by patients because of its noise and the confined structure of the imaging chamber. Also, it is less widely available.

C. Magnetic Resonance Spectroscopy

Magnetic resonance spectroscopy (MRS) is an evolving technique that allows noninvasive studies of cellular chemical mechanisms to be made *in situ*. Although MRS technology has been available for over 40 years, its stringent technical requirements previously have limited its application to *in vitro* investigations.

MRS technology is similar to MRI. Atoms with unequal numbers of protons and neutrons in their nuclei, thereby possessing a net "spin," lend themselves to MRS. An external magnetic field is imposed, aligning the nuclei. A radiofrequency energy pulse then is applied, transiently altering this alignment. As the nuclei return to their original axes (relaxation), they induce signal peaks in a recording coil. These peaks are unique for each distinct nuclear species. Because any given nucleus also is affected by the magnetic fields of its neighboring atoms, the same element in different compounds will generate distinct MRS peaks for each compound. Measuring the areas under the different peaks can identify the relative quantities of specific metabolites. MRS with phosphorus 31 (^{31}P), studying brain phospholipid metabolism and energy production, has been the most widely applied technique to date. Other promising applications include using lithium (^{7}Li) and fluorine (^{19}F) MRS to study the pharmacology of lithium and fluorinated psychopharmacological agents, using carbon (^{13}C)

and hydrogen (^{1}H) MRS to study carbohydrate, protein, and amino acid metabolism, and using sodium (^{23}Na) MRS to study electrolyte balance.

MRS is noninvasive, does not expose patients to ionizing radiation, and has no known side effects, making it ideal for longitudinal studies. The convergence of MRS and MRI technologies to yield spatial chemical images holds great promise for future research (Keshavan, Kapur, & Pettegrew, 1991; Lock, Abou-Saleh, & Edwards, 1990). Current limitations of MRS include its insensitivity, with additional time and cost required to adjust the magnetic field carefully for each measurement. A long signal acquisition time is required, preventing application of MRS to cognitive activation paradigms. MRS provides purely chemical information that conforms to a spherical "volume of interest" rather than to anatomical structures. Like MRI, MRS has a noisy and confining chamber.

D. Single-Photon Emission Computerized Tomography

Single-photon emission computerized tomography (SPECT) is an imaging technique that allows noninvasive study of brain function rather than structure. SPECT imaging was developed by combining CT imaging technology with traditional studies of cerebral blood flow using radionuclides. Radiotracers labeled with ^{123}I and ^{99}T are utilized. These compounds cross the blood–brain barrier after intravenous injection and distribute themselves in proportion to the regional cerebral blood flow (rCBF). The compounds remain trapped in the brain and emit a "single photon," enabling images of 8- to 10-mm resolution to be obtained with conventional equipment available in most nuclear medicine departments. Currently, SPECT provides an image reflecting cerebral perfusion, although new techniques are evolving that use SPECT to measure neurotransmitter receptors. SPECT is relatively inexpensive compared with other functional imaging approaches and is becoming increasingly available for clinical use (Geaney & Abou-Saleh, 1990).

E. Positron Emission Tomography

Positron emission tomography (PET) assesses and maps regional brain function. The signal measured in PET is derived from the decay of radioactive positron-emitting nuclides that are introduced into the body via intravenous injection or inhalation. Common nuclides including ^{15}O, ^{11}C, and ^{18}F, are used to label compounds such as glucose and oxygen for the purpose of evaluating cerebral metabolism or measuring rCBF. Currently DOPA can be labeled (i.e., fluorodopa) to allow PET studies of dopamine distribution in the brain; receptor ligands are being developed. The short half-life of some of these tracers (minutes) facilitates repeat studies within a single scanning session, while insuring low radiation exposure to the subject. A circumferential detector placed around the head maps the distribution of the positron-emitting nuclide with a resolution of 5 mm. The technology and labor required for PET are elaborate and expensive, requiring a cyclotron on site to produce the nuclides. Currently the prospects for general clinical application are limited (Bench, Dolan, Friston, & Frackowiak, 1990).

III. NEUROIMAGING IN DEMENTIA SYNDROMES

A. Dementia of the Alzheimer Type

The most common cause of dementia in older adults is dementia of the Alzheimer type (DAT), accounting for over 50% of all cases. DAT is characterized clinically by the insidious onset of progressive dementia; pathologically, neurofibrillary tangles, amyloid angiopathic changes, and neuritic or senile plaques are present in the cortex. The greatest pathological burden involves the medial temporal regions and the temporo-parieto-occipital junctions (Pearson & Powell, 1989).

Structural neuroimaging excludes brain diseases that can mimic the clinical symptoms of DAT. CT studies in DAT have included: (1) assessment of cerebral surface atrophy (cortical atrophy or sulcal enlargement); (2) assessment of ventricular size (subcortical or central atrophy); and (3) measurement of brain density. Atrophy is present on CT and MRI of most DAT patients, but age-related atrophy of the brain may be of similar magnitude to the atrophy noted in DAT (DeCarli, Kaye, Horwitz, & Rapoport, 1990). Cortical atrophy in DAT is usually greater than in normal aging, although some demented patients will have normal CT scans and some normal controls will have atrophic changes well into the range associated with dementia. DAT patients have significantly larger ventricles than age-matched controls. Third ventricular size has been correlated with cognitive impairment. In general, correlations between ventricular enlargement and cognitive function are more robust than those between cortical atrophy and cognition (Burns, 1990).

Methods used to evaluate these CT changes include qualitative ratings by experienced radiologists and linear, planimetric, and volumetric measurements of regions of interest (ROI). Of these, CT volumetric analysis has emerged as the best method with which to study structure–function relationships in DAT, especially when used longitudinally. This technique measures mean volumes of CSF spaces and usually has demonstrated significant differences between even mildly impaired DAT patients and normal controls. Patients with DAT have larger mean CSF and ventricular volumes than do age-matched healthy controls. Longitudinal CT studies document the progressive brain atrophy associated with DAT (DeCarli et al., 1990).

MRI has clear advantages over CT in assessment of structural changes in DAT. More precise measurements may be obtained in critical areas such as the temporal lobes and cerebral convexities in which "bone-hardening" CT artifact is most confounding. Improved differentiation of gray matter, white matter, and CSF is possible, and coronal images may be generated easily to enhance quantitative analysis (DeCarli et al., 1990). Excessive white matter ischemia is more easily detectable, identifying vascular dementia that imitates DAT or mixed vascular–DAT disorders. When patients are screened for cerebrovascular risk factors, white matter hyperintensities (WMH) on T_2-weighted MRI are uncommon and occur in DAT and age-matched controls with equal frequency; WMH severity does not correlate with cognitive decline in DAT (Kozachuk, DeCarli, Schapiro, Wagner, Rapoport, & Horwitz, 1990).

MRS in dementia has utilized ^{31}P to study membrane phospholipid alterations as well as phospholipids involved in energy metabolism. Phosphomonoesters (PME) are the precursors and phosphodiesters (PDE) are the degradation products of membrane phospholipids; these measures reflect membrane turnover. *In vivo* MRS studies of patients with DAT have shown elevations of percentage PME and PME:PDE ratio in the temporo-parietal regions. Postmortem studies using mean density of senile plaques to stage illness found PME levels in DAT to be higher than in controls early in the disease; PDE levels were higher than in controls late in the disease. These changes may be specific for DAT and could prove valuable for diagnosis as well as understanding the pathophysiological basis for the intellectual decline (Brown et al., 1989; Keshavan et al., 1991).

SPECT imaging has been used to study dementia since the mid-1980s, and DAT has been investigated most extensively. The characteristic SPECT finding in DAT is bilaterally decreased rCBF in the parietal and posterior temporal lobes, with variable frontal lobe involvement in the later stages of the illness. Primary motor, sensory, and visual cortices and basal ganglia are spared (Geaney & Abou-Saleh, 1990; Montaldi et al., 1990). Considerable asymmetry of involvement may be present in the early stages of DAT, but direct analysis of computer-generated data often reveals bilateral hypoperfusion in studies that otherwise "look" normal or appear to have lateralized changes on visual display (DeKosky, Shih, Schmitt, Coupal, & Kirkpatrick, 1990). Studies of correlations between rCBF and neuropsychological testing have documented associations of apraxia and aphasia with posterior parietal and lateral temporal perfusion, memory with left temporal lobe perfusion, and language with left hemisphere perfusion (Burns, Philpot, Costa, Ell, & Levy, 1989; Engel, Cummings, Villaneuva-Meyer, & Mena, 1993).

PET studies in DAT have demonstrated consistently a progressive decline in both CBF and global cerebral glucose metabolism that correlates with the severity of cognitive impairment. The posterior temporo-parietal association cortices are affected most, and preferential sparing of the primary motor and sensory areas is seen (Fig. 1; Benson, Kuhl, Hawkins, Phelps, Cummings, & Tsai, 1983). As the dementia progresses, frontal lobe metabolism also declines. Whether the diminished metabolic rate revealed by PET reflects a regional decrease in the number of neurons, a diminished

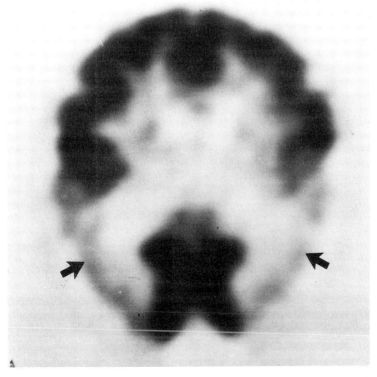

Figure 1 Fluorodeoxyglucose positron emission tomogram of a patient with Alzheimer's disease, demonstrating bilateral parietal lobe hypometabolism (arrows).

metabolism in remaining normal neurons, or both is uncertain (Bench et al., 1990). In the earliest stages of DAT, PET may demonstrate reduced metabolism in the association neocortex before the neuropsychological consequences of regional dysfunction are demonstrable (Haxby, Grady, Duara, Schlageter, Berg, & Rapoport, 1986).

B. Frontal Lobe Degenerations

A variety of degenerative diseases are recognized that preferentially affect the frontal lobes (Table 1). Neuroimaging can aid in distinguishing frontal lobe degenerations (FLD) from DAT but does not help differentiate the various types of frontal degenerative processes.

Pick's disease is the most well known FLD. Clinically the disorder is characterized by prominent personality alterations, executive function deficits, and language changes. At autopsy, lobar atrophy involving the frontal and temporal lobes is detected. Neuronal loss, gliosis, intraneuronal Pick bodies, and ballooned Pick cells are the characteristic histological findings (Cummings & Benson, 1992). CT typically demonstrates dilatation of the frontal horns, widening of the frontal sulci, and atrophy of the caudate nuclei (Fig. 2) (Knopman et al., 1989). Fluoro-deoxyglucose (FDG) PET reveals diminished frontal lobe metabolism and SPECT shows reduced frontal lobe perfusion (Kamo et al., 1987; Miller et al., 1991).

FLD of non-Alzheimer type (also known as dementia lacking distinctive histological features) is a recently recognized disorder that may be more common than Pick's disease as a cause of FLD. The patients have behavioral changes resembling those of Pick's disease, but lack Pick bodies or ballooned cells at autopsy (Brun, 1987; Gustafson, 1987; Knopman, Mastri, Frey, Sung, & Rustan, 1990). CT may reveal frontal lobe and caudate atrophy, although the atrophic changes have been less dramatic in the reported cases than typically described in Pick's disease (Knopman et al., 1990). CBF is reduced, particularly in the prefrontal regions. The reductions are less marked than in Pick's disease (Risberg, 1987).

Table 1   Principal Frontal Lobe Degenerations

Pick's disease
Non-Alzheimer type FLD
Progressive subcortical gliosis
FLD with motor neuron disease
Neuronal intranuclear hyaline inclusion disease

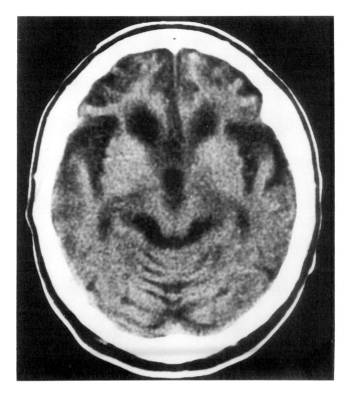

Figure 2   Computerized tomogram of a patient with Pick's disease. Marked frontal and temporal lobe atrophy is evident.

Adult-onset neuronal intraneuronal hyaline inclusion disease (NIHID) is a rare disorder that has produced a variety of clinical syndromes but most commonly shows mixtures of dementia, aphasia, personality altera- tions, psychosis, chorea, ataxia, tremor, lower motor neuron signs, dysar- thria, and dysphagia (Munoz-Garcia & Ludwin, 1986). Hyaline inclusions are seen in neurons of the cortex (particularly the frontal and temporal cortex), dorsal root ganglia, and parasympathetic neurons of the digestive tract, as well as in cells of the adrenal medulla. CT has been reported to show atrophy without a specific lobar predilection.

Progressive subcortical gliosis is a progressive degenerative dementia affecting the frontal and temporal lobes. Dementia with prominent lan- guage changes and a relatively rapidly progressive course with death in 3–7 yr are typical (Verity & Wechsler, 1987). At autopsy, neuronal loss and marked gliosis of the white matter of the frontal and temporal lobes are characteristic. Neuroimaging findings have been reported rarely but preferential anterior lobar atrophy is predicted by the postmortem findings.

FLD with motor neuron disease is a distinctive clinical syndrome combining a frontal lobe type dementia with evidence of upper and lower motor neuron dysfunction. Although few cases have been studied pathologically, preliminary reports describe frontal–temporal atrophy with regional neuronal loss, gliosis, and spongiform changes (Neary, Snowden, Mann, Northern, Goulding, & MacDermott, 1990). CT demonstrates generalized or frontally predominant atrophy and SPECT reveals diminished frontal lobe perfusion (Neary et al., 1990).

C. Vascular Dementia

Vascular dementia (VaD) is the second most common cause of dementia in the elderly, accounting for 15–30% of all cases (Cummings & Benson, 1992). Typically, more than one infarct or area of ischemic injury is visualized using brain imaging techniques; these can be large or small and are located in the cortex, white matter (Binswanger's disease), or subcortical nuclei (lacunar state). VaD develops when a threshold of total brain tissue destruction has been exceeded (Margolin & Jarema, 1989), or when critically located infarctions disrupt multiple cognitive functions.

CT in VaD initially focused on differentiation from DAT based on the presence of infarctions in VaD and atrophy in DAT. Although interobserver reliability for detection and localization of infarcts using CT has been excellent (Shinar et al., 1987), CT alone was found to detect infarctions in less than half of patients with clinical evidence of VaD (Radue, duBoulay, Harrison, & Thomas, 1978). Diagnostic studies have begun to include consideration of areas of increased lucency in the white matter (leuko-araiosis, LA). LA with or without visible infarctions is seen on CT in a majority of patients (50–90%) with VaD. Studies have noted LA to be consistently more prominent in VaD than in DAT, although the reported percentages vary from 33 to 97% for VaD and 5 to 55% for DAT. In VaD, enlargement of the lateral and third ventricles correlates significantly with severity of cognitive impairment (Aharon-Peretz, Cummings, & Hill, 1988).

MRI has emerged as the most sensitive structural imaging technique for assessing VaD (Kertesz, Black, Nicholson, & Carr, 1987). The T_2-weighted images are best for detecting the increased water content in the white matter present with ischemic changes and demyelination (Fig. 3) (Brown, Hesselink, & Rothrock, 1988; Gupta, Naheedy, Young, Ghobrial, Rubino, & Hindo, 1988). However, this increased sensitivity for detecting white

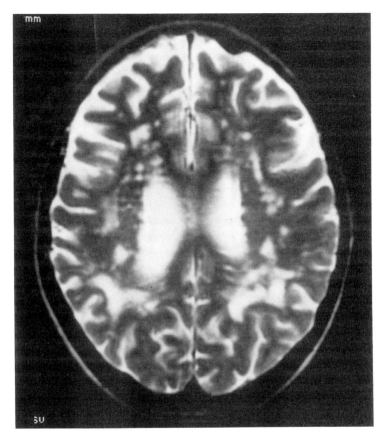

Figure 3 T$_2$-weighted magnetic resonance image of a patient with vascular dementia. Note high signal areas in the periventricular regions and deep white matter.

matter lesions has resulted in decreased specificity for diagnosis. WMH found on T$_2$-weighted MRI is seen not only in VaD patients, but in DAT patients and in healthy elderly controls, making their diagnostic interpretation difficult (Erkinjuntti, Ketonen, Sulkava, Sipponen, Vuorialho, & Iivanainen, 1987; Fein et al., 1990; Hunt et al., 1989). Commonly, WMH is located either in the periventricular region of the frontal and occipital horns of the lateral ventricles or in the deep cerebral white matter (centrum semiovale). Kertesz, Black, Tokar, Benke, Carr, and Nicholson (1988) classified periventricular WMH as "rims" or "caps" and WMH located elsewhere as "unidentified bright objects" (UBOs). These investigators found that rims were seen frequently in control subjects and correlated with increasing age; UBOs, caps, and severe rims usually signified pathology, occurring more frequently in patients with strokes and hemorrhages than in controls. Other studies confirm that WMH are correlated best with cerebrovascular disease. Preliminary studies failed to find differences in WMH occurrences among demented and nondemented patients with stroke (Hershey, Modic, Greenough, & Jaffe, 1987; Gupta et al., 1988). A recent study, however, found several factors that correlated with severity of mental status changes in VaD, including total WMH area, left WMH area, ventricular-to-brain ratio, right WMH area, age, left cortical infarction area, left parietal infarction area, and total infarction area. Location of infarction was found to be a critical factor: cortical location and left laterality strongly correlated with the development of dementia following stroke (Liu et al., 1992).

Neuropathological studies of WMH similarly demonstrate the need for caution in inferring pathology from MRI. In some cases, no neuropathological changes have been detected in regions giving rise to WMH signals. Diagnoses such as VaD or Binswanger's disease (subcortical arteriosclerotic encephalopathy) must not be made in the absence of correlative clinical information (Grafton, Sumi, Stimac, Alvord, Shaw, & Nochlin, 1991; Johnson, Davis, Buonanno, Brady, Rosen, & Growdon, 1987; Leifer, Buonanno, & Richardson, 1990).

MRS has been utilized in VaD and, although the findings must be replicated, VaD was distinguished accurately from DAT by elevation of the phosphocreatine:inorganic orthophosphate ratio in temporo-parietal and frontal regions in patients with MID. This change may reflect enhanced phosphorylation potential of brain in VaD compared with deterioration of neuronal energy mechanisms in DAT (Brown et al., 1989).

SPECT in VaD reveals diffusely diminished CBF with superimposed focal areas of more severe hypoperfusion. Theoretically, VaD could produce any pattern; however, the presence of one or more scattered perfusion defects, either unilateral or bilateral, with an asymmetric distribution is most suggestive of VaD (Geaney & Abou-Saleh, 1990).

PET studies in VaD show global reductions in cerebral metabolism with additional focal and asymmetric areas of hypometabolism that are not limited to specific cortical or subcortical brain regions (Fig. 4; Benson et al., 1983). Brain impairment demonstrated by PET is often more widespread than that shown by CT, MRI, or even neuropathological techniques, suggesting that single lesions may have extensive and distant metabolic sequelae. These remote metabolic effects have been attributed

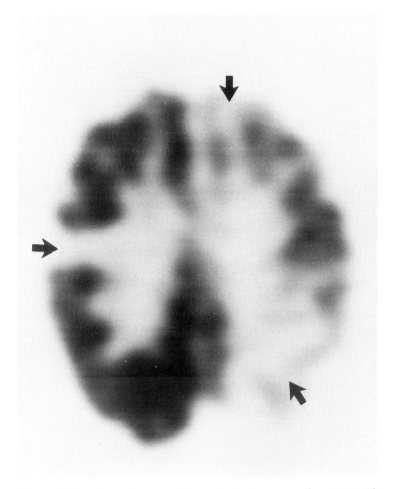

Figure 4   Fluorodeoxyglucose positron emission tomogram of a patient with vascular dementia demonstrating multifocal areas of hypometabolism (arrows). (PET courtesy of M. Mahler and D. Sultzer, West Los Angeles Veterans Affairs Medical Center.)

to degeneration of fiber tracts with disconnection of distal structures and to microscopic infarcts not apparent on gross examination but manifested as hypometabolic regions (Metter, Mazziotta, Itabashi, Mankovich, Phelps, & Kuhl, 1985). Increasing severity of dementia correlates with global hypometabolism and increasing involvement of the frontal cortex (Bench et al., 1990).

D. Basal Ganglia Disorders with Dementia

Neuroimaging can aid in the diagnosis of basal ganglia disorders associated with dementia. The most well known of these disorders with imaging abnormalities is Huntington's disease, but distinctive changes are also evident in neuroacanthocytosis, progressive supranuclear palsy, Wilson's disease, basal ganglia calcification syndromes, Hallervorden–Spatz disease, and corticobasal degeneration. In most of these diseases, imaging has been reported on well-studied cases with classical manifestations; the sensitivity, specificity, and variability of imaging abnormalities remain to be determined.

Huntington's disease is characterized clinically by progressive dementia, personality and mood alterations, and chorea. Pathologically, atrophy of the caudate nucleus and loss of the GABA-ergic interneurons of the striatum occur. CT reveals diminished volume of the caudate nuclei with loss of the convex bulge of the nucleus into the lateral aspects of the frontal horns of the lateral ventricles (Fig. 5). The bicaudate index (width of both lateral ventricles divided by distance between the outer tables of the skull at the same level) distinguishes between Huntington's disease and other disorders with cerebral atrophy (> 0.18 in Huntington's disease) (Barr, Heinze, Dobben, Valvassori, & Sugar, 1978). Similar changes are evident on MRI. FDG PET demonstrates marked hypometabolism of the caudate nuclei, and the reduced metabolism has been shown to precede detectable loss of tissue volume (Young et al., 1986).

Neuroacanthocytosis is a rare familial degenerative disease characterized by chorea, tics, peripheral neuropathy, behavioral alterations, mild dementia, and erythrocyte acanthocytosis. Postmortem examination reveals neuronal loss in the caudate and putamen (Bird, Cederbaum, Valpey, & Stahl, 1978). CT and MRI demonstrate caudate atrophy, and SPECT shows mild caudate hypoperfusion and marked frontal lobe hypoperfusion (Delecluse, Deleval, Gerard, Michotte, & Zegers de Beyl, 1991). In the few patients studied, fluorodopa PET evidenced diminished uptake in the posterior putamen; ^{11}C-labeled raclopride ratios were diminished in the caudate and putamen, indicating loss of D2 receptor binding sites in these areas; and $C^{15}O_2$ measures of CBF were reduced in the striatum and frontal lobes (Brooks et al., 1991).

Progressive supranuclear palsy (PSP) features parkinsonian-type bradykinesia, axial rigidity, supranuclear gaze palsy with marked dysarthria, and dementia (Cummings & Benson, 1992). Pathologically patients have neuronal loss, gliosis, intraneuronal neurofibrillary tangle formation, and granulovacuolar degeneration that are most marked in the midbrain. CT reveals atrophy of the midbrain, with less severe volume loss of the pons,

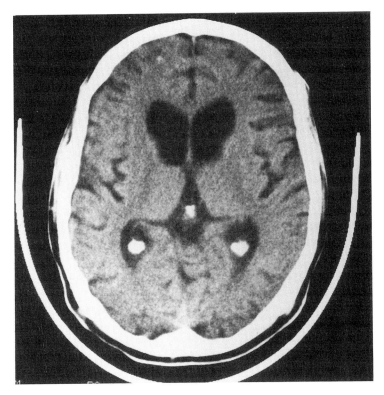

Figure 5 CT of a patient with Huntington's disease, demonstrating atrophy of the caudate nuclei. The bicaudate index (smallest width between caudate nuclei/width between outer tables of the skull at the same level) is 0.27 (abnormal is > 0.18).

cerebellum, and cerebral hemispheres (Haldeman, Goldman, Hyde, & Pribram, 1981). Preliminary MRI investigations have shown decreased T_2 signal intensity in the putamen and lateral substantia nigra that is consistent with excess iron accumulation in these structures (Drayer, Olanow, Burger, Johnson, Herfkens, & Riederer, 1986). PET studies reveal diminished blood flow and decreased oxygen extraction in the frontal lobes, as well as frontal hypometabolism (Goffinet et al., 1989; Leenders, Frackowiak, & Lees, 1988). In addition, fluorodopa PET investigations demonstrate diminished striatal uptake that is consistent with decreased receptor number (Leenders et al., 1988).

Wilson's disease is an inherited defect in the copper-carrying serum protein ceruloplasmin, resulting in abnormal copper deposition in the basal ganglia, liver, and cornea. Neurologically, the patients evidence dysarthria, dystonia, rigidity, cerebellar abnormalities, tremor, gait and postural disturbances, and mild dementia (Starosta-Rubinstein et al., 1987). In 40 patients with Wilson's disease and neurological symptoms, Williams and Walshe (1981) found ventricular dilatation in 90%, cortical atrophy in 70%, brainstem atrophy in 65%, and basal ganglia hypodensities in 57%. Similarly, MRI demonstrates hypodense lesions on T_1-weighted images and hyperintense T_2-weighted lesions in the putamen in approximately 50% of patients with neurological symptoms (Starosta-Rubinstein et al., 1987). PET reveals globally diminished cerebral metabolism with relatively more marked changes in the lenticular nuclei (Hawkins, Mazziotta, & Phelps, 1987). Fluorodopa PET demonstrates diminished uptake in the striatal nuclei in symptomatic patients with Wilson's disease (Snow, Bhatt, Martin, Li, & Calne, 1991).

Hallervorden–Spatz disease is a progressive extrapyramidal syndrome with parkinsonism, dystonia, and dementia that may present in childhood or adulthood. The characteristic pathology includes iron deposition in the globus pallidus and substantia nigra and axonal swellings and neuronal loss in the basal ganglia and cerebellum (Cummings & Benson, 1992). CT demonstrates enlarged ventricles and cortical, brainstem, and cerebellar atrophy; the subcortical structures are atrophic and the bicaudate index is abnormal (Dooling, Richardson, & Davis, 1980). MRI studies reveal markedly diminished T_2 signal intensity in the globus pallidus. Often an area of relatively normal signal intensity exists within the low intensity region, resulting in the striking "eye of the tiger" sign (Sethi, Adams, Loring, & El Gammal, 1988).

Basal ganglia calcification syndromes combine dementia with the distinctive neuroimaging finding of basal ganglia calcium deposition. Table 2 presents the syndromes to be considered in the differential diagnosis of an adult with progressive dementia and abnormal subcortical calcification. Idiopathic and hereditary dementing disorders with basal ganglia calcification include Fahr's disease (idiopathic basal ganglia calcification), membranous lipodystrophy, mitochondrial encephalopathies, and Hallervorden–Spatz disease. Metabolic disorders affecting thyroid and parathyroid function constitute another cause of dementia and basal ganglia calcification. Toxic conditions in the differential diagnosis include carbon monoxide intoxication, lead intoxication, radiation therapy, and chemotherapy with methotrexate. Infectious disorders producing dementia and basal ganglia calcification include toxoplasmosis, cysticercosis, and postencephalitic dementias. In most of these disorders, CT demonstrates speckled hyperdense areas in the basal ganglia, thalamus, and deep hemis-

Table 2 Disorders Producing Dementia and Basal Ganglia Calcification in Adults

Idiopathic and hereditary disorders
  Idiopathic basal ganglia calcification (Fahr's disease)
  Membranous lipodystrophy
  Mitochondrial encephalopathy
  Hallervorden–Spatz disease
Metabolic disorders
  Hypoparathyroidism
  Hyperparathyroidism
  Hypothyroidism
Toxic conditions
  Carbon monoxide intoxication
  Lead intoxication
  Radiation therapy
  Methotrexate therapy
Infectious disorders
  Toxoplasmosis
  Cysticercosis
  Postencephalitic dementia

pheric white matter (Cohen, Duchesneau, & Weinstein, 1980; Harrington, MacPherson, McIntosh, Allam, & Bone, 1981). MRI is less sensitive than CT in the demonstration of intracranial calcium but typically reveals areas of hypointensity in the affected areas (Fig. 6).

Olivopontocerebellar atrophy (OPCA) is a progressive degenerative disorder involving the neurons of the cerebellar cortex, pons, and inferior olivary nuclei. The existence of mental status changes in OPCA has been controversial, but studies with more sensitive neuropsychological tests reveal deficits in intelligence, memory, and executive function (Kish, El-Awar, Schut, Leach, Oscar-Berman, & Freedman, 1988). CT and MRI reveal atrophy of the cerebellum and brainstem in OPCA. PET demonstrates reduced metabolism in the cerebellar vermis, cerebellar hemispheres, and brainstem (Rosenthal et al., 1988).

Corticobasal degeneration is a progressive disorder with neuronal loss that affects the parietal lobes and the basal ganglia. Neuronal achromasia is often evident at autopsy. Clinically, the patients manifest an extrapyramidal syndrome and visuospatial deficits associated with parietal lobe dysfunction such as constructional disturbances and dressing disorders. The disorder is frequently asymmetrical; structural imaging reveals cerebral hemispheric atrophy that is more severe on one side than the other (Gibb, Luthert, & Marsden, 1989).

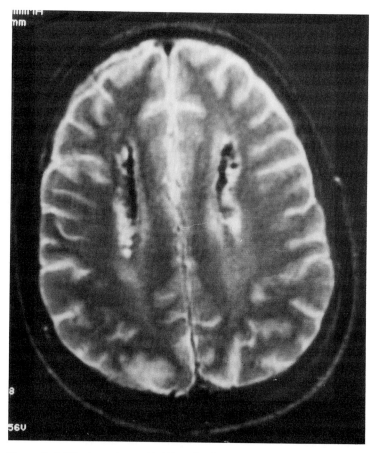

Figure 6 MRI of a patient with idiopathic basal ganglionic calcification (T_2-weighted image; calcified areas have decreased signal intensity).

E. Hydrocephalic Dementia

Normal pressure hydrocephalus (NPH) is an obstructive communicating hydrocephalus that classically produces the clinical triad of dementia, gait disturbance, and urinary incontinence. Neuroimaging reveals enlarged ventricles and is crucial to the identification and differential diagnosis of NPH. Ventricular dilatation is present in hydrocephalus *ex vacuo* with extensive tissue loss, in obstructive noncommunicating hydrocephalus, and in NPH (Cummings & Benson, 1992). Obstructive noncommunicating hydrocephalus is produced by obstruction of CSF flow within the ventricular system or at the outlet foramina between the fourth ventricle and the subarachnoid space. Noncommunicating hydrocephalus usually produces increased intracranial pressure and tends to present with headache, papilledema, cranial nerve palsies, and drowsiness or confusion rather than dementia. NPH is caused by obstruction of CSF absorption in the superior sagittal sinus. NPH may be idiopathic, may occur as a manifestation of the late decompensation of compensated congenital hydrocephalus, or may follow head trauma, subdural hematoma, subarachnoid hemorrhage, meningitis, or encephalitis.

Ventricular enlargement in NPH is characterized by greater dilatation of the anterior horns (frontal and temporal) of the lateral ventricles than of the posterior horns. Usually disproportionate enlargement of the ventricles and relatively modest sulcal dilatation, if any, are seen but atrophy of the cortical mantle occurs in the course of normal aging and the presence of some degree of peripheral atrophy does not exclude consideration of NPH. The presence of cortical atrophy does not correlate with improvement following shunting, but patients with severe peripheral atrophy usually are not considered to be suitable neurosurgical candidates (Petersen, Mokri, & Laws, 1985). Small sulci have been found to predict a favorable response to shunting (Thomsen, Mokri, & Laws, 1986).

Transependymal flow of CSF occurs in the presence of obstructive hydrocephalus. Identification of the associated periventricular lucency on CT has been found to correlate with a successful response to shunting (Thomsen et al., 1986). The transependymal flow is evident on MRI as a periventricular region of increased signal intensity on the T_2-weighted image. The abnormality may be smooth or irregular but typically does not include independent areas of hyperintensity in the corona radiata like those frequently seen in VaD.

Cisternography remains a useful tool in the imaging evaluation of NPH. Normal CSF flow, as revealed by isotope cisternography, consists of ascent of the radionuclide in the spinal subarachnoid space, flow around the cerebral convexities without ventricular reflux, and absorption in the superior sagittal sinus, usually within 24 hr of initiating the procedure. The classic pattern of NPH consists of ventricular reflux and stasis after 24–48 hr with no activity over the midline by 72 hr. A less definitive pattern consisting of persistent activity in the midline cisterns, slow flow in the Sylvian regions, and some degree of ventricular retention has been described also. Improvement after shunting has been reported with patients manifesting either cisternographic pattern (Jacobs, Conti, & Kinkel, 1976).

MRI also can be used to demonstrate CSF flow and eventually may render cisternography unnecessary. The reduced compliance of the dilated ventricular system increases flow through the cerebral aqueduct, which can be visualized on a T_1-weighted image as reduced signal intensity (flow void); reduced flow speed in hydrocephalus ex vacuo results in absence of this flow sign (Bradley et al., 1991; Kunz, Heintz, Ehrenheim, Stolke, Dietz, & Hundeshagen, 1989).

CBF is altered in NPH, and SPECT may aid in identifying patients that are likely to respond to shunting. Graff-Radford and colleagues (Graff-Radford, Godersky, & Jones, 1989; Graff-Radford, Rezai, Godersky, Eslinger, Damasio, & Kirchner, 1987) found that patients with higher anterior than posterior blood flow (measured on a slice 8 cm above the orbitomeatal line) were significantly less likely to improve following neurosurgery than patients who had equal flow anteriorly and posteriorly or greater posterior flow. Matsuda, Nakasu, Nakazawa, and Handa (1990) reported that CBF was reduced in all regions in hydrocephalus and that cerebral circulation time was increased; postoperatively, cerebral flow was unchanged but cerebral circulation time improved, particularly in the frontal lobes.

PET reveals diffusely reduced glucose metabolism in hydrocephalus. Restoration of normal metabolic rates follows successful treatment (Kay, Grady, Haxby, Moore, & Friedland, 1990).

VaD is the disorder most likely to be confused with NPH since both may present with the clinical syndrome of dementia, incontinence, and gait disturbance. Table 3 summarizes the neuroimaging changes in NPH and constrasts them with the findings of VaD.

F. Leukoencephalopathies

Hereditary leukoencephalopathies are more common in children than in adults, but white matter may be involved predominantly or exclusively in some adult-onset disorders. Structural imaging may be helpful in the identification and differential diagnosis of these disorders (Reichman & Cummings, 1990). CT often shows regions of diminished density; MRI (T_2-weighted images) reveals areas of increased signal intensity. MRI is more sensitive to white matter alterations than CT and is the superior instrument for the investigation of these dementias.

Table 4 lists the principal adult-onset dementias with marked involvement of cerebral white matter. Binswanger's disease (discussed earlier) is the most common dementia with white matter changes. Multiple sclerosis produces dementia in the later stages of the illness in at least 40% of patients, and in rare cases dementia is the presenting manifestation. Infectious diseases that may present with dementia and white matter changes include human immunodeficiency virus encephalopathy, progressive multifocal leukoencephalopathy, and Lyme disease. Toxic–metabolic

Table 3 Neuroimaging Characteristics of Normal Pressure Hydrocephalus and Vascular Dementia

Technique	Normal pressure hydrocephalus	Vascular Dementia
CT	Enlarged ventricles, particularly frontal and temporal horns; periventricular hypodensity	Diffuse ventricular dilatation; periventrical hypodensity
MRI	Enlarged ventricles, particularly frontal and temporal horns; periventricular hyperintensity; aqueductal flow void	Diffuse ventricular dilatation; periventricular hyperintensity with hyperintense regions in the corona radiata; no aqueductal flow void
Cisternography	Ventricular reflux; absence of superior perisagittal concentration	Ventricular reflux followed by superior perisagittal concentration
SPECT	Diffusely reduced cerebral blood flow; reductions more pronounced posteriorly than anteriorly	Diffuse plus multifocal reductions of cerebral blood flow
FDG PET	Diffusely reduced cerebral metabolism	Diffuse plus multifocal metabolic reductions

white matter dementias include anoxic encephalopathies, solvent-related encephalopathies, Marchiafava–Bignami disease (a disorder affecting the white matter of the corpus callosum and frontal lobes, and occurring primarily in Italian red wine drinkers), and the white matter degenerations that follow radiotherapy and chemotherapy. Hereditary adult-onset leukoencephalopathies include metachromatic leukodystrophy, adrenoleukodystrophy, cerebrotendinous xanthomatosis, Pelizaeus–Merzbacher disease, late-onset globoid cell leukodystrophy, hereditary adult-onset leukodystrophy, membranous lipodystrophy, and adult polyglucosan body disease. In hereditary leukodystrophies, CT and MRI usually demonstrate periventricular white matter abnormalities in the frontoparietal regions (Schwankhaus, Patronas, Dowart, Eldridge, Schlesinger, & McFarland, 1988). Investigation of at-risk individuals in families with hereditary leukoencephalopathies has demonstrated that MRI abnormalities may be evident prior to the onset of neurological signs and symptoms (Aubourg, Selier, Chaussain, & Kalifa, 1989).

G. Focal Atrophic Syndromes

Various disorders produce focal brain atrophy. Pick's disease and other frontal lobe degenerations, OPCA, PSP, and Huntington's disease are

Table 4 Adult-Onset Dementias with Prominent White Matter Changes

Vascular dementia
  Binswanger's disease
Demyelinating disorders
  Multiple sclerosis with dementia
Infectious illnesses
  Progressive multifocal leukoencephalopathy
  HIV encephalopathy
  Lyme disease
Toxic–metabolic disorders
  Anoxic encephalopathy
  Solvent-related encephalopathies (e.g., toluene abuse)
  Marchiafava–Bignami disease
  White matter degeneration following radiotherapy
  White matter degeneration following chemotherapy
Hereditary leukoencephalopathies
  Metachromatic leukodystrophy
  Adrenoleukodystrophy
  Cerebrotendinous xanthomatosis
  Pelizaeus–Merzbacher disease
  Late-onset globoid cell leukodystrophy
  Hereditary adult-onset leukodystrophy
  Membranous lipodystrophy
  Adult polyglucosan body disease

well-known causes of focal atrophy. Several other studied conditions also have marked local degenerative effects and produce visible focal atrophy on structural neuroimaging procedures. Table 5 summarizes conditions that may manifest dementia and focal brain atrophy. Conditions not previously discussed are described briefly here.

Primary progressive aphasia is a disorder characterized by gradually progressive linguistic deterioration (Mesulam, 1982). Dementia occurs late in the clinical course after many years of progressive language dysfunction. Structural imaging studies (CT, MRI) reveal unilateral left brain atrophy and functional imaging approaches (SPECT, PET) demonstrate unilateral left hemisphere hypoperfusion or hypometabolism (Chawluk et al., 1986). At autopsy, patients with primary progressive aphasia may have a unique localized spongiform encephalopathy (Kirshner, Tanridag, Thurman, & Whetsell, 1987) or a variety of other disorders including focal neuronal achromasia, Alzheimer's disease, Pick's disease, or Jakob–Creutzfeldt disease (Green, Morris, Sandson, McKeel, & Miller, 1990; Cohen, Smith, & Drachman).

Posterior cerebral atrophy (posterior cortical atrophy) is a clinical condition with marked visuospatial disturbances, constructional abnormali-

Table 5 Disorders That Manifest Dementia and Focal Brain Atrophy

Disorder	Focal atrophy
Pick's disease	Frontal, temporal, or combined frontotemporal lobes
Non-Alzheimer type FLD[a]	Frontal, temporal, or combined frontotemporal lobes
Motor neuron disease with dementia	Frontal lobes
NIHID[b]	Frontal lobes
Progressive subcortical gliosis	Frontal, temporal, or combined frontotemporal lobes (rarely occipital)
Alzheimer's disease	Temporal lobes, temporoparietal junction
Primary progressive aphasia	Left hemisphere
Posterior cerebral atrophy	Occipital lobes
Corticobasal degeneration	Unilateral parietal (left or right)
Huntington's disease	Caudate nuclei
Neuroacanthocytosis	Caudate nuclei
Progressive supranuclear palsy	Midbrain
Olivopontocerebellar atrophy	Pons, cerebellum

[a]Frontal lobe degeneration.
[b]Neuronal intranuclear hyaline inclusion disease.

ties, and alexia. Balint's syndrome may be present (Benson, Davis, & Snyder, 1988). The disorder is slowly progressive and dementia supervenes in the late stages. Several types of pathological alterations have been found at autopsy, including progressive subcortical gliosis and Alzheimer's disease. Figure 7 is an MRI of a patient with posterior cerebral atrophy.

H. Infectious Dementias

Jakob–Creutzfeldt disease (subacute spongiform encephalopathy) is a prion infection of the brain that is rapidly progressive and usually fatal within 1 yr. Clinically, dementia, ataxia, myoclonus, and muscle rigidity are characteristic; neuropathological alterations include spongiform changes, nerve cell loss, and marked gliosis. CT has proven of little value for diagnosis; MRI may show increased signal intensity in involved areas (Gertz, Henkes, & Cervos-Navarro, 1988). Functional imaging using SPECT and PET has revealed diffuse and multifocal reductions in gray

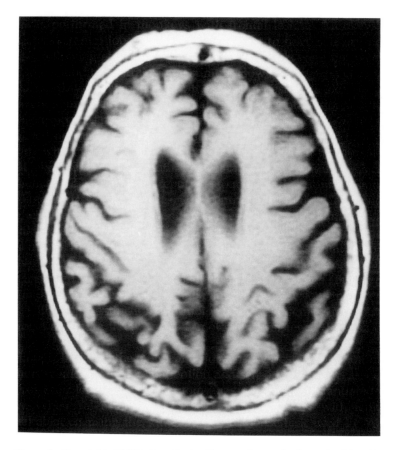

Figure 7   T₁-weighted MRI of a patient with posterior cerebral atrophy. Note the disproportionate enlargement of the sulci of the occipital lobes.

matter CBF and metabolism. These techniques may be most useful in pinpointing areas for diagnostic brain biopsy (Williams, 1991).

AIDS dementia complex is a progressive dementing illness caused by infection with the human immunodeficiency virus (HIV). Between 70 and 90% of AIDS patients develop neuropsychological impairment that is characterized by inattention, mental slowing, loss of spontaneity, reduced motor performance, and incoordination. At autopsy, mild to moderate cerebral atrophy is seen; neuropathology is found primarily in subcortical structures, particularly the basal ganglia and thalamus (Navia, 1990).

Structural imaging studies using CT or MRI show atrophy and evidence of demyelination of subcortical white matter. These techniques are helpful primarily in ruling out secondary focal neurological disease. Functional imaging studies reveal regional cortical and subcortical dysfunction. SPECT findings suggest a progression from subcortical asymmetry to cortical abnormality to more globally affected cerebral perfusion. Relative hypermetabolism of the thalamus and basal ganglia is found with PET in the early to middle stages of illness; as the dementia worsens, lowered temporal lobe metabolic function is demonstrated also (van Gorp et al., 1992).

I. Dementia Syndrome of Depression

The dementia syndrome of depression (DSD) is characterized by intellectual impairment, depressed mood, self-deprecatory delusions, appetite disturbances, prominent anxiety, agitation, a history of limited academic achievement and of previous mood disturbances or alcoholism, and a family history of affective disorder (Cummings, 1989). Several reports suggest that neuroimaging may prove useful in distinguishing this potentially reversible condition from other dementing illnesses. CT studies of DSD patients compared with age-matched control patients revealed greater central atrophy and lower CT attenuation values in individuals with DSD (Pearlson et al., 1989). Although functional imaging studies have not been reported for DSD, PET in depressed patients has shown asymmetrical frontal hypometabolism, greater on the left, that equalizes with successful treatment; significant hypofrontality as well as whole cortex hypometabolism persists in the treated state despite clinical improvement (Martinot et al., 1990). This hypofrontality of depression that is seen on functional imaging is distinct from the biparietal pattern seen with DAT and may prove helpful in distinguishing these disorders in elderly patients.

IV. PRINCIPLES OF NEUROIMAGING IN DEMENTIA

Neuroimaging is an essential part of the evaluation of dementia of unknown origin. Current algorithms directing the use of neuroimaging to specific subgroups of patients have not proven to be acceptably effective in detecting reversible intracranial pathology. Accurate diagnosis in dementia is the basis for treatment, prognosis, and genetic counseling. Restricted atrophy of cortical, basal ganglionic, brainstem, or cerebellar regions occurs in a variety of dementia syndromes; the pattern of atrophy aids in diagnosis. Frontal lobe atrophy distinguishes FLD from Alzheimer's disease but does not differentiate among the various causes of FLD. CT best demonstrates intracranial calcification. MRI is particularly sensitive to white matter changes. Functional imaging (PET, SPECT) is useful in classifying degenerative disorders that lack distinctive gross structural changes and reveals characteristic abnormalities in diseases without CT or MRI alterations. Imaging based on cerebral perfusion, glucose metabolism, neurochemical function, transmitter receptor density, and neurochemical composition eventually may generate results that are sufficiently distinctive in each dementia to be diagnostically specific.

ACKNOWLEDGMENTS

This project was supported by the Department of Veterans Affairs, a National Institute of Mental Health Training Grant, and a National Institute on Aging Alzheimer's Disease Core Center Grant (AG10123).

REFERENCES

Aharon-Peretz, J., Cummings, J. L., & Hill, M. A. (1988). Vascular dementia and dementia of the Alzheimer type: Cognition, ventricular size, and leuko-areiosis. _Archives of Neurology, 45_, 719–721.

Aubourg, P., Selier, N., Chaussain, J. L., & Kalifa, G. (1989). MRI detects cerebral involvement in neurologically asymptomatic patients with adrenoleukodystrophy. _Neurology, 39_, 1619–1621.

Barr, A. N., Heinze, W. J., Dobben, G. D., Valvassori, G. E., & Sugar, O. (1978). Bicaudate index in computerized tomography of Huntington disease and cerebral atrophy. _Neurology, 28_, 1196–1200.

Bench, C. J., Dolan, R. J., Friston, K. J., & Frackowiak, R. S. J. (1990). Positron emission tomography in the study of brain metabolism in psychiatric and neuropsychiatric disorders. _British Journal of Psychiatry, 157 (Suppl. 9)_, 82–95.

Benson, D. F., Davis, J., & Snyder, B. D. (1988). Posterior cortical atrophy. _Archives of Neurology, 45_, 789–793.

Benson, D. F., Kuhl, D. E., Hawkins, R. A., Phelps, M. E., Cummings, J. L., & Tsai, S. Y. (1983). The fluorodeoxyglucose 18F scan in Alzheimer's disease and multi-infarct dementia. _Archives of Neurology, 40_, 711–714.

Besson, J. A. O. (1990). Magnetic resonance imaging and its applications in neuropsychiatry. _British Journal of Psychiatry, 157 (Suppl. 9)_, 25–37.

Bird, T. D., Cederbaum, S., Valpey, W., & Stahl, W. L. (1978). Familial degeneration of the basal ganglia with acanthocytosis: A clinical, neuropathological, and neurochemical study. _Annals of Neurology, 3_, 253–258.

Bradley, W. G., Jr., Whittemore, A. R., Kortman, K. E., Watanabe, A. S., Homyak, M., Teresi, L. M., & Davis, S. J. (1991). Marked cerebrospinal fluid void: Indicator of successful shunt in patients with suspected normal-pressure hydrocephalus. _Radiology, 178_, 459–466.

Brooks, D. J., Ibanez, V., Playford, E. D., Sawle, G. V., Leigh, P. N., Kocen, R. S., Harding, A. E., & Marsden, C. D. (1991). Presynaptic and postsynaptic striatal dopaminergic function in neuroacanthocytosis: A positron emission tomographic study. *Annals of Neurology, 30,* 166–171.

Brown, G. G., Levine, S. R., Gorell, J. M., Pettegrew, J. W., Gdowski, B. S., Bueri, J. A., Helpern, J. A., & Welch, K. M. A. (1989). In vivo ^{31}P NMR profiles of Alzheimer's disease and multiple subcortical infarct dementia. *Neurology, 39,* 1423–1427.

Brown, J. J., Hesselink, J. R., & Rothrock, J. F. (1988). MR and CT of lacunar infarcts. *American Journal of Radiology, 151,* 367–372.

Brun, A. (1987). Frontal lobe degeneration of non-Alzheimer type. I. Neuropathology. *Archives of Gerontology and Geriatrics, 6,* 193–208.

Burns, A. (1990). Cranial computerized tomography in dementia of the Alzheimer's type. *British Journal of Psychiatry, 157 (Suppl. 9),* 10–15.

Burns, A., Philpot, M. P., Costa, D. C., Ell, P. J., & Levy, R. (1989). The investigation of Alzheimer's disease with single photon emission tomography. *Journal of Neurology, Neurosurgery and Psychiatry, 52,* 248–253.

Chawluk, J. B., Mesulam, N-M., Hurtig, H., Kushner, M., Weintraub, S., Saykin, A., Rubin, N., Alavi, A., & Reivich, M. (1986). Slowly progressive aphasia without generalized atrophy: Studies with positron emission tomography. *Annals of Neurology, 19,* 68–74.

Cohen, C. R., Duchesneau, P. M., & Weinstein, M. A. (1980). Calcification of the basal ganglia as visualized by computed tomography. *Radiology, 134,* 97–99.

Cummings, J. L., & Benson, D. F. (1992). *Dementia: A clinical approach* (2d ed.). Boston: Butterworths.

Cummings, J. L. (1989). Dementia and depression: An evolving enigma. *Journal of Neuropsychiatry, 1,* 236–242.

DeCarli, C., Kaye, J. A., Horwitz, B., & Rapoport, S. I. (1990). Critical analysis of the use of computer-assisted transverse axial tomography to study human brain in aging and dementia of the Alzheimer type. *Neurology, 40,* 872–883.

Dekosky, S. T., Shih, W.-J., Schmitt, F. A., Coupal, J., & Kirkpatrick, C. (1990). Assessing utility of single photon emission computed tomography (SPECT) scan in Alzheimer disease: Correlation with cognitive severity. *Alzheimer Disease and Associated Disorders, 4,* 14–23.

Delecluse, F., Deleval, J., Gerard, J.-M., Michotte, A., & Zegers de Beyl, D. (1991). Frontal impairment and hypoperfusion in neuroacanthocytosis. *Archives of Neurology, 48,* 232–234.

Dooling, E. C., Richardson, E. P., Jr., & Davis, K. R. (1980). Computed tomography in Hallervorden-Spatz disease. *Neurology, 30,* 1128–1130.

Drayer, B. P., Olanow, W., Burger, P., Johnson, G. A., Herfkens, R., & Riederer, S. (1986). Parkinson plus syndrome: Diagnosis using high field MR imaging of brain iron. *Radiology, 159,* 493–498.

Engel, P., Cummings, J. L., Villaneuva-Meyer, J. A. Mena, I. (1993). Single photon emission computed tomography in dementia: Relationship of perfusion to cognitive deficit. *Journal of Geriatric Psychology and Neurology, 6,* 144–151.

Erkinjuntti, T., Ketonen, L., Sulkava, R., Sipponen, J., Vuorialho, M., & Livanainen, M. (1987). Do white matter changes on MRI and CT differentiate vascular dementia from Alzheimer's disease? *Journal of Neurology, Neurosurgery and Psychiatry, 50,* 37–42.

Evans, D. A., Funkenstein, H. H., Albert, M. S., Scherr, P. A., Cook, N. R., Chown, M. J., Hebert, L. E., Hennekens, C. H., & Taylor, J. O. (1989). Prevalence of Alzheimer's disease in a community population of older persons: Higher than previously reported. *Journal of the American Medical Association, 262,* 2551–2556.

Fein, G., Van Dyke, C., Davenport, L., Turetsky, B., Brandt-Zawadzki, M., Zatz, L., Dillon, W., & Valk, P. (1990). Preservation of normal cognitive functioning in elderly subjects with extensive white-matter lesions of long duration. *Archives of General Psychiatry, 47,* 220–223.

Geaney, D. P., & Abou-Saleh, M. T. (1990). The use and applications of single-photon emission computerised tomography in dementia. *British Journal of Psychiatry, 157 (Suppl. 9)* 66–75.

Gertz, J., Henkes, H., & Cervos-Nararro, J. (1988). Creutzfeldt–Jakob disease: Correlation of MRI and neuropathologic findings. *Neurology, 38,* 1481–1482.

Gibb, W. R. G., Luthert, P. J., & Marsden, C. D. (1989). Corticobasal degeneration. *Brain, 112,* 1171–1192.

Goffinet, A. M., De Volder, A. G., Gillain, C., Rectem, D., Bol, A., Michel, C., Cogneau, M., Labar, D., & Laterre, C. (1989). Positron tomography demonstrates frontal lobe hypometabolism in progressive supranuclear palsy. *Annals of Neurology, 25,* 131–139.

Graff-Radford, N. R., Godersky, J. C., & Jones, M. P. (1989). Variables predicting surgical outcome in symptomatic hydrocephalus in the elderly. *Neurology, 39,* 1601–1604.

Graff-Radford, N. R., Rezai, K., Godersky, J. C., Eslinger, P., Damasio, H., & Kirchner, P. T. (1987). Regional cerebral blood flow in normal pressure hydrocephalus. *Journal of Neurology, Neurosurgery and Psychiatry, 50,* 1589–1596.

Grafton, S. T., Sumi, S. M., Stimac, G. K., Alvord, E. C., Jr., Shaw, C-M., & Nochlin, D. (1991). Comparison of postmortem magnetic resonance imaging and neuropathologic findings in the cerebral white matter. *Archives of Neurology, 48,* 293–298.

Green, J., Morris, J. C., Sandson, J., McKeel, D. W., Jr., & Miller, J. W. (1990). Progressive aphasia: A precursor of global dementia. *Neurology, 40,* 423–429.

Gupta, S. R., Naheedy, M. H., Young, J. C., Ghobrial, M., Rubino, F. A., & Hindo, W. (1988). Periventricular white matter changes and dementia: Clinical, neuropsychological, radiological and pathological correlation. *Archives of Neurology, 45,* 637–641.

Gustafson, L. (1987). Frontal lobe degeneration of non-Alzheimer type. II. Clinical picture and differential diagnosis. *Archives of Gerontology and Geriatrics, 6,* 209–223.

Haldeman, S., Goldman, J. W., Hyde, J., & Pribram, H. F. W. (1981). Progressive supranuclear palsy, computed tomography, and response to antiparkinsonian drugs. *Neurology, 31,* 442–445.

Harrington, M. G., MacPherson, P., McIntosh, W. B., Allam, B. F., & Bone, I. (1981). The significance of the incidental finding of basal ganglia calcification on computed tomography. *Journal of Neurology, Neurosurgery and Psychiatry, 44,* 1168–1170.

Hawkins, R. A., Mazziotta, J. C., & Phelps, M. E. (1987). Wilson's disease studied with FDG and positron emission tomography. *Neurology, 37,* 1707–1711.

Haxby, J. V., Grady, C. L., Duara, R., Schlageter, N., Berg, G., & Rapoport, S. I. (1986). Neocortical metabolic abnormalities precede nonmemory cognitive defects in early Alzheimer's-type dementia. *Archives of Neurology, 43,* 882–885.

Hershey, L. A., Modic, M. T., Greenough, P. G., & Jaffe, D. F. (1987). Magnetic resonance imaging in vascular dementia. *Neurology, 37,* 29–36.

Hunt, A. L., Orrison, W. W., Yeo, R. A., Haaland, K. Y., Rhyne, R. L., Garry, P. J., & Rosenberg, G. A. (1989). Clinical significance of MRI white matter lesions in the elderly. *Neurology, 39,* 1470–1474.

Jacobs, L., Conti, D., Kinkel, W. R., & Manning, E. J. (1976). "Normal-pressure" hydrocephalus. Relationship of clinical and radiographic findings to improvement following shunt surgery. *Journal of the American Medical Association, 235,* 510–512.

Johnson, K. A., Davis, K. R., Buonanno, F. S., Brady, T. J., Rosen, T. J., & Growdon, J. H. (1987). Comparison of magnetic resonance and roentgen ray computed tomography in dementia. *Archives of Neurology, 44,* 1075–1080.

Kamo, H., McGeer, P. L., Harrop, R., McGeer, E. G., Calne, D. B., Martin, W. R. W., & Pate, B. D. (1987). Positron emission tomography and histopathology in Pick's disease. *Neurology, 37,* 439–445.

Kaye, J. A., Grady, C. L., Haxby, J. V., Moore, A., & Friedland, R. P. (1990). Plasticity in the aging brain. Reversibility of anatomic, metabolic, and cognitive deficits in normal-pressure hydrocephalus following shunt surgery. *Archives of Neurology, 47,* 1336–1341.

Kertesz, A., Black, S. E., Nicholson, L., & Carr, T. (1987). The sensitivity and specificity of MRI in stroke. *Neurology, 37,* 1580–1585.

Kertesz, A., Black, S. E., Tokar, G., Benke, T., Carr, T., & Nicholson, L. (1988). Periventricular and subcortical hyperintensities on magnetic resonance imaging: "rims, caps, and unidentified bright objects." *Archives of Neurology, 45,* 404–408.

Keshavan, M. S., Kapur, S., & Pettegrew, J. S. (1991). Magnetic resonance spectroscopy in psychiatry: Potential, pitfalls, and promise. *American Journal of Psychiatry, 148,* 976–985.

Kirshner, S., Tanridag, O., Thurman, L., & Whetsell, W. O., Jr. (1987). Progressive aphasia without dementia: Two cases with focal spongiform degeneration. *Annals of Neurology, 22,* 527–532.

Kish, S. J., El-Awar, M., Schut, L., Leach, L., Oscar-Berman, M., & Freedman, M. (1988). Cognitive deficits in olivopontocerebellar atrophy: Implications for the cholinergic hypothesis of Alzheimer's dementia. *Annals of Neurology, 24,* 200–206.

Knopman, D. S., Christensen, K. J., Schut, L. J., Harbaugh, R. E., Reeder, T., Ngo, T., & Frey, W., II (1989). The spectrum of imaging and neuropsychological findings in Pick's disease. *Neurology, 39,* 362–368.

Knopman, D. S., Mastri, A. R., Frey, W. H., II, Sung, J. H., & Rustan, T. (1990). Dementia lacking distinctive histologic features: A common non-Alzheimer degenerative dementia. *Neurology, 40,* 251–256.

Kozachuk, W. E., DeCarli, C., Schapiro, M. B., Wagner, E. E., Rapoport, S. I., & Horwitz, B. (1990). White matter hyperintensities in dementia of Alzheimer's type and in healthy subjects without cerebrovascular risk factors: A magnetic resonance imaging study. *Archives of Neurology, 47,* 1306–1310.

Kunz, U., Heintz, P., Ehrenheim, C., Stolke, D., Dietz, H., & Hundeshagen, H. (1989). MRI as the primary diagnostic instrument in normal pressure hydrocephalus? *Psychiatry Research, 29,* 287–288.

Leenders, K. L., Frackowiak, R. S. J., & Lees, A. J. (1988). Steele–Richardson–Olszewski syndrome. Brain energy metabolism, blood flow, and fluorodopa uptake measured by positron emission tomography. *Brain, 111,* 615–630.

Leifer, D., Buonanno, F. S., & Richardson, E. P., Jr. (1990). Clinicopathologic correlations of cranial magnetic resonance imaging of periventricular white matter. *Neurology, 40,* 911–918.

Lippa, C. F., Cohen, R., Smith, T. W., & Drachman, D. A. (1991). Primary progressive aphasia with focal neuronal achromasia. *Neurology, 41,* 882–886.

Liu, C. K., Miller, B. L., Cummings, J. L., Mehringer, C. M., Goldberg, M. A., Howng, S. L., & Benson, D. F. (1992). A quantitative MRI study of vascular dementia. *Neurology, 42,* 138–173.

Lock, T., Abou-Saleh, M. T., & Edwards, R. H. T. (1990). Psychiatry and the new magnetic resonance era. *British Journal of Psychiatry, 157(Suppl. 9),* 38–55.

Margolin, R., & Jarema, M. (1989). Neuroimaging in the diagnosis of old age dementias. In E. Ban, & H. E. Lehmann (Eds.), *Modern problems in pharmacopsychiatry* (Vol. 23, pp. 28–42). Basel: Karger.

Martinot, J.-L., Hardy, P., Feline, A., Huret, J.-D., Mazoyer, B., Attar-Levy, D., Pappata, S., & Syrota, A. (1990). Left prefrontal glucose hypometabolism in the depressed state: A confirmation. *American Journal of Psychiatry, 147,* 1313–1317.

Matsuda, M., Nakasu, S., Nakazawa, T., & Handa, J. (1990). Cerebral hemodynamics in patients with normal pressure hydrocephalus: Correlation between cerebral circulation time and dementia. *Surgical Neurology, 34,* 396–401.

Metter, E. J., Mazziotta, J. C., Itabashi, H. H., Mankovich, N. J., Phelps, M. E., & Kuhl, D. E. (1985). Comparison of glucose metabolism, x-ray CT, and postmortem data in a patient with multiple cerebral infarcts. *Neurology, 35,* 1695–1701.

Miller, B. L., Cummings, J. L., Villanueva-Meyer, J., Boone, K., Mehringer, C. M., Lesser, I. M., & Mena, I. (1991). Frontal lobe degeneration: Clinical, neuropsychological, and SPECT characteristics. *Neurology, 41,* 1374–1382.

Montaldi, D., Brooks, D. N., McColl, J. H., Wyper, D., Patterson, J., Barron, E., & McCulloch, J. (1990). Measurements of regional cerebral blood flow and cognitive performance in Alzheimer's disease. *Journal of Neurology, Neurosurgery and Psychiatry, 53,* 33–38.

Munoz-Garcia, D., & Ludwin, S. K. (1986). Adult-onset neuronal intranuclear hyaline inclusion disease. *Neurology, 36,* 785–790.

Navia, B. A. (1990). The AIDS dementia complex. In J. L. Cummings (Eds.), *Subcortical dementia* (pp. 181–198). New York: Oxford University Press.

Neary, D., Snowden, J. S., Mann, D. M. A., Northern, B., Goulding, P. J., & MacDermott, N. (1990). Frontal lobe dementia and motor neuron disease. *Journal of Neurology, Neurosurgery and Psychiatry, 53,* 23–32.

Pearlson, G. D., Rabins, P. V., Kim, W. S., Speedie, L. J., Moberg, P. J., Burns, A., & Bascom, M. J. (1989). Structural brain CT changes and cognitive deficits in elderly depressives with and without reversible dementia ("pseudodementia"). *Psychological Medicine, 19,* 573–584.

Pearson, R. C. A., & Powell, T. P. S. (1989). The neuroanatomy of Alzheimer's disease. *Reviews in the Neurosciences, 2,* 101–122.

Petersen, R. C., Mokri, B., & Laws, E. R., Jr. (1985). Surgical treatment of idiopathic hydrocephalus in elderly patients. *Neurology, 35,* 307–311.

Radue, E. W., duBoulay, G. H., Harrison, M. J. G., & Thomas, D. J. (1978). Comparison of angiographic and CT findings between patients with multi-infarct dementia and those with dementia due to primary neuronal degeneration. *Neuroradiology, 16,* 113–115.

Reichman, W. E., & Cummings, J. L. (1990). Diagnosis of rare dementia syndromes: An algorithmic approach. *Journal of Geriatric Psychiatry and Neurology, 3,* 73–83.

Risberg, J. (1987). Frontal lobe degeneration of non-Alzheimer type. III. Regional cerebral blood flow. *Archives of Gerontology and Geriatrics, 6,* 225–233.

Rosenthal, G., Gilman, S., Koeppe, R. A., Kluin, K. J., Markel, D. S., Junck, L., & Gebarski, S. S. (1988). Motor dysfunction in olivopontocerebellar atrophy is related to cerebral metabolic rate studied with positron emission tomography. *Annals of Neurology, 24,* 414–419.

Schwankhaus, J. D., Patronas, N., Dowart, R., Eldridge, R., Schlesinger, S., & McFarland, H. (1988). Computed tomography and magnetic resonance imaging in adult-onset leukodystrophy. *Archives of Neurology, 45,* 1004–1008.

Sethi, K. D., Adams, R. J., Loring, D. W., & El Gammal, T. (1988). Hallervorden-Spatz syndrome: Clinical and magnetic resonance imaging correlations. *Annals of Neurology, 24,* 692–694.

Shinar, D., Gross, C. R., Hier, D. B., Caplan, L. R., Mohr, J. P., Price, T. R., Wolf, P. A., Kase, C. S., Fishman, I. G., Barwick, J. A., & Kunitz, S. C. (1987). Interobserver reliability in the interpretation of computed tomographic scans of stroke patients. *Archives of Neurology, 44,* 149–155.

Snow, B. J., Bhatt, M., Martin, W. R., Li, D., & Calne, D. B. (1991). The nigrostriatal dopaminergic pathway in Wilson's disease studied with positron emission tomography. *Journal of Neurology, Neurosurgery and Psychiatry, 54,* 12–17.

Starosta-Rubinstein, S., Young, A. B., Kluin, K., Hil, G., Aisen, S. M., Gabrielsen, T., & Brewer, G. J. (1987). Clinical assessment of 31 patients with Wilson's disease. *Archives of Neurology, 44,* 365–370.

Thomsen, A. M., Mokri, B., & Laws, E. R., Jr. (1986). Prognosis of dementia in normal-pressure hydrocephalus after a shunt operation. *Annals of Neurology, 20,* 304–310.

United States Congress Office of Technology Assessment (1987). *Losing a million minds: Confronting the tragedy of Alzheimer's disease and other dementias,* Publ. No. OTA-BA-323. Washington, D.C.: U.S. Government Printing Office.

Van Gorp, W. G., Mandelkern, M. A., Gee, M., Hinkon, C. H., Stern, C. E., Paz, D. K., Dixon, W., Evans, C., Flynn, F., Frederick, C. S., Ropchan, J. R., & Blahd, W. H. (1992). Cerebral metabolic dysfunction in AIDS: Findings in a sample with and without dementia. *Journal of Neuropsychiatry and Clinical Neuroscience, 4,* 280–287.

Verity, M. A., & Wechsler, A. F. (1987). Progressive subcortical gliosis of Neumann: A clinicopathologic study of two cases with review. *Archives of Gerontology and Geriatrics, 6,* 245–261.

Williams, F. J. B., & Walshe, J. M. (1981). Wilson's disease. An analysis of the cranial computerized tomographic appearances found in 60 patients and changes in response to treatment with chelating agents. *Brain, 104,* 735–752.

Williams, J. P. (1991). Neuroimaging. In F. O. Bastian (Ed.), *Creutzfeldt-Jakob disease and other transmissible spongiform encephalopathies* (pp. 203–212). St. Louis: Mosby Year Book.

Young, A. B., Penney, J. B., Starosta-Rubinstein, S., Markel, D. S., Berent, S., Giodani, B., Ehrenkaufer, R., Jewett, D., & Hichwa, R. (1986). PET scan investigations of Huntington's disease: Cerebral metabolic correlates of neurological features and functional decline. *Annals of Neurology, 20,* 296–303.

Index

i